The Guinness Current Car Index

Ivan Berg

GUINNESS PUBLISHING

Pictures: Courtesy of Autocar & Motor

First published in 1992 by Guinness Publishing Ltd

Cover design: Pinpoint

Design: John Mitchell

Typeset in Helvetica
by John Mitchell
set at Fraser Hamilton Associates, London

Printed and bound in Great Britain by The Bath Press, Bath

A catalogue record for this book is available from the
British Library

ISBN 0-85112-575-1

Contents

About the author

Ivan Berg joined the Merchant Navy in 1952 at the age of 16 with the romantic notion of emulating the adventurous lives of writers like Hemingway, W.W. Jacobs and Jack London. The life was sufficiently adventurous, particularly on the north Atlantic run on both of Cunard's Queens, but left him too tired to write. It was not until National Service in the Royal Air Force that he began to produce in earnest.

18 months as a medic in the ENT department at the Central Medical Establishment provided the time to create, not the adventure novel of early dreams, but the first motor racing series on British television; 'The Chequered Flag' - for Rediffusion TV in 1960.

The next 10 years were spent variously as a screenwriter, author and journalist. Apart from 'The Chequered Flag' his credits included 'Moonstrike' - BBC1, 'A Touch of Brass' - Worldwide Films, *Tackle Karting This Way* and *Tackle Motor Sport This Way* for Hutchinson Stanley Paul and *TRAD*, a reference book on the UK Trad Jazz scene for Foulsham. Weekly adventure stories and a comic strip featuring his fictional motoring hero Revs Ransome were produced for *TV Express*, and he contributed features to *TV Times*, *Daily Mail*, *Honey* Magazine and the provincial press.

In 1970 Berg started a recording studio with his wife Inge, and between 1970 and 1982 produced radio commercials and documentaries and published a very large range of spoken word cassettes. In 1982, intrigued by the publishing challenge offered by the fledgling home computer market, he bought a Commodore VIC 20 computer. The result was the publishing with Commodore Computers of a range of enormously successful home and educational software during the home computer boom years of '83 and '84.

In 1990 with both children doing well in the media (Sanchia is a BBC Television producer and reporter and Nik is a reporter with *Auto Express* magazine), Berg decided there was a case for 'following in his son's/daughter's footsteps' and returned to writing and production with 'Austins of the 1930s', a film shot in the summer.

The *Sunday Times* said it was "...an outstanding new production" featuring "four cars that, with an excellent script, electrify the small screen." Flattered and encouraged by this reception he began work on a project which combined his computer knowledge with his love of cars and the old-found writing skills. The result is this book and its much bigger sister, *The Guinness World Car Record.*

Acknowledgements and thanks

Neither this book nor its much bigger sister, *The Guinness World Car Record*, could have developed from big idea to publication without the help and support of a leading international software company, two of the world's best-known motoring journals and four individuals. They will not remain nameless.

The software company is DataEase UK Limited, part of the DataEase International group, which provided the superbly flexible and friendly DataEase 4.2 database software that made this project viable. Particular mention must be made of DataEase Senior Analyst Lee Cox for a yearful of tuition and trouble-shooting.

The first of the two of the world's best-known motoring journals is also the world's longest running. *Autocar* was first published in 1895, and introduced the 'Autocar Road Test' in 1928. Britain's other 'oldest motoring journal' was *The Motor*, first published in 1903, which merged with *Autocar* in 1988 to become *Autocar & Motor*. Thus the acknowledgement for the supply of and permission to use road-test data is to *Autocar & Motor*. Duly acknowledged.

Road & Track magazine has been giving the road-test lowdown to American enthusiasts since 1947. American cars are rarely road-tested in the UK nowadays and *Road & Track* road-test data has made it possible to include one hundred and fifty in this edition as well as three hundred and forty-five 'imports'. Thank you *Road & Track*.

Three of the individuals are Neil Eason Gibson of the RAC Motor Sports Association, Lawrence Pierce of *Auto Express* magazine and Roy Jackson-Moore. Their help on technical matters has been most generous.

The fourth is my co-author (for that is what she is) my wife Inge. She refuses, for reasons of modesty, a joint credit. I hope that she will accept acknowledgment here for the database design and the entry and husbanding of the half million or more items of data used to produce this book and its big sister The Guinness World Car Record.

Finally, a thank you to all the motor manufacturers who have provided information and photographs. Keep it coming please. There's the next edition to think about.

Ivan Berg May 1992

Introduction

There is a fascination with car facts and figures. They are the stuff of motoring magazines world-wide. Countless car books have been written around them, and they are endlessly discussed in the pubs, bars and clubs where enthusiasts meet.

Performance figures. Technical sophistication and complexity. The modern car enthusiast ranks cars according to their dynamics. It is what a car can do and how it does it that counts. Cars are compared with other cars in terms of acceleration, power-to-weight ratio, power output by engine size; by engine configuration, by number of valves, by engine revolutions per minute...

Nevertheless it is a random business. Only particular facts and figures stick in the mind. Cars once owned. Cars admired and cars coveted. Even the obsessive would be hard put to reel off figures in a particular order. Most of us would have to look them up in books, write them down and assemble the list from our notes.

Single figure ranking is easy. Take the nought to sixty miles an hour time of twenty cars and assemble them in order of least time taken. Primary school stuff. But now take 500 cars. Sort in groups by country made, with engines over 2-litres - each group to be in order of the nought to sixty time. That is a project.

This book redresses the balance. Here in an instant, you can look up the fastest Ferraris, the most accelerative Alfa Romeos, Aston Martins and Audis; Maserati, Mazda, Mercedes and Mercury maximums, and the most fuel efficient Fiats and Fords.

Being suitably superlative, (this is a Guinness publication after all) there are one thousand and six cars described in these pages, built between January 1986 and March 1992. Each car has up to forty-three dynamic and static facts and figures which are available for comparison with the up to forty-three facts and figures of every other car. The method of selection of the cars for comparison is vital.

If facts and figures, particularly performance figures, are to be compared, then not only do they have to be available, they should also be produced according to a standardised formula, if the comparisons are to be meaningful.

In addition, there is no doubt that figures obtained by driving a car according to a laid-down formula by a professional tester have more credibility in the minds of most drivers than those claimed or calculated. The selection of the cars for inclusion here was determined by the availability of relatively objective performance figures. The sources are *Autocar & Motor* in the UK and *Road & Track* in the US. Both magazines road-test cars to a standardised formula and although not identical provide results sufficiently similar for use in a computer database.

No claims are made here that the facts and figures used are complete or definitive. Simply that they are the best available for the purpose. In a very few cases, for example, the one-off Sbarro Chrono and other exotica, calculated or claimed figures have been used so that these exciting and unusual cars could be included. With the exception of modified cars, the figures for weights, dimensions and power and torque are generally as provided by the manufacturer.

Both magazines road-test a wide variety of cars, popular, interesting and exotic; as its title indicates *Road & Track* leans more than *Autocar & Motor* towards the exotic. Many models have been tested several times through their production lives and the differences, and indeed the similarities, are often surprising.British and European readers may raise an eyebrow at a detectable bias towards muscle and performance cars. Particularly those made in the US. Chevrolet Corvettes, Dodge Shelbys and Daytonas, Ford Mustangs, and Pontiac TransAms and Firebirds. Also, if it is big, if it is sporty or a foreign exotic, it seems an American performance tuning outfit will give it the treatment and provide *Road & Track* magazine with one or more examples for testing. Look out for names like Callaway, Guldstrand, Gottlieb and Lingenfelter, Norwood, Gemballa and Strosek attached to Corvette, Mustang, Ferrari and Porsche.

In the specifications section, readable type and a finite number of pages means around forty lines per car. What is fitted into those forty lines is a mix of personal preference, consultation, and what makes a good component for a comparison list.

Readers will notice that cars specified will sometimes not appear in a comparison list even though personal knowledge of the car says that it should. If you know a car gets to sixty in under seven seconds and it does not appear in the list alongside other cars which do, then it didn't when tested, or possibly, the figure is missing from the computer database because it was not available at the time the information was entered.

This book, as it deals with modern cars, has most of the required information, most of the time. The performance figures themselves may be the subject of some debate. Cars may have been faster or slower, heavier or lighter when tested by other magazines. Owners will claim better figures or complain about worse. The manufacturer's figures will differ too.

It is probably worth repeating at this point what was said earlier about the source of the performance figures: "Both magazines test cars to a standardised formula and although not identical provide results sufficiently similar for use in a computer database." The key words being standardised formula and sufficiently similar.

There is no British, American, or World Standard for the road testing of motor cars - and the least said about some of the EC bureaucrats' ideas the better. Each fact or figure included in the car specifications is discussed in more detail below, but apropos standards, times for the standing kilometre are not part of the *Road & Track* tests, neither is there compatibility between *Autocar & Motor*'s 'overall' mpg and Road & Track's 'average' mpg.

Apart from these, the performance facts and figures selected are in general, common to both the UK and US tests.

The Lists

As readers familiar with computers will know, data can be manipulated in manifold ways. Potentially, the number and variety of comparative lists which could be produced from the data used in this book is virtually limitless. More lists may be produced than there are stars in the galaxy. Not all of them would be interesting. How many of us would wish to know the number of cars with leaf springs on the rear suspension which had a ground clearance of less than five inches and where and when and by whom they were built? Or, which particular cars made in 1986 have bodies which are between twelve and sixteen inches longer than their wheelbase ?

(Readers' suggestions for lists for future editions of this book and *The Guinness World Car Record* are welcome.)

A list can be produced according to one or any number of criteria. One example should suffice. First, the query:

Make a list comprising cars manufactured between 1986 and 1992 which have engine capacities greater than four-litres, eight or more cylinders, weigh less than four thousand five hundred pounds, have a power-to-weight-ratio of 250bhp/ton or more and accelerate to sixty miles an hour in under seven seconds. Order the list by fastest acceleration time and give acceleration time, make, model, year - and because the more sceptical reader will no doubt wish to check accuracy in every detail -

the car's reference number in the *specifications section of this book.'*

Run the computer.

And the cars meeting these supercar criteria are:

1   3.5secs Jaguar XJ220 1989: 407
    3.5secs Koenig Competition Evolution
    1990: 421
2   4.0secs Koenig Competition 1989: 420
3   4.1secs Dodge Viper 1992: 241
4   4.4secs Jaguar XJS Lister Le Mans 1990:
    414
5   4.5secs Cizeta Moroder V16T 1990: 215
    4.5secs Lamborghini Diablo 1991: 432
6   4.7secs Ferrari Testa Rossa Norwood
    1989: 261
    4.7secs Lamborghini Countach 25th
    Anniversary 1989: 431
7   4.8secs Porsche 928 S4 Koenig 1989: 733
8   5.0secs Isdera Imperator 1988: 389
9   5.2secs Lamborghini Countach 5000S
    1986: 427
10 6.2secs Ferrari Pininfarina Mythos 1990:
    265

Before you put pen to paper to protest that there is not a Porsche 959 or 911 Turbo in sight, check the spec. All thirteen cars had to meet every single criterion to get a place in the list.

Size and space on the page conspire to limit the number of lists which can be fitted into this book. There are however, a great many more

in *The Guinness World Car Record*, which has the specifications of more than 5,000 cars from 1928 up to early 1992. Even so, the lid of this particular Pandora's box is being opened but a millimetre or two.

The more determined enthusiast will be delighted to know that he or she may, in the future, further explore the Virtually Limitless on the CD ROM version of *The Guinness World Car Record.*

But back to terra firma. From the limitless numbers of possible lists three of the most practical have been selected for this first edition. As a car's performance is the greatest subject of motoring debate, the reader will not be surprised to find that the three lists are maximum speed, standing start to 60mph, and fuel economy. Each of the three main lists, which include every car in the book, is then broken down into eighty-nine smaller lists by make of car.

Before calculators and computers every schoolboy and schoolgirl carried a small paperback called a 'Ready Reckoner' which provided the multiples and divisors and fractions and decimals of pages of numbers in easy to look up table form.

So why not use the lists as a 'Ready Car Comparator'. Look up the car in the makes lists, note the acceleration time, maximum speed or mpg, and see how it fares against or compares with any other car or cars you choose, in its placing in one of the main lists. Is the older Audi Quattro higher placed than the younger ? Is the Ferrari Mondial t in the fastest top twenty ? What place does your Chevrolet, Honda, Rover, or Volkswagen have in the economy stakes compared to a Ford, a Mazda or a BMW ?

Shatter some preconceptions. Confuse a pundit. Make some apparently outrageous claims. Then point to the page.

Car specifications

Reference number:

Cars are ordered alphabetically by make, then by year of manufacture and then by model name. The reference or look-up number will also appear when a car has a place in one of the comparison lists in the lists section of the book. Make: In alphabetical order. European and Japanese cars are sometimes marketed in the United States under different make names; examples are Acura (Honda), and Sterling (Rover). As the Sterling is not marketed in the US at the time of writing, it has been included with the other Rovers. All Land Rover and Range Rover vehicles are marketed as Range Rover in the US. An American tested Land Rover County will therefore appear under Range Rover.

Model:

In alphabetical order by year. The choice of models included is determined by the cars tested in that year in both the UK and the US. Model duplication will occur when a popular car has been tested regularly throughout its production life. Nevertheless the figures will generally be different year by year, and not always in the same direction! There may also be occasions where the same model appears twice in the same year. This will occur when the car has been tested both in the US and the UK. More often than not the US cars will have been differently specified, so again the figures will differ. Model names also change as cars are imported into or assembled in the US.

Many of the model name suffixes will not be familiar to European eyes. These will probably be US modified and have been included because they are based upon production cars and are available for purchase. Their performance figures will often appear outrageous.

For example, a Chevrolet Corvette Callaway Sledgehammer with a 5.7-litre turbo Vee8 which is claimed to produce 880bhp at 6250rpm and 772lbft of torque at 5250rpm. It cost $400,000 in 1990. (*Exotic Cars Quarterly* Vol.1 No.3)

Year:

The year in which the model was road tested. This is not necessarily the year of manufacture. Cars included in this book were tested between January 1986 and March 1992. The exceptions are Bentley and Rolls-Royce cars, which because of their very long model lives are included from 1980. Country: Generally, the country where the vehicle was built. Although some American/Japanese joint-venture models will be shown as built in Japan.

Maximum speed:

In miles per hour and kilometres per hour, to one decimal place.
This is the maximum speed the vehicle achieved during the road test and is not necessarily the mean of two or more runs. In a few instances a calculated or claimed figure

has been used. For example, where the maximum speed could not be measured because there was not enough straight road available !

0-50mph 80.5kmh, 0-60mph 96.5kmh, 0-1/4 mile and 0-1km:

The time in seconds to one decimal place taken by the car to accelerate from standstill to the electronically measured speed or distance from a standing start. Usually the best of several starts. A seemingly brutal test conducted by taking the engine revolutions to peak power and dropping the clutch. Wheelspin replaces clutch-slip. Tests hardiness of complete drive-train.Great fun to watch in wet weather.

Maximum power:

The maximum power produced by the engine of the car when tested by the manufacturer (usually) on a dynamometer. Shown in brake horse-power, kilowatts and the EC PS measure to one decimal place. The engine revolutions per minute, (rpm) at which the maximum power was achieved is also given. Maximum or peak power does not necessarily correlate with the engine's maximum rpm.

Maximum torque:

This figure is also measured by the manufacturer and represents the maximum twisting force or pulling power of the engine. Shown in pounds/feet (lbft) and Newton/metres (Nm). The rpm at which this maximum figure was achieved is also given. There is not necessarily any correlation between maximum rpm and maximum torque.

Power to weight ratio:

Here expressed as bhp/ton (Imperial) and bhp/tonne (Metric).Often stated in other publications as bhp/pound and bhp/kilo. This is an interesting figure, produced by dividing the bhp by the kerb weight and multiplying the result by either 2240 for bhp ton or 1000 for bhp tonne. The figure is interesting because, unlike specific output, (bhp per litre), it can bear a direct relationship to the accelerative power of the car.

Specific output:

This has to do with engine efficiency. The more power that can be squeezed from each cubic centimetre of engine capacity the better. Expressed here as bhp/litre, kW/litre and PS/litre. The figure is produced by dividing the power figure by the engine metric cubic capacity and multiplying the result by 1000 (ccs per litre).

There are many ways of improving an engine's specific output, and they tend to come in fashions. Superchargers were the bee's knees in the 1930s and have since regained favour, notably with Volkswagen. Turbocharging was in during the late 1970s and for most of the 1980s. In the 1950s the done thing was to increase compression ratios, add or change carburettors and polish ports. The latter part of the 1980s saw a trend towards more valves - up to five per cylinder, (eight per cylinder in the case of Honda's oval-piston NR750 motor-cycle engine) and the fitting of variable air intake systems.

The keen-eyed will note that Mazda RX-7s come high on the lists for specific output for normally aspirated engines. (Not turbocharged or supercharged.) This is because they have Wankel rotary engines, and as these produce three power-strokes per rotor on every revolution of the crankshaft, they need less capacity then a conventional four-stroke engine for a similar power output.

In the UK and Europe rotary engine capacity is usually doubled artificially. The accepted method of measuring the capacity of a Wankel rotary engine, used by most engineers and in the US, is to take the difference between the largest and smallest volume of one rotor chamber and multilply that by the number of chambers. Not dissimilar to the way a piston engine's capacity is measured.

Piston speed at peak power:

Expressed in feet and metres per second. In the days when very long-stroke (piston stroke greater than cylinder bore) engines were the norm piston speeds in excess of 2500 feet per minute (41.6ftsec here) were deemed to produce excessive main-bearing loads. There was a preoccupation with 'big-ends' as they were known, right up until the 1960s, the failure of which caused no end of head-shaking.

Today's high-revving engines suffer to a far lesser degree and piston speeds are very much higher. Main bearings these days fail more from lack of lubrication caused by poor maintenance than from excessive loading.

Nevertheless the figure is interesting because of its relationship with bore and stroke and engine rpm. It is produced by doubling the cylinder stroke, multiplying the result by the engine rpm at peak power and dividing that result first by 1000 and then by 60.

Fuel consumption:

Autocar & Motor and *Road & Track* are incompatible on this one. *Road & Track* quotes an 'average' or 'normal driving' figure, while *Autocar & Motor*'s 'overall' fuel consumption figure is the result of dividing the number of miles driven on the test by the amount of fuel used.

(On the basis that only like can be compared with like if comparisons are to be meaningful, there are two lists for fuel consumption, one according to *Autocar & Motor*, the other according to *Road & Track*.)

These figures are by no means definitive and they may bear no relation to manufacturer's figures, other road tests, or your own

experience. They are different from the constant speed measurements and 'Urban' figures produced using the EC criteria.

In the case of *Autocar & Motor* they are nevertheless figures achieved by the driving of the car throughout a reasonably standardised road test, one which has stayed pretty much in its present form since 1952. The figures may well be lower than figures found elsewhere, sometimes alarmingly so as the car will have been driven hard throughout.

Engine type:

With the single exception of the GM Impact, the only electrically engined car in the book, all engines are diesel or petrol, four-stroke or rotary, and may be turbocharged or supercharged.

Engine capacity:

This is the capacity of an engine cylinder chamber, measured with the piston at the bottom of its stroke, multiplied by the number of cylinders. (For Wankel rotary engines see Specific output, above) Nearly always measured in cubic centimetres, though most Americans remain loyal to the cubic inch. Both are found here. If specified, a turbocharger or supercharger is listed here.

Configuration:

Consists of cylinder configuration, number of cylinders, number of carburettors or fuel injection. An engine which has its cylinders arranged one behind the other is called

IN-LINE. If the cylinders are horizontally opposed, as in the Porsche 911, VW Beetle, and some Ferraris and Lancias, the engine is FLAT. In a VEE configuration two banks of one, two, three, four, five, six, or eight cylinders are set at an angle to each other. VEE engines cram a lot of power into a little space.

Compression ratio:

The amount by which the fuel mixture is compressed during the cylinder or rotor compression stroke prior to its ignition.

Ratios in petrol engines rarely exceed 12.5:1. Diesel engines which rely on the compression of the fuel mixture for ignition can have ratios as high as 25:1

Bore and stroke:

Normally measured in millimetres, also in inches here. Not given for rotary engines. The bore is the diameter of the cylinder, and the stroke is the distance travelled by the piston during half a revolution of the crankshaft. The terminologies are OVERSQUARE, when the bore is greater than the stroke (short-stroke), SQUARE, when the bore is equal to the stroke, and UNDERSQUARE, when the stroke is greater than the bore (long-stroke). As a rule, long-stroke engines produce high torque at low engine revolutions and short-stroke engines produce high power at high revolutions.

Valve type and No:

Virtually all modern car piston engines employ overhead-valves, to inlet the fuel charge and exhaust the spent gas. Rotary engines have inlet and exhaust ports which are opened and closed for the same purpose by the motion of the rotor.

Overhead valves can be operated indirectly by pushrods and rockers actuated by a gear driven camshaft in the engine block, or directly by overhead camshafts. High performance engines, almost always those with multi-valve cylinder heads, use one or more overhead camshafts - one camshaft to operate the inlet valves, the other the exhaust valves. Thus a twin-overhead camshaft VEE 8 engine will have 4 camshafts, 2 for each bank of cylinders.

The number of valves on most production engines used to be two per cylinder - one inlet valve, one exhaust valve. Then, in the cause of greater efficiency another inlet valve was introduced, then another exhaust valve, then another inlet valve. Currently very few engines have more than five valves per cylinder, the exception being the eight-valve Honda NR750 mentioned earlier.
The highest number of valves here is 64, comprising four valves per cylinder on a sixteen cylinder engine.

Transmission:

Categorised as manual, manual with overdrive, automatic, pre-selector, and continuously variable.

No. of forward speeds: (gears)

Where specified, the number of overdrive speeds (gears) has been added to the total. Thus, a car with a four-speed gearbox with overdrive available on third and top, is shown as having six forward speeds.

Wheels driven:

Shown as front, rear, 4-wheel drive and 4-wheel engageable.

Springs Front/Rear:

Spring types are classified as leaf, coil, torsion-bar and gas - which also covers air springs, hydro-pneumatic and hydragas systems.

Brake system:

Power-assisted (PA) and anti-lock (ABS) if specified.

Brakes Front/Rear:

Shown as Disc/Disc, Disc/Drum or Drum/Drum.

Steering:

Type of steering, ie, rack and pinion, re-circulating ball, ball and nut etc, and whether power-assisted (PA). The two most popular systems in modern cars are rack and pinion and re-circulating ball. The enthusiast will claim that rack and pinion gives the best steering feel and response.

Wheelbase:

The distance between front and rear wheels measured from the centre of the wheels.

Long wheelbase, narrow track cars tend to be good in a straight line. Short wheelbase, wide track cars can be very good at changing direction, often when you don't want them to.

Track Front/Rear:

The distance between the front/rear wheels measured from the centre of the tyre tread.

Length:

Overall. Measured from bumper to bumper. Foreign cars specified for the US market have often been longer overall in order to meet bumper regulations.

Width:

Overall. Some discrepancies may have crept in here as width overall can (in some difficult to identify cases) mean either including or excluding mirrors.

Height:

Overall. Normally measured without passengers.

Kerb weight:

Weight at the time of test, generally with a half-full fuel tank and no passengers.

Fuel tank capacity:

As specified by the manufacturer.

Maximum speed in order

1 255.0mph 410.3kmh
Chevrolet Corvette Callaway
Sledgehammer: 174
2 235.0mph 378.1kmh
Koenig C62: 422
3 230.0mph 370.1kmh
Koenig Competition Evolution:
421
4 218.0mph 350.8kmh
Vector W8 Twin Turbo: 949
5 217.0mph 349.2kmh
Koenig Competition: 420
6 216.0mph 347.5kmh
Porsche 911 Turbo RS Tuning:
744
7 214.0mph 344.3kmh
Bugatti EB110: 126
8 211.0mph 339.5kmh
Porsche 911 Turbo Ruf Twin
Turbo: 717
9 210.0mph 337.9kmh
Ferrari Testa Rossa Norwood:
261
10 208.0mph 334.7kmh
Porsche 911 Ruf CTR: 750
11 205.0mph 329.8kmh
Porsche 911 Turbo Gemballa
Mirage: 753
12 204.0mph 328.2kmh
Chevrolet Corvette
Lingenfelter: 181
Cizeta Moroder V16T: 215
13 203.0mph 326.6kmh
Chevrolet Camaro Gottlieb
1969: 159
14 202.0mph 325.0kmh
Lamborghini Diablo: 432
15 201.0mph 323.4kmh
Chevrolet Corvette Callaway
Speedster: 180
Porsche 911 Turbo Koenig RS:
716
Porsche 959 S: 709
16 200.0mph 321.8kmh
Eagle GTP: 244
Ferrari F40: 263
Jaguar XJS Lister Le Mans: 414
Jaguar XJ220: 407
Vector W2: 948
17 198.0mph 318.6kmh
Ferrari 308 Norwood
Bonneville GTO: 253
Porsche 959: 738
Porsche 959 Comfort: 724
Porsche 959 Sport: 725
18 196.1mph 315.5kmh
Porsche 911 Ruf TR2: 751
19 196.0mph 315.4kmh
Ferrari F40: 267
Porsche 911 Ruf: 749
20 193.0mph 310.5kmh
Chevrolet Camaro IROC-Z

Chevrolet Engineering: 162
Safir GT40: 839
21 192.0mph 308.9kmh
Pontiac Firebird TDM
Technologies: 692
22 191.0mph 307.3kmh
Chevrolet Corvette Callaway:
164
Chevrolet Corvette Callaway:
152
23 190.0mph 305.7kmh
Dodge Viper: 241
24 186.0mph 299.3kmh
Chevrolet Corvette ZR-1
Geiger: 184
Ford Mustang NOS/Saleen:
335
Porsche 928 S4 Koenig: 733
Zender Fact 4: 1006
25 185.0mph 297.7kmh
Chevrolet Corvette ZR-2: 185
Ferrari Testa Rossa Gemballa:
260
Ford Mustang 5.0 Cartech
Turbo: 312
Mercedes-Benz 300CE AMG
Hammer: 547
26 184.0mph 296.1kmh
Chevrolet Corvette Callaway:
179
27 183.0mph 294.4kmh
Aston Martin Zagato: 36
Mercedes-Benz 300E AMG
Hammer: 536
Porsche 911 Turbo Motorsport
Design: 731
28 182.0mph 292.8kmh
Ferrari 308 Norwood: 257
29 181.0mph 291.2kmh
Ferrari Testa Rossa Straman
Spyder: 256
Ferrari Testa Rossa: 255
30 180.0mph 289.6kmh
Ferrari Pininfarina Mythos: 265
31 179.0mph 288.0kmh
BMW 535i Alpina B10 Bi-
Turbo: 120
Lamborghini Countach 25th
Anniversary: 431
32 178.0mph 286.4kmh
Chevrolet Corvette Callaway:
140
Mazda RX-7 Cartech: 514
Pontiac Firebird T/A Turbo
Pontiac Engineering: 683
33 177.0mph 284.8kmh
Ford Mustang Cartech Turbo:
302
34 176.0mph 283.2kmh
Isdera Imperator: 389
35 175.0mph 281.6kmh
Zender Vision 3: 1005

36 174.0mph 280.0kmh
Ferrari Testa Rossa: 259
Ford Mustang Holdener: 334
Opel Omega Lotus: 641
37 173.0mph 278.4kmh
Ford Mustang SVO J Bittle
American: 313
Lamborghini Countach 5000S:
427
38 172.0mph 276.7kmh
Chevrolet Corvette Guldstrand
Grand Sport 80: 154
39 171.0mph 275.1kmh
BMW 750i Alpina B12: 114
Chevrolet Corvette Guldstrand
Grand Sport 80: 165
Ferrari 348tb: 266
Porsche 911 Turbo: 752
40 170.0mph 273.5kmh
Bertone Emotion Lotus: 79
Chevrolet Corvette Callaway:
173
Chevrolet Corvette ZR-1: 167
Mazda RX-7 Mariah Mode Six:
516
Parradine V-12: 643
41 169.0mph 271.9kmh
Chevrolet Corvette Bakeracing
SCCA Escort: 151
42 168.0mph 270.3kmh
Acura NSX: 9
De Tomaso Pantera Group 3:
227
Porsche 911 Carrera Turbo:
748
43 166.0mph 267.1kmh
Nissan 300ZX Turbo Millen
Super GTZ: 627
44 165.0mph 265.5kmh
Aston Martin Zagato Volante:
39
De Tomaso Pantera GT5-S:
226
IAD Venus: 384
Lotus Esprit Turbo SE: 469
Porsche 911 Turbo Gemballa
Avalanche: 705
Porsche 928 S4: 719
Renault Alpine A610 Turbo:
791
TVR 420 SEAC: 913
45 164.0mph 263.9kmh
Ferrari 348tb: 262
46 163.0mph 262.3kmh
Chevrolet Corvette LT1: 188
47 162.0mph 260.7kmh
Honda NSX: 373
Porsche 928 S4 SE: 734
Porsche 944 Turbo: 737
48 161.0mph 259.0kmh
BMW 850i: 121
Lotus Esprit Turbo SE: 466

Porsche 911 Carrera RS: 754
Porsche 928 S4 Automatic:
701
Porsche 911 Carrera 4: 741
Porsche 911 Carrera 4: 727
Subaru Legacy FIA Record:
853
49 160.0mph 257.4kmh
Aston Martin Virage: 40
BMW 850i Manual: 125
Ford Thunderbird Super
Coupe Ford Engineering: 318
Mercedes-Benz 500SL: 550
Porsche 911 Carrera 2: 739
50 159.0mph 255.8kmh
Dodge Stealth R/T Turbo: 240
Mercedes-Benz 500E: 554
Mitsubishi 3000GT VR-4: 588
Porsche 911 Turbo: 743
Porsche Carrera 2 Cabriolet
Tiptronic: 747
Porsche 911 Turbo: 730
Porsche 911 Carrera 2
Tiptronic: 740
51 158.0mph 254.2kmh
BMW 750i L: 93
BMW 735i Koenig: 112
Chevrolet Corvette
Convertible: 153
Chevrolet Corvette Z51: 156
Honda NSX Auto: 376
Nissan 300ZX: 621
Porsche 911 Turbo Ruf 3.4:
706
Porsche 911 3.3 Turbo: 726
Porsche 944 Turbo: 704
52 157.0mph 252.6kmh
Porsche 911 Turbo Slant-
Nose: 707
53 156.0mph 251.0kmh
Lotus Esprit Turbo: 465
Nissan Skyline GT-R: 624
54 155.0mph 249.4kmh
Aston Martin Virage: 37
BMW M5: 122
BMW M5: 108
BMW 750i L: 100
Chevrolet Corvette: 150
Chevrolet Corvette L98: 166
Ferrari 328 GTB: 247
Mercedes-Benz 190E 2.5-16
Evolution II: 545
Mercedes-Benz 500SEL: 557
Mercedes-Benz 500SL: 555
Mercedes-Benz 600SEL: 558
Nissan 300ZX Turbo: 622
Nissan 300ZX Twin Turbo: 628
Pontiac Firebird TransAm 20th
Anniversary: 684
Porsche 911 Speedster: 714
Porsche 928 Cabrio Strosek:
732

Porsche 944 Turbo: 722
Porsche 911 Turbo: 715
55 154.0mph 247.8kmh
BMW 735i Alpina B11: 105
Chevrolet Corvette ASC Geneve: 147
Chevrolet Corvette Z51: 141
Chevrolet Corvette: 146
Chevrolet Corvette ZR-1: 183
Chevrolet Corvette: 139
Ferrari Mondial t Cabriolet: 269
Ferrari Mondial: 264
Ferrari Mondial t: 268
Lamborghini Jalpa: 429
Porsche 944 Turbo SE: 723
Porsche 911 Club Sport: 729
56 153.0mph 246.2kmh
Audi Coupe S2: 67
BMW 750i: 113
Lotus Esprit Turbo: 464
MVS Venturi GT: 598
Porsche 911 Turbo: 698
57 152.0mph 244.6kmh
Audi V8 Quattro: 68
Lotus Esprit Turbo: 459
Porsche 928 S: 700
Porsche 911 Carrera Club Sport: 712
Renault GTA V6 Turbo: 772
58 151.0mph 243.0kmh
Lynx XJS Eventer: 472
Mercedes-Benz 300CE: 546
Mercedes-Benz 560SEL: 526
59 150.0mph 241.4kmh
Audi 200 Quattro: 65
BMW M635 CSi: 109
Chevrolet Corvette Convertible: 148
Chevrolet Camaro IROC-Z: 144
Infiniti Q45: 386
Irmsher GT: 388
Jaguar XJS Convertible: 408
Jaguar XJS Koenig: 404
Jaguar XJSC HE: 399
Jaguar XJRS: 412
Lexus SC400: 455
Lynx D Type: 471
Mazda RX-7 Turbo: 486
Mazda RX-7 Turbo II: 506
Pontiac Firebird TransAm Turbo: 685
Pontiac Firebird GTA: 691
TVR V8 S: 915
60 149.0mph 239.7kmh
Chevrolet Corvette L98: 175
Chevrolet Camaro IROC-Z L98: 145
Chevrolet Corvette: 163
Ferrari 328 GTS: 250
Ferrari Mondial 3.2: 248
Ford Mustang Convertible Saleen: 303
Lancia Thema 8.32: 438

Mazda RX-7 Turbo: 517
Porsche 944 S2: 745
Porsche 911 Carrera Cabriolet: 728
Porsche 911 Carrera SE: 697
Porsche 944 S2 Cabriolet: 746
Porsche 911 Carrera: 711
Porsche 911 Speedster: 742
Porsche 911 Cabriolet: 710
Porsche 911 Club Sport: 713
61 148.0mph 238.1kmh
BMW M3 Evolution: 107
Ford Mustang GT: 298
Ford Mustang GT: 304
Ford Mustang LX 5.0L: 325
Lexus LS400: 454
Mercedes-Benz 300E-24: 549
Nissan 300ZX: 615
Saab Carlsson: 831
Vauxhall Carlton GSi 3000 24v: 937
62 147.0mph 236.5kmh
BMW M6: 101
BMW 735i Automatic: 89
Chevrolet Corvette Rick Mears: 182
Ferrari 412: 251
Jaguar XJS: 417
Porsche 944 S2: 736
63 146.0mph 234.9kmh
Bentley Turbo R: 76
Ford Sierra Sapphire Cosworth 4x4: 328
Jaguar XJS V12 Convertible: 405
Lotus Esprit Turbo HC: 462
Mercedes-Benz 560SEC Cabriolet Straman: 532
Mercedes-Benz 400SE: 553
Panther Solo: 642
Toyota Supra 3.0i Turbo: 894
64 145.0mph 233.3kmh
Aston Martin Volante: 35
Bentley Turbo R: 77
BMW 535i SE: 97
Chevrolet Camaro IROC-Z Automatic: 161
Chevrolet Camaro Z/28: 171
Chevrolet Camaro IROC-Z: 160
Citroen XM V6.24: 210
Ferrari Mondial Cabriolet: 254
Ferrari 3.2 Mondial: 246
Ferrari Mondial Cabriolet 3.2: 252
Ferrari Mondial 3.2 Cabriolet: 258
Ford Thunderbird LX: 340
Lamborghini Jalpa 3500: 428
Maserati 430: 476
Mercedes-Benz 560SEC: 525
Toyota MR2 Turbo: 900
65 144.0mph 231.7kmh
Alfa Romeo 164 Cloverleaf: 33

Audi V8 Quattro Automatic: 63
Audi V8 Quattro: 62
Ferrari Mondial 3.2QV: 249
Mercedes-Benz 190E 2.5-16: 538
Nissan 300ZX Turbo: 606
Peugeot 605 SVE 24: 672
TVR 390 SE: 911
66 143.0mph 230.1kmh
BMW 535i Automatic: 104
BMW 325i SE: 116
BMW M3: 106
BMW 535i: 103
Eagle Talon TSi 4WD: 245
Eagle Talon: 243
Ford Sierra RS Cosworth: 306
Honda Legend Coupe: 375
Jaguar XJR 4.0: 411
Lancia Thema 2.0ie 16v SE Turbo: 445
Mercury Cougar XR-7: 561
Mitsubishi Eclipse: 580
Plymouth Laser RS Turbo: 675
Plymouth Laser RS: 674
Subaru SVX: 858
67 142.0mph 228.5kmh
Audi Quattro 20v: 61
Honda Legend: 372
Maserati 228: 478
Saab 9000 2.3 Turbo TCS: 835
68 141.0mph 226.9kmh
Bitter Type 3 Cabrio: 81
BMW 525i SE 24v: 119
Jaguar XJS V12 Convertible: 409
Jaguar 4.0 Sovereign: 406
Jaguar XJSC: 398
Mercedes-Benz 300CE: 529
Nissan 200SX: 614
Porsche 944 S: 703
Renault GTA V6: 779
69 140.0mph 225.3kmh
Alfa Romeo 164 3.0 Lusso: 26
Alfa Romeo 164S: 31
BMW M3: 94
Chevrolet Camaro Z28: 187
Consulier GTP LX: 216
Dodge Spirit R/T: 239
Dodge Shelby CSX: 234
Ford Taurus SHO: 316
Ford Scorpio 24v: 338
Ford Taurus SHO: 339
Ford RS200: 292
Ford Thunderbird Super Coupe: 317
Ford Mustang LX: 343
Jaguar XJ6 3.6: 396
Jaguar XJS: 403
Lancia Thema 8.32: 440
Mazda RX-7 Turbo: 499
Mazda RX-7 Turbo: 510
Mercedes-Benz 300E: 543
Mercedes-Benz 300E: 522
Mercedes-Benz 190E AMG

Baby Hammer: 540
Nissan 200SX: 634
Porsche 944 S: 721
Renault 21 Turbo Quadra: 787
Saab 9000CD: 828
Toyota Supra Turbo: 889
Toyota Supra Turbo: 895
TVR S Convertible: 914
Vauxhall Senator 3.0i 24v: 944
Volkswagen Corrado: 968
Westfield SEight: 1000
70 139.0mph 223.7kmh
BMW Z1: 110
Ford Thunderbird Bondurant 5.0: 300
Mercedes-Benz 190E 2.3-16: 520
Porsche 944: 735
Renault 21 Turbo: 778
Toyota MR2 GT: 899
Vauxhall Calibra 2.0i 16v: 935
Volkswagen Corrado G60: 976
71 138.9mph 223.5kmh
Audi 90 Quattro 20v: 56
72 138.6mph 223.0kmh
Saab 9000CS Turbo S: 837
73 138.0mph 222.0kmh
Alfa Romeo Alfa 75 V6 3.0: 20
Jaguar XJS 4.0 Auto: 418
Mercedes-Benz 420SE: 524
Porsche 924 S: 718
Rover Vitesse: 811
Toyota Supra 3.0i: 873
74 137.0mph 220.4kmh
Alfa Romeo 164 Automatic: 27
BMW 735i: 99
Lotus Elan SE: 468
Lotus Elan SE: 470
Lotus Elan: 467
Mercedes-Benz 560SL: 544
Mercedes-Benz 300CE: 534
Mercedes-Benz 300CE Cabriolet Straman: 542
Opel Calibra 2.0i 16v: 640
Toyota Celica GT4: 884
Volkswagen Corrado G60: 973
75 136.0mph 218.8kmh
Alfa Romeo Milano Verde 3.0: 21
Audi Quattro: 54
Audi 100 2.8E Auto: 64
Jaguar XJ6 3.2: 416
Jaguar XJ6 Vanden Plas: 410
Mercedes-Benz 300E 4Matic: 535
Mitsubishi Starion 2000 Turbo: 574
Pontiac Firebird TransAm GTA: 689
Porsche 924 S: 708
76 135.0mph 217.2kmh
Acura Legend LS: 10
Aston Martin Volante: 38
Bentley Mulsanne Turbo: 74

BMW 325i Convertible: 83
BMW 635 CSi: 98
Chrysler TC by Maserati: 192
Citroen XM 3.0 SEi Auto: 209
Dodge Daytona Shelby: 237
Dodge Shelby CSX: 231
Ford Mustang GT 5.0: 289
Ford Mustang GT Convertible: 324
Jaguar XJS Convertible: 419
Jaguar XJS Railton: 415
Jaguar XJS: 413
Lexus ES300: 456
Mazda RX-7: 484
Mercedes-Benz 300SL-24 5-speed Auto: 552
Pontiac Firebird Formula: 688
Saab 9000 Turbo: 830
Saab 9000 Turbo: 836
Toyota MR2 Supercharged: 880
Toyota MR2 Supercharged: 893
Vauxhall Carlton 3000 GSi: 921
Volvo 960: 999
77 134.0mph 215.6kmh
Audi 80 2.8E V6 Quattro: 69
Bentley Mulsanne Turbo: 75
Ford Mustang SVO: 290
Jaguar XJS 3.6 Automatic: 402
Lotus Excel SE: 460
Mercedes-Benz 260E Auto: 533
Mercedes-Benz 190E 2.6: 528
Peugeot 405 Mi16: 653
Rover Vitesse: 808
Subaru Legacy 2.0 4Cam Turbo Estate: 856
Toyota Celica All-Trac Turbo: 883
Toyota Celica 2.0 GT: 896
Vauxhall Cavalier GSi 2000 16v: 938
78 133.0mph 214.0kmh
Alfa Romeo GTV6: 14
Alfa Romeo 75 Evoluzione: 23
BMW 325i S: 95
Citroen BX GTi 16v: 199
Honda Legend Coupe: 356
Lancia Dedra 2000 Turbo: 447
Nissan 300ZX Turbo: 600
Peugeot 605 SV 3.0 Auto: 671
Toyota Supra: 888
Toyota Supra: 872
Toyota Celica 2.0GT: 868
79 132.0mph 212.4kmh
Audi Coupe 2.2E: 57
Audi 80 16v Sport: 66
BMW 520i SE Touring: 124
BMW 730i SE: 92
Ford Fiesta RS Turbo: 319
Honda CRX 1.6i VT: 369
Jaguar XJS: 397

Nissan Sunny 2.0 GTi: 635
Nissan Primera 2.0E ZX: 631
Porsche 924 S: 699
Porsche 944: 720
Reliant Scimitar SST 1800Ti: 766
Rover Sterling Catalyst: 810
Rover 827 SLi Auto: 807
Vauxhall Calibra 4x4: 942
Vauxhall Senator 3.0i CD: 934
Vauxhall Carlton 3.0i CDX Estate: 936
Volkswagen Corrado 16v: 969
80 131.1mph 211.0kmh
Ford Escort RS2000: 341
81 131.0mph 210.8kmh
BMW 535i: 86
BMW 635 CSi: 87
Ford Thunderbird Turbo Coupe: 301
Ford Probe GT: 305
Ford Probe GT: 326
Mazda MX-6: 495
Mazda MX-6 GT: 496
Mercedes-Benz 420SEL: 531
MG Maestro Turbo: 566
Rover Sterling Automatic: 825
82 130.0mph 209.2kmh
Acura Legend Coupe: 8
Acura Integra GS: 7
Acura Legend Coupe L: 6
Alfa Romeo Alfa 75 2.5 Green Cloverleaf: 13
Audi 90 Quattro: 60
Bitter SC: 80
BMW 325i Cabriolet: 91
BMW 520i: 117
Cadillac Allante: 130
Cadillac Allante: 132
Dodge Shadow ES VNT: 238
Ford Probe LX: 327
Ford Sierra XR 4x4: 315
Honda Prelude Si: 374
Infiniti G20: 387
Isuzu Impulse Turbo: 391
Jaguar XJ6: 400
Lancia Delta HF Integrale: 439
Lancia Delta HF 4WD: 433
Maserati Spyder: 477
Maserati 430: 474
Mazda RX-7 GTUs: 505
Mazda Protege: 509
Mercedes-Benz 560SL: 527
Merkur Scorpio: 564
Mitsubishi Diamante LS: 590
Mitsubishi Sigma: 593
Mitsubishi Starion 2.6 Turbo: 583
Mitsubishi Galant VR-4: 591
Nissan NX2000: 629
Peugeot 405 Mi16: 660
Pontiac Firebird TransAm Convertible: 693
Porsche 911 Cabriolet: 696

Renault 19 16v: 790
Rover 825i: 802
Toyota Supra: 881
Treser T1: 910
TVR S Convertible: 912
Volkswagen Jetta GTi 16v: 959
83 129.0mph 207.6kmh
Acura Legend: 2
Alfa Romeo 33 Boxer 16v: 32
Alfa Romeo Milano Platinum: 15
Autokraft AC Cobra Mk IV: 73
Lancia Delta Integrale 16v: 441
Maserati Biturbo: 475
Mercedes-Benz 190E 2.6: 539
Renault 25 V6 2.9 Auto: 789
Renault 19 16v Cabriolet: 795
Rolls-Royce Corniche: 799
Saab 9000 2.3CS: 838
Subaru Legacy Sports Sedan: 857
Vauxhall Cavalier GSi 2000 16v 4x4: 939
Volkswagen Passat GT 16v: 972
Volvo 960 24v: 998
84 128.0mph 206.0kmh
Alfa Romeo Alfa 75 2.0i Twin Spark: 19
Audi 90 Quattro: 53
Bentley Turbo R: 78
BMW 525i: 118
BMW 325i: 123
Fiat Uno Turbo ie: 283
Honda Legend Saloon Automatic: 362
Honda Civic 1.6i VT: 371
Honda Prelude 2.0i 16: 358
Infiniti M30: 385
Lotus Excel SA: 463
Maserati Biturbo Spyder: 473
Mazda RX-7 GTU: 498
Mazda RX-7 GXL: 485
Mazda 323 1.8 GT: 501
Nissan 300ZX: 599
Pontiac Bonneville SSEi: 695
Pontiac Grand Prix McLaren Turbo: 687
Renault 25 TXi: 783
Rolls-Royce Silver Spirit II: 801
Rover 220 GTi: 821
Vauxhall Calibra 2.0i: 941
Vauxhall Senator 2.5i: 933
Volvo 740 Turbo: 986
85 127.0mph 204.3kmh
Alfa Romeo 164 Twin Spark Lusso: 30
Audi 90 2.2E: 48
Audi 90 Quattro: 49
BMW 325i X: 96
Chevrolet Corvette Morrison/Baker Nelson Ledges: 155

Chrysler Conquest TSi: 191
Honda Legend: 353
Honda Prelude 2.0 Si: 357
Honda Prelude Si 4WS: 365
Mercedes-Benz 300TE: 523
Mitsubishi Starion ESI-R: 584
Peugeot 405 Mi16x4: 664
Rover Sterling Automatic: 805
Saab CDS 2.3: 834
Toyota Celica GT Cabriolet: 876
Volvo 440 Turbo: 988
86 126.8mph 204.0kmh
Renault Cleo 16v: 796
87 126.0mph 202.7kmh
Alfa Romeo 75 2.5 Auto: 22
Audi Coupe GT 2.2: 46
Caterham 7 HPC: 135
Ford Taurus LX: 344
Ford Granada Scorpio 2.9 EX: 297
Honda Prelude Si 4WS: 377
Lancia Thema 2.0ie 16v: 442
Lancia Dedra 2.0ie: 444
Mazda MX-5 Miata Millen: 507
Morgan Plus 8: 595
Renault 5 GT Turbo: 769
Renault 21 TXi: 788
Rover 820i: 824
Rover 216 GTi: 819
Toyota Camry 2.2 GL: 903
Vauxhall Cavalier SRi: 932
Vauxhall Senator 3.0i CD: 924
Volvo 460 Turbo: 993
88 125.0mph 201.1kmh
Alfa Romeo Milano 3.0: 28
BMW 318i S: 111
Buick Reatta Convertible: 128
Buick Reatta: 127
Cadillac Seville Touring Sedan: 133
Cadillac Eldorado Touring Coupe: 131
Chevrolet Cavalier Z24: 172
Chevrolet Baretta GTZ: 177
Chevrolet Beretta GTZ: 169
Chevrolet Beretta GT: 168
Citroen ZX Volcane: 212
Dodge Daytona ES: 236
Dodge Daytona Shelby: 232
Eagle Premier ES: 242
Ford Escort XR3i: 342
Ford Crown Victoria LX: 329
Ford Granada Ghia X 2.9 EFi Saloon: 322
Ford Escort RS Turbo: 287
Honda CRX Coupe 1.6i 16: 349
Honda CRX Si: 360
Honda Civic CRX: 359
Honda CRX Si Jackson: 361
Honda Accord 2.0 EXi: 367
Isuzu Impulse Turbo: 393
Lincoln Continental Mk VII: 457

Lincoln Continental Mk VII LSC: 458
Mazda 323 Turbo 4x4 Lux: 481
Mazda MX-3 V6: 513
Mazda RX-7 Convertible: 497
Mercury Tracer LTS: 563
Morgan Plus 8: 596
Nissan 300ZX: 605
Nissan 240SX: 612
Nissan 300ZX: 613
Nissan Sentra SE-R: 632
Nissan Primera 2.0 GSX: 630
Oldsmobile Cutlass Calais International Series HO: 638
Oldsmobile Toronado Trofeo: 639
Oldsmobile Calais HO Quad 4: 637
Peugeot 309 GTi: 658
Peugeot 405 Turbo 16 Pike's Peak: 665
Peugeot 505 STX: 650
Plymouth Laser RS Turbo: 676
Pontiac Fiero GT: 677
Pontiac Sunbird GT: 682
Pontiac Fiero Formula: 679
Pontiac Grand Prix GTP: 694
Saab 9000CD: 832
Saab 900SPG: 833
Toyota Paseo: 909
Toyota Celica All-Trac Turbo: 897
Toyota Celica GTS: 898
Vauxhall Astra GTE 2.0i: 919
Volkswagen Passat GL: 975
Volkswagen Jetta GLi 16v: 978
Volkswagen Scirocco 16v: 954
Volkswagen Golf GTi 16v: 951
Volkswagen Scirocco 16v: 966
Volvo 740 Turbo: 990
Volvo 740 GLE: 989

89 124.0mph 199.5kmh
BMW 325 ES: 82
Ford Sierra Ghia 4x4 Estate: 293
Lamborghini LM129: 430
Mazda MX-6 GT: 508
Mazda 626 2.0i GT 4WS: 493
Mazda MX-6 GT 4WS: 504
Mazda 323 Turbo 4x4: 491
Mercury Capri XR-2: 562
Pontiac Grand Prix STE Turbo: 690
Reliant Scimitar Ti: 765
Saab 900 Turbo S: 829
Saab 9000: 826
Saturn Sports Coupe: 840
Sbarro Chrono 3.5: 841
Seat Toledo 2.0 GTi: 845
Vauxhall Cavalier 4x4: 927
Volvo 940 SE Turbo: 997
Volvo 940 SE: 996

90 123.9mph 199.4kmh
Audi Coupe Quattro: 58
91 123.0mph 197.9kmh
Audi 100 Avant Quattro: 50
BMW 318i: 115
Mazda 323 1.8 GT: 512
Peugeot 605 SRi: 666
Porsche 944: 702
Renault 11 Turbo: 773
Toyota MR2: 906
Toyota Celica GTS: 869
Vauxhall Carlton CD 2.0i: 917
92 122.0mph 196.3kmh
Audi 80 1.8E: 52
Audi Coupe Quattro: 47
Audi 100 Sport: 51
Citroen XM 2.0SEi: 207
Ford Granada 2.4i Ghia: 296
Lancia Delta HF Turbo ie: 434
Mitsubishi Lancer Liftback 1800 GTi-16v: 587
Mitsubishi Galant GTi-16v: 577
Morgan Plus 8: 597
Nissan Maxima: 616
Peugeot 309 SR Injection: 646
Peugeot 205 GTi: 644
Peugeot 309 GTi: 649
Renault 21 Ti: 777
Renault 21 Savanna GTX: 768
Renault 5 GT Turbo: 774
Toyota MR2 T-Bar: 871
Vauxhall Carlton L 1.8i: 925
Volkswagen GTi 16v: 961
Volkswagen GTi 16v: 977
Volkswagen GTi 16v: 958
Volkswagen Passat GT: 964
93 121.0mph 194.7kmh
Audi 80 2.0E: 59
BMW 520i: 84
Fiat Croma ie Super: 273
Ford Orion 1.6i Ghia: 336
Honda Concerto 1.6i 16: 368
Mazda 929: 494
Mazda 626 2.0i 5-door: 482
Mercedes-Benz 200: 521
Mitsubishi Colt 1800 GTI-16v: 589
Peugeot 405 GR Injection Auto: 663
Rover 416 GTi: 815
Saab 9000i: 827
Subaru XT Turbo 4WD: 850
Toyota MR2: 870
Volkswagen Golf GTi: 983
Volkswagen Jetta GLi 16v: 962
94 120.0mph 193.1kmh
Acura Integra RS: 1
Alfa Romeo Alfa 75 1.8: 12
Audi 90: 55
Chevrolet Camaro IROC-Z Convertible: 170
Chevrolet Cavalier Z24 Convertible: 149

Chevrolet Camaro Z/28 Convertible: 178
Chrysler Le Baron Coupe: 190
Dodge Shelby CSX VNT: 235
Dodge Shelby GLH-S: 228
Dodge Daytona Shelby Z: 229
Fiat Tipo 1.8ie DGT SX: 284
Ford Probe GT: 337
Ford Sierra Sapphire 2000E: 314
Ford Escort GT: 333
Ford Fiesta XR2i: 320
Geo Storm GSi: 345
Isuzu Impulse XS: 394
Isuzu Impulse RS: 395
Mazda 626 GT: 483
Mazda MX-3 GS: 519
Mercedes-Benz 190E 1.8 Auto: 551
Mercedes-Benz 200E Automatic: 541
Merkur XR4Ti: 565
Mitsubishi Galant GS: 581
Nissan Maxima SE: 617
Oldsmobile Cutlass Calais International Series: 636
Pontiac Grand Am Turbo: 686
Pontiac Bonneville SE: 678
Toyota Celica GT Convertible: 904
Vauxhall Belmont 1.8 GLSi: 916
Volkswagen Scirocco GTX 16v: 955
Volkswagen Polo G40: 980
Volvo 740 Turbo: 995
95 119.0mph 191.5kmh
BMW 735i: 88
Cadillac Allante: 129
Honda Integra EX 16: 352
Lancia Prisma 1600ie LX: 435
Mazda 323 GTX: 490
Nissan 200SX SE V6: 611
Rolls-Royce Silver Spirit: 798
Subaru Legacy 2.2 GX 4WD: 855
Toyota Camry V6 GXi: 890
Vauxhall Nova GTE: 928
Volkswagen Jetta GT: 952
Volvo 740 GLT 16v Estate: 994
96 118.0mph 189.9kmh
Alfa Romeo Alfa 33 1.7 Sportwagon Veloce: 18
Alfa Romeo 33 1.7 Veloce: 17
Chrysler Laser XT: 189
Dodge Lancer Shelby: 230
Ford Escort XR3i: 288
Jaguar XJ6 2.9: 401
Lancia Dedra 1.8i: 443
Lancia Thema 2.0ie: 436
Mitsubishi Mirage Turbo: 579
Nissan Sunny ZX Coupe: 609
Nissan Bluebird 1.8 ZX: 601
Peugeot 405 SRi: 654

Peugeot 106 XSi: 668
Renault Espace V6: 794
Rover 820SE: 803
Toyota Camry 2.0 GLi: 874
97 117.0mph 188.3kmh
Ford Sierra Sapphire 2.0 Ghia: 299
Mazda MX-5 Miata: 503
Peugeot 405 GLx4: 662
Rover 216 GSi: 813
Rover 825 TD: 822
Vauxhall Cavalier SRi 130 4-door: 923
98 116.0mph 186.6kmh
Alfa Romeo Sprint 1.7 Cloverleaf: 25
BMW 318i 2-door: 90
Citroen BX GTi 4x4: 206
Citroen CX25 Ri Familiale Auto: 194
Daihatsu Charade GT ti: 219
Fiat Croma CHT: 280
Ford Escort Cabriolet: 332
Ginetta G32: 346
Honda CRX Si: 350
Mazda 323 1.6i: 500
Mitsubishi Lancer GTi 16v: 578
Nissan 100NX: 625
Peugeot 605 SR TD: 670
Renault 9 Turbo: 762
Subaru 4WD Turbo Estate: 851
Suzuki Swift 1.3 GTi: 865
99 115.0mph 185.0kmh
Audi Coupe GT: 45
Audi 90: 44
Chevrolet Beretta GT: 143
Citroen AX GTi: 213
Citroen Visa GTi: 195
Dodge Shadow ES Turbo: 233
Ford Taurus 3.8: 307
Honda Civic Si: 364
Hyundai Sonata 2.4i GLS: 380
Lotus Elan Autocrosser: 461
Mercedes-Benz 300TD: 530
Mercedes-Benz 300D: 548
Mercury Sable LS: 559
Mitsubishi Cordia Turbo: 568
Mitsubishi Space Runner 1800-16v GLXi: 594
Mitsubishi Galant Sapporo: 572
Mitsubishi Cordia Turbo: 575
Nissan 240SX: 626
Nissan Pulsar NX SE: 608
Pontiac Bonneville SE: 681
Rover 820 Fastback: 806
Rover Metro GTi 16v: 817
Subaru XT6: 854
Toyota Corolla GTS: 892
Vauxhall Cavalier 2.0i CD: 922
100 114.0mph 183.4kmh
Audi 80 1.8S: 43
BMW 316i 4-door: 102
Ford Granada Scorpio 2.0i Auto: 323

Hyundai Lantra 1.6 Cdi: 381
Renault 21 GTS: 767
Rolls-Royce Silver Shadow II: 797
Subaru 4WD 3-door Turbo Coupe: 848
Volvo 760 GLE: 987
101 113.0mph 181.8kmh
Audi 5000 S: 42
Audi 100 Turbo Diesel: 41
Chevrolet Cavalier Z24: 137
Chevrolet Lumina Z34: 186
Citroen CX25 DTR Turbo 2: 200
Fiat Tempra 1.6 SX: 281
Mazda 626 2.0 GLS Executive: 488
Mitsubishi Galant 2000 GLSi Coupe: 585
Volkswagen Quantum Synchro Wagon: 953
102 112.8mph 181.5kmh
Nissan Primera 1.6 LS: 623
103 112.0mph 180.2kmh
Acura Integra LS: 5
Acura Integra LS Special Edition: 4
Austin Montego 2.0 Vanden Plas EFI: 70
Caterham Super 7: 136
Citroen XM Turbo SD Estate: 214
Honda Civic 1.5 VEi: 370
Mazda RX-7 Infini IV: 515
Mitsubishi Galant 2000 GLSi: 576
Nissan Sunny 1.6 GS 5-door: 633
Renault 19 TXE: 781
Renault 21 GTS Hatchback: 782
Renault 5 GTX 3-DR: 775
Suzuki Swift 1.3 GTi: 860
Toyota Camry 2.0 GLi Estate: 875
Vauxhall Belmont 1.6 GL: 920
Volvo 480 ES: 985
104 111.0mph 178.6kmh
Daihatsu Applause 16 Xi: 225
Renault 19 TXE Chamade: 786
Vauxhall Cavalier 1.7 GL TD: 946
Vauxhall Cavalier 1.6L: 931
105 110.0mph 177.0kmh
Acura Integra LS: 3
Alfa Romeo Spider Quadrifoglio: 16
Chevrolet Nova Twin Cam: 157
Chevrolet Celebrity CL Eurosport: 138
Chevrolet Baretta GTU: 158
Ford Fiesta 1.6S: 311
Mazda 626 2.0 GLX Executive Estate: 492

Mazda MPV: 502
Mercedes-Benz 190: 537
Range Rover Vogue SE: 763
Renault Alliance GTA: 776
Toyota Previa: 907
Toyota Carina II 1.6 GL Liftback: 891
106 109.0mph 175.4kmh
Alfa Romeo Alfa 33 1.3 S: 11
Citroen AX GT5: 205
Peugeot 106 XT: 673
Peugeot 405 GTD Turbo: 659
Peugeot 505 GTi Family Estate: 647
Range Rover County: 762
Renault Espace TXE: 785
Toyota Carina 1.6 GL: 882
Volvo 460 GLi: 992
107 108.0mph 173.8kmh
Citroen AX GT: 196
Ford Escort 1.4LX: 330
Ford Escort 1.6 Ghia Estate: 331
Honda Civic 1.4 GL 3-door: 355
Honda Accord LXi: 348
Hyundai Sonata 2.0i GLS: 379
Peugeot 405 1.6 GL: 651
Range Rover Vogue SE: 764
Renault 19 TSE: 780
Rolls-Royce Silver Spur: 800
Rover 214 GSi: 812
Seat Ibiza SXi: 843
Vauxhall 1.4i GLS: 945
Volkswagen Jetta Syncro: 971
108 107.0mph 172.2kmh
Citroen ZX 1.4 Avantage: 211
Fiat Regata 100S ie: 274
Fiat X1/9: 279
Fiat X1/9: 277
Fiat Tipo 1.6 DGT SX: 278
Ford Granada 2.5 GL Diesel: 321
Honda Shuttle 1.6i RT4 4WD: 363
Lancia Y10 GTie: 446
Land Rover Discovery V8i 5DR: 453
Renault Clio 1.4RT: 793
Toyota Previa LE: 901
Vauxhall Cavalier 1.4L: 930
Volkswagen Golf 1.8 GL: 982
Volvo 440 GLE: 991
Volvo 340 GLE: 984
109 106.0mph 170.6kmh
Citroen BX DTR Turbo: 202
Ford Orion 1.4 GL: 291
Range Rover Vogue: 760
Range Rover Vogue: 757
Rover Montego 1.6L: 804
Rover Metro 1.4SL: 816
Volkswagen Golf Cabriolet Clipper: 957
110 105.0mph 168.9kmh
Chevrolet Lumina APV: 176
Hyundai S Coupe GSi: 382

Mercury Tracer: 560
MG B British Motor Heritage: 567
Mitsubishi Galant 2000 GLS Automatic: 571
Mitsubishi Lancer GLXi 4WD Liftback: 586
Mitsubishi Shogun V6 LWB: 592
Nissan Sunny 1.6 SLX Coupe: 610
Nissan Sunny 1.4 LS: 620
Nissan Laurel 2.4 SGL Automatic: 602
Rover 414 Si: 814
Suzuki Swift GT: 867
Toyota Corolla FX-16 GTS: 878
Toyota Tercel LE: 908
Vauxhall Carlton L 2.3D: 926
111 104.0mph 167.3kmh
Austin Metro GTa: 72
Ford Escort 1.4L: 286
Mazda 323 LX: 480
Toyota Landcruiser VX: 905
Toyota Corolla Executive: 877
Volkswagen Passat GT 5-door: 960
Volkswagen Polo GT Coupe: 981
Volkswagen Jetta Turbo Diesel: 963
112 103.0mph 165.7kmh
Alfa Romeo Spider Quadrifoglio: 29
Alfa Romeo Spider: 24
Alfa Romeo Spider Veloce: 34
Peugeot 205 D Turbo: 669
Seat Toledo 1.9 CL Diesel: 844
Suzuki Swift 1.3 GLX: 864
Volkswagen Passat CL TD Estate: 974
113 102.0mph 164.1kmh
BMW 524 TD: 85
Citroen BX14 TGE: 204
Daihatsu Charade 1.3 CX: 222
Honda Integra 1.5: 351
Mazda MX-3 1.6 Auto: 518
Mitsubishi Shogun V6 5-door: 582
Peugeot 405 GRD: 652
Range Rover Vogue SE: 761
Rover 214S: 818
Rover Montego 2.0 DSL Turbo: 809
Suzuki Swift 1.3 GLX: 862
Vauxhall Nova 1.3 GL 5-door: 918
Vauxhall Nova 1.5TD Merit: 940
114 101.0mph 162.5kmh
Citroen AX14 TRS: 198
Citroen AX11 TRE: 197
Fiat Tipo 1.4: 276
Mazda 323 1.5 GLX: 487

Peugeot 205 CJ: 656
115 100.1mph 161.0kmh
Renault Clio 1.2 RN: 792
116 100.0mph 160.9kmh
Ford Fiesta 1.4 Ghia: 310
Honda Shuttle 1.4 GL Automatic: 366
Hyundai X2 1.5 GSi: 383
Isuzu Impulse Sports Coupe: 390
Nissan Sentra Sport Coupe SE: 603
Pontiac Le Mans: 680
Proton 1.5 SE Triple Valve Aeroback: 756
Volkswagen Fox GL: 956
117 99.0mph 159.3kmh
Honda Ballade EX: 354
Land Rover Discovery V8: 452
Peugeot 309 1.6 GR: 645
Peugeot 205 1.1GL: 655
Proton 1.5 SE Aeroback: 755
Renault 5 GTL 5-door: 770
Volkswagen Polo 1.3 CL Coupe: 979
Volkswagen Golf CL Catalyst: 970
Yugo Sana 1.4: 1004
118 98.0mph 157.7kmh
Citroen AX11 TRE 5DR: 201
Mazda 323 1.5 GLX Estate: 479
Nissan Prairie: 619
Rover Mini Cooper S: 823
119 97.0mph 156.1kmh
Fiat Uno 60S: 282
Ford Escort 1.3L 3-door: 308
Mitsubishi Lancer 1.5 GLX Estate: 569
Peugeot 106 XR: 667
Rover 218 SD: 820
Suzuki Swift 1.3 GLX Executive: 863
Toyota Corolla 1.3 GL: 885
Toyota Corolla 4WD Estate: 886
Yugo 65A GLX: 1003
120 96.0mph 154.5kmh
Daihatsu Charade Turbo: 221
Fiat Regata DS: 272
Lada Samara 1300 SL: 425
Peugeot 309 Style 5-door: 661
Vauxhall Astra 1.7 DL: 929
121 95.0mph 152.9kmh
Austin Metro 1.3L 3-door: 71
Chevrolet Nova CL: 142
Citroen AX11 TZX: 208
Daihatsu Charade CLS: 223
Ford Festiva L: 294
Mitsubishi Colt 1500 GLX 5-door: 570
Peugeot 309 GLX: 657
122 94.0mph 151.2kmh
Citroen BX19 RD Estate: 193

Lada Samara 1300L 5-door: 426

Lancia Y10 Fire: 437

Seat Malaga 1.2L: 842

Vauxhall Nova 1.2 Luxe: 943

123 93.0mph 149.6kmh

Ford Fiesta 1.1LX 5-door: 309

Nissan Micra 1.2 GSX: 618

Nissan Sunny 1.3 LX 5-door: 604

Range Rover Vogue Turbo D: 759

Subaru Justy GL II 5-door 4WD: 852

Volkswagen Caravelle Carat Automatic: 967

124 92.0mph 148.0kmh

Citroen AX14 DTR: 203

Hyundai Excel GLS: 378

Land Rover Ninety County V8: 450

Land Rover Discovery TDi: 451

Mazda 121 1.3 LX Sun Top: 489

Renault 5 Campus: 784

Skoda 136 Rapide Coupe: 847

125 91.0mph 146.4kmh

Daihatsu Sportrak EL: 224

Fiat Uno Selecta: 275

Mazda 121 GSX: 511

Subaru Justy: 849

Toyota Starlet GL: 902

126 90.0mph 144.8kmh

Ford Aerostar XLT: 285

Range Rover V8: 758

Suzuki Vitara JLX: 866

Yugo GV: 1001

127 89.0mph 143.2kmh

Fiat Panda 1000 S: 270

Lada Riva 1500 Estate: 423

Nissan Micra SGL 5-DR: 607

Volkswagen Polo C Saloon: 965

128 88.0mph 141.6kmh

Daihatsu Domino: 217

Ford Fiesta 1.1 Ghia Auto: 295

Land Rover One Ten County V8: 449

Skoda Estelle 120 LSE: 846

129 87.0mph 140.0kmh

Toyota Landcruiser: 879

130 85.1mph 137.0kmh

Vauxhall Frontera 2.3 TD Estate: 947

131 85.0mph 136.8kmh

Daihatsu Fourtrak Estate DT EL: 220

132 84.0mph 135.2kmh

Mercedes-Benz G-Wagen 300 GD LWB: 556

133 83.0mph 133.5kmh

Mitsubishi Shogun Turbo Diesel 5-door: 573

Yugo 45 GLS: 1002

134 82.0mph 131.9kmh

Volkswagen Caravelle Syncro: 950

135 81.0mph 130.3kmh

Carbodies Fairway 2.7 Silver: 134

Daihatsu Charade CX Diesel Turbo: 218

Lada Niva Cossack Cabrio: 424

Toyota Landcruiser II TD: 887

136 80.0mph 128.7kmh

Fiat Panda 750 L: 271

Isuzu Trooper 3-door TD: 392

Suzuki Alto GLA: 859

137 76.0mph 122.3kmh

Land Rover Ninety County Turbo Diesel: 448

Suzuki SJ413V JX: 861

138 75.0mph 120.7kmh

GM Impact: 347

Maximum speed in make order

Acur

1   168.0mph 270.3kmh
NSX: 9
2   135.0mph 217.2kmh
Legend LS: 10
3   130.0mph 209.2kmh
Legend Coupe: 8
Integra GS: 7
Legend Coupe L: 6
4   129.0mph 207.6kmh
Legend: 2
5   120.0mph 193.1kmh
Integra RS: 1
6   112.0mph 180.2kmh
Integra LS: 5
Integra LS Special Edition: 4
7   110.0mph 177.0kmh
Integra LS: 3

Alfa Romeo

1   144.0mph 231.7kmh
164 Cloverleaf: 33
2   140.0mph 225.3kmh
164S: 31
164 3.0 Lusso: 26
3   138.0mph 222.0kmh
Alfa 75 V6 3.0: 20
4   137.0mph 220.4kmh
164 Automatic: 27
5   136.0mph 218.8kmh
Milano Verde 3.0: 21
6   133.0mph 214.0kmh
75 Evoluzione: 23
GTV6: 14
7   130.0mph 209.2kmh
Alfa 75 2.5 Green Cloverleaf:
13
8   129.0mph 207.6kmh
Milano Platinum: 15
33 Boxer 16v: 32
9   128.0mph 206.0kmh
Alfa 75 2.0i Twin Spark: 19
10   127.0mph 204.3kmh
164 Twin Spark Lusso: 30
11   126.0mph 202.7kmh
75 2.5 Auto: 22
12   125.0mph 201.1kmh
Milano 3.0: 28
13   120.0mph 193.1kmh
Alfa 75 1.8: 12
14   118.0mph 189.9kmh
Alfa 33 1.7 Sportwagon
Veloce: 18
33 1.7 Veloce: 17
15   116.0mph 186.6kmh
Sprint 1.7 Cloverleaf: 25
16   110.0mph 177.0kmh
Spider Quadrifoglio: 16
17   109.0mph 175.4kmh
Alfa 33 1.3 S: 11

18   103.0mph 165.7kmh
Spider Quadrifoglio: 29
Spider: 24
Spider Veloce: 34

Aston Martin

1   183.0mph 294.4kmh
Zagato: 36
2   165.0mph 265.5kmh
Zagato Volante: 39
3   160.0mph 257.4kmh
Virage: 40
4   155.0mph 249.4kmh
Virage: 37
5   145.0mph 233.3kmh
Volante: 35
6   135.0mph 217.2kmh
Volante: 38

Audi

1   153.0mph 246.2kmh
Coupe S2: 67
2   152.0mph 244.6kmh
V8 Quattro: 68
3   150.0mph 241.4kmh
200 Quattro: 65
4   144.0mph 231.7kmh
V8 Quattro Automatic: 63
V8 Quattro: 62
5   142.0mph 228.5kmh
Quattro 20v: 61
6   138.9mph 223.5kmh
90 Quattro 20v: 56
7   136.0mph 218.8kmh
100 2.8E Auto: 64
Quattro: 54
8   134.0mph 215.6kmh
80 2.8E V6 Quattro: 69
9   132.0mph 212.4kmh
80 16v Sport: 66
Coupe 2.2E: 57
10   130.0mph 209.2kmh
90 Quattro: 60
11   128.0mph 206.0kmh
90 Quattro: 53
12   127.0mph 204.3kmh
90 2.2E: 48
90 Quattro: 49
13   126.0mph 202.7kmh
Coupe GT 2.2: 46
14   123.9mph 199.4kmh
Coupe Quattro: 58
15   123.0mph 197.9kmh
100 Avant Quattro: 50
16   122.0mph 196.3kmh
100 Sport: 51
80 1.8E: 52
Coupe Quattro: 47
17   121.0mph 194.7kmh
80 2.0E: 59

18   120.0mph 193.1kmh
90: 55
19   115.0mph 185.0kmh
Coupe GT: 45
90: 44
20   114.0mph 183.4kmh
80 1.8S: 43
21   113.0mph 181.8kmh
5000 S: 42
100 Turbo Diesel: 41

Austin

1   112.0mph 180.2kmh
Montego 2.0 Vanden Plas EFI:
70
2   104.0mph 167.3kmh
Metro GTa: 72
395.0mph 152.9kmh
Metro 1.3L 3-door: 71

Autokraft

1   129.0mph 207.6kmh
Autokraft AC Cobra Mk IV: 73

Bentley

1   146.0mph 234.9kmh
Turbo R: 76
2   145.0mph 233.3kmh
Turbo R: 77
3   135.0mph 217.2kmh
Mulsanne Turbo: 74
4   134.0mph 215.6kmh
Mulsanne Turbo: 75
5   128.0mph 206.0kmh
Turbo R: 78

Bertone

1   170.0mph 273.5kmh
Emotion Lotus: 79

Bitter

1   141.0mph 226.9kmh
Type 3 Cabrio: 81
2   130.0mph 209.2kmh
SC: 80

BMW

1   179.0mph 288.0kmh
535i Alpina B10 Bi-Turbo: 120
2   171.0mph 275.1kmh
750i Alpina B12: 114
3   161.0mph 259.0kmh
850i: 121
4   160.0mph 257.4kmh
850i Manual: 125
5   158.0mph 254.2kmh

735i Koenig: 112
750i L: 93
6   155.0mph 249.4kmh
M5: 122
750i L: 100
M5: 108
7   154.0mph 247.8kmh
735i Alpina B11: 105
8   153.0mph 246.2kmh
750i: 113
9   150.0mph 241.4kmh
M635 CSi: 109
10   148.0mph 238.1kmh
M3 Evolution: 107
11   147.0mph 236.5kmh
735i Automatic: 89
M6: 101
12   145.0mph 233.3kmh
535i SE: 97
13   143.0mph 230.1kmh
325i SE: 116
535i Automatic: 104
535i: 103
M3: 106
14   141.0mph 226.9kmh
525i SE 24v: 119
15   140.0mph 225.3kmh
M3: 94
16   139.0mph 223.7kmh
Z1: 110
17   137.0mph 220.4kmh
735i: 99
18   135.0mph 217.2kmh
635 CSi: 98
19   133.0mph 214.0kmh
325i Convertible: 83
325i S: 95
20   132.0mph 212.4kmh
730i SE: 92
520i SE Touring: 124
21   131.0mph 210.8kmh
635 CSi: 87
535i: 86
22   130.0mph 209.2kmh
325i Cabriolet: 91
520i: 117
23   128.0mph 206.0kmh
525i: 118
325i: 123
24   127.0mph 204.3kmh
325i X: 96
25   125.0mph 201.1kmh
318i S: 111
26   124.0mph 199.5kmh
325 ES: 82
27   123.0mph 197.9kmh
318i: 115
28   121.0mph 194.7kmh
520i: 84
29   119.0mph 191.5kmh
735i: 88
30   116.0mph 186.6kmh

318i 2-door: 90
31 114.0mph 183.4kmh
316i 4-door: 102
32 102.0mph 164.1kmh
524 TD: 85

Bugatti

1 214.0mph 344.3kmh
EB110: 126

Buick. 2

1 125.0mph 201.1kmh
Reatta Convertible: 128
Reatta: 127

Cadillac

1 130.0mph 209.2kmh
Allante: 130
Allante: 132
2 125.0mph 201.1kmh
Seville Touring Sedan: 133
Eldorado Touring Coupe: 131
3 119.0mph 191.5kmh
Allante: 129

Carbodies

181.0mph 130.3kmh
Fairway 2.7 Silver: 134

Caterham

1 126.0mph 202.7kmh
7 HPC: 135
2 112.0mph 180.2kmh
Super 7: 136

Chevrolet

1 255.0mph 410.3kmh
Corvette Callaway
Sledgehammer: 174
2 204.0mph 328.2kmh
Corvette Lingenfelter: 181
3 203.0mph 326.6kmh
Camaro Gottlieb 1969: 159
4 201.0mph 323.4kmh
Corvette Callaway Speedster: 180
5 193.0mph 310.5kmh
Camaro IROC-Z Chevrolet Engineering: 162
6 191.0mph 307.3kmh
Corvette Callaway: 164
Corvette Callaway: 152
7 186.0mph 299.3kmh
Corvette ZR-1 Geiger: 184
8 185.0mph 297.7kmh
Corvette ZR-2: 185
9 184.0mph 2S5.1kmh
Corvette Callaway: 179

10 178.0mph 286.4kmh
Corvette Callaway: 140
11 172.0mph 276.7kmh
Corvette Guldstrand Grand Sport 80: 154
12 171.0mph 275.1kmh
Corvette Guldstrand Grand Sport 80: 165
13 170.0mph 273.5kmh
Corvette Callaway: 173
Corvette ZR-1: 167
14 169.0mph 271.9kmh
Corvette Bakeracing SCCA Escort: 151
15 163.0mph 262.3kmh
Corvette LT1: 188
16 158.0mph 254.2kmh
Corvette Convertible: 153
Corvette Z51: 156
17 155.0mph 249.4kmh
Corvette: 150
Corvette L98: 166
18 154.0mph 247.8kmh
Corvette ASC Geneve: 147
Corvette: 139
Corvette Z51: 141
Corvette: 146
Corvette ZR-1: 183
19 150.0mph 241.4kmh
Corvette Convertible: 148
Camaro IROC-Z: 144
20 149.0mph 239.7kmh
Corvette L98: 175
Corvette: 163
Camaro IROC-Z L98: 145
21 147.0mph 236.5kmh
Corvette Rick Mears: 182
22 145.0mph 233.3kmh
Camaro IROC-Z Automatic: 161
Camaro Z/28: 171
Camaro IROC-Z: 160
23 140.0mph 225.3kmh
Camaro Z28: 187
24 127.0mph 204.3kmh
Corvette Morrison/Baker Nelson Ledges: 155
25 125.0mph 201.1kmh
Beretta GTZ: 169
Beretta GT: 168
Cavalier Z24: 172
Baretta GTZ: 177
26 120.0mph 193.1kmh
Cavalier Z24 Convertible: 149
Camaro IROC-Z Convertible: 170
Camaro Z/28 Convertible: 178
27 115.0mph 185.0kmh
Beretta GT: 143
28 113.0mph 181.8kmh
Lumina Z34: 186
Cavalier Z24: 137
29 110.0mph 177.0kmh
Nova Twin Cam: 157

Baretta GTU: 158
Celebrity CL Eurosport: 138
30 105.0mph 168.9kmh
Lumina APV: 176
31 95.0mph 152.9kmh
Nova CL: 142

Chrysler

1 135.0mph 217.2kmh
TC by Maserati: 192
2 127.0mph 204.3kmh
Conquest TSi: 191
3 120.0mph 193.1kmh
Le Baron Coupe: 190
4 118.0mph 189.9kmh
Laser XT: 189

Citroen

1 145.0mph 233.3kmh
XM V6.24: 210
2 135.0mph 217.2kmh
XM 3.0 SEi Auto: 209
3 133.0mph 214.0kmh
BX GTi 16v: 199
4 125.0mph 201.1kmh
ZX Volcane: 212
5 122.0mph 196.3kmh
XM 2.0SEi: 207
6 116.0mph 186.6kmh
BX GTi 4x4: 206
CX25 Ri Familiale Auto: 194
7 115.0mph 185.0kmh
Visa GTi: 195
AX GTi: 213
8 113.0mph 181.8kmh
CX25 DTR Turbo 2: 200
9 112.0mph 180.2kmh
XM Turbo SD Estate: 214
10 109.0mph 175.4kmh
AX GT5: 205
11 108.0mph 173.8kmh
AX GT: 196
12 107.0mph 172.2kmh
ZX 1.4 Avantage: 211
13 106.0mph 170.6kmh
BX DTR Turbo: 202
14 102.0mph 164.1kmh
BX14 TGE: 204
15 101.0mph 162.5kmh
AX14 TRS: 198
AX11 TRE: 197
16 98.0mph 157.7kmh
AX11 TRE 5DR: 201
17 95.0mph 152.9kmh
AX11 TZX: 208
18 94.0mph 151.2kmh
BX19 RD Estate: 193
19 92.0mph 148.0kmh
AX14 DTR: 203

Cizeta

1 204.0mph 328.2kmh
Moroder V16T: 215

Consulier

1 140.0mph 225.3kmh
GTP LX: 216

Daihatsu

1 116.0mph 186.6kmh
Charade GT ti: 219
2 111.0mph 178.6kmh
Applause 16 Xi: 225
3 102.0mph 164.1kmh
Charade 1.3 CX: 222
496.0mph 154.5kmh
Charade Turbo: 221
595.0mph 152.9kmh
Charade CLS: 223
691.0mph 146.4kmh
Sportrak EL: 224
788.0mph 141.6kmh
Domino: 217
885.0mph 136.8kmh
Fourtrak Estate DT EL: 220
981.0mph 130.3kmh
Charade CX Diesel Turbo: 218

De Tomaso

1 168.0mph 270.3kmh
Pantera Group 3: 227
2 165.0mph 265.5kmh
Pantera GT5-S: 226

Dodge

1 190.0mph 305.7kmh
Viper: 241
2 159.0mph 255.8kmh
Stealth R/T Turbo: 240
3 140.0mph 225.3kmh
Spirit R/T: 239
Shelby CSX: 234
4 135.0mph 217.2kmh
Daytona Shelby: 237
Shelby CSX: 231
5 130.0mph 209.2kmh
Shadow ES VNT: 238
6 125.0mph 201.1kmh
Daytona Shelby: 232
Daytona ES: 236
7 120.0mph 193.1kmh
Shelby GLH-S: 228
Shelby CSX VNT: 235
Daytona Shelby Z: 229
8 118.0mph 189.9kmh
Lancer Shelby: 230
9 115.0mph 185.0kmh
Shadow ES Turbo: 233

Eagle

1   200.0mph 321.8kmh
GTP: 244
2   143.0mph 230.1kmh
Talon TSi 4WD: 245
Talon: 243
3   125.0mph 201.1kmh
Premier ES: 242

Ferrari

1   210.0mph 337.9kmh
Testa Rossa Norwood: 261
2   200.0mph 321.8kmh
F40: 263
3   198.0mph 318.6kmh
308 Norwood Bonneville GTO:
253
4   196.0mph 315.4kmh
F40: 267
5   185.0mph 297.7kmh
Testa Rossa Gemballa: 260
6   182.0mph 292.8kmh
308 Norwood: 257
7   181.0mph 291.2kmh
Testa Rossa: 255
Testa Rossa Straman Spyder:
256
8   180.0mph 289.6kmh
Pininfarina Mythos: 265
9   174.0mph 280.0kmh
Testa Rossa: 259
10 171.0mph 275.1kmh
348tb: 266
11 164.0mph 263.9kmh
348tb: 262
12 155.0mph 249.4kmh
328 GTB: 247
13 154.0mph 247.8kmh
Mondial: 264
Mondial t: 268
Mondial t Cabriolet: 269
14 149.0mph 239.7kmh
Mondial 3.2: 248
328 GTS: 250
15 147.0mph 236.5kmh
412: 251
16 145.0mph 233.3kmh
3.2 Mondial: 246
Mondial Cabriolet 3.2: 252
Mondial Cabriolet: 254
Mondial 3.2 Cabriolet: 258
17 144.0mph 231.7kmh
Mondial 3.2QV: 249

Fiat

1   128.0mph 206.0kmh
Uno Turbo ie: 283
2   121.0mph 194.7kmh
Croma ie Super: 273
3   120.0mph 193.0kmh
Tipo 1.8ie DGT SX: 284

4   116.0mph 186.6kmh
Croma CHT: 280
5   113.0mph 181.8kmh
Tempra 1.6 SX: 281
6   107.0mph 172.2kmh
X1/9: 279
Tipo 1.6 DGT SX: 278
Regata 100S ie: 274
X1/9: 277
7   101.0mph 162.5kmh
Tipo 1.4: 276
897.0mph 156.1kmh
Uno 60S: 282
996.0mph 154.5kmh
Regata DS: 272
10 91.0mph 146.4kmh
Uno Selecta: 275
11 89.0mph 143.2kmh
Panda 1000 S: 270
12 80.0mph 128.7kmh
Panda 750 L: 271

Ford

1   186.0mph 299.3kmh
Mustang NOS/Saleen: 335
2   185.0mph 297.7kmh
Mustang 5.0 Cartech Turbo:
312
3   177.0mph 284.8kmh
Mustang Cartech Turbo: 302
4   174.0mph 280.0kmh
Mustang Holdener: 334
5   173.0mph 278.4kmh
Mustang SVO J Bittle
American: 313
6   160.0mph 257.4kmh
Thunderbird Super Coupe
Ford Engineering: 318
7   149.0mph 239.7kmh
Mustang Convertible Saleen:
303
8   148.0mph 238.1kmh
Mustang LX 5.0L: 325
Mustang GT: 298
Mustang GT: 304
9   146.0mph 234.9kmh
Sierra Sapphire Cosworth 4x4:
328
10 145.0mph 233.3kmh
Thunderbird LX: 340
11 143.0mph 230.1kmh
Sierra RS Cosworth: 306
12 140.0mph 225.3kmh
RS200: 292
Taurus SHO: 316
Taurus SHO: 339
Scorpio 24v: 338
Thunderbird Super Coupe:
317
Mustang LX: 343
13 139.0mph 223.7kmh
Thunderbird Bondurant 5.0:
300

14 135.0mph 217.2kmh
Mustang GT Convertible: 324
Mustang GT 5.0: 289
15 134.0mph 215.6kmh
Mustang SVO: 290
16 132.0mph 212.4kmh
Fiesta RS Turbo: 319
17 131.1mph 211.0kmh
Escort RS2000: 341
18 131.0mph 210.8kmh
Probe GT: 326
Probe GT: 305
Thunderbird Turbo Coupe:
301
19 130.0mph 209.2kmh
Probe LX: 327
Sierra XR 4x4: 315
20 126.0mph 202.7kmh
Taurus LX: 344
Granada Scorpio 2.9 EX: 297
21 125.0mph 201.1kmh
Crown Victoria LX: 329
Granada Ghia X 2.9 EFi
Saloon: 322
Escort XR3i: 342
Escort RS Turbo: 287
22 124.0mph 199.5kmh
Sierra Ghia 4x4 Estate: 293
23 122.0mph 196.3kmh
Granada 2.4i Ghia: 296
24 121.0mph 194.7kmh
Orion 1.6i Ghia: 336
25 120.0mph 193.1kmh
Sierra Sapphire 2000E: 314
Fiesta XR2i: 320
Escort GT: 333
Probe GT: 337
26 118.0mph 189.9kmh
Escort XR3i: 288
27 117.0mph 188.3kmh
Sierra Sapphire 2.0 Ghia: 299
28 116.0mph 186.6kmh
Escort Cabriolet: 332
29 115.0mph 185.0kmh
Taurus 3.8: 307
30 114.0mph 183.4kmh
Granada Scorpio 2.0i Auto:
323
31 110.0mph 177.0kmh
Fiesta 1.6S: 311
32 108.0mph 173.8kmh
Escort 1.6 Ghia Estate: 331
Escort 1.4LX: 330
33 107.0mph 172.2kmh
Granada 2.5 GL Diesel: 321
34 106.0mph 170.6kmh
Orion 1.4 GL: 291
35 104.0mph 167.3kmh
Escort 1.4 GL: 286
36 100.0mph 160.9kmh
Fiesta 1.4 Ghia: 310
37 97.0mph 156.1kmh
Escort 1.3L 3-door: 308
38 95.0mph 152.9kmh

Festiva L: 294
39 93.0mph 149.6kmh
Fiesta 1.1LX 5-door: 309
40 90.0mph 144.8kmh
Aerostar XLT: 285
41 88.0mph 141.6kmh
Fiesta 1.1 Ghia Auto: 295

Geo

1   120.0mph 193.1kmh
Storm GSi: 345

Ginetta

1   116.0mph 186.6kmh
G32: 346

GM

175.0mph 120.7kmh
Impact: 347

Honda

1   162.0mph 260.7kmh
NSX: 373
2   158.0mph 254.2kmh
NSX Auto: 376
3   143.0mph 230.1kmh
Legend Coupe: 375
4   142.0mph 228.5kmh
Legend: 372
5   133.0mph 214.0kmh
Legend Coupe: 356
6   132.0mph 212.4kmh
CRX 1.6i VT: 369
7   130.0mph 209.2kmh
Prelude Si: 374
8   128.0mph 206.0kmh
Civic 1.6i VT: 371
Prelude 2.0i 16: 358
Legend Saloon Automatic: 362
9   127.0mph 204.3kmh
Prelude Si 4WS: 365
Prelude 2.0 Si: 357
Legend: 353
10 126.0mph 202.7kmh
Prelude Si 4WS: 377
11 125.0mph 201.1kmh
CRX Coupe 1.6i 16: 349
Civic CRX: 359
CRX Si Jackson: 361
CRX Si: 360
Accord 2.0 EXi: 367
12 121.0mph 194.7kmh
Concerto 1.6i 16: 368
13 119.0mph 191.5kmh
Integra EX 16: 352
14 116.0mph 186.6kmh
CRX Si: 350
15 115.0mph 185.0kmh
Civic Si: 364
16 112.0mph 180.2kmh

Civic 1.5 VEi: 370
17 108.0mph 173.8kmh
Civic 1.4 GL 3-door: 355
Accord LXi: 348
18 107.0mph 172.2kmh
Shuttle 1.6i RT4 4WD: 363
19 102.0mph 164.1kmh
Integra 1.5: 351
20 100.0mph 160.9kmh
Shuttle 1.4 GL Automatic: 366
21 99.0mph 159.3kmh
Ballade EX: 354

Hyundai

1 115.0mph 185.0kmh
Sonata 2.4i GLS: 380
2 114.0mph 183.4kmh
Lantra 1.6 Cdi: 381
3 108.0mph 173.8kmh
Sonata 2.0i GLS: 379
4 105.0mph 168.9kmh
S Coupe GSi: 382
5 100.0mph 160.9kmh
X2 1.5 GSi: 383
692.0mph 148.0kmh
Excel GLS: 378

IAD

1 165.0mph 265.5kmh
Venus: 384

Infiniti

1 150.0mph 241.4kmh
Q45: 386
2 130.0mph 209.2kmh
G20: 387
3 128.0mph 206.0kmh
M30: 385

Irmsher

1 150.0mph 241.4kmh
GT: 388

Isdera

1 176.0mph 283.2kmh
Imperator: 389

Isuzu

1 130.0mph 209.2kmh
Impulse Turbo: 391
2 125.0mph 201.1kmh
Impulse Turbo: 393
3 120.0mph 193.1kmh
Impulse XS: 394
Impulse RS: 395
4 100.0mph 160.9kmh
Impulse Sports Coupe: 390
580.0mph 128.7kmh

Trooper 3-door TD: 392

Jaguar

1 200.0mph 321.8kmh
XJ220: 407
XJS Lister Le Mans: 414
2 150.0mph 241.4kmh
XJS Convertible: 408
XJSC HE: 399
XJS Koenig: 404
XJRS: 412
3 147.0mph 236.5kmh
XJS: 417
4 146.0mph 234.9kmh
XJS V12 Convertible: 405
5 143.0mph 230.1kmh
XJR 4.0: 411
6 141.0mph 226.9kmh
4.0 Sovereign: 406
XJS V12 Convertible: 409
XJSC: 398
7 140.0mph 225.3kmh
XJ6 3.6: 396
XJS: 403
8 138.0mph 222.0kmh
XJS 4.0 Auto: 418
9 136.0mph 218.8kmh
XJ6 Vanden Plas: 410
XJ6 3.2: 416
10 135.0mph 217.2kmh
XJS: 413
XJS Railton: 415
XJS Convertible: 419
11 134.0mph 215.6kmh
XJS 3.6 Automatic: 402
12 132.0mph 212.4kmh
XJS: 397
13 130.0mph 209.2kmh
XJ6: 400
14 118.0mph 189.9kmh
XJ6 2.9: 401

Koenig

1 235.0mph 378.1kmh
C62: 422
2 230.0mph 370.1kmh
Competition Evolution: 421
3 217.0mph 349.2kmh
Competition: 420

Lada

1 96.0mph 154.5kmh
Samara 1300 SL: 425
2 94.0mph 151.2kmh
Samara 1300L 5-door: 426
3 89.0mph 143.2kmh
Riva 1500 Estate: 423
4 81.0mph 130.3kmh
Niva Cossack Cabrio: 424

Lamborghini

1 202.0mph 325.0kmh
Diablo: 432
2 179.0mph 288.0kmh
Countach 25th Anniversary:
431
3 173.0mph 278.4kmh
Countach 5000S: 427
4 154.0mph 247.8kmh
Jalpa: 429
5 145.0mph 233.3kmh
Jalpa 3500: 428
6 124.0mph 199.5kmh
LM129: 430

Lancia

1 149.0mph 239.7kmh
Thema 8.32: 438
2 143.0mph 230.1kmh
Thema 2.0ie 16v SE Turbo:
445
3 140.0mph 225.3kmh
Thema 8.32: 440
4 133.0mph 214.0kmh
Dedra 2000 Turbo: 447
5 130.0mph 209.2kmh
Delta HF Integrale: 439
Delta HF 4WD: 433
6 129.0mph 207.6kmh
Delta Integrale 16v: 441
7 126.0mph 202.7kmh
Thema 2.0ie 16v: 442
Dedra 2.0ie: 444
8 122.0mph 196.3kmh
Delta HF Turbo ie: 434
9 119.0mph 191.5kmh
Prisma 1600ie LX: 435
10 118.0mph 189.9kmh
Dedra 1.8i: 443
Thema 2.0ie: 436
11 107.0mph 172.2kmh
Y10 GTie: 446
12 94.0mph 151.2kmh
Y10 Fire: 437

Land Rover

1 107.0mph 172.2kmh
Discovery V8i 5DR: 453
2 99.0mph 159.3kmh
Discovery V8: 452
3 92.0mph 148.0kmh
Discovery TDi: 451
Ninety County V8: 450
4 88.0mph 141.6kmh
Rover One Ten County V8:
449
5 76.0mph 122.3kmh
Ninety County Turbo Diesel:
448

Lexus

1 150.0mph 241.4kmh
SC400: 455
2 148.0mph 238.1kmh
LS400: 454
3 135.0mph 217.2kmh
ES300: 456

Lincoln

1 125.0mph 201.1kmh
Continental Mk VII LSC: 458
Continental Mk VII: 457

Lotus

1 165.0mph 265.5kmh
Esprit Turbo SE: 469
2 161.0mph 259.0kmh
Esprit Turbo SE: 466
3 156.0mph 251.0kmh
Esprit Turbo: 465
4 153.0mph 246.2kmh
Esprit Turbo: 464
5 152.0mph 244.6kmh
Esprit Turbo: 459
6 146.0mph 234.9kmh
Esprit Turbo HC: 462
7 137.0mph 220.4kmh
Elan SE: 470
Elan SE: 468
Elan: 467
8 134.0mph 215.6kmh
Excel SE: 460
9 128.0mph 206.0kmh
Excel SA: 463
10 115.0mph 185.0kmh
Elan Autocrosser: 461

Lynx

1 151.0mph 243.0kmh
XJS Eventer: 472
2 150.0mph 241.4kmh
D Type: 471

Maserati

1 145.0mph 233.3kmh
430: 476
2 142.0mph 228.5kmh
228: 478
3 130.0mph 209.2kmh
430: 474
Spyder: 477
4 129.0mph 207.6kmh
Biturbo: 475
5 128.0mph 206.0kmh
Biturbo Spyder: 473

Mazda

1 178.0mph 286.4kmh

RX-7 Cartech: 514
2 170.0mph 273.5kmh
RX-7 Mariah Mode Six: 516
3 150.0mph 241.4kmh
RX-7 Turbo: 486
RX-7 Turbo II: 506
4 149.0mph 239.7kmh
RX-7 Turbo: 517
5 140.0mph 225.3kmh
RX-7 Turbo: 499
RX-7 Turbo: 510
6 135.0mph 217.2kmh
RX-7: 484
7 131.0mph 210.8kmh
MX-6 GT: 496
MX-6: 495
8 130.0mph 209.2kmh
Protege: 509
RX-7 GTUs: 505
9 128.0mph 206.0kmh
RX-7 GXL: 485
323 1.8 GT: 501
RX-7 GTU: 498
10 126.0mph 202.7kmh
MX-5 Miata Millen: 507
11 125.0mph 201.1kmh
MX-3 V6: 513
323 Turbo 4x4 Lux: 481
RX-7 Convertible: 497
12 124.0mph 199.5kmh
MX-6 GT: 508
626 2.0i GT 4WS: 493
MX-6 GT 4WS: 504
323 Turbo 4x4: 491
13 123.0mph 197.9kmh
323 1.8 GT: 512
14 121.0mph 194.7kmh
929: 494
626 2.0 5-door: 482
15 120.0mph 193.1kmh
626 GT: 483
MX-3 GS: 519
16 119.0mph 191.5kmh
323 GTX: 490
17 117.0mph 188.3kmh
MX-5 Miata: 503
18 116.0mph 186.6kmh
323 1.6i: 500
19 113.0mph 181.8kmh
626 2.0 GLS Executive: 488
20 112.0mph 180.2kmh
RX-7 Infini IV: 515
21 110.0mph 177.0kmh
626 2.0 GLX Executive Estate: 492
MPV: 502
22 104.0mph 167.3kmh
323 LX: 480
23 102.0mph 164.1kmh
MX-3 1.6 Auto: 518
24 101.0mph 162.5kmh
323 1.5 GLX: 487
25 98.0mph 157.7kmh
323 1.5 GLX Estate: 479

26 92.0mph 148.0kmh
121 1.3 LX Sun Top: 489
27 91.0mph 146.4kmh
121 GSX: 511

Mercedes-Benz

1 185.0mph 297.7kmh
300CE AMG Hammer: 547
2 183.0mph 294.4kmh
300E AMG Hammer: 536
3 160.0mph 257.4kmh
500SL: 550
4 159.0mph 255.8kmh
500E: 554
5 155.0mph 249.4kmh
500SL: 555
600SEL: 558
500SEL: 557
190E 2.5-16 Evolution II: 545
6 151.0mph 243.0kmh
560SEL: 526
300CE: 546
7 148.0mph 238.1kmh
300E-24: 549
8 146.0mph 234.9kmh
400SE: 553
560SEC Cabriolet Straman: 532
9 145.0mph 233.3kmh
560SEC: 525
10 144.0mph 231.7kmh
190E 2.5-16: 538
11 141.0mph 226.9kmh
300CE: 529
12 140.0mph 225.3kmh
300E: 543
300E: 522
190E AMG Baby Hammer: 540
13 139.0mph 223.7kmh
190E 2.3-16: 520
14 138.0mph 222.0kmh
420SE: 524
15 137.0mph 220.4kmh
300CE Cabriolet Straman: 542
560SL: 544
300CE: 534
16 136.0mph 218.8kmh
300E 4Matic: 535
17 135.0mph 217.2kmh
300SL-24 5-speed Auto: 552
18 134.0mph 215.6kmh
260E Auto: 533
190E 2.6: 528
19 131.0mph 210.8kmh
420SEL: 531
20 130.0mph 209.2kmh
560SL: 527
21 129.0mph 207.6kmh
190E 2.6: 539
22 127.0mph 204.3kmh
300TE: 523
23 121.0mph 194.7kmh
200: 521

24 120.0mph 193.1kmh
200E Automatic: 541
190E 1.8 Auto: 551
25 115.0mph 185.0kmh
300D: 548
300TD: 530
26 110.0mph 177.0kmh
190: 537
27 84.0mph 135.2kmh
G-Wagen 300 GD LWB: 556

Mercury

1 143.0mph 230.1kmh
Mercury Cougar XR-7: 561
2 125.0mph 201.1kmh
Mercury Tracer LTS: 563
3 124.0mph 199.5kmh
Mercury Capri XR-2: 562
4 115.0mph 185.0kmh
Mercury Sable LS: 559
5 105.0mph 168.9kmh
Mercury Tracer: 560

Merkur

1 130.0mph 209.2kmh
Scorpio: 564
2 120.0mph 193.1kmh
XR4Ti: 565

MG

1 131.0mph 210.8kmh
Maestro Turbo: 566
2 105.0mph 168.9kmh
B British Motor Heritage: 567

Mitsubishi

1 159.0mph 255.8kmh
3000GT VR-4: 588
2 143.0mph 230.1kmh
Eclipse: 580
3 136.0mph 218.8kmh
Starion 2000 Turbo: 574
4 130.0mph 209.2kmh
Galant VR-4: 591
Starion 2.6 Turbo: 583
Sigma: 593
Diamante LS: 590
5 127.0mph 204.3kmh
Starion ESI-R: 584
6 122.0mph 196.3kmh
Galant GTi-16v: 577
Lancer Liftback 1800 GTi-16v: 587
7 121.0mph 194.7kmh
Colt 1800 GTI-16v: 589
8 120.0mph 193.1kmh
Galant GS: 581
9 118.0mph 189.9kmh
Mirage Turbo: 579
10 116.0mph 186.6kmh

Lancer GTi 16v: 578
11 115.0mph 185.0kmh
Galant Sapporo: 572
Cordia Turbo: 568
Space Runner 1800-16v GLXi: 594
Cordia Turbo: 575
12 113.0mph 181.8kmh
Galant 2000 GLSi Coupe: 585
13 112.0mph 180.2kmh
Galant 2000 GLSi: 576
14 105.0mph 168.9kmh
Lancer GLXi 4WD Liftback: 586
Galant 2000 GLS Automatic: 571
Shogun V6 LWB: 592
15 102.0mph 164.1kmh
Shogun V6 5-door: 582
16 97.0mph 156.1kmh
Lancer 1.5 GLX Estate: 569
17 95.0mph 152.9kmh
Colt 1500 GLX 5-door: 570
18 83.0mph 133.5kmh
Shogun Turbo Diesel 5-door: 573

Morgan

1 126.0mph 202.7kmh
Plus 8: 595
2 125.0mph 201.1kmh
Plus 8: 596
3 122.0mph 196.3kmh
Plus 8: 597

MVS

1 153.0mph 246.2kmh
Venturi GT: 598

Nissan

1 166.0mph 267.1kmh
300ZX Turbo Millen Super GTZ: 627
2 158.0mph 254.2kmh
300ZX: 621
3 156.0mph 251.0kmh
Skyline GT-R: 624
4 155.0mph 249.4kmh
300ZX Turbo: 622
300ZX Twin Turbo: 628
5 148.0mph 238.1kmh
300ZX: 615
6 144.0mph 231.7kmh
300ZX Turbo: 606
7 141.0mph 226.9kmh
200SX: 614
8 140.0mph 225.3kmh
200SX: 634
9 133.0mph 214.0kmh
300ZX Turbo: 600
10 132.0mph 212.4kmh

Sunny 2.0 GTi: 635
Primera 2.0E ZX: 631
11 130.0mph 209.2kmh
NX2000: 629
12 128.0mph 206.0kmh
300ZX: 599
13 125.0mph 201.1kmh
Primera 2.0 GSX: 630
300ZX: 613
Sentra SE-R: 632
240SX: 612
300ZX: 605
14 122.0mph 196.3kmh
Maxima: 616
15 120.0mph 193.1kmh
Maxima SE: 617
16 119.0mph 191.5kmh
200SX SE V6: 611
17 118.0mph 189.9kmh
Bluebird 1.8 ZX: 601
Sunny ZX Coupe: 609
18 116.0mph 186.6kmh
100NX: 625
19 115.0mph 185.0kmh
Pulsar NX SE: 608
240SX: 626
20 112.8mph 181.5kmh
Primera 1.6 LS: 623
21 112.0mph 180.2kmh
Sunny 1.6 GS 5-door: 633
22 105.0mph 168.9kmh
Sunny 1.4 LS: 620
Sunny 1.6 SLX Coupe: 610
Laurel 2.4 SGL Automatic: 602
23 100.0mph 160.9kmh
Sentra Sport Coupe SE: 603
24 98.0mph 157.7kmh
Prairie: 619
25 93.0mph 149.6kmh
Sunny 1.3 LX 5-door: 604
Micra 1.2 GSX: 618
26 89.0mph 143.2kmh
Micra SGL 5-DR: 607

Oldsmobile

1 125.0mph 201.1kmh
Toronado Trofeo: 639
Cutlass Calais International Series HO: 638
Calais HO Quad 4: 637
2 120.0mph 193.1kmh
Cutlass Calais International Series: 636

Opel

1 174.0mph 280.0kmh
Omega: 641
2 137.0mph 220.4kmh
Calibra 2.0i 16v: 640

Panther

1 146.0mph 234.9kmh
Solo: 642

Parradine

1 170.0mph 273.5kmh
V-12: 643

Peugeot

1 144.0mph 231.7kmh
605 SVE 24: 672
2 134.0mph 215.6kmh
405 Mi16: 653
3 133.0mph 214.0kmh
605 SV 3.0 Auto: 671
4 130.0mph 209.2kmh
405 Mi16: 660
5 127.0mph 204.3kmh
405 Mi16x4: 664
6 125.0mph 201.1kmh
405 Turbo 16 Pike's Peak: 665
505 STX: 650
309 GTi: 658
7 123.0mph 197.9kmh
605 SRi: 666
205 GTi 1.9: 648
8 122.0mph 196.3kmh
309 GTi: 649
309 SR Injection: 646
205 GTi: 644
9 121.0mph 194.7kmh
405 GR Injection Auto: 663
10 118.0mph 189.9kmh
106 XSi: 668
405 SRi: 654
11 117.0mph 188.3kmh
405 GLx4: 662
12 116.0mph 186.6kmh
605 SR TD: 670
13 109.0mph 175.4kmh
106 XT: 673
405 GTD Turbo: 659
505 GTi Family Estate: 647
14 108.0mph 173.8kmh
405 1.6 GL: 651
15 103.0mph 165.7kmh
205 D Turbo: 669
16 102.0mph 164.1kmh
405 GRD: 652
17 101.0mph 162.5kmh
205 CJ: 656
18 99.0mph 159.3kmh
205 1.1GL: 655
309 1.6 GR: 645
19 97.0mph 156.1kmh
106 XR: 667
20 96.0mph 154.5kmh
309 Style 5-door: 661
21 95.0mph 152.9kmh
309 GLX: 657

Plymouth

1 143.0mph 230.1kmh
Laser RS: 674
Laser RS Turbo: 675
2 125.0mph 201.1kmh
Laser RS Turbo: 676

Pontiac

1 192.0mph 308.9kmh
Firebird TDM Technologies: 692
2 178.0mph 286.4kmh
Firebird T/A Turbo Engineering: 683
3 155.0mph 249.4kmh
Firebird TransAm 20th Anniversary: 684
4 150.0mph 241.4kmh
Firebird TransAm Turbo: 685
Firebird GTA: 691
5 136.0mph 218.8kmh
Firebird TransAm GTA: 689
6 135.0mph 217.2kmh
Firebird Formula: 688
7 130.0mph 209.2kmh
Firebird TransAm Convertible: 693
8 128.0mph 206.0kmh
Grand Prix McLaren Turbo: 687
Bonneville SSEi: 695
9 125.0mph 201.1kmh
Fiero GT: 677
Grand Prix GTP: 694
Fiero Formula: 679
Sunbird GT: 682
10 124.0mph 199.5kmh
Grand Prix STE Turbo: 690
11 120.0mph 193.1kmh
Grand Am Turbo: 686
Bonneville SE: 678
12 115.0mph 185.0kmh
Bonneville SE: 681
13 100.0mph 160.9kmh
Le Mans: 680

Porsche

1 216.0mph 347.5kmh
911 Turbo RS Tuning: 744
2 211.0mph 339.5kmh
911 Turbo Ruf Twin Turbo: 717
3 208.0mph 334.7kmh
911 Ruf CTR: 750
4 205.0mph 329.8kmh
911 Turbo Gemballa Mirage: 753
5 201.0mph 323.4kmh
911 Turbo Koenig RS: 716
959 S: 709
6 198.0mph 318.6kmh
959 Sport: 725

959 Comfort: 724
959: 738
7 196.1mph 315.5kmh
911 Ruf TR2: 751
8 196.0mph 315.4kmh
911 Ruf: 749
9 186.0mph 299.3kmh
928 S4 Koenig: 733
10 183.0mph 294.4kmh
911 Turbo Motorsport Design: 731
11 171.0mph 275.1kmh
911 Turbo: 752
12 168.0mph 270.3kmh
911 Carrera Turbo: 748
13 165.0mph 265.5kmh
928 S4: 719
911 Turbo Gemballa Avalanche: 705
14 162.0mph 260.7kmh
944 Turbo: 737
928 S4 SE: 734
15 161.0mph 259.0kmh
911 Carrera 4: 727
911 Carrera RS: 754
911 Carrera 4: 741
928 S4 Automatic: 701
16 160.0mph 257.4kmh
911 Carrera 2: 739
17 159.0mph 255.8kmh
911 Turbo: 730
Carrera 2 Cabriolet Tiptronic: 747
911 Turbo: 743
911 Carrera 2 Tiptronic: 740
18 158.0mph 254.2kmh
911 3.3 Turbo: 726
944 Turbo: 704
911 Turbo Ruf 3.4: 706
19 157.0mph 252.6kmh
911 Turbo Slant-Nose: 707
20 155.0mph 249.4kmh
928 Cabrio Strosek: 732
944 Turbo: 722
911 Turbo: 715
911 Speedster: 714
21 154.0mph 247.8kmh
911 Club Sport: 729
944 Turbo SE: 723
22 153.0mph 246.2kmh
911 Turbo: 698
23 152.0mph 244.6kmh
928 S: 700
911 Carrera Club Sport: 712
24 149.0mph 239.7kmh
911 Carrera SE: 697
911 Carrera Cabriolet: 728
944 S2 Cabriolet: 746
911 Club Sport: 713
911 Cabriolet: 710
911 Carrera: 711
944 S2: 745
911 Speedster: 742
25 147.0mph 236.5kmh

944 S2: 736
26 141.0mph 226.9kmh
944 S: 703
27 140.0mph 225.3kmh
944 S: 721
28 139.0mph 223.7kmh
944: 735
29 138.0mph 222.0kmh
924 S: 718
30 136.0mph 218.8kmh
924 S: 708
31 132.0mph 212.4kmh
924 S: 699
944: 720
32 130.0mph 209.2kmh
911 Cabriolet: 696
33 123.0mph 197.9kmh
944: 702

Proton

1 100.0mph 160.9kmh
1.5 SE Triple Valve Aeroback: 756
299.0mph 159.3kmh
1.5 SE Aeroback: 755

Range Rover

1 110.0mph 177.0kmh
Vogue SE: 763
2 109.0mph 175.4kmh
County: 762
3 108.0mph 173.8kmh
Vogue SE: 764
4 106.0mph 170.6kmh
Vogue: 757
Vogue: 760
5 102.0mph 164.1kmh
Vogue SE: 761
693.0mph 149.6kmh
Vogue Turbo D: 759
790.0mph 144.8kmh
V8: 758

Reliant

1 132.0mph 212.4kmh
Scimitar SST 1800Ti: 766
2 124.0mph 199.5kmh
Scimitar Ti: 765

Renault

1 165.0mph 265.5kmh
Alpine A610 Turbo: 791
2 152.0mph 244.6kmh
GTA V6 Turbo: 772
3 141.0mph 226.9kmh
GTA V6: 779
4 140.0mph 225.3kmh
21 Turbo Quadra: 787
5 139.0mph 223.7kmh
21 Turbo: 778

6 130.0mph 209.2kmh
19 16v: 790
7 129.0mph 207.6kmh
19 16v Cabriolet: 795
25 V6 2.9 Auto: 789
8 128.0mph 206.0kmh
25 TXi: 783
9 126.8mph 204.0kmh
Cleo 16v: 796
10 126.0mph 202.7kmh
21 TXi: 788
5 GT Turbo: 769
11 123.0mph 197.9kmh
11 Turbo: 773
12 122.0mph 196.3kmh
5 GT Turbo: 774
21 Ti: 777
21 Savanna GTX: 768
13 118.0mph 189.9kmh
Espace V6: 794
14 116.0mph 186.6kmh
9 Turbo: 771
15 114.0mph 183.4kmh
21 GTS: 767
16 112.0mph 180.2kmh
19 TXE: 781
21 GTS Hatchback: 782
5 GTX 3-DR: 775
17 111.0mph 178.6kmh
19 TXE Chamade: 786
18 110.0mph 177.0kmh
Alliance GTA: 776
19 109.0mph 175.4kmh
Espace TXE: 785
20 108.0mph 173.8kmh
19 TSE: 780
21 107.0mph 172.2kmh
Clio 1.4RT: 793
22 100.1mph 161.0kmh
Clio 1.2 RN: 792
23 99.0mph 159.3kmh
5 GTL 5-door: 770
24 92.0mph 148.0kmh
5 Campus: 784

Rolls-Royce

1 129.0mph 207.6kmh
Corniche: 799
2 128.0mph 206.0kmh
Silver Spirit II: 801
3 119.0mph 191.5kmh
Royce Silver Spirit: 798
4 114.0mph 183.4kmh
Silver Shadow II: 797
5 108.0mph 173.8kmh
Silver Spur: 800

Rover

1 138.0mph 222.0kmh
Vitesse: 811
2 134.0mph 215.6kmh
Vitesse: 808

3 132.0mph 212.4kmh
Sterling Catalyst: 810
827 SLi Auto: 807
4 131.0mph 210.8kmh
Sterling Automatic: 825
5 130.0mph 209.2kmh
825i: 802
6 128.0mph 206.0kmh
220 GTi: 821
7 127.0mph 204.3kmh
Sterling Automatic: 805
8 126.0mph 202.7kmh
820i: 824
216 GTi: 819
9 121.0mph 194.7kmh
416 GTi: 815
10 118.0mph 189.9kmh
820SE: 803
11 117.0mph 188.3kmh
216 GSi: 813
825 TD: 822
12 115.0mph 185.0kmh
Metro GTi 16v: 817
820 Fastback: 806
13 108.0mph 173.8kmh
214 GSi: 812
14 106.0mph 170.6kmh
Montego 1.6L: 804
Metro 1.4SL: 816
15 105.0mph 168.9kmh
414 Si: 814
16 102.0mph 164.1kmh
214S: 818
Montego 2.0 DSL Turbo: 809
17 98.0mph 157.7kmh
Mini Cooper S: 823
18 97.0mph 156.1kmh
218 SD: 820

Saab

1 148.0mph 238.1kmh
Carlsson: 831
2 142.0mph 228.5kmh
9000 2.3 Turbo TCS: 835
9000CD: 828
4 138.6mph 223.0kmh
9000CS Turbo S: 837
5 135.0mph 217.2kmh
9000 Turbo: 836
9000 Turbo: 830
6 129.0mph 207.6kmh
9000 2.3CS: 838
7 127.0mph 204.3kmh
CDS 2.3: 834
8 125.0mph 201.1kmh
900SPG: 833
9000CD: 832
9 124.0mph 199.5kmh
9000: 826
900 Turbo S: 829
10 121.0mph 194.7kmh
9000i: 827

Safir

1 193.0mph 310.5kmh
GT40: 839

Saturn

1 124.0mph 199.5kmh
Sports Coupe: 840

Sbarro

1 124.0mph 199.5kmh
Chrono 3.5: 841

Seat

1 124.0mph 199.5kmh
Toledo 2.0 GTi: 845
2 108.0mph 173.8kmh
Ibiza SXi: 843
3 103.0mph 165.7kmh
Toledo 1.9 CL Diesel: 844
494.0mph 151.2kmh
Malaga 1.2L: 842

Skoda

192.0mph 148.0kmh
136 Rapide Coupe: 847
288.0mph 141.6kmh
Estelle 120 LSE: 846

Subaru

1 161.0mph 259.0kmh
Legacy FIA Record: 853
2 143.0mph 230.1kmh
SVX: 858
3 134.0mph 215.6kmh
Legacy 2.0 4Cam Turbo Estate: 856
4 129.0mph 207.6kmh
Legacy Sports Sedan: 857
5 121.0mph 194.7kmh
XT Turbo 4WD: 850
6 119.0mph 191.5kmh
Legacy 2.2 GX 4WD: 855
7 116.0mph 186.6kmh
4WD Turbo Estate: 851
8 115.0mph 185.0kmh
XT6: 854
9 114.0mph 183.4kmh
4WD 3-door Turbo Coupe: 848
10 93.0mph 149.6kmh
Justy GL II 5-door 4WD: 852
11 91.0mph 146.4kmh
Justy: 849

Suzuki

1 116.0mph 186.6kmh
Swift 1.3 GTi: 865
2 112.0mph 180.2kmh

Swift 1.3 GTi: 860
3 105.0mph 168.9kmh
Swift GT: 867
4 103.0mph 165.7kmh
Swift 1.3 GLX: 864
5 102.0mph 164.1kmh
Swift 1.3 GLX: 862
697.0mph 156.1kmh
Swift 1.3 GLX Executive: 863
790.0mph 144.8kmh
Vitara JLX: 866
880.0mph 128.7kmh
Alto GLA: 859
976.0mph 122.3kmh
SJ413V JX: 861

Toyota

1 146.0mph 234.9kmh
Supra 3.0i Turbo: 894
2 145.0mph 233.3kmh
MR2 Turbo: 900
3 140.0mph 225.3kmh
Supra Turbo: 889
Supra Turbo: 895
4 139.0mph 223.7kmh
MR2 GT: 899
5 138.0mph 222.0kmh
Supra 3.0i: 873
6 137.0mph 220.4kmh
Celica GT4: 884
7 135.0mph 217.2kmh
MR2 Supercharged: 893
MR2 Supercharged: 880
8 134.0mph 215.6kmh
Celica 2.0 GT: 896
Celica All-Trac Turbo: 883
9 133.0mph 214.0kmh
Supra: 888
Celica 2.0GT: 868
Supra: 872
10 130.0mph 209.2kmh
Supra: 881
11 127.0mph 204.3kmh
Celica GT Cabriolet: 876
12 126.0mph 202.7kmh
Camry 2.2 GL: 903
13 125.0mph 201.1kmh
Celica All-Trac Turbo: 897
Celica GTS: 898
Paseo: 909
14 123.0mph 197.9kmh
Celica GTS: 869
MR2: 906
15 122.0mph 196.3kmh
MR2 T-Bar: 871
16 121.0mph 194.7kmh
MR2: 870
17 120.0mph 193.1kmh
Celica GT Convertible: 904
18 119.0mph 191.5kmh
Camry V6 GXi: 890
19 118.0mph 189.9kmh
Camry 2.0 GLi: 874

20 115.0mph 185.0kmh
Corolla GTS: 892
21 112.0mph 180.2kmh
Camry 2.0 GLi Estate: 875
22 110.0mph 177.0kmh
Carina II 1.6 GL Liftback: 891
Previa: 907
23 109.0mph 175.4kmh
Carina 1.6 GL: 882
24 107.0mph 172.2kmh
Previa LE: 901
25 105.0mph 168.9kmh
Corolla FX-16 GTS: 878
Tercel LE: 908
26 104.0mph 167.3kmh
Landcruiser VX: 905
Corolla Executive: 877
27 97.0mph 156.1kmh
Corolla 1.3 GL: 885
Corolla 4WD Estate: 886
28 91.0mph 146.4kmh
Starlet GL: 902
29 87.0mph 140.0kmh
Landcruiser: 879
30 81.0mph 130.3kmh
Landcruiser II TD: 887

Treser

1 130.0mph 209.2kmh
T1: 910

TVR

1 165.0mph 265.5kmh
420 SEAC: 913
2 150.0mph 241.4kmh
V8 S: 915
3 144.0mph 231.7kmh
390 SE: 911
4 140.0mph 225.3kmh
S Convertible: 914
5 130.0mph 209.2kmh
S Convertible: 912

Vauxhall

1 148.0mph 238.1kmh
Carlton GSi 3000 24v: 937
2 140.0mph 225.3kmh
Senator 3.0i 24v: 944
3 139.0mph 223.7kmh
Calibra 2.0i 16v: 935
4 135.0mph 217.2kmh
Carlton 3000 GSi: 921
5 134.0mph 215.6kmh
Cavalier GSi 2000 16v: 938
6 132.0mph 212.4kmh
Senator 3.0i CD: 934
Calibra 4x4: 942
Carlton 3.0i CDX Estate: 936
7 129.0mph 207.6kmh
Cavalier GSi 2000 16v 4x4: 939

8 128.0mph 206.0kmh
Senator 2.5i: 933
Calibra 2.0i: 941
9 126.0mph 202.7kmh
Cavalier SRi: 932
Senator 3.0i CD: 924
10 125.0mph 201.1kmh
Astra GTE 2.0i: 919
11 124.0mph 199.5kmh
Cavalier 4x4: 927
12 123.0mph 197.9kmh
Carlton CD 2.0i: 917
13 122.0mph 196.3kmh
Carlton L 1.8i: 925
14 120.0mph 193.1kmh
Belmont 1.8 GLSi: 916
15 119.0mph 191.5kmh
Nova GTE: 928
16 117.0mph 188.3kmh
Cavalier SRi 130 4-door: 923
17 115.0mph 185.0kmh
Cavalier 2.0i CD: 922
18 112.0mph 180.2kmh
Belmont 1.6 GL: 920
19 111.0mph 178.6kmh
Cavalier 1.6L: 931
Cavalier 1.7 GL TD: 946
1.4i GLS: 945
21 107.0mph 172.2kmh
Cavalier 1.4L: 930
22 105.0mph 168.9kmh
Carlton L 2.3D: 926
23 102.0mph 164.1kmh
Nova 1.5TD Merit: 940
Nova 1.3 GL 5-door: 918
24 96.0mph 154.5kmh
Astra 1.7 DL: 929
25 94.0mph 151.2kmh
Nova 1.2 Luxe: 943
26 85.1mph 137.0kmh
Frontera 2.3 TD Estate: 947

Vector

1 218.0mph 350.8kmh
W8 Twin Turbo: 949
2 200.0mph 321.8kmh
W2: 948

Volkswagen

1 140.0mph 225.3kmh
Corrado: 968
2 139.0mph 223.7kmh
Corrado G60: 976
3 137.0mph 220.4kmh
Corrado G60: 973
4 132.0mph 212.4kmh
Corrado 16v: 969
5 130.0mph 209.2kmh
Jetta GTi 16v: 959
6 129.0mph 207.6kmh
Passat GT 16v: 972

7 125.0mph 201.1kmh
Golf GTi 16v: 951
Scirocco 16v: 954
Passat GL: 975
Jetta GLi 16v: 978
Scirocco 16v: 966
8 122.0mph 196.3kmh
GTi 16v: 961
GTi 16v: 977
GTi 16v: 958
Passat GT: 964
9 121.0mph 194.7kmh
Golf GTi: 983
Jetta GLi 16v: 962
10 120.0mph 193.1kmh
Polo G40: 980
Scirocco GTX 16v: 955
11 119.0mph 191.5kmh
Jetta GT: 952
12 113.0mph 181.8kmh
Quantum Synchro Wagon: 953
13 108.0mph 173.8kmh
Jetta Syncro: 971
14 107.0mph 172.2kmh
Golf 1.8 GL: 982
15 106.0mph 170.6kmh
Golf Cabriolet Clipper: 957
16 104.0mph 167.3kmh
Passat GT 5-door: 960
Jetta Turbo Diesel: 963
Polo GT Coupe: 981
17 103.0mph 165.7kmh
Passat CL TD Estate: 974
18 100.0mph 160.9kmh
Fox GL: 956
19 99.0mph 159.3kmh
Golf CL Catalyst: 970
Polo 1.3 CL Coupe: 979
20 93.0mph 149.6kmh
Caravelle Carat Automatic: 967
21 89.0mph 143.2kmh
Polo C Saloon: 965
22 82.0mph 131.9kmh
Caravelle Syncro: 950

Volvo

1 135.0mph 217.2kmh
960: 999
2 129.0mph 207.6kmh
960 24v: 998
3 128.0mph 206.0kmh
740 Turbo: 986
4 127.0mph 204.3kmh
440 Turbo: 988
5 126.0mph 202.7kmh
460 Turbo: 993
6 125.0mph 201.1kmh
740 GLE: 989
740 Turbo: 990
7 124.0mph 199.5kmh
940 SE: 996
940 SE Turbo: 997

8   120.0mph 193.1kmh
740 Turbo: 995
9   119.0mph 191.5kmh
740 GLT 16v Estate: 994
10 114.0mph 183.4kmh
760 GLE: 987
11 112.0mph 180.2kmh
480 ES: 985
12 109.0mph 175.4kmh
460 GLi: 992
13 107.0mph 172.2kmh
340 GLE: 984
440 GLE: 991

Westfield

1   140.0mph 225.3kmh
SEight: 1000

Yugo

199.0mph 159.3kmh
Sana 1.4: 1004
297.0mph 156.1kmh
65A GLX: 1003
390.0mph 144.8kmh
GV: 1001
483.0mph 133.5kmh
45 GLS: 1002

Zender

1   186.0mph 299.3kmh
Fact 4: 1006
2   175.0mph 281.6kmh
Vision 3: 1005

Acceleration times in order. Standing start to 60mph, 96.5kmh.

1 3.4secs
Chevrolet Corvette Lingenfelter: 181
2 3.5secs
Jaguar XJ220: 407
Koenig Competition Evolution: 421
Koenig C62: 422
Sbarro Chrono 3.5: 841
3 3.6secs
Porsche 959 Sport: 725
Porsche 959: 738
4 3.7secs
Ferrari 308 Norwood Bonneville GTO: 253
5 3.8secs
Ferrari F40: 267
Porsche 911 Ruf: 749
Porsche 911 Ruf TR2: 751
6 3.9secs
Bugatti EB110: 126
Chevrolet Corvette Callaway Sledgehammer: 174
Porsche 911 Ruf CTR: 750
7 4.0secs
Koenig Competition: 420
Porsche 911 Turbo Ruf Twin Turbo: 717
Porsche 911 Turbo Koenig RS: 716
Porsche 959 Comfort: 724
Porsche 911 Turbo Motorsport Design: 731
Safir GT40: 839
Vector W2: 948
8 4.1secs
Dodge Viper: 241
Pontiac Firebird TDM Technologies: 692
Porsche 911 Turbo Gemballa Mirage: 753
9 4.2secs
Vector W8 Twin Turbo: 949
10 4.3secs
De Tomaso Pantera Group 3: 227
Westfield SEight: 1000
Zender Fact 4: 1006
11 4.4secs
Chevrolet Corvette Callaway Speedster: 180
Jaguar XJS Lister Le Mans: 414
12 4.5secs
Chevrolet Corvette Bakeracing SCCA Escort: 151
Chevrolet Camaro IROC-Z Chevrolet Engineering: 162
Cizeta Moroder V16T: 215
Ferrari F40: 263
Lamborghini Diablo: 432
Porsche 911 Turbo Ruf 3.4: 706
13 4.6secs
Chevrolet Corvette Callaway: 152
Porsche 911 Turbo RS Tuning: 744
Porsche 911 Carrera Turbo: 748
14 4.7secs
Ferrari Testa Rossa Norwood: 261
Ferrari 308 Norwood: 257
Lamborghini Countach 25th Anniversary: 431
Porsche 959 S: 709
Porsche 911 Turbo: 752
15 4.8secs
Aston Martin Zagato: 36
Chevrolet Camaro Gottlieb 1969: 159
Mazda RX-7 Mariah Mode Six: 516
Porsche 928 S4 Koenig: 733
Porsche 911 Turbo: 730
16 4.9secs
Chevrolet Corvette Morrison/Baker Nelson Ledges: 155
Chevrolet Corvette ZR-1: 167
Chevrolet Corvette Callaway: 164
Chevrolet Corvette Callaway: 179
Chevrolet Corvette ZR-2: 185
Chevrolet Corvette ZR-1: 183
Chevrolet Corvette ZR-1 Geiger: 184
Ford Mustang Cartech Turbo: 302
Ford Mustang Holdener: 334
Lotus Esprit Turbo SE: 466
Opel Omega Lotus: 641
Porsche 911 Turbo Gemballa Avalanche: 705
Porsche 911 3.3 Turbo: 726
Porsche 911 Carrera 4: 727
Porsche 911 Carrera RS: 754
17 5.0secs
Autokraft AC Cobra Mk IV: 73
Bertone Emotion Lotus: 79
Chevrolet Corvette Callaway: 140
Eagle GTP: 244
IAD Venus: 384
Isdera Imperator: 389
Mercedes-Benz 300CE AMG Hammer: 547
Nissan 300ZX Turbo Millen Super GTZ: 627
Porsche 911 Turbo: 698
Porsche 911 Turbo Slant-Nose: 707
Porsche 911 Turbo: 715
TVR 420 SEAC: 913
18 5.1secs
BMW 535i Alpina B10 Bi-Turbo: 120
Chevrolet Corvette Guldstrand Grand Sport 80: 165
Chevrolet Corvette Callaway: 173
Lotus Esprit Turbo SE: 469
Pontiac Firebird TransAm 20th Anniversary: 684
Porsche 911 Turbo: 743
Porsche 911 Carrera 2: 739
19 5.2secs
Caterham 7 HPC: 135
Chevrolet Corvette Guldstrand Grand Sport 80: 154
Ferrari Testa Rossa: 259
Ford Mustang SVO J Bittle American: 313
Ford Thunderbird Super Coupe Ford Engineering: 318
Lamborghini Countach 5000S: 427
Lotus Esprit Turbo: 465
Mercedes-Benz 300E AMG Hammer: 536
Porsche 911 Carrera Club Sport: 712
TVR V8 S: 915
20 5.3secs
Ferrari Testa Rossa: 255
Ferrari Testa Rossa Straman Spyder: 256
Ferrari Testa Rossa Gemballa: 260
Lotus Elan Autocrosser: 461
Lynx D Type: 471
Pontiac Firebird TransAm Turbo: 685
Porsche 911 Club Sport: 729
21 5.4secs
De Tomaso Pantera GT5-S: 226
Ford Mustang 5.0 Cartech Turbo: 312
Lotus Esprit Turbo: 464
Parradine V-12: 643
22 5.5secs
Mazda RX-7 Cartech: 514
Porsche 928 S4: 719
Porsche 944 Turbo: 737
Porsche 928 S4 SE: 734
Porsche 928 Cabrio Strosek: 732
23 5.6secs
Aston Martin Zagato Volante: 39
Caterham Super 7: 136
Ferrari 348tb: 262
Lotus Esprit Turbo: 459
Lotus Esprit Turbo HC: 462
Morgan Plus 8: 595
Nissan Skyline GT-R: 624
Nissan 300ZX: 621
Porsche 911 Carrera SE: 697
24 5.7secs
Acura NSX: 9
Chevrolet Corvette LT1: 188
Consulier GTP LX: 216
Ford Mustang NOS/Saleen: 335
Porsche 911 Cabriolet: 696
Porsche 911 Club Sport: 713
Porsche 944 Turbo SE: 723
Renault Alpine A610 Turbo: 791
TVR 390 SE: 911
25 5.8secs
Chevrolet Corvette: 139
Chevrolet Corvette Z51: 141
Chevrolet Corvette ASC Geneve: 147
Ford Sierra RS Cosworth: 306
Honda NSX: 373
Porsche 911 Carrera 4: 741
26 5.9secs
Audi Coupe S2: 67
Chevrolet Corvette: 146
Ferrari 328 GTS: 250
Mercedes-Benz 500SL: 550
27 6.0secs
Aston Martin Virage: 37
BMW M635 CSi: 109
BMW 735i Koenig: 112
Chevrolet Corvette Convertible: 153
Chevrolet Corvette Z51: 156
Chevrolet Corvette: 150
Ferrari 348tb: 266
Ford Mustang Convertible Saleen: 303
Ford Mustang GT: 304
MVS Venturi GT: 598
Porsche 944 Turbo: 704
Porsche 944 Turbo: 722
Porsche 944 S2: 736
Porsche 911 Speedster: 742
Zender Vision 3: 1005
28 6.1secs
Ford RS200: 292
Mercedes-Benz 600SEL: 558
Morgan Plus 8: 597
Porsche 911 Carrera Cabriolet: 728
29 6.2secs
Chevrolet Corvette: 163
Ferrari Pininfarina Mythos: 265
Lamborghini Jalpa 3500: 428
Peugeot 405 Turbo 16 Pike's Peak: 665
Porsche 928 S4 Automatic: 701

Porsche Carrera 2 Cabriolet Tiptronic: 747
Subaru Legacy FIA Record: 853
Toyota MR2 Turbo: 900
30 6.3secs
Audi Quattro: 54
Audi Quattro 20v: 61
BMW M5: 108
Chevrolet Corvette Convertible: 148
Chevrolet Corvette L98: 175
Chevrolet Corvette Rick Mears: 182
Dodge Stealth R/T Turbo: 240
Lancia Delta Integrale 16v: 441
Maserati 430: 476
Mercedes-Benz 500E: 554
Mitsubishi 3000GT VR-4: 588
Porsche 928 S: 700
Renault GTA V6 Turbo: 772
31 6.4secs
BMW M5: 122
Lancia Delta HF Integrale: 439
Mazda MX-5 Miata Millen: 507
Mazda RX-7 Turbo: 510
Mazda RX-7 Turbo: 517
Mercedes-Benz 500SL: 555
32 6.5secs
Dodge Spirit R/T: 239
Lotus Elan SE: 468
Maserati 430: 474
Nissan 300ZX Turbo: 622
Nissan 300ZX Twin Turbo: 628
Pontiac Firebird T/A Turbo Pontiac Engineering: 683
Porsche 911 Speedster: 714
Porsche 911 Cabriolet: 710
33 6.6secs
BMW M3 Evolution: 107
Chevrolet Camaro IROC-Z: 144
Chevrolet Corvette L98: 166
Chevrolet Camaro IROC-Z Automatic: 161
Chevrolet Camaro IROC-Z: 160
Chevrolet Camaro Z/28: 171
Ferrari 328 GTB: 247
Ferrari Mondial: 264
Ferrari Mondial t: 268
Ferrari Mondial t Cabriolet: 269
Ford Taurus SHO: 316
Ford Mustang LX 5.0L: 325
Ford Sierra Sapphire Cosworth 4x4: 328
Lancia Delta HF 4WD: 433
Lotus Elan SE: 470
Mazda RX-7 Turbo: 486
Mazda RX-7 Turbo: 499
Pontiac Firebird Formula: 688
Pontiac Firebird GTA: 691
Toyota Supra Turbo: 895

34 6.7secs
Bentley Turbo R: 77
Chevrolet Camaro Z28: 187
Dodge Shelby GLH-S: 228
Ferrari 412: 251
Ford Mustang GT: 298
Lotus Elan: 467
Maserati Spyder: 477
Maserati 228: 478
Mazda RX-7 Turbo II: 506
Porsche 944 S: 703
Porsche 944 S2: 745
Toyota MR2 GT: 899
35 6.8secs
Aston Martin Virage: 40
Chevrolet Camaro IROC-Z L98: 145
Eagle Talon: 243
Eagle Talon TSi 4WD: 245
Ferrari Mondial 3.2QV: 249
Ford Thunderbird Bondurant 5.0: 300
Honda NSX Auto: 376
Lamborghini Jalpa: 429
Lancia Thema 8.32: 438
Lancia Thema 2.0ie 16v SE Turbo: 445
Lotus Excel SE: 460
Mercedes-Benz 560SL: 544
Nissan 200SX: 634
Panther Solo: 642
Saab 9000 Turbo: 836
36 6.9secs
Bentley Mulsanne Turbo: 75
BMW 750i L: 100
BMW 750i Alpina B12: 114
Chrysler TC by Maserati: 192
Dodge Shelby CSX: 231
Ford Mustang GT 5.0: 289
Infiniti Q45: 386
Lexus SC400: 455
Maserati Biturbo: 475
Mercedes-Benz 560SEC Cabriolet Straman: 532
MG Maestro Turbo: 566
Mitsubishi Starion 2000 Turbo: 574
Plymouth Laser RS Turbo: 675
Porsche 911 Carrera 2 Tiptronic: 740
Saab 9000CD: 828
Toyota Supra 3.0i Turbo: 894
37 7.0secs
Bentley Mulsanne Turbo: 74
Bentley Turbo R: 76
Ferrari Mondial Cabriolet 3.2: 252
Ferrari Mondial Cabriolet: 254
Ferrari Mondial 3.2 Cabriolet: 258
Mazda RX-7 Infini IV: 515
Mercedes-Benz 560SEC: 525
Mercedes-Benz 190E AMG Baby Hammer: 540

Nissan 300ZX Turbo: 606
Pontiac Grand Prix McLaren Turbo: 687
Porsche 911 Carrera: 711
Porsche 944: 735
Reliant Scimitar SST 1800Ti: 766
Subaru Legacy 2.0 4Cam Turbo Estate: 856
Toyota Supra: 872
Toyota MR2 Supercharged: 880
Toyota Supra Turbo: 889
TVR S Convertible: 914
Vauxhall Carlton GSi 3000 24v: 937
38 7.1secs
Bentley Turbo R: 78
BMW M3: 94
BMW M3: 106
BMW 850i Manual: 125
Ferrari 3.2 Mondial: 246
Ford Mustang SVO: 290
Ford Mustang LX: 343
Mercedes-Benz 560SEL: 526
Mercedes-Benz 190E 2.5-16 Evolution II: 545
Nissan 300ZX: 615
Pontiac Firebird TransAm GTA: 689
Porsche 944 S2 Cabriolet: 746
Renault 5 GT Turbo: 769
Volkswagen Jetta GTi 16v: 959
39 7.2secs
Audi 200 Quattro: 65
BMW M6: 101
BMW 850i: 121
Dodge Daytona Shelby Z: 229
Dodge Shelby CSX VNT: 235
Jaguar XJSC HE: 399
Lancia Thema 8.32: 440
Maserati Biturbo Spyder: 473
Mazda MX-6 GT: 496
Mercedes-Benz 190E 2.5-16: 538
Mitsubishi Eclipse: 580
Nissan 300ZX Turbo: 600
Nissan 200SX: 614
Plymouth Laser RS: 674
Reliant Scimitar Ti: 765
40 7.3secs
BMW 750i: 113
BMW 325i SE: 116
Chrysler Conquest TSi: 191
Dodge Shadow ES VNT: 238
Ford Probe GT: 305
Renault 5 GT Turbo: 774
Saab Carlsson: 831
Subaru SVX: 858
41 7.4secs
BMW 535i SE: 97
BMW 735i Alpina B11: 105
Ferrari Mondial 3.2: 248
Ford Thunderbird Super

Coupe: 317
Ford Probe GT: 337
Lancia Dedra 2000 Turbo: 447
Mercury Cougar XR-7: 561
Morgan Plus 8: 596
Nissan 300ZX: 613
42 7.5secs
Alfa Romeo Alfa 75 V6 3.0: 20
Alfa Romeo Milano Verde 3.0: 21
Alfa Romeo 75 Evoluzione: 23
Audi V8 Quattro: 68
BMW 325i Cabriolet: 91
BMW 325i S: 95
Citroen XM V6.24: 210
Dodge Daytona Shelby: 232
Jaguar XJS Koenig: 404
Mazda MX-6: 495
Mercedes-Benz 560SL: 527
Mercedes-Benz 300E: 522
Mercedes-Benz 300CE: 529
Mercedes-Benz 560SEL: 557
Nissan Sunny 2.0 GTi: 635
Porsche 924 S: 718
Porsche 944 S: 721
Renault GTA V6: 779
Saab 9000 2.3 Turbo TCS: 835
Toyota Supra: 888
43 7.6secs
Alfa Romeo 164S: 31
Bitter Type 3 Cabrio: 81
Dodge Lancer Shelby: 230
Dodge Shelby CSX: 234
Dodge Daytona Shelby: 237
Ford Probe GT: 326
Ford Taurus SHO: 339
Honda Civic 1.6i VT: 371
Lynx XJS Eventer: 472
Mercedes-Benz 420SE: 524
Oldsmobile Cutlass Calais International Series HO: 638
Oldsmobile Calais HO Quad 4: 637
Opel Calibra 2.0i 16v: 640
Toyota Celica GT4: 884
TVR S Convertible: 912
44 7.7secs
BMW 750i L: 93
BMW 535i: 103
Dodge Shadow ES Turbo: 233
Jaguar XJRS: 412
Pontiac Fiero GT: 677
Pontiac Grand Prix STE Turbo: 690
Saab 9000 Turbo: 830
Toyota Supra 3.0i: 873
Toyota MR2 T-Bar: 871
Toyota Celica All-Trac Turbo: 883
Toyota MR2 Supercharged: 893
Volkswagen Scirocco 16v: 954
45 7.8secs
Alfa Romeo 164 Cloverleaf: 33

Aston Martin Volante: 38
Jaguar XJS 3.6 Automatic: 402
Jaguar XJS: 417
Lamborghini LM129: 430
Mazda 626 GT: 483
Mazda 323 Turbo 4x4: 491
Mazda 323 1.8 GT: 512
Mercedes-Benz 190E 2.3-16: 520
Mercedes-Benz 300E-24: 549
Mitsubishi Starion 2.6 Turbo: 583
Peugeot 205 GTi 1.9: 648
Pontiac Sunbird GT: 682
Pontiac Firebird TransAm Convertible: 693
Porsche 924 S: 699
Renault 21 Turbo Quadra: 787
Volkswagen Corrado: 968
Volvo 740 Turbo: 990

46 7.9secs
Acura Legend LS: 10
Alfa Romeo Milano 3.0: 28
Alfa Romeo 164 3.0 Lusso: 26
BMW 535i: 86
BMW 325i X: 96
BMW Z1: 110
BMW 325i: 123
Citroen BX GTi 16v: 199
Daihatsu Charade GT ti: 219
Ford Fiesta RS Turbo: 319
Honda Prelude Si 4WS: 377
Irmsher GT: 388
Mazda RX-7 GXL: 485
Mazda 323 Turbo 4x4 Lux: 481
Mitsubishi Galant VR-4: 591
Peugeot 605 SVE 24: 672
Pontiac Grand Am Turbo: 686
Renault 11 Turbo: 773
Renault 21 Turbo: 778
Toyota Supra: 881
Vauxhall Cavalier GSi 2000 16v: 938

47 8.0secs
Acura Legend Coupe L: 6
Acura Legend Coupe: 8
Chevrolet Beretta GTZ: 169
Chevrolet Baretta GTZ: 177
Chevrolet Lumina Z34: 186
Ford Mustang GT Convertible: 324
GM Impact: 347
Honda Legend: 353
Honda CRX Coupe 1.6i 16: 349
Honda Legend Coupe: 356
Honda CRX Si Jackson: 361
Honda Civic CRX: 359
Honda CRX 1.6i VT: 369
Jaguar XJS V12 Convertible: 405
Lincoln Continental Mk VII: 457
Mercedes-Benz 300CE: 534

Mitsubishi Starion ESI-R: 584
Peugeot 405 Mi16: 653
Plymouth Laser RS Turbo: 676
Pontiac Fiero Formula: 679
Porsche 924 S: 708
Rover 825i: 802
Rover Vitesse: 811
Saab 9000CS Turbo S: 837
Volkswagen Scirocco GTX 16v: 955
Volkswagen Golf GTi 16v: 951

48 8.1secs
Acura Legend: 2
BMW 325i Convertible: 83
Chevrolet Cavalier Z24: 172
Chrysler Laser XT: 189
Honda Legend Coupe: 375
Merkur XR4Ti: 565
Nissan NX2000: 629
Nissan Sentra SE-R: 632
Subaru Legacy Sports Sedan: 857
Toyota Celica 2.0 GT: 896
Vauxhall Calibra 2.0i 16v: 935

49 8.2secs
Alfa Romeo GTV6: 14
Aston Martin Volante: 35
Chevrolet Cavalier Z24 Convertible: 149
Ford Probe LX: 327
Honda CRX Si: 360
Honda Legend: 372
Lotus Excel SA: 463
Mazda MX-6 GT 4WS: 504
Mazda MX-6 GT: 508
Mitsubishi Mirage Turbo: 579
Oldsmobile Cutlass Calais International Series: 636
Toyota Celica 2.0GT: 868
Vauxhall Carlton 3000 GSi: 921
Vauxhall Cavalier SRi 130 4-door: 923

50 8.3secs
Audi 80 2.8E V6 Quattro: 69
Bitter SC: 80
Cadillac Allante: 130
Cadillac Allante: 132
Chevrolet Beretta GT: 143
Fiat Uno Turbo ie: 283
Ford Escort RS2000: 341
Jaguar 4.0 Sovereign: 406
Jaguar XJR 4.0: 411
Jaguar XJ6 3.2: 416
Lexus LS400: 454
Lincoln Continental Mk VII LSC: 458
Mercedes-Benz 190E 2.6: 528
Mercedes-Benz 300E 4Matic: 535
Mercedes-Benz 300E: 543
Mercury Capri XR-2: 562
Mitsubishi Colt 1800 GTI-16v: 589

Volkswagen Scirocco 16v: 966
Volvo 740 Turbo: 986

51 8.4secs
BMW 635 CSi: 98
Dodge Daytona ES: 236
Mercedes-Benz 400SE: 553
Mitsubishi Cordia Turbo: 575
Nissan 200SX SE V6: 611
Pontiac Grand Prix GTP: 694
Rover 827 SLi Auto: 807
Toyota MR2: 870
Vauxhall Astra GTE 2.0i: 919
Vauxhall Carlton 3.0i CDX Estate: 936
Volkswagen GTi 16v: 977
Volkswagen Polo G40: 980

52 8.5secs
Audi 90 Quattro 20v: 56
BMW 635 CSi: 87
BMW 525i: 118
Cadillac Eldorado Touring Coupe: 131
Chevrolet Cavalier Z24: 137
Ford Thunderbird Turbo Coupe: 301
Ford Scorpio 24v: 338
Honda Prelude 2.0i 16: 358
Isuzu Impulse Turbo: 391
Lancia Delta HF Turbo ie: 434
Mazda RX-7: 484
Mazda RX-7 GTU: 498
Mercedes-Benz 300CE Cabriolet Straman: 542
Mercedes-Benz 300CE: 546
Mitsubishi Cordia Turbo: 568
Mitsubishi Lancer Liftback 1800 GTi-16v: 587
Nissan Maxima SE: 617
Subaru 4WD Turbo Estate: 851
Treser T1: 910
Vauxhall Cavalier GSi 2000 16v 4x4: 939
Volkswagen Jetta GT: 952
Volkswagen GTi 16v: 958
Volkswagen GTi 16v: 961

53 8.6secs
Audi Coupe Quattro: 47
BMW 535i Automatic: 104
Cadillac Seville Touring Sedan: 133
Chevrolet Camaro IROC-Z Convertible: 170
Chevrolet Beretta GT: 168
Chevrolet Camaro Z/28 Convertible: 178
Ford Sierra XR 4x4: 315
Ford Escort XR3i: 342
Honda Integra EX 16: 352
Honda Legend Saloon Automatic: 362
Isuzu Impulse RS: 395
Jaguar XJS: 403
Mazda RX-7 GTUs: 505
Mazda 323 1.8 GT: 501

Mercedes-Benz 300SL-24 5-speed Auto: 552
Nissan Bluebird 1.8 ZX: 601
Renault Cleo 16v: 796
Saab 9000: 826
Saturn Sports Coupe: 840
Toyota Celica GTS: 869
Toyota Celica GT Cabriolet: 876
Vauxhall Cavalier 4x4: 927

54 8.7secs
Audi 80 16v Sport: 66
BMW 525i SE 24v: 119
Jaguar XJS: 397
Jaguar XJS Convertible: 408
Jaguar XJS 4.0 Auto: 418
Mazda 323 GTX: 490
Mercedes-Benz 300TE: 523
Mercedes-Benz 420SEL: 531
Mercedes-Benz 190E 2.6: 539
Mitsubishi Galant GTi-16v: 577
Nissan Primera 2.0E ZX: 631
Peugeot 205 GTi: 644
Peugeot 309 GTi: 649
Porsche 944: 720
Subaru XT Turbo 4WD: 850
Suzuki Swift 1.3 GTi: 865
Volkswagen Corrado 16v: 969

55 8.8secs
Audi Coupe GT 2.2: 46
Chrysler Le Baron Coupe: 190
Honda Prelude Si: 374
Isuzu Impulse Turbo: 393
Jaguar XJSC: 398
Lancia Thema 2.0ie 16v: 442
Mazda 323 1.6i: 500
Mercedes-Benz 260E Auto: 533
Nissan 300ZX: 605
Nissan 240SX: 612
Nissan 240SX: 626
Peugeot 309 GTi: 658
Renault 19 16v Cabriolet: 795
Rover 220 GTi: 821
Rover 216 GTi: 819
Volkswagen Jetta GLi 16v: 962
Volkswagen Passat GT 16v: 972
Volvo 740 Turbo: 995

56 8.9secs
Alfa Romeo Alfa 75 2.5 Green Cloverleaf: 13
Alfa Romeo 33 Boxer 16v: 32
Buick Reatta: 127
Buick Reatta Convertible: 128
Chevrolet Nova Twin Cam: 157
Ford Fiesta XR2i: 320
Ford Escort GT: 333
Honda CRX Si: 350
Mazda MX-3 V6: 513
Nissan Primera 2.0 GSX: 630
Porsche 944: 702
Saab 900 Turbo S: 829
Vauxhall Cavalier SRi: 932

Volkswagen Corrado G60: 973
Volkswagen Corrado G60: 976
Volvo 460 Turbo: 993
57 9.0secs
Alfa Romeo Milano Platinum: 15
Alfa Romeo 164 Automatic: 27
Audi 90 Quattro: 60
Audi V8 Quattro Automatic: 63
Audi V8 Quattro: 62
BMW 735i Automatic: 89
BMW 325 ES: 82
Chevrolet Baretta GTU: 158
Citroen AX GT: 196
Ford Thunderbird LX: 340
Ginetta G32: 346
Lexus ES300: 456
Mitsubishi Lancer GTi 16v: 578
Mitsubishi Diamante LS: 590
Peugeot 405 Mi16: 660
Pontiac Bonneville SSEi: 695
Renault 9 Turbo: 771
Toyota Camry 2.0 GLi: 874
Vauxhall Calibra 4x4: 942
58 9.1secs
Acura Integra LS Special Edition: 4
Alfa Romeo 33 1.7 Veloce: 17
BMW 735i: 88
Citroen Visa GTi: 195
Citroen ZX Volcane: 212
Mazda Protege: 509
Mercury Tracer LTS: 563
Nissan 300ZX: 599
Saab 900SPG: 833
Suzuki Swift 1.3 GTi: 860
Vauxhall Nova GTE: 928
Vauxhall Senator 3.0i 24v: 944
Volkswagen Jetta GLi 16v: 978
59 9.2secs
BMW 520i SE Touring: 124
Ford Escort RS Turbo: 287
Jaguar XJ6 Vanden Plas: 410
Mazda MX-3 GS: 519
Nissan Sunny ZX Coupe: 609
Rover 820i: 824
Toyota Corolla FX-16 GTS: 878
Vauxhall Belmont 1.8 GLSi: 916
Volvo 940 SE: 996
60 9.3secs
Acura Integra RS: 1
Acura Integra LS: 3
Alfa Romeo Alfa 75 2.0i Twin Spark: 19
Audi 90: 55
Audi 100 2.8E Auto: 64
BMW 318i 2-door: 90
BMW 735i: 99
BMW 318i S: 111
Cadillac Allante: 129
Citroen AX GTi: 213
Ford Taurus 3.8: 307
Geo Storm GSi: 345

Honda Prelude 2.0 Si: 357
Honda Concerto 1.6i 16: 368
Isuzu Impulse XS: 394
Mitsubishi Sigma: 593
Rover Sterling Catalyst: 810
Saab CDS 2.3: 834
Toyota Corolla GTS: 892
Toyota MR2: 906
Vauxhall Cavalier 2.0i CD: 922
Vauxhall Senator 3.0i CD: 934
Vauxhall Senator 2.5i: 933
Volvo 940 SE Turbo: 997
Volvo 960 24v: 998
61 9.4secs
Acura Integra GS: 7
Honda Civic Si: 364
Honda Prelude Si 4WS: 365
Mazda 626 2.0i GT 4WS: 493
Nissan Maxima: 616
Renault Espace V6: 794
Rover Sterling Automatic: 825
Volvo 740 GLT 16v Estate: 994
62 9.5secs
Alfa Romeo Alfa 33 1.7 Sportwagon Veloce: 18
Alfa Romeo Sprint 1.7 Cloverleaf: 25
Alfa Romeo 164 Twin Spark Lusso: 30
Audi 90 Quattro: 49
Audi 90 2.2E: 48
Audi 90 Quattro: 53
Audi Coupe Quattro: 58
Citroen XM 3.0 SEi Auto: 209
Ford Sierra Ghia 4x4 Estate: 293
Ford Granada 2.4i Ghia: 296
Ford Granada Ghia X 2.9 EFi Saloon: 322
Lancia Prisma 1600ie LX: 435
Mazda MX-5 Miata: 503
Peugeot 405 Mi16x4: 664
Pontiac Bonneville SE: 681
Renault 5 GTX 3-DR: 775
Renault 25 V6 2.9 Auto: 789
Renault 19 16v: 790
Saab 9000 2.3CS: 838
Suzuki Swift GT: 867
Toyota Celica All-Trac Turbo: 897
Vauxhall Calibra 2.0i: 941
63 9.6secs
Audi 90: 44
Audi Coupe GT: 45
BMW 520i: 117
Ford Escort XR3i: 288
Ford Sierra Sapphire 2.0 Ghia: 299
Saab 9000i: 827
Volkswagen Quantum Synchro Wagon: 953
Volvo 760 GLE: 987
Volvo 740 GLE: 989
Volvo 440 Turbo: 988

64 9.7secs
Acura Integra LS: 5
Audi Coupe 2.2E: 57
Honda Civic 1.4 GL 3-door: 355
Mazda RX-7 Convertible: 497
Mitsubishi Galant GS: 581
Peugeot 309 SR Injection: 646
Peugeot 106 XSi: 668
Pontiac Bonneville SE: 678
Rolls-Royce Corniche: 799
Rover Vitesse: 808
Saab 9000CD: 832
Subaru XT6: 854
Toyota Camry 2.2 GL: 903
65 9.8secs
Alfa Romeo 75 2.5 Auto: 22
Audi 80 1.8E: 52
Ford Granada Scorpio 2.9 EX: 297
Honda Accord LXi: 348
Jaguar XJ6 3.6: 396
Mazda 626 2.0i 5-door: 482
Peugeot 505 STX: 650
Renault 21 Savanna GTX: 768
Renault 21 TXi: 788
Rover Metro GTi 16v: 817
66 9.9secs
Fiat Croma ie Super: 273
Ford Crown Victoria LX: 329
Honda Accord 2.0 EXi: 367
Jaguar XJ6 2.9: 401
Jaguar XJS V12 Convertible: 409
Jaguar XJS: 413
Jaguar XJS Railton: 415
Jaguar XJS Convertible: 419
Volkswagen Golf GTi: 983
67 10.0secs
Audi 100 Sport: 51
Ford Sierra Sapphire 2000E: 314
Ford Orion 1.6i Ghia: 336
Honda Shuttle 1.6i RT4 4WD: 363
Infiniti G20: 387
Lancia Dedra 2.0ie: 444
Mitsubishi Space Runner 1800-16v GLXi: 594
Renault 25 TXi: 783
Rolls-Royce Silver Spirit: 798
Rover 416 GTi: 815
68 10.1secs
BMW 316i 4-door: 102
Citroen CX25 DTR Turbo 2: 200
Citroen AX GT5: 205
Daihatsu Applause 16 Xi: 225
Merkur Scorpio: 564
Nissan 100NX: 625
Rover Sterling Automatic: 805
Subaru 4WD 3-door Turbo Coupe: 848
Toyota Camry 2.0 GLi Estate:

875
Volvo 960: 999
69 10.2secs
BMW 318i: 115
Ford Festiva L: 294
Ford Fiesta 1.6S: 311
Ford Escort Cabriolet: 332
Rover 820SE: 803
70 10.3secs
Alfa Romeo Alfa 75 1.8: 12
Lancia Thema 2.0ie: 436
Mercedes-Benz 300TD: 530
Mercury Tracer: 560
Nissan Pulsar NX SE: 608
Peugeot 405 SRi: 654
Renault 21 Ti: 777
Seat Ibiza SXi: 843
Toyota Paseo: 909
Volkswagen Passat GT: 964
Volvo 480 ES: 985
71 10.4secs
Alfa Romeo Spider: 24
Ford Taurus LX: 344
Infiniti M30: 385
Mazda 626 2.0 GLS Executive: 488
Mitsubishi Galant Sapporo: 572
Mitsubishi Galant 2000 GLSi Coupe: 585
Oldsmobile Toronado Trofeo: 639
Rolls-Royce Silver Spirit II: 801
Vauxhall Carlton CD 2.0i: 917
72 10.5secs
BMW 520i: 84
BMW 730i SE: 92
Eagle Premier ES: 242
Fiat Tipo 1.8ie DGT SX: 284
Mazda 929: 494
Nissan Sunny 1.6 GS 5-door: 633
Peugeot 505 GTi Family Estate: 647
Toyota Carina 1.6 GL: 882
Toyota Celica GTS: 898
Volkswagen Golf Cabriolet Clipper: 957
73 10.6secs
Audi 100 Avant Quattro: 50
Audi 80 2.0E: 59
Citroen BX GTi 4x4: 206
Peugeot 605 SV 3.0 Auto: 671
Seat Toledo 2.0 GTi: 845
Vauxhall Belmont 1.6 GL: 920
74 10.7secs
Chevrolet Celebrity CL Eurosport: 138
75 10.8secs
Alfa Romeo Spider Quadrifoglio: 29
Alfa Romeo Spider Veloce: 34
Range Rover Vogue SE: 764
Renault Espace TXE: 785

35

Subaru Legacy 2.2 GX 4WD: 855

Toyota Celica GT Convertible: 904

Vauxhall Carlton L 1.8i: 925

Volkswagen Fox GL: 956

76 10.9secs

Austin Montego 2.0 Vanden Plas EFI: 70

Hyundai Lantra 1.6 Cdi: 381

Mitsubishi Galant 2000 GLSi: 576

Peugeot 405 1.6 GL: 651

Peugeot 405 GLx4: 662

Renault 21 GTS: 767

Rover 216 GSi: 813

Vauxhall Senator 3.0i CD: 924

Volkswagen Passat GT 5-door: 960

Volkswagen Passat GL: 975

77 11.0secs

Alfa Romeo Spider Quadrifoglio: 16

BMW 524 TD: 85

Citroen AX14 TRS: 198

Mazda 121 1.3 LX Sun Top: 489

Peugeot 106 XT: 673

Rover Mini Cooper S: 823

78 11.1secs

Audi 80 1.8S: 43

Honda Ballade EX: 354

Mazda 323 LX: 480

Mazda 323 1.5 GLX: 487

Renault Alliance GTA: 776

Toyota Corolla Executive: 877

Toyota Camry V6 GXi: 890

Volvo 460 GLi: 992

79 11.2secs

Citroen XM 2.0SEi: 207

Daihatsu Charade 1.3 CX: 222

Honda Civic 1.5 VEi: 370

Mercedes-Benz 190: 537

80 11.3secs

Daihatsu Charade Turbo: 221

Fiat Tipo 1.6 DGT SX: 278

Range Rover Vogue SE: 763

Rolls-Royce Silver Shadow II: 797

Suzuki Swift 1.3 GLX: 862

Volkswagen Jetta Syncro: 971

81 11.4secs

Renault 19 TXE Chamade: 786

82 11.5secs

Daihatsu Charade CLS: 223

Jaguar XJ6: 400

Lancia Y10 GTie: 446

Lancia Dedra 1.8i: 443

Rover Montego 1.6L: 804

Rover 414 Si: 814

Volkswagen Golf CL Catalyst: 970

83 11.6secs

Audi 5000 S: 42

Fiat Croma CHT: 280

Ford Granada Scorpio 2.0i Auto: 323

Nissan Laurel 2.4 SGL Automatic: 602

Nissan Sunny 1.6 SLX Coupe: 610

Renault 21 GTS Hatchback: 782

Rover 820 Fastback: 806

Suzuki Swift 1.3 GLX: 864

84 11.7secs

Alfa Romeo Alfa 33 1.3 S: 11

Hyundai Sonata 2.0i GLS: 379

Land Rover Discovery V8i 5DR: 453

Mazda 626 2.0 GLX Executive Estate: 492

Mazda MPV: 502

Mercury Sable LS: 559

Mitsubishi Lancer GLXi 4WD Liftback: 586

Peugeot 205 CJ: 656

Range Rover Vogue SE: 761

Renault 19 TXE: 781

Rover 214 GSi: 812

Rover 825 TD: 822

Volkswagen Polo GT Coupe: 981

Yugo 65A GLX: 1003

85 11.8secs

Austin Metro GTa: 72

Honda Integra 1.5: 351

Peugeot 605 SRi: 666

Vauxhall Nova 1.3 GL 5-door: 918

86 11.9secs

Citroen CX25 Ri Familiale Auto: 194

Citroen BX DTR Turbo: 202

Citroen ZX 1.4 Avantage: 211

Peugeot 405 GR Injection Auto: 663

Range Rover Vogue: 757

87 12.0secs

Fiat Regata 100S ie: 274

Fiat X1/9: 277

Nissan Sunny 1.4 LS: 620

Renault Clio 1.4RT: 793

Rover Metro 1.4SL: 816

88 12.1secs

Pontiac Le Mans: 680

Range Rover Vogue: 760

Toyota Corolla 4WD Estate: 886

Vauxhall Cavalier 1.6L: 931

89 12.2secs

Fiat Tempra 1.6 SX: 281

Honda Shuttle 1.4 GL Automatic: 366

Mitsubishi Colt 1500 GLX 5-door: 570

Peugeot 405 GTD Turbo: 659

Toyota Carina II 1.6 GL

Liftback: 891

Toyota Tercel LE: 908

Vauxhall Nova 1.5TD Merit: 940

90 12.3secs

Fiat X1/9: 279

Ford Escort 1.6 Ghia Estate: 331

Hyundai Sonata 2.4i GLS: 380

Mazda 323 1.5 GLX Estate: 479

Mercedes-Benz 200E Automatic: 541

Mercedes-Benz 190E 1.8 Auto: 551

Peugeot 205 D Turbo: 669

Range Rover County: 762

91 12.4secs

Ford Escort 1.4 GL: 286

Nissan Primera 1.6 LS: 623

Proton 1.5 SE Triple Valve Aeroback: 756

Volvo 440 GLE: 991

92 12.5secs

Hyundai S Coupe GSi: 382

Toyota Previa LE: 901

93 12.6secs

Audi 100 Turbo Diesel: 41

Chevrolet Lumina APV: 176

Citroen AX11 TRE: 197

Citroen AX11 TRE 5DR: 201

Range Rover V8: 758

Rolls-Royce Silver Spur: 800

Toyota Landcruiser VX: 905

94 12.7secs

Volkswagen Golf 1.8 GL: 982

95 12.8secs

Ford Orion 1.4 GL: 291

Ford Escort 1.4LX: 330

Land Rover Discovery V8: 452

MG B British Motor Heritage: 567

Mitsubishi Shogun V6 5-door: 582

Peugeot 605 SR TD: 670

Renault 19 TSE: 780

Toyota Corolla 1.3 GL: 885

Vauxhall 1.4i GLS: 945

Vauxhall Cavalier 1.7 GL TD: 946

96 12.9secs

Hyundai Excel GLS: 378

Isuzu Impulse Sports Coupe: 390

Peugeot 309 1.6 GR: 645

Toyota Previa: 907

Vauxhall Cavalier 1.4L: 930

Volkswagen Jetta Turbo Diesel: 963

97 13.0secs

Austin Metro 1.3L 3-door: 71

Chevrolet Nova CL: 142

Mercedes-Benz 200: 521

Subaru Justy: 849

98 13.1secs

Citroen XM Turbo SD Estate: 214

Fiat Tipo 1.4: 276

Mitsubishi Galant 2000 GLS Automatic: 571

Mitsubishi Shogun V6 LWB: 592

99 13.2secs

Rover Montego 2.0 DSL Turbo: 809

Yugo Sana 1.4: 1004

100 13.3secs

Nissan Sentra Sport Coupe SE: 603

Peugeot 205 1.1GL: 655

Rover 214S: 818

101 13.4secs

Ford Fiesta 1.4 Ghia: 310

Ford Granada 2.5 GL Diesel: 321

Lada Samara 1300L 5-door: 426

Mazda MX-3 1.6 Auto: 518

Volvo 340 GLE: 984

102 13.5secs

Citroen AX11 TZX: 208

Volkswagen Polo 1.3 CL Coupe: 979

103 13.6secs

Land Rover Ninety County V8: 450

104 13.7secs

Hyundai X2 1.5 GSi: 383

Nissan Sunny 1.3 LX 5-door: 604

105 13.8secs

Peugeot 106 XR: 667

106 13.9secs

Yugo GV: 1001

107 14.0secs

Lada Samara 1300 SL: 425

Mitsubishi Lancer 1.5 GLX Estate: 569

Nissan Micra 1.2 GSX: 618

Renault Clio 1.2 RN: 792

108 14.1secs

Ford Escort 1.3L 3-door: 308

109 14.2secs

Fiat Uno 60S: 282

110 14.3secs

Citroen BX14 TGE: 204

111 14.4secs

Ford Aerostar XLT: 285

Nissan Prairie: 619

Proton 1.5 SE Aeroback: 755

Subaru Justy GL II 5-door 4WD: 852

112 14.5secs

Nissan Micra SGL 5-DR: 607

Suzuki Swift 1.3 GLX Executive: 863

Suzuki Vitara JLX: 866

Vauxhall Nova 1.2 Luxe: 943

113 14.7secs
Daihatsu Domino: 217
Peugeot 309 Style 5-door: 661
Toyota Starlet GL: 902
114 14.8secs
Renault 5 GTL 5-door: 770
115 14.9secs
Skoda 136 Rapide Coupe:
847
116 15.1secs
Land Rover One Ten County
V8: 449
Vauxhall Astra 1.7 DL: 929
117 15.3secs
Ford Fiesta 1.1LX 5-door: 309
Mazda 121 GSX: 511
118 15.4secs
Peugeot 405 GRD: 652
119 15.5secs
Mercedes-Benz 300D: 548
120 15.6secs
Fiat Regata DS: 272
Seat Toledo 1.9 CL Diesel:
844
Volkswagen Passat CL TD
Estate: 974
121 15.7secs
Seat Malaga 1.2L: 842
122 15.8secs
Ford Fiesta 1.1 Ghia Auto:
295
123 16.0secs
Fiat Panda 1000 S: 270
124 16.1secs
Daihatsu Sportrak EL: 224
125 16.2secs
Citroen AX14 DTR: 203
Vauxhall Carlton L 2.3D: 926
126 16.3secs
Renault 5 Campus: 784
127 16.4secs
Lada Riva 1500 Estate: 423
128 16.5secs
Peugeot 309 GLX: 657
Range Rover Vogue Turbo D:
759
129 16.6secs
Fiat Uno Selecta: 275
130 16.7secs
Lancia Y10 Fire: 437
131 16.9secs
Rover 218 SD: 820
Volkswagen Polo C Saloon:
965
Volkswagen Caravelle Carat
automatic: 967
132 17.1secs
Land Rover Discovery TDi:
451
133 17.8secs
Mitsubishi Shogun Turbo
Diesel 5-door: 573
134 17.9secs
Citroen BX19 RD Estate: 193

Daihatsu Fourtrak Estate DT
EL: 220
135 18.1secs
Vauxhall Frontera 2.3 TD
Estate: 947
136 18.9secs
Lada Niva Cossack Cabrio:
424
137 19.4secs
Suzuki SJ413V JX: 861
138 19.9secs
Skoda Estelle 120 LSE: 846
139 20.6secs
Fiat Panda 750 L: 271
140 20.9secs
Daihatsu Charade CX Diesel
Turbo: 218
141 21.6secs
Toyota Landcruiser: 879
Yugo 45 GLS: 1002
142 22.3secs
Land Rover Ninety County
Turbo Diesel: 448
143 23.5secs
Toyota Landcruiser II TD: 887
144 23.6secs
Carbodies Fairway 2.7 Silver:
134
145 24.4secs
Suzuki Alto GLA: 859
146 24.7secs
Volkswagen Caravelle Syncro:
950
147 25.4secs
Mercedes-Benz G-Wagen 300
GD LWB: 556
148 26.1secs
Isuzu Trooper 3-door TD: 392

Acceleration times in make order

Acura

1 5.7secs
NSX: 9
2 7.9secs
Legend LS: 10
3 8.0secs
Legend Coupe L: 6
Legend Coupe: 8
4 8.1secs
Legend: 2
5 9.1secs
Integra LS Special Edition: 4
6 9.3secs
Integra RS: 1
Integra LS: 3
7 9.4secs
Integra GS: 7
8 9.7secs
Integra LS: 5

Alfa Romeo

1 7.5secs
Milano Verde 3.0: 21
Alfa 75 V6 3.0: 20
75 Evoluzione: 23
2 7.6secs
164S: 31
3 7.8secs
164 Cloverleaf: 33
4 7.9secs
Milano 3.0: 28
164 3.0 Lusso: 26
5 8.2secs
GTV6: 14
6 8.9secs
Alfa 75 2.5 Green Cloverleaf: 13
33 Boxer 16v: 32
7 9.0secs
Milano Platinum: 15
164 Automatic: 27
8 9.1secs
33 1.7 Veloce: 17
9 9.3secs
Alfa 75 2.0i Twin Spark: 19
10 9.5secs
Alfa 33 1.7 Sportwagon Veloce: 18
Sprint 1.7 Cloverleaf: 25
164 Twin Spark Lusso: 30
11 9.8secs
75 2.5 Auto: 22
12 10.3secs
Alfa 75 1.8: 12
13 10.4secs
Spider: 24
14 10.8secs
Spider Quadrifoglio: 29
Spider Veloce: 34
15 11.0secs

Spider Quadrifoglio: 16
16 11.7secs
Alfa 33 1.3 S: 11

Aston Martin

1 4.8secs
Zagato: 36
2 5.6secs
Zagato Volante: 39
3 6.0secs
Virage: 37
4 6.8secs
Virage: 40
5 7.8secs
Volante: 38
6 8.2secs
Volante: 35

Audi

1 5.9secs
Coupe S2: 67
2 6.3secs
Quattro: 54
Quattro 20v: 61
3 7.2secs
200 Quattro: 65
4 7.5secs
V8 Quattro: 68
5 8.3secs
80 2.8E V6 Quattro: 69
6 8.5secs
90 Quattro 20v: 56
7 8.6secs
Coupe Quattro: 47
8 8.7secs
80 16v Sport: 66
9 8.8secs
Coupe GT 2.2: 46
10 9.0secs
V8 Quattro Automatic: 63
90 Quattro: 60
V8 Quattro: 62
11 9.3secs
90: 55
100 2.8E Auto: 64
12 9.5secs
90 Quattro: 49
90 2.2E: 48
90 Quattro: 53
Coupe Quattro: 58
13 9.6secs
Coupe GT: 45
90: 44
14 9.7secs
Coupe 2.2E: 57
15 9.8secs
80 1.8E: 52
16 10.0secs
100 Sport: 51
17 10.6secs

100 Avant Quattro: 50
80 2.0E: 59
18 11.1secs
80 1.8S: 43
19 11.6secs
5000 S: 42
20 12.6secs
100 Turbo Diesel: 41

Austin

1 10.9secs
Montego 2.0 Vanden Plas EFI: 70
2 11.8secs
Metro GTa: 72
3 13.0secs
Metro 1.3L 3-door: 71

Autokraft

1 5.0secs
AC Cobra Mk IV: 73

Bentley

1 6.7secs
Turbo R: 77
2 6.9secs
Mulsanne Turbo: 75
3 7.0secs
Mulsanne Turbo: 74
Turbo R: 76
4 7.1secs
Turbo R: 78

Bertone

1 5.0secs
Emotion Lotus: 79

Bitter

1 7.6secs
Type 3 Cabrio: 81
2 8.3secs
SC: 80

BMW

1 5.1secs
535i Alpina B10 Bi-Turbo: 120
2 6.0secs
M635 CSi: 109
735i Koenig: 112
3 6.3secs
M5: 108
4 6.4secs
M5: 122
5 6.6secs
M3 Evolution: 107
6 6.9secs

750i L: 100
750i Alpina B12: 114
7 7.1secs
M3: 94
M3: 106
850i Manual: 125
8 7.2secs
M6: 101
850i: 121
9 7.3secs
750i: 113
325i SE: 116
10 7.4secs
535i SE: 97
735i Alpina B11: 105
11 7.5secs
325i Cabriolet: 91
325i S: 95
12 7.7secs
750i L: 93
535i: 103
13 7.9secs
535i: 86
325i X: 96
Z1: 110
325i: 123
14 8.1secs
325i Convertible: 83
15 8.4secs
635 CSi: 98
16 8.5secs
635 CSi: 87
525i: 118
17 8.6secs
535i Automatic: 104
18 8.7secs
525i SE 24v: 119
19 9.0secs
325 ES: 82
735i Automatic: 89
20 9.1secs
735i: 88
21 9.2secs
520i SE Touring: 124
22 9.3secs
318i 2-door: 90
735i: 99
318i S: 111
23 9.6secs
520i: 117
24 10.1secs
316i 4-door: 102
25 10.2secs
318i: 115
26 10.5secs
520i: 84
730i SE: 92
27 11.0secs
524 TD: 85

Bugatti

1 3.9secs
EB110: 126

Buick

1 8.9secs
Reatta: 127
Reatta Convertible: 128

Cadillac

1 8.3secs
Allante: 130
Allante: 132
2 8.5secs
Eldorado Touring Coupe: 131
3 8.6secs
Seville Touring Sedan: 133
4 9.3secs
Allante: 129

Carbodies

1 23.6secs
Fairway 2.7 Silver: 134

Caterham

1 5.2secs
7 HPC: 135
2 5.6secs
Super 7: 136

Chevrolet

1 3.4secs
Corvette Lingenfelter: 181
2 3.9secs
Corvette Callaway
Sledgehammer: 174
3 4.4secs
Corvette Callaway Speedster:
180
4 4.5secs
Corvette Bakeracing SCCA
Escort: 151
Camaro IROC-Z Chevrolet
Engineering: 162
5 4.6secs
Corvette Callaway: 152
6 4.8secs
Camaro Gottlieb 1969: 159
7 4.9secs
Corvette Morrison/Baker
Nelson Ledges: 155
Corvette Callaway: 164
Corvette ZR-1: 167
Corvette ZR-1 Geiger: 184
Corvette Callaway: 179
Corvette ZR-2: 185
Corvette ZR-1: 183
8 5.0secs
Corvette Callaway: 140

9 5.1secs
Corvette Guldstrand Grand
Sport 80: 165
Corvette Callaway: 173
10 5.2secs
Corvette Guldstrand Grand
Sport 80: 154
11 5.7secs
Corvette LT1: 188
12 5.8secs
Corvette Z51: 141
Corvette: 139
Corvette ASC Geneve: 147
13 5.9secs
Corvette: 146
14 6.0secs
Corvette Z51: 156
Corvette Convertible: 153
Corvette: 150
15 6.2secs
Corvette: 163
16 6.3secs
Corvette Convertible: 148
Corvette L98: 175
Corvette Rick Mears: 182
17 6.6secs
Camaro IROC-Z: 144
Corvette L98: 166
Camaro IROC-Z: 160
Camaro IROC-Z Automatic:
161
Camaro Z/28: 171
Camaro Z28: 187
19 6.8secs
Camaro IROC-Z L98: 145
20 8.0secs
Beretta GTZ: 169
Baretta GTZ: 177
Lumina Z34: 186
21 8.1secs
Cavalier Z24: 172
22 8.2secs
Cavalier Z24 Convertible: 149
23 8.3secs
Beretta GT: 143
24 8.5secs
Cavalier Z24: 137
25 8.6secs
Camaro IROC-Z Convertible:
170
Beretta GT: 168
Camaro Z/28 Convertible: 178
26 8.9secs
Nova Twin Cam: 157
27 9.0secs
Baretta GTU: 158
28 10.7secs
Celebrity CL Eurosport: 138
29 12.6secs
Lumina APV: 176
30 13.0secs
Nova CL: 142

Chrysler

1 6.9secs
TC by Maserati: 192
2 7.3secs
Conquest TSi: 191
3 8.1secs
Laser XT: 189
4 8.8secs
Le Baron Coupe: 190

Citroen

1 7.5secs
XM V6.24: 210
2 7.9secs
BX GTi 16v: 199
3 9.0secs
AX GT: 196
4 9.1secs
Visa GTi: 195
ZX Volcane: 212
5 9.3secs
AX GTi: 213
6 9.5secs
XM 3.0 SEi Auto: 209
7 10.1secs
CX25 DTR Turbo 2: 200
AX GT5: 205
8 10.6secs
BX GTi 4x4: 206
9 11.0secs
AX14 TRS: 198
10 11.2secs
XM 2.0SEi: 207
11 11.9secs
CX25 Ri Familiale Auto: 194
BX DTR Turbo: 202
ZX 1.4 Avantage: 211
12 12.6secs
AX11 TRE: 197
AX11 TRE 5DR: 201
13 13.1secs
XM Turbo SD Estate: 214
14 13.5secs
AX11 TZX: 208
15 14.3secs
BX14 TGE: 204
16 16.2secs
AX14 DTR: 203
17 17.9secs
BX19 RD Estate: 193

Cizeta

1 4.5secs
Moroder V16T: 215

Consulier

1 5.7secs
GTP LX: 216

Daihatsu

1 7.9secs
Charade GT ti: 219
2 10.1secs
Applause 16 Xi: 225
3 11.2secs
Charade 1.3 CX: 222
4 11.3secs
Charade Turbo: 221
5 11.5secs
Charade CLS: 223
6 14.7secs
Domino: 217
7 16.1secs
Sportrak EL: 224
8 17.9secs
Fourtrak Estate DT EL: 220
9 20.9secs
Charade CX Diesel Turbo: 218

De Tomaso

1 4.3secs
Pantera Group 3: 227
2 5.4secs
De Pantera GT5-S: 226

Dodge

1 4.1secs
Viper: 241
2 6.3secs
Stealth R/T Turbo: 240
3 6.5secs
Spirit R/T: 239
4 6.7secs
Shelby GLH-S: 228
5 6.9secs
Shelby CSX: 231
6 7.2secs
Daytona Shelby Z: 229
Shelby CSX VNT: 235
7 7.3secs
Shadow ES VNT: 238
8 7.5secs
Daytona Shelby: 232
9 7.6secs
Lancer Shelby: 230
Shelby CSX: 234
Daytona Shelby: 237
10 7.7secs
Shadow ES Turbo: 233
11 8.4secs
Daytona ES: 236

Eagle

1 5.0secs
GTP: 244
2 6.8secs
Talon: 243
Talon TSi 4WD: 245

3   10.5secs
Premier ES: 242

Ferrari

1   3.7secs
308 Norwood Bonneville GTO:
253
2   3.8secs
F40: 267
3   4.5secs
F40: 263
4   4.7secs
Testa Rossa Norwood: 261
308 Norwood: 257
5   5.2secs
Testa Rossa: 259
6   5.3secs
Testa Rossa Straman Spyder:
256
Testa Rossa: 255
Testa Rossa Gemballa: 260
7   5.6secs
348tb: 262
8   5.9secs
328 GTS: 250
9   6.0secs
348tb: 266
10 6.2secs
Pininfarina Mythos: 265
11 6.6secs
328 GTB: 247
Mondial: 264
Mondial t: 268
Mondial t Cabriolet: 269
12 6.7secs
412: 251
13 6.8secs
Mondial 3.2QV: 249
14 7.0secs
Mondial Cabriolet 3.2: 252
Mondial Cabriolet: 254
Mondial 3.2 Cabriolet: 258
15 7.1secs
3.2 Mondial: 246
16 7.4secs
Mondial 3.2: 248

Fiat

1   8.3secs
Uno Turbo ie: 283
2   9.9secs
Croma ie Super: 273
3   10.5secs
Tipo 1.8ie DGT SX: 284
4   11.3secs
Tipo 1.6 DGT SX: 278
5   11.6secs
Croma CHT: 280
6   12.0secs
Regata 100S ie: 274
X1/9: 277

7   12.2secs
Tempra 1.6 SX: 281
8   12.3secs
X1/9: 279
9   13.1secs
Tipo 1.4: 276
10 14.2secs
Uno 60S: 282
11 15.6secs
Regata DS: 272
12 16.0secs
Panda 1000 S: 270
13 16.6secs
Uno Selecta: 275
14 20.6secs
Panda 750 L: 271

Ford

1   4.9secs
Mustang Cartech Turbo: 302
Mustang Holdener: 334
2   5.2secs
Mustang SVO J Bittle
American: 313
Thunderbird Super Coupe
Engineering: 318
3   5.4secs
Mustang 5.0 Cartech Turbo:
312
4   5.7secs
Mustang NOS/Saleen: 335
5   5.8secs
Sierra RS Cosworth: 306
6   6.0secs
Mustang Convertible Saleen:
303
Mustang GT: 304
7   6.1secs
RS200: 292
8   6.6secs
Taurus SHO: 316
Sierra Sapphire Cosworth 4x4:
328
Mustang LX 5.0L: 325
9   6.7secs
Mustang GT: 298
10 6.8secs
Thunderbird Bondurant 5.0:
300
11 6.9secs
Mustang GT 5.0: 289
12 7.1secs
Mustang SVO: 290
Mustang LX: 343
13 7.3secs
Probe GT: 305
14 7.4secs
Thunderbird Super Coupe
317
Probe GT: 337
15 7.6secs
Probe GT: 326
Taurus SHO: 339

16 7.9secs
Fiesta RS Turbo: 319
17 8.0secs
Mustang GT Convertible: 324
18 8.2secs
Probe LX: 327
19 8.3secs
Escort RS2000: 341
20 8.5secs
Thunderbird Turbo Coupe:
301
Scorpio 24v: 338
21 8.6secs
Sierra XR 4x4: 315
Escort XR3i: 342
22 8.9secs
Fiesta XR2i: 320
Escort GT: 333
23 9.0secs
Thunderbird LX: 340
24 9.2secs
Escort RS Turbo: 287
25 9.3secs
Taurus 3.8: 307
26 9.5secs
Sierra Ghia 4x4 Estate: 293
Granada 2.4i Ghia: 296
Granada Ghia X 2.9 EFi
Saloon: 322
27 9.6secs
Escort XR3i: 288
Sierra Sapphire 2.0 Ghia: 299
28 9.8secs
Granada Scorpio 2.9 EX: 297
29 9.9secs
Crown Victoria LX: 329
30 10.0secs
Sierra Sapphire 2000E: 314
Orion 1.6i Ghia: 336
31 10.2secs
Festiva L: 294
Fiesta 1.6S: 311
Escort Cabriolet: 332
32 10.4secs
Taurus LX: 344
33 11.6secs
Granada Scorpio 2.0i Auto:
323
34 12.3secs
Escort 1.6 Ghia Estate: 331
35 12.4secs
Escort 1.4 GL: 286
36 12.8secs
Orion 1.4 GL: 291
Escort 1.4LX: 330
37 13.4secs
Fiesta 1.4 Ghia: 310
Granada 2.5 GL Diesel: 321
38 14.1secs
Escort 1.3L 3-door: 308
39 14.4secs
Aerostar XLT: 285
40 15.3secs
Fiesta 1.1LX 5-door: 309

41 15.8secs
Fiesta 1.1 Ghia Auto: 295

Geo

1   9.3secs
Storm GSi: 345

Ginetta

1   9.0secs
G32: 346

GM

1   8.0secs
Impact: 347

Honda

1   5.8secs
NSX: 373
2   6.8secs
NSX Auto: 376
3   7.6secs
Civic 1.6i VT: 371
4   7.9secs
Prelude Si 4WS: 377
5   8.0secs
Legend: 353
CRX Coupe 1.6i 16: 349
Legend Coupe: 356
Civic CRX: 359
CRX Si Jackson: 361
CRX 1.6i VT: 369
6   8.1secs
Legend Coupe: 375
7   8.2secs
CRX Si: 360
Legend: 372
8   8.5secs
Prelude 2.0i 16: 358
9   8.6secs
Integra EX 16: 352
Legend Saloon Automatic: 362
10 8.8secs
Prelude Si: 374
11 8.9secs
CRX Si: 350
12 9.3secs
Prelude 2.0 Si: 357
Concerto 1.6i 16: 368
13 9.4secs
Prelude Si 4WS: 365
Civic Si: 364
14 9.7secs
Civic 1.4 GL 3-door: 355
15 9.8secs
Accord LXi: 348
16 9.9secs
Accord 2.0 EXi: 367
17 10.0secs
Shuttle 1.6i RT4 4WD: 363
18 11.1secs

Ballade EX: 354
19 11.2secs
Civic 1.5 VEi: 370
20 11.8secs
Integra 1.5: 351
21 12.2secs
Shuttle 1.4 GL Automatic: 366

Hyundai

1 10.9secs
  Lantra 1.6 Cdi: 381
2 11.7secs
  Sonata 2.0i GLS: 379
3 12.3secs
  Sonata 2.4i GLS: 380
4 12.5secs
  S Coupe GSi: 382
5 12.9secs
  Excel GLS: 378
6 13.7secs
  X2 1.5 GSi: 383

IAD

1 5.0secs
Venus: 384

Infiniti

1 6.9secs
Q45: 386
2 10.0secs
G20: 387
3 10.4secs
M30: 385

Irmsher

1 7.9secs
GT: 388

Isdera

1 5.0secs
Imperator: 389

Isuzu

1 8.5secs
Impulse Turbo: 391
2 8.6secs
Impulse RS: 395
3 8.8secs
Impulse Turbo: 393
4 9.3secs
Impulse XS: 394
5 12.9secs
Impulse Sports Coupe: 390
6 26.1secs
Trooper 3-door TD: 392

Jaguar

1 3.5secs
XJ220: 407
2 4.4secs
XJS Lister Le Mans: 414
3 7.2secs
XJSC HE: 399
4 7.5secs
XJS Koenig: 404
5 7.7secs
XJRS: 412
6 7.8secs
XJS 3.6 Automatic: 402
XJS: 417
7 8.0secs
XJS V12 Convertible: 405
8 8.3secs
4.0 Sovereign: 406
XJR 4.0: 411
XJ6 3.2: 416
9 8.6secs
XJS: 403
10 8.7secs
XJS: 397
XJS Convertible: 408
XJS 4.0 Auto: 418
11 8.8secs
XJSC: 398
12 9.2secs
XJ6 Vanden Plas: 410
13 9.8secs
XJ6 3.6: 396
14 9.9secs
XJ6 2.9: 401
XJS V12 Convertible: 409
XJS: 413
XJS Railton: 415
XJS Convertible: 419
15 11.5secs
XJ6: 400

Koenig

1 3.5secs
Competition Evolution: 421
C62: 422
2 4.0secs
Competition: 420

Lada

1 13.4secs
Samara 1300L 5-door: 426
2 14.0secs
Samara 1300 SL: 425
3 16.4secs
Riva 1500 Estate: 423
4 18.9secs
Niva Cossack Cabrio: 424

Lamborghini

1 4.5secs

Diablo: 432
2 4.7secs
Countach 25th Anniversary:
431
3 5.2secs
Countach 5000S: 427
4 6.2secs
Jalpa 3500: 428
5 6.8secs
Jalpa: 429
6 7.8secs
LM129: 430

Lancia

1 6.3secs
Delta Integrale 16v: 441
2 6.4secs
Delta HF Integrale: 439
3 6.6secs
Delta HF 4WD: 433
4 6.8secs
Thema 8.32: 438
Thema 2.0ie 16v SE Turbo:
445
5 7.2secs
Thema 8.32: 440
6 7.4secs
Dedra 2000 Turbo: 447
7 8.5secs
Delta HF Turbo ie: 434
8 8.8secs
Thema 2.0ie 16v: 442
9 9.5secs
Prisma 1600ie LX: 435
10 10.0secs
Dedra 2.0ie: 444

11 10.3secs
Thema 2.0ie: 436
12 11.5secs
Dedra 1.8i: 443
Y10 GTie: 446
13 16.7secs
Y10 Fire: 437

Land Rover

1 11.7secs
Discovery V8i 5DR: 453
2 12.8secs
Discovery V8: 452
3 13.6secs
Ninety County V8: 450
4 15.1secs
One Ten County V8: 449
5 17.1secs
Discovery TDi: 451
6 22.3secs
Ninety County Turbo Diesel:
448

Lexus

1 6.9secs
SC400: 455
2 8.3secs
LS400: 454
3 9.0secs
ES300: 456

Lincoln

1 8.0secs
Continental Mk VII: 457
2 8.3secs
Continental Mk VII LSC: 458

Lotus

1 4.9secs
Esprit Turbo SE: 466
2 5.1secs
Esprit Turbo SE: 469
3 5.2secs
Esprit Turbo: 465
4 5.3secs
Elan Autocrosser: 461
5 5.4secs
Esprit Turbo: 464
6 5.6secs
Esprit Turbo: 459
Esprit Turbo HC: 462
7 6.5secs
Elan SE: 468
8 6.6secs
Elan SE: 470
9 6.7secs
Elan: 467
10 6.8secs
Excel SE: 460
11 8.2secs
Excel SA: 463

Lynx

1 5.3secs
D Type: 471
2 7.6secs
XJS Eventer: 472

Maserati

1 6.3secs
430: 476
2 6.5secs
430: 474
3 6.7secs
Spyder: 477
228: 478
4 6.9secs
Biturbo: 475
5 7.2secs
Biturbo Spyder: 473

Mazda

1 4.8secs
RX-7 Mariah Mode Six: 516
2 5.5secs
RX-7 Cartech: 514
3 6.4secs
MX-5 Miata Millen: 507
RX-7 Turbo: 510
RX-7 Turbo: 517
4 6.6secs
RX-7 Turbo: 486
RX-7 Turbo: 499
5 6.7secs
RX-7 Turbo II: 506
6 7.0secs
RX-7 Infini IV: 515
7 7.2secs
MX-6 GT: 496
8 7.5secs
MX-6: 495
9 7.8secs
626 GT: 483
323 Turbo 4x4: 491
323 1.8 GT: 512
10 7.9secs
323 Turbo 4x4 Lux: 481
RX-7 GXL: 485
11 8.2secs
MX-6 GT 4WS: 504
MX-6 GT: 508
12 8.5secs
RX-7: 484
RX-7 GTU: 498
13 8.6secs
RX-7 GTUs: 505
323 1.8 GT: 501
14 8.7secs
323 GTX: 490
15 8.8secs
323 1.6i: 500
16 8.9secs
MX-3 V6: 513
17 9.1secs
Protege: 509
18 9.2secs
MX-3 GS: 519
19 9.4secs
626 2.0i GT 4WS: 493
20 9.5secs
MX-5 Miata: 503
21 9.7secs
RX-7 Convertible: 497
22 9.8secs
626 2.0i 5-door: 482
23 10.4secs
626 2.0 GLS Executive: 488
24 10.5secs
929: 494
25 11.0secs
121 1.3 LX Sun Top: 489
26 11.1secs
323 LX: 480
323 1.5 GLX: 487

27 11.7secs
626 2.0 GLX Executive Estate:
492
MPV: 502
28 12.3secs
323 1.5 GLX Estate: 479
29 13.4secs
MX-3 1.6 Auto: 518
30 15.3secs
121 GSX: 511

Mercedes-Benz

1 5.0secs
300CE AMG Hammer: 547
2 5.2secs
300E AMG Hammer: 536
3 5.9secs
500SL: 550
4 6.1secs
600SEL: 558
5 6.3secs
500E: 554
6 6.4secs
500SL: 555
7 6.8secs
560SL: 544
8 6.9secs
560SEC Cabriolet Straman:
532
9 7.0secs
560SEC: 525
190E AMG Baby Hammer: 540
10 7.1secs
560SEL: 526
190E 2.5-16 Evolution II: 545
11 7.2secs
190E 2.5-16: 538
12 7.5secs
300E: 522
560SL: 527
300CE: 529
500SEL: 557
13 7.6secs
420SE: 524
14 7.8secs
190E 2.3-16: 520
300E-24: 549
15 8.0secs
300CE: 534
16 8.3secs
190E 2.6: 528
300E 4Matic: 535
300E: 543
17 8.4secs
400SE: 553
18 8.5secs
300CE Cabriolet Straman: 542
300CE: 546
19 8.6secs
300SL-24 5-speed Auto: 552
20 8.7secs
300TE: 523
420SEL: 531

190E 2.6: 539
21 8.8secs
260E Auto: 533
22 10.3secs
300TD: 530
23 11.2secs
190: 537
24 12.3secs
200E Automatic: 541
190E 1.8 Auto: 551
25 13.0secs
200: 521
26 15.5secs
300D: 548
27 25.4secs
G-Wagen 300 GD LWB: 556

Mercury

1 7.4secs
Cougar XR-7: 561
2 8.3secs
Capri XR-2: 562
3 9.1secs
Tracer LTS: 563
4 10.3secs
Tracer: 560
5 11.7secs
Sable LS: 559

Merkur

1 8.1secs
XR4Ti: 565
2 10.1secs
Scorpio: 564

MG

1 6.9secs
Maestro Turbo: 566
2 12.8secs
B British Motor Heritage: 567

Mitsubishi

1 6.3secs
3000GT VR-4: 588
2 6.9secs
Starion 2000 Turbo: 574
3 7.2secs
Eclipse: 580
4 7.8secs
Starion 2.6 Turbo: 583
5 7.9secs
Galant VR-4: 591
6 8.0secs
Starion ESI-R: 584
7 8.2secs
Mirage Turbo: 579
8 8.3secs
Colt 1800 GTI-16v: 589
9 8.4secs
Cordia Turbo: 575

10 8.5secs
Cordia Turbo: 568
Lancer Liftback 1800 GTi-16v:
587
11 8.7secs
Galant GTi-16v: 577
12 9.0secs
Lancer GTi 16v: 578
Diamante LS: 590
13 9.3secs
Sigma: 593
14 9.7secs
Galant GS: 581
15 10.0secs
Space Runner 1800-16v GLXi:
594
16 10.4secs
Galant Sapporo: 572
Galant 2000 GLSi Coupe: 585
17 10.9secs
Galant 2000 GLSi: 576
18 11.7secs
Lancer GLXi 4WD Liftback:
586
19 12.2secs
Colt 1500 GLX 5-door: 570
20 12.8secs
Shogun V6 5-door: 582
21 13.1secs
Galant 2000 GLS Automatic:
571
Shogun V6 LWB: 592
22 14.0secs
Lancer 1.5 GLX Estate: 569
23 17.8secs
Shogun Turbo Diesel 5-door:
573

Morgan

1 5.6secs
Plus 8: 595
2 6.1secs
Plus 8: 597
3 7.4secs
Plus 8: 596

MVS

1 6.0secs
Venturi GT: 598

Nissan

1 5.0secs
300ZX Turbo Millen Super
GTZ: 627
2 5.6secs
300ZX: 621
Skyline GT-R: 624
3 6.5secs
300ZX Turbo: 622
300ZX Twin Turbo: 628
4 6.8secs

200SX: 634
5 7.0secs
300ZX Turbo: 606
6 7.1secs
300ZX: 615
7 7.2secs
300ZX Turbo: 600
200SX: 614
8 7.4secs
300ZX: 613
9 7.5secs
Sunny 2.0 GTi: 635
10 8.1secs
Sentra SE-R: 632
NX2000: 629
11 8.4secs
200SX SE V6: 611
12 8.5secs
Maxima SE: 617
13 8.6secs
Bluebird 1.8 ZX: 601
14 8.7secs
Primera 2.0E ZX: 631
15 8.8secs
300ZX: 605
240SX: 612
240SX: 626
16 8.9secs
Primera 2.0 GSX: 630
17 9.1secs
300ZX: 599
18 9.2secs
Sunny ZX Coupe: 609
19 9.4secs
Maxima: 616
20 10.1secs
100NX: 625
21 10.3secs
Pulsar NX SE: 608
22 10.5secs
Sunny 1.6 GS 5-door: 633
23 11.6secs
Laurel 2.4 SGL Automatic: 602
Sunny 1.6 SLX Coupe: 610
24 12.0secs
Sunny 1.4 LS: 620
25 12.4secs
Primera 1.6 LS: 623
26 13.3secs
Sentra Sport Coupe SE: 603
27 13.7secs
Sunny 1.3 LX 5-door: 604
28 14.0secs
Micra 1.2 GSX: 618
29 14.4secs
Prairie: 619
30 14.5secs
Micra SGL 5-DR: 607

Oldsmobile

1 7.6secs
Cutlass Calais International
Series HO: 638

Calais HO Quad 4: 637
2 8.2secs
Cutlass Calais International
Series: 636
3 10.4secs
Toronado Trofeo: 639

Opel

1 4.9secs
Omega: 641
2 7.6secs
Calibra 2.0i 16v: 640

Panther

1 6.8secs
Solo: 642

Parradine

1 5.4secs
V-12: 643

Peugeot

1 6.2secs
405 Turbo 16 Pike's Peak: 665
2 7.8secs
205 GTi 1.9: 648
3 7.9secs
605 SVE 24: 672
4 8.0secs
405 Mi16: 653
5 8.7secs
205 GTi: 644
309 GTi: 649
6 8.8secs
309 GTi: 658
7 9.0secs
405 Mi16: 660
8 9.5secs
405 Mi16x4: 664
9 9.7secs
309 SR Injection: 646
106 XSi: 668
10 9.8secs
505 STX: 650
11 10.3secs
405 SRi: 654
12 10.5secs
505 GTi Family Estate: 647
13 10.6secs
605 SV 3.0 Auto: 671
14 10.9secs
405 1.6 GL: 651
405 GLx4: 662
15 11.0secs
106 XT: 673
16 11.7secs
205 CJ: 656
17 11.8secs
605 SRi: 666
18 11.9secs

405 GR Injection Auto: 663
19 12.2secs
405 GTD Turbo: 659
20 12.3secs
205 D Turbo: 669
21 12.8secs
605 SR TD: 670
22 12.9secs
309 1.6 GR: 645
23 13.3secs
205 1.1GL: 655
24 13.8secs
106 XR: 667
25 14.7secs
309 Style 5-door: 661
26 15.4secs
405 GRD: 652
27 16.5secs
309 GLX: 657

Plymouth

1 6.9secs
Laser RS Turbo: 675
2 7.2secs
Laser RS: 674
3 8.0secs
Laser RS Turbo: 676

Pontiac

1 4.1secs
Firebird TDM Technologies:
692
2 5.1secs
Firebird TransAm 20th
Anniversary: 684
3 5.3secs
Firebird TransAm Turbo: 685
4 6.5secs
Firebird T/A Turbo
Engineering: 683
5 6.6secs
Firebird Formula: 688
Firebird GTA: 691
6 7.0secs
Grand Prix McLaren Turbo:
687
7 7.1secs
Firebird TransAm GTA: 689
8 7.7secs
Fiero GT: 677
Grand Prix STE Turbo: 690
9 7.8secs
Sunbird GT: 682
Firebird TransAm Convertible:
693
10 7.9secs
Grand Am Turbo: 686
11 8.0secs
Fiero Formula: 679
12 8.4secs
Grand Prix GTP: 694
13 9.0secs

Bonneville SSEi: 695
14 9.5secs
Bonneville SE: 681
15 9.7secs
Bonneville SE: 678
16 12.1secs
Le Mans: 680

Porsche

1 3.6secs
959 Sport: 725
959: 738
2 3.8secs
911 Ruf: 749
911 Ruf TR2: 751
3 3.9secs
911 Ruf CTR: 750
4 4.0secs
911 Turbo Koenig RS: 716
911 Turbo Ruf Twin Turbo: 717
959 Comfort: 724
911 Turbo Motorsport Design:
731
5 4.1secs
911 Turbo Gemballa Mirage:
753
6 4.5secs
911 Turbo Ruf 3.4: 706
7 4.6secs
911 Turbo RS Tuning: 744
911 Carrera Turbo: 748
8 4.7secs
959 S: 709
911 Turbo: 752
9 4.8secs
928 S4 Koenig: 733
911 Turbo: 730
10 4.9secs
911 Turbo Gemballa
Avalanche: 705
911 3.3 Turbo: 726
911 Carrera 4: 727
911 Carrera RS: 754
11 5.0secs
911 Turbo: 698
911 Turbo Slant-Nose: 707
911 Turbo: 715
12 5.1secs
911 Turbo: 743
911 Carrera 2: 739
13 5.2secs
911 Carrera Club Sport: 712
14 5.3secs
911 Club Sport: 729
15 5.5secs
928 S4: 719
928 Cabrio Strosek: 732
944 Turbo: 737
928 S4 SE: 734
16 5.6secs
911 Carrera SE: 697
17 5.7secs
911 Cabriolet: 696

911 Club Sport: 713
944 Turbo SE: 723
18 5.8secs
911 Carrera 4: 741
19 6.0secs
944 Turbo: 704
944 Turbo: 722
944 S2: 736
911 Speedster: 742
20 6.1secs
911 Carrera Cabriolet: 728
21 6.2secs
928 S4 Automatic: 701
Carrera 2 Cabriolet Tiptronic: 747
22 6.3secs
928 S: 700
23 6.5secs
911 Speedster: 714
911 Cabriolet: 710
24 6.7secs
944 S: 703
944 S2: 745
25 6.9secs
911 Carrera 2 Tiptronic: 740
26 7.0secs
911 Carrera: 711
944: 735
27 7.1secs
944 S2 Cabriolet: 746
28 7.5secs
944 S: 721
924 S: 718
29 7.8secs
924 S: 699
30 8.0secs
924 S: 708
31 8.7secs
944: 720
32 8.9secs
944: 702

Proton

1 12.4secs
1.5 SE Triple Valve Aeroback: 756
2 14.4secs
1.5 SE Aeroback: 755

Range Rover

1 10.8secs
Vogue SE: 764
2 11.3secs
Vogue SE: 763
3 11.7secs
Vogue SE: 761
4 11.9secs
Vogue: 757
5 12.1secs
Vogue: 760
6 12.3secs
County: 762

7 12.6secs
V8: 758
8 16.5secs
Vogue Turbo D: 759

Reliant

1 7.0secs
Scimitar SST 1800Ti: 766
2 7.2secs
Scimitar Ti: 765

Renault

1 5.7secs
Alpine A610 Turbo: 791
2 6.3secs
GTA V6 Turbo: 772
3 7.1secs
5 GT Turbo: 769
4 7.3secs
5 GT Turbo: 774
5 7.5secs
GTA V6: 779
6 7.8secs
21 Turbo Quadra: 787
7 7.9secs
11 Turbo: 773
21 Turbo: 778
8 8.6secs
Cleo 16v: 796
9 8.8secs
19 16v Cabriolet: 795
10 9.0secs
9 Turbo: 771
11 9.4secs
Espace V6: 794
12 9.5secs
5 GTX 3-DR: 775
25 V6 2.9 Auto: 789
19 16v: 790
13 9.8secs
21 Savanna GTX: 768
21 TXi: 788
14 10.0secs
25 TXi: 783
15 10.3secs
21 Ti: 777
16 10.8secs
Espace TXE: 785
17 10.9secs
21 GTS: 767
18 11.1secs
Alliance GTA: 776
19 11.4secs
19 TXE Chamade: 786
20 11.6secs
21 GTS Hatchback: 782
21 11.7secs
19 TXE: 781
22 12.0secs
Clio 1.4RT: 793
23 12.8secs
19 TSE: 780

24 14.0secs
Clio 1.2 RN: 792
25 14.8secs
5 GTL 5-door: 770
26 16.3secs
5 Campus: 784

Rolls-Royce

1 9.7secs
Corniche: 799
2 10.0secs
Silver Spirit: 798
3 10.4secs
Silver Spirit II: 801
4 11.3secs
Silver Shadow II: 797
5 12.6secs
Silver Spur: 800

Rover

1 8.0secs
825i: 802
Vitesse: 811
2 8.4secs
827 SLi Auto: 807
3 8.8secs
220 GTi: 821
216 GTi: 819
4 9.2secs
820i: 824
5 9.3secs
Sterling Catalyst: 810
6 9.4secs
Sterling Automatic: 825
7 9.7secs
Vitesse: 808
8 9.8secs
Metro GTi 16v: 817
9 10.0secs
416 GTi: 815
10 10.1secs
Sterling Automatic: 805
11 10.2secs
820SE: 803
12 10.9secs
216 GSi: 813
13 11.0secs
Mini Cooper S: 823
14 11.5secs
Montego 1.6L: 804
414 Si: 814
15 11.6secs
820 Fastback: 806
16 11.7secs
214 GSi: 812
825 TD: 822
17 12.0secs
Metro 1.4SL: 816
18 13.2secs
Montego 2.0 DSL Turbo: 809
19 13.3secs
214S: 818

20 16.9secs
218 SD: 820

Saab

1 6.8secs
9000 Turbo: 836
2 6.9secs
9000CD: 828
3 7.3secs
Carlsson: 831
4 7.5secs
9000 2.3 Turbo TCS: 835
5 7.7secs
9000 Turbo: 830
6 8.0secs
9000CS Turbo S: 837
7 8.6secs
9000: 826
8 8.9secs
900 Turbo S: 829
9 9.1secs
900SPG: 833
10 9.3secs
CDS 2.3: 834
11 9.5secs
9000 2.3CS: 838
12 9.6secs
9000i: 827
13 9.7secs
9000CD: 832

Safir

1 4.0secs
GT40: 839

Saturn

1 8.6secs
Sports Coupe: 840

Sbarro

1 3.5secs
Chrono 3.5: 841

Seat

1 10.3secs
Ibiza SXi: 843
2 10.6secs
Toledo 2.0 GTi: 845
3 15.6secs
Toledo 1.9 CL Diesel: 844
4 15.7secs
Malaga 1.2L: 842

Skoda

1 14.9secs
136 Rapide Coupe: 847
2 19.9secs
Estelle 120 LSE: 846

Subaru

1 6.2secs
Legacy FIA Record: 853
2 7.0secs
Legacy 2.0 4Cam Turbo
Estate: 856
3 7.3secs
SVX: 858
4 8.1secs
Legacy Sports Sedan: 857
5 8.5secs
4WD Turbo Estate: 851
6 8.7secs
XT Turbo 4WD: 850
7 9.7secs
XT6: 854
8 10.1secs
4WD 3-door Turbo Coupe: 848
9 10.8secs
Legacy 2.2 GX 4WD: 855
10 13.0secs
Justy: 849
11 14.4secs
Justy GL II 5-door 4WD: 852

Suzuki

1 8.7secs
Swift 1.3 GTi: 865
2 9.1secs
Swift 1.3 GTi: 860
3 9.5secs
Swift GT: 867
4 11.3secs
Swift 1.3 GLX: 862
5 11.6secs
Swift 1.3 GLX: 864
6 14.5secs
Swift 1.3 GLX Executive: 863
Vitara JLX: 866
7 19.4secs
SJ413V JX: 861
8 24.4secs
Alto GLA: 859

Toyota

1 6.2secs
MR2 Turbo: 900
2 6.6secs
Supra Turbo: 895
3 6.7secs
MR2 GT: 899
4 6.9secs
Supra 3.0i Turbo: 894
5 7.0secs
Supra: 872
MR2 Supercharged: 880
Supra Turbo: 889
6 7.5secs
Supra: 888
7 7.6secs
Celica GT4: 884

8 7.7secs
Supra 3.0i: 873
MR2 T-Bar: 871
Celica All-Trac Turbo: 883
MR2 Supercharged: 893
9 7.9secs
Supra: 881
10 8.1secs
Celica 2.0 GT: 896
11 8.2secs
Celica 2.0GT: 868
12 8.4secs
MR2: 870
13 8.6secs
Celica GTS: 869
Celica GT Cabriolet: 876
14 9.0secs
Camry 2.0 GLi: 874
15 9.2secs
Corolla FX-16 GTS: 878
16 9.3secs
Corolla GTS: 892
MR2: 906
17 9.5secs
Celica All-Trac Turbo: 897
18 9.7secs
Camry 2.2 GL: 903
19 10.1secs
Camry 2.0 GLi Estate: 875
20 10.3secs
Paseo: 909
21 10.5secs
Carina 1.6 GL: 882
Celica GTS: 898
22 10.8secs
Celica GT Convertible: 904
23 11.1secs
Corolla Executive: 877
Camry V6 GXi: 890
24 12.1secs
Corolla 4WD Estate: 886
25 12.2secs
Carina II 1.6 GL Liftback: 891
Tercel LE: 908
26 12.5secs
Previa LE: 901
27 12.6secs
Landcruiser VX: 905
28 12.8secs
Corolla 1.3 GL: 885
29 12.9secs
Previa: 907
30 14.7secs
Starlet GL: 902
31 21.6secs
Landcruiser: 879
32 23.5secs
Landcruiser II TD: 887

Treser

1 8.5secs
T1: 910

TVR

1 5.0secs
420 SEAC: 913
2 5.2secs
V8 S: 915
3 5.7secs
390 SE: 911
4 7.0secs
S Convertible: 914
5 7.6secs
S Convertible: 912

Vauxhall

1 7.0secs
Carlton GSi 3000 24v: 937
2 7.9secs
Cavalier GSi 2000 16v: 938
3 8.1secs
Calibra 2.0i 16v: 935
4 8.2secs
Carlton 3000 GSi: 921
Cavalier SRi 130 4-door: 923
5 8.4secs
Astra GTE 2.0i: 919
Carlton 3.0i CDX Estate: 936
6 8.5secs
Cavalier GSi 2000 16v 4x4: 939
7 8.6secs
Cavalier 4x4: 927
8 8.9secs
Cavalier SRi: 932
9 9.0secs
Calibra 4x4: 942
10 9.1secs
Nova GTE: 928
Senator 3.0i 24v: 944
11 9.2secs
Belmont 1.8 GLSi: 916
12 9.3secs
Cavalier 2.0i CD: 922
Senator 3.0i CD: 934
13 9.5secs
Calibra 2.0i: 941
14 10.4secs
Carlton CD 2.0i: 917
15 10.6secs
Belmont 1.6 GL: 920
16 10.8secs
Carlton L 1.8i: 925
17 10.9secs
Senator 3.0i CD: 924
18 11.8secs
Nova 1.3 GL 5-door: 918
19 12.1secs
Cavalier 1.6L: 931
20 12.2secs
Nova 1.5TD Merit: 940
21 12.8secs
Cavalier 1.7 GL TD: 946
1.4i GLS: 945

22 12.9secs
Cavalier 1.4L: 930
23 14.5secs
Nova 1.2 Luxe: 943
24 15.1secs
Astra 1.7 DL: 929
25 16.2secs
Carlton L 2.3D: 926
26 18.1secs
Frontera 2.3 TD Estate: 947

Vector

1 4.0secs
W2: 948
2 4.2secs
W8 Twin Turbo: 949

Volkswagen

1 7.1secs
Jetta GTi 16v: 959
2 7.7secs
Scirocco 16v: 954
3 7.8secs
Corrado: 968
4 8.0secs
Scirocco GTX 16v: 955
Golf GTi 16v: 951
5 8.3secs
Scirocco 16v: 966
6 8.4secs
GTi 16v: 977
Polo G40: 980
7 8.5secs
Jetta GT: 952
GTi 16v: 958
GTi 16v: 961
8 8.7secs
Corrado 16v: 969
9 8.8secs
Jetta GLi 16v: 962
Passat GT 16v: 972
10 8.9secs
Corrado G60: 973
Corrado G60: 976
11 9.1secs
Jetta GLi 16v: 978
12 9.6secs
Quantum Synchro Wagon: 953
13 9.9secs
Golf GTi: 983
14 10.3secs
Passat GT: 964
15 10.5secs
Golf Cabriolet Clipper: 957
16 10.8secs
Fox GL: 956
17 10.9secs
Passat GT 5-door: 960
Passat GL: 975
18 11.3secs
Jetta Syncro: 971
19 11.5secs

Golf CL Catalyst: 970
20 11.7secs
Polo GT Coupe: 981
21 12.7secs
Golf 1.8 GL: 982
22 12.9secs
Jetta Turbo Diesel: 963
23 13.5secs
Polo 1.3 CL Coupe: 979
24 15.6secs
Passat CL TD Estate: 974
25 16.9secs
Polo C Saloon: 965
Caravelle Carat Automatic: 967
26 24.7secs
Caravelle Syncro: 950

45 GLS: 1002

Zender

1 4.3secs
Fact 4: 1006
2 6.0secs
Vision 3: 1005

Volvo

1 7.8secs
740 Turbo: 990
2 8.3secs
740 Turbo: 986
3 8.8secs
740 Turbo: 995
4 8.9secs
460 Turbo: 993
5 9.2secs
940 SE: 996
6 9.3secs
940 SE Turbo: 997
960 24v: 998
7 9.4secs
740 GLT 16v Estate: 994
8 9.6secs
760 GLE: 987
740 GLE: 989
440 Turbo: 988
9 10.1secs
960: 999
10 10.3secs
480 ES: 985
11 11.1secs
460 GLi: 992
12 12.4secs
440 GLE: 991
13 13.4secs
340 GLE: 984

Westfield

1 4.3secs
SEight: 1000

Yugo

1 11.7secs
65A GLX: 1003
2 13.2secs
Sana 1.4: 1004
3 13.9secs
GV: 1001
4 21.6secs

Road test fuel consumption in order, Autocar & Motor

1 54.7mpg 45.5mpg US 5.2litres/100km
Citroen AX14 DTR 1989: 203

2 45.5mpg 37.9mpg US 6.2litres/100km
Daihatsu Charade CX Diesel Turbo 1987: 218

3 44.2mpg 36.8mpg US 6.4litres/100km
Vauxhall Astra 1.7 DL 1989: 929

4 43.8mpg 36.5mpg US 6.4litres/100km
Vauxhall Nova 1.5TD Merit 1990: 940

5 41.8mpg 34.8mpg US 6.8litres/100km
Fiat Regata DS 1986: 272

6 41.0mpg 34.1mpg US 6.9litres/100km
Renault 5 Campus 1989: 784

7 40.9mpg 34.1mpg US 6.9litres/100km
Nissan Micra SGL 5-DR 1987: 607

8 40.6mpg 33.8mpg US 7.0litres/100km
Lancia Y10 Fire 1986: 437

9 39.7mpg 33.1mpg US 7.1litres/100km
Rover Montego 2.0 DSL Turbo 1989: 809

10 39.5mpg 32.9mpg US 7.2litres/100km
Peugeot 205 1.1GL 1989: 655

11 39.3mpg 32.7mpg US 7.2litres/100km
Volkswagen Jetta Turbo Diesel 1988: 963

12 39.0mpg 32.5mpg US 7.2litres/100km
Fiat Uno 60S 1990: 282

13 38.9mpg 32.4mpg US 7.3litres/100km
Citroen AX11 TRE 1987: 197

14 38.8mpg 32.3mpg US 7.3litres/100km
Citroen AX11 TZX 1991: 208

15 38.4mpg 32.0mpg US 7.4litres/100km
Citroen BX19 RD Estate 1986: 193
Volkswagen Golf CL Catalyst 1989: 970

16 38.3mpg 31.9mpg US 7.4litres/100km
Renault Clio 1.4RT 1991: 793

17 38.2mpg 31.8mpg US 7.4litres/100km
Peugeot 405 GRD 1988: 652

18 38.1mpg 31.7mpg US 7.4litres/100km
Toyota Starlet GL 1990: 902

19 37.6mpg 31.3mpg US 7.5litres/100km
Seat Toledo 1.9 CL Diesel 1992: 844

20 37.2mpg 31.0mpg US 7.6litres/100km
Daihatsu Domino 1986: 217
Peugeot 205 CJ 1989: 656

21 37.1mpg 30.9mpg US 7.6litres/100km
Vauxhall Cavalier 1.7 GL TD 1992: 946

22 37.0mpg 30.8mpg US 7.6litres/100km
Suzuki Swift 1.3 GLX 1987: 862
Volkswagen Passat CL TD Estate 1990: 974

23 36.9mpg 30.7mpg US 7.7litres/100km
Austin Metro 1.3L 3-door 1987: 71

24 36.8mpg 30.6mpg US 7.7litres/100km
Renault Clio 1.2 RN 1991: 792

25 36.2mpg 30.1mpg US 7.8litres/100km
Citroen BX DTR Turbo 1988: 202
Rover 825 TD 1991: 822

26 36.1mpg 30.1mpg US 7.8litres/100km
Ford Escort 1.3L 3-door 1989: 308

27 36.0mpg 30.0mpg US 7.8litres/100km
Suzuki Swift 1.3 GLX Executive 1987: 863

28 35.9mpg 29.9mpg US 7.9litres/100km
Citroen AX GT5 1990: 205

29 35.7mpg 29.7mpg US 7.9litres/100km
Suzuki Swift 1.3 GLX 1989: 864

30 35.6mpg 29.6mpg US 7.9litres/100km
Vauxhall Nova 1.2 Luxe 1991: 943

31 35.5mpg 29.6mpg US 8.0litres/100km
Fiat Panda 750 L 1986: 271
Renault 5 GTL 5-door 1986: 770

32 35.3mpg 29.4mpg US 8.0litres/100km
Rover 218 SD 1991: 820

33 35.2mpg 29.3mpg US 8.0litres/100km
Volkswagen Polo 1.3 CL Coupe 1991: 979

34 35.1mpg 29.2mpg US 8.0litres/100km
Vauxhall Nova 1.3 GL 5-door 1986: 918

35 35.0mpg 29.1mpg US 8.1litres/100km
Ford Fiesta 1.1 Ghia Auto 1987: 295
Suzuki Alto GLA 1986: 859

36 34.9mpg 29.1mpg US 8.1litres/100km
Fiat Uno Selecta 1987: 275
Renault 19 TSE 1989: 780
Seat Ibiza SXi 1989: 843
Volkswagen Polo GT Coupe 1991: 981

37 34.8mpg 29.0mpg US 8.1litres/100km
Vauxhall Cavalier 1.4L 1989: 930

38 34.7mpg 28.9mpg US 8.1litres/100km
Citroen AX14 TRS 1987: 198
Honda Civic 1.4 GL 3-door 1987: 355
Vauxhall 1.4i GLS 1992: 945
Yugo 45 GLS 1987: 1002

39 34.6mpg 28.8mpg US 8.2litres/100km
Ford Escort 1.4 GL 1986: 286

40 34.4mpg 28.6mpg US 8.2litres/100km
Peugeot 205 D Turbo 1991: 669

41 34.1mpg 28.4mpg US 8.3litres/100km
Vauxhall Cavalier 1.6L 1989: 931

42 34.0mpg 28.3mpg US 8.3litres/100km
Citroen BX14 TGE 1989: 204
Fiat Panda 1000 S 1986: 270
Vauxhall Belmont 1.6 GL 1987: 920

43 33.7mpg 28.1mpg US 8.4litres/100km
Rover Metro 1.4SL 1990: 816

44 33.4mpg 27.8mpg US 8.5litres/100km
Vauxhall Astra GTE 2.0i 1987: 919

45 33.3mpg 27.7mpg US 8.5litres/100km
Honda Ballade EX 1987: 354
Peugeot 605 SR TD 1991: 670

46 33.2mpg 27.6mpg US 8.5litres/100km
Ford Fiesta 1.1LX 5-door 1989: 309

47 32.9mpg 27.4mpg US 8.6litres/100km
Subaru Justy 1987: 849

48 32.7mpg 27.2mpg US 8.6litres/100km
Citroen AX11 TRE 5DR 1988: 201
Peugeot 106 XT 1992: 673
Rover 414 Si 1990: 814

49 32.5mpg 27.1mpg US 8.7litres/100km
Vauxhall Nova GTE 1988: 928

50 32.3mpg 26.9mpg US 8.7litres/100km
Nissan Micra 1.2 GSX 1989: 618

51 32.2mpg 26.8mpg US 8.8litres/100km
Fiat Tipo 1.4 1988: 276
Volkswagen Golf 1.8 GL 1992: 982

52 32.1mpg 26.7mpg US 8.8litres/100km
Peugeot 106 XR 1991: 667

53 32.0mpg 26.6mpg US 8.8litres/100km
Citroen XM Turbo SD Estate 1992: 214
Vauxhall Belmont 1.8 GLSi 1986: 916

54 31.9mpg 26.6mpg US 8.9litres/100km
Honda Civic 1.5 VEi 1991: 370

55 31.8mpg 26.5mpg US 8.9litres/100km
Seat Toledo 2.0 GTi 1992: 845

56 31.6mpg 26.3mpg US 8.9litres/100km
Nissan Sunny 1.6 GS 5-door 1991: 633
Rover Metro GTi 16v 1990: 817

57 31.4mpg 26.1mpg US 9.0litres/100km
Vauxhall Carlton L 1.8i 1988: 925

58 31.3mpg 26.1mpg US 9.0litres/100km
Peugeot 405 GTD Turbo 1989: 659
Seat Malaga 1.2L 1986: 842

59 31.2mpg 26.0mpg US 9.1litres/100km
Toyota Corolla 1.3 GL 1988: 885

60 31.1mpg 25.9mpg US 9.1litres/100km
Lada Samara 1300 SL 1987: 425
Yugo 65A GLX 1988: 1003

61 31.0mpg 25.8mpg US 9.1litres/100km
Ford Orion 1.4 GL 1986: 291
Nissan Sunny 1.3 LX 5-door 1986: 604

Subaru Justy GL II 5-door 4WD 1989: 852

Vauxhall Carlton L 2.3D 1988: 926

62 30.9mpg 25.7mpg US 9.1litres/100km

Peugeot 309 SR Injection 1986: 646

Renault 19 TXE Chamade 1990: 786

Suzuki Swift 1.3 GTi 1986: 860

Volkswagen Polo C Saloon 1988: 965

63 30.8mpg 25.6mpg US 9.2litres/100km

Audi 80 1.8S 1986: 43

Ford Escort XR3i 1986: 288

Hyundai S Coupe GSi 1991: 382

Mazda 121 1.3 LX Sun Top 1988: 489

64 30.7mpg 25.6mpg US 9.2litres/100km

Austin Metro GTa 1989: 72

Mazda 121 GSX 1991: 511

65 30.6mpg 25.5mpg US 9.2litres/100km

Ford Granada 2.5 GL Diesel 1990: 321

Skoda 136 Rapide Coupe 1989: 847

66 30.5mpg 25.4mpg US 9.3litres/100km

Lada Samara 1300L 5-door 1989: 426

Toyota Camry 2.0 GLi 1987: 874

67 30.4mpg 25.3mpg US 9.3litres/100km

Citroen AX GT 1987: 196

Fiat Tipo 1.6 DGT SX 1989: 278

Peugeot 309 Style 5-door 1990: 661

68 30.3mpg 25.2mpg US 9.3litres/100km

Volkswagen Passat GT 5-door 1987: 960

69 30.1mpg 25.1mpg US 9.4litres/100km

Citroen CX25 DTR Turbo 2 1987: 200

Lada Riva 1500 Estate 1986: 423

Mazda 323 1.5 GLX 1987: 487

Skoda Estelle 120 LSE 1986: 846

Volkswagen Polo G40 1991: 980

70 30.0mpg 25.0mpg US 9.4litres/100km

Honda CRX Coupe 1.6i 16 1986: 349

Honda Civic 1.6i VT 1991: 371

Renault 5 GTX 3-DR 1987: 775

Rover 214S 1991: 818

Toyota Carina II 1.6 GL Liftback 1989: 891

Volkswagen Golf Cabriolet Clipper 1987: 957

71 29.9mpg 24.9mpg US 9.4litres/100km

Peugeot 205 GTi 1986: 644

72 29.8mpg 24.8mpg US 9.5litres/100km

Citroen BX GTi 16v 1987: 199

Lancia Y10 GTie 1990: 446

Peugeot 309 1.6 GR 1986: 645

73 29.7mpg 24.7mpg US 9.5litres/100km

Alfa Romeo 33 Boxer 16v 1990: 32

Audi 90 2.2E 1987: 48

Renault 21 GTS 1986: 767

74 29.6mpg 24.6mpg US 9.5litres/100km

Honda Civic CRX 1988: 359

Toyota Corolla Executive 1987: 877

75 29.5mpg 24.6mpg US 9.6litres/100km

Mazda 626 2.0i 5-door 1986: 482

Rover Montego 1.6L 1987: 804

76 29.4mpg 24.5mpg US 9.6litres/100km

Renault 9 Turbo 1986: 771

77 29.3mpg 24.4mpg US 9.6litres/100km

Vauxhall Calibra 2.0i 1991: 941

78 29.2mpg 24.3mpg US 9.7litres/100km

Daihatsu Charade Turbo 1988: 221

Ford Orion 1.6i Ghia 1991: 336

Mitsubishi Colt 1500 GLX 5-door 1987: 570

Nissan Primera 1.6 LS 1990: 623

Suzuki Swift 1.3 GTi 1989: 865

Volkswagen Passat GT 1988: 964

79 29.1mpg 24.2mpg US 9.7litres/100km

Honda Integra 1.5 1986: 351

Volvo 340 GLE 1986: 984

80 29.0mpg 24.1mpg US 9.7litres/100km

Daihatsu Applause 16 Xi 1990: 225

Fiat Croma ie Super 1987: 273

Ford Fiesta 1.6S 1989: 311

Ford Escort 1.4LX 1991: 330

81 28.9mpg 24.1mpg US 9.8litres/100km

Audi 100 Turbo Diesel 1986: 41

Nissan Sunny 1.4 LS 1989: 620

Rover 214 GSi 1990: 812

82 28.8mpg 24.0mpg US 9.8litres/100km

Daihatsu Charade GT ti 1987: 219

Ford Sierra Sapphire 2000E 1989: 314

Volkswagen Jetta GTi 16v 1987: 959

Volkswagen Golf GTi 1992: 983

83 28.7mpg 23.9mpg US 9.8litres/100km

Audi 90 Quattro 1987: 49

Nissan Sunny 1.6 SLX Coupe 1987: 610

Toyota Celica GT Cabriolet 1987: 876

Toyota Camry 2.0 GLi Estate 1987: 875

84 28.6mpg 23.8mpg US 9.9litres/100km

Fiat Tipo 1.8ie DGT SX 1992: 284

Ford Escort Cabriolet 1991: 332

Ford Escort XR3i 1992: 342

Mazda 323 1.8 GT 1989: 501

Mercedes-Benz 300D 1990: 548

Toyota MR2 T-Bar 1986: 871

85 28.5mpg 23.7mpg US 9.9litres/100km

Mazda 323 1.8 GT 1991: 512

Proton 1.5 SE Triple Valve Aeroback 1991: 756

Toyota Carina 1.6 GL 1988: 882

86 28.4mpg 23.6mpg US 9.9litres/100km

Ford Escort XR2i 1990: 320

Renault 5 GT Turbo 1987: 774

87 28.3mpg 23.6mpg US 10.0litres/100km

Fiat Croma CHT 1990: 280

Peugeot 309 GTi 1987: 649

Vauxhall Cavalier 4x4 1988: 927

88 28.2mpg 23.5mpg US 10.0litres/100km

Peugeot 309 GLX 1989: 657

Renault 19 16v 1991: 790

89 28.1mpg 23.4mpg US 10.1litres/100km

Audi 80 1.8E 1988: 52

Fiat Tempra 1.6 SX 1990: 281

Ford Fiesta 1.4 Ghia 1989: 310

Honda CRX 1.6i VT 1990: 369

Peugeot 205 GTi 1.9 1987: 648

Renault 21 GTS Hatchback 1989: 782

Yugo Sana 1.4 1990: 1004

90 28.0mpg 23.3mpg US 10.1litres/100km

Audi 80 16v Sport 1991: 66

Hyundai X2 1.5 GSi 1991: 383

Mercedes-Benz 190 1989: 537

Proton 1.5 SE Aeroback 1989: 755

91 27.9mpg 23.2mpg US 10.1litres/100km

Austin Montego 2.0 Vanden Plas EFI 1986: 70

Mazda MX-3 1.6 Auto 1992: 518

Peugeot 405 1.6 GL 1988: 651

92 27.8mpg 23.1mpg US 10.2litres/100km

Ford Escort 1.6 Ghia Estate 1991: 313

Mitsubishi Lancer 1.5 GLX Estate 1986: 569

Nissan Sunny 2.0 GTi 1992: 635

Renault 21 Savanna GTX 1986: 768

93 27.6mpg 23.0mpg US 10.2litres/100km

BMW 318i 1991: 115

94 27.5mpg 22.9mpg US 10.3litres/100km

BMW 318i S 1990: 111

Honda Integra EX 16 1986: 352

Mazda 323 1.6i 1989: 500

95 27.4mpg 22.8mpg US 10.3litres/100km

Audi Coupe GT 2.2 1986: 46

Audi 100 Sport 1988: 51

BMW 318i 2-door 1987: 90

Ford Escort RS Turbo 1986: 287

Rover 820i 1992: 824

Volkswagen Passat GT 16v 1989: 972

96 27.3mpg 22.7mpg US 10.3litres/100km

Peugeot 405 GR Injection Auto 1990: 663

Renault 11 Turbo 1987: 773

TVR S Convertible 1987: 912

Vauxhall Calibra 4x4 1991: 942

97 27.2mpg 22.6mpg US 10.4litres/100km

Alfa Romeo Alfa 33 1.3 S 1986: 11

Mitsubishi Space Runner 1800-16v GLXi 1991: 594

Nissan Sunny ZX Coupe 1987: 609

Toyota MR2 GT 1990: 899

98 27.1mpg 22.6mpg US 10.4litres/100km

Hyundai Lantra 1.6 Cdi 1991: 381

Lancia Prisma 1600ie LX 1986: 435

Rover 220 GTi 1991: 821

99 27.0mpg 22.5mpg US 10.5litres/100km

Audi Coupe Quattro 1989: 58

Vauxhall Cavalier SRi 130 4-door 1987: 923

100 26.9mpg 22.4mpg US 10.5litres/100km

Mitsubishi Colt 1800 GTI-16v 1991: 589

Peugeot 106 XSi 1991: 668

Volvo 440 GLE 1990: 991

101 26.8mpg 22.3mpg US 10.5litres/100km

Ford Sierra Sapphire 2.0 Ghia 1987: 299

Nissan Primera 2.0E ZX 1991: 631

102 26.7mpg 22.2mpg US 10.6litres/100km

Renault Espace TXE 1989: 785

Toyota MR2 1991: 906

Vauxhall Carlton CD 2.0i 1986: 917

103 26.6mpg 22.1mpg US 10.6litres/100km

BMW 325i SE 1991: 116

Peugeot 405 Mi16 1988: 653

Rover 820SE 1987: 803

104 26.5mpg 22.1mpg US 10.7litres/100km

BMW 325i Convertible 1986: 83

Citroen ZX 1.4 Avantage 1991: 211

Rover Mini Cooper S 1991: 823

105 26.4mpg 22.0mpg US 10.7litres/100km

Ford Escort RS2000 1992: 341

Volvo 460 Turbo 1990: 993

Volvo 460 GLi 1990: 992

106 26.3mpg 21.9mpg US 10.7litres/100km

Citroen BX GTi 4x4 1990: 206

Rover 825i 1986: 802

107 26.2mpg 21.8mpg US 10.8litres/100km

Alfa Romeo 33 1.7 Veloce 1987: 17

Citroen Visa GTi 1986: 195

108 26.1mpg 21.7mpg US 10.8litres/100km

Audi Coupe 2.2E 1989: 57

Honda Accord 2.0 EXi 1990: 367

Mercedes-Benz 190E 1.8 Auto 1991: 551

Renault 19 TXE 1989: 781

Vauxhall Cavalier 2.0i CD 1987: 922

109 26.0mpg 21.6mpg US 10.9litres/100km

BMW M3 Evolution 1989: 107

Citroen AX GTi 1992: 213

Nissan Primera 2.0 GSX 1991: 630

Peugeot 405 Mi16x4 1990: 664

Renault Cleo 16v 1992: 796

Rover 216 GSi 1990: 813

110 25.9mpg 21.6mpg US 10.9litres/100km

Alfa Romeo 164 3.0 Lusso 1989: 26

Fiat Regata 100S ie 1987: 274

Fiat Uno Turbo ie 1990: 283

Lancia Thema 2.0ie 1986: 436

Mitsubishi Lancer GTi 16v 1988: 578

Volkswagen Corrado 16v 1989: 969

111 25.8mpg 21.5mpg US 10.9litres/100km

Alfa Romeo Alfa 33 1.7 Sportwagon Veloce 1987: 18

Citroen XM 2.0SEi 1990: 207

Daihatsu Charade 1.3 CX 1989: 222

Subaru XT Turbo 4WD 1987: 850

112 25.7mpg 21.4mpg US 11.0litres/100km

Alfa Romeo Sprint 1.7 Cloverleaf 1988: 25

Honda Concerto 1.6i 16 1990: 368

Renault 21 Ti 1988: 777

Saab 9000i 1986: 827

113 25.6mpg 21.3mpg US 11.0litres/100km

Alfa Romeo Alfa 75 2.0i Twin Spark 1987: 19

Volkswagen Jetta GT 1986: 952

114 25.5mpg 21.2mpg US 11.1litres/100km

Renault 21 TXi 1990: 788

Vauxhall Calibra 2.0i 16v 1990: 935

115 25.4mpg 21.1mpg US 11.1litres/100km

Audi 80 2.0E 1990: 59

116 25.3mpg 21.1mpg US 11.2litres/100km

BMW 316i 4-door 1989: 102

Citroen ZX Volcane 1991: 212

Mitsubishi Galant 2000 GLS Automatic 1987: 571

Vauxhall Cavalier SRi 1989: 932

Volkswagen Golf GTi 16v 1986: 951

117 25.2mpg 21.0mpg US 11.2litres/100km

Audi 90 1986: 44

Ford Granada Scorpio 2.0i Auto 1990: 323

Lancia Dedra 1.8i 1990: 443

Peugeot 309 GTi 1989: 658

118 25.1mpg 20.9mpg US 11.3litres/100km

Rover 416 GTi 1990: 815

119 25.0mpg 20.8mpg US 11.3litres/100km

Mazda MX-3 V6 1991: 513

Subaru Legacy 2.2 GX 4WD 1990: 855

Toyota Celica 2.0GT 1986: 868

Vauxhall Cavalier GSi 2000 16v 1990: 938

120 24.9mpg 20.7mpg US 11.3litres/100km

Renault 19 16v Cabriolet 1992: 795

Vauxhall Carlton 3000 GSi 1987: 921

121 24.8mpg 20.7mpg US 11.4litres/100km

Mazda 626 2.0 GLS Executive 1987: 488

Mitsubishi Galant Sapporo 1987: 572

Suzuki SJ413V JX 1987: 861

122 24.7mpg 20.6mpg US 11.4litres/100km

Peugeot 605 SRi 1990: 666

123 24.6mpg 20.5mpg US 11.5litres/100km

Mitsubishi Lancer GLXi 4WD Liftback 1990: 586

124 24.5mpg 20.4mpg US 11.5litres/100km

Alfa Romeo Alfa 75 1.8 1986: 12

Lancia Delta HF Turbo ie 1986: 434

Mazda 626 2.0i GT 4WS 1988: 493

Rover 216 GTi 1991: 819

Toyota Camry 2.2 GL 1991: 903

125 24.4mpg 20.3mpg US 11.6litres/100km

Lancia Thema 2.0ie 16v 1989: 442

Mitsubishi Galant GTi-16v 1988: 577

Peugeot 405 GLx4 1990: 662

Rover 820 Fastback 1988: 806

126 24.2mpg 20.2mpg US 11.7litres/100km

Audi Coupe Quattro 1986: 47

Suzuki Vitara JLX 1989: 866

127 24.0mpg 20.0mpg US 11.8litres/100km

BMW 21 1989: 110

Ford Granada Scorpio 2.9 EX 1987: 297

Vauxhall Cavalier GSi 2000 16v 4x4 1990: 939

128 23.9mpg 19.9mpg US 11.8litres/100km

Land Rover Discovery TDi 1990: 451

129 23.8mpg 19.8mpg US 11.9litres/100km

BMW 520i 1986: 84

Nissan Bluebird 1.8 ZX 1986: 601

Peugeot 405 SRi 1988: 654

130 23.7mpg 19.7mpg US 11.9litres/100km

Lancia Dedra 2.0ie 1990: 444

131 23.6mpg 19.7mpg US 12.0litres/100km

Ford Fiesta RS Turbo 1990: 319

Mercedes-Benz 200 1986: 521

Saab 9000 2.3CS 1992: 838

132 23.5mpg 19.6mpg US 12.0litres/100km

Lotus Esprit Turbo SE 1989: 466

Porsche 944 S 1986: 703

133 23.4mpg 19.5mpg US 12.1litres/100km

Alfa Romeo 164 Twin Spark Lusso 1990: 30

134 23.3mpg 19.4mpg US 12.1litres/100km

Ginetta G32 1990: 346

Honda Shuttle 1.4 GL Automatic 1989: 366

Mercedes-Benz 200E Automatic 1989: 541

135 23.1mpg 19.2mpg US 12.2litres/100km

Mercedes-Benz 260E Auto 1988: 533

136 23.0mpg 19.2mpg US 12.3litres/100km

Carbodies Fairway 2.7 Silver 1989: 134

Daihatsu Fourtrak Estate DT EL 1987: 220

Hyundai Sonata 2.0i GLS 1989: 379

Mitsubishi Galant 2000 GLSi Coupe 1990: 585

137 22.9mpg 19.1mpg US 12.3litres/100km

Honda Shuttle 1.6i RT4 4WD 1988: 363

Mazda 626 2.0 GLX Executive Estate 1988: 492

Mitsubishi Lancer Liftback 1800 GTi-16v 1990: 587

Porsche 944 Turbo 1986: 704

138 22.8mpg 19.0mpg US 12.4litres/100km

Alfa Romeo Alfa 75 2.5 Green Cloverleaf 1986: 13

Alfa Romeo 164 Automatic 1989: 27

Vauxhall Frontera 2.3 TD Estate 1992: 947
139 22.7mpg 18.9mpg US 12.4litres/100km
Hyundai Sonata 2.4i GLS 1990: 380
Vauxhall Senator 3.0i CD 1987: 924
140 22.6mpg 18.8mpg US 12.5litres/100km
Porsche 924 S 1987: 708
Subaru 4WD 3-door Turbo Coupe 1986: 848
141 22.5mpg 18.7mpg US 12.6litres/100km
Renault 25 TXi 1989: 783
Toyota Celica 2.0 GT 1990: 896
Volvo 440 Turbo 1989: 988
142 22.4mpg 18.7mpg US 12.6litres/100km
Alfa Romeo 75 2.5 Auto 1988: 22
143 22.3mpg 18.6mpg US 12.7litres/100km
BMW 535i SE 1988: 97
Isuzu Trooper 3-door TD 1987: 392
Volkswagen Jetta Syncro 1989: 971
144 22.2mpg 18.5mpg US 12.7litres/100km
Audi 90 Quattro 20v 1989: 56
Lotus Excel SA 1987: 463
Saab 9000CS Turbo S 1991: 837
Vauxhall Senator 2.5i 1989: 933
145 22.1mpg 18.4mpg US 12.8litres/100km
Lancia Dedra 2000 Turbo 1991: 447
Mitsubishi Shogun Turbo Diesel 5-door 1987: 573
Renault 21 Turbo Quadra 1990: 787
Rover Vitesse 1989: 811
146 22.0mpg 18.3mpg US 12.8litres/100km
BMW 520i 1991: 117
Mercedes-Benz 190E 2.5-16 1989: 538
Renault 5 GT Turbo 1986: 769
Saab CDS 2.3 1990: 834
Volkswagen Corrado G60 1991: 976
147 21.8mpg 18.2mpg US 13.0litres/100km
Subaru Legacy 2.0 4Cam Turbo Estate 1991: 856
Volvo 740 Turbo 1987: 986
148 21.7mpg 18.1mpg US 13.0litres/100km
BMW 525i SE 24v 1991: 119

Reliant Scimitar Ti 1986: 765
Reliant Scimitar SST 1800Ti 1990: 766
149 21.6mpg 18.0mpg US 13.1litres/100km
Ford Sierra Sapphire Cosworth 4x4 1990: 328
Saab Carlsson 1989: 831
Toyota Previa 1991: 907
150 21.5mpg 17.9mpg US 13.1litres/100km
Ford Granada 2.4i Ghia 1987: 296
Honda Legend 1986: 353
Honda Legend Coupe 1992: 375
Mercedes-Benz 300TE 1986: 523
151 21.4mpg 17.8mpg US 13.2litres/100km
Alfa Romeo 164 Cloverleaf 1991: 33
Audi 100 Avant Quattro 1988: 50
Daihatsu Sportrak EL 1989: 224
Mercedes-Benz 300CE 1987: 529
Range Rover Vogue Turbo D 1988: 759
Vauxhall Senator 3.0i 24v 1991: 944
152 21.3mpg 17.7mpg US 13.3litres/100km
Mazda 323 Turbo 4x4 Lux 1986: 481
Rover Sterling Catalyst 1989: 810
Toyota Corolla 4WD Estate 1988: 886
153 21.2mpg 17.7mpg US 13.3litres/100km
Mitsubishi Galant 2000 GLSi 1988: 576
Renault 25 V6 2.9 Auto 1990: 789
Volvo 940 SE Turbo 1991: 997
154 21.1mpg 17.6mpg US 13.4litres/100km
Lancia Thema 2.0ie 16v SE Turbo 1990: 445
Volvo 740 GLT 16v Estate 1990: 994
155 21.0mpg 17.5mpg US 13.5litres/100km
Ford Sierra Ghia 4x4 Estate 1986: 293
Ford Scorpio 24v 1991: 338
Honda Prelude 2.0i 16 1987: 358
Mercedes-Benz 190E 2.6 1987: 528
Rover Sterling Automatic 1987: 805

Vauxhall Carlton 3.0i CDX Estate 1990: 936
156 20.9mpg 17.4mpg US 13.5litres/100km
Lotus Esprit Turbo HC 1987: 462
MG Maestro Turbo 1989: 566
Morgan Plus 8 1987: 595
157 20.8mpg 17.3mpg US 13.6litres/100km
BMW 520i SE Touring 1992: 124
Citroen XM V6.24 1991: 210
158 20.7mpg 17.2mpg US 13.6litres/100km
Jaguar XJ6 3.6 1986: 396
Mazda 323 Turbo 4x4 1988: 491
Porsche 944 S2 1989: 736
Porsche 944 1989: 735
Volkswagen Caravelle Syncro 1986: 950
159 20.6mpg 17.2mpg US 13.7litres/100km
BMW M635 CSi 1989: 109
Nissan Prairie 1989: 619
Peugeot 505 GTi Family Estate 1986: 647
Porsche 911 Carrera Club Sport 1988: 712
Toyota Camry V6 GXi 1989: 890
160 20.5mpg 17.1mpg US 13.8litres/100km
Alfa Romeo Alfa 75 V6 3.0 1987: 20
Renault GTA V6 Turbo 1986: 772
161 20.4mpg 17.0mpg US 13.8litres/100km
Citroen CX25 Ri Familiale Auto 1986: 194
Lotus Esprit Turbo 1988: 464
Mazda 323 1.5 GLX Estate 1986: 479
Peugeot 605 SVE 24 1991: 672
Porsche 911 Carrera 2 1990: 739
162 20.3mpg 16.9mpg US 13.9litres/100km
BMW M3 1987: 94
Ford Sierra RS Cosworth 1988: 306
Honda Legend Coupe 1987: 356
Saab 9000CD 1988: 828
163 20.2mpg 16.8mpg US 14.0litres/100km
Toyota Landcruiser II TD 1988: 887
164 20.1mpg 16.7mpg US 14.1litres/100km
Lotus Elan SE 1990: 468

Morgan Plus 8 1991: 597
TVR V8 S 1991: 915
165 20.0mpg 16.7mpg US 14.1litres/100km
Audi 80 2.8E V6 Quattro 1992: 69
BMW 730i SE 1987: 92
Nissan Laurel 2.4 SGL Automatic 1986: 602
Westfield SEight 1991: 1000
166 19.9mpg 16.6mpg US 14.2litres/100km
Jaguar XJ6 3.2 1991: 416
Renault Espace V6 1991: 794
Rover Vitesse 1988: 808
Subaru 4WD Turbo Estate 1988: 851
167 19.8mpg 16.5mpg US 14.3litres/100km
Lancia Delta HF 4WD 1986: 433
Peugeot 605 SV 3.0 Auto 1991: 671
Toyota Landcruiser VX 1991: 905
168 19.7mpg 16.4mpg US 14.3litres/100km
Audi Quattro 1988: 54
Lexus LS400 1990: 454
Mitsubishi Shogun V6 5-door 1989: 582
Nissan 200SX 1992: 634
Saab 9000 2.3 Turbo TCS 1991: 835
169 19.6mpg 16.3mpg US 14.4litres/100km
Honda NSX 1991: 373
Lotus Excel SE 1986: 460
Rover Sterling Automatic 1992: 825
170 19.5mpg 16.2mpg US 14.5litres/100km
Nissan 200SX 1989: 614
Vauxhall Carlton GSi 3000 24v 1990: 937
Volvo 760 GLE 1987: 987
171 19.4mpg 16.2mpg US 14.6litres/100km
Panther Solo 1990: 642
Toyota Supra 3.0i 1986: 873
172 19.3mpg 16.1mpg US 14.6litres/100km
Jaguar XJ6 2.9 1987: 401
Vauxhall Senator 3.0i CD 1989: 934
173 19.2mpg 16.0mpg US 14.7litres/100km
Mercedes-Benz 300SL-24 5-speed Auto 1991: 552
Nissan 300ZX Turbo 1987: 606
Porsche 911 Carrera SE 1986: 697
Toyota Celica GT4 1988: 884

174 19.1mpg 15.9mpg US 14.8litres/100km

Audi Quattro 20v 1990: 61
Honda Legend 1991: 372
Lancia Delta Integrale 16v 1989: 441
Porsche 944 Turbo SE 1988: 723
Renault GTA V6 1988: 779
Saab 900 Turbo S 1989: 829

175 19.0mpg 15.8mpg US 14.9litres/100km

Audi 100 2.8E Auto 1991: 64
Mitsubishi Starion 2000 Turbo 1987: 574

176 18.9mpg 15.7mpg US 14.9litres/100km

Ford Granada Ghia X 2.9 EFi Saloon 1990: 322

177 18.8mpg 15.7mpg US 15.0litres/100km

Citroen XM 3.0 SEi Auto 1991: 209
Jaguar XJS 4.0 Auto 1991: 418
Lada Niva Cossack Cabrio 1987: 424
Mitsubishi Starion 2.6 Turbo 1989: 583
Toyota Supra 3.0i Turbo 1989: 894

178 18.7mpg 15.6mpg US 15.1litres/100km

Mazda RX-7 1986: 484
Range Rover Vogue 1989: 760

179 18.6mpg 15.5mpg US 15.2litres/100km

Mercedes-Benz 560SEL 1986: 526
Mercedes-Benz 300E 4Matic 1988: 535
Porsche 928 S4 SE 1989: 734

180 18.5mpg 15.4mpg US 15.3litres/100km

BMW 750i L 1987: 93
Jaguar 4.0 Sovereign 1989: 406
Maserati Biturbo Spyder 1987: 473
Mercedes-Benz 300E-24 1990: 549

181 18.4mpg 15.3mpg US 15.4litres/100km

Ferrari 348tb 1990: 262
Honda Legend Saloon Automatic 1988: 362
Mercedes-Benz 420SE 1986: 524
Volvo 960 24v 1991: 998

182 18.3mpg 15.2mpg US 15.4litres/100km

Ford RS200 1986: 292
Honda NSX Auto 1992: 376
Mitsubishi Sigma 1991: 593
Renault 21 Turbo 1988: 778

183 18.2mpg 15.2mpg US 15.5litres/100km

Audi Coupe S2 1991: 67

184 18.1mpg 15.1mpg US 15.6litres/100km

Volkswagen Caravelle Carat Automatic 1989: 967

185 18.0mpg 15.0mpg US 15.7litres/100km

Jaguar XJS 3.6 Automatic 1987: 402
Land Rover Ninety County Turbo Diesel 1987: 448

186 17.9mpg 14.9mpg US 15.8litres/100km

Rover 827 SLi Auto 1988: 807
Toyota Landcruiser 1987: 879

187 17.6mpg 14.7mpg US 16.1litres/100km

Lancia Delta HF Integrale 1988: 439

188 17.5mpg 14.6mpg US 16.1litres/100km

BMW 735i Automatic 1986: 89

189 17.4mpg 14.5mpg US 16.2litres/100km

Porsche Carrera 2 Cabriolet Tiptronic 1990: 747

190 17.3mpg 14.4mpg US 16.3litres/100km

Caterham 7 HPC 1990: 135

191 17.2mpg 14.3mpg US 16.4litres/100km

BMW 635 CSi 1988: 98
Jaguar XJR 4.0 1990: 411

192 17.1mpg 14.2mpg US 16.5litres/100km

Mercedes-Benz G-Wagen 300 GD LWB 1991: 556

193 17.0mpg 14.2mpg US 16.6litres/100km

Ford Sierra XR 4x4 1989: 315
Nissan 300ZX 1990: 621
Porsche 928 S4 Automatic 1986: 701

194 16.8mpg 14.0mpg US 16.8litres/100km

Ferrari Mondial 3.2QV 1986: 249
Nissan Maxima 1989: 616

195 16.6mpg 13.8mpg US 17.0litres/100km

Ferrari Testa Rossa 1989: 259
Porsche 911 3.3 Turbo 1989: 726
TVR 390 SE 1987: 911

196 16.5mpg 13.7mpg US 17.1litres/100km

Land Rover Discovery V8i 5DR 1991: 453
Mercedes-Benz 500E 1991: 554

197 16.2mpg 13.5mpg US 17.4litres/100km

BMW 750i 1990: 113

Mercedes-Benz 500SL 1990: 550

198 16.1mpg 13.4mpg US 17.5litres/100km

Mitsubishi Shogun V6 LWB 1991: 592

199 16.0mpg 13.3mpg US 17.7litres/100km

De Tomaso Pantera GT5-S 1986: 226

200 15.9mpg 13.2mpg US 17.8litres/100km

Lamborghini Jalpa 3500 1986: 428

201 15.8mpg 13.2mpg US 17.9litres/100km

Range Rover Vogue SE 1989: 761

202 15.6mpg 13.0mpg US 18.1litres/100km

Aston Martin Virage 1991: 40
Lancia Thema 8.32 1988: 440

203 15.3mpg 12.7mpg US 18.5litres/100km

Audi V8 Quattro Automatic 1990: 63
Audi V8 Quattro 1990: 62

204 15.1mpg 12.6mpg US 18.7litres/100km

Porsche 911 Turbo 1991: 752

205 15.0mpg 12.5mpg US 18.8litres/100km

Range Rover Vogue 1986: 757

206 14.7mpg 12.2mpg US 19.2litres/100km

BMW 850i Manual 1992: 125

207 14.6mpg 12.2mpg US 19.3litres/100km

Mercedes-Benz 400SE 1991: 553
Range Rover Vogue SE 1990: 763

208 14.4mpg 12.0mpg US 19.6litres/100km

BMW 850i 1991: 121

209 14.3mpg 11.9mpg US 19.8litres/100km

Mazda RX-7 Turbo II 1989: 506

210 14.2mpg 11.8mpg US 19.9litres/100km

Bentley Turbo R 1987: 76
Range Rover Vogue SE 1992: 764

211 14.0mpg 11.7mpg US 20.2litres/100km

Land Rover Discovery V8 1990: 452
Rolls-Royce Silver Spirit 1981: 798

212 13.8mpg 11.5mpg US 20.5litres/100km

Jaguar XJS V12 Convertible 1988: 405

Rolls-Royce Silver Spirit II 1991: 801

213 13.4mpg 11.2mpg US 21.1litres/100km

Bentley Mulsanne Turbo 1985: 75
Land Rover One Ten County V8 1987: 449

214 12.3mpg 10.2mpg US 23.0litres/100km

Rolls-Royce Corniche 1982: 799

215 12.1mpg 10.1mpg US 23.3litres/100km

Bentley Mulsanne Turbo 1982: 74
Land Rover Ninety County V8 1989: 450

Road test fuel consumption, Autocar & Motor. In make order

Alfa Romeo

1 29.7mpg 24.7mpg US
9.5litres/100km
33 Boxer 16v: 32
2 27.2mpg 22.6mpg US
10.4litres/100km
Alfa 33 1.3 S: 11
3 26.2mpg 21.8mpg US
10.8litres/100km
33 1.7 Veloce: 17
4 25.9mpg 21.6mpg US
10.9litres/100km
164 3.0 Lusso: 26
5 25.8mpg 21.5mpg US
10.9litres/100km
Alfa 33 1.7 Sportwagon
Veloce: 18
6 25.7mpg 21.4mpg US
11.0litres/100km
Sprint 1.7 Cloverleaf: 25
7 25.6mpg 21.3mpg US
11.0litres/100km
Alfa 75 2.0i Twin Spark: 19
8 24.5mpg 20.4mpg US
11.5litres/100km
Alfa 75 1.8: 12
9 23.4mpg 19.5mpg US
12.1litres/100km
164 Twin Spark Lusso: 30
10 22.8mpg 19.0mpg US
12.4litres/100km
Alfa 75 2.5 Green Cloverleaf:
13
164 Automatic: 27
11 22.4mpg 18.7mpg US
12.6litres/100km
75 2.5 Auto: 22
12 21.4mpg 17.8mpg US
13.2litres/100km
164 Cloverleaf: 33
13 20.5mpg 17.1mpg US
13.8litres/100km
Alfa 75 V6 3.0: 20

Aston Martin

1 15.6mpg 13.0mpg US
18.1litres/100km
Virage: 40

Audi

1 30.8mpg 25.6mpg US
9.2litres/100km
80 1.8S: 43
2 29.7mpg 24.7mpg US
9.5litres/100km
90 2.2E: 48
3 28.9mpg 24.1mpg US
9.8litres/100km
100 Turbo Diesel: 41

4 28.7mpg 23.9mpg US
9.8litres/100km
90 Quattro: 49
5 28.1mpg 23.4mpg US
10.1litres/100km
80 1.8E: 52
6 28.0mpg 23.3mpg US
10.1litres/100km
80 16v Sport: 66
7 27.4mpg 22.8mpg US
10.3litres/100km
Coupe GT 2.2: 46
100 Sport: 51
8 27.0mpg 22.5mpg US
10.5litres/100km
Coupe Quattro: 58
9 26.1mpg 21.7mpg US
10.8litres/100km
Coupe 2.2E: 57
10 25.4mpg 21.1mpg US
11.1litres/100km
80 2.0E: 59
11 25.2mpg 21.0mpg US
11.2litres/100km
90: 44
12 24.2mpg 20.2mpg US
11.7litres/100km
Coupe Quattro: 47
13 22.2mpg 18.5mpg US
12.7litres/100km
90 Quattro 20v: 56
14 21.4mpg 17.8mpg US
13.2litres/100km
100 Avant Quattro: 50
15 20.0mpg 16.7mpg US
14.1litres/100km
80 2.8E V6 Quattro: 69
16 19.7mpg 16.4mpg US
14.3litres/100km
Quattro: 54
17 19.1mpg 15.9mpg US
14.8litres/100km
Quattro 20v: 61
18 19.0mpg 15.8mpg US
14.9litres/100km
100 2.8E Auto: 64
19 18.2mpg 15.2mpg US
15.5litres/100km
Coupe S2: 67
20 15.3mpg 12.7mpg US
18.5litres/100km
V8 Quattro: 62
V8 Quattro Automatic: 63

Austin

1 36.9mpg 30.7mpg US
7.7litres/100km
Metro 1.3L 3-door: 71
2 30.7mpg 25.6mpg US
9.2litres/100km
Metro GTa: 72

3 27.9mpg 23.2mpg US
10.1litres/100km
Montego 2.0 Vanden Plas EFi:
70

Bentley

1 14.2mpg 11.8mpg US
19.9litres/100km
Turbo R: 76
2 13.4mpg 11.2mpg US
21.1litres/100km
Mulsanne Turbo: 75
3 12.1mpg 10.1mpg US
23.3litres/100km
Mulsanne Turbo: 74

BMW

1 27.6mpg 23.0mpg US
10.2litres/100km
318i: 115
2 27.5mpg 22.9mpg US
10.3litres/100km
318i S: 111
3 27.4mpg 22.8mpg US
10.3litres/100km
318i 2-door: 90
4 26.6mpg 22.1mpg US
10.6litres/100km
325i SE: 116
5 26.5mpg 22.1mpg US
10.7litres/100km
325i Convertible: 83
6 26.0mpg 21.6mpg US
10.9litres/100km
M3 Evolution: 107
7 25.3mpg 21.1mpg US
11.2litres/100km
316i 4-door: 102
8 24.0mpg 20.0mpg US
11.8litres/100km
Z1: 110
9 23.8mpg 19.8mpg US
11.9litres/100km
520i: 84
10 22.3mpg 18.6mpg US
12.7litres/100km
535i SE: 97
11 22.0mpg 18.3mpg US
12.8litres/100km
520i: 117
12 21.7mpg 18.1mpg US
13.0litres/100km
525i SE 24v: 119
13 20.8mpg 17.3mpg US
13.6litres/100km
520i SE Touring: 124
14 20.6mpg 17.2mpg US
13.7litres/100km
M635 CSi: 109
15 20.3mpg 16.9mpg US

13.9litres/100km
M3: 94
16 20.0mpg 16.7mpg US
14.1litres/100km
730i SE: 92
17 18.5mpg 15.4mpg US
15.3litres/100km
750i L: 93
18 17.5mpg 14.6mpg US
16.1litres/100km
735i Automatic: 89
19 17.2mpg 14.3mpg US
16.4litres/100km
635 CSi: 98
20 16.2mpg 13.5mpg US
17.4litres/100km
750i: 113
21 14.7mpg 12.2mpg US
19.2litres/100km
850i Manual: 125
22 14.4mpg 12.0mpg US
19.6litres/100km
850i: 121

Carbodies

1 23.0mpg 19.2mpg US
12.3litres/100km
Fairway 2.7 Silver: 134

Caterham

1 17.3mpg 14.4mpg US
16.3litres/100km
7 HPC: 135

Citroen

1 54.7mpg 45.5mpg US
5.2litres/100km
AX14 DTR: 203
2 38.9mpg 32.4mpg US
7.3litres/100km
AX11 TRE: 197
3 38.8mpg 32.3mpg US
7.3litres/100km
AX11 TZX: 208
4 38.4mpg 32.0mpg US
7.4litres/100km
BX19 RD Estate: 193
5 36.2mpg 30.1mpg US
7.8litres/100km
BX DTR Turbo: 202
6 35.9mpg 29.9mpg US
7.9litres/100km
AX GT5: 205
7 34.7mpg 28.9mpg US
8.1litres/100km
AX14 TRS: 198
8 34.0mpg 28.3mpg US
8.3litres/100km
BX14 TGE: 204

9 32.7mpg 27.2mpg US
8.6litres/100km
AX11 TRE 5DR: 201
10 32.0mpg 26.6mpg US
8.8litres/100km
XM Turbo SD Estate: 214
11 30.4mpg 25.3mpg US
9.3litres/100km
AX GT: 196
12 30.1mpg 25.1mpg US
9.4litres/100km
CX25 DTR Turbo 2: 200
13 29.8mpg 24.8mpg US
9.5litres/100km
BX GTi 16v: 199
14 26.5mpg 22.1mpg US
10.7litres/100km
ZX 1.4 Avantage: 211
15 26.3mpg 21.9mpg US
10.7litres/100km
BX GTi 4x4: 206
16 26.2mpg 21.8mpg US
10.8litres/100km
Visa GTi: 195
17 26.0mpg 21.6mpg US
10.9litres/100km
AX GTi: 213
18 25.8mpg 21.5mpg US
10.9litres/100km
XM 2.0SEi: 207
19 25.3mpg 21.1mpg US
11.2litres/100km
ZX Volcane: 212
20 20.8mpg 17.3mpg US
13.6litres/100km
XM V6.24: 210
21 20.4mpg 17.0mpg US
13.8litres/100km
CX25 Ri Familiale Auto: 194
22 18.8mpg 15.7mpg US
15.0litres/100km
XM 3.0 SEi Auto: 209

Daihatsu

1 45.5mpg 37.9mpg US
6.2litres/100km
Charade CX Diesel Turbo: 218
2 37.2mpg 31.0mpg US
7.6litres/100km
Domino: 217
3 29.2mpg 24.3mpg US
9.7litres/100km
Charade Turbo: 221
4 29.0mpg 24.1mpg US
9.7litres/100km
Applause 16 Xi: 225
5 28.8mpg 24.0mpg US
9.8litres/100km
Charade GT ti: 219
6 25.8mpg 21.5mpg US
10.9litres/100km
Charade 1.3 CX: 222
7 23.0mpg 19.2mpg US

12.3litres/100km
Fourtrak Estate DT EL: 220
8 21.4mpg 17.8mpg US
13.2litres/100km
Sportrak EL: 224

De Tomaso

1 16.0mpg 13.3mpg US
17.7litres/100km
Pantera GT5-S: 226

Ferrari

1 18.4mpg 15.3mpg US
15.4litres/100km
348tb: 262
2 16.8mpg 14.0mpg US
16.8litres/100km
Mondial 3.2QV: 249
3 16.6mpg 13.8mpg US
17.0litres/100km
Testa Rossa: 259

Fiat

1 41.8mpg 34.8mpg US
6.8litres/100km
Regata DS: 272
2 39.0mpg 32.5mpg US
7.2litres/100km
Uno 60S: 282
3 35.5mpg 29.6mpg US
8.0litres/100km
Panda 750 L: 271
4 34.9mpg 29.1mpg US
8.1litres/100km
Uno Selecta: 275
5 34.0mpg 28.3mpg US
8.3litres/100km
Panda 1000 S: 270
6 32.2mpg 26.8mpg US
8.8litres/100km
Tipo 1.4: 276
7 30.4mpg 25.3mpg US
9.3litres/100km
Tipo 1.6 DGT SX: 278
8 29.0mpg 24.1mpg US
9.7litres/100km
Croma ie Super: 273
9 28.6mpg 23.8mpg US
9.9litres/100km
Tipo 1.8ie DGT SX: 284
10 28.3mpg 23.6mpg US
10.0litres/100km
Croma CHT: 280
11 28.1mpg 23.4mpg US
10.1litres/100km
Tempra 1.6 SX: 281
12 25.9mpg 21.6mpg US
10.9litres/100km
Regata 100S ie: 274
Uno Turbo ie: 283

Ford

1 36.1mpg 30.1mpg US
7.8litres/100km
Escort 1.3L 3-door: 308
2 35.0mpg 29.1mpg US
8.1litres/100km
Fiesta 1.1 Ghia Auto: 295
3 34.6mpg 28.8mpg US
8.2litres/100km
Escort 1.4 GL: 286
4 33.2mpg 27.6mpg US
8.5litres/100km
Fiesta 1.1LX 5-door: 309
5 31.0mpg 25.8mpg US
9.1litres/100km
Orion 1.4 GL: 291
6 30.8mpg 25.6mpg US
9.2litres/100km
Escort XR3i: 288
7 30.6mpg 25.5mpg US
9.2litres/100km
Granada 2.5 GL Diesel: 321
8 29.2mpg 24.3mpg US
9.7litres/100km
Orion 1.6i Ghia: 336
9 29.0mpg 24.1mpg US
9.7litres/100km
Fiesta 1.6S: 311
10 28.8mpg 24.0mpg US
9.8litres/100km
Sierra Sapphire 2000E: 314
11 28.6mpg 23.8mpg US
9.9litres/100km
Escort Cabriolet: 332
Escort XR3i: 342
12 28.4mpg 23.6mpg US
9.9litres/100km
Fiesta XR2i: 320
13 28.1mpg 23.4mpg US
10.1litres/100km
Fiesta 1.4 Ghia: 310
14 27.8mpg 23.1mpg US
10.2litres/100km
Escort 1.6 Ghia Estate: 331
15 27.4mpg 22.8mpg US
10.3litres/100km
Escort RS Turbo: 287
16 26.8mpg 22.3mpg US
10.5litres/100km
Sierra Sapphire 2.0 Ghia: 299
17 26.4mpg 22.0mpg US
10.7litres/100km
Escort RS2000: 341
18 25.2mpg 21.0mpg US
11.2litres/100km
Granada Scorpio 2.0i Auto: 323
19 24.0mpg 20.0mpg US
11.8litres/100km
Granada Scorpio 2.9 EX: 297
20 23.6mpg 19.7mpg US
12.0litres/100km

Fiesta RS Turbo: 319
21 21.6mpg 18.0mpg US
13.1litres/100km
Sierra Sapphire Cosworth 4x4: 328
22 21.5mpg 17.9mpg US
13.1litres/100km
Granada 2.4i Ghia: 296
23 21.0mpg 17.5mpg US
13.5litres/100km
Sierra Ghia 4x4 Estate: 293
Scorpio 24v: 338
24 20.3mpg 16.9mpg US
13.9litres/100km
Sierra RS Cosworth: 306
25 18.9mpg 15.7mpg US
14.9litres/100km
Granada Ghia X 2.9 EFi Saloon: 322
26 18.3mpg 15.2mpg US
15.4litres/100km
RS200: 292
27 17.0mpg 14.2mpg US
16.6litres/100km
Sierra XR 4x4: 315

Honda

1 34.7mpg 28.9mpg US
8.1litres/100km
Civic 1.4 GL 3-door: 355
2 33.3mpg 27.7mpg US
8.5litres/100km
Ballade EX: 354
3 31.9mpg 26.6mpg US
8.9litres/100km
Civic 1.5 VEi: 370
4 30.0mpg 25.0mpg US
9.4litres/100km
CRX Coupe 1.6i 16: 349
Civic 1.6i VT: 371
5 29.6mpg 24.6mpg US
9.5litres/100km
Civic CRX: 359
6 29.1mpg 24.2mpg US
9.7litres/100km
Integra 1.5: 351
7 28.1mpg 23.4mpg US
10.1litres/100km
CRX 1.6i VT: 369
8 27.5mpg 22.9mpg US
10.3litres/100km
Integra EX 16: 352
9 26.1mpg 21.7mpg US
10.8litres/100km
Accord 2.0 EXi: 367
10 25.7mpg 21.4mpg US
11.0litres/100km
Concerto 1.6i 16: 368
11 23.3mpg 19.4mpg US
12.1litres/100km
Shuttle 1.4 GL Automatic: 366
12 22.9mpg 19.1mpg US
12.3litres/100km

Shuttle 1.6i RT4 4WD: 363
13 21.5mpg 17.9mpg US
13.1litres/100km
Legend: 353
Legend Coupe: 375
14 21.0mpg 17.5mpg US
13.5litres/100km
Prelude 2.0i 16: 358
15 20.3mpg 16.9mpg US
13.9litres/100km
Legend Coupe: 356
16 19.6mpg 16.3mpg US
14.4litres/100km
NSX: 373
17 19.1mpg 15.9mpg US
14.8litres/100km
Legend: 372
18 18.4mpg 15.3mpg US
15.4litres/100km
Legend Saloon Automatic: 362
19 18.3mpg 15.2mpg US
15.4litres/100km
NSX Auto: 376

Hyundai

1   30.8mpg 25.6mpg US
9.2litres/100km
S Coupe GSi: 382
2   28.0mpg 23.3mpg US
10.1litres/100km
X2 1.5 GSi: 383
3   27.1mpg 22.6mpg US
10.4litres/100km
Lantra 1.6 Cdi: 381
4   23.0mpg 19.2mpg US
12.3litres/100km
Sonata 2.0i GLS: 379
5   22.7mpg 18.9mpg US
12.4litres/100km
Sonata 2.4i GLS: 380

Isuzu

1   22.3mpg 18.6mpg US
12.7litres/100km
Trooper 3-door TD: 392

Jaguar

1   20.7mpg 17.2mpg US
13.6litres/100km
XJ6 3.6: 396
2   19.9mpg 16.6mpg US
14.2litres/100km
XJ6 3.2: 416
3   19.3mpg 16.1mpg US
14.6litres/100km
XJ6 2.9: 401
4   18.8mpg 15.7mpg US
15.0litres/100km
XJS 4.0 Auto: 418
5   18.5mpg 15.4mpg US
15.3litres/100km

4.0 Sovereign: 406
6   18.0mpg 15.0mpg US
15.7litres/100km
XJS 3.6 Automatic: 402
7   17.2mpg 14.3mpg US
16.4litres/100km
XJR 4.0: 411
8   13.8mpg 11.5mpg US
20.5litres/100km
XJS V12 Convertible: 405

Lada

1   31.1mpg 25.9mpg US
9.1litres/100km
Samara 1300 SL: 425
2   30.5mpg 25.4mpg US
9.3litres/100km
Samara 1300L 5-door: 426
3   30.1mpg 25.1mpg US
9.4litres/100km
Riva 1500 Estate: 423
4   18.8mpg 15.7mpg US
15.0litres/100km
Niva Cossack Cabrio: 424

Lamborghini

1   15.9mpg 13.2mpg US
17.8litres/100km
Jalpa 3500: 428

Lancia

1   40.6mpg 33.8mpg US
7.0litres/100km
Y10 Fire: 437
2   29.8mpg 24.8mpg US
9.5litres/100km
Y10 GTie: 446
3   27.1mpg 22.6mpg US
10.4litres/100km
Prisma 1600ie LX: 435
4   25.9mpg 21.6mpg US
10.9litres/100km
Thema 2.0ie: 436
5   25.2mpg 21.0mpg US
11.2litres/100km
Dedra 1.8i: 443
6   24.5mpg 20.4mpg US
11.5litres/100km
Delta HF Turbo ie: 434
7   24.4mpg 20.3mpg US
11.6litres/100km
Thema 2.0ie 16v: 442
8   23.7mpg 19.7mpg US
11.9litres/100km
Dedra 2.0ie: 444
9   22.1mpg 18.4mpg US
12.8litres/100km
Dedra 2000 Turbo: 447
10 21.1mpg 17.6mpg US
13.4litres/100km
Thema 2.0ie 16v SE Turbo: 445

11 19.8mpg 16.5mpg US
14.3litres/100km
Delta HF 4WD: 433
12 19.1mpg 15.9mpg US
14.8litres/100km
Delta Integrale 16v: 441
13 17.6mpg 14.7mpg US
16.1litres/100km
Delta HF Integrale: 439
14 15.6mpg 13.0mpg US
18.1litres/100km
Thema 8.32: 440

Land Rover

1   23.9mpg 19.9mpg US
11.8litres/100km
Discovery TDi: 451
2   18.0mpg 15.0mpg US
15.7litres/100km
Ninety County Turbo Diesel: 448
3   16.5mpg 13.7mpg US
17.1litres/100km
Discovery V8i 5DR: 453
4   14.0mpg 11.7mpg US
20.2litres/100km
Discovery V8: 452
5   13.4mpg 11.2mpg US
21.1litres/100km
One Ten County V8: 449
6   12.1mpg 10.1mpg US
23.3litres/100km
Ninety County V8: 450

Lexus

1   19.7mpg 16.4mpg US
14.3litres/100km
LS400: 454

Lotus

1   23.5mpg 19.6mpg US
12.0litres/100km
Esprit Turbo SE: 466
2   22.2mpg 18.5mpg US
12.7litres/100km
Excel SA: 463
3   20.9mpg 17.4mpg US
13.5litres/100km
Esprit Turbo HC: 462
4   20.4mpg 17.0mpg US
13.8litres/100km
Esprit Turbo: 464
5   20.1mpg 16.7mpg US
14.1litres/100km
Elan SE: 468
6   19.6mpg 16.3mpg US
14.4litres/100km
Excel SE: 460

Maserati

1   18.5mpg 15.4mpg US

15.3litres/100km
Biturbo Spyder: 473

Mazda

1   30.8mpg 25.6mpg US
9.2litres/100km
121 1.3 LX Sun Top: 489
2   30.7mpg 25.6mpg US
9.2litres/100km
121 GSX: 511
3   30.1mpg 25.1mpg US
9.4litres/100km
323 1.5 GLX: 487
4   29.5mpg 24.6mpg US
9.6litres/100km
626 2.0i 5-door: 482
5   28.6mpg 23.8mpg US
9.9litres/100km
323 1.8 GT: 501
6   28.5mpg 23.7mpg US
9.9litres/100km
323 1.8 GT: 512
7   27.9mpg 23.2mpg US
10.1litres/100km
MX-3 1.6 Auto: 518
8   27.5mpg 22.9mpg US
10.3litres/100km
323 1.6i: 500
9   25.0mpg 20.8mpg US
11.3litres/100km
MX-3 V6: 513
10 24.8mpg 20.7mpg US
11.4litres/100km
626 2.0 GLS Executive: 488
11 24.5mpg 20.4mpg US
11.5litres/100km
626 2.0i GT 4WS: 493
12 22.9mpg 19.1mpg US
12.3litres/100km
626 2.0 GLX Executive Estate:
492
13 21.3mpg 17.7mpg US
13.3litres/100km
323 Turbo 4x4 Lux: 481
14 20.7mpg 17.2mpg US
13.6litres/100km
323 Turbo 4x4: 491
15 20.4mpg 17.0mpg US
13.8litres/100km
323 1.5 GLX Estate: 479
16 18.7mpg 15.6mpg US
15.1litres/100km
RX-7: 484
17 14.3mpg 11.9mpg US
19.8litres/100km
RX-7 Turbo II: 506

Mercedes-Benz

1   28.6mpg 23.8mpg US
9.9litres/100km
300D: 548
2   28.0mpg 23.3mpg US

Road test fuel consumption, Autocar & Motor. In order by make

10.1litres/100km
190: 537
3 26.1mpg 21.7mpg US
10.8litres/100km
190E 1.8 Auto: 551
4 23.6mpg 19.7mpg US
12.0litres/100km
200: 521
5 23.3mpg 19.4mpg US
12.1litres/100km
200E Automatic: 541
6 23.1mpg 19.2mpg US
12.2litres/100km
260E Auto: 533
7 22.0mpg 18.3mpg US
12.8litres/100km
190E 2.5-16: 538
8 21.5mpg 17.9mpg US
13.1litres/100km
300TE: 523
9 21.4mpg 17.8mpg US
13.2litres/100km
300CE: 529
10 21.0mpg 17.5mpg US
13.5litres/100km
190E 2.6: 528
11 19.2mpg 16.0mpg US
14.7litres/100km
300SL-24 5-speed Auto: 552
12 18.6mpg 15.5mpg US
15.2litres/100km
560SEL: 526
300E 4Matic: 535
13 18.5mpg 15.4mpg US
15.3litres/100km
300E-24: 549
14 18.4mpg 15.3mpg US
15.4litres/100km
420SE: 524
15 17.1mpg 14.2mpg US
16.5litres/100km
G-Wagen 300 GD LWB: 556
16 16.5mpg 13.7mpg US
17.1litres/100km
500E: 554
17 16.2mpg 13.5mpg US
17.4litres/100km
500SL: 550
18 14.6mpg 12.2mpg US
19.3litres/100km
400SE: 553

MG

1 20.9mpg 17.4mpg US
13.5litres/100km
Maestro Turbo: 566

Mitsubishi

1 29.2mpg 24.3mpg US
9.7litres/100km
Colt 1500 GLX 5-door: 570
2 27.8mpg 23.1mpg US

10.2litres/100km
Lancer 1.5 GLX Estate: 569
3 27.2mpg 22.6mpg US
10.4litres/100km
Space Runner 1800-16v GLXi: 594
4 26.9mpg 22.4mpg US
10.5litres/100km
Colt 1800 GTI-16v: 589
5 25.9mpg 21.6mpg US
10.9litres/100km
Lancer GTi 16v: 578
6 25.3mpg 21.1mpg US
11.2litres/100km
Galant 2000 GLS Automatic: 571
7 24.8mpg 20.7mpg US
11.4litres/100km
Galant Sapporo: 572
8 24.6mpg 20.5mpg US
11.5litres/100km
Lancer GLXi 4WD Liftback: 586
9 24.4mpg 20.3mpg US
11.6litres/100km
Galant GTi-16v: 577
10 23.0mpg 19.2mpg US
12.3litres/100km
Galant 2000 GLSi Coupe: 585
11 22.9mpg 19.1mpg US
12.3litres/100km
Lancer Liftback 1800 GTi-16v: 587
12 22.1mpg 18.4mpg US
12.8litres/100km
Shogun Turbo Diesel 5-door: 573
13 21.2mpg 17.7mpg US
13.3litres/100km
Galant 2000 GLSi: 576
14 19.7mpg 16.4mpg US
14.3litres/100km
Shogun V6 5-door: 582
15 19.0mpg 15.8mpg US
14.9litres/100km
Starion 2000 Turbo: 574
16 18.8mpg 15.7mpg US
15.0litres/100km
Starion 2.6 Turbo: 583
17 18.3mpg 15.2mpg US
15.4litres/100km
Sigma: 593
18 16.1mpg 13.4mpg US
17.5litres/100km
Shogun V6 LWB: 592

Morgan

1 20.9mpg 17.4mpg US
13.5litres/100km
Plus 8: 595
2 20.1mpg 16.7mpg US
14.1litres/100km
Plus 8: 597

Nissan

1 40.9mpg 34.1mpg US
6.9litres/100km
Micra SGL 5-DR: 607
2 32.3mpg 26.9mpg US
8.7litres/100km
Micra 1.2 GSX: 618
3 31.6mpg 26.3mpg US
8.9litres/100km
Sunny 1.6 GS 5-door: 633
4 31.0mpg 25.8mpg US
9.1litres/100km
Sunny 1.3 LX 5-door: 604
5 29.2mpg 24.3mpg US
9.7litres/100km
Primera 1.6 LS: 623
6 28.9mpg 24.1mpg US
9.8litres/100km
Sunny 1.4 LS: 620
7 28.7mpg 23.9mpg US
9.8litres/100km
Sunny 1.6 SLX Coupe: 610
8 27.8mpg 23.1mpg US
10.2litres/100km
Sunny 2.0 GTi: 635
9 27.2mpg 22.6mpg US
10.4litres/100km
Sunny ZX Coupe: 609
100NX: 625
10 26.8mpg 22.3mpg US
10.5litres/100km
Primera 2.0E ZX: 631
11 26.0mpg 21.6mpg US
10.9litres/100km
Primera 2.0 GSX: 630
12 23.8mpg 19.8mpg US
11.9litres/100km
Bluebird 1.8 ZX: 601
13 20.6mpg 17.2mpg US
13.7litres/100km
Prairie: 619
14 20.0mpg 16.7mpg US
14.1litres/100km
Laurel 2.4 SGL Automatic: 602
15 19.7mpg 16.4mpg US
14.3litres/100km
200SX: 634
16 19.5mpg 16.2mpg US
14.5litres/100km
200SX: 614
17 19.2mpg 16.0mpg US
14.7litres/100km
300ZX Turbo: 606
18 17.0mpg 14.2mpg US
16.6litres/100km
300ZX: 621
19 16.8mpg 14.0mpg US
16.8litres/100km
Maxima: 616

Panther

1 19.4mpg 16.2mpg US

14.6litres/100km
Solo: 642

Peugeot

1 39.5mpg 32.9mpg US
7.2litres/100km
205 1.1GL: 655
2 38.2mpg 31.8mpg US
7.4litres/100km
405 GRD: 652
3 37.2mpg 31.0mpg US
7.6litres/100km
205 CJ: 656
4 34.4mpg 28.6mpg US
8.2litres/100km
205 D Turbo: 669
5 33.3mpg 27.7mpg US
8.5litres/100km
605 SR TD: 670
6 32.7mpg 27.2mpg US
8.6litres/100km
106 XT: 673
7 32.1mpg 26.7mpg US
8.8litres/100km
106 XR: 667
8 31.3mpg 26.1mpg US
9.0litres/100km
405 GTD Turbo: 659
9 30.9mpg 25.7mpg US
9.1litres/100km
309 SR Injection: 646
10 30.4mpg 25.3mpg US
9.3litres/100km
309 Style 5-door: 661
11 29.9mpg 24.9mpg US
9.4litres/100km
205 GTi: 644
12 29.8mpg 24.8mpg US
9.5litres/100km
309 1.6 GR: 645
13 28.3mpg 23.6mpg US
10.0litres/100km
309 GTi: 649
14 28.2mpg 23.5mpg US
10.0litres/100km
309 GLX: 657
15 28.1mpg 23.4mpg US
10.1litres/100km
205 GTi 1.9: 648
16 27.9mpg 23.2mpg US
10.1litres/100km
405 1.6 GL: 651
17 27.3mpg 22.7mpg US
10.3litres/100km
405 GR Injection Auto: 663
18 26.9mpg 22.4mpg US
10.5litres/100km
106 XSi: 668
19 26.6mpg 22.1mpg US
10.6litres/100km
405 Mi16: 653
20 26.0mpg 21.6mpg US
10.9litres/100km

405 Mi16x4: 664
21 25.2mpg 21.0mpg US
11.2litres/100km
309 GTi: 658
22 24.7mpg 20.6mpg US
11.4litres/100km
605 SRi: 666
23 24.4mpg 20.3mpg US
11.6litres/100km
405 GLx4: 662
24 23.8mpg 19.8mpg US
11.9litres/100km
405 SRi: 654
25 20.6mpg 17.2mpg US
13.7litres/100km
505 GTi Family Estate: 647
26 20.4mpg 17.0mpg US
13.8litres/100km
605 SVE 24: 672
27 19.8mpg 16.5mpg US
14.3litres/100km
605 SV 3.0 Auto: 671

Porsche

1 23.5mpg 19.6mpg US
12.0litres/100km
944 S: 703
2 22.9mpg 19.1mpg US
12.3litres/100km
944 Turbo: 704
3 22.6mpg 18.8mpg US
12.5litres/100km
924 S: 708
4 20.7mpg 17.2mpg US
13.6litres/100km
944 S2: 736
944: 735
5 20.6mpg 17.2mpg US
13.7litres/100km
911 Carrera Club Sport: 712
6 20.4mpg 17.0mpg US
13.8litres/100km
911 Carrera 2: 739
7 19.2mpg 16.0mpg US
14.7litres/100km
911 Carrera SE: 697
8 19.1mpg 15.9mpg US
14.8litres/100km
944 Turbo SE: 723
9 18.6mpg 15.5mpg US
15.2litres/100km
928 S4 SE: 734
10 17.4mpg 14.5mpg US
16.2litres/100km
Carrera 2 Cabriolet Tiptronic: 747
11 17.0mpg 14.2mpg US
16.6litres/100km
928 S4 Automatic: 701
12 16.6mpg 13.8mpg US
17.0litres/100km
911 3.3 Turbo: 726
13 15.1mpg 12.6mpg US

18.7litres/100km
911 Turbo: 752

Proton

1 28.5mpg 23.7mpg US
9.9litres/100km
1.5 SE Triple Valve Aeroback: 756
2 28.0mpg 23.3mpg US
10.1litres/100km
1.5 SE Aeroback: 755

Reliant

1 21.7mpg 18.1mpg US
13.0litres/100km
Scimitar Ti: 765
Scimitar SST 1800Ti: 766

Renault

1 41.0mpg 34.1mpg US
6.9litres/100km
5 Campus: 784
2 38.3mpg 31.9mpg US
7.4litres/100km
Clio 1.4RT: 793
3 36.8mpg 30.6mpg US
7.7litres/100km
Clio 1.2 RN: 792
4 35.5mpg 29.6mpg US
8.0litres/100km
5 GTL 5-door: 770
5 34.9mpg 29.1mpg US
8.1litres/100km
19 TSE: 780
6 30.9mpg 25.7mpg US
9.1litres/100km
19 TXE Chamade: 786
7 30.0mpg 25.0mpg US
9.4litres/100km
5 GTX 3-DR: 775
8 29.7mpg 24.7mpg US
9.5litres/100km
21 GTS: 767
9 29.4mpg 24.5mpg US
9.6litres/100km
9 Turbo: 771
10 28.4mpg 23.6mpg US
9.9litres/100km
5 GT Turbo: 774
11 28.2mpg 23.5mpg US
10.0litres/100km
19 16v: 790
12 28.1mpg 23.4mpg US
10.1litres/100km
21 GTS Hatchback: 782
13 27.8mpg 23.1mpg US
10.2litres/100km
21 Savanna GTX: 768
14 27.3mpg 22.7mpg US
10.3litres/100km
11 Turbo: 773

15 26.7mpg 22.2mpg US
10.6litres/100km
Espace TXE: 785
16 26.1mpg 21.7mpg US
10.8litres/100km
19 TXE: 781
17 26.0mpg 21.6mpg US
10.9litres/100km
Cleo 16v: 796
18 25.7mpg 21.4mpg US
11.0litres/100km
21 Ti: 777
19 25.5mpg 21.2mpg US
11.1litres/100km
21 TXi: 788
20 24.9mpg 20.7mpg US
11.3litres/100km
19 16v Cabriolet: 795
21 22.5mpg 18.7mpg US
12.6litres/100km
25 TXi: 783
22 22.1mpg 18.4mpg US
12.8litres/100km
21 Turbo Quadra: 787
23 22.0mpg 18.3mpg US
12.8litres/100km
5 GT Turbo: 769
24 21.2mpg 17.7mpg US
13.3litres/100km
25 V6 2.9 Auto: 789
25 20.5mpg 17.1mpg US
13.8litres/100km
GTA V6 Turbo: 772
26 19.9mpg 16.6mpg US
14.2litres/100km
Espace V6: 794
27 19.1mpg 15.9mpg US
14.8litres/100km
GTA V6: 779
28 18.3mpg 15.2mpg US
15.4litres/100km
21 Turbo: 778

Rolls-Royce

1 14.0mpg 11.7mpg US
20.2litres/100km
Silver Spirit: 798
2 13.8mpg 11.5mpg US
20.5litres/100km
Silver Spirit II: 801
3 12.3mpg 10.2mpg US
23.0litres/100km
Corniche: 799

Rover

1 39.7mpg 33.1mpg US
7.1litres/100km
Montego 2.0 DSL Turbo: 809
2 36.2mpg 30.1mpg US
7.8litres/100km
825 TD: 822
3 35.3mpg 29.4mpg US

8.0litres/100km
218 SD: 820
4 33.7mpg 28.1mpg US
8.4litres/100km
Metro 1.4SL: 816
5 32.7mpg 27.2mpg US
8.6litres/100km
414 Si: 814
6 31.6mpg 26.3mpg US
8.9litres/100km
Metro GTi 16v: 817
7 30.0mpg 25.0mpg US
9.4litres/100km
214S: 818
8 29.5mpg 24.6mpg US
9.6litres/100km
Montego 1.6L: 804
9 28.9mpg 24.1mpg US
9.8litres/100km
214 GSi: 812
10 27.4mpg 22.8mpg US
10.3litres/100km
820i: 824
11 27.1mpg 22.6mpg US
10.4litres/100km
220 GTi: 821
12 26.6mpg 22.1mpg US
10.6litres/100km
820SE: 803
13 26.5mpg 22.1mpg US
10.7litres/100km
Mini Cooper S: 823
14 26.3mpg 21.9mpg US
10.7litres/100km
825i: 802
15 26.0mpg 21.6mpg US
10.9litres/100km
216 GSi: 813
16 25.1mpg 20.9mpg US
11.3litres/100km
416 GTi: 815
17 24.5mpg 20.4mpg US
11.5litres/100km
216 GTi: 819
18 24.4mpg 20.3mpg US
11.6litres/100km
820 Fastback: 806
19 22.1mpg 18.4mpg US
12.8litres/100km
Vitesse: 811
20 21.3mpg 17.7mpg US
13.3litres/100km
Sterling Catalyst: 810
21 21.0mpg 17.5mpg US
13.5litres/100km
Sterling Automatic: 805
22 19.9mpg 16.6mpg US
14.2litres/100km
Vitesse: 808
23 19.6mpg 16.3mpg US
14.4litres/100km
Sterling Automatic: 825
24 17.9mpg 14.9mpg US
15.8litres/100km

827 SLi Auto: 807

Saab

1 25.7mpg 21.4mpg US
11.0litres/100km
9000i: 827
2 23.6mpg 19.7mpg US
12.0litres/100km
9000 2.3CS: 838
3 22.2mpg 18.5mpg US
12.7litres/100km
9000CS Turbo S: 837
4 22.0mpg 18.3mpg US
12.8litres/100km
CDS 2.3: 834
5 21.6mpg 18.0mpg US
13.1litres/100km
Carlsson: 831
6 20.3mpg 16.9mpg US
13.9litres/100km
9000CD: 828
7 19.7mpg 16.4mpg US
14.3litres/100km
9000 2.3 Turbo TCS: 835
8 19.1mpg 15.9mpg US
14.8litres/100km
900 Turbo S: 829

Seat

1 37.6mpg 31.3mpg US
7.5litres/100km
Toledo 1.9 CL Diesel: 844
2 34.9mpg 29.1mpg US
8.1litres/100km
Ibiza SXi: 843
3 31.8mpg 26.5mpg US
8.9litres/100km
Toledo 2.0 GTi: 845
4 31.3mpg 26.1mpg US
9.0litres/100km
Malaga 1.2L: 842

Skoda

1 30.6mpg 25.5mpg US
9.2litres/100km
136 Rapide Coupe: 847
2 30.1mpg 25.1mpg US
9.4litres/100km
Estelle 120 LSE: 846

Subaru

1 32.9mpg 27.4mpg US
8.6litres/100km
Justy: 849
2 31.0mpg 25.8mpg US
9.1litres/100km
Justy GL II 5-door 4WD: 852
3 25.8mpg 21.5mpg US
10.9litres/100km
XT Turbo 4WD: 850

4 25.0mpg 20.8mpg US
11.3litres/100km
Legacy 2.2 GX 4WD: 855
5 22.6mpg 18.8mpg US
12.5litres/100km
4WD 3-door Turbo Coupe: 848
6 21.8mpg 18.2mpg US
13.0litres/100km
Legacy 2.0 4Cam Turbo
Estate: 856
7 19.9mpg 16.6mpg US
14.2litres/100km
4WD Turbo Estate: 851

Suzuki

1 37.0mpg 30.8mpg US
7.6litres/100km
Swift 1.3 GLX: 862
2 36.0mpg 30.0mpg US
7.8litres/100km
Swift 1.3 GLX Executive: 863
3 35.7mpg 29.7mpg US
7.9litres/100km
Swift 1.3 GLX: 864
4 35.0mpg 29.1mpg US
8.1litres/100km
Alto GLA: 859
5 30.9mpg 25.7mpg US
9.1litres/100km
Swift 1.3 GTi: 860
6 29.2mpg 24.3mpg US
9.7litres/100km
Swift 1.3 GTi: 865
7 24.8mpg 20.7mpg US
11.4litres/100km
SJ413V JX: 861
8 24.2mpg 20.2mpg US
11.7litres/100km
Vitara JLX: 866

Toyota

1 38.1mpg 31.7mpg US
7.4litres/100km
Starlet GL: 902
2 31.2mpg 26.0mpg US
9.1litres/100km
Corolla 1.3 GL: 885
3 30.5mpg 25.4mpg US
9.3litres/100km
Camry 2.0 GLi: 874
4 30.0mpg 25.0mpg US
9.4litres/100km
Carina II 1.6 GL Liftback: 891
5 29.6mpg 24.6mpg US
9.5litres/100km
Corolla Executive: 877
6 28.7mpg 23.9mpg US
9.8litres/100km
Camry 2.0 GLi Estate: 875
7 28.6mpg 23.8mpg US
9.9litres/100km

MR2 T-Bar: 871
8 28.5mpg 23.7mpg US
9.9litres/100km
Carina 1.6 GL: 882
9 27.2mpg 22.6mpg US
10.4litres/100km
MR2 GT: 899
10 26.7mpg 22.2mpg US
10.6litres/100km
MR2: 906
11 25.0mpg 20.8mpg US
11.3litres/100km
Celica 2.0GT: 868
12 24.5mpg 20.4mpg US
11.5litres/100km
Camry 2.2 GL: 903
13 22.5mpg 18.7mpg US
12.6litres/100km
Celica 2.0 GT: 896
14 21.6mpg 18.0mpg US
13.1litres/100km
Previa: 907
15 21.3mpg 17.7mpg US
13.3litres/100km
Corolla 4WD Estate: 886
16 20.6mpg 17.2mpg US
13.7litres/100km
Camry V6 GXi: 890
17 20.2mpg 16.8mpg US
14.0litres/100km
Landcruiser II TD: 887
18 19.8mpg 16.5mpg US
14.3litres/100km
Landcruiser VX: 905
19 19.4mpg 16.2mpg US
14.6litres/100km
Supra 3.0i: 873
20 19.2mpg 16.0mpg US
14.7litres/100km
Celica GT4: 884
21 18.8mpg 15.7mpg US
15.0litres/100km
Supra 3.0i Turbo: 894
22 17.9mpg 14.9mpg US
15.8litres/100km
Landcruiser: 879

TVR

1 27.3mpg 22.7mpg US
10.3litres/100km
S Convertible: 912
2 20.1mpg 16.7mpg US
14.1litres/100km
V8 S: 915
3 16.6mpg 13.8mpg US
17.0litres/100km
390 SE: 911

Vauxhall

1 44.2mpg 36.8mpg US
6.4litres/100km
Astra 1.7 DL: 929

2 43.8mpg 36.5mpg US
6.4litres/100km
Nova 1.5TD Merit: 940
3 37.1mpg 30.9mpg US
7.6litres/100km
Cavalier 1.7 GL TD: 946
4 35.6mpg 29.6mpg US
7.9litres/100km
Nova 1.2 Luxe: 943
5 35.1mpg 29.2mpg US
8.0litres/100km
Nova 1.3 GL 5-door: 918
6 34.8mpg 29.0mpg US
8.1litres/100km
Cavalier 1.4L: 930
7 34.7mpg 28.9mpg US
8.1litres/100km
1.4i GLS: 945
8 34.1mpg 28.4mpg US
8.3litres/100km
Cavalier 1.6L: 931
9 34.0mpg 28.3mpg US
8.3litres/100km
Belmont 1.6 GL: 920
10 33.4mpg 27.8mpg US
8.5litres/100km
Astra GTE 2.0i: 919
11 32.5mpg 27.1mpg US
8.7litres/100km
Nova GTE: 928
12 32.0mpg 26.6mpg US
8.8litres/100km
Belmont 1.8 GLSi: 916
13 31.4mpg 26.1mpg US
9.0litres/100km
Carlton L 1.8i: 925
14 31.0mpg 25.8mpg US
9.1litres/100km
Carlton L 2.3D: 926
15 29.3mpg 24.4mpg US
9.6litres/100km
Calibra 2.0i: 941
16 28.3mpg 23.6mpg US
10.0litres/100km
Cavalier 4x4: 927
17 27.3mpg 22.7mpg US
10.3litres/100km
Calibra 4x4: 942
18 27.0mpg 22.5mpg US
10.5litres/100km
Cavalier SRi 130 4-door: 923
19 26.7mpg 22.2mpg US
10.6litres/100km
Carlton CD 2.0i: 917
20 26.1mpg 21.7mpg US
10.8litres/100km
Cavalier 2.0i CD: 922
21 25.5mpg 21.2mpg US
11.1litres/100km
Calibra 2.0i 16v: 935
22 25.3mpg 21.1mpg US
11.2litres/100km
Cavalier SRi: 932
23 25.0mpg 20.8mpg US

11.3litres/100km
Cavalier GSi 2000 16v: 938
24 24.9mpg 20.7mpg US
11.3litres/100km
Carlton 3000 GSi: 921
25 24.0mpg 20.0mpg US
11.8litres/100km
Cavalier GSi 2000 16v 4x4:
939
26 22.8mpg 19.0mpg US
12.4litres/100km
Frontera 2.3 TD Estate: 947
27 22.7mpg 18.9mpg US
12.4litres/100km
Senator 3.0i CD: 924
28 22.2mpg 18.5mpg US
12.7litres/100km
Senator 2.5i: 933
29 21.4mpg 17.8mpg US
13.2litres/100km
Senator 3.0i 24v: 944
30 21.0mpg 17.5mpg US
13.5litres/100km
Carlton 3.0i CDX Estate: 936
31 19.5mpg 16.2mpg US
14.5litres/100km
Carlton GSi 3000 24v: 937
32 19.3mpg 16.1mpg US
14.6litres/100km
Senator 3.0i CD: 934

Volkswagen

1 39.3mpg 32.7mpg US
7.2litres/100km
Jetta Turbo Diesel: 963
2 38.4mpg 32.0mpg US
7.4litres/100km
Golf CL Catalyst: 970
3 37.0mpg 30.8mpg US
7.6litres/100km
Passat CL TD Estate: 974
4 35.2mpg 29.3mpg US
8.0litres/100km
Polo 1.3 CL Coupe: 979
5 34.9mpg 29.1mpg US
8.1litres/100km
Polo GT Coupe: 981
6 32.2mpg 26.8mpg US
8.8litres/100km
Golf 1.8 GL: 982
7 30.9mpg 25.7mpg US
9.1litres/100km
Polo C Saloon: 965
8 30.3mpg 25.2mpg US
9.3litres/100km
Passat GT 5-door: 960
9 30.1mpg 25.1mpg US
9.4litres/100km
Polo G40: 980
10 30.0mpg 25.0mpg US
9.4litres/100km
Golf Cabriolet Clipper: 957
11 29.2mpg 24.3mpg US

9.7litres/100km
Passat GT: 964
12 28.8mpg 24.0mpg US
9.8litres/100km
Jetta GTi 16v: 959
Golf GTi: 983
13 27.4mpg 22.8mpg US
10.3litres/100km
Passat GT 16v: 972
14 25.9mpg 21.6mpg US
10.9litres/100km
Corrado 16v: 969
15 25.6mpg 21.3mpg US
11.0litres/100km
Jetta GT: 952
16 25.3mpg 21.1mpg US
11.2litres/100km
Golf GTi 16v: 951
17 22.3mpg 18.6mpg US
12.7litres/100km
Jetta Syncro: 971
18 22.0mpg 18.3mpg US
12.8litres/100km
Corrado G60: 976
19 20.7mpg 17.2mpg US
13.6litres/100km
Caravelle Syncro: 950
20 18.1mpg 15.1mpg US
15.6litres/100km
Caravelle Carat Automatic: 967

Volvo

1 29.1mpg 24.2mpg US
9.7litres/100km
340 GLE: 984
2 28.8mpg 24.0mpg US
9.8litres/100km
480 ES: 985
3 26.9mpg 22.4mpg US
10.5litres/100km
440 GLE: 991
4 26.4mpg 22.0mpg US
10.7litres/100km
460 Turbo: 993
460 GLi: 992
5 22.5mpg 18.7mpg US
12.6litres/100km
440 Turbo: 988
6 21.8mpg 18.2mpg US
13.0litres/100km
740 Turbo: 986
7 21.2mpg 17.7mpg US
13.3litres/100km
940 SE Turbo: 997
8 21.1mpg 17.6mpg US
13.4litres/100km
740 GLT 16v Estate: 994
9 19.5mpg 16.2mpg US
14.5litres/100km
760 GLE: 987
10 18.4mpg 15.3mpg US
15.4litres/100km
960 24v: 998

Westfield

1 20.0mpg 16.7mpg US
14.1litres/100km
SEight: 1000

Yugo

1 34.7mpg 28.9mpg US
8.1litres/100km
45 GLS: 1002
2 31.1mpg 25.9mpg US
9.1litres/100km
65A GLX: 1003
3 28.1mpg 23.4mpg US
10.1litres/100km
Sana 1.4: 1004

Road test fuel consumption. All cars in order, Road & Track

**1 44.4mpg 37.0mpg US
6.4litres/100km**
Ford Festiva L: 294
**2 39.6mpg 33.0mpg US
7.1litres/100km**
Mercury Tracer: 560
**3 37.8mpg 31.5mpg US
7.5litres/100km**
Honda CRX Si: 350
Hyundai Excel GLS: 378
Suzuki Swift GT: 867
**4 37.2mpg 31.0mpg US
7.6litres/100km**
Honda CRX Si: 360
**5 36.6mpg 30.5mpg US
7.7litres/100km**
Pontiac Le Mans: 680
**6 36.0mpg 30.0mpg US
7.8litres/100km**
Chevrolet Nova CL: 142
Nissan NX2000: 629
Toyota Tercel LE: 908
Toyota Paseo: 909
Yugo GV: 1001
**7 35.4mpg 29.5mpg US
8.0litres/100km**
Daihatsu Charade CLS: 223
**8 34.8mpg 29.0mpg US
8.1litres/100km**
Honda CRX Si Jackson: 361
Isuzu Impulse Sports Coupe:
390
**9 34.2mpg 28.5mpg US
8.3litres/100km**
Honda Civic Si: 364
**10 33.6mpg 28.0mpg US
8.4litres/100km**
BMW 325 ES: 82
Ford Escort GT: 333
Toyota Corolla GTS: 892
**11 32.4mpg 27.0mpg US
8.7litres/100km**
Geo Storm GSi: 345
Mercury Tracer LTS: 563
Nissan Sentra Sport Coupe
SE: 603
Oldsmobile Cutlass Calais
International Series HO: 638
Oldsmobile Calais HO Quad 4:
637
**12 32.3mpg 26.9mpg US
8.7litres/100km**
Acura Integra GS: 7
**13 31.8mpg 26.5mpg US
8.9litres/100km**
Chevrolet Nova Twin Cam: 157
Volvo 740 Turbo: 995
**14 31.2mpg 26.0mpg US
9.0litres/100km**
Isuzu Impulse Turbo: 391
Lotus Elan SE: 470
Mazda Protege: 509

Pontiac Sunbird GT: 682
Saturn Sports Coupe: 840
**15 30.6mpg 25.5mpg US
9.2litres/100km**
BMW 524 TD: 85
Isuzu Impulse Turbo: 393
Mazda MX-5 Miata: 503
Renault Alliance GTA: 776
Toyota Corolla FX-16 GTS: 878
**16 30.0mpg 25.0mpg US
9.4litres/100km**
Acura Integra LS Special
Edition: 4
Acura Integra LS: 5
Chevrolet Cavalier Z24
Convertible: 149
Chevrolet Baretta GTZ: 177
Mercury Capri XR-2: 562
MG B British Motor Heritage:
567
Mitsubishi Mirage Turbo: 579
Nissan 240SX: 612
Opel Calibra 2.0i 16v: 640
Toyota MR2: 870
Volkswagen Scirocco 16v: 954
Volkswagen Fox GL: 956
Volkswagen Scirocco 16v: 966
Volkswagen Passat GL: 975
Volkswagen GTi 16v: 977
**17 29.9mpg 24.9mpg US
9.4litres/100km**
Isuzu Impulse RS: 395
**18 29.8mpg 24.8mpg US
9.5litres/100km**
Toyota Celica GTS: 898
**19 29.5mpg 24.6mpg US
9.6litres/100km**
Plymouth Laser RS Turbo: 675
**20 29.4mpg 24.5mpg US
9.6litres/100km**
Acura Integra RS: 1
Audi 5000 S: 42
Mazda 323 LX: 480
**21 29.3mpg 24.4mpg US
9.6litres/100km**
Subaru XT6: 854
**22 29.1mpg 24.2mpg US
9.7litres/100km**
Cadillac Eldorado Touring
Coupe: 131
**23 28.8mpg 24.0mpg US
9.8litres/100km**
Honda Accord LXi: 348
Honda Prelude Si: 374
Honda Prelude Si 4WS: 377
Infiniti G20: 387
Isuzu Impulse XS: 394
Mazda 626 GT: 483
Mazda MX-6 GT: 496
Mazda MX-3 GS: 519
Merkur XR4Ti: 565
Nissan Pulsar NX SE: 608

Peugeot 505 STX: 650
Volkswagen GTi 16v: 958
Volkswagen Jetta GLi 16v: 962
Volkswagen GTi 16v: 961
Volkswagen Jetta GLi 16v: 978
**24 28.6mpg 23.8mpg US
9.9litres/100km**
Nissan 240SX: 626
**25 28.2mpg 23.5mpg US
10.0litres/100km**
Fiat X1/9: 277
Fiat X1/9: 279
Nissan 300ZX: 615
**26 28.1mpg 23.4mpg US
10.1litres/100km**
Alfa Romeo Spider Veloce: 34
Ford Probe GT: 326
**27 27.6mpg 23.0mpg US
10.2litres/100km**
Acura NSX: 9
Alfa Romeo 164S: 31
Dodge Daytona Shelby Z: 229
Dodge Daytona Shelby: 237
Mitsubishi Eclipse: 580
Oldsmobile Cutlass Calais
International Series: 636
Plymouth Laser RS: 674
Saab 9000 Turbo: 836
Toyota MR2 Supercharged:
893
Toyota Celica GT Convertible:
904
**28 27.5mpg 22.9mpg US
10.3litres/100km**
Mitsubishi Galant GS: 581
**29 27.0mpg 22.5mpg US
10.5litres/100km**
Acura Integra LS: 3
Alfa Romeo Spider: 24
Chevrolet Lumina APV: 176
Ford Crown Victoria LX: 329
Pontiac Fiero GT: 677
Porsche 911 Cabriolet: 710
**30 26.9mpg 22.4mpg US
10.5litres/100km**
Cadillac Allante: 129
**31 26.7mpg 22.2mpg US
10.6litres/100km**
Volvo 740 GLE: 989
**32 26.5mpg 22.1mpg US
10.6litres/100km**
Porsche 944: 702
**33 26.4mpg 22.0mpg US
10.7litres/100km**
Audi Coupe GT: 45
BMW 325i: 123
Chevrolet Cavalier Z24: 137
Chevrolet Baretta GTU: 158
Ford Taurus SHO: 316
Ford Taurus SHO: 339
Honda Prelude 2.0 Si: 357
Mercedes-Benz 300E: 522

Mercedes-Benz 300TD: 530
Peugeot 405 Mi16: 660
Porsche 924 S: 699
Porsche 924 S: 718
Toyota Supra Turbo: 895
**34 26.3mpg 21.9mpg US
10.7litres/100km**
Alfa Romeo GTV6: 14
Dodge Shadow ES VNT: 238
**35 26.2mpg 21.8mpg US
10.8litres/100km**
Alfa Romeo Milano Platinum:
15
**36 25.8mpg 21.5mpg US
10.9litres/100km**
Chevrolet Lumina Z34: 186

Dodge Lancer Shelby: 230
Dodge Shelby CSX: 231
Dodge Spirit R/T: 239
Mazda 323 GTX: 490
Volkswagen Corrado G60: 973
**37 25.7mpg 21.4mpg US
11.0litres/100km**
Saab 900SPG: 833
**38 25.3mpg 21.1mpg US
11.1litres/100km**
Chevrolet Beretta GTZ: 169
Toyota MR2 Supercharged:
880
**39 25.2mpg 21.0mpg US
11.2litres/100km**
Alfa Romeo Spider
Quadrifoglio: 16
Alfa Romeo Milano Verde 3.0:
21
Audi 90 Quattro: 53
Audi 90: 55
Chrysler Laser XT: 189
Ford Probe LX: 327
Lexus ES300: 456
Mazda RX-7 GTU: 498
Mazda MPV: 502
Mercedes-Benz 190E 2.3-16:
520
Merkur Scorpio: 564
Mitsubishi Cordia Turbo: 575
Mitsubishi Galant VR-4: 591
Nissan 300ZX: 613
Pontiac Grand Prix McLaren
Turbo: 687
Pontiac Grand Prix GTP: 694
Porsche 944: 720
Saab 9000: 826
Saab 9000 Turbo: 830
Saab 9000CD: 832
**40 24.6mpg 20.5mpg US
11.5litres/100km**
Acura Legend: 2
BMW 325i S: 95
Chrysler Le Baron Coupe: 190
Dodge Shelby CSX: 234

Dodge Daytona ES: 236
Ford Probe GT: 337
Honda Prelude Si 4WS: 365
Mercedes-Benz 190E 2.6: 539
Mercedes-Benz 300E: 543
Mitsubishi Cordia Turbo: 568
Nissan Maxima SE: 617

41 24.5mpg 20.4mpg US 11.5litres/100km
Pontiac Grand Prix STE Turbo: 690

42 24.3mpg 20.2mpg US 11.6litres/100km
Toyota Celica GTS: 869

43 24.1mpg 20.1mpg US 11.7litres/100km
Dodge Shelby CSX VNT: 235

44 24.0mpg 20.0mpg US 11.8litres/100km
Alfa Romeo Milano 3.0: 28
BMW 325i X: 96
BMW 525i: 118
Chevrolet Beretta GT: 143
Chevrolet Beretta GT: 168
Chrysler Conquest TSi: 191
Chrysler TC by Maserati: 192
Dodge Shelby GLH-S: 228
Ford Probe GT: 305
Infiniti M30: 385
Lexus SC400: 455
Mazda MX-6: 495
Mitsubishi Starion ESI-R: 584
Plymouth Laser RS Turbo: 676
Pontiac Bonneville SSEi: 695
Porsche 911 Carrera Cabriolet: 728
Porsche 911 Speedster: 742
Subaru Legacy Sports Sedan: 857
Toyota Supra Turbo: 889
Toyota Celica All-Trac Turbo: 883
Volvo 940 SE: 996
Volvo 960: 999

45 23.8mpg 19.8mpg US 11.9litres/100km
Mercury Sable LS: 559
Nissan 300ZX: 605

46 23.5mpg 19.6mpg US 12.0litres/100km
Nissan 300ZX Twin Turbo: 628

47 23.4mpg 19.5mpg US 12.1litres/100km
Acura Legend Coupe L: 6
Acura Legend Coupe: 8
Audi 200 Quattro: 65
Chevrolet Corvette L98: 166
Mercedes-Benz 300CE: 546
Morgan Plus 8: 599
Nissan 200SX SE V6: 611
Porsche 944 S: 721
Volkswagen Quantum Synchro Wagon: 953

48 23.2mpg 19.3mpg US 12.2litres/100km
Chevrolet Celebrity CL Eurosport: 138
Chevrolet Corvette Convertible: 153

49 23.1mpg 19.2mpg US 12.3litres/100km
Chevrolet Cavalier Z24: 172
Mazda RX-7 Turbo: 486

50 22.8mpg 19.0mpg US 12.4litres/100km
Acura Legend LS: 10
Chevrolet Corvette Z51: 141
Dodge Shadow ES Turbo: 233
Eagle Premier ES: 242
Ford Mustang LX: 343
Mazda RX-7 GXL: 485
Mazda RX-7 Turbo: 499
Mazda RX-7 GTUs: 505
Mazda MX-6 GT 4WS: 504
Mazda MX-6 GT: 508
Mitsubishi Diamante LS: 590
Pontiac Firebird TransAm Convertible: 693
Porsche 944 Turbo: 722
Toyota Supra: 881
Toyota Previa LE: 901

51 22.6mpg 18.8mpg US 12.5litres/100km
Volvo 740 Turbo: 990

52 22.5mpg 18.7mpg US 12.6litres/100km
Jaguar XJS: 413
Jaguar XJS Convertible: 419

53 22.3mpg 18.6mpg US 12.6litres/100km
Chevrolet Camaro IROC-Z Convertible: 170
Chevrolet Camaro Z/28 Convertible: 178
Porsche 911 Cabriolet: 696

54 22.2mpg 18.5mpg US 12.7litres/100km
BMW 535i Automatic: 104
Cadillac Seville Touring Sedan: 133
Ford Aerostar XLT: 285
Mazda 929: 494
Pontiac Bonneville SE: 678
Porsche 944 Turbo: 737
Porsche 944 S2: 745
Porsche 944 S2 Cabriolet: 746
Toyota MR2 Turbo: 900

55 21.9mpg 18.2mpg US 12.9litres/100km
Consulier GTP LX: 216

56 21.7mpg 18.1mpg US 13.0litres/100km
Chevrolet Corvette ZR-1: 183

57 21.6mpg 18.0mpg US 13.1litres/100km
Audi 90 Quattro: 60
BMW 735i: 99

Chevrolet Corvette Callaway: 173
Chevrolet Camaro Z28: 187
Dodge Stealth R/T Turbo: 240
Jaguar XJ6 Vanden Plas: 410
Mitsubishi 3000GT VR-4: 588
Nissan 300ZX: 599
Pontiac Bonneville SE: 681
Porsche 911 Carrera: 711
Porsche 911 Club Sport: 713
Toyota Supra: 872
Toyota Supra: 888

58 21.5mpg 17.9mpg US 13.1litres/100km
Chevrolet Corvette: 163

59 21.4mpg 17.8mpg US 13.2litres/100km
Oldsmobile Toronado Trofeo: 639

60 21.3mpg 17.7mpg US 13.3litres/100km
Pontiac Firebird TransAm GTA: 689
Pontiac Firebird GTA: 691

61 21.0mpg 17.5mpg US 13.4litres/100km
BMW 635 CSi: 87
Chevrolet Corvette: 146
Chevrolet Corvette: 150
Chevrolet Corvette Z51: 156
Ford Taurus 3.8: 307
Ford Taurus LX: 344
Jaguar XJ6: 400
Porsche 911 Carrera 2 Tiptronic: 740
Subaru SVX: 858

62 20.9mpg 17.4mpg US 13.5litres/100km
Chevrolet Corvette L98: 175

63 20.7mpg 17.2mpg US 13.7litres/100km
Infiniti Q45: 386
Pontiac Fiero Formula: 679

64 20.4mpg 17.0mpg US 13.8litres/100km
BMW 535i: 86
BMW 735i: 88
BMW 535i: 103
Chevrolet Corvette Convertible: 148
Chevrolet Corvette LT1: 188
Ford Thunderbird Turbo Coupe: 301
Ford Thunderbird Super Coupe: 317
Ford Mustang LX 5.0L: 325
Ford Thunderbird LX: 340
Lotus Esprit Turbo: 465
Maserati 430: 476
Mercury Cougar XR-7: 561
Nissan 300ZX Turbo: 600
Porsche 911 Turbo: 698
Porsche 911 Carrera 4: 727
Porsche 911 Carrera 4: 741

65 20.3mpg 16.9mpg US 13.9litres/100km
Dodge Daytona Shelby: 232

66 20.1mpg 16.7mpg US 14.1litres/100km
Chevrolet Camaro IROC-Z: 144

67 19.9mpg 16.6mpg US 14.2litres/100km
Cadillac Allante: 132

68 19.8mpg 16.5mpg US 14.3litres/100km
Buick Reatta: 127
Buick Reatta Convertible: 128
Eagle Talon: 243
Eagle Talon TSi 4WD: 245
Ferrari 328 GTS: 250
Lotus Esprit Turbo SE: 469
Mercedes-Benz 560SL: 544
Porsche 928 S4: 719
Porsche 911 Turbo: 743

69 19.6mpg 16.3mpg US 14.4litres/100km
Porsche 928 S: 700

70 19.2mpg 16.0mpg US 14.7litres/100km
Audi V8 Quattro: 68
BMW 750i L: 100
Chevrolet Corvette: 139
Chevrolet Camaro IROC-Z Automatic: 161
Chevrolet Camaro IROC-Z: 160
Chevrolet Camaro Z/28: 171
Ferrari Mondial Cabriolet 3.2: 252
Ferrari Mondial Cabriolet: 254
Ferrari Mondial 3.2 Cabriolet: 258
Ferrari 348tb: 266
Ford Mustang GT: 304
Lincoln Continental Mk VII: 457
Lincoln Continental Mk VII LSC: 458
Mazda RX-7 Convertible: 497
Mazda RX-7 Turbo: 510
Mazda RX-7 Turbo: 517
Mercedes-Benz 420SEL: 531
Mercedes-Benz 500SL: 555
Nissan Sentra SE-R: 632
Nissan 300ZX Turbo Millen Super GTZ: 627
Porsche 911 Turbo: 715
Porsche 911 Turbo: 730

71 18.6mpg 15.5mpg US 15.2litres/100km
BMW M5: 122
Lotus Esprit Turbo: 459
Maserati Spyder: 477
Maserati 228: 478

72 18.0mpg 15.0mpg US 15.7litres/100km
Ferrari Mondial t Cabriolet: 269
Mazda MX-5 Miata Millen: 507

Mercedes-Benz 560SL: 527

Porsche 911 Carrera Turbo: 748

73 17.4mpg 14.5mpg US
16.2litres/100km

Mazda RX-7 Infini IV: 515

Mercedes-Benz 560SEC: 525

74 17.2mpg 14.3mpg US
16.4litres/100km

Ford Mustang GT Convertible: 324

75 16.8mpg 14.0mpg US
16.8litres/100km

Bitter SC: 80

Jaguar XJS: 403

Pontiac Firebird TransAm Turbo: 685

Porsche 911 Ruf CTR: 750

76 16.2mpg 13.5mpg US
17.4litres/100km

Jaguar XJSC HE: 399

Lamborghini Jalpa: 429

Maserati Biturbo: 475

Range Rover V8: 758

Range Rover County: 762

Vector W8 Twin Turbo: 949

77 15.9mpg 13.2mpg US
17.8litres/100km

Jaguar XJS: 397

78 15.6mpg 13.0mpg US
18.1litres/100km

Ferrari F40: 267

Jaguar XJSC: 398

Jaguar XJS V12 Convertible: 409

79 15.0mpg 12.5mpg US
18.8litres/100km

BMW M6: 101

80 14.4mpg 12.0mpg US
19.6litres/100km

Ferrari Testa Rossa: 255

81 13.8mpg 11.5mpg US
20.5litres/100km

Bentley Turbo R: 78

Mercedes-Benz 500SEL: 557

82 13.2mpg 11.0mpg US
21.4litres/100km

Lamborghini Diablo: 432

83 12.0mpg 10.0mpg US
23.5litres/100km

Chevrolet Corvette Morrison/Baker Nelson Ledges: 155

Chevrolet Corvette Bakeracing SCCA Escort: 151

Lamborghini Countach 5000S: 427

Lamborghini Countach 25th Anniversary: 431

Rolls-Royce Silver Spur: 800

84 11.4mpg 9.5mpg US
24.8litres/100km

Rolls-Royce Silver Shadow II: 797

85 10.8mpg 9.0mpg US
26.1litres/100km

Ferrari 412: 251

86 9.0mpg 7.5mpg US
31.4litres/100km

Lamborghini LM129: 430

87 7.2mpg 6.0mpg US
39.2litres/100km

Lotus Elan Autocrosser: 461

Road test fuel consumption, Road & Track. In make order

Acura

1 32.3mpg 26.9mpg US
8.7litres/100km
Integra GS: 7
2 30.0mpg 25.0mpg US
9.4litres/100km
Integra LS Special Edition: 4
Integra LS: 5
3 29.4mpg 24.5mpg US
9.6litres/100km
Integra RS: 1
4 27.6mpg 23.0mpg US
10.2litres/100km
NSX: 9
5 27.0mpg 22.5mpg US
10.5litres/100km
Integra LS: 3
6 24.6mpg 20.5mpg US
11.5litres/100km
Legend: 2
7 23.4mpg 19.5mpg US
12.1litres/100km
Legend Coupe L: 6
Legend Coupe: 8
8 22.8mpg 19.0mpg US
12.4litres/100km
Legend LS: 10

Alfa Romeo

1 28.1mpg 23.4mpg US
10.1litres/100km
Spider Veloce: 34
2 27.6mpg 23.0mpg US
10.2litres/100km
164S: 31
3 27.0mpg 22.5mpg US
10.5litres/100km
Spider: 24
4 26.3mpg 21.9mpg US
10.7litres/100km
GTV6: 14
5 26.2mpg 21.8mpg US
10.8litres/100km
Milano Platinum: 15
6 25.2mpg 21.0mpg US
11.2litres/100km
Spider Quadrifoglio: 16
Milano Verde 3.0: 21
7 24.0mpg 20.0mpg US
11.8litres/100km
Milano 3.0: 28

Audi

1 29.4mpg 24.5mpg US
9.6litres/100km
5000 S: 42
2 26.4mpg 22.0mpg US
10.7litres/100km
Coupe GT: 45
3 25.2mpg 21.0mpg US
11.2litres/100km
90 Quattro: 53
90: 55
4 23.4mpg 19.5mpg US
12.1litres/100km
200 Quattro: 65
5 21.6mpg 18.0mpg US
13.1litres/100km
90 Quattro: 60
6 19.2mpg 16.0mpg US
14.7litres/100km
V8 Quattro: 68

Bentley

1 13.8mpg 11.5mpg US
20.5litres/100km
Turbo R: 78

Bitter

1 16.8mpg 14.0mpg US
16.8litres/100km
SC: 80

BMW

1 33.6mpg 28.0mpg US
8.4litres/100km
325 ES: 82
2 30.6mpg 25.5mpg US
9.2litres/100km
524 TD: 85
3 26.4mpg 22.0mpg US
10.7litres/100km
325i: 123
4 24.6mpg 20.5mpg US
11.5litres/100km
325i S: 95
5 24.0mpg 20.0mpg US
11.8litres/100km
325i X: 96
525i: 118
6 22.2mpg 18.5mpg US
12.7litres/100km
535i Automatic: 104
7 21.6mpg 18.0mpg US
13.1litres/100km
735i: 99
8 21.0mpg 17.5mpg US
13.4litres/100km
635 CSi: 87
9 20.4mpg 17.0mpg US
13.8litres/100km
535i: 86
735i: 88
535i: 103
10 19.2mpg 16.0mpg US
14.7litres/100km
750i L: 100
11 18.6mpg 15.5mpg US
15.2litres/100km
M5: 122
12 15.0mpg 12.5mpg US
18.8litres/100km
M6: 101

Buick

1 19.8mpg 16.5mpg US
14.3litres/100km
Reatta: 127
Reatta Convertible: 128

Cadillac

1 29.1mpg 24.2mpg US
9.7litres/100km
Eldorado Touring Coupe: 131
2 26.9mpg 22.4mpg US
10.5litres/100km
Allante: 129
3 22.2mpg 18.5mpg US
12.7litres/100km
Seville Touring Sedan: 133
4 19.9mpg 16.6mpg US
14.2litres/100km
Allante: 132

Chevrolet

1 36.0mpg 30.0mpg US
7.8litres/100km
Nova CL: 142
2 31.8mpg 26.5mpg US
8.9litres/100km
Nova Twin Cam: 157
3 30.0mpg 25.0mpg US
9.4litres/100km
Cavalier Z24 Convertible: 149
Baretta GTZ: 177
4 27.0mpg 22.5mpg US
10.5litres/100km
Lumina APV: 176
5 26.4mpg 22.0mpg US
10.7litres/100km
Cavalier Z24: 137
Baretta GTU: 158
6 25.8mpg 21.5mpg US
10.9litres/100km
Lumina Z34: 186
7 25.3mpg 21.1mpg US
11.1litres/100km
Beretta GTZ: 169
8 24.0mpg 20.0mpg US
11.8litres/100km
Beretta GT: 143
Beretta GT: 168
9 23.4mpg 19.5mpg US
12.1litres/100km
Corvette L98: 166
10 23.2mpg 19.3mpg US
12.2litres/100km
Celebrity CL Eurosport: 138
Corvette Convertible: 153
11 23.1mpg 19.2mpg US
12.3litres/100km
Cavalier Z24: 172
12 22.8mpg 19.0mpg US
12.4litres/100km
Corvette Z51: 141
13 22.3mpg 18.6mpg US
12.6litres/100km
Camaro IROC-Z Convertible:
170
Camaro Z/28 Convertible: 178
14 21.7mpg 18.1mpg US
13.0litres/100km
Corvette ZR-1: 183
15 21.6mpg 18.0mpg US
13.1litres/100km
Corvette Callaway: 173
Camaro Z28: 187
16 21.5mpg 17.9mpg US
13.1litres/100km
Corvette: 163
17 21.0mpg 17.5mpg US
13.4litres/100km
Corvette: 146
Corvette Z51: 156
Corvette: 150
18 20.9mpg 17.4mpg US
13.5litres/100km
Corvette L98: 175
19 20.4mpg 17.0mpg US
13.8litres/100km
Corvette Convertible: 148
Corvette LT1: 188
20 20.1mpg 16.7mpg US
14.1litres/100km
Camaro IROC-Z: 144
21 19.2mpg 16.0mpg US
14.7litres/100km
Corvette: 139
Camaro IROC-Z Automatic: 161
Camaro IROC-Z: 160
Camaro Z/28: 171
22 12.0mpg 10.0mpg US
23.5litres/100km
Corvette Morrison/Baker
Nelson Ledges: 155
Corvette Bakeracing SCCA
Escort: 151

Consulier

1 21.9mpg 18.2mpg US
12.9litres/100km
GTP LX: 216

Daihatsu

1 35.4mpg 29.5mpg US

8.0litres/100km
Charade CLS: 223

Dodge

1 27.6mpg 23.0mpg US
10.2litres/100km
Daytona Shelby Z: 229
Daytona Shelby: 237
2 26.3mpg 21.9mpg US
10.7litres/100km
Shadow ES VNT: 238
3 25.8mpg
21.5mpg US
10.9litres/100km
Lancer Shelby: 230
Shelby CSX: 231
Spirit R/T: 239
4 24.6mpg 20.5mpg US
11.5litres/100km
Shelby CSX: 234
Daytona ES: 236
5 24.1mpg 20.1mpg US
11.7litres/100km
Shelby CSX VNT: 235
6 24.0mpg 20.0mpg US
11.8litres/100km
Shelby GLH-S: 228
7 22.8mpg 19.0mpg US
12.4litres/100km
Shadow ES Turbo: 233
8 21.6mpg 18.0mpg US
13.1litres/100km
Stealth R/T Turbo: 240
9 20.3mpg 16.9mpg US
13.9litres/100km
Daytona Shelby: 232

Eagle

1 22.8mpg 19.0mpg US
12.4litres/100km
Premier ES: 242
2 19.8mpg 16.5mpg US
14.3litres/100km
Talon: 243
Talon TSi 4WD: 245

Ferrari

1 19.8mpg 16.5mpg US
14.3litres/100km
328 GTS: 250
2 19.2mpg 16.0mpg US
14.7litres/100km
Mondial Cabriolet 3.2: 252
Mondial Cabriolet: 254
Mondial 3.2 Cabriolet: 258
348tb: 266
3 18.0mpg 15.0mpg US
15.7litres/100km
Mondial t Cabriolet: 269
4 15.6mpg 13.0mpg US
18.1litres/100km

F40: 267
5 14.4mpg 12.0mpg US
19.6litres/100km
Testa Rossa: 255
6 10.8mpg 9.0mpg US
26.1litres/100km
412: 251

Fiat

1 28.2mpg 23.5mpg US
10.0litres/100km
X1/9: 277
X1/9: 279

Ford

1 44.4mpg 37.0mpg US
6.4litres/100km
Festiva L: 294
2 33.6mpg 28.0mpg US
8.4litres/100km
Escort GT: 333
3 28.1mpg 23.4mpg US
10.1litres/100km
Probe GT: 326
4 27.0mpg 22.5mpg US
10.5litres/100km
Crown Victoria LX: 329
5 26.4mpg 22.0mpg US
10.7litres/100km
Taurus SHO: 316
Taurus SHO: 339
6 25.2mpg 21.0mpg US
11.2litres/100km
Probe LX: 327
7 24.6mpg 20.5mpg US
11.5litres/100km
Probe GT: 337
8 24.0mpg 20.0mpg US
11.8litres/100km
Probe GT: 305
9 22.8mpg 19.0mpg US
12.4litres/100km
Mustang LX: 343
10 22.2mpg 18.5mpg US
12.7litres/100km
Aerostar XLT: 285
11 21.0mpg 17.5mpg US
13.4litres/100km
Taurus 3.8: 307
Taurus LX: 344
12 20.4mpg 17.0mpg US
13.8litres/100km
Thunderbird Turbo Coupe:
301
Thunderbird Super Coupe:
317
Mustang LX 5.0L: 325
Thunderbird LX: 340
13 19.2mpg 16.0mpg US
14.7litres/100km
Mustang GT: 304
14 17.2mpg 14.3mpg US

16.4litres/100km
Mustang GT Convertible: 324

Geo

1 32.4mpg 27.0mpg US
8.7litres/100km
Storm GSi: 345

Honda

1 37.8mpg 31.5mpg US
7.5litres/100km
CRX Si: 350
2 37.2mpg 31.0mpg US
7.6litres/100km
CRX Si: 360
3 34.8mpg 29.0mpg US
8.1litres/100km
CRX Si Jackson: 361
4 34.2mpg 28.5mpg US
8.3litres/100km
Civic Si: 364
5 28.8mpg 24.0mpg US
9.8litres/100km
Accord LXi: 348
Prelude Si: 374
Prelude Si 4WS: 377
6 26.4mpg 22.0mpg US
10.7litres/100km
Prelude 2.0 Si: 357
7 24.6mpg 20.5mpg US
11.5litres/100km
Prelude Si 4WS: 365

Hyundai

1 37.8mpg 31.5mpg US
7.5litres/100km
Excel GLS: 378

Infiniti

1 28.8mpg 24.0mpg US
9.8litres/100km
G20: 387
2 24.0mpg 20.0mpg US
11.8litres/100km
M30: 385
3 20.7mpg 17.2mpg US
13.7litres/100km
Q45: 386

Isuzu

1 34.8mpg 29.0mpg US
8.1litres/100km
Impulse Sports Coupe: 390
2 31.2mpg 26.0mpg US
9.0litres/100km
Impulse Turbo: 391
3 30.6mpg 25.5mpg US
9.2litres/100km
Impulse Turbo: 393

4 29.9mpg 24.9mpg US
9.4litres/100km
Impulse RS: 395
5 28.8mpg 24.0mpg US
9.8litres/100km
Impulse XS: 394

Jaguar

1 22.5mpg 18.7mpg US
12.6litres/100km
XJS: 413
XJS Convertible: 419
2 21.6mpg 18.0mpg US
13.1litres/100km
XJ6 Vanden Plas: 410
3 21.0mpg 17.5mpg US
13.4litres/100km
XJ6: 400
4 16.8mpg 14.0mpg US
16.8litres/100km
XJS: 403
5 16.2mpg 13.5mpg US
17.4litres/100km
XJSC HE: 399
6 15.9mpg 13.2mpg US
17.8litres/100km
XJS: 397
7 15.6mpg 13.0mpg US
18.1litres/100km
XJSC: 398
XJS V12 Convertible: 409

Lamborghini

1 16.2mpg 13.5mpg US
17.4litres/100km
Jalpa: 429
2 13.2mpg 11.0mpg US
21.4litres/100km
Diablo: 432
3 12.0mpg 10.0mpg US
23.5litres/100km
Countach 5000S: 427
Countach 25th Anniversary:
431
49.0mpg 7.5mpg US
31.4litres/100km
LM129: 430

Lexus

1 25.2mpg 21.0mpg US
11.2litres/100km
ES300: 456
2 24.0mpg 20.0mpg US
11.8litres/100km
SC400: 455

Lincoln

1 19.2mpg 16.0mpg US
14.7litres/100km
Continental Mk VII: 457

Continental Mk VII LSC: 458

Lotus

1 31.2mpg 26.0mpg US
9.0litres/100km
Elan SE: 470
2 20.4mpg 17.0mpg US
13.8litres/100km
Esprit Turbo: 465
3 19.8mpg 16.5mpg US
14.3litres/100km
Esprit Turbo SE: 469
4 18.6mpg 15.5mpg US
15.2litres/100km
Esprit Turbo: 459
57.2mpg 6.0mpg US
39.2litres/100km
Elan Autocrosser: 461

Maserati

1 20.4mpg 17.0mpg US
13.8litres/100km
430: 476
2 18.6mpg 15.5mpg US
15.2litres/100km
Spyder: 477
228: 478
3 16.2mpg 13.5mpg US
17.4litres/100km
Biturbo: 475

Mazda

1 31.2mpg 26.0mpg US
9.0litres/100km
Protege: 509
2 30.6mpg 25.5mpg US
9.2litres/100km
MX-5 Miata: 503
3 29.4mpg 24.5mpg US
9.6litres/100km
323 LX: 480
4 28.8mpg 24.0mpg US
9.8litres/100km
626 GT: 483
MX-6 GT: 496
MX-3 GS: 519
5 25.8mpg 21.5mpg US
10.9litres/100km
323 GTX: 490
6 25.2mpg 21.0mpg US
11.2litres/100km
RX-7 GTU: 498
MPV: 502
7 24.0mpg 20.0mpg US
11.8litres/100km
MX-6: 495
8 23.1mpg 19.2mpg US
12.3litres/100km
RX-7 Turbo: 486
9 22.8mpg 19.0mpg US
12.4litres/100km

RX-7 GXL: 485
RX-7 Turbo: 499
MX-6 GT 4WS: 504
RX-7 GTUs: 505
MX-6 GT: 508
10 22.2mpg 18.5mpg US
12.7litres/100km
929: 494
11 19.2mpg 16.0mpg US
14.7litres/100km
RX-7 Convertible: 497
RX-7 Turbo: 510
RX-7 Turbo: 517
12 18.0mpg 15.0mpg US
15.7litres/100km
MX-5 Miata Millen: 507
13 17.4mpg 14.5mpg US
16.2litres/100km
RX-7 Infini IV: 515

Mercedes-Benz

1 26.4mpg 22.0mpg US
10.7litres/100km
300E: 522
300TD: 530
2 25.2mpg 21.0mpg US
11.2litres/100km
190E 2.3-16: 520
3 24.6mpg 20.5mpg US
11.5litres/100km
190E 2.6: 539
300E: 543
4 23.4mpg 19.5mpg US
12.1litres/100km
300CE: 546
5 19.8mpg 16.5mpg US
14.3litres/100km
560SL: 544
6 19.2mpg 16.0mpg US
14.7litres/100km
420SEL: 531
500SL: 555
7 18.0mpg 15.0mpg US
15.7litres/100km
560SL: 527
8 17.4mpg 14.5mpg US
16.2litres/100km
560SEC: 525
9 13.8mpg 11.5mpg US
20.5litres/100km
500SEL: 557

Mercury

1 39.6mpg 33.0mpg US
7.1litres/100km
Tracer: 560
2 32.4mpg 27.0mpg US
8.7litres/100km
Tracer LTS: 563
3 30.0mpg 25.0mpg US
9.4litrcs/100km
Capri XR-2: 562

4 23.8mpg 19.8mpg US
11.9litres/100km
Sable LS: 559
5 20.4mpg 17.0mpg US
13.8litres/100km
Cougar XR-7: 561

Merkur

1 28.8mpg 24.0mpg US
9.8litres/100km
XR4Ti: 565
2 25.2mpg 21.0mpg US
11.2litres/100km
Scorpio: 564

MG

1 30.0mpg 25.0mpg US
9.4litres/100km
B British Motor Heritage: 567

Mitsubishi

1 30.0mpg 25.0mpg US
9.4litres/100km
Mirage Turbo: 579
2 27.6mpg 23.0mpg US
10.2litres/100km
Eclipse: 580
3 27.5mpg 22.9mpg US
10.3litres/100km
Galant GS: 581
4 25.2mpg 21.0mpg US
11.2litres/100km
Cordia Turbo: 575
Galant VR-4: 591
5 24.6mpg 20.5mpg US
11.5litres/100km
Cordia Turbo: 568
6 24.0mpg 20.0mpg US
11.8litres/100km
Starion ESI-R: 584
7 22.8mpg 19.0mpg US
12.4litres/100km
Diamante LS: 590
8 21.6mpg 18.0mpg US
13.1litres/100km
3000GT VR-4: 588

Morgan

1 23.4mpg 19.5mpg US
12.1litres/100km
Plus 8: 596

Nissan

1 36.0mpg 30.0mpg US
7.8litres/100km
NX2000: 629
2 32.4mpg 27.0mpg US
8.7litres/100km
Sentra Sport Coupe SE: 603

3 30.0mpg 25.0mpg US
9.4litres/100km
240SX: 612
4 28.8mpg 24.0mpg US
9.8litres/100km
Pulsar NX SE: 608
5 28.6mpg 23.8mpg US
9.9litres/100km
240SX: 626
6 28.2mpg 23.5mpg US
10.0litres/100km
300ZX: 615
7 25.2mpg 21.0mpg US
11.2litres/100km
300ZX: 613
8 24.6mpg 20.5mpg US
11.5litres/100km
Maxima SE: 617
9 23.8mpg 19.8mpg US
11.9litres/100km
300ZX: 605
10 23.5mpg 19.6mpg US
12.0litres/100km
300ZX Twin Turbo: 628
11 23.4mpg 19.5mpg US
12.1litres/100km
200SX SE V6: 611
12 21.6mpg 18.0mpg US
13.1litres/100km
300ZX: 599
13 20.4mpg 17.0mpg US
13.8litres/100km
300ZX Turbo: 600
14 19.2mpg 16.0mpg US
14.7litres/100km
300ZX Turbo Millen Super GTZ: 627
Sentra SE-R: 632

Oldsmobile

1 32.4mpg 27.0mpg US
8.7litres/100km
Calais HO Quad 4: 637
Cutlass Calais International Series HO: 638
2 27.6mpg 23.0mpg US
10.2litres/100km
Cutlass Calais International Series: 636
3 21.4mpg 17.8mpg US
13.2litres/100km
Toronado Trofeo: 639

Opel

1 30.0mpg 25.0mpg US
9.4litres/100km
Calibra 2.0i 16v: 640

Peugeot

1 28.8mpg 24.0mpg US
9.8litres/100km

505 STX: 650
2 26.4mpg 22.0mpg US
10.7litres/100km
405 Mi16: 660

Plymouth

1 29.5mpg 24.6mpg US
9.6litres/100km
Laser RS Turbo: 675
2 27.6mpg 23.0mpg US
10.2litres/100km
Laser RS: 674
3 24.0mpg 20.0mpg US
11.8litres/100km
Laser RS Turbo: 676

Pontiac

1 36.6mpg 30.5mpg US
7.7litres/100km
Le Mans: 680
2 31.2mpg 26.0mpg US
9.0litres/100km
Sunbird GT: 682
3 27.0mpg 22.5mpg US
10.5litres/100km
Fiero GT: 677
4 25.2mpg 21.0mpg US
11.2litres/100km
Grand Prix McLaren Turbo: 687
Grand Prix GTP: 694
5 24.5mpg 20.4mpg US
11.5litres/100km
Grand Prix STE Turbo: 690
6 24.0mpg 20.0mpg US
11.8litres/100km
Bonneville SSEi: 695
7 22.8mpg 19.0mpg US
12.4litres/100km
Firebird TransAm Convertible: 693
8 22.2mpg 18.5mpg US
12.7litres/100km
Bonneville SE: 678
9 21.6mpg 18.0mpg US
13.1litres/100km
Bonneville SE: 681
10 21.3mpg 17.7mpg US
13.3litres/100km
Firebird TransAm GTA: 689
Firebird GTA: 691
11 20.7mpg 17.2mpg US
13.7litres/100km
Fiero Formula: 679
12 16.8mpg 14.0mpg US
16.8litres/100km
Firebird TransAm Turbo: 685

Porsche

1 27.0mpg 22.5mpg US
10.5litres/100km

911 Cabriolet: 710
2 26.5mpg 22.1mpg US
10.6litres/100km
944: 702
3 26.4mpg 22.0mpg US
10.7litres/100km
924 S: 699
924 S: 718
4 25.2mpg 21.0mpg US
11.2litres/100km
944: 720
5 24.0mpg 20.0mpg US
11.8litres/100km
911 Carrera Cabriolet: 728
911 Speedster: 742
6 23.4mpg 19.5mpg US
12.1litres/100km
944 S: 721
7 22.8mpg 19.0mpg US
12.4litres/100km
944 Turbo: 722
8 22.3mpg 18.6mpg US
12.6litres/100km
911 Cabriolet: 696
9 22.2mpg 18.5mpg US
12.7litres/100km
944 Turbo: 737
944 S2 Cabriolet: 746
944 S2: 745
10 21.6mpg 18.0mpg US
13.1litres/100km
911 Club Sport: 713
911 Carrera: 711
11 21.0mpg 17.5mpg US
13.4litres/100km
911 Carrera 2 Tiptronic: 740
12 20.4mpg 17.0mpg US
13.8litres/100km
911 Turbo: 698
911 Carrera 4: 727
911 Carrera 4: 741
13 19.8mpg 16.5mpg US
14.3litres/100km
928 S4: 719
911 Turbo: 743
14 19.6mpg 16.3mpg US
14.4litres/100km
928 S: 700
15 19.2mpg 16.0mpg US
14.7litres/100km
911 Turbo: 715
911 Turbo: 730
16 18.0mpg 15.0mpg US
15.7litres/100km
911 Carrera Turbo: 748
17 16.8mpg 14.0mpg US
16.8litres/100km
911 Ruf CTR: 750

Range Rover

1 16.2mpg 13.5mpg US
17.4litres/100km
V8: 758

County: 762   Suzuki

Renault

1 30.6mpg 25.5mpg US
9.2litres/100km
Alliance GTA: 776

Rolls-Royce

1 12.0mpg 10.0mpg US
23.5litres/100km
Silver Spur: 800
2 11.4mpg 9.5mpg US
24.8litres/100km
Silver Shadow II: 797

Saab

1 27.6mpg 23.0mpg US
10.2litres/100km
9000 Turbo: 836
2 25.7mpg 21.4mpg US
11.0litres/100km
900SPG: 833
3 25.2mpg 21.0mpg US
11.2litres/100km
9000: 826
9000 Turbo: 830
9000CD: 832

Saturn

1 31.2mpg 26.0mpg US
9.0litres/100km
Sports Coupe: 840

Subaru

1 29.3mpg 24.4mpg US
9.6litres/100km
XT6: 854
2 24.0mpg 20.0mpg US
11.8litres/100km
Legacy Sports Sedan: 857
3 21.0mpg 17.5mpg US
13.4litres/100km
SVX: 858

Suzuki

1 37.8mpg 31.5mpg US
7.5litres/100km
Swift GT: 867

Toyota

1 36.0mpg 30.0mpg US
7.8litres/100km
Tercel LE: 908
Paseo: 909
2 33.6mpg 28.0mpg US
8.4litres/100km
Corolla GTS: 892
3 30.6mpg 25.5mpg US

9.2litres/100km
Corolla FX-16 GTS: 878
4 30.0mpg 25.0mpg US
9.4litres/100km
MR2: 870
5 29.8mpg 24.8mpg US
9.5litres/100km
Celica GTS: 898
6 27.6mpg 23.0mpg US
10.2litres/100km
MR2 Supercharged: 893
Celica GT Convertible: 904
7 26.4mpg 22.0mpg US
10.7litres/100km
Supra Turbo: 895
8 25.3mpg 21.1mpg US
11.1litres/100km
MR2 Supercharged: 880
9 24.3mpg 20.2mpg US
11.6litres/100km
Celica GTS: 869
10 24.0mpg 20.0mpg US
11.8litres/100km
Celica All-Trac Turbo: 883
Supra Turbo: 889
11 22.8mpg 19.0mpg US
12.4litres/100km
Supra: 881
Previa LE: 901
12 22.2mpg 18.5mpg US
12.7litres/100km
MR2 Turbo: 900
13 21.6mpg 18.0mpg US
13.1litres/100km
Supra: 872
Supra: 888

Vector

1 16.2mpg 13.5mpg US
17.4litres/100km
W8 Twin Turbo: 949

Volkswagen

1 30.0mpg 25.0mpg US
9.4litres/100km
Scirocco 16v: 954
Fox GL: 956
Scirocco 16v: 966
Passat GL: 975
GTi 16v: 977
2 28.8mpg 24.0mpg US
9.8litres/100km
GTi 16v: 958
GTi 16v: 961
Jetta GLi 16v: 962
Jetta GLi 16v: 978
3 25.8mpg 21.5mpg US
10.9litres/100km
Corrado G60: 973
4 23.4mpg 19.5mpg US
12.1litres/100km
Quantum Synchro Wagon: 953

Volvo

1   31.8mpg 26.5mpg US
8.9litres/100km
740 Turbo: 995
2   26.7mpg 22.2mpg US
10.6litres/100km
740 GLE: 989
3   24.0mpg 20.0mpg US
11.8litres/100km
940 SE: 996
960: 999
4   22.6mpg 18.8mpg US
12.5litres/100km
740 Turbo: 990

Yugo

1   36.0mpg 30.0mpg US
7.8litres/100km
GV: 1001

Car Specifications

In alphabetical order

Car specifications A - Z

1 Acura
Integra RS
1986 Japan
120.0mph 193.1kmh
0-60mph 96.5kmh: 9.3secs
0-1/4 mile: 17.0secs
113.0bhp 84.3kW 114.6PS
@ 6250rpm
99.0lbft 134.1Nm @ 5500rpm
102.3bhp/ton 100.6bhp/tonne
71.1bhp/L 53.0kW/L 72.1PS/L
61.5ft/sec 18.8m/sec
29.4mpg 24.5mpgUS
9.6L/100km
Petrol 4-stroke piston
1590cc 97.0cu in
In-line 4 fuel injection
Compression ratio: 9.3:1
Bore: 75.0mm 2.9in
Stroke: 90.0mm 3.5in
Valve type/No: Overhead 16
Transmission: Manual
No. of forward speeds: 5
Wheels driven: Front
Springs F/R: Torsion bar/Coil
Brake system: PA
Brakes F/R: Disc/Disc
Steering: Rack & pinion PA
Wheelbase: 252.0cm 99.2in
Track F: 142.0cm 55.9in
Track R: 143.5cm 56.5in
Length: 435.1cm 171.3in
Width: 166.6cm 65.6in
Height: 134.6cm 53.0in
Ground clearance: 16.0cm
6.3in
Kerb weight: 1123.6kg
2475.0lb
Fuel: 50.0L 11.0gal 13.2galUS

2 Acura
Legend
1986 Japan
129.0mph 207.6kmh
0-60mph 96.5kmh: 8.1secs
0-1/4 mile: 16.4secs
151.0bhp 112.6kW 153.1PS
@ 5800rpm
154.0lbft 208.7Nm @ 4500rpm
107.2bhp/ton 105.4bhp/tonne
60.5bhp/L 45.1kW/L 61.4PS/L
47.5ft/sec 14.5m/sec
24.6mpg 20.5mpgUS
11.5L/100km
Petrol 4-stroke piston
2494cc 152.2cu in
Vee 6 fuel injection
Compression ratio: 9.0:1
Bore: 84.0mm 3.3in
Stroke: 75.0mm 2.9in
Valve type/No: Overhead 24
Transmission: Manual

No. of forward speeds: 5
Wheels driven: Front
Springs F/R: Coil/Coil
Brake.system: PA
Brakes F/R: Disc/Disc
Steering: Rack & pinion PA
Wheelbase: 275.8cm 108.6in
Track F: 149.9cm 59.0in
Track R: 146.1cm 57.5in
Length: 503.9cm 198.4in
Width: 173.5cm 68.3in
Height: 138.9cm 54.7in
Ground clearance: 15.0cm
5.9in
Kerb weight: 1432.4kg
3155.0lb
Fuel: 68.1L 15.0gal 18.0galUS

3 Acura
Integra LS
1987 Japan
110.0mph 177.0kmh
0-60mph 96.5kmh: 9.3secs
0-1/4 mile: 17.0secs
113.0bhp 84.3kW 114.6PS
@ 6250rpm
99.0lbft 134.1Nm @ 5500rpm
102.3bhp/ton 100.6bhp/tonne
71.1bhp/L 53.0kW/L 72.1PS/L
61.5ft/sec 18.8m/sec
27.0mpg 22.5mpgUS
10.5L/100km
Petrol 4-stroke piston
1590cc 97.0cu in
In-line 4 fuel injection
Compression ratio: 9.3:1
Bore: 75.0mm 2.9in
Stroke: 90.0mm 3.5in
Valve type/No: Overhead 16
Transmission: Manual
No. of forward speeds: 5
Wheels driven: Front
Springs F/R: Torsion bar/Coil
Brake system: PA
Brakes F/R: Disc/Disc
Steering: Rack & pinion PA
Wheelbase: 252.0cm 99.2in
Track F: 142.0cm 55.9in
Track R: 143.5cm 56.5in
Length: 435.1cm 171.3in
Width: 166.6cm 65.6in
Height: 134.6cm 53.0in
Kerb weight: 1123.6kg
2475.0lb
Fuel: 50.0L 11.0gal 13.2galUS

4 Acura
Integra LS Special Edition
1988 Japan
112.0mph 180.2kmh
0-50mph 80.5kmh: 6.4secs
0-60mph 96.5kmh: 9.1secs

0-1/4 mile: 16.8secs
113.0bhp 84.3kW 114.6PS
@ 6250rpm
99.0lbft 134.1Nm @ 5500rpm
104.4bhp/ton 102.6bhp/tonne
71.1bhp/L 53.0kW/L 72.1PS/L
61.5ft/sec 18.8m/sec
30.0mpg 25.0mpgUS
9.4L/100km
Petrol 4-stroke piston
1590cc 97.0cu in
In-line 4 fuel injection
Compression ratio: 9.3:1
Bore: 75.0mm 2.9in
Stroke: 90.0mm 3.5in
Valve type/No: Overhead 16
Transmission: Manual
No. of forward speeds: 5
Wheels driven: Front
Springs F/R: Torsion bar/Coil
Brake system: PA
Brakes F/R: Disc/Disc
Steering: Rack & pinion PA
Wheelbase: 245.1cm 96.5in
Track F: 142.0cm 55.9in
Track R: 143.5cm 56.5in
Length: 428.0cm 168.5in
Width: 163.8cm 64.9in
Height: 134.4cm 52.9in
Kerb weight: 1100.9kg
2425.0lb
Fuel: 50.0L 11.0gal 13.2galUS

5 Acura
Integra LS
1989 Japan
112.0mph 180.2kmh
0-50mph 80.5kmh: 6.9secs
0-60mph 96.5kmh: 9.7secs
0-1/4 mile: 17.3secs
118.0bhp 88.0kW 119.6PS
@ 6500rpm
103.0lbft 139.6Nm @ 5500rpm
108.5bhp/ton 106.7bhp/tonne
74.2bhp/L 55.3kW/L 75.2PS/L
63.9ft/sec 19.5m/sec
30.0mpg 25.0mpgUS
9.4L/100km
Petrol 4-stroke piston
1590cc 97.0cu in
In-line 4 fuel injection
Compression ratio: 9.5:1
Bore: 75.0mm 2.9in
Stroke: 90.0mm 3.5in
Valve type/No: Overhead 16
Transmission: Manual
No. of forward speeds: 5
Wheels driven: Front
Springs F/R: Torsion bar/Coil
Brake system: PA
Brakes F/R: Disc/Disc
Steering: Rack & pinion PA

Wheelbase: 245.1cm 96.5in
Track F: 142.0cm 55.9in
Track R: 143.5cm 56.5in
Length: 428.0cm 168.5in
Width: 163.8cm 64.5in
Height: 134.4cm 52.9in
Kerb weight: 1105.5kg
2435.0lb
Fuel: 50.0L 11.0gal 13.2galUS

6 Acura
Legend Coupe L
1989 Japan
130.0mph 209.2kmh
0-50mph 80.5kmh: 5.9secs
0-60mph 96.5kmh: 8.0secs
0-1/4 mile: 16.2secs
161.0bhp 120.1kW 163.2PS
@ 5900rpm
162.0lbft 219.5Nm @ 4500rpm
112.0bhp/ton 110.1bhp/tonne
60.2bhp/L 44.9kW/L 61.0PS/L
48.3ft/sec 14.7m/sec
23.4mpg 19.5mpgUS
12.1L/100km
Petrol 4-stroke piston .
2675cc 163.2cu in
Vee 6 fuel injection
Compression ratio: 9.0:1
Bore: 87.0mm 3.4in
Stroke: 75.0mm 2.9in
Valve type/No: Overhead 24
Transmission: Manual
No. of forward speeds: 5
Wheels driven: Front
Springs F/R: Coil/Coil
Brake system: PA ABS
Brakes F/R: Disc/Disc
Steering: Rack & pinion PA
Wheelbase: 270.5cm 106.5in
Track F: 150.1cm 59.1in
Track R: 150.1cm 59.1in
Length: 477.5cm 188.0in
Width: 174.5cm 68.7in
Height: 136.9cm 53.9in
Ground clearance: 17.3cm
6.8in
Kerb weight: 1461.9kg
3220.0lb
Fuel: 68.1L 15.0gal 18.0galUS

7 Acura
Integra GS
1990 Japan
130.0mph 209.2kmh
0-50mph 80.5kmh: 6.6secs
0-60mph 96.5kmh: 9.4secs
0-1/4 mile: 16.6secs
130.0bhp 96.9kW 131.8PS
@ 6000rpm
121.0lbft 164.0Nm @ 5000rpm
108.2bhp/ton 106.4bhp/tonne

70.9bhp/L 52.9kW/L 71.9PS/L
58.3ft/sec 17.8m/sec
32.3mpg 26.9mpgUS
8.7L/100km
Petrol 4-stroke piston
1834cc 111.9cu in
In-line 4 fuel injection
Compression ratio: 9.2:1
Bore: 81.0mm 3.2in
Stroke: 89.0mm 3.5in
Valve type/No: Overhead 16
Transmission: Manual
No. of forward speeds: 5
Wheels driven: Front
Springs F/R: Coil/Coil
Brake system: PA ABS
Brakes F/R: Disc/Disc
Steering: Rack & pinion PA
Wheelbase: 255.0cm 100.4in
Track F: 147.6cm 58.1in
Track R: 147.6cm 58.1in
Length: 439.2cm 172.9in
Width: 171.2cm 67.4in
Height: 132.6cm 52.2in
Ground clearance: 15.0cm
5.9in
Kerb weight: 1221.3kg
2690.0lb
Fuel: 50.0L 11.0gal 13.2galUS

8 Acura
Legend Coupe
1990 Japan
130.0mph 209.2kmh
0-50mph 80.5kmh: 5.9secs
0-60mph 96.5kmh: 8.0secs
0-1/4 mile: 16.2secs
161.0bhp 120.1kW 163.2PS
@ 5900rpm
162.0lbft 219.5Nm @ 4500rpm
112.0bhp/ton 110.1bhp/tonne
60.2bhp/L 44.9kW/L 61.0PS/L
48.3ft/sec 14.7m/sec
23.4mpg 19.5mpgUS
12.1L/100km
Petrol 4-stroke piston
2675cc 163.2cu in
Vee 6 fuel injection
Compression ratio: 9.0:1
Bore: 87.0mm 3.4in
Stroke: 75.0mm 2.9in
Valve type/No: Overhead 24
Transmission: Manual
No. of forward speeds: 5
Wheels driven: Front
Springs F/R: Coil/Coil
Brake system: PA ABS
Brakes F/R: Disc/Disc
Steering: Rack & pinion PA
Wheelbase: 270.5cm 106.5in
Track F: 150.1cm 59.1in
Track R: 150.1cm 59.1in
Length: 477.5cm 188.0in
Width: 174.5cm 68.7in

Height: 136.9cm 53.9in
Kerb weight: 1461.9kg
3220.0lb
Fuel: 68.1L 15.0gal 18.0galUS

9 Acura
NSX
1990 Japan
168.0mph 270.3kmh
0-50mph 80.5kmh: 4.4secs
0-60mph 96.5kmh: 5.7secs
0-1/4 mile: 14.0secs
270.0bhp 201.3kW 273.7PS
@ 7100rpm
210.0lbft 284.6Nm @ 5300rpm
202.6bhp/ton 199.2bhp/tonne
90.7bhp/L 67.6kW/L 91.9PS/L
60.5ft/sec 18.5m/sec
27.6mpg 23.0mpgUS
10.2L/100km
Petrol 4-stroke piston
2977cc 181.6cu in
Vee 6 fuel injection
Compression ratio: 10.2:1
Bore: 90.0mm 3.5in
Stroke: 78.0mm 3.1in
Valve type/No: Overhead 24
Transmission: Manual
No. of forward speeds: 5
Wheels driven: Rear
Springs F/R: Coil/Coil
Brake system: PA ABS
Brakes F/R: Disc/Disc
Steering: Rack & pinion PA
Wheelbase: 253.0cm 99.6in
Track F: 150.9cm 59.4in
Track R: 152.9cm 60.2in
Length: 440.4cm 173.4in
Width: 181.1cm 71.3in
Height: 117.1cm 46.1in
Ground clearance: 13.5cm
5.3in
Kerb weight: 1355.2kg
2985.0lb
Fuel: 70.0L 15.4gal 18.5galUS

10 Acura
Legend LS
1991 Japan
135.0mph 217.2kmh
0-50mph 80.5kmh: 6.0secs
0-60mph 96.5kmh: 7.9secs
0-1/4 mile: 16.1secs
200.0bhp 149.1kW 202.8PS
@ 5500rpm
210.0lbft 284.6Nm @ 4500rpm
128.0bhp/ton 125.9bhp/tonne
62.4bhp/L 46.5kW/L 63.2PS/L
50.6ft/sec 15.4m/sec
22.8mpg 19.0mpgUS
12.4L/100km
Petrol 4-stroke piston
3206cc 195.6cu in
Vee 6 fuel injection

Compression ratio: 9.6:1
Bore: 90.0mm 3.5in
Stroke: 84.0mm 3.3in
Valve type/No: Overhead 24
Transmission: Automatic
No. of forward speeds: 4
Wheels driven: Front
Springs F/R: Coil/Coil
Brake system: PA ABS
Brakes F/R: Disc/Disc
Steering: Rack & pinion PA
Wheelbase: 291.1cm 114.6in
Track F: 154.9cm 61.0in
Track R: 153.9cm 60.6in
Length: 495.0cm 194.9in
Width: 181.1cm 71.3in
Height: 140.0cm 55.1in
Ground clearance: 15.2cm
6.0in
Kerb weight: 1589.0kg
3500.0lb
Fuel: 68.1L 15.0gal 18.0galUS

11 Alfa Romeo
Alfa 33 1.3 S
1986 Italy
109.0mph 175.4kmh
0-50mph 80.5kmh: 7.9secs
0-60mph 96.5kmh: 11.7secs
0-1/4 mile: 18.0secs
0-1km: 33.8secs
85.0bhp 63.4kW 86.2PS
@ 5800rpm
88.0lbft 119.2Nm @ 4000rpm
95.3bhp/ton 93.7bhp/tonne
62.9bhp/L 46.9kW/L 63.8PS/L
42.7ft/sec 13.0m/sec
21.8mph 35.1kmh/1000rpm
27.2mpg 22.6mpgUS
10.4L/100km
Petrol 4-stroke piston
1351cc 82.0cu in
Flat 4 2 Carburettor
Compression ratio: 9.5:1
Bore: 80.0mm 3.1in
Stroke: 67.2mm 2.6in
Valve type/No: Overhead 8
Transmission: Manual
No. of forward speeds: 5
Wheels driven: Front
Springs F/R: Coil/Coil
Brake system: PA
Brakes F/R: Disc/Drum
Steering: Rack & pinion
Wheelbase: 245.5cm 96.7in
Track F: 139.2cm 54.8in
Track R: 135.9cm 53.5in
Length: 401.5cm 158.1in
Width: 161.2cm 63.5in
Height: 134.0cm 52.8in
Ground clearance: 15.0cm
5.9in
Kerb weight: 907.0kg 1997.8lb
Fuel: 50.0L 11.0gal 13.2galUS

12 Alfa Romeo
Alfa 75 1.8
1986 Italy
120.0mph 193.1kmh
0-50mph 80.5kmh: 7.4secs
0-60mph 96.5kmh: 10.3secs
0-1/4 mile: 17.5secs
0-1km: 32.4secs
120.0bhp 89.5kW 121.7PS
@ 5300rpm
123.0lbft 166.7Nm @ 4000rpm
107.0bhp/ton 105.3bhp/tonne
67.4bhp/L 50.3kW/L 68.4PS/L
51.2ft/sec 15.6m/sec
20.1mph 32.3kmh/1000rpm
24.5mpg 20.4mpgUS
11.5L/100km
Petrol 4-stroke piston
1779cc 109.0cu in
In-line 4 2 Carburettor
Compression ratio: 9.5:1
Bore: 80.0mm 3.1in
Stroke: 88.5mm 3.5in
Valve type/No: Overhead 8
Transmission: Manual
No. of forward speeds: 5
Wheels driven: Rear
Springs F/R: Torsion bar/Coil
Brake system: PA
Brakes F/R: Disc/Disc
Steering: Rack & pinion
Wheelbase: 251.0cm 98.8in
Track F: 136.8cm 53.9in
Track R: 135.8cm 53.5in
Length: 433.0cm 170.5in
Width: 163.0cm 64.2in
Height: 135.0cm 53.1in
Ground clearance: 20.3cm
8.0in
Kerb weight: 1140.0kg
2511.0lb
Fuel: 49.1L 10.8gal 13.0galUS

13 Alfa Romeo
Alfa 75 2.5 Green Cloverleaf
1986 Italy
130.0mph 209.2kmh
0-50mph 80.5kmh: 6.4secs
0-60mph 96.5kmh: 8.9secs
0-1/4 mile: 16.7secs
0-1km: 30.4secs
156.0bhp 116.3kW 158.2PS
@ 5600rpm
155.0lbft 210.0Nm @ 4000rpm
130.0bhp/ton 127.9bhp/tonne
62.6bhp/L 46.7kW/L 63.5PS/L
41.8ft/sec 12.7m/sec
20.8mph 33.5kmh/1000rpm
22.8mpg 19.0mpgUS
12.4L/100km
Petrol 4-stroke piston
2492cc 152.0cu in
Vee 6 fuel injection
Compression ratio: 9.5:1

17 Alfa Romeo
33 1.7 Veloce

Bore: 88.0mm 3.5in
Stroke: 68.3mm 2.7in
Valve type/No: Overhead 12
Transmission: Manual
No. of forward speeds: 5
Wheels driven: Rear
Springs F/R: Torsion bar/Coil
Brake system: PA
Brakes F/R: Disc/Disc
Steering: Rack & pinion PA
Wheelbase: 251.0cm 98.8in
Track F: 136.8cm 53.9in
Track R: 135.8cm 53.5in
Length: 433.0cm 170.5in
Width: 163.0cm 64.2in
Height: 135.0cm 53.1in
Ground clearance: 20.3cm
8.0in
Kerb weight: 1220.0kg
2687.2lb
Fuel: 56.9L 12.5gal 15.0galUS

14 Alfa Romeo
GTV6
1986 Italy
133.0mph 214.0kmh
0-60mph 96.5kmh: 8.2secs
0-1/4 mile: 16.2secs
160.0bhp 119.3kW 162.2PS
@ 5800rpm
157.0lbft 212.7Nm @ 4000rpm
134.2bhp/ton 132.0bhp/tonne
64.2bhp/L 47.9kW/L 65.1PS/L
43.3ft/sec 13.2m/sec
21.8mph 35.1kmh/1000rpm
26.3mpg 21.9mpgUS
10.7L/100km
Petrol 4-stroke piston
2492cc 152.0cu in
Vee 6 fuel injection
Compression ratio: 9.0:1
Bore: 88.0mm 3.5in
Stroke: 68.3mm 2.7in

Valve type/No: Overhead 12
Transmission: Manual
No. of forward speeds: 5
Wheels driven: Rear
Springs F/R: Torsion bar/Coil
Brakes F/R: Disc/Disc
Steering: Rack & pinion
Wheelbase: 240.0cm 94.5in
Track F: 137.4cm 54.1in
Track R: 135.1cm 53.2in
Length: 426.0cm 167.7in
Width: 166.4cm 65.5in
Height: 130.0cm 51.2in
Kerb weight: 1212.2kg
2670.0lb

15 Alfa Romeo
Milano Platinum
1986 Italy
129.0mph 207.6kmh
0-50mph 80.5kmh: 6.7secs
0-60mph 96.5kmh: 9.0secs
0-1/4 mile: 16.8secs
154.0bhp 114.8kW 156.1PS
@ 5600rpm
152.0lbft 206.0Nm @ 3200rpm
115.6bhp/ton 113.6bhp/tonne
61.8bhp/L 46.1kW/L 62.7PS/L
41.8ft/sec 12.7m/sec
26.2mpg 21.8mpgUS
10.8L/100km
Petrol 4-stroke piston
2492cc 152.0cu in
Vee 6 fuel injection
Compression ratio: 9.0:1
Bore: 88.0mm 3.5in
Stroke: 68.3mm 2.7in
Valve type/No: Overhead 12
Transmission: Manual
No. of forward speeds: 5
Wheels driven: Rear
Springs F/R: Gas/Coil
Brake system: PA ABS

Brakes F/R: Disc/Disc
Steering: Rack & pinion PA
Wheelbase: 251.0cm 98.8in
Track F: 137.9cm 54.3in
Track R: 136.9cm 53.9in
Length: 433.1cm 170.5in
Width: 163.1cm 64.2in
Height: 134.9cm 53.1in
Ground clearance: 10.2cm
4.0in
Kerb weight: 1355.2kg
2985.0lb
Fuel: 66.6L 14.6gal 17.6galUS

16 Alfa Romeo
Spider Quadrifoglio
1986 Italy
110.0mph 177.0kmh
0-50mph 80.5kmh: 7.5secs
0-60mph 96.5kmh: 11.0secs
0-1/4 mile: 17.5secs
115.0bhp 85.8kW 116.6PS
@ 5500rpm
119.0lbft 161.2Nm @ 2750rpm
99.5bhp/ton 97.8bhp/tonne
58.6bhp/L 43.7kW/L 59.4PS/L
53.2ft/sec 16.2m/sec
25.2mpg 21.0mpgUS
11.2L/100km
Petrol 4-stroke piston
1962cc 119.7cu in
In-line 4 fuel injection
Compression ratio: 9.0:1
Bore: 84.0mm 3.3in
Stroke: 88.5mm 3.5in
Valve type/No: Overhead 8
Transmission: Manual
No. of forward speeds: 5
Wheels driven: Rear
Springs F/R: Coil/Coil
Brake system: PA
Brakes F/R: Disc/Disc
Steering: Worm & roller

Wheelbase: 225.0cm 88.6in
Track F: 132.3cm 52.1in
Track R: 127.3cm 50.1in
Length: 428.8cm 168.8in
Width: 162.8cm 64.1in
Height: 124.0cm 48.8in
Kerb weight: 1175.9kg
2590.0lb
Fuel: 46.2L 10.1gal 12.2galUS

17 Alfa Romeo
33 1.7 Veloce
1987 Italy
118.0mph 189.9kmh
0-50mph 80.5kmh: 6.8secs
0-60mph 96.5kmh: 9.1secs
0-1/4 mile: 16.5secs
0-1km: 31.2secs
118.0bhp 88.0kW 119.6PS
@ 6000rpm
108.5lbft 147.0Nm @ 3500rpm
128.8bhp/ton 126.6bhp/tonne
68.9bhp/L 51.4kW/L 69.9PS/L
47.3ft/sec 14.4m/sec
20.7mph 33.3kmh/1000rpm
26.2mpg 21.8mpgUS
10.8L/100km
Petrol 4-stroke piston
1712cc 104.0cu in
Flat 4 2 Carburettor
Compression ratio: 9.5:1
Bore: 87.0mm 3.4in
Stroke: 72.2mm 2.8in
Valve type/No: Overhead 8
Transmission: Manual
No. of forward speeds: 5
Wheels driven: Front
Springs F/R: Coil/Coil
Brake system: PA
Brakes F/R: Disc/Drum
Steering: Rack & pinion
Wheelbase: 245.5cm 96.7in
Track F: 139.2cm 54.8in

Track R: 135.9cm 53.5in
Length: 401.5cm 158.1in
Width: 161.2cm 63.5in
Height: 134.0cm 52.8in
Ground clearance: 15.0cm
5.9in
Kerb weight: 932.0kg 2052.9lb
Fuel: 50.0L 11.0gal 13.2galUS

18 Alfa Romeo

Alfa 33 1.7 Sportwagon Veloce
1987 Italy
118.0mph 189.9kmh
0-50mph 80.5kmh: 7.0secs
0-60mph 96.5kmh: 9.5secs
0-1/4 mile: 17.5secs
0-1km: 31.9secs
118.0bhp 88.0kW 119.6PS
@ 6000rpm
109.0lbft 147.7Nm @ 3900rpm
123.1bhp/ton 121.0bhp/tonne
68.9bhp/L 51.4kW/L 69.9PS/L
47.3ft/sec 14.4m/sec
20.7mph 33.3kmh/1000rpm
25.8mpg 21.5mpgUS
10.9L/100km
Petrol 4-stroke piston
1712cc 104.0cu in
Flat 4 2 Carburettor
Compression ratio: 9.5:1
Bore: 87.0mm 3.4in
Stroke: 72.2mm 2.8in
Valve type/No: Overhead 8
Transmission: Manual
No. of forward speeds: 5
Wheels driven: Front
Springs F/R: Coil/Coil
Brake system: PA
Brakes F/R: Disc/Drum
Steering: Rack & pinion
Wheelbase: 245.5cm 96.7in
Track F: 139.7cm 55.0in
Track R: 137.5cm 54.1in
Length: 414.0cm 163.0in
Width: 161.2cm 63.5in
Height: 132.5cm 52.2in
Ground clearance: 13.1cm
5.2in
Kerb weight: 975.0kg 2147.6lb
Fuel: 50.0L 11.0gal 13.2galUS

19 Alfa Romeo

Alfa 75 2.0i Twin Spark
1987 Italy
128.0mph 206.0kmh
'0-50mph 80.5kmh: 6.7secs
0-60mph 96.5kmh: 9.3secs
0-1/4 mile: 16.3secs
0-1km: 30.5secs
148.0bhp 110.4kW 150.0PS
@ 5800rpm
137.0lbft 185.6Nm @ 4700rpm
131.2bhp/ton 129.0bhp/tonne
75.4bhp/L 56.3kW/L 76.5PS/L

56.1ft/sec 17.1m/sec
20.8mph 33.5kmh/1000rpm
25.6mpg 21.3mpgUS
11.0L/100km
Petrol 4-stroke piston
1962cc 120.0cu in
In-line 4 fuel injection
Compression ratio: 10.0:1
Bore: 84.0mm 3.3in
Stroke: 88.5mm 3.5in
Valve type/No: Overhead 8
Transmission: Manual
No. of forward speeds: 5
Wheels driven: Rear
Springs F/R: Coil/Coil
Brake system: PA
Brakes F/R: Disc/Disc
Steering: Rack & pinion PA
Wheelbase: 251.0cm 98.8in
Track F: 136.8cm 53.9in
Track R: 135.8cm 53.5in
Length: 433.0cm 170.5in
Width: 163.0cm 64.2in
Height: 135.0cm 53.1in
Ground clearance: 20.3cm
8.0in
Kerb weight: 1147.0kg
2526.4lb
Fuel: 49.1L 10.8gal 13.0galUS

20 Alfa Romeo

Alfa 75 V6 3.0
1987 Italy
138.0mph 222.0kmh
0-50mph 80.5kmh: 5.7secs
0-60mph 96.5kmh: 7.5secs
0-1/4 mile: 16.1secs
0-1km: 24.9secs
188.0bhp 140.2kW 190.6PS
@ 5800rpm
180.0lbft 243.9Nm @ 4000rpm
158.0bhp/ton 155.4bhp/tonne
63.5bhp/L 47.4kW/L 64.4PS/L
45.6ft/sec 13.9m/sec
24.2mph 38.9kmh/1000rpm
20.5mpg 17.1mpgUS
13.8L/100km
Petrol 4-stroke piston
2959cc 181.0cu in
Vee 6 fuel injection
Compression ratio: 9.5:1
Bore: 93.0mm 3.7in
Stroke: 72.0mm 2.8in
Valve type/No: Overhead 12
Transmission: Manual
No. of forward speeds: 5
Wheels driven: Rear
Springs F/R: Coil/Coil
Brake system: PA
Brakes F/R: Disc/Disc
Steering: Rack & pinion PA
Wheelbase: 251.0cm 98.8in
Track F: 136.8cm 53.9in
Track R: 135.8cm 53.5in

Length: 433.0cm 170.5in
Width: 163.0cm 64.2in
Height: 135.0cm 53.1in
Ground clearance: 20.3cm
8.0in
Kerb weight: 1210.0kg
2665.2lb
Fuel: 68.2L 15.0gal 18.0galUS

21 Alfa Romeo

Milano Verde 3.0
1987 Italy
136.0mph 218.8kmh
0-50mph 80.5kmh: 5.7secs
0-60mph 96.5kmh: 7.5secs
0-1/4 mile: 15.8secs
183.0bhp 136.5kW 185.5PS
@ 5800rpm
181.0lbft 245.3Nm @ 3000rpm
136.0bhp/ton 133.7bhp/tonne
61.8bhp/L 46.1kW/L 62.7PS/L
46.1ft/sec 14.0m/sec
25.2mpg 21.0mpgUS
11.2L/100km
Petrol 4-stroke piston
2959cc 180.5cu in
Vee 6 fuel injection
Compression ratio: 9.0:1
Bore: 93.0mm 3.7in
Stroke: 72.6mm 2.9in
Valve type/No: Overhead 12
Transmission: Manual
No. of forward speeds: 5
Wheels driven: Rear
Springs F/R: Torsion bar/Coil
Brake system: PA ABS
Brakes F/R: Disc/Disc
Steering: Rack & pinion PA
Wheelbase: 251.0cm 98.8in
Track F: 137.9cm 54.3in
Track R: 137.4cm 54.1in
Length: 433.1cm 170.5in
Width: 163.1cm 64.2in
Height: 134.9cm 53.1in
Kerb weight: 1368.8kg
3015.0lb
Fuel: 66.6L 14.6gal 17.6galUS

22 Alfa Romeo

75 2.5 Auto
1988 Italy
126.0mph 202.7kmh
0-50mph 80.5kmh: 7.4secs
0-60mph 96.5kmh: 9.8secs
0-1/4 mile: 17.7secs
0-1km: 32.0secs
156.0bhp 116.3kW 158.2PS
@ 5600rpm
155.0lbft 210.0Nm @ 4000rpm
122.9bhp/ton 120.9bhp/tonne
62.6bhp/L 46.7kW/L 63.5PS/L
41.8ft/sec 12.7m/sec
21.6mph 34.8kmh/1000rpm
22.4mpg 18.7mpgUS

12.6L/100km
Petrol 4-stroke piston
2492cc 152.0cu in
Vee 6 fuel injection
Compression ratio: 9.5:1
Bore: 88.0mm 3.5in
Stroke: 68.3mm 2.7in
Valve type/No: Overhead 12
Transmission: Automatic
No. of forward speeds: 3
Wheels driven: Rear
Springs F/R: Torsion bar/Coil
Brake system: PA
Brakes F/R: Disc/Disc
Steering: Rack & pinion PA
Wheelbase: 251.0cm 98.8in
Track F: 136.7cm 53.8in
Track R: 135.6cm 53.4in
Length: 432.8cm 170.4in
Width: 162.8cm 64.1in
Height: 134.9cm 53.1in
Ground clearance: 20.3cm 8.0in
Kerb weight: 1290.7kg
2843.0lb
Fuel: 49.1L 10.8gal 13.0galUS

23 Alfa Romeo

75 Evoluzione
1988 Italy
133.0mph 214.0kmh
0-60mph 96.5kmh: 7.5secs
155.0bhp 115.6kW 157.1PS
@ 5800rpm
166.0lbft 224.9Nm @ 2600rpm
139.7bhp/ton 137.4bhp/tonne
87.1bhp/L 65.0kW/L 88.3PS/L
56.1ft/sec 17.1m/sec
Petrol 4-stroke piston
1779cc 108.5cu in
turbocharged
In-line 4 fuel injection
Compression ratio: 7.5:1
Bore: 80.0mm 3.1in
Stroke: 88.5mm 3.5in
Valve type/No: Overhead 16
Transmission: Manual
No. of forward speeds: 5
Wheels driven: Rear
Springs F/R: Torsion bar/Coil
Brake system: PA ABS
Brakes F/R: Disc/Disc
Steering: Rack & pinion PA
Wheelbase: 251.0cm 98.8in
Track F: 139.7cm 55.0in
Track R: 138.7cm 54.6in
Length: 433.1cm 170.5in
Width: 163.1cm 64.2in
Height: 134.9cm 53.1in
Kerb weight: 1128.2kg 2485.0lb
Fuel: 48.8L 10.7gal 12.9galUS

24 Alfa Romeo

Spider
1988 Italy

103.0mph 165.7kmh
0-50mph 80.5kmh: 7.3secs
0-60mph 96.5kmh: 10.4secs
0-1/4 mile: 17.6secs
125.0bhp 93.2kW 126.7PS
@ 5500rpm
119.0lbft 161.2Nm @ 2750rpm
109.4bhp/ton 107.5bhp/tonne
63.7bhp/L 47.5kW/L 64.6PS/L
53.2ft/sec 16.2m/sec
27.0mpg 22.5mpgUS
10.5L/100km
Petrol 4-stroke piston
1962cc 119.7cu in
In-line 4 fuel injection
Compression ratio: 9.0:1
Bore: 84.0mm 3.3in
Stroke: 88.5mm 3.5in
Valve type/No: Overhead 8
Transmission: Manual
No. of forward speeds: 5
Wheels driven: Rear
Springs F/R: Coil/Coil
Brake system: PA
Brakes F/R: Disc/Disc
Steering: Worm & sector
Wheelbase: 225.0cm 88.6in
Track F: 134.1cm 52.8in
Track R: 129.0cm 50.8in
Length: 426.7cm 168.0in
Width: 163.1cm 64.2in
Height: 129.0cm 50.8in
Kerb weight: 1162.2kg
2560.0lb
Fuel: 46.2L 10.1gal 12.2galUS

25 Alfa Romeo
Sprint 1.7 Cloverleaf
1988 Italy
116.0mph 186.6kmh
0-50mph 80.5kmh: 7.2secs
0-60mph 96.5kmh: 9.5secs
0-1/4 mile: 17.6secs

0-1km: 33.5secs
118.0bhp 88.0kW 119.6PS
@ 5800rpm
109.0lbft 147.7Nm @ 3500rpm
123.0bhp/ton 121.0bhp/tonne
68.9bhp/L 51.4kW/L 69.9PS/L
45.8ft/sec 14.0m/sec
20.7mph 33.3kmh/1000rpm
25.7mpg 21.4mpgUS
11.0L/100km
Petrol 4-stroke piston
1712cc 104.4cu in
Flat 4 2 Carburettor
Compression ratio: 9.5:1
Bore: 87.0mm 3.4in
Stroke: 72.2mm 2.8in
Valve type/No: Overhead 8
Transmission: Manual
No. of forward speeds: 5
Wheels driven: Front
Springs F/R: Coil/Coil
Brake system: PA
Brakes F/R: Disc/Drum
Steering: Rack & pinion
Wheelbase: 245.6cm 96.7in
Track F: 139.7cm 55.0in
Track R: 136.4cm 53.7in
Length: 402.3cm 158.4in
Width: 162.1cm 63.8in
Height: 130.6cm 51.4in
Ground clearance: 16.5cm
6.5in
Kerb weight: 975.2kg 2148.0lb
Fuel: 50.0L 11.0gal 13.2galUS

26 Alfa Romeo
164 3.0 Lusso
1989 Italy
140.0mph 225.3kmh
0-50mph 80.5kmh: 6.1secs
0-60mph 96.5kmh: 7.9secs
0-1/4 mile: 16.2secs
0-1km: 29.3secs

192.0bhp 143.2kW 194.7PS
@ 5600rpm
181.5lbft 246.0Nm @ 3000rpm
143.0bhp/ton 140.7bhp/tonne
64.9bhp/L 48.4kW/L 65.8PS/L
44.5ft/sec 13.5m/sec
25.8mph 41.5kmh/1000rpm
25.9mpg 21.6mpgUS
10.9L/100km
Petrol 4-stroke piston
2959cc 181.0cu in
Vee 6 fuel injection
Compression ratio: 9.5:1
Bore: 93.0mm 3.7in
Stroke: 72.6mm 2.9in
Valve type/No: Overhead 12
Transmission: Manual
No. of forward speeds: 5
Wheels driven: Front
Springs F/R: Coil/Coil
Brake system: PA ABS
Brakes F/R: Disc/Disc
Steering: Rack & pinion PA
Wheelbase: 266.0cm 104.7in
Track F: 151.5cm 59.6in
Track R: 148.8cm 58.6in
Length: 455.5cm 179.3in
Width: 176.0cm 69.3in
Height: 140.0cm 55.1in
Ground clearance: 16.0cm
6.3in
Kerb weight: 1365.0kg
3006.6lb
Fuel: 70.5L 15.5gal 18.6galUS

27 Alfa Romeo
164 Automatic
1989 Italy
137.0mph 220.4kmh
0-50mph 80.5kmh: 6.8secs
0-60mph 96.5kmh: 9.0secs
0-1/4 mile: 17.3secs
0-1km: 30.6secs

185.0bhp 137.9kW 187.6PS
@ 5600rpm
191.1lbft 259.0Nm @ 4400rpm
132.5bhp/ton 130.3bhp/tonne
62.5bhp/L 46.6kW/L 63.4PS/L
44.6ft/sec 13.6m/sec
24.2mph 38.9kmh/1000rpm
22.8mpg 19.0mpgUS
12.4L/100km
Petrol 4-stroke piston
2959cc 181.0cu in
Vee 6 fuel injection
Compression ratio: 9.5:1
Bore: 93.0mm 3.7in
Stroke: 73.0mm 2.9in
Valve type/No: Overhead 12
Transmission: Automatic
No. of forward speeds: 4
Wheels driven: Front
Springs F/R: Coil/Coil
Brake system: PA ABS
Brakes F/R: Disc/Disc
Steering: Rack & pinion PA
Wheelbase: 266.0cm 104.7in
Track F: 151.5cm 59.6in
Track R: 148.8cm 58.6in
Length: 455.5cm 179.3in
Width: 176.0cm 69.3in
Height: 140.0cm 55.1in
Ground clearance: 16.0cm
6.3in
Kerb weight: 1420.0kg
3127.7lb
Fuel: 70.5L 15.5gal 18.6galUS

28 Alfa Romeo
Milano 3.0
1989 Italy
125.0mph 201.1kmh
0-50mph 80.5kmh: 5.9secs
0-60mph 96.5kmh: 7.9secs
0-1/4 mile: 16.0secs
183.0bhp 136.5kW 185.5PS

26 Alfa Romeo
164 3.0 Lusso

@ 5800rpm
181.0lbft 245.3Nm @ 3000rpm
136.0bhp/ton 133.7bhp/tonne
61.8bhp/L 46.1kW/L 62.7PS/L
46.1ft/sec 14.0m/sec
24.0mpg 20.0mpgUS
11.8L/100km
Petrol 4-stroke piston
2959cc 180.5cu in
Vee 6 fuel injection
Compression ratio: 9.0:1
Bore: 93.0mm 3.7in
Stroke: 72.6mm 2.9in
Valve type/No: Overhead 12
Transmission: Manual
No. of forward speeds: 5
Wheels driven: Rear
Springs F/R: Torsion bar/Coil
Brake system: PA ABS
Brakes F/R: Disc/Disc
Steering: Rack & pinion PA
Wheelbase: 251.0cm 98.8in
Track F: 137.9cm 54.3in
Track R: 137.4cm 54.1in
Length: 433.1cm 170.5in
Width: 163.1cm 64.2in
Height: 134.9cm 53.1in
Kerb weight: 1368.8kg
3015.0lb
Fuel: 66.6L 14.6gal 17.6galUS

29 Alfa Romeo

Spider Quadrifoglio
1989 Italy
103.0mph 165.7kmh
0-50mph 80.5kmh: 7.7secs
0-60mph 96.5kmh: 10.8secs
0-1/4 mile: 17.8secs
120.0bhp 89.5kW 121.7PS
@ 5800rpm
117.0lbft 158.5Nm @ 2600rpm
106.0bhp/ton 104.3bhp/tonne
61.2bhp/L 45.6kW/L 62.0PS/L
56.1ft/sec 17.1m/sec
Petrol 4-stroke piston
1962cc 119.7cu in
In-line 4 fuel injection
Compression ratio: 9.0:1
Bore: 84.0mm 3.3in
Stroke: 88.5mm 3.5in
Valve type/No: Overhead 8
Transmission: Manual
No. of forward speeds: 5
Wheels driven: Rear
Springs F/R: Coil/Coil
Brake system: PA
Brakes F/R: Disc/Disc
Steering: Worm & sector
Wheelbase: 225.0cm 88.6in
Track F: 132.3cm 52.1in
Track R: 127.5cm 50.2in
Length: 424.9cm 167.3in
Width: 163.1cm 64.2in
Height: 142.0cm 55.9in

Kerb weight: 1150.9kg
2535.0lb
Fuel: 46.2L 10.1gal 12.2galUS

30 Alfa Romeo

164 Twin Spark Lusso
1990 Italy
127.0mph 204.3kmh
0-50mph 80.5kmh: 7.2secs
0-60mph 96.5kmh: 9.5secs
0-1/4 mile: 17.7secs
0-1km: 31.4secs
148.0bhp 110.4kW 150.0PS
@ 5800rpm
137.3lbft 186.0Nm @ 4000rpm
112.7bhp/ton 110.9bhp/tonne
75.4bhp/L 56.3kW/L 76.5PS/L
56.1ft/sec 17.1m/sec
23.8mph 38.3kmh/1000rpm
23.4mpg 19.5mpgUS
12.1L/100km
Petrol 4-stroke piston
1962cc 120.0cu in
In-line 4 fuel injection
Compression ratio: 10.0:1
Bore: 84.0mm 3.3in
Stroke: 88.5mm 3.5in
Valve type/No: Overhead 8
Transmission: Manual
No. of forward speeds: 5
Wheels driven: Front
Springs F/R: Coil/Coil
Brake system: PA ABS
Brakes F/R: Disc/Disc
Steering: Rack & pinion PA
Wheelbase: 265.9cm 104.7in
Track F: 141.2cm 55.6in
Track R: 148.6cm 58.5in
Length: 455.4cm 179.3in
Width: 175.8cm 69.2in
Height: 140.0cm 55.1in
Ground clearance: 16.0cm
6.3in
Kerb weight: 1335.0kg
2940.5lb
Fuel: 70.5L 15.5gal 18.6galUS

31 Alfa Romeo

164S
1990 Italy
140.0mph 225.3kmh
0-50mph 80.5kmh: 5.7secs
0-60mph 96.5kmh: 7.6secs
0-1/4 mile: 15.8secs
200.0bhp 149.1kW 202.8PS
@ 6000rpm
189.0lbft 256.1Nm @ 4400rpm
134.3bhp/ton 132.1bhp/tonne
67.6bhp/L 50.4kW/L 68.5PS/L
47.7ft/sec 14.5m/sec
27.6mpg 23.0mpgUS
10.2L/100km
Petrol 4-stroke piston
2959cc 180.5cu in

Vee 6 fuel injection
Compression ratio: 10.0:1
Bore: 93.0mm 3.7in
Stroke: 72.6mm 2.9in
Valve type/No: Overhead 12
Transmission: Manual
No. of forward speeds: 5
Wheels driven: Front
Springs F/R: Coil/Coil
Brake system: PA ABS
Brakes F/R: Disc/Disc
Steering: Rack & pinion PA
Wheelbase: 265.9cm 104.7in
Track F: 151.4cm 59.6in
Track R: 148.8cm 58.6in
Length: 455.7cm 179.4in
Width: 176.0cm 69.3in
Height: 139.2cm 54.8in
Ground clearance: 15.7cm
6.2in
Kerb weight: 1514.1kg
3335.0lb
Fuel: 65.1L 14.3gal 17.2galUS

32 Alfa Romeo

33 Boxer 16v
1990 Italy
129.0mph 207.6kmh
0-50mph 80.5kmh: 6.8secs
0-60mph 96.5kmh: 8.9secs
0-1/4 mile: 16.7secs
0-1km: 29.9secs
137.0bhp 102.2kW 138.9PS
@ 6500rpm
118.8lbft 161.0Nm @ 4600rpm
137.9bhp/ton 135.6bhp/tonne
80.0bhp/L 59.7kW/L 81.1PS/L
51.1ft/sec 15.6m/sec
19.9mph 32.0kmh/1000rpm
29.7mpg 24.7mpgUS
9.5L/100km
Petrol 4-stroke piston
1712cc 104.0cu in
Flat 4 fuel injection
Compression ratio: 9.5:1
Bore: 87.0mm 3.4in
Stroke: 72.0mm 2.8in
Valve type/No: Overhead 16
Transmission: Manual
No. of forward speeds: 5
Wheels driven: Front
Springs F/R: Coil/Coil
Brake system: PA
Brakes F/R: Disc/Drum
Steering: Rack & pinion PA
Wheelbase: 247.4cm 97.4in
Track F: 136.4cm 53.7in
Track R: 136.4cm 53.7in
Length: 407.4cm 160.4in
Width: 161.3cm 63.5in
Height: 134.9cm 53.1in
Kerb weight: 1010.0kg
2224.7lb
Fuel: 49.6L 10.9gal 13.1galUS

33 Alfa Romeo

164 Cloverleaf
1991 Italy
144.0mph 231.7kmh
0-50mph 80.5kmh: 6.2secs
0-60mph 96.5kmh: 7.8secs
0-1/4 mile: 16.4secs
0-1km: 29.0secs
200.0bhp 149.1kW 202.8PS
@ 5800rpm
198.0lbft 268.3Nm @ 4400rpm
140.8bhp/ton 138.5bhp/tonne
67.6bhp/L 50.4kW/L 68.5PS/L
46.1ft/sec 14.0m/sec
22.9mph 36.8kmh/1000rpm
21.4mpg 17.8mpgUS
13.2L/100km
Petrol 4-stroke piston
2959cc 180.5cu in
Vee 6 fuel injection
Compression ratio: 10.0:1
Bore: 93.0mm 3.7in
Stroke: 72.6mm 2.9in
Valve type/No: Overhead 12
Transmission: Manual
No. of forward speeds: 5
Wheels driven: Front
Springs F/R: Coil/Coil
Brake system: PA ABS
Brakes F/R: Disc/Disc
Steering: Rack & pinion PA
Wheelbase: 266.0cm 104.7in
Track F: 151.5cm 59.6in
Track R: 148.8cm 58.6in
Length: 455.5cm 179.3in
Width: 176.0cm 69.3in
Height: 140.0cm 55.1in
Ground clearance: 16.0cm
6.3in
Kerb weight: 1444.2kg
3181.0lb
Fuel: 70.5L 15.5gal 18.6galUS

34 Alfa Romeo

Spider Veloce
1991 Italy
103.0mph 165.7kmh
0-50mph 80.5kmh: 7.7secs
0-60mph 96.5kmh: 10.8secs
0-1/4 mile: 17.8secs
120.0bhp 89.5kW 121.7PS
@ 5800rpm
117.0lbft 158.5Nm @ 2700rpm
105.4bhp/ton 103.6bhp/tonne
61.2bhp/L 45.6kW/L 62.0PS/L
56.1ft/sec 17.1m/sec
28.1mpg 23.4mpgUS
10.1L/100km
Petrol 4-stroke piston
1962cc 119.7cu in
In-line 4 fuel injection
Compression ratio: 9.0:1
Bore: 84.0mm 3.3in
Stroke: 88.5mm 3.5in

37 Aston Martin
Virage

Valve type/No: Overhead 8
Transmission: Manual
No. of forward speeds: 5
Wheels driven: Rear
Springs F/R: Coil/Coil
Brake system: PA
Brakes F/R: Disc/Disc
Steering: Worm & sector PA
Wheelbase: 225.0cm 88.6in
Track F: 132.3cm 52.1in
Track R: 127.3cm 50.1in
Length: 426.0cm 167.7in
Width: 162.3cm 63.9in
Height: 126.2cm 49.7in
Kerb weight: 1157.7kg
2550.0lb
Fuel: 46.2L 10.1gal 12.2galUS

35 Aston Martin
Volante
1986 UK
145.0mph 233.3kmh
0-60mph 96.5kmh: 8.2secs
0-1/4 mile: 16.3secs
350.0bhp 261.0kW 354.8PS
@ 5500rpm
400.0lbft 542.0Nm @ 4250rpm
195.5bhp/ton 192.2bhp/tonne
65.5bhp/L 48.9kW/L 66.4PS/L
Petrol 4-stroke piston
5340cc 325.8cu in
Vee 8 fuel injection
Valve type/No: Overhead 16
Transmission: Automatic
No. of forward speeds: 3
Wheels driven: Rear
Springs F/R: Coil/Coil
Brakes F/R: Disc/Disc
Steering: Rack & pinion PA
Wheelbase: 261.1cm 102.8in
Length: 467.4cm 184.0in
Width: 182.9cm 72.0in
Height: 137.2cm 54.0in

Kerb weight: 1820.5kg
4010.0lb
Fuel: 97.6L 21.5gal 25.8galUS

36 Aston Martin
Zagato
1988 UK
183.0mph 294.4kmh
0-60mph 96.5kmh: 4.8secs
432.0bhp 322.1kW 438.0PS
@ 6200rpm
395.0lbft 535.2Nm @ 5100rpm
266.6bhp/ton 262.1bhp/tonne
80.9bhp/L 60.3kW/L 82.0PS/L
57.7ft/sec 17.6m/sec
Petrol 4-stroke piston
5340cc 325.8cu in
Vee 8 4 Carburettor
Compression ratio: 10.2:1
Bore: 100.0mm 3.9in
Stroke: 85.0mm 3.3in
Valve type/No: Overhead 16
Transmission: Manual
No. of forward speeds: 5
Wheels driven: Rear
Springs F/R: Coil/Coil
Brake system: PA
Brakes F/R: Disc/Disc
Steering: Rack & pinion PA
Wheelbase: 262.1cm 103.2in
Track F: 152.7cm 60.1in
Track R: 154.4cm 60.8in
Length: 440.7cm 173.5in
Width: 186.7cm 73.5in
Height: 129.8cm 51.1in
Kerb weight: 1648.0kg
3630.0lb
Fuel: 97.6L 21.5gal 25.8galUS

37 Aston Martin
Virage
1989 UK
155.0mph 249.4kmh

0-60mph 96.5kmh: 6.0secs
330.0bhp 246.1kW 334.6PS
@ 6000rpm
350.0lbft 474.3Nm @ 4000rpm
187.6bhp/ton 184.5bhp/tonne
61.8bhp/L 46.1kW/L 62.7PS/L
Petrol 4-stroke piston
5340cc 325.8cu in
Vee 8 fuel injection
Compression ratio: 9.5:1
Valve type/No: Overhead 32
Transmission: Manual
No. of forward speeds: 5
Wheels driven: Rear
Springs F/R: Coil/Coil
Brake system: PA
Brakes F/R: Disc/Disc
Wheelbase: 261.1cm 102.8in
Length: 473.7cm 186.5in
Width: 185.4cm 73.0in
Height: 132.1cm 52.0in
Kerb weight: 1788.8kg
3940.0lb
Fuel: 90.5L 19.9gal 23.9galUS

38 Aston Martin
Volante
1989 UK
135.0mph 217.2kmh
0-60mph 96.5kmh: 7.8secs
240.0bhp 179.0kW 243.3PS
@ 5500rpm
288.0lbft 390.2Nm @ 5500rpm
134.1bhp/ton 131.8bhp/tonne
44.9bhp/L 33.5kW/L 45.6PS/L
51.2ft/sec 15.6m/sec
Petrol 4-stroke piston
5340cc 325.8cu in
Vee 8 fuel injection
Compression ratio: 8.0:1
Bore: 100.0mm 3.9in
Stroke: 85.0mm 3.3in
Valve type/No: Overhead 16

Transmission: Automatic
No. of forward speeds: 3
Wheels driven: Rear
Springs F/R: Coil/Coil
Brake system: PA
Brakes F/R: Disc/Disc
Steering: Rack & pinion PA
Wheelbase: 261.1cm 102.8in
Track F: 155.2cm 61.1in
Track R: 158.5cm 62.4in
Length: 467.1cm 183.9in
Width: 191.5cm 75.4in
Height: 136.9cm 53.9in
Kerb weight: 1820.5kg
4010.0lb
Fuel: 81.4L 17.9gal 21.5galUS

39 Aston Martin
Zagato Volante
1989 UK
165.0mph 265.5kmh
0-60mph 96.5kmh: 5.6secs
305.0bhp 227.4kW 309.2PS
@ 5500rpm
330.0lbft 447.2Nm @ 4750rpm
184.4bhp/ton 181.3bhp/tonne
57.1bhp/L 42.6kW/L 57.9PS/L
51.2ft/sec 15.6m/sec
Petrol 4-stroke piston
5340cc 325.8cu in
Vee 8 fuel injection
Compression ratio: 10.2:1
Bore: 100.0mm 3.9in
Stroke: 85.0mm 3.3in
Valve type/No: Overhead 16
Transmission: Manual
No. of forward speeds: 5
Wheels driven: Rear
Springs F/R: Coil/Coil
Brake system: PA
Brakes F/R: Disc/Disc
Steering: Rack & pinion PA
Wheelbase: 262.1cm 103.2in

Track F: 152.7cm 60.1in
Track R: 154.4cm 60.8in
Length: 440.7cm 173.5in
Width: 186.7cm 73.5in
Kerb weight: 1682.1kg
3705.0lb
Fuel: 97.6L 21.5gal 25.8galUS

40 Aston Martin

Virage
1991 UK
160.0mph 257.4kmh
0-50mph 80.5kmh: 5.2secs
0-60mph 96.5kmh: 6.8secs
0-1/4 mile: 14.7secs
0-1km: 26.7secs
330.0bhp 246.1kW 334.6PS
@ 6000rpm
339.5lbft 460.0Nm @ 3700rpm
172.1bhp/ton 169.2bhp/tonne
61.8bhp/L 46.1kW/L 62.7PS/L
55.8ft/sec 17.0m/sec
27.1mph 43.6kmh/1000rpm
15.6mpg 13.0mpgUS
18.1L/100km
Petrol 4-stroke piston
5340cc 326.0cu in
Vee 8 fuel injection
Compression ratio: 9.5:1
Bore: 100.0mm 3.9in
Stroke: 85.0mm 3.3in
Valve type/No: Overhead 32
Transmission: Manual
No. of forward speeds: 5
Wheels driven: Rear
Springs F/R: Coil/Coil
Brake system: PA
Brakes F/R: Disc/Disc
Steering: Rack & pinion PA
Wheelbase: 261.1cm 102.8in
Track F: 151.0cm 59.4in
Track R: 152.2cm 59.9in
Length: 474.5cm 186.8in
Width: 185.6cm 73.1in
Height: 132.0cm 52.0in
Kerb weight: 1950.0kg
4295.1lb
Fuel: 118.3L 26.0gal
31.3galUS

41 Audi

100 Turbo Diesel
1986 Germany
113.0mph 181.8kmh
0-50mph 80.5kmh: 9.2secs
0-60mph 96.5kmh: 12.6secs
0-1/4 mile: 18.6secs
0-1km: 34.7secs
87.0bhp 64.9kW 88.2PS
@ 4500rpm
126.0lbft 170.7Nm @ 2750rpm
70.5bhp/ton 69.3bhp/tonne
43.8bhp/L 32.7kW/L 44.4PS/L
42.5ft/sec 13.0m/sec

26.6mph 42.8kmh/1000rpm
28.9mpg 24.1mpgUS
9.8L/100km
Diesel 4-stroke piston
1986cc 121.2cu in
turbocharged
In-line 5 fuel injection
Compression ratio: 23.0:1
Bore: 76.4mm 3.0in
Stroke: 86.4mm 3.4in
Valve type/No: Overhead 10
Transmission: Manual
No. of forward speeds: 5
Wheels driven: Front
Springs F/R: Coil/Coil
Brake system: PA
Brakes F/R: Disc/Drum
Steering: Rack & pinion PA
Wheelbase: 268.7cm 105.8in
Track F: 146.8cm 57.8in
Track R: 146.6cm 57.7in
Length: 479.3cm 188.7in
Width: 181.4cm 71.4in
Height: 142.2cm 56.0in
Ground clearance: 13.7cm
5.4in
Kerb weight: 1254.9kg
2764.0lb
Fuel: 80.1L 17.6gal 21.1galUS

42 Audi

5000 S
1986 Germany
113.0mph 181.8kmh
0-60mph 96.5kmh: 11.6secs
0-1/4 mile: 18.4secs
110.0bhp 82.0kW 111.5PS
@ 5500rpm
122.0lbft 165.3Nm @ 2500rpm
85.8bhp/ton 84.4bhp/tonne
49.4bhp/L 36.8kW/L 50.1PS/L
51.9ft/sec 15.8m/sec
29.4mpg 24.5mpgUS
9.6L/100km
Petrol 4-stroke piston
2226cc 135.8cu in
In-line 5 fuel injection
Compression ratio: 8.5:1
Bore: 81.0mm 3.2in
Stroke: 86.4mm 3.4in
Valve type/No: Overhead 10
Transmission: Automatic
No. of forward speeds: 3
Wheels driven: Front
Springs F/R: Coil/Coil
Brakes F/R: Disc/Drum
Steering: Rack & pinion PA
Wheelbase: 268.7cm 105.8in
Track F: 146.8cm 57.8in
Track R: 146.8cm 57.8in
Length: 489.5cm 192.7in
Width: 181.4cm 71.4in
Height: 141.5cm 55.7in
Kerb weight: 1303.0kg

2870.0lb
Fuel: 79.9L 17.6gal 21.1galUS

43 Audi

80 1.8S
1986 Germany
114.0mph 183.4kmh
0-50mph 80.5kmh: 7.8secs
0-60mph 96.5kmh: 11.1secs
0-1/4 mile: 18.2secs
0-1km: 33.2secs
90.0bhp 67.1kW 91.2PS
@ 5200rpm
150.0lbft 203.3Nm @ 3300rpm
86.5bhp/ton 85.1bhp/tonne
50.5bhp/L 37.7kW/L 51.2PS/L
49.1ft/sec 15.0m/sec
24.2mph 38.9kmh/1000rpm
30.8mpg 25.6mpgUS
9.2L/100km
Petrol 4-stroke piston
1781cc 109.0cu in
In-line 4 1 Carburettor
Compression ratio: 10.0:1
Bore: 81.0mm 3.2in
Stroke: 86.4mm 3.4in
Valve type/No: Overhead 8
Transmission: Manual
No. of forward speeds: 5
Wheels driven: Front
Springs F/R: Coil/Coil
Brake system: PA
Brakes F/R: Disc/Drum
Steering: Rack & pinion
Wheelbase: 254.6cm 100.2in
Track F: 141.1cm 55.6in
Track R: 143.1cm 56.3in
Length: 439.3cm 173.0in
Width: 169.5cm 66.7in
Height: 139.7cm 55.0in
Ground clearance: 12.0cm
4.7in
Kerb weight: 1058.0kg
2330.4lb
Fuel: 67.8L 14.9gal 17.9galUS

44 Audi

90
1986 Germany
115.0mph 185.0kmh
0-50mph 80.5kmh: 6.9secs
0-60mph 96.5kmh: 9.6secs
0-1/4 mile: 17.2secs
0-1km: 32.0secs
115.0bhp 85.8kW 116.6PS
@ 5400rpm
121.0lbft 164.0Nm @ 3200rpm
106.0bhp/ton 104.3bhp/tonne
57.7bhp/L 43.0kW/L 58.5PS/L
45.7ft/sec 13.9m/sec
20.9mph 33.6kmh/1000rpm
25.2mpg 21.0mpgUS
11.2L/100km
Petrol 4-stroke piston

1994cc 121.7cu in
In-line 5 fuel injection
Compression ratio: 10.0:1
Bore: 81.0mm 3.2in
Stroke: 77.4mm 3.0in
Valve type/No: Overhead 10
Transmission: Manual
No. of forward speeds: 5
Wheels driven: Front
Springs F/R: Coil/Coil
Brake system: PA
Brakes F/R: Disc/Drum
Steering: Rack & pinion PA
Wheelbase: 253.7cm 99.9in
Track F: 140.0cm 55.1in
Track R: 142.0cm 55.9in
Length: 440.7cm 173.5in
Width: 168.1cm 66.2in
Height: 138.9cm 54.7in
Ground clearance: 11.9cm
4.7in
Kerb weight: 1102.8kg
2429.0lb
Fuel: 68.2L 15.0gal 18.0galUS

45 Audi

Coupe GT
1986 Germany
115.0mph 185.0kmh
0-50mph 80.5kmh: 6.9secs
0-60mph 96.5kmh: 9.6secs
0-1/4 mile: 17.2secs
110.0bhp 82.0kW 111.5PS
@ 5500rpm
122.0lbft 165.3Nm @ 2500rpm
98.2bhp/ton 96.5bhp/tonne
49.4bhp/L 36.8kW/L 50.1PS/L
51.9ft/sec 15.8m/sec
26.4mpg 22.0mpgUS
10.7L/100km
Petrol 4-stroke piston
2226cc 135.8cu in
In-line 5 fuel injection
Compression ratio: 8.5:1
Bore: 81.0mm 3.2in
Stroke: 86.4mm 3.4in
Valve type/No: Overhead 10
Transmission: Manual
No. of forward speeds: 5
Wheels driven: Front
Springs F/R: Coil/Coil
Brake system: PA
Brakes F/R: Disc/Drum
Steering: Rack & pinion PA
Wheelbase: 253.5cm 99.8in
Track F: 140.2cm 55.2in
Track R: 142.0cm 55.9in
Length: 450.3cm 177.3in
Width: 168.4cm 66.3in
Height: 134.9cm 53.1in
Kerb weight: 1139.5kg
2510.0lb
Fuel: 60.2L 13.2gal 15.9galUS

46 Audi

Coupe GT 2.2
1986 Germany
126.0mph 202.7kmh
0-50mph 80.5kmh: 6.4secs
0-60mph 96.5kmh: 8.8secs
0-1/4 mile: 16.5secs
0-1km: 30.8secs
136.0bhp 101.4kW 137.9PS
@ 5700rpm
137.0lbft 185.6Nm @ 3500rpm
133.6bhp/ton 131.4bhp/tonne
61.1bhp/L 45.6kW/L 61.9PS/L
53.8ft/sec 16.4m/sec
21.8mph 35.1kmh/1000rpm
27.4mpg 22.8mpgUS
10.3L/100km
Petrol 4-stroke piston
2226cc 136.0cu in
In-line 5 fuel injection
Compression ratio: 10.0:1
Bore: 81.0mm 3.2in
Stroke: 86.4mm 3.4in
Valve type/No: Overhead 10
Transmission: Manual
No. of forward speeds: 5
Wheels driven: Front
Springs F/R: Coil/Coil
Brake system: PA
Brakes F/R: Disc/Disc
Steering: Rack & pinion PA
Wheelbase: 253.8cm 99.9in
Track F: 140.0cm 55.1in
Track R: 142.0cm 55.9in
Length: 442.1cm 174.1in
Width: 168.2cm 66.2in
Height: 135.0cm 53.1in
Ground clearance: 12.0cm
4.7in
Kerb weight: 1035.0kg
2279.7lb
Fuel: 68.2L 15.0gal 18.0galUS

47 Audi

Coupe Quattro
1986 Germany
122.0mph 196.3kmh
0-50mph 80.5kmh: 6.2secs
0-60mph 96.5kmh: 8.6secs
0-1/4 mile: 16.9secs
0-1km: 31.9secs
136.0bhp 101.4kW 137.9PS
@ 5700rpm
137.0lbft 185.6Nm @ 3500rpm
113.9bhp/ton 112.0bhp/tonne
61.1bhp/L 45.6kW/L 61.9PS/L
53.8ft/sec 16.4m/sec
19.6mph 31.5kmh/1000rpm
24.2mpg 20.2mpgUS
11.7L/100km
Petrol 4-stroke piston
2226cc 135.8cu in
In-line 5 fuel injection
Compression ratio: 10.0:1

Bore: 81.0mm 3.2in
Stroke: 86.4mm 3.4in
Valve type/No: Overhead 10
Transmission: Manual
No. of forward speeds: 5
Wheels driven: 4-wheel drive
Springs F/R: Coil/Coil
Brake system: PA
Brakes F/R: Disc/Disc
Steering: Rack & pinion PA
Wheelbase: 254.0cm 100.0in
Track F: 140.0cm 55.1in
Track R: 142.0cm 55.9in
Length: 434.8cm 171.2in
Width: 168.1cm 66.2in
Height: 134.9cm 53.1in
Ground clearance: 11.9cm
4.7in
Kerb weight: 1214.0kg
2674.0lb
Fuel: 70.1L 15.4gal 18.5galUS

48 Audi

90 2.2E
1987 Germany
127.0mph 204.3kmh
0-50mph 80.5kmh: 6.7secs
0-60mph 96.5kmh: 9.5secs
0-1/4 mile: 17.2secs
0-1km: 31.6secs
136.0bhp 101.4kW 137.9PS
@ 5700rpm
137.0lbft 185.6Nm @ 3500rpm
122.2bhp/ton 120.1bhp/tonne
61.1bhp/L 45.6kW/L 61.9PS/L
53.8ft/sec 16.4m/sec
21.9mph 35.2kmh/1000rpm
29.7mpg 24.7mpgUS
9.5L/100km
Petrol 4-stroke piston
2226cc 136.0cu in
In-line 5 fuel injection
Compression ratio: 10.0:1
Bore: 81.0mm 3.2in
Stroke: 86.4mm 3.4in
Valve type/No: Overhead 10
Transmission: Manual
No. of forward speeds: 5
Wheels driven: Front
Springs F/R: Coil/Coil
Brake system: PA
Brakes F/R: Disc/Disc
Steering: Rack & pinion PA
Wheelbase: 254.6cm 100.2in
Track F: 141.1cm 55.6in
Track R: 143.2cm 56.4in
Length: 439.3cm 173.0in
Width: 169.5cm 66.7in
Height: 139.7cm 55.0in
Ground clearance: 12.0cm
4.7in
Kerb weight: 1132.0kg
2493.4lb
Fuel: 70.1L 15.4gal 18.5galUS

49 Audi

90 Quattro
1987 Germany
127.0mph 204.3kmh
0-50mph 80.5kmh: 6.6secs
0-60mph 96.5kmh: 9.5secs
0-1/4 mile: 16.7secs
0-1km: 31.6secs
136.0bhp 101.4kW 137.9PS
@ 5700rpm
137.0lbft 185.6Nm @ 3500rpm
108.6bhp/ton 106.8bhp/tonne
61.1bhp/L 45.6kW/L 61.9PS/L
53.8ft/sec 16.4m/sec
21.2mph 34.1kmh/1000rpm
28.7mpg 23.9mpgUS
9.8L/100km
Petrol 4-stroke piston
2226cc 136.0cu in
In-line 5 fuel injection
Compression ratio: 10.0:1
Bore: 81.0mm 3.2in
Stroke: 86.4mm 3.4in
Valve type/No: Overhead 10
Transmission: Manual
No. of forward speeds: 5
Wheels driven: 4-wheel drive
Springs F/R: Coil/Coil
Brake system: PA
Brakes F/R: Disc/Disc
Steering: Rack & pinion PA
Wheelbase: 254.6cm 100.2in
Track F: 141.1cm 55.6in
Track R: 143.1cm 56.3in
Length: 439.3cm 173.0in
Width: 169.5cm 66.7in
Height: 139.7cm 55.0in
Ground clearance: 12.0cm
4.7in
Kerb weight: 1273.0kg
2804.0lb
Fuel: 70.1L 15.4gal 18.5galUS

50 Audi

100 Avant Quattro
1988 Germany
123.0mph 197.9kmh
0-50mph 80.5kmh: 7.6secs
0-60mph 96.5kmh: 10.6secs
0-1/4 mile: 17.2secs
0-1km: 31.9secs
138.0bhp 102.9kW 139.9PS
@ 5700rpm
138.0lbft 187.0Nm @ 3500rpm
96.9bhp/ton 95.3bhp/tonne
62.0bhp/L 46.2kW/L 62.8PS/L
53.8ft/sec 16.4m/sec
21.3mph 34.3kmh/1000rpm
21.4mpg 17.8mpgUS
13.2L/100km
Petrol 4-stroke piston
2226cc 136.0cu in
In-line 5 fuel injection
Compression ratio: 10.0:1

Bore: 81.0mm 3.2in
Stroke: 86.4mm 3.4in
Valve type/No: Overhead 10
Transmission: Manual
No. of forward speeds: 5
Wheels driven: 4-wheel drive
Springs F/R: Coil/Coil
Brake system: PA
Brakes F/R: Disc/Disc
Steering: Rack & pinion PA
Wheelbase: 268.7cm 105.8in
Track F: 146.8cm 57.8in
Track R: 146.9cm 57.8in
Length: 480.8cm 189.3in
Width: 181.4cm 71.4in
Height: 142.2cm 56.0in
Ground clearance: 13.6cm
5.4in
Kerb weight: 1448.0kg
3189.4lb
Fuel: 80.1L 17.6gal 21.1galUS

51 Audi

100 Sport
1988 Germany
122.0mph 196.3kmh
0-50mph 80.5kmh: 7.2secs
0-60mph 96.5kmh: 10.0secs
0-1/4 mile: 17.5secs
0-1km: 32.5secs
138.0bhp 102.9kW 139.9PS
@ 5700rpm
138.0lbft 187.0Nm @ 3500rpm
110.6bhp/ton 108.7bhp/tonne
62.0bhp/L 46.2kW/L 62.8PS/L
53.8ft/sec 16.4m/sec
21.5mph 34.6kmh/1000rpm
27.4mpg 22.8mpgUS
10.3L/100km
Petrol 4-stroke piston
2226cc 136.0cu in
In-line 5 fuel injection
Compression ratio: 10.0:1
Bore: 81.0mm 3.2in
Stroke: 86.4mm 3.4in
Valve type/No: Overhead 10
Transmission: Manual
No. of forward speeds: 5
Wheels driven: Front
Springs F/R: Coil/Coil
Brake system: PA ABS
Brakes F/R: Disc/Disc
Steering: Rack & pinion PA
Wheelbase: 268.7cm 105.8in
Track F: 146.8cm 57.8in
Track R: 146.9cm 57.8in
Length: 480.7cm 189.3in
Width: 181.4cm 71.4in
Height: 142.2cm 56.0in
Ground clearance: 13.6cm
5.4in
Kerb weight: 1269.0kg
2795.1lb
Fuel: 80.1L 17.6gal 21.1galUS

52 Audi
80 1.8E
1988 Germany
122.0mph 196.3kmh
0-50mph 80.5kmh: 6.9secs
0-60mph 96.5kmh: 9.8secs
0-1/4 mile: 17.7secs
0-1km: 32.4secs
112.0bhp 83.5kW 113.5PS
@ 5800rpm
118.1lbft 160.0Nm @ 3400rpm
103.1bhp/ton 101.4bhp/tonne
62.9bhp/L 46.9kW/L 63.8PS/L
54.6ft/sec 16.6m/sec
20.6mph 33.1kmh/1000rpm
28.1mpg 23.4mpgUS
10.1L/100km
Petrol 4-stroke piston
1781cc 109.0cu in
In-line 4 fuel injection
Compression ratio: 10.0:1
Bore: 81.0mm 3.2in
Stroke: 86.0mm 3.4in
Valve type/No: Overhead 8
Transmission: Manual
No. of forward speeds: 5
Wheels driven: Front
Springs F/R: Coil/Torsion bar
Brake system: PA
Brakes F/R: Disc/Disc
Steering: Rack & pinion PA
Wheelbase: 254.6cm 100.2in
Track F: 141.1cm 55.6in
Track R: 143.2cm 56.4in
Length: 439.3cm 173.0in
Width: 169.5cm 66.7in
Height: 139.7cm 55.0in
Ground clearance: 12.0cm 4.7in
Kerb weight: 1105.0kg
2433.9lb
Fuel: 67.8L 14.9gal 17.9galUS

53 Audi
90 Quattro
1988 Germany
128.0mph 206.0kmh
0-50mph 80.5kmh: 6.9secs
0-60mph 96.5kmh: 9.5secs
0-1/4 mile: 17.0secs
130.0bhp 96.9kW 131.8PS
@ 5700rpm
140.0lbft 189.7Nm @ 4500rpm
97.1bhp/ton 95.4bhp/tonne
56.3bhp/L 42.0kW/L 57.1PS/L
53.8ft/sec 16.4m/sec
25.2mpg 21.0mpgUS
11.2L/100km
Petrol 4-stroke piston
2309cc 140.9cu in
In-line 5 fuel injection
Compression ratio: 10.0:1
Bore: 82.5mm 3.2in
Stroke: 86.4mm 3.4in
Valve type/No: Overhead 10

Transmission: Manual
No. of forward speeds: 5
Wheels driven: 4-wheel drive
Springs F/R: Coil/Coil
Brake system: PA ABS
Brakes F/R: Disc/Disc
Steering: Rack & pinion PA
Wheelbase: 253.7cm 99.9in
Track F: 141.2cm 55.6in
Track R: 143.0cm 56.3in
Length: 447.8cm 176.3in
Width: 169.4cm 66.7in
Height: 139.2cm 54.8in
Ground clearance: 15.2cm
6.0in
Kerb weight: 1362.0kg
3000.0lb
Fuel: 70.0L 15.4gal
18.5galUS

54 Audi
Quattro
1988 Germany
136.0mph 218.8kmh
0-50mph 80.5kmh: 4.4secs
0-60mph 96.5kmh: 6.3secs
0-1/4 mile: 14.7secs
0-1km: 27.5secs
200.0bhp 149.1kW 202.8PS
@ 5500rpm
199.3lbft 270.0Nm @ 3500rpm
152.8bhp/ton 150.3bhp/tonne
89.8bhp/L 67.0kW/L 91.1PS/L
51.9ft/sec 15.8m/sec
22.3mph 35.9kmh/1000rpm
19.7mpg 16.4mpgUS
14.3L/100km
Petrol 4-stroke piston
2226cc 136.0cu in
turbocharged
In-line 5 fuel injection
Compression ratio: 8.6:1
Bore: 81.0mm 3.2in
Stroke: 86.4mm 3.4in
Valve type/No: Overhead 10
Transmission: Manual
No. of forward speeds: 5
Wheels driven: 4-wheel drive
Springs F/R: Coil/Coil
Brake system: PA ABS
Brakes F/R: Disc/Disc
Steering: Rack & pinion PA
Wheelbase: 252.4cm 99.4in
Track F: 146.1cm 57.5in
Track R: 149.4cm 58.8in
Length: 440.4cm 173.4in
Width: 172.3cm 67.8in
Height: 134.4cm 52.9in
Ground clearance: 15.3cm
6.0in
Kerb weight: 1331.0kg
2931.7lb
Fuel: 90.1L 19.8gal
23.8galUS

55 Audi
90
1989 Germany
120.0mph 193.1kmh
0-50mph 80.5kmh: 6.6secs
0-60mph 96.5kmh: 9.3secs
0-1/4 mile: 17.0secs
130.0bhp 96.9kW 131.8PS
@ 5700rpm
140.0lbft 189.7Nm @ 4500rpm
104.6bhp/ton 102.8bhp/tonne
56.3bhp/L 42.0kW/L 57.1PS/L
53.8ft/sec 16.4m/sec
25.2mpg 21.0mpgUS
11.2L/100km
Petrol 4-stroke piston
2309cc 140.9cu in
In-line 4 fuel injection
Compression ratio: 10.0:1
Bore: 82.5mm 3.2in
Stroke: 86.4mm 3.4in
Valve type/No: Overhead 8
Transmission: Manual
No. of forward speeds: 5
Wheels driven: Front
Springs F/R: Coil/Coil
Brake system: PA ABS
Brakes F/R: Disc/Disc
Steering: Rack & pinion PA
Wheelbase: 254.5cm 100.2in
Track F: 141.2cm 55.6in
Track R: 143.0cm 56.3in
Length: 447.8cm 176.3in
Width: 169.4cm 66.7in
Height: 139.2cm 54.8in
Kerb weight: 1264.4kg
2785.0lb
Fuel: 67.7L 14.9gal 17.9galUS

56 Audi
90 Quattro 20v
1989 Germany
138.9mph 223.5kmh
0-50mph 80.5kmh: 6.1secs
0-60mph 96.5kmh: 8.5secs
0-1/4 mile: 16.9secs
0-1km: 29.8secs
170.0bhp 126.8kW 172.4PS
@ 6000rpm
162.4lbft 220.0Nm @ 4500rpm
125.6bhp/ton 123.5bhp/tonne
73.6bhp/L 54.9kW/L 74.6PS/L
56.7ft/sec 17.3m/sec
22.1mph 35.6kmh/1000rpm
22.2mpg 18.5mpgUS
12.7L/100km
Petrol 4-stroke piston
2309cc 141.0cu in
In-line 5 fuel injection
Compression ratio: 10.3:1
Bore: 82.5mm 3.2in
Stroke: 86.4mm 3.4in
Valve type/No: Overhead 20
Transmission: Manual

No. of forward speeds: 5
Wheels driven: 4-wheel drive
Springs F/R: Coil/Coil
Brake system: PA ABS
Brakes F/R: Disc/Disc
Steering: Rack & pinion PA
Wheelbase: 254.6cm 100.2in
Track F: 141.1cm 55.6in
Track R: 143.2cm 56.4in
Length: 439.3cm 173.0in
Width: 169.5cm 66.7in
Height: 139.7cm 55.0in
Ground clearance: 12.0cm
4.7in
Kerb weight: 1376.0kg
3030.8lb
Fuel: 68.2L 15.0gal 18.0galUS

57 Audi
Coupe 2.2E
1989 Germany
132.0mph 212.4kmh
0-50mph 80.5kmh: 7.1secs
0-60mph 96.5kmh: 9.7secs
0-1/4 mile: 17.3secs
0-1km: 31.6secs
136.0bhp 101.4kW 137.9PS
@ 5700rpm
137.3lbft 186.0Nm @ 3500rpm
114.1bhp/ton 112.2bhp/tonne
61.1bhp/L 45.6kW/L 61.9PS/L
53.8ft/sec 16.4m/sec
21.6mph 34.8kmh/1000rpm
26.1mpg 21.7mpgUS
10.8L/100km
Petrol 4-stroke piston
2226cc 136.0cu in
In-line 5 fuel injection
Compression ratio: 10.0:1
Bore: 81.0mm 3.2in
Stroke: 86.4mm 3.4in
Valve type/No: Overhead 10
Transmission: Manual
No. of forward speeds: 5
Wheels driven: Front
Springs F/R: Coil/Coil
Brake system: PA
Brakes F/R: Disc/Disc
Steering: Rack & pinion PA
Wheelbase: 254.8cm 100.3in
Track F: 145.3cm 57.2in
Track R: 143.7cm 56.6in
Length: 436.6cm 171.9in
Width: 171.6cm 67.6in
Height: 136.5cm 53.7in
Ground clearance: 12.9cm 5.1in
Kerb weight: 1212.0kg
2669.6lb
Fuel: 70.1L 15.4gal 18.5galUS

58 Audi
Coupe Quattro
1989 Germany
123.9mph 199.4kmh

0-50mph 80.5kmh: 6.6secs
0-60mph 96.5kmh: 9.5secs
0-1/4 mile: 17.0secs
0-1km: 31.7secs
136.0bhp 101.4kW 137.9PS
@ 5700rpm
137.3lbft 186.0Nm @ 3500rpm
103.8bhp/ton 102.1bhp/tonne
61.1bhp/L 45.6kW/L 61.9PS/L
53.8ft/sec 16.4m/sec
20.5mph 33.0kmh/1000rpm
27.0mpg 22.5mpgUS
10.5L/100km
Petrol 4-stroke piston
2226cc 136.0cu in
In-line 5 fuel injection
Compression ratio: 10.0:1
Bore: 81.0mm 3.2in
Stroke: 86.4mm 3.4in
Valve type/No: Overhead 10
Transmission: Manual
No. of forward speeds: 5
Wheels driven: 4-wheel drive
Springs F/R: Coil/Coil
Brake system: PA ABS
Brakes F/R: Disc/Disc
Steering: Rack & pinion PA
Wheelbase: 254.8cm 100.3in
Track F: 145.3cm 57.2in
Track R: 143.7cm 56.6in
Length: 436.6cm 171.9in
Width: 171.6cm 67.6in
Height: 136.5cm 53.7in
Ground clearance: 12.9cm
5.1in
Kerb weight: 1332.0kg
2933.9lb
Fuel: 68.2L 15.0gal 18.0galUS

59 Audi
80 2.0E
1990 Germany

121.0mph 194.7kmh
0-50mph 80.5kmh: 7.7secs
0-60mph 96.5kmh: 10.6secs
0-1/4 mile: 18.4secs
0-1km: 33.6secs
113.0bhp 84.3kW 114.6PS
@ 5300rpm
125.5lbft 170.0Nm @ 3250rpm
100.4bhp/ton 98.7bhp/tonne
57.0bhp/L 42.5kW/L 57.7PS/L
53.7ft/sec 16.4m/sec
21.1mph 33.9kmh/1000rpm
25.4mpg 21.1mpgUS
11.1L/100km
Petrol 4-stroke piston
1984cc 121.0cu in
In-line 4 fuel injection
Compression ratio: 10.5:1
Bore: 82.5mm 3.2in
Stroke: 92.8mm 3.6in
Valve type/No: Overhead 8
Transmission: Manual
No. of forward speeds: 5
Wheels driven: Front
Springs F/R: Coil/Coil
Brake system: PA ABS
Brakes F/R: Disc/Disc
Steering: Rack & pinion PA
Wheelbase: 254.5cm 100.2in
Track F: 141.0cm 55.5in
Track R: 143.0cm 56.3in
Length: 439.2cm 172.9in
Width: 169.4cm 66.7in
Height: 139.7cm 55.0in
Ground clearance: 11.9cm
4.7in
Kerb weight: 1145.0kg
2522.0lb
Fuel: 68.2L 15.0gal 18.0galUS

60 Audi
90 Quattro

1990 Germany
130.0mph 209.2kmh
0-50mph 80.5kmh: 6.4secs
0-60mph 96.5kmh: 9.0secs
0-1/4 mile: 16.6secs
164.0bhp 122.3kW 166.3PS
@ 6000rpm
157.0lbft 212.7Nm @ 4500rpm
115.5bhp/ton 113.6bhp/tonne
71.0bhp/L 53.0kW/L 72.0PS/L
56.7ft/sec 17.3m/sec
21.6mpg 18.0mpgUS
13.1L/100km
Petrol 4-stroke piston
2309cc 140.9cu in
In-line 5 fuel injection
Compression ratio: 10.3:1
Bore: 82.5mm 3.2in
Stroke: 86.4mm 3.4in
Valve type/No: Overhead 20
Transmission: Manual
No. of forward speeds: 5
Wheels driven: Front
Springs F/R: Coil/Coil
Brake system: PA ABS
Brakes F/R: Disc/Disc
Steering: Rack & pinion PA
Wheelbase: 254.0cm 100.0in
Track F: 141.2cm 55.6in
Track R: 143.0cm 56.3in
Length: 447.8cm 176.3in
Width: 169.4cm 66.7in
Height: 139.2cm 54.8in
Kerb weight: 1443.7kg
3180.0lb
Fuel: 70.0L 15.4gal 18.5galUS

61 Audi
Quattro 20v
1990 Germany
142.0mph 228.5kmh
0-50mph 80.5kmh: 4.3secs

0-60mph 96.5kmh: 6.3secs
0-1/4 mile: 14.5secs
0-1km: 26.9secs
220.0bhp 164.0kW 223.0PS
@ 5900rpm
228.0lbft 309.0Nm @ 1950rpm
160.4bhp/ton 157.7bhp/tonne
98.8bhp/L 73.7kW/L 100.2PS/L
55.6ft/sec 16.9m/sec
22.1mph 35.6kmh/1000rpm
19.1mpg 15.9mpgUS
14.8L/100km
Petrol 4-stroke piston
2226cc 136.0cu in
turbocharged
In-line 5 fuel injection
Compression ratio: 9.3:1
Bore: 81.0mm 3.2in
Stroke: 86.0mm 3.4in
Valve type/No: Overhead 20
Transmission: Manual
No. of forward speeds: 5
Wheels driven: 4-wheel drive
Springs F/R: Coil/Coil
Brake system: PA ABS
Brakes F/R: Disc/Disc
Steering: Rack & pinion PA
Wheelbase: 252.4cm 99.4in
Track F: 146.1cm 57.5in
Track R: 149.4cm 58.8in
Length: 440.4cm 173.4in
Width: 172.3cm 67.8in
Height: 134.4cm 52.9in
Ground clearance: 15.3cm
6.0in
Kerb weight: 1395.0kg
3072.7lb
Fuel: 90.1L 19.8gal 23.8galUS

62 Audi
V8 Quattro
1990 Germany

61 Audi
Quattro 20v

144.0mph 231.7kmh
0-50mph 80.5kmh: 6.8secs
0-60mph 96.5kmh: 9.0secs
0-1/4 mile: 16.8secs
250.0bhp 186.4kW 253.5PS
@ 5800rpm
215.0lbft 291.3Nm @ 4000rpm
139.7bhp/ton 137.4bhp/tonne
70.2bhp/L 52.3kW/L 71.2PS/L
54.8ft/sec 16.7m/sec
24.5mph 39.4kmh/1000rpm
15.3mpg 12.7mpgUS
18.5L/100km
Petrol 4-stroke piston
3562cc 217.3cu in
Vee 8 fuel injection
Compression ratio: 10.6:1
Bore: 81.0mm 3.2in
Stroke: 86.4mm 3.4in
Valve type/No: Overhead 32
Transmission: Automatic
No. of forward speeds: 4
Wheels driven: 4-wheel drive
Springs F/R: Coil/Coil
Brake system: PA ABS
Brakes F/R: Disc/Disc
Steering: Rack & pinion PA
Wheelbase: 269.2cm 106.0in
Track F: 151.4cm 59.6in
Track R: 152.9cm 60.2in
Length: 487.2cm 191.8in
Width: 181.4cm 71.4in
Height: 142.0cm 55.9in
Ground clearance: 10.2cm
4.0in
Kerb weight: 1820.1kg
4009.0lb
Fuel: 80.1L 17.6gal 21.1galUS

63 Audi

V8 Quattro Automatic
1990 Germany
144.0mph 231.7kmh
0-50mph 80.5kmh: 6.8secs
0-60mph 96.5kmh: 9.0secs
0-1/4 mile: 16.8secs
250.0bhp 186.4kW 253.5PS
@ 5800rpm
250.9lbft 340.0Nm @ 4000rpm
139.7bhp/ton 137.4bhp/tonne
70.2bhp/L 52.3kW/L 71.2PS/L
54.8ft/sec 16.7m/sec
24.5mph 39.4kmh/1000rpm
15.3mpg 12.7mpgUS
18.5L/100km
Petrol 4-stroke piston
3562cc 217.0cu in
Vee 8 fuel injection
Compression ratio: 10.6:1
Bore: 81.0mm 3.2in
Stroke: 86.4mm 3.4in
Valve type/No: Overhead 32
Transmission: Automatic

No. of forward speeds: 4
Wheels driven: 4-wheel drive
Springs F/R: Coil/Coil
Brake system: PA ABS
Brakes F/R: Disc/Disc
Steering: Rack & pinion PA
Wheelbase: 270.2cm 106.4in
Track F: 151.4cm 59.6in
Track R: 153.1cm 60.3in
Length: 487.4cm 191.9in
Width: 181.4cm 71.4in
Height: 142.0cm 55.9in
Ground clearance: 10.2cm
4.0in
Kerb weight: 1820.0kg
4008.8lb
Fuel: 80.1L 17.6gal 21.1galUS

64 Audi

100 2.8E Auto
1991 Germany
136.0mph 218.8kmh
0-50mph 80.5kmh: 7.0secs
0-60mph 96.5kmh: 9.3secs
0-1/4 mile: 17.2secs
0-1km: 31.1secs
174.0bhp 129.7kW 176.4PS
@ 5500rpm
184.5lbft 250.0Nm @ 4000rpm
117.6bhp/ton 115.6bhp/tonne
62.8bhp/L 46.8kW/L 63.7PS/L
51.9ft/sec 15.8m/sec
24.8mph 39.9kmh/1000rpm
19.0mpg 15.8mpgUS
14.9L/100km
Petrol 4-stroke piston
2771cc 169.0cu in
Vee 6 fuel injection
Compression ratio: 10.0:1
Bore: 82.5mm 3.2in
Stroke: 86.4mm 3.4in
Valve type/No: Overhead 12
Transmission: Automatic
No. of forward speeds: 4
Wheels driven: Front
Springs F/R: Coil/Coil
Brake system: PA ABS
Brakes F/R: Disc/Disc
Steering: Rack & pinion PA
Wheelbase: 268.7cm 105.8in
Track F: 152.8cm 60.2in
Track R: 152.4cm 60.0in
Length: 479.0cm 188.6in
Width: 177.7cm 70.0in
Height: 142.0cm 55.9in
Ground clearance: 10.0cm
3.9in
Kerb weight: 1505.0kg
3315.0lb
Fuel: 80.0L 17.6gal 21.1galUS

65 Audi

200 Quattro
1991 Germany

150.0mph 241.4kmh
0-60mph 96.5kmh: 7.2secs
0-1/4 mile: 15.4secs
217.0bhp 161.8kW 220.0PS
@ 5700rpm
228.0lbft 308.9Nm @ 1950rpm
133.4bhp/ton 131.1bhp/tonne
97.5bhp/L 72.7kW/L 98.8PS/L
53.8ft/sec 16.4m/sec
23.4mpg 19.5mpgUS
12.1L/100km
Petrol 4-stroke piston
2226cc 135.8cu in
turbocharged
In-line 5 fuel injection
Compression ratio: 9.3:1
Bore: 81.0mm 3.2in
Stroke: 86.4mm 3.4in
Valve type/No: Overhead 20
Transmission: Manual
No. of forward speeds: 5
Wheels driven: 4-wheel drive
Springs F/R: Coil/Coil
Brake system: PA ABS
Brakes F/R: Disc/Disc
Steering: Rack & pinion PA
Wheelbase: 269.5cm 106.1in
Track F: 151.4cm 59.6in
Track R: 151.1cm 59.5in
Length: 491.2cm 193.4in
Width: 181.4cm 71.4in
Height: 142.5cm 56.1in
Kerb weight: 1654.8kg
3645.0lb
Fuel: 79.9L 17.6gal 21.1galUS

66 Audi

80 16v Sport
1991 Germany
132.0mph 212.4kmh
0-50mph 80.5kmh: 6.3secs
0-60mph 96.5kmh: 8.7secs
0-1/4 mile: 16.7secs
0-1km: 30.2secs
137.0bhp 102.2kW 138.9PS
@ 5800rpm
132.8lbft 180.0Nm @ 4500rpm
115.1bhp/ton 113.2bhp/tonne
69.0bhp/L 51.5kW/L 70.0PS/L
59.0ft/sec 18.0m/sec
21.1mph 33.9kmh/1000rpm
28.0mpg 23.3mpgUS
10.1L/100km
Petrol 4-stroke piston
1984cc 121.0cu in
In-line 4 fuel injection
Compression ratio: 10.8:1
Bore: 83.0mm 3.3in
Stroke: 93.0mm 3.7in
Valve type/No: Overhead 16
Transmission: Manual
No. of forward speeds: 5
Wheels driven: Front
Springs F/R: Coil/Coil

Brake system: PA ABS
Brakes F/R: Disc/Disc
Steering: Rack & pinion PA
Wheelbase: 254.5cm 100.2in
Track F: 141.0cm 55.5in
Track R: 143.0cm 56.3in
Length: 439.2cm 172.9in
Width: 169.4cm 66.7in
Height: 139.7cm 55.0in
Ground clearance: 11.9cm
4.7in
Kerb weight: 1210.0kg
2665.2lb
Fuel: 68.2L 15.0gal 18.0galUS

67 Audi

Coupe S2
1991 Germany
153.0mph 246.2kmh
0-50mph 80.5kmh: 4.6secs
0-60mph 96.5kmh: 5.9secs
0-1/4 mile: 14.5secs
0-1km: 26.7secs
220.0bhp 164.0kW 223.0PS
@ 5900rpm
228.0lbft 309.0Nm @ 1950rpm
151.7bhp/ton 149.1bhp/tonne
98.8bhp/L 73.7kW/L 100.2PS/L
55.6ft/sec 16.9m/sec
24.7mph 39.7kmh/1000rpm
18.2mpg 15.2mpgUS
15.5L/100km
Petrol 4-stroke piston
2226cc 136.0cu in
turbocharged
In-line 5 fuel injection
Compression ratio: 9.3:1
Bore: 81.0mm 3.2in
Stroke: 86.0mm 3.4in
Valve type/No: Overhead 20
Transmission: Manual
No. of forward speeds: 5
Wheels driven: 4-wheel drive
Springs F/R: Coil/Coil
Brake system: PA ABS
Brakes F/R: Disc/Disc
Steering: Rack & pinion PA
Wheelbase: 252.5cm 99.4in
Track F: 146.3cm 57.6in
Track R: 142.7cm 56.2in
Length: 436.6cm 171.9in
Width: 171.7cm 67.6in
Height: 127.0cm 50.0in
Ground clearance: 12.7cm
5.0in
Kerb weight: 1475.0kg
3248.9lb
Fuel: 54.6L 12.0gal 14.4galUS

68 Audi

V8 Quattro
1991 Germany
152.0mph 244.6kmh
0-50mph 80.5kmh: 5.8secs

0-60mph 96.5kmh: 7.5secs
0-1/4 mile: 15.6secs
240.0bhp 179.0kW 243.3PS
@ 5800rpm
245.0lbft 332.0Nm @ 4000rpm
137.7bhp/ton 135.4bhp/tonne
67.4bhp/L 50.2kW/L 68.3PS/L
54.8ft/sec 16.7m/sec
19.2mpg 16.0mpgUS
14.7L/100km
Petrol 4-stroke piston
3562cc 217.3cu in
Vee 8 fuel injection
Compression ratio: 10.6:1
Bore: 81.0mm 3.2in
Stroke: 86.4mm 3.4in
Valve type/No: Overhead 32
Transmission: Manual
No. of forward speeds: 5
Wheels driven: 4-wheel drive
Springs F/R: Coil/Coil
Brake system: ABS
Brakes F/R: Disc/Disc
Steering: Rack & pinion PA
Wheelbase: 270.3cm 106.4in
Track F: 151.4cm 59.6in
Track R: 153.2cm 60.3in
Length: 487.4cm 191.9in
Width: 181.4cm 71.4in
Height: 142.0cm 55.9in
Kerb weight: 1772.9kg
3905.0lb
Fuel: 79.9L 17.6gal 21.1galUS

69 Audi
80 2.8E V6 Quattro
1992 Germany
134.0mph 215.6kmh
0-50mph 80.5kmh: 6.2secs
0-60mph 96.5kmh: 8.3secs
0-1/4 mile: 16.5secs
0-1km: 30.2secs

174.0bhp 129.7kW 176.4PS
@ 5500rpm
184.0lbft 249.3Nm @ 3000rpm
121.6bhp/ton 119.6bhp/tonne
62.8bhp/L 46.8kW/L 63.7PS/L
51.9ft/sec 15.8m/sec
22.7mph 36.5kmh/1000rpm
20.0mpg 16.7mpgUS
14.1L/100km
Petrol 4-stroke piston
2771cc 169.1cu in
Vee 6 fuel injection
Compression ratio: 10.0:1
Bore: 82.5mm 3.2in
Stroke: 86.4mm 3.4in
Valve type/No: Overhead 12
Transmission: Manual
No. of forward speeds: 5
Wheels driven: 4-wheel drive
Springs F/R: Coil/Coil
Brake system: PA
Brakes F/R: Disc/Disc
Steering: Rack & pinion PA
Wheelbase: 261.4cm 102.9in
Track F: 144.8cm 57.0in
Track R: 147.3cm 58.0in
Length: 448.3cm 176.5in
Width: 193.8cm 76.3in
Height: 140.7cm 55.4in
Ground clearance: 13.5cm
5.3in
Kerb weight: 1455.1kg
3205.0lb
Fuel: 64.2L 14.1gal 16.9galUS

70 Austin
Montego 2.0 Vanden Plas EFI
1986 UK
112.0mph 180.2kmh
0-50mph 80.5kmh: 7.7secs
0-60mph 96.5kmh: 10.9secs
0-1/4 mile: 17.6secs

0-1km: 32.7secs
115.0bhp 85.8kW 116.6PS
@ 5500rpm
134.0lbft 181.6Nm @ 2800rpm
108.8bhp/ton 107.0bhp/tonne
57.7bhp/L 43.0kW/L 58.5PS/L
53.5ft/sec 16.3m/sec
26.8mph 43.1kmh/1000rpm
27.9mpg 23.2mpgUS
10.1L/100km
Petrol 4-stroke piston
1994cc 121.7cu in
In-line 4 fuel injection
Compression ratio: 9.0:1
Bore: 84.5mm 3.3in
Stroke: 89.0mm 3.5in
Valve type/No: Overhead 8
Transmission: Manual
No. of forward speeds: 5
Wheels driven: Front
Springs F/R: Coil/Coil
Brake system: PA
Brakes F/R: Disc/Drum
Steering: Rack & pinion PA
Wheelbase: 256.5cm 101.0in
Track F: 144.0cm 56.7in
Track R: 145.8cm 57.4in
Length: 446.8cm 175.9in
Width: 170.9cm 67.3in
Height: 142.0cm 55.9in
Ground clearance: 15.7cm
6.2in
Kerb weight: 1074.6kg
2367.0lb
Fuel: 50.0L 11.0gal 13.2galUS

71 Austin
Metro 1.3L 3-door
1987 UK
95.0mph 152.9kmh
0-50mph 80.5kmh: 8.5secs
0-60mph 96.5kmh: 13.0secs

0-1/4 mile: 19.4secs
0-1km: 36.5secs
62.0bhp 46.2kW 62.9PS
@ 5300rpm
72.0lbft 97.6Nm @ 3200rpm
78.8bhp/ton 77.5bhp/tonne
48.6bhp/L 36.3kW/L 49.3PS/L
47.1ft/sec 14.4m/sec
18.5mph 29.8kmh/1000rpm
36.9mpg 30.7mpgUS
7.7L/100km
Petrol 4-stroke piston
1275cc 78.0cu in
In-line 4 1 Carburettor
Compression ratio: 9.7:1
Bore: 70.6mm 2.8in
Stroke: 81.3mm 3.2in
Valve type/No: Overhead 8
Transmission: Manual
No. of forward speeds: 4
Wheels driven: Front
Springs F/R: Gas/Gas
Brake system: PA
Brakes F/R: Disc/Drum
Steering: Rack & pinion
Wheelbase: 225.0cm 88.6in
Track F: 127.3cm 50.1in
Track R: 127.5cm 50.2in
Length: 340.4cm 134.0in
Width: 154.7cm 60.9in
Height: 135.9cm 53.5in
Ground clearance: 16.5cm
6.5in
Kerb weight: 800.0kg 1762.1lb
Fuel: 35.5L 7.8gal 9.4galUS

72 Austin
Metro GTa
1989 UK
104.0mph 167.3kmh
0-50mph 80.5kmh: 8.2secs
0-60mph 96.5kmh: 11.8secs

71 Austin
Metro 1.3L 3-door

73 Autokraft
AC Cobra Mk IV

74 Bentley
Mulsanne Turbo

0-1/4 mile: 18.6secs
0-1km: 34.7secs
73.0bhp 54.4kW 74.0PS
@ 6000rpm
73.1lbft 99.0Nm @ 4000rpm
89.4bhp/ton 87.9bhp/tonne
57.2bhp/L 42.7kW/L 58.0PS/L
53.3ft/sec 16.3m/sec
17.6mph 28.3kmh/1000rpm
30.7mpg 25.6mpgUS
9.2L/100km
Petrol 4-stroke piston
1275cc 78.0cu in
In-line 4 1 Carburettor
Compression ratio: 10.5:1
Bore: 70.6mm 2.8in
Stroke: 81.3mm 3.2in
Valve type/No: Overhead 8
Transmission: Manual
No. of forward speeds: 4
Wheels driven: Front
Springs F/R: Gas/Gas

Brake system: PA
Brakes F/R: Disc/Drum
Steering: Rack & pinion
Wheelbase: 225.0cm 88.6in
Track F: 127.3cm 50.1in
Track R: 127.5cm 50.2in
Length: 340.6cm 134.1in
Width: 156.0cm 61.4in
Height: 138.0cm 54.3in
Ground clearance: 16.5cm
6.5in
Kerb weight: 830.0kg 1828.2lb
Fuel: 35.5L 7.8gal 9.4galUS

73 Autokraft
AC Cobra Mk IV
1987 UK
129.0mph 207.6kmh
0-60mph 96.5kmh: 5.0secs
0-1/4 mile: 13.5secs
300.0bhp 223.7kW 304.2PS
@ 6000rpm

285.0lbft 386.2Nm @ 3800rpm
274.3bhp/ton 269.7bhp/tonne
52.1bhp/L 38.8kW/L 52.8PS/L
20.9mph 33.6kmh/1000rpm
Petrol 4-stroke piston
5763cc 351.6cu in
Vee 8 1 Carburettor
Compression ratio: 10.6:1
Valve type/No: Overhead 16
Transmission: Manual
No. of forward speeds: 5
Wheels driven: Rear
Springs F/R: Coil/Coil
Brake system: PA
Brakes F/R: Disc/Disc
Steering: Rack & pinion
Wheelbase: 228.6cm 90.0in
Track F: 142.2cm 56.0in
Track R: 152.4cm 60.0in
Length: 411.5cm 162.0in
Width: 172.7cm 68.0in
Height: 124.5cm 49.0in

Kerb weight: 1112.3kg
2450.0lb

74 Bentley
Mulsanne Turbo
1982 UK
135.0mph 217.2kmh
0-50mph 80.5kmh: 5.1secs
0-60mph 96.5kmh: 7.0secs
0-1/4 mile: 15.1secs
0-1km: 27.8secs
298.0bhp 222.2kW 302.1PS
@ 3800rpm
135.5bhp/ton 133.2bhp/tonne
44.1bhp/L 32.9kW/L 44.8PS/L
41.2ft/sec 12.5m/sec
29.9mph 48.1kmh/1000rpm
12.1mpg 10.1mpgUS
23.3L/100km
Petrol 4-stroke piston
6750cc 411.8cu in
turbocharged

Vee 8 1 Carburettor
Compression ratio: 8.0:1
Bore: 104.1mm 4.1in
Stroke: 99.1mm 3.9in
Valve type/No: Overhead 16
Transmission: Automatic
No. of forward speeds: 3
Wheels driven: Rear
Springs F/R: Coil/Coil
Brake system: PA
Brakes F/R: Disc/Disc
Steering: Rack & pinion PA
Wheelbase: 306.1cm 120.5in
Track F: 153.7cm 60.5in
Track R: 153.7cm 60.5in
Length: 526.8cm 207.4in
Width: 188.7cm 74.3in
Height: 148.6cm 58.5in
Ground clearance: 16.5cm
6.5in
Kerb weight: 2236.4kg
4926.0lb
Fuel: 106.9L 23.5gal
28.2galUS

75 Bentley
Mulsanne Turbo
1985 UK
134.0mph 215.6kmh
0-50mph 80.5kmh: 5.2secs
0-60mph 96.5kmh: 6.9secs
0-1/4 mile: 15.0secs
0-1km: 27.8secs
298.0bhp 222.2kW 302.1PS
@ 3800rpm
136.3bhp/ton 134.1bhp/tonne
44.1bhp/L 32.9kW/L 44.8PS/L
41.2ft/sec 12.5m/sec
28.7mph 46.2kmh/1000rpm
13.4mpg 11.2mpgUS
21.1L/100km
Petrol 4-stroke piston
6750cc 411.8cu in
turbocharged
Vee 8 1 Carburettor
Compression ratio: 8.0:1
Bore: 104.1mm 4.1in
Stroke: 99.1mm 3.9in
Valve type/No: Overhead 16
Transmission: Automatic
No. of forward speeds: 3
Wheels driven: Rear
Springs F/R: Coil/Coil
Brake system: PA
Brakes F/R: Disc/Disc
Steering: Rack & pinion PA
Wheelbase: 306.1cm 120.5in
Track F: 153.7cm 60.5in
Track R: 153.7cm 60.5in
Length: 526.8cm 207.4in
Width: 188.7cm 74.3in
Height: 148.6cm 58.5in
Ground clearance: 16.5cm
6.5in

Kerb weight: 2222.8kg
4896.0lb
Fuel: 108.1L 23.7gal
28.5galUS

76 Bentley
Turbo R
1987 UK
146.0mph 234.9kmh
0-50mph 80.5kmh: 5.2secs
0-60mph 96.5kmh: 7.0secs
0-1/4 mile: 15.4secs
0-1km: 28.3secs
328.0bhp 244.6kW 332.5PS
148.1bhp/ton 145.6bhp/tonne
48.6bhp/L 36.2kW/L 49.3PS/L
35.6mph 57.3kmh/1000rpm
14.2mpg 11.8mpgUS
19.9L/100km
Petrol 4-stroke piston
6750cc 412.0cu in
turbocharged
Vee 8 fuel injection
Compression ratio: 8.0:1
Bore: 104.1mm 4.1in
Stroke: 99.1mm 3.9in
Valve type/No: Overhead 16
Transmission: Automatic
No. of forward speeds: 3
Wheels driven: Rear
Springs F/R: Coil/Coil
Brake system: PA ABS
Brakes F/R: Disc/Disc
Steering: Rack & pinion PA
Wheelbase: 306.1cm 120.5in
Track F: 153.8cm 60.6in
Track R: 153.8cm 60.6in
Length: 526.8cm 207.4in
Width: 188.7cm 74.3in
Height: 148.6cm 58.5in
Ground clearance: 16.5cm
6.5in
Kerb weight: 2252.0kg
4960.3lb
Fuel: 108.3L 23.8gal
28.6galUS

77 Bentley
Turbo R
1990 UK
145.0mph 233.3kmh
0-60mph 96.5kmh: 6.7secs
Petrol 4-stroke piston
6750cc 411.8cu in
turbocharged
Vee 8 fuel injection
Compression ratio: 8.0:1
Bore: 104.1mm 4.1in
Stroke: 99.1mm 3.9in
Valve type/No: Overhead 16
Transmission: Automatic
No. of forward speeds: 3
Wheels driven: Rear
Springs F/R: Coil/Coil

Brake system: ABS
Brakes F/R: Disc/Disc
Steering: Rack & pinion PA
Wheelbase: 306.1cm 120.5in
Track F: 154.9cm 61.0in
Track R: 154.9cm 61.0in
Length: 526.8cm 207.4in
Width: 200.7cm 79.0in
Height: 148.6cm 58.5in
Kerb weight: 2392.6kg
5270.0lb
Fuel: 107.9L 23.7gal
28.5galUS

78 Bentley
Turbo R
1991 UK
128.0mph 206.0kmh
0-60mph 96.5kmh: 7.1secs
0-1/4 mile: 15.5secs
315.0bhp 234.9kW 319.4PS
@ 4300rpm
485.0lbft 657.2Nm @ 2250rpm
135.3bhp/ton 133.0bhp/tonne
46.7bhp/L 34.8kW/L 47.3PS/L
46.6ft/sec 14.2m/sec
13.8mpg 11.5mpgUS
20.5L/100km
Petrol 4-stroke piston
6750cc 411.8cu in
turbocharged
Vee 8 fuel injection
Compression ratio: 8.0:1
Bore: 104.0mm 4.1in
Stroke: 99.1mm 3.9in
Valve type/No: Overhead 16
Transmission: Automatic
No. of forward speeds: 3
Wheels driven: Rear
Springs F/R: Coil/Coil
Brake system: PA ABS
Brakes F/R: Disc/Disc
Steering: Rack & pinion PA
Wheelbase: 306.1cm 120.5in
Track F: 154.9cm 61.0in
Track R: 154.9cm 61.0in
Length: 526.8cm 207.4in
Width: 189.0cm 74.4in
Height: 148.6cm 58.5in
Kerb weight: 2367.6kg
5215.0lb
Fuel: 107.9L 23.7gal
28.5galUS

79 Bertone
Emotion Lotus
1991 Italy
170.0mph 273.5kmh
0-60mph 96.5kmh: 5.0secs
264.0bhp 196.9kW 267.7PS
@ 6500rpm
261.0lbft 353.7Nm @ 3900rpm
121.4bhp/L 90.6kW/L
123.1PS/L

54.2ft/sec 16.5m/sec
Petrol 4-stroke piston
2174cc 132.6cu in
turbocharged
In-line 4 fuel injection
Compression ratio: 8.0:1
Bore: 95.3mm 3.8in
Stroke: 76.2mm 3.0in
Valve type/No: Overhead 16
Transmission: Manual
No. of forward speeds: 5
Wheels driven: Rear
Springs F/R: Coil/Coil
Brake system: PA ABS
Brakes F/R: Disc/Disc
Steering: Rack & pinion
Wheelbase: 245.1cm 96.5in
Track F: 157.0cm 61.8in
Track R: 159.0cm 62.6in
Length: 406.4cm 160.0in
Width: 190.0cm 74.8in
Height: 109.7cm 43.2in
Fuel: 70.0L 15.4gal 18.5galUS

80 Bitter
SC
1986 Germany
130.0mph 209.2kmh
0-50mph 80.5kmh: 6.1secs
0-60mph 96.5kmh: 8.3secs
0-1/4 mile: 16.4secs
200.0bhp 149.1kW 202.8PS
@ 5100rpm
236.0lbft 319.8Nm @ 3400rpm
128.4bhp/ton 126.2bhp/tonne
52.0bhp/L 38.8kW/L 52.7PS/L
50.4ft/sec 15.4m/sec
25.2mph 40.5kmh/1000rpm
16.8mpg 14.0mpgUS
16.8L/100km
Petrol 4-stroke piston
3848cc 234.8cu in
In-line 6 fuel injection
Compression ratio: 9.5:1
Bore: 95.0mm 3.7in
Stroke: 90.5mm 3.6in
Valve type/No: Overhead 12
Transmission: Manual
No. of forward speeds: 5
Wheels driven: Rear
Springs F/R: Coil/Coil
Brake system: PA
Brakes F/R: Disc/Disc
Steering: Recirculating ball PA
Wheelbase: 268.2cm 105.6in
Track F: 144.3cm 56.8in
Track R: 147.3cm 58.0in
Length: 491.0cm 193.3in
Width: 182.1cm 71.7in
Height: 134.9cm 53.1in
Ground clearance: 10.9cm
4.3in
Kerb weight: 1584.5kg
3490.0lb

Fuel: 73.0L 16.1gal 19.3galUS

81 Bitter

Type 3 Cabrio
1990 Germany
141.0mph 226.9kmh
0-60mph 96.5kmh: 7.6secs
240.0bhp 179.0kW 243.3PS
@ 5800rpm
235.0lbft 318.4Nm @ 4200rpm
180.1bhp/ton 177.1bhp/tonne
80.8bhp/L 60.3kW/L 82.0PS/L
44.3ft/sec 13.5m/sec
Petrol 4-stroke piston
2969cc 181.1cu in
In-line 6 fuel injection
Compression ratio: 10.0:1
Bore: 95.0mm 3.7in
Stroke: 69.8mm 2.7in
Valve type/No: Overhead 24
Transmission: Manual
No. of forward speeds: 5
Wheels driven: Rear
Springs F/R: Coil/Coil
Brake system: PA ABS
Brakes F/R: Disc/Disc
Steering: Rack & pinion PA
Wheelbase: 236.2cm 93.0in
Track F: 148.3cm 58.4in
Track R: 150.6cm 59.3in
Length: 452.1cm 178.0in
Width: 176.8cm 69.6in
Height: 141.7cm 55.8in
Kerb weight: 1355.2kg
2985.0lb
Fuel: 74.9L 16.5gal 19.8galUS

82 BMW

325 ES
1986 Germany
124.0mph 199.5kmh
0-50mph 80.5kmh: 6.0secs
0-60mph 96.5kmh: 9.0secs
0-1/4 mile: 16.7secs
121.0bhp 90.2kW 122.7PS
@ 4250rpm
170.0lbft 230.4Nm @ 3250rpm
97.3bhp/ton 95.7bhp/tonne
44.9bhp/L 33.5kW/L 45.6PS/L
37.7ft/sec 11.5m/sec
33.6mpg 28.0mpgUS
8.4L/100km
Petrol 4-stroke piston
2693cc 164.3cu in
In-line 6 fuel injection
Compression ratio: 9.0:1
Bore: 84.0mm 3.3in
Stroke: 81.0mm 3.2in
Valve type/No: Overhead 12
Transmission: Manual
No. of forward speeds: 5
Wheels driven: Rear
Springs F/R: Coil/Coil
Brake system: PA ABS

Brakes F/R: Disc/Disc
Steering: Rack & pinion PA
Wheelbase: 257.0cm 101.2in
Track F: 140.7cm 55.4in
Track R: 141.5cm 55.7in
Length: 446.0cm 175.6in
Width: 164.6cm 64.8in
Height: 137.9cm 54.3in
Kerb weight: 1264.4kg
2785.0lb
Fuel: 54.9L 12.1gal 14.5galUS

83 BMW

325i Convertible
1986 Germany
135.0mph 217.2kmh
0-50mph 80.5kmh: 5.8secs
0-60mph 96.5kmh: 8.1secs
0-1/4 mile: 16.1secs
0-1km: 29.4secs
171.0bhp 127.5kW 173.4PS
@ 5800rpm
167.0lbft 226.3Nm @ 4000rpm
148.8bhp/ton 146.3bhp/tonne
68.6bhp/L 51.1kW/L 69.5PS/L
47.5ft/sec 14.5m/sec
23.0mph 37.0kmh/1000rpm
26.5mpg 22.1mpgUS
10.7L/100km
Petrol 4-stroke piston
2494cc 152.0cu in
In-line 6 fuel injection
Compression ratio: 9.7:1
Bore: 84.0mm 3.3in
Stroke: 75.0mm 2.9in
Valve type/No: Side 12
Transmission: Manual
No. of forward speeds: 5
Wheels driven: Rear
Springs F/R: Coil/Coil
Brake system: PA
Brakes F/R: Disc/Disc
Steering: Rack & pinion PA
Wheelbase: 257.0cm 101.2in
Track F: 140.7cm 55.4in
Track R: 141.5cm 55.7in
Length: 432.5cm 170.3in
Width: 164.5cm 64.8in
Height: 138.0cm 54.3in
Ground clearance: 17.8cm
7.0in
Kerb weight: 1169.0kg
2574.9lb
Fuel: 55.1L 12.1gal 14.5galUS

84 BMW

520i
1986 Germany
121.0mph 194.7kmh
0-50mph 80.5kmh: 7.2secs
0-60mph 96.5kmh: 10.5secs
0-1/4 mile: 17.6secs
0-1km: 32.7secs
129.0bhp 96.2kW 130.8PS

@ 5800rpm
128.0lbft 173.4Nm @ 4500rpm
99.8bhp/ton 98.1bhp/tonne
64.8bhp/L 48.3kW/L 65.7PS/L
41.9ft/sec 12.8m/sec
21.5mph 34.6kmh/1000rpm
23.8mpg 19.8mpgUS
11.9L/100km
Petrol 4-stroke piston
1990cc 121.0cu in
In-line 6 fuel injection
Compression ratio: 9.8:1
Bore: 80.0mm 3.1in
Stroke: 66.0mm 2.6in
Valve type/No: Overhead 12
Transmission: Manual
No. of forward speeds: 5
Wheels driven: Rear
Springs F/R: Coil/Coil
Brake system: PA
Brakes F/R: Disc/Drum
Steering: Recirculating ball PA
Wheelbase: 262.5cm 103.3in
Track F: 143.0cm 56.3in
Track R: 146.0cm 57.5in
Length: 462.0cm 181.9in
Width: 170.0cm 66.9in
Height: 141.5cm 55.7in
Ground clearance: 15.2cm
6.0in
Kerb weight: 1315.0kg
2896.5lb
Fuel: 70.1L 15.4gal 18.5galUS

85 BMW

524 TD
1986 Germany
102.0mph 164.1kmh
0-60mph 96.5kmh: 11.0secs
0-1/4 mile: 18.1secs
114.0bhp 85.0kW 115.6PS
@ 4800rpm
155.0lbft 210.0Nm @ 2400rpm
78.7bhp/ton 77.4bhp/tonne
46.7bhp/L 34.8kW/L 47.3PS/L
42.5ft/sec 13.0m/sec
30.6mpg 25.5mpgUS
9.2L/100km
Diesel 4-stroke piston
2443cc 149.0cu in
turbocharged
In-line 6 fuel injection
Compression ratio: 22.0:1
Bore: 80.0mm 3.1in
Stroke: 81.0mm 3.2in
Valve type/No: Overhead 12
Transmission: Automatic
No. of forward speeds: 4
Wheels driven: Rear
Springs F/R: Coil/Coil
Brake system: PA
Brakes F/R: Disc/Disc
Steering: Recirculating ball PA
Wheelbase: 262.4cm 103.3in

Track F: 143.0cm 56.3in
Track R: 146.1cm 57.5in
Length: 480.1cm 189.0in
Width: 169.9cm 66.9in
Height: 141.5cm 55.7in
Kerb weight: 1473.2kg
3245.0lb
Fuel: 62.8L 13.8gal 16.6galUS

86 BMW

535i
1986 Germany
131.0mph 210.8kmh
0-60mph 96.5kmh: 7.9secs
0-1/4 mile: 16.1secs
182.0bhp 135.7kW 184.5PS
@ 5400rpm
214.0lbft 290.0Nm @ 4000rpm
126.0bhp/ton 123.9bhp/tonne
53.1bhp/L 39.6kW/L 53.8PS/L
50.8ft/sec 15.5m/sec
20.4mpg 17.0mpgUS
13.8L/100km
Petrol 4-stroke piston
3430cc 209.3cu in
In-line 6 fuel injection
Compression ratio: 8.0:1
Bore: 92.0mm 3.6in
Stroke: 86.0mm 3.4in
Valve type/No: Overhead 12
Transmission: Manual
No. of forward speeds: 5
Wheels driven: Rear
Springs F/R: Coil/Coil
Brake system: ABS
Brakes F/R: Disc/Disc
Steering: Recirculating ball PA
Wheelbase: 262.4cm 103.3in
Track F: 143.0cm 56.3in
Track R: 146.1cm 57.5in
Length: 480.1cm 189.0in
Width: 169.9cm 66.9in
Height: 141.5cm 55.7in
Kerb weight: 1468.7kg
3235.0lb
Fuel: 62.8L 13.8gal 16.6galUS

87 BMW

635 CSi
1986 Germany
131.0mph 210.8kmh
0-60mph 96.5kmh: 8.5secs
0-1/4 mile: 16.3secs
182.0bhp 135.7kW 184.5PS
@ 5400rpm
214.0lbft 290.0Nm @ 4000rpm
119.4bhp/ton 117.4bhp/tonne
53.1bhp/L 39.6kW/L 53.8PS/L
50.8ft/sec 15.5m/sec
21.0mpg 17.5mpgUS
13.4L/100km
Petrol 4-stroke piston
3430cc 209.3cu in
In-line 6 fuel injection

Bore: 92.0mm 3.6in
Stroke: 86.0mm 3.4in
Valve type/No: Overhead 12
Transmission: Manual
No. of forward speeds: 5
Wheels driven: Rear
Springs F/R: Coil/Coil
Brake system: ABS
Brakes F/R: Disc/Disc
Steering: Recirculating ball PA
Wheelbase: 262.9cm 103.5in
Track F: 143.0cm 56.3in
Track R: 147.1cm 57.9in
Length: 492.3cm 193.8in
Width: 172.5cm 67.9in
Height: 136.4cm 53.7in
Kerb weight: 1550.4kg
3415.0lb
Fuel: 62.8L 13.8gal 16.6galUS

88 BMW

735i
1986 Germany
119.0mph 191.5kmh
0-50mph 80.5kmh: 6.7secs
0-60mph 96.5kmh: 9.1secs
0-1/4 mile: 16.7secs
182.0bhp 135.7kW 184.5PS
@ 5400rpm
214.0lbft 290.0Nm @ 4000rpm
115.0bhp/ton 113.1bhp/tonne
53.1bhp/L 39.6kW/L 53.8PS/L
50.8ft/sec 15.5m/sec
20.4mpg 17.0mpgUS
13.8L/100km
Petrol 4-stroke piston
3430cc 209.3cu in
In-line 6 fuel injection
Compression ratio: 8.0:1
Bore: 92.0mm 3.6in
Stroke: 86.0mm 3.4in
Valve type/No: Overhead 12
Transmission: Manual
No. of forward speeds: 5
Wheels driven: Rear
Springs F/R: Coil/Coil
Brake system: ABS
Brakes F/R: Disc/Disc
Steering: Recirculating ball PA
Wheelbase: 279.4cm 110.0in
Track F: 150.1cm 59.1in
Track R: 151.6cm 59.7in
Length: 501.4cm 197.4in
Width: 180.1cm 70.9in
Height: 143.0cm 56.3in
Kerb weight: 1609.4kg
3545.0lb
Fuel: 85.2L 18.7gal 22.5galUS

89 BMW

735i Automatic
1986 Germany
147.0mph 236.5kmh
0-50mph 80.5kmh: 6.7secs

0-60mph 96.5kmh: 9.0secs
0-1/4 mile: 16.7secs
0-1km: 30.1secs
217.0bhp 161.8kW 220.0PS
@ 5700rpm
232.0lbft 314.4Nm @ 4000rpm
125.3bhp/ton 123.2bhp/tonne
63.3bhp/L 47.2kW/L 64.1PS/L
53.7ft/sec 16.3m/sec
28.3mph 45.5kmh/1000rpm
17.5mpg 14.6mpgUS
16.1L/100km
Petrol 4-stroke piston
3430cc 209.0cu in
In-line 6 fuel injection
Compression ratio: 9.2:1
Bore: 92.0mm 3.6in
Stroke: 86.0mm 3.4in
Valve type/No: Overhead 12
Transmission: Automatic
No. of forward speeds: 4
Wheels driven: Rear
Springs F/R: Coil/Coil
Brake system: PA ABS
Brakes F/R: Disc/Disc
Steering: Recirculating ball PA
Wheelbase: 283.2cm 111.5in
Track F: 152.7cm 60.1in
Track R: 155.0cm 61.0in
Length: 491.0cm 193.3in
Width: 184.5cm 72.6in
Height: 141.1cm 55.6in
Ground clearance: 20.3cm
8.0in
Kerb weight: 1761.0kg
3878.8lb
Fuel: 90.1L 19.8gal 23.8galUS

90 BMW

318i 2-door
1987 Germany
116.0mph 186.6kmh
0-50mph 80.5kmh: 6.6secs
0-60mph 96.5kmh: 9.3secs
0-1/4 mile: 16.9secs
0-1km: 31.8secs
115.0bhp 85.8kW 116.6PS
@ 5500rpm
122.0lbft 165.3Nm @ 4250rpm
107.8bhp/ton 106.0bhp/tonne
64.1bhp/L 47.8kW/L 65.0PS/L
48.7ft/sec 14.8m/sec
20.8mph 33.5kmh/1000rpm
27.4mpg 22.8mpgUS
10.3L/100km
Petrol 4-stroke piston
1795cc 110.0cu in
In-line 4 fuel injection
Compression ratio: 8.8:1
Bore: 84.0mm 3.3in
Stroke: 81.0mm 3.2in
Valve type/No: Overhead 8
Transmission: Manual
No. of forward speeds: 5

Wheels driven: Rear
Springs F/R: Coil/Coil
Brake system: PA
Brakes F/R: Disc/Drum
Steering: Rack & pinion PA
Wheelbase: 257.0cm 101.2in
Track F: 140.7cm 55.4in
Track R: 141.5cm 55.7in
Length: 432.5cm 170.3in
Width: 164.5cm 64.8in
Height: 138.0cm 54.3in
Ground clearance: 15.5cm
6.1in
Kerb weight: 1085.0kg
2389.9lb
Fuel: 55.1L 12.1gal 14.5galUS

91 BMW

325i Cabriolet
1987 Germany
130.0mph 209.2kmh
0-60mph 96.5kmh: 7.5secs
168.0bhp 125.3kW 170.3PS
@ 5800rpm
164.0lbft 222.2Nm @ 4300rpm
124.8bhp/ton 122.7bhp/tonne
67.4bhp/L 50.2kW/L 68.3PS/L
47.5ft/sec 14.5m/sec
Petrol 4-stroke piston
2494cc 152.2cu in
In-line 6 fuel injection
Compression ratio: 8.8:1
Bore: 84.0mm 3.3in
Stroke: 75.0mm 2.9in
Valve type/No: Overhead 12
Transmission: Manual
No. of forward speeds: 5
Wheels driven: Rear
Springs F/R: Coil/Coil
Brake system: PA ABS
Brakes F/R: Disc/Disc
Steering: Rack & pinion PA
Wheelbase: 257.0cm 101.2in
Track F: 140.7cm 55.4in
Track R: 141.5cm 55.7in
Length: 446.0cm 175.6in
Width: 164.6cm 64.8in
Height: 136.9cm 53.9in
Kerb weight: 1368.8kg
3015.0lb
Fuel: 54.9L 12.1gal 14.5galUS

92 BMW

730i SE
1987 Germany
132.0mph 212.4kmh
0-50mph 80.5kmh: 7.9secs
0-60mph 96.5kmh: 10.5secs
0-1/4 mile: 17.7secs
0-1km: 32.0secs
197.0bhp 146.9kW 199.7PS
@ 5800rpm
199.0lbft 269.6Nm @ 4000rpm
117.8bhp/ton 115.9bhp/tonne

66.0bhp/L 49.2kW/L 66.9PS/L
50.7ft/sec 15.5m/sec
27.6mph 44.4kmh/1000rpm
20.0mpg 16.7mpgUS
14.1L/100km
Petrol 4-stroke piston
2986cc 182.0cu in
In-line 6 fuel injection
Compression ratio: 9.2:1
Bore: 89.0mm 3.5in
Stroke: 80.0mm 3.1in
Valve type/No: Overhead 12
Transmission: Automatic
No. of forward speeds: 4
Wheels driven: Rear
Springs F/R: Coil/Coil
Brake system: PA ABS
Brakes F/R: Disc/Disc
Steering: Recirculating ball PA
Wheelbase: 283.2cm 111.5in
Track F: 152.7cm 60.1in
Track R: 155.0cm 61.0in
Length: 491.0cm 193.3in
Width: 184.5cm 72.6in
Height: 141.1cm 55.6in
Ground clearance: 20.3cm
8.0in
Kerb weight: 1700.0kg
3744.5lb
Fuel: 90.1L 19.8gal 23.8galUS

93 BMW

750i L
1987 Germany
158.0mph 254.2kmh
0-50mph 80.5kmh: 5.8secs
0-60mph 96.5kmh: 7.7secs
0-1/4 mile: 15.9secs
0-1km: 28.1secs
300.0bhp 223.7kW 304.2PS
@ 5200rpm
332.0lbft 449.9Nm @ 4100rpm
159.3bhp/ton 156.7bhp/tonne
60.1bhp/L 44.8kW/L 61.0PS/L
42.6ft/sec 13.0m/sec
32.1mph 51.6kmh/1000rpm
18.5mpg 15.4mpgUS
15.3L/100km
Petrol 4-stroke piston
4988cc 304.0cu in
Vee 12 fuel injection
Compression ratio: 8.8:1
Bore: 84.0mm 3.3in
Stroke: 75.0mm 2.9in
Valve type/No: Overhead 48
Transmission: Automatic
No. of forward speeds: 4
Wheels driven: Rear
Springs F/R: Coil/Coil
Brake system: PA
Brakes F/R: Disc/Disc
Steering: Recirculating ball PA
Wheelbase: 294.7cm 116.0in
Track F: 152.8cm 60.2in

94 BMW
M3

Track R: 155.6cm 61.3in
Length: 502.4cm 197.8in
Width: 184.5cm 72.6in
Height: 140.0cm 55.1in
Ground clearance: 20.3cm
8.0in
Kerb weight: 1915.0kg
4218.1lb
Fuel: 102.4L 22.5gal
27.0galUS

94 BMW
M3
1987 Germany
140.0mph 225.3kmh
0-50mph 80.5kmh: 5.5secs
0-60mph 96.5kmh: 7.1secs
0-1/4 mile: 15.7secs
0-1km: 28.3secs
200.0bhp 149.1kW 202.8PS
@ 6750rpm
177.0lbft 239.8Nm @ 4750rpm
162.4bhp/ton 159.7bhp/tonne
86.9bhp/L 64.8kW/L 88.1PS/L
62.1ft/sec 18.9m/sec
21.2mph 34.1kmh/1000rpm
20.3mpg 16.9mpgUS
13.9L/100km
Petrol 4-stroke piston
2302cc 140.0cu in
In-line 4 fuel injection
Compression ratio: 10.5:1
Bore: 93.4mm 3.7in
Stroke: 84.0mm 3.3in
Valve type/No: Overhead 16
Transmission: Manual
No. of forward speeds: 5
Wheels driven: Rear
Springs F/R: Coil/Coil
Brake system: PA
Brakes F/R: Disc/Disc
Steering: Rack & pinion PA
Wheelbase: 256.2cm 100.9in

Track F: 141.2cm 55.6in
Track R: 143.3cm 56.4in
Length: 434.5cm 171.1in
Width: 168.0cm 66.1in
Height: 136.5cm 53.7in
Ground clearance: 15.3cm
6.0in
Kerb weight: 1252.0kg
2757.7lb
Fuel: 70.1L 15.4gal 18.5galUS

95 BMW
325i S
1988 Germany
133.0mph 214.0kmh
0-50mph 80.5kmh: 5.6secs
0-60mph 96.5kmh: 7.5secs
0-1/4 mile: 15.7secs
168.0bhp 125.3kW 170.3PS
@ 5800rpm
164.0lbft 222.2Nm @ 4300rpm
133.2bhp/ton 131.0bhp/tonne
67.4bhp/L 50.2kW/L 68.3PS/L
47.5ft/sec 14.5m/sec
24.6mpg 20.5mpgUS
11.5L/100km
Petrol 4-stroke piston
2494cc 152.2cu in
In-line 6 fuel injection
Compression ratio: 8.8:1
Bore: 84.0mm 3.3in
Stroke: 75.0mm 2.9in
Valve type/No: Overhead 12
Transmission: Manual
No. of forward speeds: 5
Wheels driven: Rear
Springs F/R: Coil/Coil
Brake system: PA ABS
Brakes F/R: Disc/Disc
Steering: Rack & pinion PA
Wheelbase: 257.0cm 101.2in
Track F: 140.7cm 55.4in
Track R: 141.5cm 55.7in

Length: 445.0cm 175.2in
Width: 164.6cm 64.8in
Height: 137.9cm 54.3in
Kerb weight: 1282.5kg
2825.0lb
Fuel: 62.1L 13.6gal 16.4galUS

96 BMW
325i X
1988 Germany
127.0mph 204.3kmh
0-50mph 80.5kmh: 5.7secs
0-60mph 96.5kmh: 7.9secs
0-1/4 mile: 15.9secs
168.0bhp 125.3kW 170.3PS
@ 5800rpm
164.0lbft 222.2Nm @ 4300rpm
126.3bhp/ton 124.2bhp/tonne
67.4bhp/L 50.2kW/L 68.3PS/L
47.5ft/sec 14.5m/sec
24.0mpg 20.0mpgUS
11.8L/100km
Petrol 4-stroke piston
2494cc 152.2cu in
In-line 6 fuel injection
Compression ratio: 8.8:1
Bore: 84.0mm 3.3in
Stroke: 75.0mm 2.9in
Valve type/No: Overhead 12
Transmission: Manual
No. of forward speeds: 5
Wheels driven: 4-wheel drive
Springs F/R: Coil/Coil
Brake system: PA ABS
Brakes F/R: Disc/Disc
Steering: Rack & pinion PA
Wheelbase: 257.0cm 101.2in
Track F: 142.0cm 55.9in
Track R: 141.5cm 55.7in
Length: 445.0cm 175.2in
Width: 166.1cm 65.4in
Height: 140.0cm 55.1in
Kerb weight: 1352.9kg

2980.0lb
Fuel: 62.1L 13.6gal 16.4galUS

97 BMW
535i SE
1988 Germany
145.0mph 233.3kmh
0-50mph 80.5kmh: 5.6secs
0-60mph 96.5kmh: 7.4secs
0-1/4 mile: 15.9secs
0-1km: 28.9secs
211.0bhp 157.3kW 213.9PS
@ 5700rpm
225.1lbft 305.0Nm @ 4000rpm
135.2bhp/ton 133.0bhp/tonne
61.5bhp/L 45.9kW/L 62.4PS/L
53.7ft/sec 16.3m/sec
25.1mph 40.4kmh/1000rpm
22.3mpg 18.6mpgUS
12.7L/100km
Petrol 4-stroke piston
3430cc 209.0cu in
In-line 6 fuel injection
Compression ratio: 9.0:1
Bore: 92.0mm 3.6in
Stroke: 86.0mm 3.4in
Valve type/No: Overhead 12
Transmission: Manual
No. of forward speeds: 5
Wheels driven: Rear
Springs F/R: Coil/Coil
Brake system: PA ABS
Brakes F/R: Disc/Disc
Steering: Recirculating ball PA
Wheelbase: 276.1cm 108.7in
Track F: 146.6cm 57.7in
Track R: 148.7cm 58.5in
Length: 472.0cm 185.8in
Width: 175.1cm 68.9in
Height: 141.2cm 55.6in
Ground clearance: 17.9cm
7.0in
Kerb weight: 1587.0kg

3495.6lb
Fuel: 80.1L 17.6gal 21.1galUS

98 BMW

635 CSi
1988 Germany
135.0mph 217.2kmh
0-50mph 80.5kmh: 6.3secs
0-60mph 96.5kmh: 8.4secs
0-1/4 mile: 16.7secs
0-1km: 29.4secs
220.0bhp 164.0kW 223.0PS
@ 5700rpm
232.5lbft 315.0Nm @ 4000rpm
142.0bhp/ton 139.6bhp/tonne
64.1bhp/L 47.8kW/L 65.0PS/L
53.7ft/sec 16.3m/sec
27.3mph 43.9kmh/1000rpm
17.2mpg 14.3mpgUS
16.4L/100km
Petrol 4-stroke piston
3430cc 209.0cu in
In-line 6 fuel injection
Compression ratio: 9.2:1
Bore: 92.0mm 3.6in
Stroke: 86.0mm 3.4in
Valve type/No: Overhead 12
Transmission: Automatic
No. of forward speeds: 4
Wheels driven: Rear
Springs F/R: Coil/Coil
Brake system: PA ABS
Brakes F/R: Disc/Disc
Steering: Ball & nut PA
Wheelbase: 262.5cm 103.3in
Track F: 143.0cm 56.3in
Track R: 146.0cm 57.5in
Length: 481.5cm 189.6in
Width: 172.5cm 67.9in
Height: 136.5cm 53.7in
Ground clearance: 9.5cm
3.7in
Kerb weight: 1576.0kg 3471.4lb
Fuel: 70.1L 15.4gal 18.5galUS

99 BMW

735i
1988 Germany
137.0mph 220.4kmh
0-50mph 80.5kmh: 7.2secs
0-60mph 96.5kmh: 9.3secs
0-1/4 mile: 17.0secs
208.0bhp 155.1kW 210.9PS
@ 5700rpm
225.0lbft 304.9Nm @ 4000rpm
120.1bhp/ton 118.1bhp/tonne
60.6bhp/L 45.2kW/L 61.5PS/L
53.7ft/sec 16.3m/sec
21.6mpg 18.0mpgUS
13.1L/100km
Petrol 4-stroke piston
3430cc 209.3cu in
In-line 6 fuel injection
Compression ratio: 9.0:1

Bore: 92.0mm 3.6in
Stroke: 86.0mm 3.4in
Valve type/No: Overhead 12
Transmission: Automatic
No. of forward speeds: 4
Wheels driven: Rear
Springs F/R: Coil/Coil
Brake system: ABS
Brakes F/R: Disc/Disc
Steering: Recirculating ball PA
Wheelbase: 283.2cm 111.5in
Track F: 152.9cm 60.2in
Track R: 156.0cm 61.4in
Length: 491.0cm 193.3in
Width: 184.4cm 72.6in
Height: 141.2cm 55.6in
Ground clearance: 17.8cm
7.0in
Kerb weight: 1761.5kg
3880.0lb
Fuel: 81.4L 17.9gal 21.5galUS

100 BMW

750i L
1988 Germany
155.0mph 249.4kmh
0-50mph 80.5kmh: 5.2secs
0-60mph 96.5kmh: 6.9secs
0-1/4 mile: 15.2secs
300.0bhp 223.7kW 304.2PS
@ 5200rpm
332.0lbft 449.9Nm @ 4100rpm
158.7bhp/ton 156.0bhp/tonne
60.1bhp/L 44.8kW/L 61.0PS/L
42.6ft/sec 13.0m/sec
19.2mpg 16.0mpgUS
14.7L/100km
Petrol 4-stroke piston
4988cc 304.3cu in
Vee 12 fuel injection
Compression ratio: 8.8:1
Bore: 84.0mm 3.3in
Stroke: 75.0mm 2.9in
Valve type/No: Overhead 24
Transmission: Automatic
No. of forward speeds: 4
Wheels driven: Rear
Springs F/R: Coil/Coil
Brake system: ABS
Brakes F/R: Disc/Disc
Steering: Recirculating ball PA
Wheelbase: 294.6cm 116.0in
Track F: 152.9cm 60.2in
Track R: 155.7cm 61.3in
Length: 502.4cm 197.8in
Width: 184.4cm 72.6in
Height: 140.0cm 55.1in
Kerb weight: 1922.7kg
4235.0lb
Fuel: 90.8L 20.0gal 24.0galUS

101 BMW

M6
1988 Germany

147.0mph 236.5kmh
0-50mph 80.5kmh: 5.3secs
0-60mph 96.5kmh: 7.2secs
0-1/4 mile: 15.3secs
256.0bhp 190.9kW 259.5PS
@ 6500rpm
243.0lbft 329.3Nm @ 4500rpm
160.8bhp/ton 158.2bhp/tonne
74.1bhp/L 55.3kW/L 75.2PS/L
59.8ft/sec 18.2m/sec
15.0mpg 12.5mpgUS
18.8L/100km
Petrol 4-stroke piston
3453cc 210.7cu in
In-line 6 fuel injection
Compression ratio: 9.8:1
Bore: 93.4mm 3.7in
Stroke: 84.0mm 3.3in
Valve type/No: Overhead 24
Transmission: Manual
No. of forward speeds: 5
Wheels driven: Rear
Springs F/R: Coil/Coil
Brake system: ABS
Brakes F/R: Disc/Disc
Steering: Recirculating ball PA
Wheelbase: 262.4cm 103.3in
Track F: 143.0cm 56.3in
Track R: 146.3cm 57.6in
Length: 492.3cm 193.8in
Width: 172.5cm 67.9in
Height: 135.4cm 53.3in
Kerb weight: 1618.5kg
3565.0lb
Fuel: 62.8L 13.8gal 16.6galUS

102 BMW

316i 4-door
1989 Germany
114.0mph 183.4kmh
0-50mph 80.5kmh: 7.1secs
0-60mph 96.5kmh: 10.1secs
0-1/4 mile: 17.4secs
0-1km: 32.4secs
102.0bhp 76.1kW 103.4PS
@ 5500rpm
105.5lbft 143.0Nm @ 4250rpm
93.5bhp/ton 92.0bhp/tonne
63.9bhp/L 47.7kW/L 64.8PS/L
43.2ft/sec 13.2m/sec
20.8mph 33.5kmh/1000rpm
25.3mpg 21.1mpgUS
11.2L/100km
Petrol 4-stroke piston
1596cc 97.0cu in
In-line 4 fuel injection
Compression ratio: 9.0:1
Bore: 84.0mm 3.3in
Stroke: 72.0mm 2.8in
Valve type/No: Overhead 8
Transmission: Manual
No. of forward speeds: 5
Wheels driven: Rear
Springs F/R: Coil/Coil

Brake system: PA ABS
Brakes F/R: Disc/Drum
Steering: Rack & pinion PA
Wheelbase: 257.0cm 101.2in
Track F: 140.7cm 55.4in
Track R: 141.5cm 55.7in
Length: 432.5cm 170.3in
Width: 164.5cm 64.8in
Height: 138.0cm 54.3in
Ground clearance: 15.3cm
6.0in
Kerb weight: 1109.0kg
2442.7lb
Fuel: 55.1L 12.1gal 14.5galUS

103 BMW

535i
1989 Germany
143.0mph 230.1kmh
0-50mph 80.5kmh: 5.8secs
0-60mph 96.5kmh: 7.7secs
0-1/4 mile: 15.9secs
208.0bhp 155.1kW 210.9PS
@ 5700rpm
225.0lbft 304.9Nm @ 4000rpm
128.0bhp/ton 125.9bhp/tonne
60.6bhp/L 45.2kW/L 61.5PS/L
53.7ft/sec 16.3m/sec
20.4mpg 17.0mpgUS
13.8L/100km
Petrol 4-stroke piston
3430cc 209.3cu in
In-line 6 fuel injection
Compression ratio: 9.0:1
Bore: 92.0mm 3.6in
Stroke: 86.0mm 3.4in
Valve type/No: Overhead 12
Transmission: Manual
No. of forward speeds: 5
Wheels driven: Rear
Springs F/R: Coil/Coil
Brake system: PA ABS
Brakes F/R: Disc/Disc
Steering: Recirculating ball PA
Wheelbase: 276.1cm 108.7in
Track F: 146.6cm 57.7in
Track R: 148.6cm 58.5in
Length: 471.9cm 185.8in
Width: 175.0cm 68.9in
Height: 141.2cm 55.6in
Ground clearance: 17.0cm
6.7in
Kerb weight: 1652.6kg
3640.0lb
Fuel: 79.9L 17.6gal 21.1galUS

104 BMW

535i Automatic
1989 Germany
143.0mph 230.1kmh
0-60mph 96.5kmh: 8.6secs
0-1/4 mile: 16.5secs
208.0bhp 155.1kW 210.9PS
@ 5700rpm

225.0lbft 304.9Nm @ 4000rpm
126.6bhp/ton 124.5bhp/tonne
60.6bhp/L 45.2kW/L 61.5PS/L
53.7ft/sec 16.3m/sec
22.2mpg 18.5mpgUS
12.7L/100km
Petrol 4-stroke piston
3430cc 209.3cu in
In-line 6 fuel injection
Compression ratio: 9.0:1
Bore: 92.0mm 3.6in
Stroke: 86.0mm 3.4in
Valve type/No: Overhead 12
Transmission: Automatic
No. of forward speeds: 4
Wheels driven: Rear
Springs F/R: Coil/Coil
Brake system: PA ABS
Brakes F/R: Disc/Disc
Steering: Recirculating ball PA
Wheelbase: 276.1cm 108.7in
Track F: 146.6cm 57.7in
Track R: 148.6cm 58.5in
Length: 471.9cm 185.8in
Width: 175.0cm 68.9in
Height: 141.2cm 55.6in
Kerb weight: 1670.7kg
3680.0lb
Fuel: 79.9L 17.6gal 21.1galUS

105 BMW

735i Alpina B11
1989 Germany
154.0mph 247.8kmh
0-60mph 96.5kmh: 7.4secs
0-1/4 mile: 15.5secs
250.0bhp 186.4kW 253.5PS
@ 5700rpm
243.0lbft 329.3Nm @ 4000rpm
146.0bhp/ton 143.6bhp/tonne
72.9bhp/L 54.3kW/L 73.9PS/L
53.7ft/sec 16.3m/sec
Petrol 4-stroke piston
3430cc 209.3cu in
In-line 6 fuel injection
Compression ratio: 10.4:1
Bore: 92.0mm 3.6in
Stroke: 86.0mm 3.4in
Valve type/No: Overhead 12
Transmission: Manual
No. of forward speeds: 5
Wheels driven: Rear
Springs F/R: Coil/Coil
Brake system: ABS
Brakes F/R: Disc/Disc
Steering: Recirculating ball PA
Wheelbase: 279.4cm 110.0in
Track F: 150.1cm 59.1in
Track R: 151.6cm 59.7in
Length: 501.4cm 197.4in
Width: 180.1cm 70.9in
Height: 140.5cm 55.3in
Kerb weight: 1741.1kg 3835.0lb
Fuel: 85.2L 18.7gal 22.5galUS

106 BMW

M3
1989 Germany
143.0mph 230.1kmh
0-60mph 96.5kmh: 7.1secs
0-1/4 mile: 15.4secs
192.0bhp 143.2kW 194.7PS
@ 6750rpm
170.0lbft 230.4Nm @ 4750rpm
150.1bhp/ton 147.6bhp/tonne
83.4bhp/L 62.2kW/L 84.6PS/L
62.1ft/sec 18.9m/sec
Petrol 4-stroke piston
2302cc 140.4cu in
In-line 4 fuel injection
Compression ratio: 10.5:1
Bore: 93.4mm 3.7in
Stroke: 84.0mm 3.3in
Valve type/No: Overhead 16
Transmission: Manual
No. of forward speeds: 5
Wheels driven: Rear
Springs F/R: Coil/Coil
Brake system: PA ABS
Brakes F/R: Disc/Disc
Steering: Recirculating ball PA
Wheelbase: 256.5cm 101.0in
Track F: 141.2cm 55.6in
Track R: 142.5cm 56.1in
Length: 434.6cm 171.1in
Width: 167.9cm 66.1in
Height: 136.9cm 53.9in
Kerb weight: 1300.7kg
2865.0lb
Fuel: 54.9L 12.1gal 14.5galUS

107 BMW

M3 Evolution
1989 Germany
148.0mph 238.1kmh
0-50mph 80.5kmh: 5.1secs
0-60mph 96.5kmh: 6.6secs
0-1/4 mile: 15.2secs
0-1km: 27.6secs
220.0bhp 164.0kW 223.0PS
@ 6750rpm
180.8lbft 245.0Nm @ 4750rpm
175.6bhp/ton 172.7bhp/tonne
95.6bhp/L 71.3kW/L 96.9PS/L
62.1ft/sec 18.9m/sec
22.2mpg 35.7kmh/1000rpm
26.0mpg 21.6mpgUS
10.9L/100km
Petrol 4-stroke piston
2302cc 140.0cu in
In-line 4 fuel injection
Compression ratio: 11.0:1
Bore: 93.4mm 3.7in
Stroke: 84.0mm 3.3in
Valve type/No: Overhead 16
Transmission: Manual
No. of forward speeds: 5
Wheels driven: Rear
Springs F/R: Coil/Coil

Brake system: PA
Brakes F/R: Disc/Disc
Steering: Rack & pinion PA
Wheelbase: 257.0cm 101.2in
Track F: 140.7cm 55.4in
Track R: 141.5cm 55.7in
Length: 432.5cm 170.3in
Width: 164.5cm 64.8in
Height: 138.0cm 54.3in
Ground clearance: 15.3cm
6.0in
Kerb weight: 1274.0kg
2806.2lb
Fuel: 70.1L 15.4gal 18.5galUS

108 BMW

M5
1989 Germany
155.0mph 249.4kmh
0-60mph 96.5kmh: 6.3secs
315.0bhp 234.9kW 319.4PS
@ 6900rpm
266.0lbft 360.4Nm @ 4750rpm
184.7bhp/ton 181.6bhp/tonne
89.1bhp/L 66.4kW/L 90.3PS/L
65.0ft/sec 19.8m/sec
Petrol 4-stroke piston
3535cc 215.7cu in
In-line 6 fuel injection
Compression ratio: 10.0:1
Bore: 93.4mm 3.7in
Stroke: 86.0mm 3.4in
Valve type/No: Overhead 24
Transmission: Manual
No. of forward speeds: 5
Wheels driven: Rear
Springs F/R: Coil/Coil
Brake system: ABS
Brakes F/R: Disc/Disc
Steering: Recirculating ball PA
Wheelbase: 276.1cm 108.7in
Track F: 147.3cm 58.0in
Track R: 149.6cm 58.9in
Length: 471.9cm 185.8in
Width: 175.0cm 68.9in
Height: 139.2cm 54.8in
Kerb weight: 1734.3kg
3820.0lb
Fuel: 90.1L 19.8gal 23.8galUS

109 BMW

M635 CSi
1989 Germany
150.0mph 241.4kmh
0-50mph 80.5kmh: 4.6secs
0-60mph 96.5kmh: 6.0secs
0-1/4 mile: 14.6secs
0-1km: 23.6secs
286.0bhp 213.3kW 290.0PS
@ 6500rpm
250.9lbft 340.0Nm @ 4500rpm
185.3bhp/ton 182.2bhp/tonne
82.8bhp/L 61.8kW/L 84.0PS/L
59.8ft/sec 18.2m/sec

24.0mph 38.6kmh/1000rpm
20.6mpg 17.2mpgUS
13.7L/100km
Petrol 4-stroke piston
3453cc 211.0cu in
In-line 6 fuel injection
Compression ratio: 10.5:1
Bore: 93.4mm 3.7in
Stroke: 84.0mm 3.3in
Valve type/No: Overhead 24
Transmission: Manual
No. of forward speeds: 5
Wheels driven: Rear
Springs F/R: Coil/Coil
Brake system: PA
Brakes F/R: Disc/Disc
Steering: Recirculating ball PA
Wheelbase: 262.5cm 103.3in
Track F: 143.0cm 56.3in
Track R: 146.0cm 57.5in
Length: 481.5cm 189.6in
Width: 172.5cm 67.9in
Height: 133.8cm 52.7in
Ground clearance: 9.5cm
3.7in
Kerb weight: 1570.0kg
3458.1lb
Fuel: 70.1L 15.4gal 18.5galUS

110 BMW

Z1
1989 Germany
139.0mph 223.7kmh
0-50mph 80.5kmh: 5.8secs
0-60mph 96.5kmh: 7.9secs
0-1/4 mile: 15.9secs
0-1km: 29.4secs
170.0bhp 126.8kW 172.4PS
@ 5800rpm
163.8lbft 222.0Nm @ 4300rpm
129.3bhp/ton 127.1bhp/tonne
68.2bhp/L 50.8kW/L 69.1PS/L
47.5ft/sec 14.5m/sec
24.8mph 39.9kmh/1000rpm
24.0mpg 20.0mpgUS
11.8L/100km
Petrol 4-stroke piston
2494cc 152.0cu in
In-line 6 fuel injection
Compression ratio: 8.8:1
Bore: 84.0mm 3.3in
Stroke: 75.0mm 2.9in
Valve type/No: Overhead 12
Transmission: Manual
No. of forward speeds: 5
Wheels driven: Rear
Springs F/R: Coil/Coil
Brake system: PA ABS
Brakes F/R: Disc/Drum
Steering: Rack & pinion PA
Wheelbase: 243.8cm 96.0in
Track F: 142.2cm 56.0in
Track R: 144.7cm 57.0in
Length: 392.4cm 154.5in

Width: 170.1cm 67.0in
Height: 124.4cm 49.0in
Ground clearance: 16.5cm
6.5in
Kerb weight: 1337.0kg
2944.9lb
Fuel: 58.2L 12.8gal 15.4galUS

111 BMW
318i S
1990 Germany
125.0mph 201.1kmh
0-50mph 80.5kmh: 6.5secs
0-60mph 96.5kmh: 9.3secs
0-1/4 mile: 16.8secs
0-1km: 30.8secs
136.0bhp 101.4kW 137.9PS
@ 6000rpm
126.9lbft 172.0Nm @ 4600rpm
119.2bhp/ton 117.2bhp/tonne
75.7bhp/L 56.5kW/L 76.8PS/L
53.2ft/sec 16.2m/sec
20.8mph 33.5kmh/1000rpm
27.5mpg 22.9mpgUS
10.3L/100km
Petrol 4-stroke piston
1796cc 110.0cu in
In-line 4 fuel injection
Compression ratio: 10.0:1
Bore: 84.0mm 3.3in
Stroke: 81.0mm 3.2in
Valve type/No: Overhead 16
Transmission: Manual
No. of forward speeds: 5
Wheels driven: Rear
Springs F/R: Coil/Coil
Brake system: PA ABS
Brakes F/R: Disc/Disc
Steering: Rack & pinion PA
Wheelbase: 257.0cm 101.2in
Track F: 140.7cm 55.4in
Track R: 141.5cm 55.7in
Length: 432.5cm 170.3in
Width: 164.5cm 64.8in
Height: 138.0cm 54.3in
Ground clearance: 15.5cm
6.1in
Kerb weight: 1160.0kg
2555.1lb
Fuel: 55.1L 12.1gal 14.5galUS

112 BMW
735i Koenig
1990 Germany
158.0mph 254.2kmh
0-60mph 96.5kmh: 6.0secs
290.0bhp 216.2kW 294.0PS
@ 5200rpm
299.0lbft 405.1Nm
164.5bhp/ton 161.7bhp/tonne
84.5bhp/L 63.0kW/L 85.7PS/L
49.0ft/sec 14.9m/sec
Petrol 4-stroke piston
3430cc 209.3cu in

supercharged
In-line 6 fuel injection
Compression ratio: 8.0:1
Bore: 92.0mm 3.6in
Stroke: 86.0mm 3.4in
Valve type/No: Overhead 12
Transmission: Automatic
No. of forward speeds: 4
Wheels driven: Rear
Springs F/R: Coil/Coil
Brake system: ABS
Brakes F/R: Disc/Disc
Steering: Recirculating ball PA
Wheelbase: 279.4cm 110.0in
Track F: 150.1cm 59.1in
Track R: 151.6cm 59.7in
Length: 515.6cm 203.0in
Width: 203.2cm 80.0in
Height: 139.7cm 55.0in
Kerb weight: 1793.3kg
3950.0lb
Fuel: 85.2L 18.7gal 22.5galUS

113 BMW
750i
1990 Germany
153.0mph 246.2kmh
0-50mph 80.5kmh: 5.4secs
0-60mph 96.5kmh: 7.3secs
0-1/4 mile: 15.2secs
0-1km: 27.7secs
300.0bhp 223.7kW 304.2PS
@ 5200rpm
332.1lbft 450.0Nm @ 4100rpm
164.9bhp/ton 162.2bhp/tonne
60.1bhp/L 44.8kW/L 61.0PS/L
42.6ft/sec 13.0m/sec
32.1mph 51.6kmh/1000rpm
16.2mpg 13.5mpgUS
17.4L/100km
Petrol 4-stroke piston
4988cc 304.0cu in
Vee 12 fuel injection
Compression ratio: 8.8:1
Bore: 84.0mm 3.3in
Stroke: 75.0mm 2.9in
Valve type/No: Overhead 48
Transmission: Automatic
No. of forward speeds: 4
Wheels driven: Rear
Springs F/R: Coil/Coil
Brake system: PA ABS
Brakes F/R: Disc/Disc
Steering: Recirculating ball PA
Wheelbase: 283.2cm 111.5in
Track F: 152.7cm 60.1in
Track R: 155.0cm 61.0in
Length: 491.0cm 193.3in
Width: 184.5cm 72.6in
Height: 141.1cm 55.6in
Ground clearance: 20.3cm
8.0in
Kerb weight: 1850.0kg
4074.9lb

Fuel: 101.9L 22.4gal
26.9galUS

114 BMW
750i Alpina B12
1990 Germany
171.0mph 275.1kmh
0-60mph 96.5kmh: 6.9secs
350.0bhp 261.0kW 354.8PS
@ 5300rpm
347.0lbft 470.2Nm @ 4000rpm
191.7bhp/ton 188.5bhp/tonne
70.2bhp/L 52.3kW/L 71.1PS/L
43.4ft/sec 13.2m/sec
Petrol 4-stroke piston
4988cc 304.3cu in
Vee 12 fuel injection
Compression ratio: 9.5:1
Bore: 84.0mm 3.3in
Stroke: 75.0mm 2.9in
Valve type/No: Overhead 24
Transmission: Automatic
No. of forward speeds: 4
Wheels driven: Rear
Springs F/R: Coil/Coil
Brake system: ABS
Brakes F/R: Disc/Disc
Steering: Recirculating ball PA
Wheelbase: 283.2cm 111.5in
Track F: 153.9cm 60.6in
Track R: 153.4cm 60.4in
Length: 491.0cm 193.3in
Width: 184.4cm 72.6in
Height: 138.9cm 54.7in
Kerb weight: 1856.9kg
4090.0lb
Fuel: 101.8L 22.4gal
26.9galUS

115 BMW
318i
1991 Germany
123.0mph 197.9kmh
0-50mph 80.5kmh: 7.4secs
0-60mph 96.5kmh: 10.2secs
0-1/4 mile: 17.5secs
0-1km: 32.0secs
113.0bhp 84.3kW 114.6PS
@ 5900rpm
119.6lbft 162.0Nm @ 3000rpm
91.2bhp/ton 89.7bhp/tonne
62.9bhp/L 46.9kW/L 63.8PS/L
52.3ft/sec 15.9m/sec
20.5mph 33.0kmh/1000rpm
27.6mpg 23.0mpgUS
10.2L/100km
Petrol 4-stroke piston
1796cc 110.0cu in
In-line 4 fuel injection
Compression ratio: 8.8:1
Bore: 84.0mm 3.3in
Stroke: 81.0mm 3.2in
Valve type/No: Overhead 8
Transmission: Manual

No. of forward speeds: 5
Wheels driven: Rear
Springs F/R: Coil/Coil
Brake system: PA ABS
Brakes F/R: Disc/Drum
Steering: Rack & pinion PA
Wheelbase: 269.7cm 106.2in
Track F: 140.7cm 55.4in
Track R: 142.0cm 55.9in
Length: 443.2cm 174.5in
Width: 169.7cm 66.8in
Height: 139.2cm 54.8in
Kerb weight: 1260.0kg
2775.3lb
Fuel: 65.1L 14.3gal 17.2galUS

116 BMW
325i SE
1991 Germany
143.0mph 230.1kmh
0-50mph 80.5kmh: 5.3secs
0-60mph 96.5kmh: 7.3secs
0-1/4 mile: 15.8secs
0-1km: 28.3secs
189.0bhp 140.9kW 191.6PS
@ 5900rpm
180.8lbft 245.0Nm @ 4700rpm
144.5bhp/ton 142.1bhp/tonne
75.8bhp/L 56.5kW/L 76.8PS/L
51.0ft/sec 15.5m/sec
22.5mph 36.2kmh/1000rpm
26.6mpg 22.1mpgUS
10.6L/100km
Petrol 4-stroke piston
2494cc 152.0cu in
In-line 6 fuel injection
Compression ratio: 10.0:1
Bore: 84.0mm 3.3in
Stroke: 79.0mm 3.1in
Valve type/No: Overhead 24
Transmission: Manual
No. of forward speeds: 5
Wheels driven: Rear
Springs F/R: Coil/Coil
Brake system: PA ABS
Brakes F/R: Disc/Disc
Steering: Rack & pinion PA
Wheelbase: 269.7cm 106.2in
Track F: 140.7cm 55.4in
Track R: 142.0cm 55.9in
Length: 443.2cm 174.5in
Width: 169.7cm 66.8in
Height: 139.2cm 54.8in
Kerb weight: 1330.0kg
2929.5lb
Fuel: 65.1L 14.3gal 17.2galUS

117 BMW
520i
1991 Germany
130.0mph 209.2kmh
0-50mph 80.5kmh: 6.7secs
0-60mph 96.5kmh: 9.6secs
0-1/4 mile: 17.1secs

0-1km: 31.4secs
150.0bhp 111.9kW 152.1PS @ 5900rpm
140.2lbft 190.0Nm @ 4700rpm
102.4bhp/ton 100.7bhp/tonne
75.3bhp/L 56.2kW/L 76.4PS/L
42.6ft/sec 13.0m/sec
24.4mph 39.3kmh/1000rpm
22.0mpg 18.3mpgUS
12.8L/100km
Petrol 4-stroke piston
1991cc 121.0cu in
In-line 6 fuel injection
Compression ratio: 10.5:1
Bore: 80.0mm 3.1in
Stroke: 66.0mm 2.6in
Valve type/No: Overhead 24
Transmission: Manual
No. of forward speeds: 5
Wheels driven: Rear
Springs F/R: Coil/Coil
Brake system: PA
Brakes F/R: Disc/Disc
Steering: Recirculating ball PA
Wheelbase: 276.1cm 108.7in
Track F: 146.6cm 57.7in
Track R: 154.9cm 61.0in
Length: 471.9cm 185.8in
Width: 175.0cm 68.9in
Height: 141.2cm 55.6in
Ground clearance: 17.8cm 7.0in
Kerb weight: 1490.0kg 3281.9lb
Fuel: 80.1L 17.6gal 21.1galUS

118 BMW

525i
1991 Germany
128.0mph 206.0kmh
0-50mph 80.5kmh: 6.2secs
0-60mph 96.5kmh: 8.5secs
0-1/4 mile: 16.3secs
189.0bhp 140.9kW 191.6PS @ 5900rpm
181.0lbft 245.3Nm @ 4700rpm
121.5bhp/ton 119.4bhp/tonne
75.8bhp/L 56.5kW/L 76.8PS/L
48.3ft/sec 14.7m/sec
24.0mpg 20.0mpgUS
11.8L/100km
Petrol 4-stroke piston
2494cc 152.2cu in
In-line 6 fuel injection
Compression ratio: 10.0:1
Bore: 84.0mm 3.3in
Stroke: 75.0mm 2.9in
Valve type/No: Overhead 24
Transmission: Manual
No. of forward speeds: 5
Wheels driven: Rear
Springs F/R: Coil/Coil
Brake system: PA ABS
Brakes F/R: Disc/Disc

Steering: Recirculating ball PA
Wheelbase: 276.1cm 108.7in
Track F: 147.1cm 57.9in
Track R: 149.6cm 58.9in
Length: 471.9cm 185.8in
Width: 175.0cm 68.9in
Height: 141.2cm 55.6in
Kerb weight: 1582.2kg 3485.0lb
Fuel: 79.9L 17.6gal 21.1galUS

119 BMW

525i SE 24v
1991 Germany
141.0mph 226.9kmh
0-50mph 80.5kmh: 6.3secs
0-60mph 96.5kmh: 8.7secs
0-1/4 mile: 16.5secs
0-1km: 29.9secs
192.0bhp 143.2kW 194.7PS @ 5900rpm
180.8lbft 245.0Nm @ 4700rpm
127.2bhp/ton 125.1bhp/tonne
77.0bhp/L 57.4kW/L 78.0PS/L
48.3ft/sec 14.7m/sec
22.2mph 35.7kmh/1000rpm
21.7mpg 18.1mpgUS
13.0L/100km
Petrol 4-stroke piston
2494cc 152.0cu in
In-line 6 fuel injection
Compression ratio: 10.0:1
Bore: 84.0mm 3.3in
Stroke: 75.0mm 2.9in
Valve type/No: Overhead 24
Transmission: Manual
No. of forward speeds: 5
Wheels driven: Rear
Springs F/R: Coil/Coil
Brake system: PA ABS
Brakes F/R: Disc/Disc
Steering: Recirculating ball PA
Wheelbase: 276.1cm 108.7in
Track F: 146.6cm 57.7in
Track R: 154.9cm 61.0in
Length: 471.9cm 185.8in
Width: 175.0cm 68.9in
Height: 141.2cm 55.6in
Ground clearance: 17.8cm 7.0in
Kerb weight: 1535.0kg 3381.1lb
Fuel: 80.1L 17.6gal 21.1galUS

120 BMW

535i Alpina B10 Bi-Turbo
1991 Germany
179.0mph 288.0kmh
0-60mph 96.5kmh: 5.1secs
0-1/4 mile: 13.6secs
360.0bhp 268.4kW 365.0PS @ 6000rpm
382.0lbft 517.6Nm @ 4000rpm
216.2bhp/ton 212.6bhp/tonne

105.0bhp/L 78.3kW/L
106.4PS/L
56.5ft/sec 17.2m/sec
Petrol 4-stroke piston
3430cc 209.3cu in
turbocharged
In-line 6 fuel injection
Compression ratio: 7.2:1
Bore: 92.0mm 3.6in
Stroke: 86.0mm 3.4in
Valve type/No: Overhead 12
Transmission: Manual
No. of forward speeds: 5
Wheels driven: Rear
Springs F/R: Coil/Coil
Brake system: PA ABS
Brakes F/R: Disc/Disc
Steering: Recirculating ball PA
Wheelbase: 278.1cm 109.5in
Track F: 149.4cm 58.8in
Track R: 148.8cm 58.6in
Length: 475.5cm 187.2in
Width: 176.3cm 69.4in
Height: 140.2cm 55.2in
Kerb weight: 1693.4kg 3730.0lb
Fuel: 109.8L 24.1gal 29.0galUS

121 BMW

850i
1991 Germany
161.0mph 259.0kmh
0-50mph 80.5kmh: 5.3secs
0-60mph 96.5kmh: 7.2secs
0-1/4 mile: 15.3secs
0-1km: 27.2secs
300.0bhp 223.7kW 304.2PS @ 5200rpm
332.0lbft 449.9Nm @ 4100rpm
162.3bhp/ton 159.6bhp/tonne
60.1bhp/L 44.8kW/L 61.0PS/L
42.6ft/sec 13.0m/sec
31.8mph 51.2kmh/1000rpm
14.4mpg 12.0mpgUS
19.6L/100km
Petrol 4-stroke piston
4988cc 304.3cu in
Vee 12 fuel injection
Compression ratio: 8.8:1
Bore: 84.0mm 3.3in
Stroke: 75.0mm 2.9in
Valve type/No: Overhead 24
Transmission: Automatic
No. of forward speeds: 4
Wheels driven: Rear
Springs F/R: Coil/Coil
Brake system: PA ABS
Brakes F/R: Disc/Disc
Steering: Recirculating ball PA
Wheelbase: 268.2cm 105.6in
Track F: 154.9cm 61.0in
Track R: 154.9cm 61.0in
Length: 477.5cm 188.0in

Width: 185.4cm 73.0in
Height: 131.3cm 51.7in
Ground clearance: 15.2cm 6.0in
Kerb weight: 1880.0kg 4141.0lb
Fuel: 90.1L 19.8gal 23.8galUS

122 BMW

M5
1991 Germany
155.0mph 249.4kmh
0-60mph 96.5kmh: 6.4secs
0-1/4 mile: 15.0secs
310.0bhp 231.2kW 314.3PS @ 6900rpm
265.0lbft 359.1Nm @ 4750rpm
172.9bhp/ton 172.9bhp/tonne
87.7bhp/L 65.4kW/L 88.9PS/L
65.0ft/sec 19.8m/sec
18.6mpg 15.5mpgUS
15.2L/100km
Petrol 4-stroke piston
3535cc 215.7cu in
In-line 6 fuel injection
Compression ratio: 10.0:1
Bore: 93.4mm 3.7in
Stroke: 86.0mm 3.4in
Valve type/No: Overhead 24
Transmission: Manual
No. of forward speeds: 5
Wheels driven: Rear
Springs F/R: Coil/Coil
Brake system: PA ABS
Brakes F/R: Disc/Disc
Steering: Recirculating ball PA
Wheelbase: 276.1cm 108.7in
Track F: 147.3cm 58.0in
Track R: 149.6cm 58.9in
Length: 471.9cm 185.8in
Width: 175.0cm 68.9in
Height: 140.7cm 55.4in
Kerb weight: 1793.3kg 3950.0lb
Fuel: 90.1L 19.8gal 23.8galUS

123 BMW

325i
1992 Germany
128.0mph 206.0kmh
0-50mph 80.5kmh: 5.9secs
0-60mph 96.5kmh: 7.9secs
0-1/4 mile: 16.1secs
189.0bhp 140.9kW 191.6PS @ 5900rpm
181.0lbft 245.3Nm @ 4700rpm
137.4bhp/ton 135.2bhp/tonne
75.8bhp/L 56.5kW/L 76.8PS/L
48.3ft/sec 14.7m/sec
26.4mpg 22.0mpgUS
10.7L/100km
Petrol 4-stroke piston
2494cc 152.2cu in
In-line 6 fuel injection

Compression ratio: 10.0:1
Bore: 84.0mm 3.3in
Stroke: 75.0mm 2.9in
Valve type/No: Overhead 24
Transmission: Manual
No. of forward speeds: 5
Wheels driven: Rear
Springs F/R: Coil/Coil
Brake system: PA ABS
Brakes F/R: Disc/Disc
Steering: Rack & pinion PA
Wheelbase: 270.0cm 106.3in
Track F: 140.7cm 55.4in
Track R: 142.0cm 55.9in
Length: 443.2cm 174.5in
Width: 169.7cm 66.8in
Height: 139.2cm 54.8in
Ground clearance: 16.5cm
6.5in
Kerb weight: 1398.3kg
3080.0lb
Fuel: 65.1L 14.3gal 17.2galUS

124 BMW

520i SE Touring
1992 Germany
132.0mph 212.4kmh
0-50mph 80.5kmh: 6.6secs
0-60mph 96.5kmh: 9.2secs
0-1/4 mile: 17.0secs
0-1km: 31.0secs
192.0bhp 143.2kW 194.7PS
@ 5900rpm
180.0lbft 243.9Nm @ 4700rpm
122.8bhp/ton 120.7bhp/tonne
77.0bhp/L 57.4kW/L 78.0PS/L
48.3ft/sec 14.7m/sec
22.9mph 36.8kmh/1000rpm
20.8mpg 17.3mpgUS
13.6L/100km
Petrol 4-stroke piston
2494cc 152.2cu in
In-line 6 fuel injection
Compression ratio: 10.1:1
Bore: 84.0mm 3.3in
Stroke: 75.0mm 2.9in
Valve type/No: Overhead 24
Transmission: Manual
No. of forward speeds: 5
Wheels driven: Rear
Springs F/R: Coil/Coil
Brake system: PA ABS
Brakes F/R: Disc/Disc
Steering: Ball & nut PA
Wheelbase: 276.1cm 108.7in
Track F: 147.0cm 57.9in
Track R: 149.5cm 58.9in
Length: 472.0cm 185.8in
Width: 175.1cm 68.9in
Height: 141.7cm 55.8in
Ground clearance: 17.8cm 7.0in
Kerb weight: 1590.0kg
3502.2lb
Fuel: 80.1L 17.6gal 21.1galUS

125 BMW

850i Manual
1992 Germany
160.0mph 257.4kmh
0-50mph 80.5kmh: 5.3secs
0-60mph 96.5kmh: 7.1secs
0-1/4 mile: 15.5secs
0-1km: 27.5secs
300.0bhp 223.7kW 304.2PS
@ 5200rpm
332.1lbft 450.0Nm @ 4100rpm
161.8bhp/ton 159.1bhp/tonne
60.1bhp/L 44.8kW/L 61.0PS/L
42.6ft/sec 13.0m/sec
30.4mph 48.9kmh/1000rpm
14.7mpg 12.2mpgUS
19.2L/100km
Petrol 4-stroke piston
4988cc 304.0cu in
Vee 12 fuel injection
Compression ratio: 8.8:1
Bore: 84.0mm 3.3in
Stroke: 75.0mm 2.9in
Valve type/No: Overhead 24
Transmission: Manual
No. of forward speeds: 6
Wheels driven: Rear
Springs F/R: Coil/Coil
Brake system: PA ABS
Brakes F/R: Disc/Disc
Steering: Recirculating ball PA
Wheelbase: 268.4cm 105.7in
Track F: 155.2cm 61.1in
Track R: 156.2cm 61.5in
Length: 478.0cm 188.2in
Width: 185.5cm 73.0in
Height: 134.0cm 52.8in
Kerb weight: 1885.0kg
4152.0lb
Fuel: 90.0L 19.8gal 23.8galUS

126 Bugatti

EB110
1991 Italy
214.0mph 344.3kmh
0-60mph 96.5kmh: 3.9secs
600.0bhp 447.4kW 608.3PS
@ 9000rpm
419.0lbft 567.8Nm @ 4200rpm
454.8bhp/ton 447.2bhp/tonne
171.5bhp/L 127.9kW/L
173.9PS/L
55.7ft/sec 17.0m/sec
Petrol 4-stroke piston
3498cc 213.4cu in
turbocharged
Vee 12 fuel injection
Compression ratio: 7.5:1
Bore: 81.0mm 3.2in
Stroke: 56.6mm 2.2in
Valve type/No: Overhead 48
Transmission: Manual
No. of forward speeds: 6
Wheels driven: 4-wheel drive

Brake system: ABS
Brakes F/R: Disc/Disc
Steering: Rack & pinion PA
Wheelbase: 255.0cm 100.4in
Track F: 153.9cm 60.6in
Track R: 153.9cm 60.6in
Length: 407.9cm 160.6in
Width: 198.9cm 78.3in
Height: 105.9cm 41.7in
Kerb weight: 1341.6kg
2955.0lb
Fuel: 120.0L 26.4gal
31.7galUS

127 Buick

Reatta
1988 USA
125.0mph 201.1kmh
0-50mph 80.5kmh: 6.3secs
0-60mph 96.5kmh: 8.9secs
0-1/4 mile: 16.8secs
165.0bhp 123.0kW 167.3PS
@ 4800rpm
210.0lbft 284.6Nm @ 2000rpm
109.3bhp/ton 107.5bhp/tonne
43.5bhp/L 32.5kW/L 44.1PS/L
45.3ft/sec 13.8m/sec
19.8mpg 16.5mpgUS
14.3L/100km
Petrol 4-stroke piston
3791cc 231.3cu in
Vee 6 fuel injection
Compression ratio: 8.5:1
Bore: 96.5mm 3.8in
Stroke: 86.4mm 3.4in
Valve type/No: Overhead 12
Transmission: Automatic
No. of forward speeds: 4
Wheels driven: Front
Springs F/R: Coil/Leaf
Brake system: PA ABS
Brakes F/R: Disc/Disc
Steering: Rack & pinion PA
Wheelbase: 250.2cm 98.5in
Track F: 153.2cm 60.3in
Track R: 153.2cm 60.3in
Length: 464.3cm 182.8in
Width: 185.4cm 73.0in
Height: 130.0cm 51.2in
Kerb weight: 1534.5kg
3380.0lb
Fuel: 68.9L 15.1gal
18.2galUS

128 Buick

Reatta Convertible
1991 USA
125.0mph 201.1kmh
0-50mph 80.5kmh: 6.3secs
0-60mph 96.5kmh: 8.9secs
0-1/4 mile: 16.8secs
170.0bhp 126.8kW 172.4PS
@ 4800rpm
220.0lbft 298.1Nm @ 3200rpm

105.9bhp/ton 104.2bhp/tonne
44.8bhp/L 33.4kW/L 45.5PS/L
45.3ft/sec 13.8m/sec
19.8mpg 16.5mpgUS
14.3L/100km
Petrol 4-stroke piston
3791cc 231.3cu in
Vee 6 fuel injection
Compression ratio: 8.5:1
Bore: 96.5mm 3.8in
Stroke: 86.4mm 3.4in
Valve type/No: Overhead 12
Transmission: Automatic
No. of forward speeds: 4
Wheels driven: Front
Springs F/R: Coil/Leaf
Brake system: PA ABS
Brakes F/R: Disc/Disc
Steering: Rack & pinion PA
Wheelbase: 250.2cm 98.5in
Track F: 153.2cm 60.3in
Track R: 153.2cm 60.3in
Length: 466.6cm 183.7in
Width: 185.4cm 73.0in
Height: 132.8cm 52.3in
Kerb weight: 1632.1kg
3595.0lb
Fuel: 68.9L 15.1gal
18.2galUS

129 Cadillac

Allante
1988 USA
119.0mph 191.5kmh
0-60mph 96.5kmh: 9.3secs
0-1/4 mile: 17.1secs
170.0bhp 126.8kW 172.4PS
@ 4300rpm
230.0lbft 311.7Nm @ 3200rpm
109.0bhp/ton 107.1bhp/tonne
41.6bhp/L 31.0kW/L 42.2PS/L
39.5ft/sec 12.0m/sec
26.9mpg 22.4mpgUS
10.5L/100km
Petrol 4-stroke piston
4087cc 249.4cu in
Vee 8 fuel injection
Compression ratio: 8.5:1
Bore: 88.0mm 3.5in
Stroke: 84.0mm 3.3in
Valve type/No: Overhead 16
Transmission: Automatic
No. of forward speeds: 4
Wheels driven: Front
Springs F/R: Coil/Leaf
Brake system: ABS
Brakes F/R: Disc/Disc
Steering: Rack & pinion PA
Wheelbase: 252.5cm 99.4in
Track F: 153.7cm 60.5in
Track R: 153.7cm 60.5in
Length: 453.6cm 178.6in
Width: 186.4cm 73.4in
Height: 132.1cm 52.0in

129 Cadillac
Allante

Kerb weight: 1586.7kg 3495.0lb
Fuel: 83.3L 18.3gal 22.0galUS

130 Cadillac
Allante
1990 USA
130.0mph 209.2kmh
0-50mph 80.5kmh: 6.2secs
0-60mph 96.5kmh: 8.3secs
0-1/4 mile: 16.4secs
200.0bhp 149.1kW 202.8PS @ 4400rpm
270.0lbft 365.9Nm @ 3200rpm
129.3bhp/ton 127.1bhp/tonne
44.8bhp/L 33.4kW/L 45.4PS/L
40.5ft/sec 12.3m/sec
Petrol 4-stroke piston
4467cc 272.5cu in
Vee 8 fuel injection
Compression ratio: 9.0:1
Bore: 92.0mm 3.6in
Stroke: 84.0mm 3.3in
Valve type/No: Overhead 16
Transmission: Automatic
No. of forward speeds: 4
Wheels driven: Front
Springs F/R: Coil/Leaf
Brake system: PA ABS
Brakes F/R: Disc/Disc
Steering: Rack & pinion PA
Wheelbase: 252.5cm 99.4in
Track F: 153.4cm 60.4in
Track R: 153.4cm 60.4in
Length: 478.7cm 178.7in
Width: 186.7cm 73.5in
Height: 132.6cm 52.2in
Kerb weight: 1573.1kg 3465.0lb
Fuel: 83.3L 18.3gal 22.0galUS

131 Cadillac
Eldorado Touring Coupe
1990 USA

125.0mph 201.1kmh
0-50mph 80.5kmh: 6.3secs
0-60mph 96.5kmh: 8.5secs
0-1/4 mile: 16.6secs
180.0bhp 134.2kW 182.5PS @ 4300rpm
240.0lbft 325.2Nm @ 2600rpm
114.7bhp/ton 112.8bhp/tonne
40.3bhp/L 30.0kW/L 40.9PS/L
39.5ft/sec 12.0m/sec
29.1mpg 24.2mpgUS
9.7L/100km
Petrol 4-stroke piston
4467cc 272.5cu in
Vee 8 fuel injection
Compression ratio: 9.5:1
Bore: 91.9mm 3.6in
Stroke: 84.1mm 3.3in
Valve type/No: Overhead 16
Transmission: Automatic
No. of forward speeds: 4
Wheels driven: Front
Springs F/R: Coil/Leaf
Brake system: PA ABS
Brakes F/R: Disc/Disc
Steering: Rack & pinion PA
Wheelbase: 274.3cm 108.0in
Track F: 152.1cm 59.9in
Track R: 152.1cm 59.9in
Length: 486.2cm 191.4in
Width: 183.9cm 72.4in
Height: 136.4cm 53.7in
Kerb weight: 1595.8kg 3515.0lb
Fuel: 71.2L 15.6gal 18.8galUS

132 Cadillac
Allante
1991 USA
130.0mph 209.2kmh
0-50mph 80.5kmh: 6.2secs
0-60mph 96.5kmh: 8.3secs
0-1/4 mile: 16.4secs

200.0bhp 149.1kW 202.8PS @ 4400rpm
270.0lbft 365.9Nm @ 3200rpm
126.7bhp/ton 124.6bhp/tonne
44.8bhp/L 33.4kW/L 45.4PS/L
40.5ft/sec 12.3m/sec
19.9mpg 16.6mpgUS
14.2L/100km
Petrol 4-stroke piston
4467cc 272.5cu in
Vee 8 fuel injection
Compression ratio: 9.0:1
Bore: 92.0mm 3.6in
Stroke: 84.0mm 3.3in
Valve type/No: Overhead 16
Transmission: Automatic
No. of forward speeds: 4
Wheels driven: Front
Springs F/R: Coil/Leaf
Brake system: PA ABS
Brakes F/R: Disc/Disc
Steering: Rack & pinion PA
Wheelbase: 252.5cm 99.4in
Track F: 153.4cm 60.4in
Track R: 153.4cm 60.4in
Length: 453.9cm 178.7in
Width: 186.7cm 73.5in
Height: 130.0cm 51.2in
Kerb weight: 1604.9kg 3535.0lb
Fuel: 83.3L 18.3gal 22.0galUS

133 Cadillac
Seville Touring Sedan
1991 USA
125.0mph 201.1kmh
0-60mph 96.5kmh: 8.6secs
0-1/4 mile: 16.6secs
200.0bhp 149.1kW 202.8PS @ 4100rpm
275.0lbft 372.6Nm @ 3000rpm
125.5bhp/ton 123.4bhp/tonne
40.9bhp/L 30.5kW/L 41.4PS/L

41.2ft/sec 12.6m/sec
22.2mpg 18.5mpgUS
12.7L/100km
Petrol 4-stroke piston
4894cc 298.6cu in
Vee 8 fuel injection
Compression ratio: 9.5:1
Bore: 92.0mm 3.6in
Stroke: 92.0mm 3.6in
Valve type/No: Overhead 16
Transmission: Automatic
No. of forward speeds: 4
Wheels driven: Front
Springs F/R: Coil/Leaf
Brake system: PA ABS
Brakes F/R: Disc/Disc
Steering: Rack & pinion PA
Wheelbase: 274.3cm 108.0in
Track F: 152.1cm 59.9in
Track R: 152.1cm 59.9in
Length: 484.6cm 190.8in
Width: 182.9cm 72.0in
Height: 135.1cm 53.2in
Kerb weight: 1620.8kg 3570.0lb
Fuel: 71.2L 15.6gal 18.8galUS

134 Carbodies
Fairway 2.7 Silver
1989 UK
81.0mph 130.3kmh
0-50mph 80.5kmh: 14.6secs
0-60mph 96.5kmh: 23.6secs
0-1/4 mile: 22.0secs
0-1km: 43.1secs
78.0bhp 58.2kW 79.1PS @ 4300rpm
127.7lbft 173.0Nm @ 2200rpm
51.4bhp/ton 50.6bhp/tonne
29.3bhp/L 21.8kW/L 29.7PS/L
43.2ft/sec 13.2m/sec
29.5mph 47.5kmh/1000rpm
23.0mpg 19.2mpgUS

12.3L/100km
Diesel 4-stroke piston
2663cc 162.0cu in
In-line 4 fuel injection
Compression ratio: 22.8:1
Bore: 96.0mm 3.8in
Stroke: 92.0mm 3.6in
Valve type/No: Overhead 8
Transmission: Manual
No. of forward speeds: 4
Wheels driven: Rear
Springs F/R. Coil/Leaf
Brake system: PA
Brakes F/R: Drum/Drum
Steering: Worm & roller PA
Wheelbase: 281.0cm 110.6in
Track F: 142.0cm 55.9in
Track R: 142.0cm 55.9in
Length: 458.0cm 180.3in
Width: 175.0cm 68.9in
Height: 177.0cm 69.7in
Ground clearance: 20.4cm
8.0in
Kerb weight: 1542.0kg
3396.5lb
Fuel: 52.3L 11.5gal 13.8galUS

135 Caterham
7 HPC
1990 UK
126.0mph 202.7kmh
0-50mph 80.5kmh: 4.1secs
0-60mph 96.5kmh: 5.2secs
0-1/4 mile: 13.6secs
0-1km: 25.7secs
175.0bhp 130.5kW 177.4PS
@ 6000rpm
155.0lbft 210.0Nm @ 4800rpm
283.4bhp/ton 278.7bhp/tonne
87.6bhp/L 65.3kW/L 88.8PS/L
56.5ft/sec 17.2m/sec
20.2mph 32.5kmh/1000rpm

17.3mpg 14.4mpgUS
16.3L/100km
Petrol 4-stroke piston
1998cc 121.9cu in
In-line 4 2 Carburettor
Compression ratio: 10.5:1
Bore: 86.0mm 3.4in
Stroke: 86.0mm 3.4in
Valve type/No: Overhead 16
Transmission: Manual
No. of forward speeds: 5
Wheels driven: Rear
Springs F/R: Coil/Coil
Brakes F/R: Disc/Disc
Steering: Rack & pinion
Wheelbase: 223.5cm 88.0in
Track F: 124.5cm 49.0in
Track R: 132.1cm 52.0in
Length: 339.1cm 133.5in
Width: 157.5cm 62.0in
Height: 108.0cm 42.5in
Ground clearance: 12.7cm
5.0in
Kerb weight: 627.9kg 1383.0lb
Fuel: 45.5L 10.0gal 12.0galUS

136 Caterham
Super 7
1991 UK
112.0mph 180.2kmh
0-60mph 96.5kmh: 5.6secs
135.0bhp 100.7kW 136.9PS
@ 6000rpm
122.0lbft 165.3Nm @ 4500rpm
236.2bhp/ton 232.3bhp/tonne
79.9bhp/L 59.6kW/L 81.0PS/L
51.0ft/sec 15.5m/sec
Petrol 4-stroke piston
1690cc 103.1cu in
In-line 4 2 Carburettor
Compression ratio: 9.7:1
Bore: 88.3mm 3.5in

Stroke: 77.6mm 3.1in
Valve type/No: Overhead 8
Transmission: Manual
No. of forward speeds: 5
Wheels driven: Rear
Springs F/R: Coil/Coil
Brakes F/R: Disc/Disc
Steering: Rack & pinion
Wheelbase: 224.8cm 88.5in
Track F: 127.0cm 50.0in
Track R: 132.1cm 52.0in
Length: 337.8cm 133.0in
Width: 158.0cm 62.2in
Height: 109.2cm 43.0in
Kerb weight: 581.1kg 1280.0lb
Fuel: 30.3L 6.7gal 8.0galUS

137 Chevrolet
Cavalier Z24
1986 USA
113.0mph 181.8kmh
0-60mph 96.5kmh: 8.5secs
0-1/4 mile: 16.7secs
120.0bhp 89.5kW 121.7PS
@ 4800rpm
155.0lbft 210.0Nm @ 3600rpm
101.4bhp/ton 99.7bhp/tonne
42.3bhp/L 31.5kW/L 42.9PS/L
39.9ft/sec 12.2m/sec
26.4mpg 22.0mpgUS
10.7L/100km
Petrol 4-stroke piston
2837cc 173.1cu in
Vee 6 fuel injection
Compression ratio: 8.9:1
Bore: 89.0mm 3.5in
Stroke: 76.0mm 3.0in
Valve type/No: Overhead 12
Transmission: Manual
No. of forward speeds: 4
Wheels driven: Front
Springs F/R: Coil/Coil

Brake system: PA
Brakes F/R: Disc/Drum
Steering: Rack & pinion PA
Wheelbase: 257.0cm 101.2in
Track F: 140.7cm 55.4in
Track R: 140.2cm 55.2in
Length: 437.9cm 172.4in
Width: 167.6cm 66.0in
Height: 131.8cm 51.9in
Kerb weight: 1203.1kg
2650.0lb
Fuel: 51.5L 11.3gal 13.6galUS

138 Chevrolet
Celebrity CL Eurosport
1986 USA
110.0mph 177.0kmh
0-60mph 96.5kmh: 10.7secs
0-1/4 mile: 17.9secs
125.0bhp 93.2kW 126.7PS
@ 4800rpm
160.0lbft 216.8Nm @ 3600rpm
92.3bhp/ton 90.7bhp/tonne
44.1bhp/L 32.9kW/L 44.7PS/L
39.9ft/sec 12.2m/sec
23.2mpg 19.3mpgUS
12.2L/100km
Petrol 4-stroke piston
2837cc 173.1cu in
Vee 6 fuel injection
Compression ratio: 8.5:1
Bore: 89.0mm 3.5in
Stroke: 76.0mm 3.0in
Valve type/No: Overhead 12
Transmission: Automatic
No. of forward speeds: 4
Wheels driven: Front
Springs F/R: Coil/Coil
Brake system: PA
Brakes F/R: Disc/Drum
Steering: Rack & pinion PA
Wheelbase: 266.4cm 104.9in

136 Caterham
Super 7

Track F: 149.1cm 58.7in
Track R: 144.8cm 57.0in
Length: 478.3cm 188.3in
Width: 176.0cm 69.3in
Height: 137.4cm 54.1in
Kerb weight: 1377.9kg
3035.0lb
Fuel: 59.4L 13.1gal 15.7galUS

139 Chevrolet
Corvette
1986 USA
154.0mph 247.8kmh
0-60mph 96.5kmh: 5.8secs
0-1/4 mile: 14.4secs
230.0bhp 171.5kW 233.2PS
@ 4000rpm
330.0lbft 447.2Nm @ 3200rpm
157.1bhp/ton 154.4bhp/tonne
40.1bhp/L 29.9kW/L 40.7PS/L
38.7ft/sec 11.8m/sec
36.4mph 58.6kmh/1000rpm
19.2mpg 16.0mpgUS
14.7L/100km
Petrol 4-stroke piston
5733cc 349.8cu in
Vee 8 fuel injection
Compression ratio: 9.5:1
Bore: 101.6mm 4.0in
Stroke: 88.4mm 3.5in
Valve type/No: Overhead 16
Transmission: Manual with
overdrive
Wheels driven: Rear
Springs F/R: Leaf/Leaf
Brake system: ABS
Brakes F/R: Disc/Disc
Steering: Rack & pinion PA
Wheelbase: 244.3cm 96.2in
Track F: 151.4cm 59.6in
Track R: 153.4cm 60.4in
Length: 448.3cm 176.5in

Width: 180.3cm 71.0in
Height: 118.6cm 46.7in
Kerb weight: 1489.1kg
3280.0lb

140 Chevrolet
Corvette Callaway
1986 USA
178.0mph 286.4kmh
0-50mph 80.5kmh: 3.8secs
0-60mph 96.5kmh: 5.0secs
0-1/4 mile: 13.7secs
345.0bhp 257.3kW 349.8PS
@ 4000rpm
465.0lbft 630.1Nm @ 2800rpm
225.3bhp/ton 221.5bhp/tonne
60.2bhp/L 44.9kW/L 61.0PS/L
38.7ft/sec 11.8m/sec
Petrol 4-stroke piston
5733cc 349.8cu in
turbocharged
Vee 8 fuel injection
Compression ratio: 7.5:1
Bore: 101.6mm 4.0in
Stroke: 88.4mm 3.5in
Valve type/No: Overhead 16
Transmission: Manual with
overdrive
Wheels driven: Rear
Springs F/R: Leaf/Leaf
Brake system: PA ABS
Brakes F/R: Disc/Disc
Steering: Rack & pinion PA
Wheelbase: 244.3cm 96.2in
Track F: 151.4cm 59.6in
Track R: 153.4cm 60.4in
Length: 448.3cm 176.5in
Width: 180.3cm 71.0in
Height: 118.6cm 46.7in
Kerb weight: 1557.2kg
3430.0lb
Fuel: 75.7L 16.6gal 20.0galUS

141 Chevrolet
Corvette Z51
1986 USA
154.0mph 247.8kmh
0-50mph 80.5kmh: 4.4secs
0-60mph 96.5kmh: 5.8secs
0-1/4 mile: 14.4secs
230.0bhp 171.5kW 233.2PS
@ 4000rpm
330.0lbft 447.2Nm @ 3200rpm
157.1bhp/ton 154.4bhp/tonne
40.1bhp/L 29.9kW/L 40.7PS/L
38.7ft/sec 11.8m/sec
22.8mpg 19.0mpgUS
12.4L/100km
Petrol 4-stroke piston
5733cc 349.8cu in
Vee 8 fuel injection
Compression ratio: 9.5:1
Bore: 101.6mm 4.0in
Stroke: 88.4mm 3.5in
Valve type/No: Overhead 16
Transmission: Manual with
overdrive
Wheels driven: Rear
Springs F/R: Leaf/Leaf
Brake system: PA ABS
Brakes F/R: Disc/Disc
Steering: Rack & pinion PA
Wheelbase: 244.3cm 96.2in
Track F: 151.4cm 59.6in
Track R: 153.4cm 60.4in
Length: 448.3cm 176.5in
Width: 180.3cm 71.0in
Height: 118.6cm 46.7in
Kerb weight: 1489.1kg
3280.0lb
Fuel: 75.7L 16.6gal 20.0galUS

142 Chevrolet
Nova CL
1986 USA

95.0mph 152.9kmh
0-50mph 80.5kmh: 9.3secs
0-60mph 96.5kmh: 13.0secs
0-1/4 mile: 18.8secs
70.0bhp 52.2kW 71.0PS
@ 4800rpm
85.0lbft 115.2Nm @ 2800rpm
69.7bhp/ton 68.5bhp/tonne
44.1bhp/L 32.9kW/L 44.7PS/L
40.4ft/sec 12.3m/sec
36.0mpg 30.0mpgUS
7.8L/100km
Petrol 4-stroke piston
1587cc 96.8cu in
In-line 4 1 Carburettor
Compression ratio: 9.0:1
Bore: 81.0mm 3.2in
Stroke: 77.0mm 3.0in
Valve type/No: Overhead 8
Transmission: Manual
No. of forward speeds: 5
Wheels driven: Front
Springs F/R: Coil/Coil
Brake system: PA
Brakes F/R: Disc/Drum
Steering: Rack & pinion PA
Wheelbase: 243.1cm 95.7in
Track F: 142.5cm 56.1in
Track R: 140.5cm 55.3in
Length: 422.4cm 166.3in
Width: 163.6cm 64.4in
Height: 134.1cm 52.8in
Kerb weight: 1021.5kg
2250.0lb
Fuel: 50.0L 11.0gal 13.2galUS

143 Chevrolet
Beretta GT
1987 USA
115.0mph 185.0kmh
0-50mph 80.5kmh: 5.9secs
0-60mph 96.5kmh: 8.3secs

143 Chevrolet
Beretta GT

0-1/4 mile: 16.3secs
125.0bhp 93.2kW 126.7PS
@ 4500rpm
160.0lbft 216.8Nm @ 3600rpm
98.1bhp/ton 96.4bhp/tonne
44.1bhp/L 32.9kW/L 44.7PS/L
37.4ft/sec 11.4m/sec
24.0mpg 20.0mpgUS
11.8L/100km
Petrol 4-stroke piston
2837cc 173.1cu in
Vee 6 fuel injection
Compression ratio: 8.9:1
Bore: 89.0mm 3.5in
Stroke: 76.0mm 3.0in
Valve type/No: Overhead 12
Transmission: Manual
No. of forward speeds: 5
Wheels driven: Front
Springs F/R: Coil/Coil
Brake system: PA
Brakes F/R: Disc/Drum
Steering: Rack & pinion PA
Wheelbase: 262.6cm 103.4in
Track F: 141.2cm 55.6in
Track R: 143.5cm 56.5in
Length: 475.5cm 187.2in
Width: 173.2cm 68.2in
Height: 133.6cm 52.6in
Ground clearance: 16.5cm
6.5in
Kerb weight: 1296.2kg
2855.0lb
Fuel: 51.5L 11.3gal 13.6galUS

144 Chevrolet
Camaro IROC-Z
1987 USA
150.0mph 241.4kmh
0-60mph 96.5kmh: 6.6secs
0-1/4 mile: 14.9secs
215.0bhp 160.3kW 218.0PS
@ 4400rpm
295.0lbft 399.7Nm @ 3200rpm
140.4bhp/ton 138.1bhp/tonne
43.0bhp/L 32.1kW/L 43.6PS/L
40.1ft/sec 12.2m/sec
20.1mpg 16.7mpgUS
14.1L/100km
Petrol 4-stroke piston
4999cc 305.0cu in
Vee 8 fuel injection
Compression ratio: 9.3:1
Bore: 94.9mm 3.7in
Stroke: 83.4mm 3.3in
Valve type/No: Overhead 16
Transmission: Manual
No. of forward speeds: 5
Wheels driven: Rear
Springs F/R: Coil/Coil
Brakes F/R: Disc/Disc
Steering: Recirculating ball PA
Wheelbase: 256.5cm 101.0in
Track F: 154.2cm 60.7in

Track R: 153.9cm 60.6in
Length: 487.7cm 192.0in
Width: 184.9cm 72.8in
Height: 127.8cm 50.3in
Kerb weight: 1557.2kg
3430.0lb
Fuel: 58.7L 12.9gal 15.5galUS

145 Chevrolet
Camaro IROC-Z L98
1987 USA
149.0mph 239.7kmh
0-60mph 96.5kmh: 6.8secc
0-1/4 mile: 15.3secs
220.0bhp 164.0kW 223.0PS
@ 4200rpm
320.0lbft 433.6Nm @ 3200rpm
141.2bhp/ton 138.8bhp/tonne
38.4bhp/L 28.6kW/L 38.9PS/L
39.7ft/sec 12.1m/sec
33.1mph 53.3kmh/1000rpm
Petrol 4-stroke piston
5733cc 349.8cu in
Vee 8 fuel injection
Compression ratio: 9.0:1
Bore: 101.6mm 4.0in
Stroke: 86.4mm 3.4in
Valve type/No: Overhead 16
Transmission: Automatic
No. of forward speeds: 4
Wheels driven: Rear
Springs F/R: Coil/Coil
Brakes F/R: Disc/Disc
Steering: Recirculating ball PA
Wheelbase: 256.5cm 101.0in
Track F: 154.2cm 60.7in
Track R: 153.9cm 60.6in
Length: 487.7cm 192.0in
Width: 184.9cm 72.8in
Height: 127.8cm 50.3in
Kerb weight: 1584.5kg
3490.0lb
Fuel: 58.7L 12.9gal 15.5galUS

146 Chevrolet
Corvette
1987 USA
154.0mph 247.8kmh
0-50mph 80.5kmh: 4.3secs
0-60mph 96.5kmh: 5.9secs
0-1/4 mile: 14.5secs
240.0bhp 179.0kW 243.3PS
@ 4000rpm
345.0lbft 467.5Nm @ 3200rpm
163.9bhp/ton 161.2bhp/tonne
41.9bhp/L 31.2kW/L 42.4PS/L
38.7ft/sec 11.8m/sec
21.0mpg 17.5mpgUS
13.4L/100km
Petrol 4-stroke piston
5733cc 349.8cu in
Vee 8 fuel injection
Compression ratio: 9.5:1
Bore: 101.6mm 4.0in

Stroke: 88.4mm 3.5in
Valve type/No: Overhead 16
Transmission: Manual with
overdrive
Wheels driven: Rear
Springs F/R: Leaf/Leaf
Brake system: PA ABS
Brakes F/R: Disc/Disc
Steering: Rack & pinion PA
Wheelbase: 244.3cm 96.2in
Track F: 151.4cm 59.6in
Track R: 153.4cm 60.4in
Length: 448.3cm 176.5in
Width: 180.3cm 71.0in
Height: 118.6cm 46.7in
Ground clearance: 11.9cm
4.7in
Kerb weight: 1489.1kg
3280.0lb
Fuel: 75.7L 16.6gal 20.0galUS

147 Chevrolet
Corvette ASC Geneve
1987 USA
154.0mph 247.8kmh
0-60mph 96.5kmh: 5.8secs
0-1/4 mile: 14.4secs
230.0bhp 171.5kW 233.2PS
@ 4000rpm
330.0lbft 447.2Nm @ 3200rpm
153.8bhp/ton 151.2bhp/tonne
40.1bhp/L 29.9kW/L 40.7PS/L
38.7ft/sec 11.8m/sec
Petrol 4-stroke piston
5733cc 349.8cu in
Vee 8 fuel injection
Compression ratio: 9.5:1
Bore: 101.6mm 4.0in
Stroke: 88.4mm 3.5in
Valve type/No: Overhead 16
Transmission: Manual with
overdrive
Wheels driven: Rear
Springs F/R: Leaf/Leaf
Brake system: PA ABS
Brakes F/R: Disc/Disc
Steering: Rack & pinion PA
Wheelbase: 244.3cm 96.2in
Track F: 169.7cm 66.8in
Track R: 184.2cm 72.5in
Length: 467.4cm 184.0in
Width: 193.0cm 76.0in
Height: 121.9cm 48.0in
Kerb weight: 1520.9kg
3350.0lb
Fuel: 75.7L 16.6gal 20.0galUS

148 Chevrolet
Corvette Convertible
1987 USA
150.0mph 241.4kmh
0-50mph 80.5kmh: 4.7secs
0-60mph 96.5kmh: 6.3secs
0-1/4 mile: 14.8secs

240.0bhp 179.0kW 243.3PS
@ 4000rpm
345.0lbft 467.5Nm @ 3200rpm
161.9bhp/ton 159.2bhp/tonne
41.9bhp/L 31.2kW/L 42.4PS/L
38.7ft/sec 11.8m/sec
20.4mpg 17.0mpgUS
13.8L/100km
Petrol 4-stroke piston
5733cc 349.8cu in
Vee 8 fuel injection
Compression ratio: 9.5:1
Bore: 101.6mm 4.0in
Stroke: 88.4mm 3.5in
Valve type/No: Overhead 16
Transmission: Manual with
overdrive
Wheels driven: Rear
Springs F/R: Leaf/Leaf
Brake system: PA ABS
Brakes F/R: Disc/Disc
Steering: Rack & pinion PA
Wheelbase: 244.3cm 96.2in
Track F: 151.4cm 59.6in
Track R: 153.4cm 60.4in
Length: 448.3cm 176.5in
Width: 180.3cm 71.0in
Height: 117.9cm 46.4in
Ground clearance: 12.7cm
5.0in
Kerb weight: 1507.3kg
3320.0lb
Fuel: 75.7L 16.6gal 20.0galUS

149 Chevrolet
Cavalier Z24 Convertible
1988 USA
120.0mph 193.1kmh
0-50mph 80.5kmh: 5.8secs
0-60mph 96.5kmh: 8.2secs
0-1/4 mile: 16.2secs
125.0bhp 93.2kW 126.7PS
@ 4500rpm
160.0lbft 216.8Nm @ 3600rpm
104.7bhp/ton 102.9bhp/tonne
44.1bhp/L 32.9kW/L 44.7PS/L
37.4ft/sec 11.4m/sec
30.0mpg 25.0mpgUS
9.4L/100km
Petrol 4-stroke piston
2837cc 173.1cu in
Vee 6 fuel injection
Compression ratio: 8.9:1
Bore: 89.0mm 3.5in
Stroke: 76.0mm 3.0in
Valve type/No: Overhead 12
Transmission: Manual
No. of forward speeds: 5
Wheels driven: Front
Springs F/R: Coil/Coil
Brake system: PA
Brakes F/R: Disc/Drum
Steering: Rack & pinion PA
Wheelbase: 257.0cm 101.2in

Track F: 140.7cm 55.4in
Track R: 140.2cm 55.2in
Length: 437.9cm 172.4in
Width: 167.6cm 66.0in
Height: 131.8cm 51.9in
Kerb weight: 1214.4kg
2675.0lb
Fuel: 51.5L 11.3gal 13.6galUS

150 Chevrolet
Corvette
1988 USA
155.0mph 249.4kmh
0-50mph 80.5kmh: 4.3secs
0-60mph 96.5kmh: 6.0secs
0-1/4 mile: 14.6secs
245.0bhp 182.7kW 248.4PS
@ 4300rpm
340.0lbft 460.7Nm @ 3200rpm
164.8bhp/ton 162.1bhp/tonne
42.7bhp/L 31.8kW/L 43.2PS/L
41.6ft/sec 12.7m/sec
21.0mpg 17.5mpgUS
13.4L/100km
Petrol 4-stroke piston
5743cc 350.4cu in
Vee 8 fuel injection
Compression ratio: 9.5:1
Bore: 101.6mm 4.0in
Stroke: 88.4mm 3.5in
Valve type/No: Overhead 16
Transmission: Manual with
overdrive
Wheels driven: Rear
Springs F/R: Leaf/Leaf
Brake system: PA ABS
Brakes F/R: Disc/Disc
Steering: Rack & pinion PA
Wheelbase: 244.3cm 96.2in
Track F: 151.4cm 59.6in
Track R: 153.4cm 60.4in
Length: 448.3cm 176.5in
Width: 180.3cm 71.0in
Height: 118.6cm 46.7in
Kerb weight: 1511.8kg
3330.0lb
Fuel: 75.7L 16.6gal 20.0galUS

151 Chevrolet
Corvette Bakeracing SCCA
Escort
1988 USA
169.0mph 271.9kmh
0-50mph 80.5kmh: 3.3secs
0-60mph 96.5kmh: 4.5secs
0-1/4 mile: 13.2secs
325.0bhp 242.3kW 329.5PS
@ 4400rpm
410.0lbft 555.6Nm @ 3250rpm
227.5bhp/ton 223.7bhp/tonne
56.6bhp/L 42.2kW/L 57.4PS/L
42.5ft/sec 13.0m/sec
12.0mpg 10.0mpgUS
23.5L/100km

Petrol 4-stroke piston
5743cc 350.4cu in
Vee 8 fuel injection
Compression ratio: 10.0:1
Bore: 101.6mm 4.0in
Stroke: 88.4mm 3.5in
Valve type/No: Overhead 16
Transmission: Manual with
overdrive
Wheels driven: Rear
Springs F/R: Leaf/Leaf
Brake system: PA ABS
Brakes F/R: Disc/Disc
Steering: Rack & pinion PA
Wheelbase: 244.3cm 96.2in
Track F: 151.4cm 59.6in
Track R: 153.4cm 60.4in
Length: 448.3cm 176.5in
Width: 180.3cm 71.0in
Height: 116.1cm 45.7in
Kerb weight: 1452.8kg
3200.0lb
Fuel: 75.7L 16.6gal 20.0galUS

152 Chevrolet
Corvette Callaway
1988 USA
191.0mph 307.3kmh
0-60mph 96.5kmh: 4.6secs
0-1/4 mile: 13.0secs
382.0bhp 284.9kW 387.3PS
@ 4250rpm
545.0lbft 738.5Nm @ 2750rpm
248.0bhp/ton 243.9bhp/tonne
66.5bhp/L 49.6kW/L 67.4PS/L
41.1ft/sec 12.5m/sec
Petrol 4-stroke piston
5743cc 350.4cu in
turbocharged
Vee 8 fuel injection
Compression ratio: 7.5:1
Bore: 101.6mm 4.0in
Stroke: 88.4mm 3.5in
Valve type/No: Overhead 16
Transmission: Manual with
overdrive
Wheels driven: Rear
Springs F/R: Leaf/Leaf
Brake system: PA ABS
Brakes F/R: Disc/Disc
Steering: Rack & pinion PA
Wheelbase: 244.3cm 96.2in
Track F: 151.4cm 59.6in
Track R: 153.4cm 60.4in
Length: 448.3cm 176.5in
Width: 180.3cm 71.0in
Height: 118.6cm 46.7in
Kerb weight: 1566.3kg
3450.0lb
Fuel: 75.7L 16.6gal 20.0galUS

153 Chevrolet
Corvette Convertible
1988 USA

158.0mph 254.2kmh
0-60mph 96.5kmh: 6.0secs
0-1/4 mile: 14.6secs
245.0bhp 182.7kW 248.4PS
@ 4300rpm
340.0lbft 460.7Nm @ 3200rpm
164.3bhp/ton 161.6bhp/tonne
42.7bhp/L 31.9kW/L 43.3PS/L
41.6ft/sec 12.7m/sec
23.2mpg 19.3mpgUS
12.2L/100km
Petrol 4-stroke piston
5733cc 349.8cu in
Vee 8 fuel injection
Compression ratio: 9.5:1
Bore: 101.6mm 4.0in
Stroke: 88.4mm 3.5in
Valve type/No: Overhead 16
Transmission: Manual with
overdrive
Wheels driven: Rear
Springs F/R: Leaf/Leaf
Brake system: PA ABS
Brakes F/R: Disc/Disc
Steering: Rack & pinion PA
Wheelbase: 244.3cm 96.2in
Track F: 151.4cm 59.6in
Track R: 153.4cm 60.4in
Length: 448.3cm 176.5in
Width: 180.3cm 71.0in
Height: 117.9cm 46.4in
Kerb weight: 1516.4kg
3340.0lb
Fuel: 75.7L 16.6gal 20.0galUS

154 Chevrolet
Corvette Guldstrand Grand
Sport 80
1988 USA
172.0mph 276.7kmh
0-60mph 96.5kmh: 5.2secs
0-1/4 mile: 13.6secs
374.0bhp 278.9kW 379.2PS
@ 5600rpm
442.0lbft 598.9Nm @ 3600rpm
254.2bhp/ton 250.0bhp/tonne
61.3bhp/L 45.7kW/L 62.2PS/L
54.1ft/sec 16.5m/sec
Petrol 4-stroke piston
6096cc 371.9cu in
Vee 8 fuel injection
Compression ratio: 10.0:1
Bore: 104.7mm 4.1in
Stroke: 88.4mm 3.5in
Valve type/No: Overhead 16
Transmission: Manual with
overdrive
Wheels driven: Rear
Springs F/R: Leaf/Leaf
Brake system: PA ABS
Brakes F/R: Disc/Disc
Steering: Rack & pinion PA
Wheelbase: 244.3cm 96.2in
Track F: 151.4cm 59.6in

Track R: 153.4cm 60.4in
Length: 448.3cm 176.5in
Width: 180.3cm 71.0in
Height: 117.9cm 46.4in
Kerb weight: 1495.9kg
3295.0lb
Fuel: 75.7L 16.6gal 20.0galUS

155 Chevrolet
Corvette Morrison/Baker
Nelson Ledges
1988 USA
127.0mph 204.3kmh
0-50mph 80.5kmh: 3.6secs
0-60mph 96.5kmh: 4.9secs
0-1/4 mile: 13.7secs
360.0bhp 268.4kW 365.0PS
@ 5000rpm
437.0lbft 592.1Nm @ 3500rpm
252.0bhp/ton 247.8bhp/tonne
60.9bhp/L 45.4kW/L 61.8PS/L
48.3ft/sec 14.7m/sec
12.0mpg 10.0mpgUS
23.5L/100km
Petrol 4-stroke piston
5907cc 360.4cu in
Vee 8 fuel injection
Compression ratio: 10.5:1
Bore: 103.1mm 4.1in
Stroke: 88.4mm 3.5in
Valve type/No: Overhead 16
Transmission: Manual
No. of forward speeds: 5
Wheels driven: Rear
Springs F/R: Leaf/Leaf
Brake system: PA ABS
Brakes F/R: Disc/Disc
Steering: Rack & pinion PA
Wheelbase: 244.3cm 96.2in
Track F: 151.4cm 59.6in
Track R: 153.4cm 60.4in
Length: 452.1cm 178.0in
Width: 180.3cm 71.0in
Height: 116.1cm 45.7in
Kerb weight: 1452.8kg
3200.0lb
Fuel: 75.7L 16.6gal 20.0galUS

156 Chevrolet
Corvette Z51
1988 USA
158.0mph 254.2kmh
0-50mph 80.5kmh: 4.3secs
0-60mph 96.5kmh: 6.0secs
0-1/4 mile: 14.6secs
245.0bhp 182.7kW 248.4PS
@ 4300rpm
340.0lbft 460.7Nm @ 3200rpm
164.8bhp/ton 162.1bhp/tonne
42.7bhp/L 31.8kW/L 43:2PS/L
41.6ft/sec 12.7m/sec
21.0mpg 17.5mpgUS
13.4L/100km
Petrol 4-stroke piston

5743cc 350.4cu in
Vee 8 fuel injection
Compression ratio: 9.5:1
Bore: 101.6mm 4.0in
Stroke: 88.4mm 3.5in
Valve type/No: Overhead 16
Transmission: Manual with overdrive
Wheels driven: Rear
Springs F/R: Leaf/Leaf
Brake system: PA ABS
Brakes F/R: Disc/Disc
Steering: Rack & pinion PA
Wheelbase: 244.3cm 96.2in
Track F: 151.4cm 59.6in
Track R: 153.4cm 60.4in
Length: 448.3cm 176.5in
Width: 180.3cm 71.0in
Height: 118.6cm 46.7in
Kerb weight: 1511.8kg 3330.0lb
Fuel: 75.7L 16.6gal 20.0galUS

157 Chevrolet

Nova Twin Cam
1988 Japan
110.0mph 177.0kmh
0-50mph 80.5kmh: 6.3secs
0-60mph 96.5kmh: 8.9secs
0-1/4 mile: 16.7secs
110.0bhp 82.0kW 111.5PS @ 6600rpm
98.0lbft 132.8Nm @ 4800rpm
103.1bhp/ton 101.4bhp/tonne
69.3bhp/L 51.7kW/L 70.3PS/L
55.5ft/sec 16.9m/sec
31.8mpg 26.5mpgUS
8.9L/100km
Petrol 4-stroke piston
1587cc 96.8cu in
In-line 4 fuel injection
Compression ratio: 9.4:1
Bore: 81.0mm 3.2in
Stroke: 77.0mm 3.0in
Valve type/No: Overhead 16
Transmission: Manual
No. of forward speeds: 5
Wheels driven: Front
Springs F/R: Coil/Coil
Brake system: PA
Brakes F/R: Disc/Disc
Steering: Rack & pinion PA
Wheelbase: 243.1cm 95.7in
Track F: 142.5cm 56.1in
Track R: 140.5cm 55.3in
Length: 422.4cm 166.3in
Width: 163.6cm 64.4in
Height: 134.1cm 52.8in
Kerb weight: 1085.1kg 2390.0lb
Fuel: 50.0L 11.0gal 13.2galUS

158 Chevrolet

Baretta GTU
1989 USA
110.0mph 177.0kmh
0-50mph 80.5kmh: 6.3secs
0-60mph 96.5kmh: 9.0secs
0-1/4 mile: 16.7secs
130.0bhp 96.9kW 131.8PS @ 4700rpm
160.0lbft 216.8Nm @ 3600rpm
103.3bhp/ton 101.5bhp/tonne
45.8bhp/L 34.2kW/L 46.5PS/L
39.0ft/sec 11.9m/sec
26.4mpg 22.0mpgUS
10.7L/100km
Petrol 4-stroke piston
2837cc 173.1cu in
Vee 6 fuel injection
Compression ratio: 8.9:1
Bore: 89.0mm 3.5in
Stroke: 76.0mm 3.0in
Valve type/No: Overhead 12
Transmission: Manual
No. of forward speeds: 5
Wheels driven: Front
Springs F/R: Coil/Coil
Brake system: PA
Brakes F/R: Disc/Drum
Steering: Rack & pinion PA
Wheelbase: 262.6cm 103.4in
Track F: 141.2cm 55.6in
Track R: 143.5cm 56.5in
Length: 475.7cm 187.2in
Width: 172.7cm 68.0in
Height: 133.6cm 52.6in
Kerb weight: 1280.3kg 2820.0lb
Fuel: 51.5L 11.3gal 13.6galUS

159 Chevrolet

Camaro Gottlieb 1969
1989 USA
203.0mph 326.6kmh
0-50mph 80.5kmh: 3.6secs
0-60mph 96.5kmh: 4.8secs
0-1/4 mile: 12.2secs
798.0bhp 595.1kW 809.1PS @ 7000rpm
638.0lbft 864.5Nm @ 6500rpm
595.8bhp/ton 585.9bhp/tonne
90.0bhp/L 67.1kW/L 91.3PS/L
82.6ft/sec 25.2m/sec
Petrol 4-stroke piston
8865cc 540.9cu in
Vee 8 1 Carburettor
Compression ratio: 12.8:1
Bore: 114.3mm 4.5in
Stroke: 108.0mm 4.2in
Valve type/No: Overhead 16
Transmission: Manual
No. of forward speeds: 4
Wheels driven: Rear
Springs F/R: Coil/Coil
Brakes F/R: Disc/Disc
Steering: Rack & pinion
Wheelbase: 274.3cm 108.0in
Track F: 144.8cm 57.0in
Track R: 144.8cm 57.0in
Length: 469.1cm 184.7in
Width: 184.2cm 72.5in
Height: 129.5cm 51.0in
Kerb weight: 1362.0kg 3000.0lb

160 Chevrolet

Camaro IROC-Z
1989 USA
145.0mph 233.3kmh
0-50mph 80.5kmh: 4.7secs
0-60mph 96.5kmh: 6.6secs
0-1/4 mile: 14.9secs
230.0bhp 171.5kW 233.2PS @ 4600rpm
300.0lbft 406.5Nm @ 3200rpm
155.2bhp/ton 152.6bhp/tonne
46.0bhp/L 34.3kW/L 46.6PS/L
44.5ft/sec 13.5m/sec
19.2mpg 16.0mpgUS
14.7L/100km
Petrol 4-stroke piston
5001cc 305.1cu in
Vee 8 fuel injection
Compression ratio: 9.3:1
Bore: 94.9mm 3.7in
Stroke: 88.4mm 3.5in
Valve type/No: Overhead 16
Transmission: Manual
No. of forward speeds: 5
Wheels driven: Rear
Springs F/R: Coil/Coil
Brake system: PA
Brakes F/R: Disc/Disc
Steering: Recirculating ball PA
Wheelbase: 256.5cm 101.0in
Track F: 152.4cm 60.0in
Track R: 154.7cm 60.9in
Length: 487.7cm 192.0in
Width: 184.9cm 72.8in
Height: 127.8cm 50.3in
Kerb weight: 1507.3kg 3320.0lb
Fuel: 58.7L 12.9gal 15.5galUS

161 Chevrolet

Camaro IROC-Z Automatic
1989 USA
145.0mph 233.3kmh
0-50mph 80.5kmh: 4.8secs
0-60mph 96.5kmh: 6.6secs
0-1/4 mile: 15.2secs
240.0bhp 179.0kW 243.3PS @ 4400rpm
345.0lbft 467.5Nm @ 3200rpm
154.5bhp/ton 151.9bhp/tonne
41.9bhp/L 31.2kW/L 42.4PS/L
42.5ft/sec 13.0m/sec
19.2mpg 16.0mpgUS
14.7L/100km
Petrol 4-stroke piston
5733cc 349.8cu in
Vee 8 fuel injection
Compression ratio: 9.3:1
Bore: 101.6mm 4.0in
Stroke: 88.4mm 3.5in
Valve type/No: Overhead 16
Transmission: Automatic
No. of forward speeds: 4
Wheels driven: Rear
Springs F/R: Coil/Coil
Brake system: PA
Brakes F/R: Disc/Disc
Steering: Recirculating ball PA
Wheelbase: 256.5cm 101.0in
Track F: 152.4cm 60.0in
Track R: 154.7cm 60.9in
Length: 487.7cm 192.0in
Width: 184.9cm 72.8in
Height: 127.8cm 50.3in
Kerb weight: 1579.9kg 3480.0lb
Fuel: 58.7L 12.9gal 15.5galUS

162 Chevrolet

Camaro IROC-Z Chevrolet Engineering
1989 USA
193.0mph 310.5kmh
0-50mph 80.5kmh: 3.7secs
0-60mph 96.5kmh: 4.5secs
0-1/4 mile: 12.8secs
530.0bhp 395.2kW 537.3PS @ 5000rpm
672.0lbft 910.6Nm @ 4500rpm
320.9bhp/ton 315.5bhp/tonne
63.9bhp/L 47.7kW/L 64.8PS/L
55.6ft/sec 16.9m/sec
Petrol 4-stroke piston
8292cc 505.9cu in
Vee 8 fuel injection
Compression ratio: 9.7:1
Bore: 114.3mm 4.5in
Stroke: 101.6mm 4.0in
Valve type/No: Overhead 16
Transmission: Manual
No. of forward speeds: 6
Wheels driven: Rear
Springs F/R: Coil/Coil
Brake system: PA
Brakes F/R: Disc/Disc
Steering: Recirculating ball PA
Wheelbase: 256.5cm 101.0in
Track F: 152.4cm 60.0in
Track R: 154.7cm 60.9in
Length: 487.7cm 192.0in
Width: 184.9cm 72.8in
Height: 127.8cm 50.3in
Kerb weight: 1679.8kg 3700.0lb

163 Chevrolet

Corvette
1989 USA
149.0mph 239.7kmh
0-50mph 80.5kmh: 4.8secs
0-60mph 96.5kmh: 6.2secs

167 Chevrolet
Corvette ZR-1

0-1/4 mile: 14.8secs
245.0bhp 182.7kW 248.4PS
@ 4300rpm
340.0lbft 460.7Nm @ 3200rpm
164.8bhp/ton 162.1bhp/tonne
42.7bhp/L 31.9kW/L 43.3PS/L
41.6ft/sec 12.7m/sec
21.5mpg 17.9mpgUS
13.1L/100km
Petrol 4-stroke piston
5733cc 349.8cu in
Vee 8 fuel injection
Compression ratio: 9.5:1
Bore: 101.6mm 4.0in
Stroke: 88.4mm 3.5in
Valve type/No: Overhead 16
Transmission: Manual
No. of forward speeds: 6
Wheels driven: Rear
Springs F/R: Leaf/Leaf
Brake system: PA ABS
Brakes F/R: Disc/Disc
Steering: Rack & pinion PA
Wheelbase: 244.3cm 96.2in
Track F: 151.4cm 59.6in
Track R: 153.4cm 60.4in
Length: 448.3cm 176.5in
Width: 180.3cm 71.0in
Height: 118.6cm 46.7in
Kerb weight: 1511.8kg
3330.0lb
Fuel: 75.7L 16.6gal 20.0galUS

164 Chevrolet
Corvette Callaway
1989 USA
191.0mph 307.3kmh
0-50mph 80.5kmh: 3.9secs
0-60mph 96.5kmh: 4.9secs
0-1/4 mile: 13.4secs
382.0bhp 284.9kW 387.3PS
@ 4250rpm
562.0lbft 761.5Nm @ 2500rpm

244.5bhp/ton 240.4bhp/tonne
66.6bhp/L 49.7kW/L 67.6PS/L
41.1ft/sec 12.5m/sec
Petrol 4-stroke piston
5733cc 349.8cu in
turbocharged
Vee 8 fuel injection
Compression ratio: 7.5:1
Bore: 101.6mm 4.0in
Stroke: 88.4mm 3.5in
Valve type/No: Overhead 16
Transmission: Manual
No. of forward speeds: 6
Wheels driven: Rear
Springs F/R: Leaf/Leaf
Brake system: PA ABS
Brakes F/R: Disc/Disc
Steering: Rack & pinion PA
Wheelbase: 244.3cm 96.2in
Track F: 151.4cm 59.6in
Track R: 153.4cm 60.4in
Length: 448.3cm 176.5in
Width: 180.3cm 71.0in
Height: 118.6cm 46.7in
Kerb weight: 1589.0kg
3500.0lb
Fuel: 75.7L 16.6gal 20.0galUS

165 Chevrolet
Corvette Guldstrand Grand
Sport 80
1989 USA
171.0mph 275.1kmh
0-50mph 80.5kmh: 4.0secs
0-60mph 96.5kmh: 5.1secs
0-1/4 mile: 13.5secs
355.0bhp 264.7kW 359.9PS
@ 5200rpm
425.0lbft 575.9Nm @ 4000rpm
238.8bhp/ton 234.8bhp/tonne
58.2bhp/L 43.4kW/L 59.0PS/L
50.3ft/sec 15.3m/sec
Petrol 4-stroke piston

6097cc 372.0cu in
Vee 8 fuel injection
Compression ratio: 9.5:1
Bore: 104.8mm 4.1in
Stroke: 88.4mm 3.5in
Valve type/No: Overhead 16
Transmission: Manual with
overdrive
Wheels driven: Rear
Springs F/R: Leaf/Leaf
Brake system: PA ABS
Brakes F/R: Disc/Disc
Steering: Rack & pinion PA
Wheelbase: 244.3cm 96.2in
Track F: 151.4cm 59.6in
Track R: 153.4cm 60.4in
Length: 448.3cm 176.5in
Width: 180.3cm 71.0in
Height: 118.6cm 46.7in
Kerb weight: 1511.8kg
3330.0lb
Fuel: 75.7L 16.6gal 20.0galUS

166 Chevrolet
Corvette L98
1989 USA
155.0mph 249.4kmh
0-60mph 96.5kmh: 6.6secs
0-1/4 mile: 14.8secs
245.0bhp 182.7kW 248.4PS
@ 4300rpm
340.0lbft 460.7Nm @ 3200rpm
164.3bhp/ton 161.6bhp/tonne
42.7bhp/L 31.9kW/L 43.3PS/L
41.6ft/sec 12.7m/sec
23.4mpg 19.5mpgUS
12.1L/100km
Petrol 4-stroke piston
5733cc 349.8cu in
Vee 8 fuel injection
Compression ratio: 9.5:1
Bore: 101.6mm 4.0in
Stroke: 88.4mm 3.5in

Valve type/No: Overhead 16
Transmission: Manual
No. of forward speeds: 6
Wheels driven: Rear
Springs F/R: Leaf/Leaf
Brake system: PA ABS
Brakes F/R: Disc/Disc
Steering: Rack & pinion PA
Wheelbase: 244.3cm 96.2in
Track F: 151.4cm 59.6in
Track R: 153.4cm 60.4in
Length: 446.0cm 175.6in
Width: 180.3cm 71.0in
Height: 117.9cm 46.4in
Kerb weight: 1516.4kg
3340.0lb
Fuel: 75.7L 16.6gal 20.0galUS

167 Chevrolet
Corvette ZR-1
1989 USA
170.0mph 273.5kmh
0-50mph 80.5kmh: 3.6secs
0-60mph 96.5kmh: 4.9secs
0-1/4 mile: 13.4secs
380.0bhp 283.4kW 385.3PS
@ 6200rpm
370.0lbft 501.4Nm @ 4500rpm
241.1bhp/ton 237.1bhp/tonne
66.3bhp/L 49.5kW/L 67.3PS/L
63.0ft/sec 19.2m/sec
Petrol 4-stroke piston
5727cc 349.4cu in
Vee 8 fuel injection
Compression ratio: 11.3:1
Bore: 99.0mm 3.9in
Stroke: 93.0mm 3.7in
Valve type/No: Overhead 32
Transmission: Manual
No. of forward speeds: 6
Wheels driven: Rear
Springs F/R: Leaf/Leaf
Brake system: PA ABS

Brakes F/R: Disc/Disc
Steering: Rack & pinion PA
Wheelbase: 244.3cm 96.2in
Track F: 151.4cm 59.6in
Track R: 153.4cm 60.4in
Length: 448.3cm 176.5in
Width: 180.3cm 71.0in
Height: 118.6cm 46.7in
Kerb weight: 1602.6kg
3530.0lb
Fuel: 75.7l 16.6qal 20.0galUS

168 Chevrolet
Beretta GT
1990 USA
125.0mph 201.1kmh
0-60mph 96.5kmh: 8.6secs
0-1/4 mile: 16.4secs
135.0bhp 100.7kW 136.9PS
@ 4200rpm
180.0lbft 243.9Nm @ 3600rpm
105.7bhp/ton 104.0bhp/tonne
43.1bhp/L 32.1kW/L 43.7PS/L
38.6ft/sec 11.8m/sec
24.0mpg 20.0mpgUS
11.8L/100km
Petrol 4-stroke piston
3135cc 191.3cu in
Vee 6 fuel injection
Compression ratio: 8.8:1
Bore: 89.0mm 3.5in
Stroke: 84.0mm 3.3in
Valve type/No: Overhead 12
Transmission: Manual
No. of forward speeds: 5
Wheels driven: Front
Springs F/R: Coil/Coil
Brake system: PA
Brakes F/R: Disc/Drum
Steering: Rack & pinion PA
Wheelbase: 262.6cm 103.4in
Track F: 141.2cm 55.6in
Track R: 140.0cm 55.1in
Length: 465.8cm 183.4in
Width: 173.2cm 68.2in
Height: 142.7cm 56.2in
Kerb weight: 1298.4kg
2860.0lb
Fuel: 59.0L 13.0gal 15.6galUS

169 Chevrolet
Beretta GTZ
1990 USA
125.0mph 201.1kmh
0-50mph 80.5kmh: 6.3secs
0-60mph 96.5kmh: 8.0secs
0-1/4 mile: 16.4secs
180.0bhp 134.2kW 182.5PS
@ 6200rpm
160.0lbft 216.8Nm @ 5200rpm
140.5bhp/ton 138.1bhp/tonne
59.2bhp/L 59.1kW/L 80.3PS/L
7.7ft/sec 17.6m/sec
25.3mpg 21.1mpgUS

11.1L/100km
Petrol 4-stroke piston
2272cc 138.6cu in
In-line 4 fuel injection
Compression ratio: 10.0:1
Bore: 92.2mm 3.6in
Stroke: 85.1mm 3.3in
Valve type/No: Overhead 16
Transmission: Manual
No. of forward speeds: 5
Wheels driven: Front
Springs F/R: Coil/Coil
Brake system: PA
Brakes F/R: Disc/Drum
Steering: Rack & pinion PA
Wheelbase: 262.6cm 103.4in
Track F: 141.2cm 55.6in
Track R: 143.5cm 56.5in
Length: 475.5cm 187.2in
Width: 172.7cm 68.0in
Height: 133.6cm 52.6in
Kerb weight: 1303.0kg
2870.0lb
Fuel: 51.5L 11.3gal 13.6galUS

170 Chevrolet
Camaro IROC-Z Convertible
1990 USA
120.0mph 193.1kmh
0-50mph 80.5kmh: 6.2secs
0-60mph 96.5kmh: 8.6secs
0-1/4 mile: 16.6secs
195.0bhp 145.4kW 197.7PS
@ 4000rpm
295.0lbft 399.7Nm @ 2800rpm
124.1bhp/ton 122.0bhp/tonne
39.0bhp/L 29.1kW/L 39.5PS/L
38.7ft/sec 11.8m/sec
22.3mpg 18.6mpgUS
12.6L/100km
Petrol 4-stroke piston
5002cc 305.2cu in
Vee 8 fuel injection
Compression ratio: 9.3:1
Bore: 94.9mm 3.7in
Stroke: 88.4mm 3.5in
Valve type/No: Overhead 16
Transmission: Automatic
No. of forward speeds: 4
Wheels driven: Rear
Springs F/R: Coil/Coil
Brake system: PA
Brakes F/R: Disc/Disc
Steering: Recirculating ball PA
Wheelbase: 256.5cm 101.0in
Track F: 152.4cm 60.0in
Track R: 154.7cm 60.9in
Length: 487.7cm 192.0in
Width: 184.9cm 72.8in
Height: 127.8cm 50.3in
Kerb weight: 1598.1kg
3520.0lb
Fuel: 58.7L 12.9gal 15.5galUS

171 Chevrolet
Camaro Z/28
1990 USA
145.0mph 233.3kmh
0-50mph 80.5kmh: 4.8secs
0-60mph 96.5kmh: 6.6secs
0-1/4 mile: 15.2secs
240.0bhp 179.0kW 243.3PS
@ 4400rpm
345.0lbft 467.5Nm @ 3200rpm
153.6bhp/ton 151.0bhp/tonne
41.9bhp/L 31.2kW/L 42.4PS/L
42.5ft/sec 13.0m/sec
19.2mpg 16.0mpgUS
14.7L/100km
Petrol 4-stroke piston
5733cc 349.8cu in
Vee 8 fuel injection
Compression ratio: 9.3:1
Bore: 101.6mm 4.0in
Stroke: 88.4mm 3.5in
Valve type/No: Overhead 16
Transmission: Automatic
No. of forward speeds: 4
Wheels driven: Rear
Springs F/R: Coil/Coil
Brake system: PA
Brakes F/R: Disc/Disc
Steering: Recirculating ball PA
Wheelbase: 256.5cm 101.0in
Track F: 152.4cm 60.0in
Track R: 154.7cm 60.9in
Length: 487.7cm 192.0in
Width: 184.9cm 72.8in
Height: 127.8cm 50.3in
Kerb weight: 1589.0kg
3500.0lb
Fuel: 58.7L 12.9gal 15.5galUS

172 Chevrolet
Cavalier Z24
1990 USA
125.0mph 201.1kmh
0-50mph 80.5kmh: 5.8secs
0-60mph 96.5kmh: 8.1secs
0-1/4 mile: 16.1secs
140.0bhp 104.4kW 141.9PS
@ 4500rpm
185.0lbft 250.7Nm @ 3600rpm
128.8bhp/ton 126.6bhp/tonne
44.7bhp/L 33.3kW/L 45.3PS/L
41.4ft/sec 12.6m/sec
23.1mpg 19.2mpgUS
12.3L/100km
Petrol 4-stroke piston
3135cc 191.3cu in
Vee 6 fuel injection
Compression ratio: 9.0:1
Bore: 89.0mm 3.5in
Stroke: 84.0mm 3.3in
Valve type/No: Overhead 12
Transmission: Manual
No. of forward speeds: 5
Wheels driven: Front

Springs F/R: Coil/Coil
Brake system: PA
Brakes F/R: Disc/Drum
Steering: Rack & pinion PA
Wheelbase: 257.0cm 101.2in
Track F: 141.7cm 55.8in
Track R: 140.2cm 55.2in
Length: 453.6cm 178.6in
Width: 167.6cm 66.0in
Height: 132.1cm 52.0in
Kerb weight: 1105.5kg
2435.0lb
Fuel: 51.5L 11.3gal 13.6galUS

173 Chevrolet
Corvette Callaway
1990 USA
170.0mph 273.5kmh
0-50mph 80.5kmh: 3.9secs
0-60mph 96.5kmh: 5.1secs
0-1/4 mile: 13.4secs
390.0bhp 290.8kW 395.4PS
@ 4250rpm
562.0lbft 761.5Nm @ 2500rpm
256.9bhp/ton 252.7bhp/tonne
68.0bhp/L 50.7kW/L 68.9PS/L
41.1ft/sec 12.5m/sec
21.6mpg 18.0mpgUS
13.1L/100km
Petrol 4-stroke piston
5735cc 349.9cu in
turbocharged
Vee 8 fuel injection
Compression ratio: 7.5:1
Bore: 101.6mm 4.0in
Stroke: 88.4mm 3.5in
Valve type/No: Overhead 16
Transmission: Manual
No. of forward speeds: 6
Wheels driven: Rear
Springs F/R: Leaf/Leaf
Brake system: PA
Brakes F/R: Disc/Disc
Steering: Rack & pinion PA
Wheelbase: 244.3cm 96.2in
Track F: 151.4cm 59.6in
Track R: 153.4cm 60.4in
Length: 448.3cm 176.5in
Width: 180.3cm 71.0in
Height: 118.6cm 46.7in
Kerb weight: 1543.6kg
3400.0lb
Fuel: 75.7L 16.6gal 20.0galUS

174 Chevrolet
Corvette Callaway
Sledgehammer
1990 USA
255.0mph 410.3kmh
0-60mph 96.5kmh: 3.9secs
0-1/4 mile: 10.6secs
880.0bhp 656.2kW 892.2PS
@ 6250rpm
772.0lbft 1046.1Nm @

99

5250rpm
537.1bhp/ton 528.1bhp/tonne
153.5bhp/L 114.5kW/L
155.6PS/L
60.4ft/sec 18.4m/sec
Petrol 4-stroke piston
5733cc 349.8cu in
turbocharged
Vee 8 fuel injection
Compression ratio: 7.5:1
Bore: 101.6mm 4.0in
Stroke: 88.4mm 3.5in
Valve type/No: Overhead 16
Transmission: Manual
No. of forward speeds: 6
Wheels driven: Rear
Springs F/R: Leaf/Leaf
Brake system: PA ABS
Brakes F/R: Disc/Disc
Steering: Rack & pinion PA
Wheelbase: 244.3cm 96.2in
Track F: 151.4cm 59.6in
Track R: 153.4cm 60.4in
Length: 448.3cm 176.5in
Width: 180.3cm 71.0in
Height: 118.6cm 46.7in
Kerb weight: 1666.2kg
3670.0lb
Fuel: 75.7L 16.6gal 20.0galUS

175 Chevrolet
Corvette L98
1990 USA
149.0mph 239.7kmh
0-50mph 80.5kmh: 4.8secs
0-60mph 96.5kmh: 6.3secs
0-1/4 mile: 14.8secs
245.0bhp 182.7kW 248.4PS
@ 4300rpm
340.0lbft 460.7Nm @ 3200rpm
164.8bhp/ton 162.1bhp/tonne
42.7bhp/L 31.9kW/L 43.3PS/L
41.6ft/sec 12.7m/sec
20.9mpg 17.4mpgUS
13.5L/100km
Petrol 4-stroke piston
5735cc 349.9cu in
Vee 8 fuel injection
Compression ratio: 9.5:1
Bore: 101.6mm 4.0in
Stroke: 88.4mm 3.5in
Valve type/No: Overhead 16
Transmission: Manual
No. of forward speeds: 6
Wheels driven: Rear
Springs F/R: Leaf/Leaf
Brake system: PA ABS
Brakes F/R: Disc/Disc
Steering: Rack & pinion PA
Wheelbase: 244.3cm 96.2in
Track F: 151.4cm 59.6in
Track R: 153.4cm 60.4in
Length: 448.3cm 176.5in
Width: 180.3cm 71.0in

Height: 118.6cm 46.7in
Kerb weight: 1511.8kg
3330.0lb
Fuel: 75.7L 16.6gal 20.0galUS

176 Chevrolet
Lumina APV
1990 USA
105.0mph 168.9kmh
0-50mph 80.5kmh: 8.7secs
0-60mph 96.5kmh: 12.6secs
0-1/4 mile: 18.8secs
120.0bhp 89.5kW 121.7PS
@ 4200rpm
175.0lbft 237.1Nm @ 2200rpm
76.8bhp/ton 75.5bhp/tonne
38.3bhp/L 28.5kW/L 38.8PS/L
38.6ft/sec 11.8m/sec
27.0mpg 22.5mpgUS
10.5L/100km
Petrol 4-stroke piston
3135cc 191.3cu in
Vee 6 fuel injection
Compression ratio: 8.5:1
Bore: 89.0mm 3.5in
Stroke: 84.0mm 3.3in
Valve type/No: Overhead 12
Transmission: Automatic
No. of forward speeds: 3
Wheels driven: Front
Springs F/R: Coil/Coil
Brake system: PA
Steering: Rack & pinion PA
Wheelbase: 278.9cm 109.8in
Track F: 149.1cm 58.7in
Track R: 154.7cm 60.9in
Length: 492.5cm 193.9in
Width: 188.5cm 74.2in
Height: 166.4cm 65.5in
Ground clearance: 18.3cm
7.2in
Kerb weight: 1589.0kg
3500.0lb
Fuel: 75.7L 16.6gal 20.0galUS

177 Chevrolet
Baretta GTZ
1991 USA
125.0mph 201.1kmh
0-50mph 80.5kmh: 6.3secs
0-60mph 96.5kmh: 8.0secs
0-1/4 mile: 16.4secs
180.0bhp 134.2kW 182.5PS
@ 6200rpm
160.0lbft 216.8Nm @ 5200rpm
140.5bhp/ton 138.1bhp/tonne
79.6bhp/L 59.4kW/L 80.7PS/L
57.7ft/sec 17.6m/sec
30.0mpg 25.0mpgUS
9.4L/100km
Petrol 4-stroke piston
2260cc 137.9cu in
In-line 4 fuel injection
Compression ratio: 10.0:1

Bore: 92.0mm 3.6in
Stroke: 85.0mm 3.3in
Valve type/No: Overhead 16
Transmission: Manual
No. of forward speeds: 5
Wheels driven: Front
Springs F/R: Coil/Coil
Brake system: PA
Brakes F/R: Disc/Drum
Steering: Rack & pinion PA
Wheelbase: 262.6cm 103.4in
Track F: 141.2cm 55.6in
Track R: 143.8cm 56.6in
Length: 465.8cm 183.4in
Width: 173.2cm 68.2in
Height: 134.4cm 52.9in
Kerb weight: 1303.0kg
2870.0lb
Fuel: 59.0L 13.0gal 15.6galUS

178 Chevrolet
Camaro Z/28 Convertible
1991 USA
120.0mph 193.1kmh
0-50mph 80.5kmh: 6.2secs
0-60mph 96.5kmh: 8.6secs
0-1/4 mile: 16.6secs
195.0bhp 145.4kW 197.7PS
@ 4000rpm
295.0lbft 399.7Nm @ 2800rpm
128.5bhp/ton 126.3bhp/tonne
39.0bhp/L 29.1kW/L 39.5PS/L
38.7ft/sec 11.8m/sec
22.3mpg 18.6mpgUS
12.6L/100km
Petrol 4-stroke piston
5002cc 305.2cu in
Vee 8 fuel injection
Compression ratio: 9.3:1
Bore: 94.9mm 3.7in
Stroke: 88.4mm 3.5in
Valve type/No: Overhead 16
Transmission: Automatic
No. of forward speeds: 4
Wheels driven: Rear
Springs F/R: Coil/Coil
Brake system: PA
Brakes F/R: Disc/Disc
Steering: Recirculating ball PA
Wheelbase: 256.5cm 101.0in
Track F: 152.4cm 60.0in
Track R: 154.7cm 60.9in
Length: 487.7cm 192.0in
Width: 184.9cm 72.8in
Height: 127.8cm 50.3in
Kerb weight: 1543.6kg
3400.0lb
Fuel: 58.7L 12.9gal 15.5galUS

179 Chevrolet
Corvette Callaway
1991 USA
184.0mph 296.1kmh
0-60mph 96.5kmh: 4.9secs

0-1/4 mile: 13.3secs
403.0bhp 300.5kW 408.6PS
@ 4500rpm
575.0lbft 779.1Nm @ 3000rpm
262.4bhp/ton 258.0bhp/tonne
70.3bhp/L 52.4kW/L 71.2PS/L
43.5ft/sec 13.3m/sec
Petrol 4-stroke piston
5735cc 349.9cu in
turbocharged
Vee 8 fuel injection
Compression ratio: 7.5:1
Bore: 101.6mm 4.0in
Stroke: 88.4mm 3.5in
Valve type/No: Overhead 16
Transmission: Manual
No. of forward speeds: 6
Wheels driven: Rear
Springs F/R: Leaf/Leaf
Brake system: PA ABS
Brakes F/R: Disc/Disc
Steering: Rack & pinion PA
Wheelbase: 244.3cm 96.2in
Track F: 151.4cm 59.6in
Track R: 153.4cm 60.4in
Length: 448.3cm 176.5in
Width: 180.3cm 71.0in
Height: 118.6cm 46.7in
Kerb weight: 1561.8kg 3440.0lb

180 Chevrolet
Corvette Callaway Speedster
1991 USA
201.0mph 323.4kmh
0-60mph 96.5kmh: 4.4secs
403.0bhp 300.5kW 408.6PS
@ 4500rpm
575.0lbft 779.1Nm @ 3000rpm
282.1bhp/ton 277.4bhp/tonne
70.3bhp/L 52.4kW/L 71.2PS/L
43.5ft/sec 13.3m/sec
Petrol 4-stroke piston
5735cc 349.9cu in
turbocharged
Vee 8 fuel injection
Compression ratio: 7.5:1
Bore: 101.6mm 4.0in
Stroke: 88.4mm 3.5in
Valve type/No: Overhead 16
Transmission: Manual
No. of forward speeds: 6
Wheels driven: Rear
Springs F/R: Leaf/Leaf
Brake system: PA ABS
Brakes F/R: Disc/Disc
Steering: Rack & pinion PA
Wheelbase: 234.2cm 92.2in
Track F: 151.4cm 59.6in
Track R: 153.4cm 60.4in
Length: 448.3cm 176.5in
Width: 180.3cm 71.0in
Kerb weight: 1452.8kg
3200.0lb
Fuel: 75.7L 16.6gal 20.0galUS

181 Chevrolet

Corvette Lingenfelter
1991 USA
204.0mph 328.2kmh
0-60mph 96.5kmh: 3.4secs
0-1/4 mile: 11.8secs
530.0bhp 395.2kW 537.3PS
@ 5500rpm
570.0lbft 772.4Nm @ 4300rpm
341.6bhp/ton 335.9bhp/tonne
79.7bhp/L 59.4kW/L 80.8PS/L
57.3ft/sec 17.5m/sec
Petrol 4-stroke piston
6653cc 405.9cu in
Vee 8 fuel injection
Compression ratio: 10.5:1
Bore: 105.5mm 4.1in
Stroke: 95.3mm 3.8in
Valve type/No: Overhead 16
Transmission: Automatic
No. of forward speeds: 4
Wheels driven: Rear
Springs F/R: Coil/Leaf
Brake system: PA ABS
Brakes F/R: Disc/Disc
Steering: Rack & pinion PA
Wheelbase: 244.3cm 96.2in
Track F: 151.4cm 59.6in
Track R: 153.4cm 60.4in
Length: 453.6cm 178.6in
Width: 181.4cm 71.4in
Height: 118.6cm 46.7in
Kerb weight: 1577.6kg
3475.0lb

182 Chevrolet

Corvette Rick Mears
1991 USA
147.0mph 236.5kmh
0-60mph 96.5kmh: 6.3secs
0-1/4 mile: 14.8secs
245.0bhp 182.7kW 248.4PS
@ 4000rpm
340.0lbft 460.7Nm @ 3200rpm
164.8bhp/ton 162.1bhp/tonne
42.7bhp/L 31.9kW/L 43.3PS/L
38.7ft/sec 11.8m/sec
Petrol 4-stroke piston
5735cc 349.9cu in
Vee 8 fuel injection
Compression ratio: 10.0:1
Bore: 101.6mm 4.0in
Stroke: 88.4mm 3.5in
Valve type/No: Overhead 16
Transmission: Manual
No. of forward speeds: 6
Wheels driven: Rear
Springs F/R: Leaf/Leaf
Brake system: PA ABS
Brakes F/R: Disc/Disc
Steering: Rack & pinion PA
Wheelbase: 244.3cm 96.2in
Track F: 151.4cm 59.6in
Track R: 153.4cm 60.4in

Length: 461.0cm 181.5in
Width: 182.9cm 72.0in
Height: 118.6cm 46.7in
Kerb weight: 1511.8kg
3330.0lb

183 Chevrolet

Corvette ZR-1
1991 USA
154.0mph 247.8kmh
0-50mph 80.5kmh: 3.6secs
0-60mph 96.5kmh: 4.9secs
0-1/4 mile: 13.4secs
375.0bhp 279.6kW 380.2PS
@ 5800rpm
370.0lbft 501.4Nm @ 4800rpm
242.4bhp/ton 238.4bhp/tonne
65.5bhp/L 48.8kW/L 66.4PS/L
59.0ft/sec 18.0m/sec
21.7mpg 18.1mpgUS
13.0L/100km
Petrol 4-stroke piston
5727cc 349.4cu in
Vee 8 fuel injection
Compression ratio: 11.0:1
Bore: 99.0mm 3.9in
Stroke: 93.0mm 3.7in
Valve type/No: Overhead 32
Transmission: Manual
No. of forward speeds: 6
Wheels driven: Rear
Springs F/R: Leaf/Leaf
Brake system: PA ABS
Brakes F/R: Disc/Disc
Steering: Rack & pinion PA
Wheelbase: 244.3cm 96.2in
Track F: 151.4cm 59.6in
Track R: 157.2cm 61.9in
Length: 450.6cm 177.4in
Width: 188.0cm 74.0in
Height: 118.6cm 46.7in
Kerb weight: 1573.1kg
3465.0lb
Fuel: 75.7L 16.6gal 20.0galUS

184 Chevrolet

Corvette ZR-1 Geiger
1991 USA
186.0mph 299.3kmh
0-60mph 96.5kmh: 4.9secs
410.0bhp 305.7kW 415.7PS
@ 5800rpm
370.0lbft 501.4Nm @ 4800rpm
71.6bhp/L 53.4kW/L 72.6PS/L
59.0ft/sec 18.0m/sec
Petrol 4-stroke piston
5727cc 349.4cu in
Vee 8 fuel injection
Compression ratio: 11.0:1
Bore: 99.0mm 3.9in
Stroke: 93.0mm 3.7in
Valve type/No: Overhead 32
Transmission: Manual
No. of forward speeds: 6

Wheels driven: Rear
Springs F/R: Leaf/Leaf
Brake system: PA ABS
Brakes F/R: Disc/Disc
Steering: Rack & pinion PA
Wheelbase: 244.3cm 96.2in
Track F: 151.4cm 59.6in
Length: 473.7cm 186.5in
Width: 180.6cm 71.1in
Height: 118.6cm 46.7in
Fuel: 75.7L 16.6gal 20.0galUS

185 Chevrolet

Corvette ZR-2
1991 USA
185.0mph 297.7kmh
0-60mph 96.5kmh: 4.9secs
0-1/4 mile: 13.3secs
385.0bhp 287.1kW 390.3PS
@ 5200rpm
445.0lbft 603.0Nm @ 3400rpm
244.3bhp/ton 240.2bhp/tonne
51.7bhp/L 38.6kW/L 52.5PS/L
57.8ft/sec 17.6m/sec
Petrol 4-stroke piston
7439cc 453.9cu in
Vee 8 fuel injection
Compression ratio: 9.0:1
Bore: 107.9mm 4.2in
Stroke: 101.6mm 4.0in
Valve type/No: Overhead 16
Transmission: Manual
No. of forward speeds: 6
Wheels driven: Rear
Springs F/R: Leaf/Leaf
Brake system: PA ABS
Brakes F/R: Disc/Disc
Steering: Rack & pinion PA
Wheelbase: 244.3cm 96.2in
Track F: 151.4cm 59.6in
Track R: 157.2cm 61.9in
Length: 453.4cm 178.5in
Width: 188.0cm 74.0in
Height: 115.3cm 45.4in
Kerb weight: 1602.6kg
3530.0lb
Fuel: 75.7L 16.6gal 20.0galUS

186 Chevrolet

Lumina Z34
1991 USA
113.0mph 181.8kmh
0-60mph 96.5kmh: 8.0secs
0-1/4 mile: 15.8secs
210.0bhp 156.6kW 212.9PS
@ 5200rpm
215.0lbft 291.3Nm @ 4000rpm
137.5bhp/ton 135.2bhp/tonne
62.6bhp/L 46.7kW/L 63.5PS/L
47.8ft/sec 14.6m/sec
25.8mpg 21.5mpgUS
10.9L/100km
Petrol 4-stroke piston
3352cc 204.5cu in

Vee 6 fuel injection
Compression ratio: 9.3:1
Bore: 92.0mm 3.6in
Stroke: 84.0mm 3.3in
Valve type/No: Overhead 24
Transmission: Manual
No. of forward speeds: 5
Wheels driven: Front
Springs F/R: Coil/Leaf
Brake system: PA
Brakes F/R: Disc/Disc
Steering: Rack & pinion PA
Wheelbase: 273.1cm 107.5in
Track F: 151.1cm 59.5in
Track R: 147.3cm 58.0in
Length: 506.2cm 199.3in
Width: 182.1cm 71.7in
Height: 135.4cm 53.3in
Kerb weight: 1552.7kg
3420.0lb
Fuel: 62.4L 13.7gal 16.5galUS

187 Chevrolet

Camaro Z28
1992 USA
140.0mph 225.3kmh
0-50mph 80.5kmh: 4.9secs
0-60mph 96.5kmh: 6.7secs
0-1/4 mile: 15.2secs
230.0bhp 171.5kW 233.2PS
@ 4400rpm
300.0lbft 406.5Nm @ 3200rpm
155.2bhp/ton 152.6bhp/tonne
46.0bhp/L 34.3kW/L 46.6PS/L
42.5ft/sec 13.0m/sec
21.6mpg 18.0mpgUS
13.1L/100km
Petrol 4-stroke piston
5002cc 305.2cu in
Vee 8 fuel injection
Compression ratio: 9.3:1
Bore: 94.9mm 3.7in
Stroke: 88.4mm 3.5in
Valve type/No: Overhead 16
Transmission: Manual
No. of forward speeds: 5
Wheels driven: Rear
Springs F/R: Coil/Coil
Brake system: PA
Brakes F/R: Disc/Drum
Steering: Recirculating ball PA
Wheelbase: 256.5cm 101.0in
Track F: 152.4cm 60.0in
Track R: 154.7cm 60.9in
Length: 489.2cm 192.6in
Width: 183.9cm 72.4in
Height: 128.0cm 50.4in
Kerb weight: 1507.3kg
3320.0lb
Fuel: 58.7L 12.9gal 15.5galUS

188 Chevrolet

Corvette LT1
1992 USA

189 Chrysler
Laser XT

163.0mph 262.3kmh
0-60mph 96.5kmh: 5.7secs
0-1/4 mile: 14.1secs
300.0bhp 223.7kW 304.2PS
@ 5000rpm
330.0lbft 447.2Nm @ 4000rpm
201.8bhp/ton 198.4bhp/tonne
52.3bhp/L 39.0kW/L 53.0PS/L
48.3ft/sec 14.7m/sec
20.4mpg 17.0mpgUS
13.8L/100km
Petrol 4-stroke piston
5733cc 349.8cu in
Vee 8 fuel injection
Compression ratio: 10.5:1
Bore: 101.6mm 4.0in
Stroke: 88.4mm 3.5in
Valve type/No: Overhead 16
Transmission: Manual
No. of forward speeds: 6
Wheels driven: Rear
Springs F/R: Leaf/Leaf
Brake system: PA ABS
Brakes F/R: Disc/Disc
Steering: Rack & pinion PA
Wheelbase: 244.3cm 96.2in
Track F: 146.6cm 57.7in
Track R: 149.9cm 59.0in
Length: 453.6cm 178.6in
Width: 180.3cm 71.0in
Height: 118.6cm 46.7in
Kerb weight: 1511.8kg
3330.0lb
Fuel: 75.7L 16.6gal 20.0galUS

189 Chrysler
Laser XT
1986 USA
118.0mph 189.9kmh
0-60mph 96.5kmh: 8.1secs
0-1/4 mile: 16.2secs
146.0bhp 108.9kW 148.0PS
@ 5200rpm

170.0lbft 230.4Nm @ 3600rpm
115.1bhp/ton 113.2bhp/tonne
66.0bhp/L 49.2kW/L 66.9PS/L
52.3ft/sec 15.9m/sec
25.2mpg 21.0mpgUS
11.2L/100km
Petrol 4-stroke piston
2213cc 135.0cu in
turbocharged
In-line 4 fuel injection
Compression ratio: 8.1:1
Bore: 87.5mm 3.4in
Stroke: 92.0mm 3.6in
Valve type/No: Overhead 8
Transmission: Manual
No. of forward speeds: 5
Wheels driven: Front
Springs F/R: Coil/Coil
Brake system: PA
Brakes F/R: Disc/Drum
Steering: Rack & pinion PA
Wheelbase: 246.6cm 97.1in
Track F: 146.3cm 57.6in
Track R: 146.3cm 57.6in
Length: 444.5cm 175.0in
Width: 176.0cm 69.3in
Height: 128.0cm 50.4in
Kerb weight: 1289.4kg
2840.0lb
Fuel: 53.0L 11.6gal 14.0galUS

190 Chrysler
Le Baron Coupe
1987 USA
120.0mph 193.1kmh
0-50mph 80.5kmh: 6.4secs
0-60mph 96.5kmh: 8.8secs
0-1/4 mile: 16.8secs
146.0bhp 108.9kW 148.0PS
@ 5200rpm
170.0lbft 230.4Nm @ 3600rpm
113.4bhp/ton 111.5bhp/tonne
66.0bhp/L 49.2kW/L 66.9PS/L

52.3ft/sec 15.9m/sec
24.6mpg 20.5mpgUS
11.5L/100km
Petrol 4-stroke piston
2213cc 135.0cu in
turbocharged
In-line 4 fuel injection
Compression ratio: 8.1:1
Bore: 87.5mm 3.4in
Stroke: 92.0mm 3.6in
Valve type/No: Overhead 8
Transmission: Manual
No. of forward speeds: 5
Wheels driven: Front
Springs F/R: Coil/Coil
Brake system: PA
Brakes F/R: Disc/Drum
Steering: Rack & pinion PA
Wheelbase: 254.8cm 100.3in
Track F: 146.1cm 57.5in
Track R: 146.3cm 57.6in
Length: 468.4cm 184.9in
Width: 173.7cm 68.4in
Height: 129.3cm 50.9in
Ground clearance: 11.7cm
4.6in
Kerb weight: 1309.8kg
2885.0lb
Fuel: 53.0L 11.6gal 14.0galUS

191 Chrysler
Conquest TSi
1988 Japan
127.0mph 204.3kmh
0-50mph 80.5kmh: 5.1secs
0-60mph 96.5kmh: 7.3secs
0-1/4 mile: 15.4secs
188.0bhp 140.2kW 190.6PS
@ 5000rpm
234.0lbft 317.1Nm @ 2500rpm
135.8bhp/ton 133.6bhp/tonne
73.6bhp/L 54.9kW/L 74.6PS/L
53.6ft/sec 16.3m/sec

24.0mpg 20.0mpgUS
11.8L/100km
Petrol 4-stroke piston
2555cc 155.9cu in
turbocharged
In-line 4 fuel injection
Compression ratio: 7.0:1
Bore: 91.1mm 3.6in
Stroke: 98.0mm 3.9in
Valve type/No: Overhead 8
Transmission: Manual
No. of forward speeds: 5
Wheels driven: Rear
Springs F/R: Coil/Coil
Brake system: PA ABS
Brakes F/R: Disc/Disc
Steering: Recirculating ball PA
Wheelbase: 243.6cm 95.9in
Track F: 146.6cm 57.7in
Track R: 145.5cm 57.3in
Length: 439.9cm 173.2in
Width: 173.5cm 68.3in
Height: 127.5cm 50.2in
Kerb weight: 1407.4kg
3100.0lb
Fuel: 74.9L 16.5gal 19.8galUS

192 Chrysler
TC by Maserati
1989 Italy
135.0mph 217.2kmh
0-60mph 96.5kmh: 6.9secs
0-1/4 mile: 15.4secs
200.0bhp 149.1kW 202.8PS
@ 5500rpm
220.0lbft 298.1Nm @ 3400rpm
139.3bhp/ton 137.0bhp/tonne
90.4bhp/L 67.4kW/L 91.6PS/L
55.3ft/sec 16.9m/sec
24.0mpg 20.0mpgUS
11.8L/100km
Petrol 4-stroke piston
2213cc 135.0cu in

turbocharged
In-line 4 fuel injection
Compression ratio: 7.3:1
Bore: 87.5mm 3.4in
Stroke: 92.0mm 3.6in
Valve type/No: Overhead 16
Transmission: Manual
No. of forward speeds: 5
Wheels driven: Front
Springs F/R: Coil/Coil
Brake system: PA ABS
Brakes F/R: Disc/Disc
Steering: Rack & pinion PA
Wheelbase: 237.0cm 93.3in
Track F: 146.3cm 57.6in
Track R: 146.3cm 57.6in
Length: 446.5cm 175.8in
Width: 174.0cm 68.5in
Height: 131.8cm 51.9in
Kerb weight: 1459.6kg 3215.0lb
Fuel: 53.0L 11.6gal 14.0galUS

193 Citroen

BX19 RD Estate
1986 France
94.0mph 151.2kmh
0-50mph 80.5kmh: 12.2secs
0-60mph 96.5kmh: 17.9secs
0-1/4 mile: 20.8secs
0-1km: 38.7secs
65.0bhp 48.5kW 65.9PS
@ 4600rpm
88.0lbft 119.2Nm @ 2000rpm
62.1bhp/ton 61.0bhp/tonne
34.1bhp/L 25.4kW/L 34.6PS/L
44.2ft/sec 13.5m/sec
21.8mph 35.1kmh/1000rpm
38.4mpg 32.0mpgUS
7.4L/100km
Diesel 4-stroke piston
1905cc 116.2cu in
In-line 4 fuel injection
Compression ratio: 23.5:1
Bore: 83.0mm 3.3in
Stroke: 88.0mm 3.5in
Valve type/No: Overhead 8
Transmission: Manual
No. of forward speeds: 5
Wheels driven: Front
Springs F/R: Gas/Gas
Brake system: PA
Brakes F/R: Disc/Disc
Steering: Rack & pinion PA
Wheelbase: 265.4cm 104.5in
Track F: 141.0cm 55.5in
Track R: 135.4cm 53.3in
Length: 439.9cm 173.2in
Width: 165.6cm 65.2in
Height: 142.7cm 56.2in
Ground clearance: 16.3cm
6.4in
Kerb weight: 1064.6kg
2345.0lb
Fuel: 51.9L 11.4gal 13.7galUS

194 Citroen

CX25 Ri Familiale Auto
1986 France
116.0mph 186.6kmh
0-50mph 80.5kmh: 8.7secs
0-60mph 96.5kmh: 11.9secs
0-1/4 mile: 18.7secs
0-1km: 34.9secs
138.0bhp 102.9kW 139.9PS
@ 5000rpm
155.0lbft 210.0Nm @ 4000rpm
97.5bhp/ton 95.8bhp/tonne
55.8bhp/L 41.6kW/L 56.6PS/L
50.3ft/sec 15.3m/sec
21.7mph 34.9kmh/1000rpm
20.4mpg 17.0mpgUS
13.8L/100km
Petrol 4-stroke piston
2473cc 151.0cu in
In-line 4 fuel injection
Compression ratio: 8.7:1
Bore: 93.0mm 3.7in
Stroke: 92.0mm 3.6in
Valve type/No: Overhead 8
Transmission: Automatic
No. of forward speeds: 3
Wheels driven: Front
Springs F/R: Gas/Gas
Brake system: PA
Brakes F/R: Disc/Disc
Steering: Rack & pinion PA
Wheelbase: 307.3cm 121.0in
Track F: 147.3cm 58.0in
Track R: 135.9cm 53.5in
Length: 495.2cm 195.0in
Width: 177.0cm 69.7in
Height: 146.5cm 57.7in
Ground clearance: 15.5cm
6.1in
Kerb weight: 1440.0kg
3171.8lb
Fuel: 68.2L 15.0gal 18.0galUS

195 Citroen

Visa GTi
1986 France
115.0mph 185.0kmh
0-50mph 80.5kmh: 6.5secs
0-60mph 96.5kmh: 9.1secs
0-1/4 mile: 17.1secs
0-1km: 31.7secs
115.0bhp 85.8kW 116.6PS
@ 6250rpm
97.0lbft 131.4Nm @ 4000rpm
133.3bhp/ton 131.1bhp/tonne
72.8bhp/L 54.3kW/L 73.8PS/L
49.8ft/sec 15.2m/sec
18.3mph 29.4kmh/1000rpm
26.2mpg 21.8mpgUS
10.8L/100km
Petrol 4-stroke piston
1580cc 96.0cu in
In-line 4 fuel injection
Compression ratio: 9.8:1

Bore: 83.0mm 3.3in
Stroke: 73.0mm 2.9in
Valve type/No: Overhead 8
Transmission: Manual
No. of forward speeds: 5
Wheels driven: Front
Springs F/R: Coil/Coil
Brake system: PA
Brakes F/R: Disc/Drum
Steering: Rack & pinion
Wheelbase: 242.1cm 95.3in
Track F: 129.2cm 50.9in
Track R: 124.1cm 48.9in
Length: 369.0cm 145.3in
Width: 153.4cm 60.4in
Height: 141.0cm 55.5in
Ground clearance: 13.1cm
5.2in
Kerb weight: 877.0kg 1931.7lb
Fuel: 43.2L 9.5gal 11.4galUS

196 Citroen

AX GT
1987 France
108.0mph 173.8kmh
0-50mph 80.5kmh: 6.3secs
0-60mph 96.5kmh: 9.0secs
0-1/4 mile: 16.8secs
0-1km: 31.6secs
85.0bhp 63.4kW 86.2PS
@ 6400rpm
85.3lbft 115.6Nm @ 4000rpm
121.3bhp/ton 119.3bhp/tonne
62.5bhp/L 46.6kW/L 63.4PS/L
53.9ft/sec 16.4m/sec
18.2mph 29.3kmh/1000rpm
30.4mpg 25.3mpgUS
9.3L/100km
Petrol 4-stroke piston
1360cc 83.0cu in
In-line 4 1 Carburettor
Compression ratio: 9.3:1
Bore: 75.0mm 2.9in
Stroke: 77.0mm 3.0in
Valve type/No: Overhead 8
Transmission: Manual
No. of forward speeds: 5
Wheels driven: Front
Springs F/R: Coil/Torsion bar
Brake system: PA
Brakes F/R: Disc/Drum
Steering: Rack & pinion
Wheelbase: 228.5cm 90.0in
Track F: 138.0cm 54.3in
Track R: 130.0cm 51.2in
Length: 349.5cm 137.6in
Width: 155.5cm 61.2in
Height: 135.5cm 53.3in
Ground clearance: 11.0cm
4.3in
Kerb weight: 712.3kg 1569.0lb
Fuel: 43.2L 9.5gal 11.4galUS

197 Citroen

AX11 TRE
1987 France
101.0mph 162.5kmh
0-50mph 80.5kmh: 8.3secs
0-60mph 96.5kmh: 12.6secs
0-1/4 mile: 17.9secs
0-1km: 34.6secs
54.0bhp 40.3kW 54.7PS
@ 5800rpm
65.7lbft 89.0Nm @ 3200rpm
86.9bhp/ton 85.4bhp/tonne
48.0bhp/L 35.8kW/L 48.7PS/L
43.8ft/sec 13.3m/sec
20.2mph 32.5kmh/1000rpm
38.9mpg 32.4mpgUS
7.3L/100km
Petrol 4-stroke piston
1124cc 69.0cu in
In-line 4 1 Carburettor
Compression ratio: 9.4:1
Bore: 72.0mm 2.8in
Stroke: 69.0mm 2.7in
Valve type/No: Overhead 8
Transmission: Manual
No. of forward speeds: 5
Wheels driven: Front
Springs F/R: Coil/Torsion bar
Brake system: PA
Brakes F/R: Disc/Drum
Steering: Rack & pinion
Wheelbase: 228.5cm 90.0in
Track F: 138.0cm 54.3in
Track R: 130.0cm 51.2in
Length: 349.5cm 137.6in
Width: 155.5cm 61.2in
Height: 135.5cm 53.3in
Ground clearance: 11.0cm
4.3in
Kerb weight: 632.0kg 1392.1lb
Fuel: 35.9L 7.9gal 9.5galUS

198 Citroen

AX14 TRS
1987 France
101.0mph 162.5kmh
0-50mph 80.5kmh: 7.7secs
0-60mph 96.5kmh: 11.0secs
0-1/4 mile: 18.1secs
0-1km: 34.1secs
65.0bhp 48.5kW 65.9PS
@ 5400rpm
83.2lbft 112.7Nm @ 3000rpm
96.9bhp/ton 95.3bhp/tonne
47.8bhp/L 35.6kW/L 48.5PS/L
45.4ft/sec 13.9m/sec
21.6mph 34.8kmh/1000rpm
34.7mpg 28.9mpgUS
8.1L/100km
Petrol 4-stroke piston
1360cc 83.0cu in
In-line 4 1 Carburettor
Compression ratio: 9.3:1
Bore: 75.0mm 2.9in
Stroke: 77.0mm 3.0in

Valve type/No: Overhead 8
Transmission: Manual
No. of forward speeds: 5
Wheels driven: Front
Springs F/R: Coil/Torsion bar
Brake system: PA
Brakes F/R: Disc/Drum
Steering: Rack & pinion
Wheelbase: 228.5cm 90.0in
Track F: 138.0cm 54.3in
Track R: 130.0cm 51.2in
Length: 349.5cm 137.6in
Width: 145.5cm 57.3in
Height: 135.5cm 53.3in
Ground clearance: 11.0cm 4.3in
Kerb weight: 682.0kg 1502.2lb
Fuel: 43.2L 9.5gal 11.4galUS

199 Citroen
BX GTi 16v
1987 France
133.0mph 214.0kmh
0-50mph 80.5kmh: 5.9secs
0-60mph 96.5kmh: 7.9secs
0-1/4 mile: 16.2secs
0-1km: 29.1secs
155.0bhp 115.6kW 157.1PS @ 6500rpm
133.0lbft 180.2Nm @ 5000rpm
145.5bhp/ton 143.1bhp/tonne
81.4bhp/L 60.7kW/L 82.5PS/L
62.5ft/sec 19.1m/sec
19.9mph 32.0kmh/1000rpm
29.8mpg 24.8mpgUS
9.5L/100km
Petrol 4-stroke piston
1905cc 116.0cu in
In-line 4 fuel injection
Compression ratio: 10.4:1
Bore: 83.0mm 3.3in
Stroke: 88.0mm 3.5in
Valve type/No: Overhead 16
Transmission: Manual
No. of forward speeds: 5
Wheels driven: Front
Springs F/R: Gas/Gas
Brake system: PA ABS
Brakes F/R: Disc/Disc
Steering: Rack & pinion PA
Wheelbase: 265.5cm 104.5in
Track F: 141.0cm 55.5in
Track R: 135.4cm 53.3in
Length: 422.9cm 166.5in
Width: 165.7cm 65.2in
Height: 136.5cm 53.7in
Ground clearance: 16.2cm 6.4in
Kerb weight: 1083.0kg 2385.5lb
Fuel: 66.0L 14.5gal 17.4galUS

200 Citroen
CX25 DTR Turbo 2
1987 France

113.0mph 181.8kmh
0-50mph 80.5kmh: 7.3secs
0-60mph 96.5kmh: 10.1secs
0-1/4 mile: 17.8secs
0-1km: 32.8secs
120.0bhp 89.5kW 121.7PS @ 3900rpm
188.0lbft 254.7Nm @ 2000rpm
90.3bhp/ton 88.8bhp/tonne
48.5bhp/L 36.2kW/L 49.2PS/L
39.2ft/sec 12.0m/sec
28.9mph 46.5kmh/1000rpm
30.1mpg 25.1mpgUS
9.4L/100km
Diesel 4-stroke piston
2473cc 151.0cu in
turbocharged
In-line 4 fuel injection
Compression ratio: 21.0:1
Bore: 93.0mm 3.7in
Stroke: 92.0mm 3.6in
Valve type/No: Overhead 8
Transmission: Manual
No. of forward speeds: 5
Wheels driven: Front
Springs F/R: Gas/Gas
Brake system: PA ABS
Brakes F/R: Disc/Disc
Steering: Rack & pinion PA
Wheelbase: 284.5cm 112.0in
Track F: 152.2cm 59.9in
Track R: 136.8cm 53.9in
Length: 465.0cm 183.1in
Width: 177.0cm 69.7in
Height: 136.0cm 53.5in
Ground clearance: 15.3cm 6.0in
Kerb weight: 1351.0kg 2975.8lb
Fuel: 68.2L 15.0gal 18.0galUS

201 Citroen
AX11 TRE 5DR
1988 France
98.0mph 157.7kmh
0-50mph 80.5kmh: 8.8secs
0-60mph 96.5kmh: 12.6secs
0-1/4 mile: 19.1secs
0-1km: 35.4secs
55.0bhp 41.0kW 55.8PS @ 5800rpm
65.7lbft 89.0Nm @ 3200rpm
76.6bhp/ton 75.3bhp/tonne
48.9bhp/L 36.5kW/L 49.6PS/L
43.8ft/sec 13.3m/sec
20.1mph 32.3kmh/1000rpm
32.7mpg 27.2mpgUS
8.6L/100km
Petrol 4-stroke piston
1124cc 69.0cu in
In-line 4 1 Carburettor
Compression ratio: 9.4:1
Bore: 72.0mm 2.8in
Stroke: 69.0mm 2.7in

Valve type/No: Overhead 8
Transmission: Manual
No. of forward speeds: 5
Wheels driven: Front
Springs F/R: Coil/Torsion bar
Brake system: PA
Brakes F/R: Disc/Drum
Steering: Rack & pinion
Wheelbase: 228.5cm 90.0in
Track F: 138.0cm 54.3in
Track R: 130.0cm 51.2in
Length: 349.5cm 137.6in
Width: 155.5cm 61.2in
Height: 135.5cm 53.3in
Ground clearance: 11.0cm 4.3in
Kerb weight: 730.0kg 1607.9lb
Fuel: 35.9L 7.9gal 9.5galUS

202 Citroen
BX DTR Turbo
1988 France
106.0mph 170.6kmh
0-50mph 80.5kmh: 8.7secs
0-60mph 96.5kmh: 11.9secs
0-1/4 mile: 18.3secs
0-1km: 34.0secs
90.0bhp 67.1kW 91.2PS @ 4300rpm
134.3lbft 182.0Nm @ 2100rpm
85.9bhp/ton 84.5bhp/tonne
50.9bhp/L 37.9kW/L 51.6PS/L
41.3ft/sec 12.6m/sec
25.6mph 41.2kmh/1000rpm
36.2mpg 30.1mpgUS
7.8L/100km
Diesel 4-stroke piston
1769cc 108.0cu in
turbocharged
In-line 4 fuel injection
Compression ratio: 22.1:1
Bore: 80.0mm 3.1in
Stroke: 88.0mm 3.5in
Valve type/No: Overhead 8
Transmission: Manual
No. of forward speeds: 5
Wheels driven: Front
Springs F/R: Gas/Gas
Brake system: PA ABS
Brakes F/R: Disc/Disc
Steering: Rack & pinion PA
Wheelbase: 265.5cm 104.5in
Track F: 141.0cm 55.5in
Track R: 135.4cm 53.3in
Length: 422.9cm 166.5in
Width: 165.7cm 65.2in
Height: 136.5cm 53.7in
Ground clearance: 16.2cm 6.4in
Kerb weight: 1065.0kg 2345.8lb
Fuel: 66.0L 14.5gal 17.4galUS

203 Citroen
AX14 DTR

1989 France
92.0mph 148.0kmh
0-50mph 80.5kmh: 11.4secs
0-60mph 96.5kmh: 16.2secs
0-1/4 mile: 20.3secs
0-1km: 37.5secs
53.7bhp 40.0kW 54.4PS @ 5000rpm
62.4lbft 84.6Nm @ 2500rpm
73.8bhp/ton 72.6bhp/tonne
39.5bhp/L 29.4kW/L 40.0PS/L
42.1ft/sec 12.8m/sec
20.3mph 32.7kmh/1000rpm
54.7mpg 45.5mpgUS
5.2L/100km
Diesel 4-stroke piston
1360cc 83.0cu in
In-line 4 fuel injection
Compression ratio: 22.0:1
Bore: 75.0mm 2.9in
Stroke: 77.0mm 3.0in
Valve type/No: Overhead 8
Transmission: Manual
No. of forward speeds: 5
Wheels driven: Front
Springs F/R: Coil/Torsion bar
Brakes F/R: Disc/Drum
Steering: Rack & pinion
Wheelbase: 228.5cm 90.0in
Track F: 138.0cm 54.3in
Track R: 130.0cm 51.2in
Length: 349.5cm 137.6in
Width: 155.5cm 61.2in
Height: 135.5cm 53.3in
Ground clearance: 11.0cm 4.3in
Kerb weight: 740.0kg 1630.0lb
Fuel: 43.2L 9.5gal 11.4galUS

204 Citroen
BX14 TGE
1989 France
102.0mph 164.1kmh
0-50mph 80.5kmh: 9.8secs
0-60mph 96.5kmh: 14.3secs
0-1/4 mile: 19.3secs
0-1km: 36.3secs
73.0bhp 54.4kW 74.0PS @ 5600rpm
81.9lbft 111.0Nm @ 3000rpm
78.1bhp/ton 76.8bhp/tonne
53.7bhp/L 40.0kW/L 54.4PS/L
47.1ft/sec 14.4m/sec
19.3mph 31.1kmh/1000rpm
34.0mpg 28.3mpgUS
8.3L/100km
Petrol 4-stroke piston
1360cc 83.0cu in
In-line 4 1 Carburettor
Compression ratio: 9.3:1
Bore: 75.0mm 2.9in
Stroke: 77.0mm 3.0in
Valve type/No: Overhead 8
Transmission: Manual

No. of forward speeds: 5
Wheels driven: Front
Springs F/R: Gas/Gas
Brake system: PA
Brakes F/R: Disc/Disc
Steering: Rack & pinion
Wheelbase: 265.5cm 104.5in
Track F: 141.0cm 55.5in
Track R: 135.4cm 53.3in
Length: 422.9cm 166.5in
Width: 165.7cm 65.2in
Height: 136.5cm 53.7in
Ground clearance: 16.2cm
6.4in
Kerb weight: 950.0kg 2092.5lb
Fuel: 44.1L 9.7gal 11.7galUS

205 Citroen
AX GT5
1990 France
109.0mph 175.4kmh
0-50mph 80.5kmh: 7.4secs
0-60mph 96.5kmh: 10.1secs
0-1/4 mile: 17.6secs
0-1km: 31.9secs
85.0bhp 63.4kW 86.2PS
@ 6400rpm
85.6lbft 116.0Nm @ 4000rpm
106.2bhp/ton 104.4bhp/tonne
62.5bhp/L 46.6kW/L 63.4PS/L
53.9ft/sec 16.4m/sec
18.3mph 29.4kmh/1000rpm
35.9mpg 29.9mpgUS
7.9L/100km
Petrol 4-stroke piston
1360cc 83.0cu in
In-line 4 1 Carburettor
Compression ratio: 9.3:1
Bore: 75.0mm 2.9in
Stroke: 77.0mm 3.0in
Valve type/No: Overhead 8
Transmission: Manual
No. of forward speeds: 5
Wheels driven: Front
Springs F/R: Coil/Torsion bar
Brake system: PA
Brakes F/R: Disc/Drum
Steering: Rack & pinion
Wheelbase: 228.5cm 90.0in
Track F: 138.0cm 54.3in
Track R: 130.0cm 51.2in
Length: 349.5cm 137.6in
Width: 155.5cm 61.2in
Height: 135.5cm 53.3in
Ground clearance: 11.0cm
4.3in
Kerb weight: 814.0kg 1792.9lb
Fuel: 43.2L 9.5gal 11.4galUS

206 Citroen
BX GTi 4x4
1990 France
116.0mph 186.6kmh
0-50mph 80.5kmh: 7.4secs

0-60mph 96.5kmh: 10.6secs
0-1/4 mile: 17.8secs
0-1km: 32.7secs
125.0bhp 93.2kW 126.7PS
@ 5500rpm
124.7lbft 169.0Nm @ 4500rpm
110.5bhp/ton 108.7bhp/tonne
65.6bhp/L 48.9kW/L 66.5PS/L
52.9ft/sec 16.1m/sec
20.7mph 33.3kmh/1000rpm
26.3mpg 21.9mpgUS
10.7L/100km
Petrol 4-stroke piston
1905cc 116.0cu in
In-line 4 fuel injection
Compression ratio: 9.3:1
Bore: 83.0mm 3.3in
Stroke: 88.0mm 3.5in
Valve type/No: Overhead 8
Transmission: Manual
No. of forward speeds: 5
Wheels driven: 4-wheel drive
Springs F/R: Gas/Gas
Brake system: PA ABS
Brakes F/R: Disc/Disc
Steering: Rack & pinion PA
Wheelbase: 265.4cm 104.5in
Track F: 141.0cm 55.5in
Track R: 135.4cm 53.3in
Length: 422.9cm 166.5in
Width: 168.9cm 66.5in
Height: 136.4cm 53.7in
Ground clearance: 16.3cm
6.4in
Kerb weight: 1150.0kg
2533.0lb
Fuel: 66.0L 14.5gal 17.4galUS

207 Citroen
XM 2.0SEi
1990 France
122.0mph 196.3kmh
0-50mph 80.5kmh: 8.0secs
0-60mph 96.5kmh: 11.2secs
0-1/4 mile: 19.0secs
0-1km: 33.0secs
130.0bhp 96.9kW 131.8PS
@ 5600rpm

129.2lbft 175.0Nm @ 4800rpm
92.4bhp/ton 90.9bhp/tonne
65.1bhp/L 48.5kW/L 66.0PS/L
52.7ft/sec 16.0m/sec
20.9mph 33.6kmh/1000rpm
25.8mpg 21.5mpgUS
10.9L/100km
Petrol 4-stroke piston
1998cc 122.0cu in
In-line 4 fuel injection
Compression ratio: 8.8:1
Bore: 86.0mm 3.4in
Stroke: 86.0mm 3.4in
Valve type/No: Overhead 8
Transmission: Manual
No. of forward speeds: 5
Wheels driven: Front
Springs F/R: Gas/Gas
Brake system: PA ABS
Brakes F/R: Disc/Disc
Steering: Rack & pinion PA
Wheelbase: 285.0cm 112.2in
Track F: 153.0cm 60.2in
Track R: 145.7cm 57.4in
Length: 470.8cm 185.4in
Width: 179.4cm 70.6in
Height: 138.2cm 54.4in
Ground clearance: 17.9cm
7.0in
Kerb weight: 1430.0kg
3149.8lb
Fuel: 80.1L 17.6gal 21.1galUS

208 Citroen
AX11 TZX
1991 France
95.0mph 152.9kmh
0-50mph 80.5kmh: 9.4secs
0-60mph 96.5kmh: 13.5secs
0-1/4 mile: 19.5secs
0-1km: 36.2secs
55.0bhp 41.0kW 55.8PS
@ 5800rpm
66.0lbft 89.4Nm @ 3200rpm
75.6bhp/ton 74.3bhp/tonne
48.9bhp/L 36.5kW/L 49.6PS/L
43.8ft/sec 13.3m/sec
20.3mph 32.7kmh/1000rpm

38.8mpg 32.3mpgUS
7.3L/100km
Petrol 4-stroke piston
1124cc 69.0cu in
In-line 4 1 Carburettor
Compression ratio: 9.4:1
Bore: 72.0mm 2.8in
Stroke: 69.0mm 2.7in
Valve type/No: Overhead 8
Transmission: Manual
No. of forward speeds: 5
Wheels driven: Front
Springs F/R: Coil/Coil
Brake system: PA
Brakes F/R: Disc/Drum
Steering: Rack & pinion
Track F: 137.9cm 54.3in
Track R: 130.0cm 51.2in
Length: 350.5cm 138.0in
Width: 177.8cm 70.0in
Height: 134.6cm 53.0in
Ground clearance: 10.9cm
4.3in
Kerb weight: 740.0kg 1630.0lb
Fuel: 36.4L 8.0gal 9.6galUS

209 Citroen
XM 3.0 SEi Auto
1991 France
135.0mph 217.2kmh
0-50mph 80.5kmh: 7.2secs
0-60mph 96.5kmh: 9.5secs
0-1/4 mile: 17.3secs
0-1km: 30.9secs
170.0bhp 126.8kW 172.4PS
@ 5600rpm
173.4lbft 235.0Nm @ 4600rpm
115.0bhp/ton 113.1bhp/tonne
57.1bhp/L 42.6kW/L 57.9PS/L
44.6ft/sec 13.6m/sec
25.1mph 40.4kmh/1000rpm
18.8mpg 15.7mpgUS
15.0L/100km
Petrol 4-stroke piston
2975cc 182.0cu in
Vee 6 fuel injection
Compression ratio: 9.5:1
Bore: 93.0mm 3.7in

209 Citroen
XM 3.0 SEi Auto

Stroke: 73.0mm 2.9in
Valve type/No: Overhead 12
Transmission: Automatic
No. of forward speeds: 4
Wheels driven: Front
Springs F/R: Gas/Gas
Brake system: PA ABS
Brakes F/R: Disc/Disc
Steering: Rack & pinion PA
Wheelbase: 285.0cm 112.2in
Track F: 153.0cm 60.2in
Track R: 145.7cm 57.4in
Length: 470.8cm 185.4in
Width: 179.4cm 70.6in
Height: 138.2cm 54.4in
Ground clearance: 17.9cm
7.0in
Kerb weight: 1503.0kg
3310.6lb
Fuel: 80.1L 17.6gal 21.1galUS

210 Citroen

XM V6.24
1991 France
145.0mph 233.3kmh
0-50mph 80.5kmh: 5.8secs
0-60mph 96.5kmh: 7.5secs
0-1/4 mile: 15.5secs
0-1km: 28.7secs
200.0bhp 149.1kW 202.8PS
@ 6000rpm
191.9lbft 260.0Nm @ 3600rpm
133.4bhp/ton 131.1bhp/tonne
67.2bhp/L 50.1kW/L 68.2PS/L
47.8ft/sec 14.6m/sec
22.9mph 36.8kmh/1000rpm
20.8mpg 17.3mpgUS
13.6L/100km
Petrol 4-stroke piston
2975cc 182.0cu in
Vee 6 fuel injection
Compression ratio: 9.5:1

Bore: 93.0mm 3.7in
Stroke: 73.0mm 2.9in
Valve type/No: Overhead 24
Transmission: Manual
No. of forward speeds: 5
Wheels driven: Front
Springs F/R: Gas/Gas
Brake system: PA ABS
Brakes F/R: Disc/Disc
Steering: Rack & pinion PA
Wheelbase: 285.0cm 112.2in
Track F: 152.9cm 60.2in
Track R: 145.5cm 57.3in
Length: 470.7cm 185.3in
Width: 179.3cm 70.6in
Height: 138.2cm 54.4in
Ground clearance: 17.8cm
7.0in
Kerb weight: 1525.0kg
3359.0lb
Fuel: 80.1L 17.6gal 21.1galUS

211 Citroen

ZX 1.4 Avantage
1991 France
107.0mph 172.2kmh
0-50mph 80.5kmh: 8.3secs
0-60mph 96.5kmh: 11.9secs
0-1/4 mile: 18.7secs
0-1km: 29.6secs
75.0bhp 55.9kW 76.0PS
@ 5800rpm
84.9lbft 115.0Nm @ 3800rpm
75.5bhp/ton 74.3bhp/tonne
55.1bhp/L 41.1kW/L 55.9PS/L
48.8ft/sec 14.9m/sec
20.6mph 33.1kmh/1000rpm
26.5mpg 22.1mpgUS
10.7L/100km
Petrol 4-stroke piston
1360cc 83.0cu in
In-line 4 1 Carburettor
Compression ratio: 9.3:1

Bore: 75.0mm 2.9in
Stroke: 77.0mm 3.0in
Valve type/No: Overhead 8
Transmission: Manual
No. of forward speeds: 5
Wheels driven: Front
Springs F/R: Coil/Torsion bar
Brake system: PA
Brakes F/R: Disc/Drum
Steering: Rack & pinion
Wheelbase: 254.0cm 100.0in
Length: 407.0cm 160.2in
Width: 169.0cm 66.5in
Height: 140.0cm 55.1in
Kerb weight: 1010.0kg
2224.7lb
Fuel: 56.0L 12.3gal 14.8galUS

212 Citroen

ZX Volcane
1991 France
125.0mph 201.1kmh
0-50mph 80.5kmh: 6.9secs
0-60mph 96.5kmh: 9.1secs
0-1/4 mile: 17.1secs
0-1km: 31.1secs
130.0bhp 96.9kW 131.8PS
@ 6000rpm
118.1lbft 160.0Nm @ 3250rpm
120.2bhp/ton 118.2bhp/tonne
68.2bhp/L 50.9kW/L 69.2PS/L
57.7ft/sec 17.6m/sec
19.9mph 32.0kmh/1000rpm
25.3mpg 21.1mpgUS
11.2L/100km
Petrol 4-stroke piston
1905cc 116.0cu in
In-line 4 fuel injection
Compression ratio: 9.2:1
Bore: 83.0mm 3.3in
Stroke: 88.0mm 3.5in
Valve type/No: Overhead 8

Transmission: Manual
No. of forward speeds: 5
Wheels driven: Front
Springs F/R: Coil/Torsion bar
Brake system: PA
Brakes F/R: Disc/Disc
Steering: Rack & pinion PA
Wheelbase: 254.0cm 100.0in
Length: 407.9cm 160.6in
Width: 170.9cm 67.3in
Height: 138.9cm 54.7in
Ground clearance: 19.1cm
7.5in
Kerb weight: 1100.0kg
2422.9lb
Fuel: 56.0L 12.3gal 14.8galUS

213 Citroen

AX GTi
1992 France
115.0mph 185.0kmh
0-50mph 80.5kmh: 6.6secs
0-60mph 96.5kmh: 9.3secs
0-1/4 mile: 17.1secs
0-1km: 31.5secs
100.0bhp 74.6kW 101.4PS
@ 6800rpm
90.0lbft 122.0Nm @ 4200rpm
124.0bhp/ton 121.9bhp/tonne
73.5bhp/L 54.8kW/L 74.5PS/L
57.2ft/sec 17.4m/sec
18.5mph 29.8kmh/1000rpm
26.0mpg 21.6mpgUS
10.9L/100km
Petrol 4-stroke piston
1360cc 83.0cu in
In-line 4 fuel injection
Compression ratio: 9.9:1
Bore: 75.0mm 2.9in
Stroke: 77.0mm 3.0in
Valve type/No: Overhead 8
Transmission: Manual

213 Citroen
AX GTi

No. of forward speeds: 5
Wheels driven: Front
Springs F/R: Coil/Torsion bar
Brake system: PA ABS
Brakes F/R: Disc/Drum
Steering: Rack & pinion
Wheelbase: 228.5cm 90.0in
Track F: 138.0cm 54.3in
Track R: 130.0cm 51.2in
Length: 350.0cm 137.8in
Width: 156.0cm 61.4in
Height: 135.5cm 53.3in
Ground clearance: 11.0cm
4.3in
Kerb weight: 820.0kg 1806.2lb
Fuel: 43.1L 9.5gal 11.4galUS

214 Citroen
XM Turbo SD Estate
1992 France
112.0mph 180.2kmh
0-50mph 80.5kmh: 9.6secs
0-60mph 96.5kmh: 13.1secs
0-1/4 mile: 19.5secs
0-1km: 33.9secs
110.0bhp 82.0kW 111.5PS
@ 4300rpm
183.0lbft 248.0Nm @ 2000rpm
65.1bhp/ton 64.1bhp/tonne
52.7bhp/L 39.3kW/L 53.4PS/L
43.2ft/sec 13.2m/sec
32.0mpg 26.6mpgUS
8.8L/100km
Diesel 4-stroke piston
2088cc 127.4cu in
turbocharged
In-line 4 fuel injection
Compression ratio: 21.5:1
Bore: 85.0mm 3.3in
Stroke: 92.0mm 3.6in
Valve type/No: Overhead 12
Transmission: Manual
No. of forward speeds: 5
Wheels driven: Front
Springs F/R: Gas/Gas
Brake system: PA
Brakes F/R: Disc/Disc
Steering: Rack & pinion PA
Kerb weight: 1716.9kg
3781.8lb
Fuel: 80.1L 17.6gal 21.1galUS

215 Cizeta
Moroder V16T
1990 Italy
204.0mph 328.2kmh
0-60mph 96.5kmh: 4.5secs
560.0bhp 417.6kW 567.8PS
@ 8000rpm
398.0lbft 539.3Nm @ 6000rpm
334.5bhp/ton 328.9bhp/tonne
93.4bhp/L 69.7kW/L 94.7PS/L
56.4ft/sec 17.2m/sec
Petrol 4-stroke piston

5995cc 365.8cu in
Vee 16 fuel injection
Compression ratio: 9.3:1
Bore: 86.0mm 3.4in
Stroke: 64.5mm 2.5in
Valve type/No: Overhead 64
Transmission: Manual
No. of forward speeds: 5
Wheels driven: Rear
Springs F/R: Coil/Coil
Brake system: PA
Brakes F/R: Disc/Disc
Steering: Rack & pinion PA
Wheelbase: 269.0cm 105.9in
Track F: 161.0cm 63.4in
Track R: 166.6cm 65.6in
Length: 449.3cm 176.9in
Width: 206.0cm 81.1in
Height: 111.5cm 43.9in
Kerb weight: 1702.5kg
3750.0lb
Fuel: 120.4L 26.5gal
31.8galUS

216 Consulier
GTP LX
1990 USA
140.0mph 225.3kmh
0-50mph 80.5kmh: 4.2secs
0-60mph 96.5kmh: 5.7secs
0-1/4 mile: 14.2secs
174.0bhp 129.7kW 176.4PS
@ 5200rpm
200.0lbft 271.0Nm @ 2400rpm
166.6bhp/ton 163.8bhp/tonne
78.6bhp/L 58.6kW/L 79.7PS/L
52.3ft/sec 15.9m/sec
21.9mpg 18.2mpgUS
12.9L/100km
Petrol 4-stroke piston
2213cc 135.0cu in
turbocharged
In-line 4 fuel injection
Compression ratio: 8.0:1
Bore: 87.5mm 3.4in
Stroke: 92.0mm 3.6in
Valve type/No: Overhead 8
Transmission: Manual
No. of forward speeds: 5
Wheels driven: Rear
Springs F/R: Coil/Coil
Brake system: PA
Brakes F/R: Disc/Disc
Steering: Rack & pinion
Wheelbase: 254.0cm 100.0in
Track F: 153.7cm 60.5in
Track R: 153.7cm 60.5in
Length: 436.9cm 172.0in
Width: 182.9cm 72.0in
Height: 114.3cm 45.0in
Kerb weight: 1062.4kg
2340.0lb
Fuel: 64.3L 14.1gal 17.0galUS

217 Daihatsu

Domino
1986 Japan
88.0mph 141.6kmh
0-50mph 80.5kmh: 10.1secs
0-60mph 96.5kmh: 14.7secs
0-1/4 mile: 19.7secs
0-1km: 37.8secs
44.0bhp 32.8kW 44.6PS
@ 5500rpm
50.0lbft 67.8Nm @ 3200rpm
72.6bhp/ton 71.4bhp/tonne
52.0bhp/L 38.8kW/L 52.7PS/L
48.7ft/sec 14.0m/sec
17.8mph 28.6kmh/1000rpm
37.2mpg 31.0mpgUS
7.6L/100km
Petrol 4-stroke piston
846cc 52.0cu in
In-line 3 1 Carburettor
Compression ratio: 9.5:1
Bore: 66.6mm 2.6in
Stroke: 81.0mm 3.2in
Valve type/No: Overhead 6
Transmission: Manual
No. of forward speeds: 5
Wheels driven: Front
Springs F/R: Coil/Coil
Brake system: PA
Brakes F/R: Disc/Drum
Steering: Rack & pinion
Wheelbase: 225.0cm 88.6in
Track F: 121.5cm 47.8in
Track R: 120.5cm 47.4in
Length: 319.5cm 125.8in
Width: 139.5cm 54.9in
Height: 140.0cm 55.1in
Ground clearance: 12.0cm
4.7in
Kerb weight: 616.0kg 1356.8lb
Fuel: 28.2L 6.2gal 7.4galUS

218 Daihatsu
Charade CX Diesel Turbo
1987 Japan
81.0mph 130.3kmh
0-50mph 80.5kmh: 13.9secs
0-60mph 96.5kmh: 20.9secs
0-1/4 mile: 22.0secs
0-1km: 41.3secs
47.0bhp 35.0kW 47.6PS
@ 4800rpm
63.0lbft 85.4Nm @ 2300rpm
58.1bhp/ton 57.2bhp/tonne
47.3bhp/L 35.3kW/L 48.0PS/L
38.3ft/sec 11.7m/sec
18.5mph 29.8kmh/1000rpm
45.5mpg 37.9mpgUS
6.2L/100km
Diesel 4-stroke piston
993cc 61.0cu in turbocharged
In-line 3 fuel injection
Compression ratio: 21.5:1
Bore: 76.0mm 3.0in
Stroke: 73.0mm 2.9in

Valve type/No: Overhead 6
Transmission: Manual
No. of forward speeds: 5
Wheels driven: Front
Springs F/R: Coil/Coil
Brake system: PA
Brakes F/R: Disc/Drum
Steering: Rack & pinion
Wheelbase: 234.0cm 92.1in
Track F: 138.5cm 54.5in
Track R: 136.5cm 53.7in
Length: 361.0cm 142.1in
Width: 160.0cm 63.0in
Height: 138.5cm 54.5in
Ground clearance: 16.0cm 6.3in
Kerb weight: 822.0kg 1810.6lb
Fuel: 36.9L 8.1gal 9.7galUS

219 Daihatsu
Charade GT ti
1987 Japan
116.0mph 186.6kmh
0-50mph 80.5kmh: 5.8secs
0-60mph 96.5kmh: 7.9secs
0-1/4 mile: 16.4secs
0-1km: 30.4secs
99.0bhp 73.8kW 100.4PS
@ 6500rpm
95.8lbft 129.8Nm @ 3500rpm
123.1bhp/ton 121.0bhp/tonne
99.7bhp/L 74.3kW/L 101.1PS/L
51.8ft/sec 15.8m/sec
18.4mph 29.6kmh/1000rpm
28.8mpg 24.0mpgUS
9.8L/100km
Petrol 4-stroke piston
993cc 61.0cu in turbocharged
In-line 3 fuel injection
Compression ratio: 7.8:1
Bore: 76.0mm 3.0in
Stroke: 73.0mm 2.9in
Valve type/No: Overhead 12
Transmission: Manual
No. of forward speeds: 5
Wheels driven: Front
Springs F/R: Coil/Coil
Brake system: PA
Brakes F/R: Disc/Disc
Steering: Rack & pinion
Wheelbase: 234.0cm 92.1in
Track F: 138.5cm 54.5in
Track R: 137.5cm 54.1in
Length: 361.0cm 142.1in
Width: 161.5cm 63.6in
Height: 138.5cm 54.5in
Ground clearance: 16.0cm
6.3in
Kerb weight: 818.0kg 1801.8lb
Fuel: 40.0L 8.8gal 10.6galUS

220 Daihatsu
Fourtrak Estate DT EL
1987 Japan
85.0mph 136.8kmh

0-50mph 80.5kmh: 12.0secs
0-60mph 96.5kmh: 17.9secs
0-1/4 mile: 20.7secs
0-1km: 38.7secs
87.0bhp 64.9kW 88.2PS
@ 3500rpm
155.0lbft 210.0Nm @ 2200rpm
53.3bhp/ton 52.4bhp/tonne
31.5bhp/L 23.5kW/L 31.9PS/L
39.8ft/sec 12.1m/sec
21.5mph 34.6kmh/1000rpm
23.0mpg 19.2mpgUS
12.3L/100km
Diesel 4-stroke piston
2765cc 169.0cu in
turbocharged
In-line 4 fuel injection
Compression ratio: 21.5:1
Bore: 92.0mm 3.6in
Stroke: 104.0mm 4.1in
Valve type/No: Overhead 8
Transmission: Manual
No. of forward speeds: 5
Wheels driven: 4-wheel
engageable
Springs F/R: Leaf/Leaf
Brake system: PA
Brakes F/R: Disc/Drum
Steering: Ball & nut PA
Wheelbase: 210.0cm 82.7in
Track F: 132.0cm 52.0in
Track R: 130.0cm 51.2in
Length: 412.5cm 162.4in
Width: 158.0cm 62.2in
Height: 191.5cm 75.4in
Ground clearance: 21.0cm
8.3in
Kerb weight: 1660.0kg
3656.4lb
Fuel: 60.1L 13.2gal 15.9galUS

221 Daihatsu
Charade Turbo
1988 Japan
96.0mph 154.5kmh
0-50mph 80.5kmh: 8.1secs
0-60mph 96.5kmh: 11.3secs
0-1/4 mile: 18.2secs
0-1km: 34.3secs
67.0bhp 50.0kW 67.9PS
@ 5500rpm
78.2lbft 106.0Nm @ 3500rpm
85.7bhp/ton 84.3bhp/tonne
67.5bhp/L 50.3kW/L 68.4PS/L
43.8ft/sec 13.4m/sec
19.8mph 31.9kmh/1000rpm
29.2mpg 24.3mpgUS
9.7L/100km
Petrol 4-stroke piston
993cc 61.0cu in turbocharged
In-line 3 1 Carburettor
Compression ratio: 8.0:1
Bore: 76.0mm 3.0in
Stroke: 73.0mm 2.9in

Valve type/No: Overhead 6
Transmission: Manual
No. of forward speeds: 5
Wheels driven: Front
Springs F/R: Coil/Coil
Brake system: PA
Brakes F/R: Disc/Drum
Steering: Rack & pinion
Wheelbase: 234.0cm 92.1in
Track F: 138.5cm 54.5in
Track R: 136.5cm 53.7in
Length: 361.0cm 142.1in
Width: 160.0cm 63.0in
Height: 138.5cm 54.5in
Ground clearance: 16.0cm
6.3in
Kerb weight: 795.0kg 1751.1lb
Fuel: 40.0L 8.8gal 10.6galUS

222 Daihatsu
Charade 1.3 CX
1989 Japan
102.0mph 164.1kmh
0-50mph 80.5kmh: 8.0secs
0-60mph 96.5kmh: 11.2secs
0-1/4 mile: 18.3secs
0-1km: 34.4secs
75.0bhp 55.9kW 76.0PS
@ 6500rpm
75.3lbft 102.0Nm @ 3900rpm
90.8bhp/ton 89.3bhp/tonne
57.9bhp/L 43.2kW/L 58.7PS/L
50.6ft/sec 15.4m/sec
18.4mph 29.6kmh/1000rpm
25.8mpg 21.5mpgUS
10.9L/100km
Petrol 4-stroke piston
1295cc 79.0cu in
In-line 4 1 Carburettor
Compression ratio: 9.5:1
Bore: 76.0mm 3.0in
Stroke: 71.0mm 2.8in
Valve type/No: Overhead 16
Transmission: Manual
No. of forward speeds: 5
Wheels driven: Front
Springs F/R: Coil/Coil
Brake system: PA
Brakes F/R: Disc/Drum
Steering: Rack & pinion
Wheelbase: 234.0cm 92.1in
Track F: 138.5cm 54.5in
Track R: 136.5cm 53.7in
Length: 361.0cm 142.1in
Width: 160.0cm 63.0in
Height: 138.5cm 54.5in
Ground clearance: 16.0cm
6.3in
Kerb weight: 840.0kg 1850.2lb
Fuel: 44.6L 9.8gal 11.8galUS

223 Daihatsu
Charade CLS
1989 Japan

95.0mph 152.9kmh
0-60mph 96.5kmh: 11.5secs
0-1/4 mile: 18.2secs
80.0bhp 59.7kW 81.1PS
@ 6000rpm
74.0lbft 100.3Nm @ 4400rpm
88.9bhp/ton 87.4bhp/tonne
61.7bhp/L 46.0kW/L 62.6PS/L
46.8ft/sec 14.3m/sec
35.4mpg 29.5mpgUS
8.0L/100km
Petrol 4-stroke piston
1296cc 79.1cu in
In-line 4 fuel injection
Compression ratio: 9.5:1
Bore: 76.0mm 3.0in
Stroke: 71.4mm 2.8in
Valve type/No: Overhead 16
Transmission: Manual
No. of forward speeds: 5
Wheels driven: Front
Springs F/R: Coil/Coil
Brake system: PA
Brakes F/R: Disc/Drum
Steering: Rack & pinion PA
Wheelbase: 233.9cm 92.1in
Track F: 138.4cm 54.5in
Track R: 136.4cm 53.7in
Length: 368.0cm 144.9in
Width: 161.5cm 63.6in
Height: 138.4cm 54.5in
Kerb weight: 914.8kg 2015.0lb
Fuel: 40.1L 8.8gal 10.6galUS

224 Daihatsu
Sportrak EL
1989 Japan
91.0mph 146.4kmh
0-50mph 80.5kmh: 11.1secs
0-60mph 96.5kmh: 16.1secs
0-1/4 mile: 20.3secs
0-1km: 40.1secs
85.0bhp 63.4kW 86.2PS
@ 6000rpm
93.0lbft 126.0Nm @ 3500rpm
68.3bhp/ton 67.1bhp/tonne
53.5bhp/L 39.9kW/L 54.2PS/L
57.5ft/sec 17.5m/sec
16.9mph 27.2kmh/1000rpm
21.4mpg 17.8mpgUS
13.2L/100km
Petrol 4-stroke piston
1589cc 97.0cu in
In-line 4 1 Carburettor
Compression ratio: 9.5:1
Bore: 76.0mm 3.0in
Stroke: 87.6mm 3.4in
Valve type/No: Overhead 16
Transmission: Manual
No. of forward speeds: 5
Wheels driven: 4-wheel
engageable
Springs F/R: Torsion bar/Leaf
Brake system: PA

Brakes F/R: Disc/Drum
Steering: Recirculating ball PA
Wheelbase: 217.5cm 85.6in
Track F: 132.0cm 52.0in
Track R: 132.0cm 52.0in
Length: 368.5cm 145.1in
Width: 158.0cm 62.2in
Height: 172.0cm 67.7in
Ground clearance: 21.0cm
8.3in
Kerb weight: 1266.0kg
2788.5lb
Fuel: 60.1L 13.2gal 15.9galUS

225 Daihatsu
Applause 16 Xi
1990 Japan
111.0mph 178.6kmh
0-50mph 80.5kmh: 7.2secs
0-60mph 96.5kmh: 10.1secs
0-1/4 mile: 17.6secs
0-1km: 32.6secs
105.0bhp 78.3kW 106.5PS
@ 6000rpm
98.9lbft 134.0Nm @ 4800rpm
107.3bhp/ton 105.5bhp/tonne
66.1bhp/L 49.3kW/L 67.0PS/L
57.7ft/sec 17.6m/sec
20.4mph 32.8kmh/1000rpm
29.0mpg 24.1mpgUS
9.7L/100km
Petrol 4-stroke piston
1589cc 97.0cu in
In-line 4 fuel injection
Compression ratio: 9.5:1
Bore: 76.0mm 3.0in
Stroke: 88.0mm 3.5in
Valve type/No: Overhead 16
Transmission: Manual
No. of forward speeds: 5
Wheels driven: Front
Springs F/R: Coil/Coil
Brake system: PA
Brakes F/R: Disc/Disc
Steering: Rack & pinion PA
Wheelbase: 247.0cm 97.2in
Track F: 142.5cm 56.1in
Track R: 141.5cm 55.7in
Length: 426.0cm 167.7in
Width: 166.0cm 65.4in
Height: 137.5cm 54.1in
Ground clearance: 14.9cm
5.9in
Kerb weight: 995.0kg 2191.6lb
Fuel: 60.1L 13.2gal 15.9galUS

226 De Tomaso
Pantera GT5-S
1986 Italy
165.0mph 265.5kmh
0-50mph 80.5kmh: 4.1secs
0-60mph 96.5kmh: 5.4secs
0-1/4 mile: 13.6secs
0-1km: 25.2secs

226 De Tomaso
Pantera GT5-S

300.0bhp 223.7kW 304.2PS
@ 6500rpm
209.0bhp/ton 205.5bhp/tonne
52.1bhp/L 38.8kW/L 52.8PS/L
63.2ft/sec 19.3m/sec
27.0mph 43.4kmh/1000rpm
16.0mpg 13.3mpgUS
17.7L/100km
Petrol 4-stroke piston
5763cc 352.0cu in
Vee 8 1 Carburettor
Compression ratio: 10.5:1
Bore: 101.6mm 4.0in
Stroke: 88.9mm 3.5in
Valve type/No: Overhead 16
Transmission: Manual
No. of forward speeds: 5
Wheels driven: Rear
Springs F/R: Coil/Coil
Brake system: PA
Brakes F/R: Disc/Disc
Steering: Rack & pinion
Wheelbase: 251.4cm 99.0in
Track F: 151.1cm 59.5in
Track R: 157.7cm 62.1in
Length: 426.9cm 168.1in
Width: 196.8cm 77.5in
Height: 109.9cm 43.3in
Ground clearance: 14.5cm
5.7in
Kerb weight: 1460.0kg
3215.9lb
Fuel: 79.6L 17.5gal 21.0galUS

227 De Tomaso
Pantera Group 3
1989 Italy
168.0mph 270.3kmh
0-50mph 80.5kmh: 3.4secs
0-60mph 96.5kmh: 4.3secs
0-1/4 mile: 12.7secs
465.0bhp 346.7kW 471.4PS
@ 5800rpm

440.0lbft 596.2Nm @ 3800rpm
315.2bhp/ton 309.9bhp/tonne
73.9bhp/L 55.1kW/L 74.9PS/L
60.4ft/sec 18.4m/sec
Petrol 4-stroke piston
6291cc 383.8cu in
Vee 8 4 Carburettor
Compression ratio: 10.1:1
Bore: 102.5mm 4.0in
Stroke: 95.3mm 3.8in
Valve type/No: Overhead 16
Transmission: Manual
No. of forward speeds: 5
Wheels driven: Rear
Springs F/R: Coil/Coil
Brakes F/R: Disc/Disc
Steering: Rack & pinion
Wheelbase: 249.9cm 98.4in
Track F: 157.5cm 62.0in
Track R: 162.6cm 64.0in
Length: 457.2cm 180.0in
Width: 198.1cm 78.0in
Height: 109.2cm 43.0in
Kerb weight: 1500.5kg
3305.0lb

228 Dodge
Shelby GLH-S
1986 USA
120.0mph 193.1kmh
0-50mph 80.5kmh: 4.9secs
0-60mph 96.5kmh: 6.7secs
0-1/4 mile: 15.3secs
175.0bhp 130.5kW 177.4PS
@ 5300rpm
175.0lbft 237.1Nm @ 2200rpm
154.3bhp/ton 151.8bhp/tonne
79.1bhp/L 59.0kW/L 80.2PS/L
53.3ft/sec 16.2m/sec
24.0mpg 20.0mpgUS
11.8L/100km
Petrol 4-stroke piston
2213cc 135.0cu in

turbocharged
In-line 4 fuel injection
Compression ratio: 8.5:1
Bore: 87.5mm 3.4in
Stroke: 92.0mm 3.6in
Valve type/No: Overhead 8
Transmission: Manual
No. of forward speeds: 5
Wheels driven: Front
Springs F/R: Coil/Coil
Brake system: PA
Brakes F/R: Disc/Drum
Steering: Rack & pinion PA
Wheelbase: 251.7cm 99.1in
Track F: 142.5cm 56.1in
Track R: 142.5cm 56.1in
Length: 414.5cm 163.2in
Width: 162.1cm 63.8in
Height: 134.6cm 53.0in
Kerb weight: 1153.2kg
2540.0lb
Fuel: 49.2L 10.8gal 13.0galUS

229 Dodge
Daytona Shelby Z
1988 USA
120.0mph 193.1kmh
0-50mph 80.5kmh: 5.2secs
0-60mph 96.5kmh: 7.2secs
0-1/4 mile: 15.5secs
174.0bhp 129.7kW 176.4PS
@ 5200rpm
200.0lbft 271.0Nm @ 2400rpm
137.2bhp/ton 134.9bhp/tonne
78.6bhp/L 58.6kW/L 79.7PS/L
52.3ft/sec 15.9m/sec
27.6mpg 23.0mpgUS
10.2L/100km
Petrol 4-stroke piston
2213cc 135.0cu in
turbocharged
In-line 4 fuel injection
Compression ratio: 8.0:1

Bore: 87.5mm 3.4in
Stroke: 92.0mm 3.6in
Valve type/No: Overhead 8
Transmission: Manual
No. of forward speeds: 5
Wheels driven: Rear
Springs F/R: Coil/Coil
Brake system: PA
Brakes F/R: Disc/Drum
Steering: Rack & pinion PA
Wheelbase: 246.6cm 97.1in
Track F: 146.3cm 57.6in
Track R: 146.3cm 57.6in
Length: 446.8cm 175.9in
Width: 177.3cm 69.8in
Height: 129.0cm 50.8in
Kerb weight: 1289.4kg
2840.0lb
Fuel: 53.0L 11.6gal 14.0galUS

230 Dodge
Lancer Shelby
1988 USA
118.0mph 189.9kmh
0-50mph 80.5kmh: 5.5secs
0-60mph 96.5kmh: 7.6secs
0-1/4 mile: 15.8secs
174.0bhp 129.7kW 176.4PS
@ 5200rpm
200.0lbft 271.0Nm @ 2400rpm
140.2bhp/ton 137.9bhp/tonne
78.6bhp/L 58.6kW/L 79.7PS/L
52.3ft/sec 15.9m/sec
25.8mpg 21.5mpgUS
10.9L/100km
Petrol 4-stroke piston
2213cc 135.0cu in
turbocharged
In-line 4 fuel injection
Compression ratio: 8.0:1
Bore: 87.5mm 3.4in
Stroke: 92.0mm 3.6in
Valve type/No: Overhead 8

Transmission: Manual
No. of forward speeds: 5
Wheels driven: Front
Springs F/R: Coil/Coil
Brake system: PA
Brakes F/R: Disc/Drum
Steering: Rack & pinion PA
Wheelbase: 261.9cm 103.1in
Track F: 146.3cm 57.6in
Track R: 145.3cm 57.2in
Length: 458.0cm 180.3in
Width: 173.5cm 68.3in
Height: 134.6cm 53.0in
Kerb weight: 1262.1kg
2780.0lb
Fuel: 53.0L 11.6gal 14.0galUS

231 Dodge

Shelby CSX
1988 USA
135.0mph 217.2kmh
0-50mph 80.5kmh: 5.0secs
0-60mph 96.5kmh: 6.9secs
0-1/4 mile: 14.9secs
175.0bhp 130.5kW 177.4PS
@ 5300rpm
200.0lbft 271.0Nm @ 2200rpm
155.6bhp/ton 153.0bhp/tonne
79.1bhp/L 59.0kW/L 80.2PS/L
53.3ft/sec 16.2m/sec
25.8mpg 21.5mpgUS
10.9L/100km
Petrol 4-stroke piston
2213cc 135.0cu in
turbocharged
In-line 4 fuel injection
Compression ratio: 8.0:1
Bore: 87.5mm 3.4in
Stroke: 92.0mm 3.6in
Valve type/No: Overhead 8
Transmission: Manual
No. of forward speeds: 5
Wheels driven: Front
Springs F/R: Coil/Coil
Brake system: PA
Brakes F/R: Disc/Disc
Steering: Rack & pinion PA
Wheelbase: 246.4cm 97.0in
Track F: 148.8cm 58.6in
Track R: 147.8cm 58.2in
Length: 436.1cm 171.7in
Width: 170.9cm 67.3in
Height: 132.1cm 52.0in
Kerb weight: 1144.1kg
2520.0lb
Fuel: 53.0L 11.6gal 14.0galUS

232 Dodge

Daytona Shelby
1989 USA
125.0mph 201.1kmh
0-50mph 80.5kmh: 5.6secs
0-60mph 96.5kmh: 7.5secs
0-1/4 mile: 15.9secs

174.0bhp 129.7kW 176.4PS
@ 5200rpm
200.0lbft 271.0Nm @ 2400rpm
128.8bhp/ton 126.7bhp/tonne
78.6bhp/L 58.6kW/L 79.7PS/L
52.3ft/sec 15.9m/sec
20.3mpg 16.9mpgUS
13.9L/100km
Petrol 4-stroke piston
2213cc 135.0cu in
turbocharged
In-line 4 fuel injection
Compression ratio: 8.1:1
Bore: 87.5mm 3.4in
Stroke: 92.0mm 3.6in
Valve type/No: Overhead 8
Transmission: Manual
No. of forward speeds: 5
Wheels driven: Front
Springs F/R: Coil/Coil
Brake system: PA
Brakes F/R: Disc/Disc
Steering: Rack & pinion PA
Wheelbase: 246.4cm 97.0in
Track F: 146.3cm 57.6in
Track R: 146.3cm 57.6in
Length: 455.4cm 179.3in
Width: 176.0cm 69.3in
Height: 130.3cm 51.3in
Kerb weight: 1373.3kg
3025.0lb
Fuel: 53.0L 11.6gal 14.0galUS

233 Dodge

Shadow ES Turbo
1989 USA
115.0mph 185.0kmh
0-50mph 80.5kmh: 5.5secs
0-60mph 96.5kmh: 7.7secs
0-1/4 mile: 16.0secs
150.0bhp 111.9kW 152.1PS
@ 4800rpm
180.0lbft 243.9Nm @ 2000rpm
121.7bhp/ton 119.7bhp/tonne
60.0bhp/L 44.7kW/L 60.8PS/L
54.5ft/sec 16.6m/sec
22.8mpg 19.0mpgUS
12.4L/100km
Petrol 4-stroke piston
2501cc 152.6cu in
turbocharged
In-line 4 fuel injection
Compression ratio: 7.8:1
Bore: 87.5mm 3.4in
Stroke: 104.0mm 4.1in
Valve type/No: Overhead 8
Transmission: Manual
No. of forward speeds: 5
Wheels driven: Front
Springs F/R: Coil/Coil
Brake system: PA
Brakes F/R: Disc/Drum
Steering: Rack & pinion PA
Wheelbase: 246.4cm 97.0in

Track F: 146.3cm 57.6in
Track R: 145.3cm 57.2in
Length: 434.6cm 171.1in
Width: 170.9cm 67.3in
Height: 133.9cm 52.7in
Kerb weight: 1253.0kg
2760.0lb
Fuel: 53.0L 11.6gal 14.0galUS

234 Dodge

Shelby CSX
1989 USA
140.0mph 225.3kmh
0-60mph 96.5kmh: 7.6secs
0-1/4 mile: 15.7secs
175.0bhp 130.5kW 177.4PS
@ 5200rpm
210.0lbft 284.6Nm @ 2400rpm
140.7bhp/ton 138.4bhp/tonne
79.1bhp/L 59.0kW/L 80.2PS/L
52.3ft/sec 15.9m/sec
24.6mpg 20.5mpgUS
11.5L/100km
Petrol 4-stroke piston
2213cc 135.0cu in
turbocharged
In-line 4 fuel injection
Compression ratio: 8.1:1
Bore: 87.5mm 3.4in
Stroke: 92.0mm 3.6in
Valve type/No: Overhead 8
Transmission: Manual
No. of forward speeds: 5
Wheels driven: Front
Springs F/R: Coil/Coil
Brake system: PA
Brakes F/R: Disc/Disc
Steering: Rack & pinion PA
Wheelbase: 246.4cm 97.0in
Track F: 147.6cm 58.1in
Track R: 146.6cm 57.7in
Length: 436.1cm 171.7in
Width: 170.9cm 67.3in
Height: 133.9cm 52.7in
Kerb weight: 1264.4kg
2785.0lb
Fuel: 53.0L 11.6gal 14.0galUS

235 Dodge

Shelby CSX VNT
1989 USA
120.0mph 193.1kmh
0-50mph 80.5kmh: 5.2secs
0-60mph 96.5kmh: 7.2secs
0-1/4 mile: 15.6secs
175.0bhp 130.5kW 177.4PS
@ 5300rpm
200.0lbft 271.0Nm @ 3700rpm
141.3bhp/ton 138.9bhp/tonne
79.1bhp/L 59.0kW/L 80.2PS/L
53.3ft/sec 16.2m/sec
24.1mpg 20.1mpgUS
11.7L/100km
Petrol 4-stroke piston

2213cc 135.0cu in
turbocharged
In-line 4 fuel injection
Compression ratio: 8.1:1
Bore: 87.5mm 3.4in
Stroke: 92.0mm 3.6in
Valve type/No: Overhead 8
Transmission: Manual
No. of forward speeds: 5
Wheels driven: Front
Springs F/R: Coil/Coil
Brake system: PA
Brakes F/R: Disc/Disc
Steering: Rack & pinion PA
Wheelbase: 246.4cm 97.0in
Track F: 147.6cm 58.1in
Track R: 146.6cm 57.7in
Length: 434.6cm 171.1in
Width: 170.9cm 67.3in
Height: 133.9cm 52.7in
Kerb weight: 1259.8kg
2775.0lb
Fuel: 53.0L 11.6gal
14.0galUS

236 Dodge

Daytona ES
1990 USA
125.0mph 201.1kmh
0-60mph 96.5kmh: 8.4secs
0-1/4 mile: 16.2secs
141.0bhp 105.1kW 143.0PS
@ 5000rpm
171.0lbft 231.7Nm @ 2800rpm
104.2bhp/ton 102.5bhp/tonne
47.4bhp/L 35.4kW/L 48.1PS/L
41.5ft/sec 12.7m/sec
24.6mpg 20.5mpgUS
11.5L/100km
Petrol 4-stroke piston
2972cc 181.3cu in
Vee 6 fuel injection
Compression ratio: 8.9:1
Bore: 91.1mm 3.6in
Stroke: 76.0mm 3.0in
Valve type/No: Overhead 12
Transmission: Manual
No. of forward speeds: 5
Wheels driven: Front
Springs F/R: Coil/Coil
Brake system: PA
Brakes F/R: Disc/Disc
Steering: Rack & pinion PA
Wheelbase: 246.4cm 97.0in
Track F: 146.1cm 57.5in
Track R: 146.3cm 57.6in
Length: 455.2cm 179.2in
Width: 176.0cm 69.3in
Height: 127.3cm 50.1in
Kerb weight: 1375.6kg
3030.0lb
Fuel: 53.0L 11.6gal
14.0galUS

237 Dodge

237 Dodge
Daytona
Shelby

Daytona Shelby
1990 USA
135.0mph 217.2kmh
0-60mph 96.5kmh: 7.6secs
0-1/4 mile: 16.0secs
174.0bhp 129.7kW 176.4PS
@ 5200rpm
210.0lbft 284.6Nm @ 2400rpm
128.2bhp/ton 126.1bhp/tonne
78.6bhp/L 58.6kW/L 79.7PS/L
52.3ft/sec 15.9m/sec
27.6mpg 23.0mpgUS
10.2L/100km
Petrol 4-stroke piston
2213cc 135.0cu in
turbocharged
In-line 4 fuel injection
Compression ratio: 8.0:1
Bore: 87.5mm 3.4in
Stroke: 92.0mm 3.6in
Valve type/No: Overhead 8
Transmission: Manual
No. of forward speeds: 5
Wheels driven: Front
Springs F/R: Coil/Coil
Brake system: PA
Brakes F/R: Disc/Disc
Steering: Rack & pinion PA
Wheelbase: 246.4cm 97.0in
Track F: 146.1cm 57.5in
Track R: 146.3cm 57.6in
Length: 455.2cm 179.2in
Width: 176.0cm 69.3in
Height: 127.3cm 50.1in
Kerb weight: 1380.2kg
3040.0lb
Fuel: 53.0L 11.6gal 14.0galUS

238 Dodge
Shadow ES VNT
1990 USA
130.0mph 209.2kmh
0-50mph 80.5kmh: 5.4secs

0-60mph 96.5kmh: 7.3secs
0-1/4 mile: 15.7secs
174.0bhp 129.7kW 176.4PS
@ 5200rpm
210.0lbft 284.6Nm @ 2400rpm
152.8bhp/ton 150.3bhp/tonne
78.6bhp/L 58.6kW/L 79.7PS/L
52.3ft/sec 15.9m/sec
26.3mpg 21.9mpgUS
10.7L/100km
Petrol 4-stroke piston
2213cc 135.0cu in
turbocharged
In-line 4 fuel injection
Compression ratio: 8.0:1
Bore: 87.5mm 3.4in
Stroke: 92.0mm 3.6in
Valve type/No: Overhead 8
Transmission: Manual
No. of forward speeds: 5
Wheels driven: Front
Springs F/R: Coil/Coil
Brake system: PA
Brakes F/R: Disc/Disc
Steering: Rack & pinion PA
Wheelbase: 246.4cm 97.0in
Track F: 146.3cm 57.6in
Track R: 145.3cm 57.2in
Length: 436.1cm 171.7in
Width: 170.9cm 67.3in
Height: 128.0cm 50.4in
Kerb weight: 1157.7kg
2550.0lb
Fuel: 53.0L 11.6gal 14.0galUS

239 Dodge
Spirit R/T
1991 USA
140.0mph 225.3kmh
0-60mph 96.5kmh: 6.5secs
0-1/4 mile: 15.0secs
224.0bhp 167.0kW 227.1PS
@ 6000rpm

217.0lbft 294.0Nm @ 2800rpm
160.6bhp/ton 157.9bhp/tonne
101.6bhp/L 75.8kW/L
103.0PS/L
60.3ft/sec 18.4m/sec
25.8mpg 21.5mpgUS
10.9L/100km
Petrol 4-stroke piston
2205cc 134.5cu in
turbocharged
In-line 4 fuel injection
Compression ratio: 8.5:1
Bore: 87.4mm 3.4in
Stroke: 92.0mm 3.6in
Valve type/No: Overhead 16
Transmission: Manual
No. of forward speeds: 5
Wheels driven: Front
Springs F/R: Coil/Coil
Brake system: PA
Brakes F/R: Disc/Disc
Steering: Rack & pinion PA
Wheelbase: 262.4cm 103.3in
Track F: 146.3cm 57.6in
Track R: 145.3cm 57.2in
Length: 460.2cm 181.2in
Width: 173.0cm 68.1in
Height: 135.9cm 53.5in
Kerb weight: 1418.7kg
3125.0lb
Fuel: 60.6L 13.3gal 16.0galUS

240 Dodge
Stealth R/T Turbo
1991 Japan
159.0mph 255.8kmh
0-50mph 80.5kmh: 4.8secs
0-60mph 96.5kmh: 6.3secs
0-1/4 mile: 14.7secs
300.0bhp 223.7kW 304.2PS
@ 6000rpm
307.0lbft 416.0Nm @ 2500rpm
177.1bhp/ton 174.1bhp/tonne

100.9bhp/L 75.3kW/L
102.3PS/L
49.8ft/sec 15.2m/sec
21.6mpg 18.0mpgUS
13.1L/100km
Petrol 4-stroke piston
2972cc 181.3cu in
turbocharged
Vee 6 fuel injection
Compression ratio: 8.0:1
Bore: 91.1mm 3.6in
Stroke: 76.0mm 3.0in
Valve type/No: Overhead 24
Transmission: Manual
No. of forward speeds: 5
Wheels driven: 4-wheel drive
Springs F/R: Coil/Coil
Brake system: PA ABS
Brakes F/R: Disc/Disc
Steering: Rack & pinion PA
Wheelbase: 246.9cm 97.2in
Track F: 156.0cm 61.4in
Track R: 158.0cm 62.2in
Length: 458.5cm 180.5in
Width: 183.9cm 72.4in
Height: 124.7cm 49.1in
Kerb weight: 1722.9kg
3795.0lb
Fuel: 74.9L 16.5gal 19.8galUS

241 Dodge
Viper
1992 USA
190.0mph 305.7kmh
0-60mph 96.5kmh: 4.1secs
0-1/4 mile: 12.5secs
400.0bhp 298.3kW 405.5PS
@ 5500rpm
450.0lbft 609.8Nm @ 2200rpm
299.7bhp/ton 294.7bhp/tonne
50.1bhp/L 37.3kW/L 50.8PS/L
59.3ft/sec 18.1m/sec
Petrol 4-stroke piston

7989cc 487.4cu in
Vee 10 fuel injection
Bore: 101.6mm 4.0in
Stroke: 98.6mm 3.9in
Valve type/No: Overhead 20
Transmission: Manual
No. of forward speeds: 6
Wheels driven: Rear
Springs F/R: Coil/Coil
Brakes F/R: Disc/Disc
Steering: Rack & pinion
Wheelbase: 244.3cm 96.2in
Track F: 151.4cm 59.6in
Track R: 154.4cm 60.8in
Length: 436.9cm 172.0in
Width: 192.0cm 75.6in
Height: 117.3cm 46.2in
Kerb weight: 1357.5kg
2990.0lb

242 Eagle

Premier ES
1988 USA
125.0mph 201.1kmh
0-60mph 96.5kmh: 10.5secs
0-1/4 mile: 17.7secs
150.0bhp 111.9kW 152.1PS
@ 5000rpm
171.0lbft 231.7Nm @ 3750rpm
110.5bhp/ton 108.7bhp/tonne
50.4bhp/L 37.6kW/L 51.1PS/L
39.9ft/sec 12.2m/sec
22.8mpg 19.0mpgUS
12.4L/100km
Petrol 4-stroke piston
2975cc 181.5cu in
Vee 6 fuel injection
Compression ratio: 9.3:1
Bore: 93.0mm 3.7in
Stroke: 73.0mm 2.9in
Valve type/No: Overhead 12
Transmission: Automatic
No. of forward speeds: 4
Wheels driven: Front
Springs F/R: Coil/Torsion bar
Brake system: PA
Brakes F/R: Disc/Drum
Steering: Rack & pinion PA
Wheelbase: 268.0cm 105.5in
Track F: 147.8cm 58.2in
Track R: 145.0cm 57.1in
Length: 489.5cm 192.7in
Width: 177.3cm 69.8in
Height: 142.0cm 55.9in
Kerb weight: 1380.2kg
3040.0lb
Fuel: 64.3L 14.1gal 17.0galUS

243 Eagle

Talon
1990 USA
143.0mph 230.1kmh
0-50mph 80.5kmh: 4.9secs
0-60mph 96.5kmh: 6.8secs

0-1/4 mile: 15.3secs
195.0bhp 145.4kW 197.7PS
@ 6000rpm
203.0lbft 275.1Nm @ 3000rpm
138.9bhp/ton 136.6bhp/tonne
97.6bhp/L 72.8kW/L 99.0PS/L
57.7ft/sec 17.6m/sec
19.8mpg 16.5mpgUS
14.3L/100km
Petrol 4-stroke piston
1997cc 121.8cu in
turbocharged
In-line 4 fuel injection
Compression ratio: 7.8:1
Bore: 85.0mm 3.3in
Stroke: 88.0mm 3.5in
Valve type/No: Overhead 16
Transmission: Manual
No. of forward speeds: 5
Wheels driven: 4-wheel drive
Springs F/R: Coil/Coil
Brake system: PA
Brakes F/R: Disc/Disc
Steering: Rack & pinion PA
Wheelbase: 246.9cm 97.2in
Track F: 146.6cm 57.7in
Track R: 145.0cm 57.1in
Length: 433.1cm 170.5in
Width: 169.4cm 66.7in
Height: 130.6cm 51.4in
Kerb weight: 1427.8kg
3145.0lb
Fuel: 60.2L 13.2gal 15.9galUS

244 Eagle

GTP
1991 USA
200.0mph 321.8kmh
0-60mph 96.5kmh: 5.0secs
0-1/4 mile: 13.8secs
345.0bhp 257.3kW 349.8PS
@ 5600rpm
340.0lbft 460.7Nm @ 3200rpm
396.3bhp/ton 389.7bhp/tonne
60.2bhp/L 44.9kW/L 61.0PS/L
54.1ft/sec 16.5m/sec
Petrol 4-stroke piston
5733cc 349.8cu in
Vee 8 1 Carburettor
Compression ratio: 9.0:1
Bore: 101.6mm 4.0in
Stroke: 88.4mm 3.5in
Valve type/No: Overhead 16
Transmission: Manual
No. of forward speeds: 4
Wheels driven: Rear
Springs F/R: Coil/Coil
Brakes F/R: Disc/Disc
Steering: Rack & pinion
Wheelbase: 270.5cm 106.5in
Track F: 160.0cm 63.0in
Track R: 154.9cm 61.0in
Length: 477.5cm 188.0in
Width: 200.7cm 79.0in

Height: 104.1cm 41.0in
Kerb weight: 885.3kg 1950.0lb
Fuel: 60.6L 13.3gal 16.0galUS

245 Eagle

Talon TSi 4WD
1991 USA
143.0mph 230.1kmh
0-50mph 80.5kmh: 4.9secs
0-60mph 96.5kmh: 6.8secs
0-1/4 mile: 15.3secs
195.0bhp 145.4kW 197.7PS
@ 6000rpm
203.0lbft 275.1Nm @ 3000rpm
140.9bhp/ton 138.5bhp/tonne
97.6bhp/L 72.8kW/L 99.0PS/L
57.7ft/sec 17.6m/sec
19.8mpg 16.5mpgUS
14.3L/100km
Petrol 4-stroke piston
1997cc 121.8cu in
turbocharged
In-line 4 fuel injection
Compression ratio: 7.8:1
Bore: 85.0mm 3.3in
Stroke: 88.0mm 3.5in
Valve type/No: Overhead 16
Transmission: Manual
No. of forward speeds: 5
Wheels driven: 4-wheel drive
Springs F/R: Coil/Coil
Brake system: PA ABS
Brakes F/R: Disc/Disc
Steering: Rack & pinion PA
Wheelbase: 246.9cm 97.2in
Track F: 146.6cm 57.7in
Track R: 145.5cm 57.3in
Length: 437.9cm 172.4in
Width: 169.9cm 66.9in
Height: 132.1cm 52.0in
Kerb weight: 1407.4kg
3100.0lb
Fuel: 60.2L 13.2gal 15.9galUS

246 Ferrari

3.2 Mondial
1986 Italy
145.0mph 233.3kmh
0-60mph 96.5kmh: 7.1secs
0-1/4 mile: 15.3secs
260.0bhp 193.9kW 263.6PS
@ 7000rpm
213.0lbft 288.6Nm @ 5500rpm
164.3bhp/ton 161.5bhp/tonne
81.6bhp/L 60.9kW/L 82.8PS/L
56.4ft/sec 17.2m/sec
19.4mph 31.2kmh/1000rpm
Petrol 4-stroke piston
3185cc 194.3cu in
Vee 8 fuel injection
Compression ratio: 9.2:1
Bore: 83.0mm 3.3in
Stroke: 73.6mm 2.9in
Valve type/No: Overhead 32

Transmission: Manual
No. of forward speeds: 5
Wheels driven: Rear
Springs F/R: Coil/Coil
Brakes F/R: Disc/Disc
Steering: Rack & pinion
Wheelbase: 264.9cm 104.3in
Track F: 151.4cm 59.6in
Track R: 153.4cm 60.4in
Length: 464.1cm 182.7in
Width: 179.1cm 70.5in
Height: 126.0cm 49.6in
Kerb weight: 1609.4kg
3545.0lb

247 Ferrari

328 GTB
1986 Italy
155.0mph 249.4kmh
0-60mph 96.5kmh: 6.6secs
0-1/4 mile: 14.6secs
260.0bhp 193.9kW 263.6PS
@ 7000rpm
213.0lbft 288.6Nm @ 5500rpm
185.5bhp/ton 182.4bhp/tonne
81.6bhp/L 60.9kW/L 82.8PS/L
Petrol 4-stroke piston
3185cc 194.3cu in
Vee 8 fuel injection
Valve type/No: Overhead 32
Transmission: Manual
No. of forward speeds: 5
Wheels driven: Rear
Springs F/R: Coil/Coil
Brakes F/R: Disc/Disc
Steering: Rack & pinion
Wheelbase: 235.0cm 92.5in
Length: 429.5cm 169.1in
Width: 172.0cm 67.7in
Height: 112.0cm 44.1in
Kerb weight: 1425.6kg
3140.0lb
Fuel: 51.5L 11.3gal 13.6galUS

248 Ferrari

Mondial 3.2
1986 Italy
149.0mph 239.7kmh
0-60mph 96.5kmh: 7.4secs
0-1/4 mile: 15.0secs
260.0bhp 193.9kW 263.6PS
@ 7000rpm
213.0lbft 288.6Nm @ 5500rpm
171.3bhp/ton 168.4bhp/tonne
81.6bhp/L 60.9kW/L 82.8PS/L
Petrol 4-stroke piston
3185cc 194.3cu in
Vee 8 fuel injection
Valve type/No: Overhead 32
Transmission: Manual
No. of forward speeds: 5
Wheels driven: Rear
Springs F/R: Coil/Coil
Brakes F/R: Disc/Disc

Steering: Rack & pinion
Wheelbase: 264.9cm 104.3in
Length: 453.4cm 178.5in
Width: 179.1cm 70.5in
Height: 125.0cm 49.2in
Kerb weight: 1543.6kg
3400.0lb
Fuel: 60.2L 13.2gal 15.9galUS

249 Ferrari
Mondial 3.2QV
1986 Italy
144.0mph 231.7kmh
0-50mph 80.5kmh: 5.3secs
0-60mph 96.5kmh: 6.8secs
0-1/4 mile: 14.9secs
0-1km: 27.4secs
270.0bhp 201.3kW 273.7PS
@ 7000rpm
224.0lbft 303.5Nm @ 5500rpm
185.9bhp/ton 182.8bhp/tonne
84.7bhp/L 63.2kW/L 85.9PS/L
56.4ft/sec 17.2m/sec
20.9mph 33.6kmh/1000rpm
16.8mpg 14.0mpgUS
16.8L/100km
Petrol 4-stroke piston
3186cc 194.0cu in
Vee 8 1 Carburettor
Compression ratio: 9.8:1
Bore: 83.0mm 3.3in
Stroke: 73.6mm 2.9in
Valve type/No: Overhead 32
Transmission: Manual
No. of forward speeds: 5
Wheels driven: Rear
Springs F/R: Coil/Coil
Brake system: PA
Brakes F/R: Disc/Disc
Steering: Rack & pinion
Wheelbase: 264.9cm 104.3in
Track F: 151.8cm 59.8in
Track R: 150.8cm 59.4in
Length: 453.3cm 178.5in
Width: 179.5cm 70.7in
Height: 123.4cm 48.6in
Ground clearance: 15.3cm
6.0in
Kerb weight: 1477.0kg
3253.3lb
Fuel: 80.1L 17.6gal 21.1galUS

250 Ferrari
328 GTS
1987 Italy
149.0mph 239.7kmh
0-60mph 96.5kmh: 5.9secs
0-1/4 mile: 14.4secs
260.0bhp 193.9kW 263.6PS
@ 7000rpm
213.0lbft 288.6Nm @ 5500rpm
183.7bhp/ton 180.7bhp/tonne
81.6bhp/L 60.9kW/L 82.8PS/L
56.4ft/sec 17.2m/sec

19.9mph 32.0kmh/1000rpm
19.8mpg 16.5mpgUS
14.3L/100km
Petrol 4-stroke piston
3185cc 194.3cu in
Vee 8 fuel injection
Compression ratio: 9.2:1
Bore: 83.0mm 3.3in
Stroke: 73.6mm 2.9in
Valve type/No: Overhead 32
Transmission: Manual
No. of forward speeds: 5
Wheels driven: Rear
Springs F/R: Coil/Coil
Brake system: PA
Brakes F/R: Disc/Disc
Steering: Rack & pinion
Wheelbase: 235.0cm 92.5in
Track F: 147.3cm 58.0in
Track R: 146.8cm 57.8in
Length: 428.5cm 168.7in
Width: 173.0cm 68.1in
Height: 112.8cm 44.4in
Ground clearance: 10.2cm
4.0in
Kerb weight: 1439.2kg
3170.0lb
Fuel: 70.0L 15.4gal 18.5galUS

251 Ferrari
412
1987 Italy
147.0mph 236.5kmh
0-60mph 96.5kmh: 6.7secs
0-1/4 mile: 14.9secs
340.0bhp 253.5kW 344.7PS
@ 6000rpm
333.0lbft 451.2Nm @ 4200rpm
191.4bhp/ton 188.2bhp/tonne
68.8bhp/L 51.3kW/L 69.7PS/L
51.2ft/sec 15.6m/sec
23.5mph 37.8kmh/1000rpm
10.8mpg 9.0mpgUS
26.1L/100km
Petrol 4-stroke piston
4942cc 301.5cu in
Vee 12 fuel injection
Compression ratio: 9.6:1
Bore: 82.0mm 3.2in
Stroke: 78.0mm 3.1in
Valve type/No: Overhead 24
Transmission: Manual
No. of forward speeds: 5
Wheels driven: Rear
Springs F/R: Coil/Coil
Brake system: PA ABS
Brakes F/R: Disc/Disc
Steering: Rack & pinion PA
Wheelbase: 270.0cm 106.3in
Track F: 148.1cm 58.3in
Track R: 149.9cm 59.0in
Length: 481.1cm 189.4in
Width: 179.8cm 70.8in

Height: 131.3cm 51.7in
Kerb weight: 1806.9kg
3980.0lb
Fuel: 109.8L 24.1gal
29.0galUS

252 Ferrari
Mondial Cabriolet 3.2
1987 Italy
145.0mph 233.3kmh
0-60mph 96.5kmh: 7.0secs
0-1/4 mile: 15.2secs
260.0bhp 193.9kW 263.6PS
@ 7000rpm
213.0lbft 288.6Nm @ 5500rpm
164.3bhp/ton 161.5bhp/tonne
81.6bhp/L 60.9kW/L 82.8PS/L
56.4ft/sec 17.2m/sec
19.4mph 31.2kmh/1000rpm
19.2mpg 16.0mpgUS
14.7L/100km
Petrol 4-stroke piston
3185cc 194.3cu in
Vee 8 fuel injection
Compression ratio: 9.2:1
Bore: 83.0mm 3.3in
Stroke: 73.6mm 2.9in
Valve type/No: Overhead 32
Transmission: Manual
No. of forward speeds: 5
Wheels driven: Rear
Springs F/R: Coil/Coil
Brake system: PA
Brakes F/R: Disc/Disc
Steering: Rack & pinion
Wheelbase: 264.9cm 104.3in
Track F: 151.4cm 59.6in
Track R: 153.4cm 60.4in
Length: 464.1cm 182.7in
Width: 179.1cm 70.5in
Height: 126.0cm 49.6in
Ground clearance: 11.7cm
4.6in
Kerb weight: 1609.4kg
3545.0lb
Fuel: 70.0L 15.4gal 18.5galUS

253 Ferrari
308 Norwood Bonneville GTO
1988 Italy
198.0mph 318.6kmh
0-60mph 96.5kmh: 3.7secs
0-1/4 mile: 12.3secs
671.0bhp 500.4kW 680.3PS
@ 7850rpm
502.0lbft 680.2Nm @ 5250rpm
589.4bhp/ton 579.6bhp/tonne
111.8bhp/L 83.4kW/L
113.4PS/L
77.6ft/sec 23.6m/sec
Petrol 4-stroke piston
6000cc 366.1cu in
Vee 8 fuel injection
Compression ratio: 13.2:1

Bore: 104.9mm 4.1in
Stroke: 90.4mm 3.6in
Valve type/No: Overhead 16
Transmission: Manual
No. of forward speeds: 5
Wheels driven: Rear
Springs F/R: Coil/Coil
Brake system: PA
Brakes F/R: Disc/Disc
Steering: Rack & pinion
Wheelbase: 245.1cm 96.5in
Track F: 149.4cm 58.8in
Track R: 158.8cm 62.5in
Length: 429.0cm 168.9in
Width: 191.0cm 75.2in
Height: 112.0cm 44.1in
Kerb weight: 1157.7kg
2550.0lb
Fuel: 56.8L 12.5gal 15.0galUS

254 Ferrari
Mondial Cabriolet
1988 Italy
145.0mph 233.3kmh
0-50mph 80.5kmh: 5.4secs
0-60mph 96.5kmh: 7.0secs
0-1/4 mile: 15.2secs
260.0bhp 193.9kW 263.6PS
@ 7000rpm
213.0lbft 288.6Nm @ 5500rpm
164.3bhp/ton 161.5bhp/tonne
81.6bhp/L 60.9kW/L 82.8PS/L
56.4ft/sec 17.2m/sec
19.2mpg 16.0mpgUS
14.7L/100km
Petrol 4-stroke piston
3185cc 194.3cu in
Vee 8 fuel injection
Compression ratio: 9.2:1
Bore: 83.0mm 3.3in
Stroke: 73.6mm 2.9in
Valve type/No: Overhead 32
Transmission: Manual
No. of forward speeds: 5
Wheels driven: Rear
Springs F/R: Coil/Coil
Brake system: PA
Brakes F/R: Disc/Disc
Steering: Rack & pinion
Wheelbase: 264.9cm 104.3in
Track F: 151.4cm 59.6in
Track R: 153.4cm 60.4in
Length: 464.1cm 182.7in
Width: 179.1cm 70.5in
Height: 126.0cm 49.6in
Kerb weight: 1609.4kg
3545.0lb
Fuel: 70.0L 15.4gal 18.5galUS

255 Ferrari
Testa Rossa
1988 Italy
181.0mph 291.2kmh

0-50mph 80.5kmh: 3.9secs
0-60mph 96.5kmh: 5.3secs
0-1/4 mile: 13.3secs
380.0bhp 283.4kW 385.3PS
@ 5750rpm
354.0lbft 479.7Nm @ 4500rpm
232.6bhp/ton 228.7bhp/tonne
76.9bhp/L 57.3kW/L 78.0PS/L
49.0ft/sec 14.9m/sec
14.4mpg 12.0mpgUS
19.6L/100km
Petrol 4-stroke piston
4942cc 301.5cu in
Flat 12 fuel injection
Compression ratio: 8.7:1
Bore: 82.0mm 3.2in
Stroke: 78.0mm 3.1in
Valve type/No: Overhead 48
Transmission: Manual
No. of forward speeds: 5
Wheels driven: Rear
Springs F/R: Coil/Coil
Brake system: PA
Brakes F/R: Disc/Disc
Steering: Rack & pinion
Wheelbase: 255.0cm 100.4in
Track F: 151.9cm 59.8in
Track R: 166.1cm 65.4in
Length: 448.6cm 176.6in
Width: 197.1cm 77.6in
Height: 113.0cm 44.5in
Kerb weight: 1661.6kg
3660.0lb
Fuel: 104.8L 23.0gal
27.7galUS

256 Ferrari

Testa Rossa Straman Spyder
1988 Italy
181.0mph 291.2kmh
0-60mph 96.5kmh: 5.3secs
0-1/4 mile: 13.3secs
380.0bhp 283.4kW 385.3PS
@ 5750rpm
354.0lbft 479.7Nm @ 4500rpm
221.7bhp/ton 218.0bhp/tonne
76.9bhp/L 57.3kW/L 78.0PS/L
49.0ft/sec 14.9m/sec
Petrol 4-stroke piston
4942cc 301.5cu in
Flat 12 fuel injection
Compression ratio: 8.7:1
Bore: 82.0mm 3.2in
Stroke: 78.0mm 3.1in
Valve type/No: Overhead 48
Transmission: Manual
No. of forward speeds: 5
Wheels driven: Rear
Springs F/R: Coil/Coil
Brake system: PA
Brakes F/R: Disc/Disc
Steering: Rack & pinion
Wheelbase: 255.0cm 100.4in
Track F: 151.9cm 59.8in

Track R: 166.1cm 65.4in
Length: 448.6cm 176.6in
Width: 197.1cm 77.6in
Height: 114.3cm 45.0in
Kerb weight: 1743.4kg
3840.0lb
Fuel: 104.8L 23.0gal
27.7galUS

257 Ferrari

308 Norwood
1989 Italy
182.0mph 292.8kmh
0-50mph 80.5kmh: 3.8secs
0-60mph 96.5kmh: 4.7secs
0-1/4 mile: 12.7secs
450.0bhp 335.6kW 456.2PS
@ 6700rpm
311.1bhp/ton 305.9bhp/tonne
153.8bhp/L 114.7kW/L
155.9PS/L
52.1ft/sec 15.9m/sec
Petrol 4-stroke piston
2926cc 178.5cu in
turbocharged
Vee 8 fuel injection
Compression ratio: 8.0:1
Bore: 81.0mm 3.2in
Stroke: 71.0mm 2.8in
Valve type/No: Overhead 16
Transmission: Manual
No. of forward speeds: 5
Wheels driven: Rear
Springs F/R: Coil/Coil
Brake system: PA
Brakes F/R: Disc/Disc
Steering: Rack & pinion
Wheelbase: 233.9cm 92.1in
Track F: 146.8cm 57.8in
Track R: 146.8cm 57.8in
Length: 442.5cm 174.2in
Width: 172.0cm 67.7in
Height: 112.0cm 44.1in
Kerb weight: 1471.0kg
3240.0lb

258 Ferrari

Mondial 3.2 Cabriolet
1989 Italy
145.0mph 233.3kmh
0-50mph 80.5kmh: 5.4secs
0-60mph 96.5kmh: 7.0secs
0-1/4 mile: 15.2secs
260.0bhp 193.9kW 263.6PS
@ 7000rpm
213.0lbft 288.6Nm @ 5500rpm
164.3bhp/ton 161.5bhp/tonne
81.6bhp/L 60.9kW/L 82.8PS/L
56.4ft/sec 17.2m/sec
19.2mpg 16.0mpgUS
14.7L/100km
Petrol 4-stroke piston
3185cc 194.3cu in
Vee 8 fuel injection

Compression ratio: 9.2:1
Bore: 83.0mm 3.3in
Stroke: 73.6mm 2.9in
Valve type/No: Overhead 32
Transmission: Manual
No. of forward speeds: 5
Wheels driven: Rear
Springs F/R: Coil/Coil
Brake system: PA
Brakes F/R: Disc/Disc
Steering: Rack & pinion
Wheelbase: 264.9cm 104.3in
Track F: 151.4cm 59.6in
Track R: 153.4cm 60.4in
Length: 464.1cm 182.7in
Width: 179.1cm 70.5in
Height: 126.0cm 49.6in
Kerb weight: 1609.4kg
3545.0lb
Fuel: 70.0L 15.4gal 18.5galUS

259 Ferrari

Testa Rossa
1989 Italy
174.0mph 280.0kmh
0-50mph 80.5kmh: 4.1secs
0-60mph 96.5kmh: 5.2secs
0-1/4 mile: 13.5secs
0-1km: 23.8secs
390.0bhp 290.8kW 395.4PS
@ 6300rpm
363.1lbft 492.0Nm @ 4500rpm
237.6bhp/ton 233.7bhp/tonne
78.9bhp/L 58.8kW/L 80.0PS/L
53.7ft/sec 16.4m/sec
26.5mph 42.6kmh/1000rpm
16.6mpg 13.8mpgUS
17.0L/100km
Petrol 4-stroke piston
4942cc 302.0cu in
Flat 12 fuel injection
Compression ratio: 9.3:1
Bore: 82.0mm 3.2in
Stroke: 78.0mm 3.1in
Valve type/No: Overhead 48
Transmission: Manual
No. of forward speeds: 5
Wheels driven: Rear
Springs F/R: Coil/Coil
Brake system: PA
Brakes F/R: Disc/Disc
Steering: Rack & pinion
Wheelbase: 255.0cm 100.4in
Track F: 151.8cm 59.8in
Track R: 166.1cm 65.4in
Length: 448.5cm 176.6in
Width: 196.6cm 77.4in
Height: 113.0cm 44.5in
Ground clearance: 12.6cm
5.0in
Kerb weight: 1669.0kg
3676.2lb
Fuel: 113.7L 25.0gal
30.0galUS

260 Ferrari

Testa Rossa Gemballa
1989 Italy
185.0mph 297.7kmh
0-60mph 96.5kmh: 5.3secs
0-1/4 mile: 13.3secs
380.0bhp 283.4kW 385.3PS
@ 5750rpm
354.0lbft 479.7Nm @ 4500rpm
231.3bhp/ton 227.4bhp/tonne
76.9bhp/L 57.3kW/L 78.0PS/L
49.0ft/sec 14.9m/sec
Petrol 4-stroke piston
4942cc 301.5cu in
Vee 12 fuel injection
Compression ratio: 8.7:1
Bore: 82.0mm 3.2in
Stroke: 78.0mm 3.1in
Valve type/No: Overhead 48
Transmission: Manual
No. of forward speeds: 5
Wheels driven: Rear
Springs F/R: Coil/Coil
Brake system: PA
Brakes F/R: Disc/Disc
Steering: Rack & pinion PA
Wheelbase: 255.0cm 100.4in
Track F: 151.9cm 59.8in
Track R: 168.7cm 66.4in
Length: 460.0cm 181.1in
Width: 210.1cm 82.7in
Height: 113.0cm 44.5in
Kerb weight: 1670.7kg
3680.0lb
Fuel: 104.8L 23.0gal
27.7galUS

261 Ferrari

Testa Rossa Norwood
1989 Italy
210.0mph 337.9kmh
0-50mph 80.5kmh: 4.0secs
0-60mph 96.5kmh: 4.7secs
0-1/4 mile: 12.1secs
943.0bhp 703.2kW 956.1PS
@ 10000rpm
570.9bhp/ton 561.4bhp/tonne
190.8bhp/L 142.3kW/L
193.5PS/L
85.3ft/sec 26.0m/sec
Petrol 4-stroke piston
4942cc 301.5cu in
turbocharged
Flat 12 fuel injection
Compression ratio: 8.2:1
Bore: 82.0mm 3.2in
Stroke: 78.0mm 3.1in
Valve type/No: Overhead 48
Transmission: Manual
No. of forward speeds: 5
Wheels driven: Rear
Springs F/R: Coil/Coil
Brake system: PA
Brakes F/R: Disc/Disc

262 Ferrari
348tb

Steering: Rack & pinion
Wheelbase: 255.0cm 100.4in
Track F: 151.9cm 59.8in
Track R: 166.1cm 65.4in
Length: 448.6cm 176.6in
Width: 197.1cm 77.6in
Height: 113.0cm 44.5in
Kerb weight: 1679.8kg
3700.0lb

262 Ferrari
348tb
1990 Italy
164.0mph 263.9kmh
0-50mph 80.5kmh: 4.4secs
0-60mph 96.5kmh: 5.6secs
0-1/4 mile: 13.8secs
0-1km: 25.3secs
300.0bhp 223.7kW 304.2PS
@ 7200rpm
238.4lbft 323.0Nm @ 4200rpm
208.2bhp/ton 204.8bhp/tonne
88.1bhp/L 65.7kW/L 89.3PS/L
59.0ft/sec 18.0m/sec
23.3mph 37.5kmh/1000rpm
18.4mpg 15.3mpgUS
15.4L/100km
Petrol 4-stroke piston
3405cc 208.0cu in
Vee 8 fuel injection
Compression ratio: 10.4:1
Bore: 85.0mm 3.3in
Stroke: 75.0mm 2.9in
Valve type/No: Overhead 32
Transmission: Manual
No. of forward speeds: 5
Wheels driven: Rear
Springs F/R: Coil/Coil
Brake system: PA ABS
Brakes F/R: Disc/Disc
Steering: Rack & pinion
Wheelbase: 245.0cm 96.5in
Track F: 150.2cm 59.1in

Track R: 157.8cm 62.1in
Length: 423.0cm 166.5in
Width: 189.4cm 74.6in
Height: 117.0cm 46.1in
Kerb weight: 1465.0kg
3226.9lb
Fuel: 95.0L 20.9gal 25.1galUS

263 Ferrari
F40
1990 Italy
200.0mph 321.8kmh
0-60mph 96.5kmh: 4.5secs
0-1/4 mile: 11.6secs
478.0bhp 356.4kW 484.6PS
@ 7000rpm
425.0lbft 575.9Nm @ 4000rpm
441.5bhp/ton 434.2bhp/tonne
162.8bhp/L 121.4kW/L
165.1PS/L
53.3ft/sec 16.2m/sec
Petrol 4-stroke piston
2936cc 179.1cu in
turbocharged
Vee 8 fuel injection
Compression ratio: 7.8:1
Bore: 82.0mm 3.2in
Stroke: 69.5mm 2.7in
Valve type/No: Overhead 32
Transmission: Manual
No. of forward speeds: 5
Wheels driven: Rear
Springs F/R: Coil/Coil
Brakes F/R: Disc/Disc
Steering: Rack & pinion
Wheelbase: 245.1cm 96.5in
Track F: 159.5cm 62.8in
Track R: 161.0cm 63.4in
Length: 443.0cm 174.4in
Width: 198.1cm 78.0in
Height: 113.0cm 44.5in
Kerb weight: 1100.9kg
2425.0lb

Fuel: 117.3L 25.8gal
31.0galUS

264 Ferrari
Mondial
1990 Italy
154.0mph 247.8kmh
0-50mph 80.5kmh: 5.1secs
0-60mph 96.5kmh: 6.6secs
0-1/4 mile: 15.0secs
300.0bhp 223.7kW 304.2PS
@ 7000rpm
229.0lbft 310.3Nm @ 4000rpm
194.2bhp/ton 191.0bhp/tonne
88.1bhp/L 65.7kW/L 89.3PS/L
57.4ft/sec 17.5m/sec
Petrol 4-stroke piston
3405cc 207.7cu in
Vee 8 fuel injection
Compression ratio: 10.4:1
Bore: 85.0mm 3.3in
Stroke: 75.0mm 2.9in
Valve type/No: Overhead 32
Transmission: Manual
No. of forward speeds: 5
Wheels driven: Rear
Springs F/R: Coil/Coil
Brake system: ABS
Brakes F/R: Disc/Disc
Steering: Rack & pinion PA
Wheelbase: 264.9cm 104.3in
Track F: 152.1cm 59.9in
Track R: 156.0cm 61.4in
Length: 453.4cm 178.5in
Width: 181.1cm 71.3in
Height: 123.4cm 48.6in
Kerb weight: 1570.8kg
3460.0lb
Fuel: 85.2L 18.7gal 22.5galUS

265 Ferrari
Pininfarina Mythos
1990 Italy

180.0mph 289.6kmh
0-60mph 96.5kmh: 6.2secs
380.0bhp 283.4kW 385.3PS
@ 5750rpm
354.0lbft 479.7Nm @ 4500rpm
309.0bhp/ton 303.8bhp/tonne
76.9bhp/L 57.3kW/L 78.0PS/L
49.0ft/sec 14.9m/sec
Petrol 4-stroke piston
4942cc 301.5cu in
Flat 12 fuel injection
Compression ratio: 8.7:1
Bore: 82.0mm 3.2in
Stroke: 78.0mm 3.1in
Valve type/No: Overhead 48
Transmission: Manual
No. of forward speeds: 5
Wheels driven: Rear
Springs F/R: Coil/Coil
Brake system: PA
Brakes F/R: Disc/Disc
Steering: Rack & pinion
Wheelbase: 255.0cm 100.4in
Track F: 151.9cm 59.8in
Track R: 172.7cm 68.0in
Length: 430.5cm 169.5in
Width: 209.8cm 82.6in
Height: 106.4cm 41.9in
Kerb weight: 1250.8kg
2755.0lb
Fuel: 115.1L 25.3gal
30.4galUS

266 Ferrari
348tb
1991 Italy
171.0mph 275.1kmh
0-50mph 80.5kmh: 4.7secs
0-60mph 96.5kmh: 6.0secs
0-1/4 mile: 14.3secs
300.0bhp 223.7kW 304.2PS
@ 7000rpm
229.0lbft 310.3Nm @ 4000rpm

267 Ferrari
F40

205.5bhp/ton 202.1bhp/tonne
88.1bhp/L 65.7kW/L 89.3PS/L
57.4ft/sec 17.5m/sec
19.2mpg 16.0mpgUS
14.7L/100km
Petrol 4-stroke piston
3405cc 207.7cu in
Vee 8 fuel injection
Compression ratio: 10.4:1
Bore: 85.0mm 3.3in
Stroke: 75.0mm 2.9in
Valve type/No: Overhead 32
Transmission: Manual
No. of forward speeds: 5
Wheels driven: Rear
Springs F/R: Coil/Coil
Brake system: PA ABS
Brakes F/R: Disc/Disc
Steering: Rack & pinion
Wheelbase: 245.1cm 96.5in
Track F: 150.4cm 59.2in
Track R: 158.0cm 62.2in
Length: 423.4cm 166.7in
Width: 189.5cm 74.6in
Height: 117.1cm 46.1in
Ground clearance: 14.0cm
5.5in
Kerb weight: 1484.6kg
3270.0lb
Fuel: 87.8L 19.3gal 23.2galUS

267 Ferrari
F40
1991 Italy
196.0mph 315.4kmh
0-50mph 80.5kmh: 2.8secs
0-60mph 96.5kmh: 3.8secs
0-1/4 mile: 11.8secs
478.0bhp 356.4kW 484.6PS
@ 7000rpm
425.0lbft 575.9Nm @ 4500rpm
359.3bhp/ton 353.3bhp/tonne
162.8bhp/L 121.4kW/L

165.1PS/L
53.3ft/sec 16.2m/sec
15.6mpg 13.0mpgUS
18.1L/100km
Petrol 4-stroke piston
2936cc 179.1cu in
turbocharged
Vee 8 fuel injection
Compression ratio: 7.7:1
Bore: 82.0mm 3.2in
Stroke: 69.5mm 2.7in
Valve type/No: Overhead 32
Transmission: Manual
No. of forward speeds: 5
Wheels driven: Rear
Springs F/R: Coil/Coil
Brake system: PA
Brakes F/R: Disc/Disc
Steering: Rack & pinion
Wheelbase: 245.1cm 96.5in
Track F: 159.5cm 62.8in
Track R: 161.0cm 63.4in
Length: 443.0cm 174.4in
Width: 198.1cm 78.0in
Height: 113.0cm 44.5in
Ground clearance: 10.9cm
4.3in
Kerb weight: 1352.9kg
2980.0lb
Fuel: 120.0L 26.4gal
31.7galUS

268 Ferrari
Mondial t
1991 Italy
154.0mph 247.8kmh
0-50mph 80.5kmh: 5.1secs
0-60mph 96.5kmh: 6.6secs
0-1/4 mile: 15.0secs
300.0bhp 223.7kW 304.2PS
@ 7200rpm
237.0lbft 321.1Nm @ 4200rpm
207.7bhp/ton 204.3bhp/tonne

88.1bhp/L 65.7kW/L 89.3PS/L
59.0ft/sec 18.0m/sec
Petrol 4-stroke piston
3405cc 207.7cu in
Vee 8 fuel injection
Compression ratio: 10.4:1
Bore: 85.0mm 3.3in
Stroke: 75.0mm 2.9in
Valve type/No: Overhead 32
Transmission: Manual
No. of forward speeds: 5
Wheels driven: Rear
Springs F/R: Coil/Coil
Brake system: ABS
Brakes F/R: Disc/Disc
Steering: Rack & pinion
Wheelbase: 264.9cm 104.3in
Track F: 152.1cm 59.9in
Track R: 156.0cm 61.4in
Length: 453.4cm 178.5in
Width: 180.8cm 71.2in
Height: 123.4cm 48.6in
Kerb weight: 1468.7kg
3235.0lb
Fuel: 85.2L 18.7gal 22.5galUS

269 Ferrari
Mondial t Cabriolet
1991 Italy
154.0mph 247.8kmh
0-60mph 96.5kmh: 6.6secs
0-1/4 mile: 15.0secs
300.0bhp 223.7kW 304.2PS
@ 7000rpm
229.0lbft 310.3Nm @ 4000rpm
184.6bhp/ton 181.5bhp/tonne
88.1bhp/L 65.7kW/L 89.3PS/L
57.4ft/sec 17.5m/sec
18.0mpg 15.0mpgUS
15.7L/100km
Petrol 4-stroke piston
3405cc 207.7cu in
Vee 8 fuel injection

Compression ratio: 10.4:1
Bore: 85.0mm 3.3in
Stroke: 75.0mm 2.9in
Valve type/No: Overhead 32
Transmission: Manual
No. of forward speeds: 5
Wheels driven: Rear
Springs F/R: Coil/Coil
Brake system: PA ABS
Brakes F/R: Disc/Disc
Steering: Rack & pinion PA
Wheelbase: 264.9cm 104.3in
Track F: 152.1cm 59.9in
Track R: 156.0cm 61.4in
Length: 453.4cm 178.5in
Width: 181.1cm 71.3in
Height: 123.4cm 48.6in
Kerb weight: 1652.6kg
3640.0lb
Fuel: 85.2L 18.7gal 22.5galUS

270 Fiat
Panda 1000 S
1986 Italy
89.0mph 143.2kmh
0-50mph 80.5kmh: 11.1secs
0-60mph 96.5kmh: 16.0secs
0-1/4 mile: 20.2secs
0-1km: 38.2secs
45.0bhp 33.6kW 45.6PS
@ 5000rpm
59.0lbft 79.9Nm @ 2750rpm
60.7bhp/ton 59.7bhp/tonne
45.0bhp/L 33.6kW/L 45.7PS/L
35.6ft/sec 10.8m/sec
18.8mpg 30.2kmh/1000rpm
34.0mpg 28.3mpgUS
8.3L/100km
Petrol 4-stroke piston
999cc 61.0cu in
In-line 4 1 Carburettor
Compression ratio: 9.8:1
Bore: 70.0mm 2.8in

Stroke: 64.9mm 2.6in
Valve type/No: Overhead 8
Transmission: Manual
No. of forward speeds: 5
Wheels driven: Front
Springs F/R: Coil/Coil
Brakes F/R: Disc/Drum
Steering: Rack & pinion
Wheelbase: 215.9cm 85.0in
Track F: 125.1cm 49.3in
Track R: 125.1cm 49.3in
Length: 337.8cm 133.0in
Width: 146.1cm 57.5in
Height: 144.2cm 56.8in
Ground clearance: 16.5cm
6.5in
Kerb weight: 754.0kg 1660.8lb
Fuel: 40.0L 8.8gal 10.6galUS

271 Fiat

Panda 750 L
1986 Italy
80.0mph 128.7kmh
0-50mph 80.5kmh: 13.2secs
0-60mph 96.5kmh: 20.6secs
0-1/4 mile: 22.1secs
0-1km: 42.4secs
34.0bhp 25.3kW 34.5PS
@ 2250rpm
42.0lbft 56.9Nm @ 3000rpm
45.9bhp/ton 45.1bhp/tonne
44.2bhp/L 33.0kW/L 44.8PS/L
14.2ft/sec 4.3m/sec
18.8mph 30.2kmh/1000rpm
35.5mpg 29.6mpgUS
8.0L/100km
Petrol 4-stroke piston
769cc 47.0cu in
In-line 4 1 Carburettor
Compression ratio: 9.4:1
Bore: 65.0mm 2.6in
Stroke: 58.0mm 2.3in
Valve type/No: Overhead 8
Transmission: Manual
No. of forward speeds: 4
Wheels driven: Front
Springs F/R: Coil/Coil
Brakes F/R: Disc/Drum
Steering: Rack & pinion
Wheelbase: 215.9cm 85.0in
Track F: 125.1cm 49.3in
Track R: 125.1cm 49.3in
Length: 337.8cm 133.0in
Width: 146.1cm 57.5in
Height: 144.2cm 56.8in
Ground clearance: 16.5cm
6.5in
Kerb weight: 754.0kg 1660.8lb
Fuel: 40.0L 8.8gal 10.6galUS

272 Fiat

Regata DS
1986 Italy
96.0mph 154.5kmh

0-50mph 80.5kmh: 10.6secs
0-60mph 96.5kmh: 15.6secs
0-1/4 mile: 19.9secs
0-1km: 37.7secs
65.0bhp 48.5kW 65.9PS
@ 4600rpm
88.0lbft 119.2Nm @ 2000rpm
63.9bhp/ton 62.8bhp/tonne
33.7bhp/L 25.1kW/L 34.2PS/L
45.2ft/sec 13.8m/sec
21.6mph 34.8kmh/1000rpm
41.8mpg 34.8mpgUS
6.8L/100km
Diesel 4-stroke piston
1929cc 118.0cu in
In-line 4 fuel injection
Compression ratio: 21.0:1
Bore: 82.6mm 3.2in
Stroke: 90.0mm 3.5in
Valve type/No: Overhead 8
Transmission: Manual
No. of forward speeds: 5
Wheels driven: Front
Springs F/R: Coil/Leaf
Brake system: PA
Brakes F/R: Disc/Drum
Steering: Rack & pinion PA
Wheelbase: 244.8cm 96.4in
Track F: 141.4cm 55.7in
Track R: 141.2cm 55.6in
Length: 426.0cm 167.7in
Width: 165.0cm 65.0in
Height: 142.0cm 55.9in
Ground clearance: 15.2cm
6.0in
Kerb weight: 1035.0kg
2279.7lb
Fuel: 55.1L 12.1gal 14.5galUS

273 Fiat

Croma ie Super
1987 Italy
121.0mph 194.7kmh
0-50mph 80.5kmh: 7.2secs
0-60mph 96.5kmh: 9.9secs
0-1/4 mile: 17.3secs
0-1km: 31.7secs
120.0bhp 89.5kW 121.7PS
@ 5250rpm
123.0lbft 166.7Nm @ 3300rpm
101.8bhp/ton 100.1bhp/tonne
60.1bhp/L 44.8kW/L 61.0PS/L
51.6ft/sec 15.7m/sec
24.9mph 40.1kmh/1000rpm
29.0mpg 24.1mpgUS
9.7L/100km
Petrol 4-stroke piston
1995cc 122.0cu in
In-line 4 fuel injection
Compression ratio: 9.8:1
Bore: 84.0mm 3.3in
Stroke: 90.0mm 3.5in
Valve type/No: Overhead 8
Transmission: Manual

No. of forward speeds: 5
Wheels driven: Front
Springs F/R: Coil/Coil
Brake system: PA
Brakes F/R: Disc/Drum
Steering: Rack & pinion PA
Wheelbase: 266.0cm 104.7in
Track F: 148.2cm 58.3in
Track R: 147.2cm 58.0in
Length: 449.5cm 177.0in
Width: 176.0cm 69.3in
Height: 143.3cm 56.4in
Ground clearance: 17.9cm
7.0in
Kerb weight: 1199.0kg
2641.0lb
Fuel: 70.1L 15.4gal 18.5galUS

274 Fiat

Regata 100S ie
1987 Italy
107.0mph 172.2kmh
0-50mph 80.5kmh: 8.4secs
0-60mph 96.5kmh: 12.0secs
0-1/4 mile: 18.6secs
0-1km: 34.7secs
100.0bhp 74.6kW 101.4PS
@ 6000rpm
94.0lbft 127.4Nm @ 4000rpm
98.7bhp/ton 97.1bhp/tonne
63.1bhp/L 47.0kW/L 64.0PS/L
46.8ft/sec 14.3m/sec
21.9mph 35.2kmh/1000rpm
25.9mpg 21.6mpgUS
10.9L/100km
Petrol 4-stroke piston
1585cc 97.0cu in
In-line 4 fuel injection
Compression ratio: 9.7:1
Bore: 84.0mm 3.3in
Stroke: 71.5mm 2.8in
Valve type/No: Overhead 8
Transmission: Manual
No. of forward speeds: 5
Wheels driven: Front
Springs F/R: Coil/Leaf
Brake system: PA
Brakes F/R: Disc/Drum
Steering: Rack & pinion PA
Wheelbase: 245.5cm 96.7in
Track F: 140.0cm 55.1in
Track R: 139.5cm 54.9in
Length: 426.0cm 167.7in
Width: 165.0cm 65.0in
Height: 141.2cm 55.6in
Ground clearance: 15.2cm
6.0in
Kerb weight: 1030.0kg
2268.7lb
Fuel: 55.1L 12.1gal 14.5galUS

275 Fiat

Uno Selecta
1987 Italy

91.0mph 146.4kmh
0-50mph 80.5kmh: 11.7secs
0-60mph 96.5kmh: 16.6secs
0-1/4 mile: 21.1secs
0-1km: 41.0secs
59.0bhp 44.0kW 59.8PS
@ 5700rpm
64.0lbft 86.7Nm @ 3000rpm
77.9bhp/ton 76.6bhp/tonne
52.9bhp/L 39.4kW/L 53.6PS/L
34.7ft/sec 10.5m/sec
25.7mph 41.4kmh/1000rpm
34.9mpg 29.1mpgUS
8.1L/100km
Petrol 4-stroke piston
1116cc 68.0cu in
In-line 4 1 Carburettor
Compression ratio: 9.2:1
Bore: 80.0mm 3.1in
Stroke: 55.5mm 2.2in
Valve type/No: Overhead 8
Transmission: Continuously
variable
Wheels driven: Front
Springs F/R: Coil/Torsion bar
Brake system: PA
Brakes F/R: Disc/Drum
Steering: Rack & pinion
Wheelbase: 236.2cm 93.0in
Track F: 133.9cm 52.7in
Track R: 130.9cm 51.5in
Length: 364.4cm 143.5in
Width: 155.5cm 61.2in
Height: 142.0cm 55.9in
Ground clearance: 15.0cm
5.9in
Kerb weight: 770.0kg 1696.0lb
Fuel: 42.0L 9.2gal 11.1galUS

276 Fiat

Tipo 1.4
1988 Italy
101.0mph 162.5kmh
0-50mph 80.5kmh: 9.2secs
0-60mph 96.5kmh: 13.1secs
0-1/4 mile: 19.1secs
0-1km: 35.5secs
72.0bhp 53.7kW 73.0PS
@ 6000rpm
78.2lbft 106.0Nm @ 2900rpm
73.4bhp/ton 72.1bhp/tonne
52.5bhp/L 39.1kW/L 53.2PS/L
44.5ft/sec 13.5m/sec
20.6mph 33.1kmh/1000rpm
32.2mpg 26.8mpgUS
8.8L/100km
Petrol 4-stroke piston
1372cc 84.0cu in
In-line 4 1 Carburettor
Compression ratio: 9.2:1
Bore: 80.5mm 3.2in
Stroke: 67.7mm 2.7in
Valve type/No: Overhead 8
Transmission: Manual

No. of forward speeds: 5
Wheels driven: Front
Springs F/R: Coil/Coil
Brake system: PA
Brakes F/R: Disc/Drum
Steering: Rack & pinion
Wheelbase: 254.0cm 100.0in
Track F: 142.9cm 56.3in
Track R: 141.5cm 55.7in
Length: 395.8cm 155.8in
Width: 170.0cm 66.9in
Height: 144.5cm 56.9in
Ground clearance: 15.2cm
6.0in
Kerb weight: 998.0kg 2198.2lb
Fuel: 60.0L 13.2gal 15.8galUS

277 Fiat

X1/9
1988 Italy
107.0mph 172.2kmh
0-50mph 80.5kmh: 8.6secs
0-60mph 96.5kmh: 12.0secs
0-1/4 mile: 18.4secs
75.0bhp 55.9kW 76.0PS
@ 5500rpm
79.0lbft 107.0Nm @ 3000rpm
76.0bhp/ton 74.7bhp/tonne
50.1bhp/L 37.3kW/L 50.8PS/L
38.5ft/sec 11.7m/sec
28.2mpg 23.5mpgUS
10.0L/100km
Petrol 4-stroke piston
1498cc 91.4cu in
In-line 4 fuel injection
Compression ratio: 8.5:1
Bore: 86.4mm 3.4in
Stroke: 63.9mm 2.5in
Valve type/No: Overhead 8
Transmission: Manual
No. of forward speeds: 5
Wheels driven: Rear
Springs F/R: Coil/Coil
Brake system: PA
Brakes F/R: Disc/Disc
Steering: Rack & pinion
Wheelbase: 220.2cm 86.7in
Track F: 135.4cm 53.3in
Track R: 136.1cm 53.6in
Length: 397.0cm 156.3in
Width: 157.0cm 61.8in
Height: 118.9cm 46.8in
Kerb weight: 1003.3kg
2210.0lb
Fuel: 46.2L 10.1gal 12.2galUS

278 Fiat

Tipo 1.6 DGT SX
1989 Italy
107.0mph 172.2kmh
0-50mph 80.5kmh: 8.1secs
0-60mph 96.5kmh: 11.3secs
0-1/4 mile: 18.4secs
0-1km: 33.9secs

83.0bhp 61.9kW 84.1PS
@ 6000rpm
95.9lbft 130.0Nm @ 2900rpm
80.4bhp/ton 79.0bhp/tonne
52.5bhp/L 39.2kW/L 53.3PS/L
44.2ft/sec 13.5m/sec
21.8mph 35.1kmh/1000rpm
30.4mpg 25.3mpgUS
9.3L/100km
Petrol 4-stroke piston
1580cc 96.0cu in
In-line 4 1 Carburettor
Compression ratio: 9.2:1
Bore: 86.4mm 3.4in
Stroke: 67.4mm 2.6in
Valve type/No: Overhead 8
Transmission: Manual
No. of forward speeds: 5
Wheels driven: Front
Springs F/R: Coil/Coil
Brake system: PA
Brakes F/R: Disc/Drum
Steering: Rack & pinion
Wheelbase: 254.0cm 100.0in
Track F: 142.9cm 56.3in
Track R: 141.5cm 55.7in
Length: 395.8cm 155.8in
Width: 170.0cm 66.9in
Height: 144.5cm 56.9in
Ground clearance: 15.2cm
6.0in
Kerb weight: 1050.0kg
2312.8lb
Fuel: 55.1L 12.1gal 14.5galUS

279 Fiat

X1/9
1989 Italy
107.0mph 172.2kmh
0-50mph 80.5kmh: 8.7secs
0-60mph 96.5kmh: 12.3secs
0-1/4 mile: 18.6secs
75.0bhp 55.9kW 76.0PS
@ 5500rpm
79.0lbft 107.0Nm @ 3000rpm
76.0bhp/ton 74.7bhp/tonne
50.1bhp/L 37.3kW/L 50.8PS/L
38.5ft/sec 11.7m/sec
28.2mpg 23.5mpgUS
10.0L/100km
Petrol 4-stroke piston
1498cc 91.4cu in
In-line 4 fuel injection
Compression ratio: 8.5:1
Bore: 86.4mm 3.4in
Stroke: 63.9mm 2.5in
Valve type/No: Overhead 8
Transmission: Manual
No. of forward speeds: 5
Wheels driven: Rear
Springs F/R: Coil/Coil
Brake system: PA
Brakes F/R: Disc/Disc
Steering: Rack & pinion

Wheelbase: 220.2cm 86.7in
Track F: 135.4cm 53.3in
Track R: 136.1cm 53.6in
Length: 397.0cm 156.3in
Width: 157.0cm 61.8in
Height: 118.9cm 46.8in
Kerb weight: 1003.3kg
2210.0lb
Fuel: 46.2L 10.1gal 12.2galUS

280 Fiat

Croma CHT
1990 Italy
116.0mph 186.6kmh
0-50mph 80.5kmh: 8.2secs
0-60mph 96.5kmh: 11.6secs
0-1/4 mile: 18.2secs
0-1km: 33.4secs
98.0bhp 73.1kW 99.4PS
@ 5250rpm
123.2lbft 167.0Nm @ 2750rpm
84.7bhp/ton 83.3bhp/tonne
49.1bhp/L 36.6kW/L 49.8PS/L
51.6ft/sec 15.7m/sec
24.5mph 39.4kmh/1000rpm
28.3mpg 23.6mpgUS
10.0L/100km
Petrol 4-stroke piston
1995cc 122.0cu in
In-line 4 1 Carburettor
Compression ratio: 9.5:1
Bore: 84.0mm 3.3in
Stroke: 90.0mm 3.5in
Valve type/No: Overhead 8
Transmission: Manual
No. of forward speeds: 5
Wheels driven: Front
Springs F/R: Coil/Coil
Brake system: PA
Brakes F/R: Disc/Disc
Steering: Rack & pinion PA
Wheelbase: 266.0cm 104.7in
Track F: 149.0cm 58.7in
Track R: 148.2cm 58.3in
Length: 449.5cm 177.0in
Width: 176.0cm 69.3in
Height: 143.3cm 56.4in
Ground clearance: 17.9cm
7.0in
Kerb weight: 1177.0kg
2592.5lb
Fuel: 70.1L 15.4gal 18.5galUS

281 Fiat

Tempra 1.6 SX
1990 Italy
113.0mph 181.8kmh
0-50mph 80.5kmh: 8.6secs
0-60mph 96.5kmh: 12.2secs
0-1/4 mile: 18.7secs
0-1km: 34.6secs
86.0bhp 64.1kW 87.2PS
@ 6000rpm
95.9lbft 130.0Nm @ 2900rpm

75.4bhp/ton 74.1bhp/tonne
54.4bhp/L 40.6kW/L 55.1PS/L
44.0ft/sec 13.4m/sec
18.8mph 30.2kmh/1000rpm
28.1mpg 23.4mpgUS
10.1L/100km
Petrol 4-stroke piston
1581cc 96.0cu in
In-line 4 1 Carburettor
Compression ratio: 9.2:1
Bore: 86.0mm 3.4in
Stroke: 67.0mm 2.6in
Valve type/No: Overhead 8
Transmission: Manual
No. of forward speeds: 5
Wheels driven: Front
Springs F/R: Coil/Coil
Brake system: PA
Brakes F/R: Disc/Drum
Steering: Rack & pinion PA
Wheelbase: 254.0cm 100.0in
Track F: 142.5cm 56.1in
Track R: 141.5cm 55.7in
Length: 435.4cm 171.4in
Width: 169.5cm 66.7in
Height: 144.5cm 56.9in
Kerb weight: 1160.0kg
2555.1lb
Fuel: 65.1L 14.3gal 17.2galUS

282 Fiat

Uno 60S
1990 Italy
97.0mph 156.1kmh
0-50mph 80.5kmh: 9.3secs
0-60mph 96.5kmh: 14.2secs
0-1/4 mile: 19.6secs
0-1km: 36.5secs
57.0bhp 42.5kW 57.8PS
@ 5500rpm
64.2lbft 87.0Nm @ 2900rpm
72.3bhp/ton 71.1bhp/tonne
51.4bhp/L 38.4kW/L 52.2PS/L
43.2ft/sec 13.2m/sec
21.7mph 34.9kmh/1000rpm
39.0mpg 32.5mpgUS
7.2L/100km
Petrol 4-stroke piston
1108cc 68.0cu in
In-line 4 1 Carburettor
Compression ratio: 9.6:1
Bore: 70.0mm 2.8in
Stroke: 72.0mm 2.8in
Valve type/No: Overhead 8
Transmission: Manual
No. of forward speeds: 5
Wheels driven: Front
Springs F/R: Coil/Torsion bar
Brake system: PA
Brakes F/R: Disc/Drum
Steering: Rack & pinion
Wheelbase: 236.2cm 93.0in
Track F: 134.4cm 52.9in
Track R: 130.0cm 51.2in

282 Fiat
Uno 60S

Length: 368.8cm 145.2in
Width: 155.7cm 61.3in
Height: 141.5cm 55.7in
Ground clearance: 15.0cm
5.9in
Kerb weight: 802.0kg 1766.5lb
Fuel: 41.9L 9.2gal 11.1galUS

283 Fiat
Uno Turbo ie
1990 Italy
128.0mph 206.0kmh
0-50mph 80.5kmh: 6.0secs
0-60mph 96.5kmh: 8.3secs
0-1/4 mile: 16.4secs
0-1km: 30.1secs
118.0bhp 88.0kW 119.6PS
@ 6000rpm
118.8lbft 161.0Nm @ 3500rpm
127.1bhp/ton 125.0bhp/tonne
86.0bhp/L 64.1kW/L 87.2PS/L
44.2ft/sec 13.5m/sec
20.9mph 33.6kmh/1000rpm
25.9mpg 21.6mpgUS
10.9L/100km
Petrol 4-stroke piston
1372cc 84.0cu in
turbocharged
In-line 4 fuel injection
Compression ratio: 7.7:1
Bore: 80.5mm 3.2in
Stroke: 67.4mm 2.6in
Valve type/No: Overhead 8
Transmission: Manual
No. of forward speeds: 5
Wheels driven: Front
Springs F/R: Coil/Coil
Brake system: PA
Brakes F/R: Disc/Drum
Steering: Rack & pinion
Wheelbase: 236.2cm 93.0in
Track F: 135.1cm 53.2in
Track R: 130.6cm 51.4in

Length: 368.9cm 145.2in
Width: 156.2cm 61.5in
Height: 140.5cm 55.3in
Ground clearance: 17.9cm
7.0in
Kerb weight: 944.0kg 2079.3lb
Fuel: 50.0L 11.0gal 13.2galUS

284 Fiat
Tipo 1.8ie DGT SX
1992 Italy
120.0mph 193.0kmh
0-50mph 80.5kmh: 7.6secs
0-60mph 96.5kmh: 10.5secs
0-1/4 mile: 17.9secs
0-1km: 32.7secs
110.0bhp 82.0kW 111.5PS
@ 6000rpm
104.8lbft 142.0Nm @ 2500rpm
94.0bhp/ton 92.4bhp/tonne
62.6bhp/L 46.7kW/L 63.5PS/L
52.0ft/sec 15.8m/sec
19.5mph 31.4kmh/1000rpm
28.6mpg 23.8mpgUS
9.9L/100km
Petrol 4-stroke piston
1756cc 107.0cu in
In-line 4 fuel injection
Compression ratio: 9.5:1
Bore: 84.0mm 3.3in
Stroke: 79.2mm 3.1in
Valve type/No: Overhead 8
Transmission: Manual
No. of forward speeds: 5
Wheels driven: Front
Springs F/R: Coil/Coil
Brake system: PA ABS
Brakes F/R: Disc/Drum
Steering: Rack & pinion PA
Wheelbase: 254.0cm 100.0in
Track F: 142.9cm 56.3in
Track R: 141.5cm 55.7in
Length: 395.8cm 155.8in

Width: 170.0cm 66.9in
Height: 144.5cm 56.9in
Ground clearance: 15.2cm
6.0in
Kerb weight: 1190.0kg
2621.1lb
Fuel: 54.9L 12.1gal 14.5galUS

285 Ford
Aerostar XLT
1986 USA
90.0mph 144.8kmh
0-60mph 96.5kmh: 14.4secs
0-1/4 mile: 19.5secs
115.0bhp 85.8kW 116.6PS
@ 4600rpm
150.0lbft 203.3Nm @ 2600rpm
97.6bhp/ton 95.9bhp/tonne
41.0bhp/L 30.6kW/L 41.6PS/L
34.5ft/sec 10.5m/sec
22.2mpg 18.5mpgUS
12.7L/100km
Petrol 4-stroke piston
2802cc 171.0cu in
Vee 6 1 Carburettor
Compression ratio: 8.7:1
Bore: 93.0mm 3.7in
Stroke: 68.6mm 2.7in
Valve type/No: Overhead 12
Transmission: Manual
No. of forward speeds: 5
Wheels driven: Rear
Springs F/R: Coil/Coil
Brake system: PA
Brakes F/R: Disc/Drum
Steering: Rack & pinion PA
Wheelbase: 302.0cm 118.9in
Track F: 156.2cm 61.5in
Track R: 152.4cm 60.0in
Length: 444.2cm 174.9in
Width: 182.1cm 71.7in
Height: 184.4cm 72.6in
Ground clearance: 16.0cm

6.3in
Kerb weight: 1198.6kg
2640.0lb
Fuel: 64.3L 14.1gal 17.0galUS

286 Ford
Escort 1.4 GL
1986 UK
104.0mph 167.3kmh
0-50mph 80.5kmh: 8.4secs
0-60mph 96.5kmh: 12.4secs
0-1/4 mile: 18.5secs
0-1km: 35.1secs
75.0bhp 55.9kW 76.0PS
@ 5600rpm
80.0lbft 108.4Nm @ 4000rpm
87.5bhp/ton 86.0bhp/tonne
53.9bhp/L 40.2kW/L 54.6PS/L
45.6ft/sec 13.9m/sec
22.5mph 36.2kmh/1000rpm
34.6mpg 28.8mpgUS
8.2L/100km
Petrol 4-stroke piston
1392cc 85.0cu in
In-line 4 1 Carburettor
Compression ratio: 9.5:1
Bore: 77.2mm 3.0in
Stroke: 74.3mm 2.9in
Valve type/No: Overhead 8
Transmission: Manual
No. of forward speeds: 5
Wheels driven: Front
Springs F/R: Coil/Coil
Brake system: PA ABS
Brakes F/R: Disc/Drum
Steering: Rack & pinion
Wheelbase: 239.8cm 94.4in
Track F: 138.4cm 54.5in
Track R: 143.0cm 56.3in
Length: 402.0cm 158.3in
Width: 158.8cm 62.5in
Height: 136.9cm 53.9in
Ground clearance: 17.9cm

7.0in
Kerb weight: 872.0kg 1920.7lb
Fuel: 48.2L 10.6gal 12.7galUS

287 Ford
Escort RS Turbo
1986 UK
125.0mph 201.1kmh
0-50mph 80.5kmh: 6.1secs
0-60mph 96.5kmh: 9.2secs
0-1/4 mile: 16.8secs
0-1km: 30.2secs
132.0bhp 98.4kW 133.8PS
@ 5750rpm
132.7lbft 179.8Nm @ 2750rpm
132.0bhp/ton 129.8bhp/tonne
82.6bhp/L 61.6kW/L 83.8PS/L
50.0ft/sec 15.2m/sec
22.5mph 36.2kmh/1000rpm
27.4mpg 22.8mpgUS
10.3L/100km
Petrol 4-stroke piston
1597cc 97.0cu in
turbocharged
In-line 4 fuel injection
Compression ratio: 8.2:1
Bore: 79.6mm 3.1in
Stroke: 79.5mm 3.1in
Valve type/No: Overhead 8
Transmission: Manual
No. of forward speeds: 5
Wheels driven: Front
Springs F/R: Coil/Coil
Brake system: PA ABS
Brakes F/R: Disc/Drum
Steering: Rack & pinion PA
Wheelbase: 239.8cm 94.4in
Track F: 138.4cm 54.5in
Track R: 143.0cm 56.3in
Length: 404.6cm 159.3in
Width: 183.3cm 72.2in
Height: 134.8cm 53.1in
Ground clearance: 17.9cm
7.0in
Kerb weight: 1017.0kg
2240.1lb
Fuel: 48.2L 10.6gal 12.7galUS

288 Ford
Escort XR3i
1986 UK
118.0mph 189.9kmh
0-50mph 80.5kmh: 7.0secs
0-60mph 96.5kmh: 9.6secs
0-1/4 mile: 17.0secs
0-1km: 31.3secs
105.0bhp 78.3kW 106.5PS
@ 6000rpm
104.0lbft 140.9Nm @ 4800rpm
110.6bhp/ton 108.8bhp/tonne
65.7bhp/L 49.0kW/L 66.7PS/L
52.2ft/sec 15.9m/sec
20.2mph 32.5kmh/1000rpm
30.8mpg 25.6mpgUS

9.2L/100km
Petrol 4-stroke piston
1597cc 97.0cu in
In-line 4 fuel injection
Compression ratio: 9.5:1
Bore: 79.9mm 3.1in
Stroke: 79.5mm 3.1in
Valve type/No: Overhead 8
Transmission: Manual
No. of forward speeds: 5
Wheels driven: Front
Springs F/R: Coil/Coil
Brake system: PA
Brakes F/R: Disc/Drum
Steering: Rack & pinion
Wheelbase: 179.0cm 70.5in
Track F: 138.4cm 54.5in
Track R: 143.0cm 56.3in
Length: 404.6cm 159.3in
Width: 158.8cm 62.5in
Height: 134.8cm 53.1in
Ground clearance: 17.9cm
7.0in
Kerb weight: 965.0kg 2125.5lb
Fuel: 48.2L 10.6gal 12.7galUS

289 Ford
Mustang GT 5.0
1986 USA
135.0mph 217.2kmh
0-60mph 96.5kmh: 6.9secs
0-1/4 mile: 15.0secs
200.0bhp 149.1kW 202.8PS
@ 4000rpm
285.0lbft 386.2Nm @ 3500rpm
142.7bhp/ton 140.3bhp/tonne
40.5bhp/L 30.2kW/L 41.0PS/L
Petrol 4-stroke piston
4942cc 301.5cu in
Vee 8 fuel injection
Valve type/No: Overhead 16
Transmission: Manual
No. of forward speeds: 5
Wheels driven: Rear
Springs F/R: Coil/Coil
Brakes F/R: Disc/Drum
Steering: Rack & pinion PA
Wheelbase: 255.3cm 100.5in
Length: 456.9cm 179.9in
Width: 175.5cm 69.1in
Height: 132.3cm 52.1in
Kerb weight: 1425.6kg
3140.0lb
Fuel: 58.3L 12.8gal 15.4galUS

290 Ford
Mustang SVO
1986 USA
134.0mph 215.6kmh
0-60mph 96.5kmh: 7.1secs
0-1/4 mile: 15.2secs
200.0bhp 149.1kW 202.8PS
@ 5000rpm
240.0lbft 325.2Nm @ 3200rpm

142.7bhp/ton 140.3bhp/tonne
87.0bhp/L 64.8kW/L 88.2PS/L
Petrol 4-stroke piston
2300cc 140.3cu in
turbocharged
In-line 4 fuel injection
Valve type/No: Overhead 8
Transmission: Manual
No. of forward speeds: 5
Wheels driven: Rear
Springs F/R: Coil/Coil
Brakes F/R: Disc/Disc
Steering: Rack & pinion PA
Wheelbase: 255.3cm 100.5in
Length: 459.2cm 180.8in
Width: 175.5cm 69.1in
Height: 132.3cm 52.1in
Kerb weight: 1425.6kg
3140.0lb
Fuel: 58.3L 12.8gal
15.4galUS

291 Ford
Orion 1.4 GL
1986 UK
106.0mph 170.6kmh
0-50mph 80.5kmh: 8.8secs
0-60mph 96.5kmh: 12.8secs
0-1/4 mile: 19.1secs
0-1km: 42.9secs
75.0bhp 55.9kW 76.0PS
@ 5600rpm
80.0lbft 108.4Nm @ 4000rpm
80.4bhp/ton 79.0bhp/tonne
53.9bhp/L 40.2kW/L 54.6PS/L
45.6ft/sec 13.9m/sec
22.4mph 36.0kmh/1000rpm
31.0mpg 25.8mpgUS
9.1L/100km
Petrol 4-stroke piston
1392cc 85.0cu in
In-line 4 1 Carburettor
Compression ratio: 9.5:1
Bore: 77.2mm 3.0in
Stroke: 74.3mm 2.9in
Valve type/No: Overhead 8
Transmission: Manual
No. of forward speeds: 5
Wheels driven: Front
Springs F/R: Coil/Coil
Brake system: PA
Brakes F/R: Disc/Drum
Steering: Rack & pinion
Wheelbase: 240.0cm 94.5in
Track F: 140.0cm 55.1in
Track R: 140.0cm 55.1in
Length: 421.3cm 165.9in
Width: 183.3cm 72.2in
Height: 138.9cm 54.7in
Ground clearance: 17.9cm
7.0in
Kerb weight: 949.0kg 2090.3lb
Fuel: 46.9L 10.3gal
12.4galUS

292 Ford
RS200
1986 UK
140.0mph 225.3kmh
0-50mph 80.5kmh: 4.8secs
0-60mph 96.5kmh: 6.1secs
0-1/4 mile: 15.0secs
0-1km: 27.4secs
230.0bhp 171.5kW 233.2PS
@ 6000rpm
280.0lbft 379.4Nm @ 4500rpm
181.9bhp/ton 178.8bhp/tonne
127.8bhp/L 95.3kW/L
129.5PS/L
51.0ft/sec 15.5m/sec
20.1mph 32.3kmh/1000rpm
18.3mpg 15.2mpgUS
15.4L/100km
Petrol 4-stroke piston
1800cc 110.0cu in
turbocharged
In-line 4 fuel injection
Compression ratio: 8.2:1
Bore: 86.0mm 3.4in
Stroke: 77.6mm 3.1in
Valve type/No: Overhead 16
Transmission: Manual
No. of forward speeds: 5
Wheels driven: 4-wheel drive
Springs F/R: Coil/Coil
Brakes F/R: Disc/Disc
Steering: Rack & pinion PA
Wheelbase: 253.0cm 99.6in
Track F: 147.3cm 58.0in
Track R: 149.9cm 59.0in
Length: 399.8cm 157.4in
Width: 175.3cm 69.0in
Height: 132.1cm 52.0in
Ground clearance: 17.8cm
7.0in
Kerb weight: 1286.0kg
2832.6lb
Fuel: 90.1L 19.8gal 23.8galUS

293 Ford
Sierra Ghia 4x4 Estate
1986 UK
124.0mph 199.5kmh
0-50mph 80.5kmh: 7.0secs
0-60mph 96.5kmh: 9.5secs
0-1/4 mile: 17.6secs
0-1km: 32.1secs
150.0bhp 111.9kW 152.1PS
@ 5700rpm
216.0lbft 292.7Nm @ 3800rpm
114.7bhp/ton 112.8bhp/tonne
53.7bhp/L 40.1kW/L 54.5PS/L
42.7ft/sec 13.0m/sec
22.9mph 36.8kmh/1000rpm
21.0mpg 17.5mpgUS
13.5L/100km
Petrol 4-stroke piston
2792cc 170.0cu in
Vee 6 fuel injection

Compression ratio: 9.2:1
Bore: 93.0mm 3.7in
Stroke: 68.5mm 2.7in
Valve type/No: Overhead 12
Transmission: Manual
No. of forward speeds: 5
Wheels driven: 4-wheel drive
Springs F/R: Coil/Coil
Brake system: PA ABS
Brakes F/R: Disc/Disc
Steering: Rack & pinion PA
Wheelbase: 260.8cm 102.7in
Track F: 146.8cm 57.8in
Track R: 146.6cm 57.7in
Length: 452.1cm 178.0in
Width: 192.0cm 75.6in
Height: 150.6cm 59.3in
Ground clearance: 17.9cm
7.0in
Kerb weight: 1330.0kg
2929.5lb
Fuel: 60.1L 13.2gal 15.9galUS

294 Ford

Festiva L
1987 Japan
95.0mph 152.9kmh
0-50mph 80.5kmh: 7.2secs
0-60mph 96.5kmh: 10.2secs
0-1/4 mile: 17.8secs
58.0bhp 43.2kW 58.8PS
@ 5000rpm
73.0lbft 98.9Nm @ 3500rpm
75.5bhp/ton 74.3bhp/tonne
43.8bhp/L 32.7kW/L 44.4PS/L
45.7ft/sec 13.9m/sec
44.4mpg 37.0mpgUS
6.4L/100km
Petrol 4-stroke piston
1324cc 80.8cu in
In-line 4 1 Carburettor
Compression ratio: 9.0:1
Bore: 71.0mm 2.8in
Stroke: 83.6mm 3.3in
Valve type/No: Overhead 8
Transmission: Manual
No. of forward speeds: 4
Wheels driven: Front
Springs F/R: Coil/Coil
Brake system: PA
Brakes F/R: Disc/Drum
Steering: Rack & pinion
Wheelbase: 229.1cm 90.2in
Track F: 140.0cm 55.1in
Track R: 138.4cm 54.5in
Length: 356.9cm 140.5in
Width: 160.5cm 63.2in
Height: 140.5cm 55.3in
Kerb weight: 780.9kg 1720.0lb
Fuel: 37.8L 8.3gal 10.0galUS

295 Ford

Fiesta 1.1 Ghia Auto
1987 UK
88.0mph 141.6kmh
0-50mph 80.5kmh: 10.9secs
0-60mph 96.5kmh: 15.8secs
0-1/4 mile: 20.4secs
0-1km: 38.8secs
50.0bhp 37.3kW 50.7PS
@ 5000rpm
61.0lbft 82.7Nm @ 2700rpm
62.4bhp/ton 61.3bhp/tonne
44.8bhp/L 33.4kW/L 45.4PS/L
35.6ft/sec 10.8m/sec
25.7mph 41.4kmh/1000rpm
35.0mpg 29.1mpgUS
8.1L/100km
Petrol 4-stroke piston
1117cc 68.0cu in
In-line 4 1 Carburettor
Compression ratio: 9.5:1
Bore: 74.0mm 2.9in
Stroke: 64.9mm 2.6in
Valve type/No: Overhead 8
Transmission: Automatic
Wheels driven: Front
Springs F/R: Coil/Coil
Brake system: PA
Brakes F/R: Disc/Drum
Steering: Rack & pinion
Wheelbase: 228.8cm 90.1in
Track F: 136.7cm 53.8in
Track R: 132.1cm 52.0in
Length: 364.8cm 143.6in
Width: 172.4cm 67.9in
Height: 133.5cm 52.6in
Ground clearance: 15.2cm
6.0in
Kerb weight: 815.0kg 1795.1lb
Fuel: 40.0L 8.8gal 10.6galUS

296 Ford

Granada 2.4i Ghia
1987 UK
122.0mph 196.3kmh
0-50mph 80.5kmh: 7.2secs
0-60mph 96.5kmh: 9.5secs
0-1/4 mile: 17.4secs
0-1km: 32.0secs
130.0bhp 96.9kW 131.8PS
@ 5800rpm
142.0lbft 192.4Nm @ 3000rpm
98.7bhp/ton 97.0bhp/tonne
54.3bhp/L 40.5kW/L 55.1PS/L
45.6ft/sec 13.9m/sec
23.6mph 38.0kmh/1000rpm
21.5mpg 17.9mpgUS
13.1L/100km
Petrol 4-stroke piston
2393cc 146.0cu in
Vee 6 fuel injection
Compression ratio: 9.5:1
Bore: 84.0mm 3.3in
Stroke: 72.0mm 2.8in
Valve type/No: Overhead 12
Transmission: Manual
No. of forward speeds: 5

Wheels driven: Rear
Springs F/R: Coil/Coil
Brake system: PA
Brakes F/R: Disc/Disc
Steering: Rack & pinion PA
Wheelbase: 276.1cm 108.7in
Track F: 147.7cm 58.1in
Track R: 149.2cm 58.7in
Length: 466.9cm 183.8in
Width: 176.6cm 69.5in
Height: 145.0cm 57.1in
Ground clearance: 17.9cm
7.0in
Kerb weight: 1340.0kg
2951.5lb
Fuel: 70.1L 15.4gal 18.5galUS

297 Ford

Granada Scorpio 2.9 EX
1987 UK
126.0mph 202.7kmh
0-50mph 80.5kmh: 6.2secs
0-60mph 96.5kmh: 9.8secs
0-1/4 mile: 17.6secs
0-1km: 31.5secs
150.0bhp 111.9kW 152.1PS
@ 5700rpm
172.0lbft 233.1Nm @ 3000rpm
106.3bhp/ton 104.5bhp/tonne
51.1bhp/L 38.1kW/L 51.8PS/L
44.8ft/sec 13.7m/sec
26.2mph 42.2kmh/1000rpm
24.0mpg 20.0mpgUS
11.8L/100km
Petrol 4-stroke piston
2933cc 179.0cu in
Vee 6 fuel injection
Compression ratio: 9.5:1
Bore: 93.0mm 3.7in
Stroke: 72.0mm 2.8in
Valve type/No: Overhead 12
Transmission: Automatic
No. of forward speeds: 4
Wheels driven: Rear
Springs F/R: Coil/Coil
Brake system: PA
Brakes F/R: Disc/Disc
Steering: Rack & pinion PA
Wheelbase: 276.1cm 108.7in
Track F: 147.7cm 58.1in
Track R: 149.2cm 58.7in
Length: 466.9cm 183.8in
Width: 176.6cm 69.5in
Height: 145.0cm 57.1in
Ground clearance: 17.9cm
7.0in
Kerb weight: 1435.0kg
3160.8lb
Fuel: 70.1L 15.4gal 18.5galUS

298 Ford

Mustang GT
1987 USA
148.0mph 238.1kmh

0-60mph 96.5kmh: 6.7secs
0-1/4 mile: 15.3secs
225.0bhp 167.8kW 228.1PS
@ 4400rpm
300.0lbft 406.5Nm @ 3000rpm
154.1bhp/ton 151.6bhp/tonne
45.5bhp/L 33.9kW/L 46.2PS/L
36.7ft/sec 11.2m/sec
37.0mph 59.5kmh/1000rpm
Petrol 4-stroke piston
4942cc 301.5cu in
Vee 8 fuel injection
Compression ratio: 9.0:1
Bore: 101.6mm 4.0in
Stroke: 76.2mm 3.0in
Valve type/No: Overhead 16
Transmission: Manual
No. of forward speeds: 5
Wheels driven: Rear
Springs F/R: Coil/Coil
Brakes F/R: Disc/Drum
Steering: Rack & pinion PA
Wheelbase: 255.3cm 100.5in
Track F: 143.8cm 56.6in
Track R: 144.8cm 57.0in
Length: 455.4cm 179.3in
Width: 175.5cm 69.1in
Height: 132.3cm 52.1in
Kerb weight: 1484.6kg
3270.0lb
Fuel: 58.3L 12.8gal 15.4galUS

299 Ford

Sierra Sapphire 2.0 Ghia
1987 UK
117.0mph 188.3kmh
0-50mph 80.5kmh: 6.7secs
0-60mph 96.5kmh: 9.6secs
0-1/4 mile: 17.3secs
0-1km: 31.8secs
115.0bhp 85.8kW 116.6PS
@ 5500rpm
118.0lbft 159.9Nm @ 4000rpm
102.6bhp/ton 100.9bhp/tonne
57.7bhp/L 43.0kW/L 58.5PS/L
46.3ft/sec 14.1m/sec
21.9mph 35.2kmh/1000rpm
26.8mpg 22.3mpgUS
10.5L/100km
Petrol 4-stroke piston
1993cc 122.0cu in
In-line 4 fuel injection
Compression ratio: 9.2:1
Bore: 90.8mm 3.6in
Stroke: 77.0mm 3.0in
Valve type/No: Overhead 8
Transmission: Manual
No. of forward speeds: 5
Wheels driven: Rear
Springs F/R: Coil/Coil
Brake system: PA
Brakes F/R: Disc/Drum
Steering: Rack & pinion
Wheelbase: 260.8cm 102.7in

Track F: 145.0cm 57.1in
Track R: 146.6cm 57.7in
Length: 446.7cm 175.9in
Width: 192.0cm 75.6in
Height: 135.8cm 53.5in
Ground clearance: 17.9cm
7.0in
Kerb weight: 1140.0kg 2511.0lb
Fuel: 60.1L 13.2gal 15.9galUS

300 Ford
Thunderbird Bondurant 5.0
1987 USA
139.0mph 223.7kmh
0-50mph 80.5kmh: 5.3secs
0-60mph 96.5kmh: 6.8secs
0-1/4 mile: 15.2secs
225.0bhp 167.8kW 228.1PS
@ 4000rpm
300.0lbft 406.5Nm @ 3200rpm
140.0bhp/ton 137.7bhp/tonne
45.4bhp/L 33.9kW/L 46.1PS/L
33.3ft/sec 10.2m/sec
Petrol 4-stroke piston
4950cc 302.0cu in
Vee 8 fuel injection
Compression ratio: 9.2:1
Bore: 101.6mm 4.0in
Stroke: 76.2mm 3.0in
Valve type/No: Overhead 16
Transmission: Manual
No. of forward speeds: 5
Wheels driven: Rear
Springs F/R: Coil/Coil
Brake system: ABS
Brakes F/R: Disc/Disc
Steering: Rack & pinion PA
Wheelbase: 264.7cm 104.2in
Track F: 147.6cm 58.1in
Track R: 148.6cm 58.5in
Length: 513.3cm 202.1in
Width: 180.6cm 71.1in

Height: 135.6cm 53.4in
Kerb weight: 1634.4kg
3600.0lb
Fuel: 68.9L 15.1gal 18.2galUS

301 Ford
Thunderbird Turbo Coupe
1987 USA
131.0mph 210.8kmh
0-50mph 80.5kmh: 5.9secs
0-60mph 96.5kmh: 8.5secs
0-1/4 mile: 16.3secs
190.0bhp 141.7kW 192.6PS
@ 4600rpm
240.0lbft 325.2Nm @ 3400rpm
122.1bhp/ton 120.1bhp/tonne
82.8bhp/L 61.7kW/L 83.9PS/L
39.9ft/sec 12.1m/sec
20.4mpg 17.0mpgUS
13.8L/100km
Petrol 4-stroke piston
2295cc 140.0cu in
turbocharged
In-line 4 fuel injection
Compression ratio: 8.0:1
Bore: 96.0mm 3.8in
Stroke: 79.2mm 3.1in
Valve type/No: Overhead 8
Transmission: Manual
No. of forward speeds: 5
Wheels driven: Rear
Springs F/R: Coil/Coil
Brake system: ABS
Brakes F/R: Disc/Disc
Steering: Rack & pinion PA
Wheelbase: 264.7cm 104.2in
Track F: 147.6cm 58.1in
Track R: 148.6cm 58.5in
Length: 513.3cm 202.1in
Width: 180.6cm 71.1in
Height: 135.6cm 53.4in
Kerb weight: 1582.2kg 3485.0lb

Fuel: 68.9L 15.1gal 18.2galUS

302 Ford
Mustang Cartech Turbo
1988 USA
177.0mph 284.8kmh
0-60mph 96.5kmh: 4.9secs
0-1/4 mile: 13.3secs
392.0bhp 292.3kW 397.4PS
@ 5000rpm
550.0lbft 745.3Nm @ 4200rpm
310.3bhp/ton 305.1bhp/tonne
79.3bhp/L 59.1kW/L 80.4PS/L
41.7ft/sec 12.7m/sec
Petrol 4-stroke piston
4942cc 301.5cu in
turbocharged
Vee 8 fuel injection
Compression ratio: 9.2:1
Bore: 101.6mm 4.0in
Stroke: 76.2mm 3.0in
Valve type/No: Overhead 16
Transmission: Manual
No. of forward speeds: 5
Wheels driven: Rear
Springs F/R: Coil/Coil
Brake system: PA
Brakes F/R: Disc/Drum
Steering: Rack & pinion PA
Wheelbase: 255.3cm 100.5in
Track F: 145.3cm 57.2in
Track R: 147.3cm 58.0in
Length: 456.2cm 179.6in
Width: 175.5cm 69.1in
Height: 128.5cm 50.6in
Kerb weight: 1284.8kg
2830.0lb
Fuel: 96.1L 21.1gal 25.4galUS

303 Ford
Mustang Convertible Saleen
1988 USA

149.0mph 239.7kmh
0-60mph 96.5kmh: 6.0secs
0-1/4 mile: 14.5secs
225.0bhp 167.8kW 228.1PS
@ 4000rpm
300.0lbft 406.5Nm @ 3200rpm
168.8bhp/ton 166.0bhp/tonne
45.4bhp/L 33.9kW/L 46.1PS/L
33.3ft/sec 10.2m/sec
Petrol 4-stroke piston
4950cc 302.0cu in
Vee 8 fuel injection
Compression ratio: 9.2:1
Bore: 101.6mm 4.0in
Stroke: 76.2mm 3.0in
Valve type/No: Overhead 16
Transmission: Manual
No. of forward speeds: 5
Wheels driven: Rear
Springs F/R: Coil/Coil
Brake system: PA
Brakes F/R: Disc/Disc
Steering: Rack & pinion PA
Wheelbase: 255.3cm 100.5in
Track F: 147.1cm 57.9in
Track R: 149.1cm 58.7in
Length: 456.2cm 179.6in
Width: 175.5cm 69.1in
Height: 127.0cm 50.0in
Kerb weight: 1355.2kg
2985.0lb
Fuel: 58.3L 12.8gal 15.4galUS

304 Ford
Mustang GT
1988 USA
148.0mph 238.1kmh
0-50mph 80.5kmh: 4.3secs
0-60mph 96.5kmh: 6.0secs
0-1/4 mile: 14.6secs
225.0bhp 167.8kW 228.1PS
@ 4000rpm

304 Ford
Mustang GT

300.0lbft 406.5Nm @ 3200rpm
154.1bhp/ton 151.6bhp/tonne
45.4bhp/L 33.9kW/L 46.1PS/L
33.3ft/sec 10.2m/sec
19.2mpg 16.0mpgUS
14.7L/100km
Petrol 4-stroke piston
4950cc 302.0cu in
Vee 8 fuel injection
Compression ratio: 9.2:1
Bore: 101.6mm 4.0in
Stroke: 76.2mm 3.0in
Valve type/No: Overhead 16
Transmission: Manual
No. of forward speeds: 5
Wheels driven: Rear
Springs F/R: Coil/Coil
Brake system: PA
Brakes F/R: Disc/Drum
Steering: Rack & pinion PA
Wheelbase: 255.3cm 100.5in
Track F: 143.8cm 56.6in
Track R: 144.8cm 57.0in
Length: 455.4cm 179.3in
Width: 175.5cm 69.1in
Height: 132.3cm 52.1in
Kerb weight: 1484.6kg
3270.0lb
Fuel: 58.3L 12.8gal 15.4galUS

305 Ford
Probe GT
1988 USA
131.0mph 210.8kmh
0-60mph 96.5kmh: 7.3secs
0-1/4 mile: 15.6secs
145.0bhp 108.1kW 147.0PS
@ 4300rpm
190.0lbft 257.5Nm @ 3500rpm
110.7bhp/ton 108.8bhp/tonne
66.4bhp/L 49.5kW/L 67.3PS/L
44.2ft/sec 13.5m/sec
24.0mpg 20.0mpgUS
11.8L/100km
Petrol 4-stroke piston
2184cc 133.2cu in
turbocharged
In-line 4 fuel injection
Compression ratio: 7.8:1
Bore: 86.0mm 3.4in
Stroke: 94.0mm 3.7in
Valve type/No: Overhead 12
Transmission: Manual
No. of forward speeds: 5
Wheels driven: Front
Springs F/R: Coil/Coil
Brake system: PA ABS
Brakes F/R: Disc/Disc
Steering: Rack & pinion PA
Wheelbase: 251.5cm 99.0in
Track F: 145.5cm 57.3in
Track R: 146.6cm 57.7in
Length: 449.6cm 177.0in
Width: 174.0cm 68.5in

Height: 131.6cm 51.8in
Kerb weight: 1332.5kg
2935.0lb
Fuel: 57.1L 12.6gal 15.1galUS

306 Ford
Sierra RS Cosworth
1988 UK
143.0mph 230.1kmh
0-50mph 80.5kmh: 4.4secs
0-60mph 96.5kmh: 5.8secs
0-1/4 mile: 14.4secs
0-1km: 26.8secs
204.0bhp 152.1kW 206.8PS
@ 6000rpm
205.0lbft 277.8Nm @ 4500rpm
171.8bhp/ton 168.9bhp/tonne
102.4bhp/L 76.3kW/L
103.8PS/L
50.5ft/sec 15.4m/sec
22.9mph 36.8kmh/1000rpm
20.3mpg 16.9mpgUS
13.9L/100km
Petrol 4-stroke piston
1993cc 121.6cu in
turbocharged
In-line 4 fuel injection
Compression ratio: 8.0:1
Bore: 90.8mm 3.6in
Stroke: 76.9mm 3.0in
Valve type/No: Overhead 16
Transmission: Manual
No. of forward speeds: 5
Wheels driven: Rear
Springs F/R: Coil/Coil
Brake system: PA ABS
Brakes F/R: Disc/Disc
Steering: Rack & pinion PA
Wheelbase: 260.9cm 102.7in
Track F: 144.3cm 56.8in
Track R: 146.1cm 57.5in
Length: 449.3cm 176.9in
Width: 169.7cm 66.8in
Height: 137.7cm 54.2in
Ground clearance: 17.8cm
7.0in
Kerb weight: 1207.6kg
2660.0lb
Fuel: 60.1L 13.2gal 15.9galUS

307 Ford
Taurus 3.8
1988 USA
115.0mph 185.0kmh
0-50mph 80.5kmh: 6.8secs
0-60mph 96.5kmh: 9.3secs
0-1/4 mile: 17.2secs
140.0bhp 104.4kW 141.9PS
@ 3800rpm
215.0lbft 291.3Nm @ 2200rpm
95.6bhp/ton 94.0bhp/tonne
36.9bhp/L 27.5kW/L 37.4PS/L
35.8ft/sec 10.9m/sec
21.0mpg 17.5mpgUS

13.4L/100km
Petrol 4-stroke piston
3797cc 231.7cu in
Vee 6 fuel injection
Compression ratio: 9.0:1
Bore: 96.8mm 3.8in
Stroke: 86.0mm 3.4in
Valve type/No: Overhead 12
Transmission: Automatic
No. of forward speeds: 4
Wheels driven: Front
Springs F/R: Coil/Coil
Brake system: PA
Brakes F/R: Disc/Drum
Steering: Rack & pinion PA
Wheelbase: 269.2cm 106.0in
Track F: 156.5cm 61.6in
Track R: 153.7cm 60.5in
Length: 478.5cm 188.4in
Width: 179.8cm 70.8in
Height: 137.9cm 54.3in
Kerb weight: 1489.1kg
3280.0lb
Fuel: 60.6L 13.3gal 16.0galUS

308 Ford
Escort 1.3L 3-door
1989 UK
97.0mph 156.1kmh
0-50mph 80.5kmh: 9.8secs
0-60mph 96.5kmh: 14.1secs
0-1/4 mile: 19.8secs
0-1km: 35.1secs
63.0bhp 47.0kW 63.9PS
@ 5000rpm
74.5lbft 101.0Nm @ 3000rpm
76.9bhp/ton 75.6bhp/tonne
48.6bhp/L 36.2kW/L 49.2PS/L
41.2ft/sec 12.6m/sec
22.1mph 35.6kmh/1000rpm
36.1mpg 30.1mpgUS
7.8L/100km
Petrol 4-stroke piston
1297cc 79.0cu in
In-line 4 1 Carburettor
Compression ratio: 9.5:1
Bore: 74.0mm 2.9in
Stroke: 75.5mm 3.0in
Valve type/No: Overhead 8
Transmission: Manual
No. of forward speeds: 5
Wheels driven: Front
Springs F/R: Coil/Coil
Brake system: PA
Brakes F/R: Disc/Drum
Steering: Rack & pinion
Wheelbase: 239.8cm 94.4in
Track F: 138.4cm 54.5in
Track R: 143.0cm 56.3in
Length: 402.0cm 158.3in
Width: 158.8cm 62.5in
Height: 136.9cm 53.9in
Ground clearance: 17.9cm
7.0in

Kerb weight: 833.0kg 1834.8lb
Fuel: 48.2L 10.6gal 12.7galUS

309 Ford
Fiesta 1.1LX 5-door
1989 UK
93.0mph 149.6kmh
0-50mph 80.5kmh: 10.5secs
0-60mph 96.5kmh: 15.3secs
0-1/4 mile: 20.0secs
0-1km: 37.7secs
55.0bhp 41.0kW 55.8PS
@ 5200rpm
63.5lbft 86.0Nm @ 2700rpm
63.9bhp/ton 62.9bhp/tonne
49.2bhp/L 36.7kW/L 49.9PS/L
42.9ft/sec 13.1m/sec
20.3mph 32.7kmh/1000rpm
33.2mpg 27.6mpgUS
8.5L/100km
Petrol 4-stroke piston
1118cc 68.0cu in
In-line 4 1 Carburettor
Compression ratio: 9.5:1
Bore: 68.7mm 2.7in
Stroke: 75.5mm 3.0in
Valve type/No: Overhead 8
Transmission: Manual
No. of forward speeds: 5
Wheels driven: Front
Springs F/R: Coil/Coil
Brake system: PA
Brakes F/R: Disc/Disc
Steering: Rack & pinion
Wheelbase: 244.6cm 96.3in
Length: 374.3cm 147.4in
Width: 185.4cm 73.0in
Height: 132.0cm 52.0in
Ground clearance: 18.5cm
7.3in
Kerb weight: 875.0kg 1927.3lb
Fuel: 41.9L 9.2gal 11.1galUS

310 Ford
Fiesta 1.4 Ghia
1989 UK
100.0mph 160.9kmh
0-50mph 80.5kmh: 9.4secs
0-60mph 96.5kmh: 13.4secs
0-1/4 mile: 19.1secs
0-1km: 35.9secs
75.0bhp 55.9kW 76.0PS
@ 5600rpm
80.4lbft 109.0Nm @ 4000rpm
85.7bhp/ton 84.3bhp/tonne
53.9bhp/L 40.2kW/L 54.6PS/L
45.3ft/sec 13.8m/sec
20.6mph 33.1kmh/1000rpm
28.1mpg 23.4mpgUS
10.1L/100km
Petrol 4-stroke piston
1392cc 85.0cu in
In-line 4 1 Carburettor
Compression ratio: 9.5:1

Bore: 77.0mm 3.0in
Stroke: 74.0mm 2.9in
Valve type/No: Overhead 8
Transmission: Manual
No. of forward speeds: 5
Wheels driven: Front
Springs F/R: Coil/Coil
Brake system: PA
Brakes F/R: Disc/Drum
Steering: Rack & pinion
Wheelbase: 244.6cm 96.3in
Length: 374.3cm 147.4in
Width: 185.4cm 73.0in
Height: 132.0cm 52.0in
Ground clearance: 18.5cm
7.3in
Kerb weight: 890.0kg 1960.3lb
Fuel: 41.9L 9.2gal 11.1galUS

311 Ford
Fiesta 1.6S
1989 UK
110.0mph 177.0kmh
0-50mph 80.5kmh: 7.3secs
0-60mph 96.5kmh: 10.2secs
0-1/4 mile: 18.0secs
0-1km: 33.0secs
90.0bhp 67.1kW 91.2PS
@ 5800rpm
98.2lbft 133.0Nm @ 4000rpm
103.6bhp/ton 101.9bhp/tonne
56.4bhp/L 42.0kW/L 57.2PS/L
50.4ft/sec 15.4m/sec
21.4mph 34.4kmh/1000rpm
29.0mpg 24.1mpgUS
9.7L/100km
Petrol 4-stroke piston
1596cc 97.0cu in
In-line 4 1 Carburettor
Compression ratio: 9.5:1
Bore: 80.0mm 3.1in
Stroke: 79.5mm 3.1in
Valve type/No: Overhead 8
Transmission: Manual
No. of forward speeds: 5
Wheels driven: Front
Springs F/R: Coil/Coil
Brake system: PA ABS
Brakes F/R: Disc/Drum
Steering: Rack & pinion
Wheelbase: 244.6cm 96.3in
Length: 374.3cm 147.4in
Width: 185.4cm 73.0in
Height: 132.0cm 52.0in
Ground clearance: 18.5cm
7.3in
Kerb weight: 883.0kg 1944.9lb
Fuel: 42.1L 9.2gal 11.1galUS

312 Ford
Mustang 5.0 Cartech Turbo
1989 USA
185.0mph 297.7kmh
0-50mph 80.5kmh: 4.4secs

0-60mph 96.5kmh: 5.4secs
0-1/4 mile: 13.4secs
470.0bhp 350.5kW 476.5PS
@ 5500rpm
445.0lbft 603.0Nm @ 5200rpm
319.0bhp/ton 313.7bhp/tonne
95.1bhp/L 70.9kW/L 96.4PS/L
45.8ft/sec 14.0m/sec
Petrol 4-stroke piston
4942cc 301.5cu in
turbocharged
Vee 8 fuel injection
Compression ratio: 9.2:1
Bore: 101.6mm 4.0in
Stroke: 76.2mm 3.0in
Valve type/No: Overhead 16
Transmission: Manual
No. of forward speeds: 5
Wheels driven: Rear
Springs F/R: Coil/Coil
Brake system: PA
Brakes F/R: Disc/Drum
Steering: Rack & pinion PA
Wheelbase: 255.3cm 100.5in
Track F: 144.0cm 56.7in
Track R: 144.8cm 57.0in
Length: 456.2cm 179.6in
Width: 175.5cm 69.1in
Height: 132.3cm 52.1in
Kerb weight: 1498.2kg
3300.0lb

313 Ford
Mustang SVO J Bittle
American
1989 USA
173.0mph 278.4kmh
0-50mph 80.5kmh: 4.2secs
0-60mph 96.5kmh: 5.2secs
0-1/4 mile: 13.4secs
510.0bhp 380.3kW 517.1PS
@ 7900rpm
368.0lbft 498.6Nm @ 6400rpm
355.9bhp/ton 349.9bhp/tonne
101.2bhp/L 75.5kW/L
102.6PS/L
65.8ft/sec 20.1m/sec
Petrol 4-stroke piston
5040cc 307.5cu in
Vee 8 1 Carburettor
Compression ratio: 13.0:1
Bore: 102.6mm 4.0in
Stroke: 76.2mm 3.0in
Valve type/No: Overhead 16
Transmission: Manual
No. of forward speeds: 4
Wheels driven: Rear
Springs F/R: Coil/Coil
Brakes F/R: Disc/Disc
Steering: Rack & pinion
Wheelbase: 255.3cm 100.5in
Track F: 160.0cm 63.0in
Track R: 160.0cm 63.0in
Length: 456.2cm 179.6in

Width: 196.9cm 77.5in
Height: 129.5cm 51.0in
Kerb weight: 1457.3kg
3210.0lb

314 Ford
Sierra Sapphire 2000E
1989 UK
120.0mph 193.1kmh
0-50mph 80.5kmh: 7.1secs
0-60mph 96.5kmh: 10.0secs
0-1/4 mile: 17.3secs
0-1km: 32.2secs
125.0bhp 93.2kW 126.7PS
@ 5600rpm
128.4lbft 174.0Nm @ 2500rpm
104.0bhp/ton 102.3bhp/tonne
62.6bhp/L 46.6kW/L 63.4PS/L
52.7ft/sec 16.0m/sec
21.7mph 34.9kmh/1000rpm
28.8mpg 24.0mpgUS
9.8L/100km
Petrol 4-stroke piston
1998cc 122.0cu in
In-line 4 fuel injection
Compression ratio: 10.3:1
Bore: 86.0mm 3.4in
Stroke: 86.0mm 3.4in
Valve type/No: Overhead 8
Transmission: Manual
No. of forward speeds: 5
Wheels driven: Rear
Springs F/R: Coil/Coil
Brake system: PA ABS
Brakes F/R: Disc/Drum
Steering: Rack & pinion PA
Wheelbase: 260.8cm 102.7in
Track F: 145.0cm 57.1in
Track R: 146.6cm 57.7in
Length: 446.7cm 175.9in
Width: 192.0cm 75.6in
Height: 135.8cm 53.5in
Ground clearance: 17.9cm
7.0in
Kerb weight: 1222.0kg
2691.6lb
Fuel: 60.1L 13.2gal 15.9galUS

315 Ford
Sierra XR 4x4
1989 UK
130.0mph 209.2kmh
0-50mph 80.5kmh: 6.2secs
0-60mph 96.5kmh: 8.6secs
0-1/4 mile: 16.6secs
0-1km: 30.4secs
150.0bhp 111.9kW 152.1PS
@ 5700rpm
172.7lbft 234.0Nm @ 3000rpm
126.1bhp/ton 124.0bhp/tonne
51.1bhp/L 38.1kW/L 51.8PS/L
44.8ft/sec 13.7m/sec
22.4mph 36.0kmh/1000rpm
17.0mpg 14.2mpgUS

16.6L/100km
Petrol 4-stroke piston
2933cc 179.0cu in
Vee 6 fuel injection
Compression ratio: 9.5:1
Bore: 93.0mm 3.7in
Stroke: 72.0mm 2.8in
Valve type/No: Overhead 12
Transmission: Manual
No. of forward speeds: 5
Wheels driven: 4-wheel drive
Springs F/R: Coil/Coil
Brake system: PA
Brakes F/R: Disc/Disc
Steering: Rack & pinion PA
Wheelbase: 260.6cm 102.6in
Track F: 146.8cm 57.8in
Track R: 146.6cm 57.7in
Length: 445.8cm 175.5in
Width: 172.8cm 68.0in
Height: 137.8cm 54.3in
Ground clearance: 17.9cm
7.0in
Kerb weight: 1210.0kg
2665.2lb
Fuel: 60.1L 13.2gal 15.9galUS

316 Ford
Taurus SHO
1989 USA
140.0mph 225.3kmh
0-50mph 80.5kmh: 5.0secs
0-60mph 96.5kmh: 6.6secs
0-1/4 mile: 15.2secs
220.0bhp 164.0kW 223.0PS
@ 6000rpm
200.0lbft 271.0Nm @ 4800rpm
148.9bhp/ton 146.4bhp/tonne
73.7bhp/L 54.9kW/L 74.7PS/L
52.5ft/sec 16.0m/sec
26.4mpg 22.0mpgUS
10.7L/100km
Petrol 4-stroke piston
2986cc 182.2cu in
Vee 6 fuel injection
Compression ratio: 9.8:1
Bore: 89.0mm 3.5in
Stroke: 80.0mm 3.1in
Valve type/No: Overhead 24
Transmission: Manual
No. of forward speeds: 5
Wheels driven: Front
Springs F/R: Coil/Coil
Brake system: PA
Brakes F/R: Disc/Disc
Steering: Rack & pinion PA
Wheelbase: 269.2cm 106.0in
Track F: 156.5cm 61.6in
Track R: 153.7cm 60.5in
Length: 478.5cm 188.4in
Width: 179.8cm 70.8in
Height: 137.9cm 54.3in
Ground clearance: 14.0cm
5.5in

319 Ford
Fiesta RS Turbo

Kerb weight: 1502.7kg
3310.0lb
Fuel: 70.4L 15.5gal
18.6galUS

317 Ford
Thunderbird Super Coupe
1989 USA
140.0mph 225.3kmh
0-50mph 80.5kmh: 5.5secs
0-60mph 96.5kmh: 7.4secs
0-1/4 mile: 15.9secs
210.0bhp 156.6kW 212.9PS
@ 4000rpm
315.0lbft 426.8Nm @ 2600rpm
124.8bhp/ton 122.7bhp/tonne
55.3bhp/L 41.2kW/L 56.1PS/L
37.7ft/sec 11.5m/sec
20.4mpg 17.0mpgUS
13.8L/100km
Petrol 4-stroke piston
3797cc 231.7cu in
supercharged
Vee 6 fuel injection
Compression ratio: 8.2:1
Bore: 96.8mm 3.8in
Stroke: 86.0mm 3.4in
Valve type/No: Overhead 12
Transmission: Manual
No. of forward speeds: 5
Wheels driven: Rear
Springs F/R: Coil/Coil
Brake system: PA ABS
Brakes F/R: Disc/Disc
Steering: Rack & pinion PA
Wheelbase: 287.0cm 113.0in
Track F: 156.0cm 61.4in
Track R: 152.9cm 60.2in
Length: 504.7cm 198.7in
Width: 184.7cm 72.7in
Height: 133.9cm 52.7in
Ground clearance: 13.7cm
5.4in

Kerb weight: 1711.6kg
3770.0lb
Fuel: 71.9L 15.8gal
19.0galUS

318 Ford
Thunderbird Super Coupe
Ford Engineering
1989 USA
160.0mph 257.4kmh
0-50mph 80.5kmh: 4.0secs
0-60mph 96.5kmh: 5.2secs
0-1/4 mile: 14.1secs
330.0bhp 246.1kW 334.6PS
@ 5500rpm
403.0lbft 546.1Nm @ 4000rpm
201.6bhp/ton 198.2bhp/tonne
86.9bhp/L 64.8kW/L 88.1PS/L
51.8ft/sec 15.8m/sec
Petrol 4-stroke piston
3797cc 231.7cu in
supercharged
Vee 6 fuel injection
Compression ratio: 8.0:1
Bore: 96.8mm 3.8in
Stroke: 86.0mm 3.4in
Valve type/No: Overhead 12
Transmission: Manual
No. of forward speeds: 5
Wheels driven: Rear
Springs F/R: Coil/Coil
Brake system: PA ABS
Brakes F/R: Disc/Disc
Steering: Rack & pinion PA
Wheelbase: 287.0cm 113.0in
Track F: 156.0cm 61.4in
Track R: 152.9cm 60.2in
Length: 504.7cm 198.7in
Width: 184.7cm 72.7in
Height: 132.1cm 52.0in
Kerb weight: 1664.8kg
3667.0lb

319 Ford
Fiesta RS Turbo
1990 UK
132.0mph 212.4kmh
0-50mph 80.5kmh: 5.9secs
0-60mph 96.5kmh: 7.9secs
0-1/4 mile: 16.1secs
0-1km: 28.9secs
133.0bhp 99.2kW 134.8PS
@ 5500rpm
135.1lbft 183.0Nm @ 2400rpm
148.6bhp/ton 146.1bhp/tonne
83.3bhp/L 62.1kW/L 84.5PS/L
47.8ft/sec 14.6m/sec
20.8mph 33.5kmh/1000rpm
23.6mpg 19.7mpgUS
12.0L/100km
Petrol 4-stroke piston
1596cc 97.0cu in
turbocharged
In-line 4 fuel injection
Compression ratio: 8.2:1
Bore: 80.0mm 3.1in
Stroke: 79.5mm 3.1in
Valve type/No: Overhead 8
Transmission: Manual
No. of forward speeds: 5
Wheels driven: Front
Springs F/R: Coil/Coil
Brake system: PA ABS
Brakes F/R: Disc/Drum
Steering: Rack & pinion
Wheelbase: 244.6cm 96.3in
Track F: 143.0cm 56.3in
Track R: 137.7cm 54.2in
Length: 380.0cm 149.6in
Width: 163.1cm 64.2in
Height: 132.1cm 52.0in
Ground clearance: 18.5cm
7.3in
Kerb weight: 910.0kg 2004.4lb
Fuel: 42.1L 9.2gal
11.1galUS

320 Ford
Fiesta XR2i
1990 UK
120.0mph 193.1kmh
0-50mph 80.5kmh: 6.3secs
0-60mph 96.5kmh: 8.9secs
0-1/4 mile: 16.7secs
0-1km: 30.8secs
110.0bhp 82.0kW 111.5PS
@ 6000rpm
101.8lbft 138.0Nm @ 2800rpm
121.9bhp/ton 119.8bhp/tonne
68.9bhp/L 51.4kW/L 69.9PS/L
52.2ft/sec 15.9m/sec
20.3mph 32.7kmh/1000rpm
28.4mpg 23.6mpgUS
9.9L/100km
Petrol 4-stroke piston
1596cc 97.0cu in
In-line 4 fuel injection
Compression ratio: 9.7:1
Bore: 80.0mm 3.1in
Stroke: 79.5mm 3.1in
Valve type/No: Overhead 8
Transmission: Manual
No. of forward speeds: 5
Wheels driven: Front
Springs F/R: Coil/Coil
Brake system: PA ABS
Brakes F/R: Disc/Disc
Steering: Rack & pinion
Wheelbase: 244.6cm 96.3in
Length: 374.4cm 147.4in
Width: 185.4cm 73.0in
Height: 132.1cm 52.0in
Ground clearance: 18.5cm
7.3in
Kerb weight: 918.0kg 2022.0lb
Fuel: 41.9L 9.2gal 11.1galUS

321 Ford
Granada 2.5 GL Diesel
1990 UK

107.0mph 172.2kmh
0-50mph 80.5kmh: 9.4secs
0-60mph 96.5kmh: 13.4secs
0-1/4 mile: 19.2secs
0-1km: 35.5secs
92.0bhp 68.6kW 93.3PS
@ 4150rpm
150.6lbft 204.0Nm @ 2250rpm
67.3bhp/ton 66.2bhp/tonne
36.8bhp/L 27.5kW/L 37.3PS/L
40.8ft/sec 12.4m/sec
24.3mph 39.1kmh/1000rpm
30.6mpg 25.5mpgUS
9.2L/100km
Diesel 4-stroke piston
2498cc 152.0cu in
turbocharged
In-line 4 fuel injection
Compression ratio: 21.1:1
Bore: 94.0mm 3.7in
Stroke: 90.0mm 3.5in
Valve type/No: Overhead 8
Transmission: Manual
No. of forward speeds: 5
Wheels driven: Rear
Springs F/R: Coil/Coil
Brake system: PA ABS
Brakes F/R: Disc/Disc
Steering: Rack & pinion PA
Wheelbase: 276.1cm 108.7in
Track F: 147.7cm 58.1in
Track R: 149.2cm 58.7in
Length: 466.9cm 183.8in
Width: 176.6cm 69.5in
Height: 145.0cm 57.1in
Ground clearance: 17.9cm
7.0in
Kerb weight: 1390.0kg
3061.7lb
Fuel: 70.1L 15.4gal
18.5galUS

322 Ford

Granada Ghia X 2.9 EFi
Saloon
1990 UK
125.0mph 201.1kmh
0-50mph 80.5kmh: 7.0secs
0-60mph 96.5kmh: 9.5secs
0-1/4 mile: 17.3secs
0-1km: 31.8secs
150.0bhp 111.9kW 152.1PS
@ 5700rpm
172.0lbft 233.0Nm @ 5900rpm
113.8bhp/ton 111.9bhp/tonne
51.1bhp/L 38.1kW/L 51.8PS/L
44.8ft/sec 13.7m/sec
26.4mph 42.5kmh/1000rpm
18.9mpg 15.7mpgUS
14.9L/100km
Petrol 4-stroke piston
2933cc 179.0cu in
Vee 6 fuel injection
Compression ratio: 9.5:1

Bore: 93.0mm 3.7in
Stroke: 72.0mm 2.8in
Valve type/No: Overhead 12
Transmission: Automatic
No. of forward speeds: 4
Wheels driven: Rear
Springs F/R: Coil/Coil
Brake system: PA ABS
Brakes F/R: Disc/Disc
Steering: Rack & pinion PA
Wheelbase: 226.1cm 89.0in
Track F: 147.7cm 58.1in
Track R: 149.2cm 58.7in
Length: 466.9cm 183.8in
Width: 176.6cm 69.5in
Height: 145.0cm 57.1in
Ground clearance: 17.9cm
7.0in
Kerb weight: 1340.0kg
2951.5lb
Fuel: 70.1L 15.4gal 18.5galUS

323 Ford

Granada Scorpio 2.0i Auto
1990 UK
114.0mph 183.4kmh
0-50mph 80.5kmh: 8.5secs
0-60mph 96.5kmh: 11.6secs
0-1/4 mile: 18.6secs
0-1km: 33.9secs
125.0bhp 93.2kW 126.7PS
@ 5600rpm
128.4lbft 174.0Nm @ 2500rpm
92.4bhp/ton 90.9bhp/tonne
62.6bhp/L 46.6kW/L 63.4PS/L
52.7ft/sec 16.0m/sec
24.4mph 39.3kmh/1000rpm
25.2mpg 21.0mpgUS
11.2L/100km
Petrol 4-stroke piston
1998cc 122.0cu in
In-line 4 fuel injection
Compression ratio: 10.3:1
Bore: 86.0mm 3.4in
Stroke: 86.0mm 3.4in
Valve type/No: Overhead 8
Transmission: Automatic
No. of forward speeds: 4
Wheels driven: Rear
Springs F/R: Coil/Coil
Brake system: PA ABS
Brakes F/R: Disc/Disc
Steering: Rack & pinion PA
Wheelbase: 276.1cm 108.7in
Track F: 147.6cm 58.1in
Track R: 149.1cm 58.7in
Length: 466.3cm 183.6in
Width: 196.3cm 77.3in
Height: 139.2cm 54.8in
Ground clearance: 17.8cm
7.0in
Kerb weight: 1375.0kg
3028.6lb
Fuel: 70.1L 15.4gal 18.5galUS

324 Ford

Mustang GT Convertible
1990 USA
135.0mph 217.2kmh
0-50mph 80.5kmh: 6.0secs
0-60mph 96.5kmh: 8.0secs
0-1/4 mile: 16.1secs
225.0bhp 167.8kW 228.1PS
@ 4000rpm
300.0lbft 406.5Nm @ 3200rpm
144.4bhp/ton 142.0bhp/tonne
45.5bhp/L 33.9kW/L 46.2PS/L
33.3ft/sec 10.2m/sec
17.2mpg 14.3mpgUS
16.4L/100km
Petrol 4-stroke piston
4942cc 301.5cu in
Vee 8 fuel injection
Compression ratio: 9.2:1
Bore: 101.1mm 4.0in
Stroke: 76.2mm 3.0in
Valve type/No: Overhead 16
Transmission: Automatic
No. of forward speeds: 4
Wheels driven: Rear
Springs F/R: Coil/Coil
Brake system: PA
Brakes F/R: Disc/Drum
Steering: Rack & pinion PA
Wheelbase: 255.3cm 100.5in
Track F: 143.8cm 56.6in
Track R: 144.8cm 57.0in
Length: 456.2cm 179.6in
Width: 175.5cm 69.1in
Height: 132.3cm 52.1in
Kerb weight: 1584.5kg
3490.0lb
Fuel: 58.3L 12.8gal 15.4galUS

325 Ford

Mustang LX 5.0L
1990 USA
148.0mph 238.1kmh
0-50mph 80.5kmh: 4.9secs
0-60mph 96.5kmh: 6.6secs
0-1/4 mile: 15.3secs
225.0bhp 167.8kW 228.1PS
@ 4000rpm
300.0lbft 406.5Nm @ 3200rpm
153.7bhp/ton 151.1bhp/tonne
45.5bhp/L 33.9kW/L 46.2PS/L
33.3ft/sec 10.2m/sec
20.4mpg 17.0mpgUS
13.8L/100km
Petrol 4-stroke piston
4942cc 301.5cu in
Vee 8 fuel injection
Compression ratio: 9.2:1
Bore: 101.1mm 4.0in
Stroke: 76.2mm 3.0in
Valve type/No: Overhead 16
Transmission: Manual
No. of forward speeds: 5
Wheels driven: Rear

Springs F/R: Coil/Coil
Brake system: PA
Brakes F/R: Disc/Drum
Steering: Rack & pinion PA
Wheelbase: 255.3cm 100.5in
Track F: 143.8cm 56.6in
Track R: 144.8cm 57.0in
Length: 456.2cm 179.6in
Width: 175.5cm 69.1in
Height: 132.3cm 52.1in
Kerb weight: 1489.1kg
3280.0lb
Fuel: 58.3L 12.8gal 15.4galUS

326 Ford

Probe GT
1990 USA
131.0mph 210.8kmh
0-60mph 96.5kmh: 7.6secs
0-1/4 mile: 16.0secs
145.0bhp 108.1kW 147.0PS
@ 4300rpm
190.0lbft 257.5Nm @ 3500rpm
110.7bhp/ton 108.8bhp/tonne
66.4bhp/L 49.5kW/L 67.3PS/L
44.2ft/sec 13.5m/sec
28.1mpg 23.4mpgUS
10.1L/100km
Petrol 4-stroke piston
2184cc 133.2cu in
turbocharged
In-line 4 fuel injection
Compression ratio: 7.8:1
Bore: 86.0mm 3.4in
Stroke: 94.0mm 3.7in
Valve type/No: Overhead 12
Transmission: Manual
No. of forward speeds: 5
Wheels driven: Front
Springs F/R: Coil/Coil
Brake system: PA ABS
Brakes F/R: Disc/Disc
Steering: Rack & pinion PA
Wheelbase: 251.5cm 99.0in
Track F: 145.5cm 57.3in
Track R: 146.6cm 57.7in
Length: 449.6cm 177.0in
Width: 174.0cm 68.5in
Height: 131.6cm 51.8in
Kerb weight: 1332.5kg
2935.0lb
Fuel: 57.1L 12.6gal 15.1galUS

327 Ford

Probe LX
1990 USA
130.0mph 209.2kmh
0-60mph 96.5kmh: 8.2secs
0-1/4 mile: 16.2secs
140.0bhp 104.4kW 141.9PS
@ 4800rpm
160.0lbft 216.8Nm @ 3000rpm
108.1bhp/ton 106.3bhp/tonne
48.5bhp/L 36.2kW/L 49.2PS/L

42.0ft/sec 12.8m/sec
25.2mpg 21.0mpgUS
11.2L/100km
Petrol 4-stroke piston
2886cc 176.1cu in
Vee 6 fuel injection
Compression ratio: 9.3:1
Bore: 89.0mm 3.5in
Stroke: 80.0mm 3.1in
Valve type/No: Overhead 12
Transmission: Manual
No. of forward speeds: 5
Wheels driven: Front
Springs F/R: Coil/Coil
Brake system: PA ABS
Brakes F/R: Disc/Disc
Steering: Rack & pinion PA
Wheelbase: 251.5cm 99.0in
Track F: 145.5cm 57.3in
Track R: 146.6cm 57.7in
Length: 449.6cm 177.0in
Width: 172.5cm 67.9in
Height: 131.8cm 51.9in
Kerb weight: 1316.6kg
2900.0lb
Fuel: 57.1L 12.6gal 15.1galUS

328 Ford

Sierra Sapphire Cosworth 4x4
1990 UK
146.0mph 234.9kmh
0-50mph 80.5kmh: 4.8secs
0-60mph 96.5kmh: 6.6secs
0-1/4 mile: 14.3secs
0-1km: 26.8secs
220.0bhp 164.0kW 223.0PS
@ 6250rpm
214.0lbft 290.0Nm @ 3500rpm
171.4bhp/ton 168.6bhp/tonne
110.4bhp/L 82.3kW/L
111.9PS/L
52.6ft/sec 16.0m/sec
22.2mph 35.7kmh/1000rpm
21.6mpg 18.0mpgUS
13.1L/100km
Petrol 4-stroke piston
1993cc 122.0cu in
Turbocharged
In-line 4 fuel injection
Compression ratio: 8.0:1
Bore: 91.0mm 3.6in
Stroke: 77.0mm 3.0in
Valve type/No: Overhead 16
Transmission: Manual
No. of forward speeds: 5
Wheels driven: 4-wheel drive
Springs F/R: Coil/Coil
Brake system: PA ABS
Brakes F/R: Disc/Disc
Steering: Rack & pinion PA
Wheelbase: 260.9cm 102.7in
Track F: 144.3cm 56.8in
Track R: 146.1cm 57.5in
Length: 423.9cm 166.9in

Width: 169.7cm 66.8in
Height: 137.7cm 54.2in
Ground clearance: 17.8cm
7.0in
Kerb weight: 1305.0kg
2874.4lb
Fuel: 59.1L 13.0gal 15.6galUS

329 Ford

Crown Victoria LX
1991 USA
125.0mph 201.1kmh
0-50mph 80.5kmh: 7.2secs
0-60mph 96.5kmh: 9.9secs
0-1/4 mile: 17.4secs
210.0bhp 156.6kW 212.9PS
@ 4600rpm
270.0lbft 365.9Nm @ 3400rpm
117.2bhp/ton 115.2bhp/tonne
45.6bhp/L 34.0kW/L 46.3PS/L
45.2ft/sec 13.8m/sec
27.0mpg 22.5mpgUS
10.5L/100km
Petrol 4-stroke piston
4601cc 280.7cu in
Vee 8 fuel injection
Compression ratio: 9.0:1
Bore: 90.2mm 3.5in
Stroke: 90.0mm 3.5in
Valve type/No: Overhead 16
Transmission: Automatic
No. of forward speeds: 4
Wheels driven: Rear
Springs F/R: Coil/Coil
Brake system: PA ABS
Brakes F/R: Disc/Disc
Steering: Recirculating ball PA
Wheelbase: 290.6cm 114.4in
Track F: 159.5cm 62.8in
Track R: 160.8cm 63.3in
Length: 539.5cm 212.4in
Width: 197.6cm 77.8in
Height: 144.0cm 56.7in
Ground clearance: 15.5cm
6.1in
Kerb weight: 1822.8kg
4015.0lb
Fuel: 75.7L 16.6gal 20.0galUS

330 Ford

Escort 1.4LX
1991 UK
108.0mph 173.8kmh
0-50mph 80.5kmh: 9.2secs
0-60mph 96.5kmh: 12.8secs
0-1/4 mile: 18.8secs
0-1km: 34.2secs
73.0bhp 54.4kW 74.0PS
@ 5500rpm
79.7lbft 108.0Nm @ 4000rpm
73.5bhp/ton 72.3bhp/tonne
52.4bhp/L 39.1kW/L 53.2PS/L
44.8ft/sec 13.6m/sec
22.8mph 36.7kmh/1000rpm

29.0mpg 24.1mpgUS
9.7L/100km
Petrol 4-stroke piston
1392cc 85.0cu in
In-line 4 1 Carburettor
Compression ratio: 8.5:1
Bore: 77.2mm 3.0in
Stroke: 74.3mm 2.9in
Valve type/No: Overhead 8
Transmission: Manual
No. of forward speeds: 5
Wheels driven: Front
Springs F/R: Coil/Torsion bar
Brake system: PA ABS
Brakes F/R: Disc/Drum
Steering: Rack & pinion
Wheelbase: 252.5cm 99.4in
Track F: 143.8cm 56.6in
Track R: 146.1cm 57.5in
Length: 403.6cm 158.9in
Width: 168.4cm 66.3in
Height: 135.1cm 53.2in
Kerb weight: 1010.0kg
2224.7lb
Fuel: 55.1L 12.1gal 14.5galUS

331 Ford

Escort 1.6 Ghia Estate
1991 UK
108.0mph 173.8kmh
0-50mph 80.5kmh: 8.7secs
0-60mph 96.5kmh: 12.3secs
0-1/4 mile: 18.9secs
0-1km: 34.7secs
90.0bhp 67.1kW 91.2PS
@ 5800rpm
95.9lbft 130.0Nm @ 4000rpm
82.1bhp/ton 80.7bhp/tonne
56.4bhp/L 42.0kW/L 57.1PS/L
50.7ft/sec 15.5m/sec
22.7mph 36.5kmh/1000rpm
27.8mpg 23.1mpgUS
10.2L/100km
Petrol 4-stroke piston
1597cc 97.0cu in
In-line 4 1 Carburettor
Compression ratio: 9.5:1
Bore: 80.0mm 3.1in
Stroke: 80.0mm 3.1in
Valve type/No: Overhead 8
Transmission: Manual
No. of forward speeds: 5
Wheels driven: Front
Springs F/R: Coil/Coil
Brake system: PA ABS
Brakes F/R: Disc/Drum
Steering: Rack & pinion PA
Wheelbase: 252.5cm 99.4in
Track F: 143.8cm 56.6in
Track R: 146.1cm 57.5in
Length: 428.8cm 168.8in
Width: 168.4cm 66.3in
Height: 135.1cm 53.2in
Kerb weight: 1115.0kg

2455.9lb
Fuel: 55.1L 12.1gal 14.5galUS

332 Ford

Escort Cabriolet
1991 UK
116.0mph 186.6kmh
0-50mph 80.5kmh: 7.3secs
0-60mph 96.5kmh: 10.2secs
0-1/4 mile: 17.7secs
0-1km: 32.6secs
108.0bhp 80.5kW 109.5PS
@ 6000rpm
104.1lbft 141.0Nm @ 4500rpm
98.3bhp/ton 96.7bhp/tonne
67.7bhp/L 50.5kW/L 68.6PS/L
52.2ft/sec 15.9m/sec
22.7mph 36.5kmh/1000rpm
28.6mpg 23.8mpgUS
9.9L/100km
Petrol 4-stroke piston
1596cc 97.0cu in
In-line 4 fuel injection
Compression ratio: 9.7:1
Bore: 80.0mm 3.1in
Stroke: 79.5mm 3.1in
Valve type/No: Overhead 8
Transmission: Manual
No. of forward speeds: 5
Wheels driven: Front
Springs F/R: Coil/Coil
Brake system: PA
Brakes F/R: Disc/Drum
Steering: Rack & pinion PA
Wheelbase: 251.5cm 99.0in
Track F: 144.8cm 57.0in
Track R: 144.8cm 57.0in
Length: 403.9cm 159.0in
Width: 170.2cm 67.0in
Height: 139.7cm 55.0in
Kerb weight: 1117.0kg
2460.3lb
Fuel: 66.0L 14.5gal 17.4galUS

333 Ford

Escort GT
1991 USA
120.0mph 193.1kmh
0-50mph 80.5kmh: 6.6secs
0-60mph 96.5kmh: 8.9secs
0-1/4 mile: 16.8secs
127.0bhp 94.7kW 128.8PS
@ 6500rpm
114.0lbft 154.5Nm @ 4500rpm
113.6bhp/ton 111.7bhp/tonne
69.0bhp/L 51.5kW/L 70.0PS/L
60.5ft/sec 18.4m/sec
33.6mpg 28.0mpgUS
8.4L/100km
Petrol 4-stroke piston
1840cc 112.3cu in
In-line 4 fuel injection
Compression ratio: 9.0:1
Bore: 83.0mm 3.3in

Stroke: 85.0mm 3.3in
Valve type/No: Overhead 16
Transmission: Manual
No. of forward speeds: 5
Wheels driven: Front
Springs F/R: Coil/Coil
Brake system: PA
Brakes F/R: Disc/Disc
Steering: Rack & pinion PA
Wheelbase: 249.9cm 98.4in
Track F: 143.5cm 56.5in
Track R: 143.5cm 56.5in
Length: 434.1cm 170.9in
Width: 169.4cm 66.7in
Height: 133.4cm 52.5in
Ground clearance: 13.0cm
5.1in
Kerb weight: 1137.3kg
2505.0lb
Fuel: 45.0L 9.9gal 11.9galUS

334 Ford

Mustang Holdener
1991 USA
174.0mph 280.0kmh
0-60mph 96.5kmh: 4.9secs
0-1/4 mile: 13.3secs
450.0bhp 335.6kW 456.2PS
@ 5600rpm
420.0lbft 569.1Nm @ 4300rpm
362.6bhp/ton 356.5bhp/tonne
91.1bhp/L 67.9kW/L 92.3PS/L
46.7ft/sec 14.2m/sec
Petrol 4-stroke piston
4942cc 301.5cu in
supercharged
Vee 8
Compression ratio: 8.0:1
Bore: 101.0mm 4.0in
Stroke: 76.2mm 3.0in
Valve type/No: Overhead 16
Transmission: Manual
No. of forward speeds: 5
Wheels driven: Rear
Springs F/R: Coil/Coil
Brake system: PA
Brakes F/R: Disc/Drum
Steering: Rack & pinion PA
Wheelbase: 255.3cm 100.5in
Track F: 143.8cm 56.6in
Track R: 144.8cm 57.0in
Length: 456.2cm 179.6in
Width: 175.5cm 69.1in
Height: 132.3cm 52.1in
Kerb weight: 1262.1kg
2780.0lb

335 Ford

Mustang NOS/Saleen
1991 USA
186.0mph 299.3kmh
0-60mph 96.5kmh: 5.7secs
0-1/4 mile: 14.1secs
310.0bhp 231.2kW 314.3PS

@ 6000rpm
323.0lbft 437.7Nm @ 4500rpm
231.8bhp/ton 228.0bhp/tonne
62.7bhp/L 46.8kW/L 63.6PS/L
50.0ft/sec 15.2m/sec
Petrol 4-stroke piston
4942cc 301.5cu in
Vee 8 fuel injection
Compression ratio: 9.2:1
Bore: 101.6mm 4.0in
Stroke: 76.2mm 3.0in
Valve type/No: Overhead 16
Transmission: Manual
No. of forward speeds: 5
Wheels driven: Rear
Springs F/R: Coil/Coil
Brake system: PA
Brakes F/R: Disc/Disc
Steering: Rack & pinion PA
Wheelbase: 255.3cm 100.5in
Track F: 147.1cm 57.9in
Track R: 149.1cm 58.7in
Length: 456.2cm 179.6in
Width: 175.5cm 69.1in
Height: 127.0cm 50.0in
Kerb weight: 1359.7kg
2995.0lb

336 Ford

Orion 1.6i Ghia
1991 UK
121.0mph 194.7kmh
0-50mph 80.5kmh: 7.2secs
0-60mph 96.5kmh: 10.0secs
0-1/4 mile: 17.6secs
0-1km: 32.2secs
108.0bhp 80.5kW 109.5PS
@ 6000rpm
101.8lbft 138.0Nm @ 4500rpm
102.6bhp/ton 100.9bhp/tonne
67.7bhp/L 50.5kW/L 68.6PS/L
52.2ft/sec 15.9m/sec
22.8mph 36.7kmh/1000rpm
29.2mpg 24.3mpgUS
9.7L/100km
Petrol 4-stroke piston
1596cc 97.0cu in
In-line 4 fuel injection
Compression ratio: 9.7:1
Bore: 80.0mm 3.1in
Stroke: 79.5mm 3.1in
Valve type/No: Overhead 8
Transmission: Manual
No. of forward speeds: 5
Wheels driven: Front
Springs F/R: Coil/Coil
Brake system: PA ABS
Brakes F/R: Disc/Drum
Steering: Rack & pinion PA
Wheelbase: 252.5cm 99.4in
Track F: 144.0cm 56.7in
Track R: 143.9cm 56.7in
Length: 422.9cm 166.5in
Width: 187.5cm 73.8in

Height: 134.5cm 53.0in
Kerb weight: 1070.0kg
2356.8lb
Fuel: 55.1L 12.1gal 14.5galUS

337 Ford

Probe GT
1991 USA
120.0mph 193.1kmh
0-60mph 96.5kmh: 7.4secs
0-1/4 mile: 15.4secs
145.0bhp 108.1kW 147.0PS
@ 4300rpm
190.0lbft 257.5Nm @ 3500rpm
112.0bhp/ton 110.1bhp/tonne
66.4bhp/L 49.5kW/L 67.3PS/L
44.2ft/sec 13.5m/sec
24.6mpg 20.5mpgUS
11.5L/100km
Petrol 4-stroke piston
2184cc 133.2cu in
turbocharged
In-line 4 fuel injection
Compression ratio: 7.8:1
Bore: 86.0mm 3.4in
Stroke: 94.0mm 3.7in
Valve type/No: Overhead 12
Transmission: Automatic
No. of forward speeds: 4
Wheels driven: Front
Springs F/R: Coil/Coil
Brake system: PA ABS
Brakes F/R: Disc/Disc
Steering: Rack & pinion PA
Wheelbase: 251.5cm 99.0in
Track F: 145.5cm 57.3in
Track R: 146.6cm 57.7in
Length: 449.6cm 177.0in
Width: 174.0cm 68.5in
Height: 131.6cm 51.8in
Kerb weight: 1316.6kg
2900.0lb
Fuel: 57.1L 12.6gal 15.1galUS

338 Ford

Scorpio 24v
1991 UK
140.0mph 225.3kmh
0-50mph 80.5kmh: 6.5secs
0-60mph 96.5kmh: 8.5secs
0-1/4 mile: 16.6secs
0-1km: 30.1secs
195.0bhp 145.4kW 197.7PS
@ 5750rpm
203.0lbft 275.0Nm @ 4500rpm
133.1bhp/ton 130.9bhp/tonne
66.4bhp/L 49.5kW/L 67.4PS/L
45.2ft/sec 13.8m/sec
25.5mph 41.0kmh/1000rpm
21.0mpg 17.5mpgUS
13.5L/100km
Petrol 4-stroke piston
2935cc 179.0cu in
Vee 6 fuel injection

Compression ratio: 9.7:1
Bore: 93.0mm 3.7in
Stroke: 72.0mm 2.8in
Valve type/No: Overhead 24
Transmission: Automatic
No. of forward speeds: 4
Wheels driven: Rear
Springs F/R: Coil/Coil
Brake system: PA ABS
Brakes F/R: Disc/Disc
Steering: Rack & pinion PA
Wheelbase: 276.1cm 108.7in
Track F: 147.7cm 58.1in
Track R: 150.0cm 59.1in
Length: 474.4cm 186.8in
Width: 176.6cm 69.5in
Height: 145.0cm 57.1in
Ground clearance: 17.9cm
7.0in
Kerb weight: 1490.0kg
3281.9lb
Fuel: 70.0L 15.4gal 18.5galUS

339 Ford

Taurus SHO
1991 USA
140.0mph 225.3kmh
0-60mph 96.5kmh: 7.6secs
0-1/4 mile: 15.9secs
220.0bhp 164.0kW 223.0PS
@ 6200rpm
200.0lbft 271.0Nm @ 4800rpm
144.5bhp/ton 142.1bhp/tonne
73.7bhp/L 54.9kW/L 74.7PS/L
54.2ft/sec 16.5m/sec
26.4mpg 22.0mpgUS
10.7L/100km
Petrol 4-stroke piston
2986cc 182.2cu in
Vee 6 fuel injection
Compression ratio: 9.8:1
Bore: 89.0mm 3.5in
Stroke: 80.0mm 3.1in
Valve type/No: Overhead 24
Transmission: Manual
No. of forward speeds: 5
Wheels driven: Front
Springs F/R: Coil/Coil
Brake system: PA ABS
Brakes F/R: Disc/Disc
Steering: Rack & pinion PA
Wheelbase: 269.2cm 106.0in
Track F: 156.5cm 61.6in
Track R: 153.7cm 60.5in
Length: 478.5cm 188.4in
Width: 179.8cm 70.8in
Height: 137.9cm 54.3in
Kerb weight: 1548.1kg
3410.0lb
Fuel: 60.6L 13.3gal 16.0galUS

340 Ford

Thunderbird LX
1991 USA

145.0mph 233.3kmh
0-60mph 96.5kmh: 9.0secs
0-1/4 mile: 16.7secs
200.0bhp 149.1kW 202.8PS
@ 4000rpm
275.0lbft 372.6Nm @ 3000rpm
116.1bhp/ton 114.1bhp/tonne
40.5bhp/L 30.2kW/L 41.0PS/L
33.3ft/sec 10.2m/sec
20.4mpg 17.0mpgUS
13.8L/100km
Petrol 4-stroke piston
4942cc 301.5cu in
Vee 8 fuel injection
Compression ratio: 9.0:1
Bore: 101.6mm 4.0in
Stroke: 76.2mm 3.0in
Valve type/No: Overhead 16
Transmission: Automatic
No. of forward speeds: 4
Wheels driven: Rear
Springs F/R: Coil/Coil
Brake system: PA ABS
Brakes F/R: Disc/Disc
Steering: Rack & pinion PA
Wheelbase: 287.0cm 113.0in
Track F: 156.5cm 61.6in
Track R: 152.9cm 60.2in
Length: 504.7cm 198.7in
Width: 184.7cm 72.7in
Height: 133.9cm 52.7in
Kerb weight: 1752.4kg
3860.0lb
Fuel: 71.9L 15.8gal 19.0galUS

341 Ford
Escort RS2000
1992 UK
131.1mph 211.0kmh
0-50mph 80.5kmh: 6.0secs
0-60mph 96.5kmh: 8.3secs
0-1/4 mile: 16.4secs

0-1km: 29.7secs
150.0bhp 111.9kW 152.1PS
@ 6000rpm
140.2lbft 190.0Nm @ 4500rpm
135.6bhp/ton 133.3bhp/tonne
75.1bhp/L 56.0kW/L 76.1PS/L
56.5ft/sec 17.2m/sec
20.2mph 32.5kmh/1000rpm
26.4mpg 22.0mpgUS
10.7L/100km
Petrol 4-stroke piston
1998cc 122.0cu in
In-line 4 fuel injection
Compression ratio: 10.3:1
Bore: 86.0mm 3.4in
Stroke: 86.0mm 3.4in
Valve type/No: Overhead 16
Transmission: Manual
No. of forward speeds: 5
Wheels driven: Front
Springs F/R: Coil/Coil
Brake system: PA ABS
Brakes F/R: Disc/Disc
Steering: Rack & pinion PA
Wheelbase: 251.5cm 99.0in
Track F: 144.8cm 57.0in
Track R: 144.8cm 57.0in
Length: 403.9cm 159.0in
Width: 170.2cm 67.0in
Height: 139.7cm 55.0in
Kerb weight: 1125.0kg
2478.0lb
Fuel: 54.6L 12.0gal 14.4galUS

342 Ford
Escort XR3i
1992 UK
125.0mph 201.1kmh
0-50mph 80.5kmh: 6.3secs
0-60mph 96.5kmh: 8.6secs
0-1/4 mile: 16.7secs
0-1km: 30.5secs

130.0bhp 96.9kW 131.8PS
@ 6250rpm
119.0lbft 161.2Nm @ 4500rpm
72.4bhp/L 54.0kW/L 73.4PS/L
60.1ft/sec 18.3m/sec
20.2mph 32.5kmh/1000rpm
28.6mpg 23.8mpgUS
9.9L/100km
Petrol 4-stroke piston
1796cc 109.6cu in
In-line 4 fuel injection
Compression ratio: 10.0:1
Bore: 80.6mm 3.2in
Stroke: 88.0mm 3.5in
Valve type/No: Overhead 16
Transmission: Manual
No. of forward speeds: 5
Wheels driven: Front
Springs F/R: Coil/Coil
Brake system: PA
Brakes F/R: Disc/Disc
Steering: Rack & pinion PA
Wheelbase: 252.5cm 99.4in
Track F: 144.3cm 56.8in
Track R: 143.9cm 56.7in
Length: 404.0cm 159.1in
Width: 169.2cm 66.6in
Height: 139.7cm 55.0in
Fuel: 68.2L 15.0gal 18.0galUS

343 Ford
Mustang LX
1992 USA
140.0mph 225.3kmh
0-50mph 80.5kmh: 5.4secs
0-60mph 96.5kmh: 7.1secs
0-1/4 mile: 15.5secs
225.0bhp 167.8kW 228.1PS
@ 4200rpm
300.0lbft 406.5Nm @ 3200rpm
156.8bhp/ton 154.1bhp/tonne
45.5bhp/L 33.9kW/L 46.2PS/L

35.0ft/sec 10.7m/sec
22.8mpg 19.0mpgUS
12.4L/100km
Petrol 4-stroke piston
4942cc 301.5cu in
Vee 8 fuel injection
Compression ratio: 9.0:1
Bore: 101.6mm 4.0in
Stroke: 76.2mm 3.0in
Valve type/No: Overhead 16
Transmission: Manual
No. of forward speeds: 5
Wheels driven: Rear
Springs F/R: Coil/Coil
Brake system: PA
Brakes F/R: Disc/Drum
Steering: Rack & pinion PA
Wheelbase: 255.3cm 100.5in
Track F: 143.8cm 56.6in
Track R: 144.8cm 57.0in
Length: 456.2cm 179.6in
Width: 173.5cm 68.3in
Height: 132.3cm 52.1in
Kerb weight: 1459.6kg
3215.0lb
Fuel: 58.3L 12.8gal 15.4galUS

344 Ford
Taurus LX
1992 USA
126.0mph 202.7kmh
0-60mph 96.5kmh: 10.4secs
0-1/4 mile: 17.7secs
140.0bhp 104.4kW 141.9PS
@ 3800rpm
215.0lbft 291.3Nm @ 2200rpm
94.7bhp/ton 93.2bhp/tonne
36.9bhp/L 27.5kW/L 37.4PS/L
35.8ft/sec 10.9m/sec
21.0mpg 17.5mpgUS
13.4L/100km
Petrol 4-stroke piston

342 Ford
Escort XR3i

3791cc 231.3cu in
Vee 6 fuel injection
Compression ratio: 9.0:1
Bore: 96.8mm 3.8in
Stroke: 86.0mm 3.4in
Valve type/No: Overhead 12
Transmission: Automatic
No. of forward speeds: 4
Wheels driven: Front
Springs F/R: Coil/Coil
Brake system: PA ABS
Brakes F/R: Disc/Disc
Steering: Rack & pinion PA
Wheelbase: 269.2cm 106.0in
Track F: 156.5cm 61.6in
Track R: 153.7cm 60.5in
Length: 487.7cm 192.0in
Width: 180.8cm 71.2in
Height: 137.4cm 54.1in
Kerb weight: 1502.7kg
3310.0lb
Fuel: 60.6L
13.3gal 16.0galUS

345 Geo
Storm GSi
1991 Japan
120.0mph 193.1kmh
0-60mph 96.5kmh: 9.3secs
0-1/4 mile: 16.9secs
130.0bhp 96.9kW 131.8PS
@ 7000rpm
102.0lbft 138.2Nm @ 5800rpm
118.9bhp/ton 116.9bhp/tonne
81.9bhp/L 61.0kW/L 83.0PS/L
60.5ft/sec 18.4m/sec
32.4mpg 27.0mpgUS
8.7L/100km
Petrol 4-stroke piston
1588cc 96.9cu in
In-line 4 fuel injection
Compression ratio: 9.8:1

Bore: 80.0mm 3.1in
Stroke: 79.0mm 3.1in
Valve type/No: Overhead 16
Transmission: Manual
No. of forward speeds: 5
Wheels driven: Front
Springs F/R: Coil/Coil
Brake system: PA
Brakes F/R: Disc/Disc
Steering: Rack & pinion PA
Wheelbase: 245.1cm 96.5in
Track F: 143.0cm 56.3in
Track R: 140.2cm 55.2in
Length: 415.0cm 163.4in
Width: 169.4cm 66.7in
Height: 129.8cm 51.1in
Kerb weight: 1112.3kg
2450.0lb
Fuel: 46.9L 10.3gal
12.4galUS

346 Ginetta
G32
1990 UK
116.0mph 186.6kmh
0-50mph 80.5kmh: 6.4secs
0-60mph 96.5kmh: 9.0secs
0-1/4 mile: 16.8secs
0-1km: 31.4secs
110.0bhp 82.0kW 111.5PS
@ 6000rpm
104.1lbft 141.0Nm @ 2800rpm
127.1bhp/ton 125.0bhp/tonne
68.8bhp/L 51.3kW/L 69.8PS/L
52.2ft/sec 15.9m/sec
22.6mph 36.4kmh/1000rpm
23.3mpg 19.4mpgUS
12.1L/100km
Petrol 4-stroke piston
1598cc 97.0cu in
In-line 4 fuel injection
Compression ratio: 9.5:1

Bore: 80.0mm 3.1in
Stroke: 79.5mm 3.1in
Valve type/No: Overhead 8
Transmission: Manual
No. of forward speeds: 5
Wheels driven: Rear
Springs F/R: Coil/Coil
Brake system: PA
Brakes F/R: Disc/Disc
Steering: Rack & pinion PA
Wheelbase: 221.0cm 87.0in
Track F: 139.7cm 55.0in
Track R: 139.7cm 55.0in
Length: 375.9cm 148.0in
Width: 165.1cm 65.0in
Height: 116.8cm 46.0in
Kerb weight: 880.0kg 1938.3lb
Fuel: 44.1L 9.7gal 11.7galUS

347 GM
Impact
1990 USA
75.0mph 120.7kmh
0-60mph 96.5kmh: 8.0secs
0-1/4 mile: 16.7secs
57.0bhp 42.5kW 57.8PS
@ 6600rpm
47.0lbft 63.7Nm @ 6000rpm
58.0bhp/ton 57.1bhp/tonne
AC electric
No. of forward speeds: 1
Wheels driven: Front
Springs F/R: Coil/Coil
Brakes F/R: Disc/Drum
Steering: Rack & pinion
Wheelbase: 241.3cm 95.0in
Track F: 147.1cm 57.9in
Track R: 122.9cm 48.4in
Length: 464.8cm 183.0in
Width: 173.2cm 68.2in
Height: 120.7cm 47.5in
Kerb weight: 998.8kg 2200.0lb

348 Honda
Accord LXi
1986 Japan
108.0mph 173.8kmh
0-60mph 96.5kmh: 9.8secs
0-1/4 mile: 17.3secs
110.0bhp 82.0kW 111.5PS
@ 5500rpm
114.0lbft 154.5Nm @ 4500rpm
87.7bhp/ton 86.2bhp/tonne
56.3bhp/L 42.0kW/L 57.0PS/L
54.7ft/sec 16.7m/sec
28.8mpg 24.0mpgUS
9.8L/100km
Petrol 4-stroke piston
1955cc 119.3cu in
In-line 4 fuel injection
Compression ratio: 8.8:1
Bore: 82.7mm 3.3in
Stroke: 91.0mm 3.6in
Valve type/No: Overhead 8
Transmission: Manual
No. of forward speeds: 5
Wheels driven: Front
Springs F/R: Coil/Coil
Steering: Rack & pinion PA
Wheelbase: 260.1cm 102.4in
Track F: 148.1cm 58.3in
Track R: 147.6cm 58.1in
Length: 453.4cm 178.5in
Width: 169.4cm 66.7in
Height: 135.4cm 53.3in
Ground clearance: 16.0cm
6.3in
Kerb weight: 1275.7kg
2810.0lb
Fuel: 60.2L 13.2gal 15.9galUS

349 Honda
CRX Coupe 1.6i 16
1986 Japan

346 Ginetta
G32

125.0mph 201.1kmh
0-50mph 80.5kmh: 5.9secs
0-60mph 96.5kmh: 8.0secs
0-1/4 mile: 16.2secs
0-1km: 29.7secs
125.0bhp 93.2kW 126.7PS
@ 6500rpm
103.0lbft 139.6Nm @ 5500rpm
141.2bhp/ton 138.9bhp/tonne
78.6bhp/L 58.6kW/L 79.7PS/L
63.9ft/sec 19.5m/sec
19.4mph 31.2kmh/1000rpm
30.0mpg 25.0mpgUS
9.4L/100km
Petrol 4-stroke piston
1590cc 97.0cu in
In-line 4 fuel injection
Compression ratio: 9.3:1
Bore: 75.0mm 2.9in
Stroke: 90.0mm 3.5in
Valve type/No: Overhead 16
Transmission: Manual
No. of forward speeds: 5
Wheels driven: Front
Springs F/R: Torsion bar/Coil
Brake system: PA
Brakes F/R: Disc/Drum
Steering: Rack & pinion
Wheelbase: 220.0cm 86.6in
Track F: 140.0cm 55.1in
Track R: 141.5cm 55.7in
Length: 375.5cm 147.8in
Width: 163.0cm 64.2in
Height: 129.0cm 50.8in
Ground clearance: 15.3cm
6.0in
Kerb weight: 900.0kg 1982.4lb
Fuel: 40.9L 9.0gal 10.8galUS

350 Honda
CRX Si
1986 Japan
116.0mph 186.6kmh
0-50mph 80.5kmh: 6.3secs
0-60mph 96.5kmh: 8.9secs
0-1/4 mile: 16.7secs
91.0bhp 67.9kW 92.3PS
@ 5500rpm
93.0lbft 126.0Nm @ 4500rpm
104.3bhp/ton 102.5bhp/tonne
61.2bhp/L 45.6kW/L 62.0PS/L
52.1ft/sec 15.9m/sec
37.8mpg 31.5mpgUS
7.5L/100km
Petrol 4-stroke piston
1488cc 90.8cu in
In-line 4 fuel injection
Compression ratio: 8.7:1
Bore: 73.9mm 2.9in
Stroke: 86.5mm 3.4in
Valve type/No: Overhead 8
Transmission: Manual
No. of forward speeds: 5
Wheels driven: Front

Springs F/R: Torsion bar/Coil
Brake system: PA
Brakes F/R: Disc/Drum
Steering: Rack & pinion
Wheelbase: 220.0cm 86.6in
Track F: 140.0cm 55.1in
Track R: 141.5cm 55.7in
Length: 375.4cm 147.8in
Width: 162.3cm 63.9in
Height: 129.0cm 50.8in
Kerb weight: 887.6kg 1955.0lb
Fuel: 40.9L 9.0gal 10.8galUS

351 Honda
Integra 1.5
1986 Japan
102.0mph 164.1kmh
0-50mph 80.5kmh: 8.2secs
0-60mph 96.5kmh: 11.8secs
0-1/4 mile: 18.2secs
0-1km: 34.5secs
85.0bhp 63.4kW 86.2PS
@ 6000rpm
87.0lbft 117.9Nm @ 3750rpm
85.6bhp/ton 84.2bhp/tonne
57.1bhp/L 42.6kW/L 57.9PS/L
56.8ft/sec 17.3m/sec
21.2mph 34.1kmh/1000rpm
29.1mpg 24.2mpgUS
9.7L/100km
Petrol 4-stroke piston
1488cc 91.0cu in
In-line 4 1 Carburettor
Compression ratio: 8.7:1
Bore: 74.0mm 2.9in
Stroke: 86.5mm 3.4in
Valve type/No: Overhead 12
Transmission: Manual
No. of forward speeds: 5
Wheels driven: Front
Springs F/R: Torsion bar/Coil
Brake system: PA
Brakes F/R: Disc/Drum
Steering: Rack & pinion
Wheelbase: 252.0cm 99.2in
Track F: 142.0cm 55.9in
Track R: 141.5cm 55.7in
Length: 435.0cm 171.3in
Width: 166.5cm 65.6in
Height: 134.5cm 53.0in
Ground clearance: 15.3cm
6.0in
Kerb weight: 1010.0kg
2224.7lb
Fuel: 50.0L 11.0gal 13.2galUS

352 Honda
Integra EX 16
1986 Japan
119.0mph 191.5kmh
0-50mph 80.5kmh: 6.4secs
0-60mph 96.5kmh: 8.6secs
0-1/4 mile: 17.0secs
0-1km: 30.8secs

125.0bhp 93.2kW 126.7PS
@ 6500rpm
103.0lbft 139.6Nm @ 5500rpm
124.6bhp/ton 122.5bhp/tonne
78.6bhp/L 58.6kW/L 79.7PS/L
63.9ft/sec 19.5m/sec
18.4mph 29.6kmh/1000rpm
27.5mpg 22.9mpgUS
10.3L/100km
Petrol 4-stroke piston
1590cc 97.0cu in
In-line 4 fuel injection
Compression ratio: 9.3:1
Bore: 75.0mm 2.9in
Stroke: 90.0mm 3.5in
Valve type/No: Overhead 16
Transmission: Manual
No. of forward speeds: 5
Wheels driven: Front
Springs F/R: Torsion bar/Coil
Brake system: PA
Brakes F/R: Disc/Disc
Steering: Rack & pinion PA
Wheelbase: 252.0cm 99.2in
Track F: 142.0cm 55.9in
Track R: 141.5cm 55.7in
Length: 435.0cm 171.3in
Width: 166.5cm 65.6in
Height: 134.5cm 53.0in
Ground clearance: 15.3cm
6.0in
Kerb weight: 1020.0kg
2246.7lb
Fuel: 50.0L 11.0gal 13.2galUS

353 Honda
Legend
1986 Japan
127.0mph 204.3kmh
0-50mph 80.5kmh: 6.0secs
0-60mph 96.5kmh: 8.0secs
0-1/4 mile: 16.5secs
0-1km: 30.0secs
173.0bhp 129.0kW 175.4PS
@ 6000rpm
160.0lbft 216.8Nm @ 5000rpm
127.5bhp/ton 125.4bhp/tonne
69.4bhp/L 51.7kW/L 70.4PS/L
49.2ft/sec 15.0m/sec
24.2mph 38.9kmh/1000rpm
21.5mpg 17.9mpgUS
13.1L/100km
Petrol 4-stroke piston
2493cc 152.0cu in
Vee 6 fuel injection
Compression ratio: 9.6:1
Bore: 84.0mm 3.3in
Stroke: 75.0mm 2.9in
Valve type/No: Overhead 24
Transmission: Manual
No. of forward speeds: 5
Wheels driven: Front
Springs F/R: Coil/Coil
Brake system: PA

Brakes F/R: Disc/Disc
Steering: Rack & pinion PA
Wheelbase: 276.0cm 108.7in
Track F: 149.0cm 58.7in
Track R: 145.0cm 57.1in
Length: 481.0cm 189.4in
Width: 173.5cm 68.3in
Height: 139.0cm 54.7in
Ground clearance: 15.0cm
5.9in
Kerb weight: 1380.0kg
3039.6lb
Fuel: 68.2L 15.0gal 18.0galUS

354 Honda
Ballade EX
1987 Japan
99.0mph 159.3kmh
0-50mph 80.5kmh: 7.9secs
0-60mph 96.5kmh: 11.1secs
0-1/4 mile: 18.5secs
0-1km: 34.8secs
85.0bhp 63.4kW 86.2PS
@ 6000rpm
93.0lbft 126.0Nm @ 3500rpm
93.4bhp/ton 91.9bhp/tonne
57.1bhp/L 42.6kW/L 57.9PS/L
56.8ft/sec 17.3m/sec
22.0mph 35.4kmh/1000rpm
33.3mpg 27.7mpgUS
8.5L/100km
Petrol 4-stroke piston
1488cc 91.0cu in
In-line 4 1 Carburettor
Compression ratio: 8.7:1
Bore: 74.0mm 2.9in
Stroke: 86.5mm 3.4in
Valve type/No: Overhead 12
Transmission: Manual
No. of forward speeds: 5
Wheels driven: Front
Springs F/R: Torsion bar/Coil
Brake system: PA
Brakes F/R: Disc/Drum
Steering: Rack & pinion PA
Wheelbase: 245.0cm 96.5in
Track F: 141.5cm 55.7in
Track R: 140.0cm 55.1in
Length: 414.5cm 163.2in
Width: 163.5cm 64.4in
Height: 139.5cm 54.9in
Ground clearance: 17.9cm
7.0in
Kerb weight: 925.0kg 2037.4lb
Fuel: 45.5L 10.0gal 12.0galUS

355 Honda
Civic 1.4 GL 3-door
1987 Japan
108.0mph 173.8kmh
0-50mph 80.5kmh: 7.0secs
0-60mph 96.5kmh: 9.7secs
0-1/4 mile: 17.4secs
0-1km: 32.2secs

90.0bhp 67.1kW 91.2PS
@ 6300rpm
82.5lbft 111.8Nm @ 4500rpm
101.5bhp/ton 99.8bhp/tonne
64.5bhp/L 48.1kW/L 65.4PS/L
54.4ft/sec 16.6m/sec
20.1mph 32.3kmh/1000rpm
34.7mpg 28.9mpgUS
8.1L/100km
Petrol 4-stroke piston
1396cc 85.0cu in
In-line 4 2 Carburettor
Compression ratio: 9.3:1
Bore: 75.0mm 2.9in
Stroke: 79.0mm 3.1in
Valve type/No: Overhead 16
Transmission: Manual
No. of forward speeds: 5
Wheels driven: Front
Springs F/R: Coil/Coil
Brake system: PA
Brakes F/R: Disc/Drum
Steering: Rack & pinion PA
Wheelbase: 250.0cm 98.4in
Track F: 145.0cm 57.1in
Track R: 145.5cm 57.3in
Length: 396.5cm 156.1in
Width: 168.0cm 66.1in
Height: 133.0cm 52.4in
Ground clearance: 15.3cm
6.0in
Kerb weight: 902.0kg 1986.8lb
Fuel: 43.2L 9.5gal 11.4galUS

356 Honda

Legend Coupe
1987 Japan
133.0mph 214.0kmh
0-50mph 80.5kmh: 6.0secs
0-60mph 96.5kmh: 8.0secs
0-1/4 mile: 16.4secs
0-1km: 29.8secs
177.0bhp 132.0kW 179.4PS
@ 6000rpm
165.0lbft 223.6Nm @ 4500rpm
126.8bhp/ton 124.6bhp/tonne
66.2bhp/L 49.3kW/L 67.1PS/L
49.2ft/sec 15.0m/sec
22.6mph 36.4kmh/1000rpm
20.3mpg 16.9mpgUS
13.9L/100km
Petrol 4-stroke piston
2675cc 163.0cu in
Vee 6 fuel injection
Compression ratio: 9.0:1
Bore: 87.0mm 3.4in
Stroke: 75.0mm 2.9in
Valve type/No: Overhead 24
Transmission: Manual
No. of forward speeds: 5
Wheels driven: Front
Springs F/R: Coil/Coil
Brake system: PA ABS
Brakes F/R: Disc/Disc

Steering: Rack & pinion PA
Wheelbase: 270.5cm 106.5in
Track F: 150.0cm 59.1in
Track R: 150.0cm 59.1in
Length: 477.5cm 188.0in
Width: 174.5cm 68.7in
Height: 137.0cm 53.9in
Ground clearance: 17.9cm
7.0in
Kerb weight: 1420.0kg
3127.7lb
Fuel: 68.2L 15.0gal 18.0galUS

357 Honda

Prelude 2.0 Si
1987 Japan
127.0mph 204.3kmh
0-50mph 80.5kmh: 6.7secs
0-60mph 96.5kmh: 9.3secs
0-1/4 mile: 16.8secs
135.0bhp 100.7kW 136.9PS
@ 6200rpm
127.0lbft 172.1Nm @ 4500rpm
111.8bhp/ton 109.9bhp/tonne
68.9bhp/L 51.4kW/L 69.9PS/L
64.4ft/sec 19.6m/sec
26.4mpg 22.0mpgUS
10.7L/100km
Petrol 4-stroke piston
1958cc 119.5cu in
In-line 4 fuel injection
Compression ratio: 9.0:1
Bore: 81.0mm 3.2in
Stroke: 95.0mm 3.7in
Valve type/No: Overhead 16
Transmission: Manual
No. of forward speeds: 5
Wheels driven: Front
Springs F/R: Coil/Coil
Brake system: PA
Brakes F/R: Disc/Disc
Steering: Rack & pinion PA
Wheelbase: 256.5cm 101.0in
Track F: 148.1cm 58.3in
Track R: 147.1cm 57.9in
Length: 446.0cm 175.6in
Width: 170.9cm 67.3in
Height: 129.5cm 51.0in
Ground clearance: 14.5cm
5.7in
Kerb weight: 1228.1kg
2705.0lb
Fuel: 60.2L 13.2gal 15.9galUS

358 Honda

Prelude 2.0i 16
1987 Japan
128.0mph 206.0kmh
0-50mph 80.5kmh: 6.2secs
0-60mph 96.5kmh: 8.5secs
0-1/4 mile: 16.4secs
0-1km: 30.0secs
150.0bhp 111.9kW 152.1PS
@ 6000rpm

133.0lbft 180.2Nm @ 5500rpm
133.6bhp/ton 131.3bhp/tonne
76.6bhp/L 57.1kW/L 77.7PS/L
62.3ft/sec 19.0m/sec
20.7mph 33.3kmh/1000rpm
21.0mpg 17.5mpgUS
13.5L/100km
Petrol 4-stroke piston
1958cc 119.0cu in
In-line 4 fuel injection
Compression ratio: 10.5:1
Bore: 81.0mm 3.2in
Stroke: 95.0mm 3.7in
Valve type/No: Overhead 16
Transmission: Manual
No. of forward speeds: 5
Wheels driven: Front
Springs F/R: Coil/Coil
Brake system: PA
Brakes F/R: Disc/Disc
Steering: Rack & pinion PA
Wheelbase: 256.5cm 101.0in
Track F: 148.0cm 58.3in
Track R: 147.0cm 57.9in
Length: 446.0cm 175.6in
Width: 169.5cm 66.7in
Height: 129.5cm 51.0in
Ground clearance: 14.5cm
5.7in
Kerb weight: 1142.0kg
2515.4lb
Fuel: 59.6L 13.1gal 15.7galUS

359 Honda

Civic CRX
1988 Japan
125.0mph 201.1kmh
0-50mph 80.5kmh: 6.0secs
0-60mph 96.5kmh: 8.0secs
0-1/4 mile: 16.2secs
0-1km: 29.9secs
130.0bhp 96.9kW 131.8PS
@ 6800rpm
105.0lbft 142.3Nm @ 5700rpm
146.5bhp/ton 144.1bhp/tonne
81.8bhp/L 61.0kW/L 82.9PS/L
66.9ft/sec 20.4m/sec
19.2mph 30.9kmh/1000rpm
29.6mpg 24.6mpgUS
9.5L/100km
Petrol 4-stroke piston
1590cc 97.0cu in
In-line 4 fuel injection
Compression ratio: 9.5:1
Bore: 75.0mm 2.9in
Stroke: 90.0mm 3.5in
Valve type/No: Overhead 16
Transmission: Manual
No. of forward speeds: 5
Wheels driven: Front
Springs F/R: Coil/Coil
Brake system: PA
Brakes F/R: Disc/Disc
Steering: Rack & pinion

Wheelbase: 229.9cm 90.5in
Track F: 144.8cm 57.0in
Track R: 145.3cm 57.2in
Length: 377.4cm 148.6in
Width: 167.4cm 65.9in
Height: 127.0cm 50.0in
Ground clearance: 15.7cm
6.2in
Kerb weight: 902.1kg 1987.0lb
Fuel: 45.0L 9.9gal 11.9galUS

360 Honda

CRX Si
1988 Japan
125.0mph 201.1kmh
0-50mph 80.5kmh: 5.9secs
0-60mph 96.5kmh: 8.2secs
0-1/4 mile: 16.4secs
105.0bhp 78.3kW 106.5PS
@ 6000rpm
98.0lbft 132.8Nm @ 5000rpm
111.2bhp/ton 109.3bhp/tonne
66.0bhp/L 49.2kW/L 67.0PS/L
59.0ft/sec 18.0m/sec
37.2mpg 31.0mpgUS
7.6L/100km
Petrol 4-stroke piston
1590cc 97.0cu in
In-line 4 fuel injection
Compression ratio: 9.1:1
Bore: 75.0mm 2.9in
Stroke: 90.0mm 3.5in
Valve type/No: Overhead 16
Transmission: Manual
No. of forward speeds: 5
Wheels driven: Front
Springs F/R: Coil/Coil
Brake system: PA
Brakes F/R: Disc/Drum
Steering: Rack & pinion PA
Wheelbase: 230.1cm 90.6in
Track F: 145.0cm 57.1in
Track R: 145.0cm 57.1in
Length: 375.4cm 147.8in
Width: 166.9cm 65.7in
Height: 127.0cm 50.0in
Ground clearance: 16.5cm
6.5in
Kerb weight: 960.2kg 2115.0lb
Fuel: 45.0L 9.9gal 11.9galUS

361 Honda

CRX Si Jackson
1988 Japan
125.0mph 201.1kmh
0-60mph 96.5kmh: 8.0secs
0-1/4 mile: 16.2secs
131.0bhp 97.7kW 132.8PS
@ 6500rpm
134.9bhp/ton 132.7bhp/tonne
82.4bhp/L 61.4kW/L 83.5PS/L
63.9ft/sec 19.5m/sec
34.8mpg 29.0mpgUS
8.1L/100km

Petrol 4-stroke piston
1590cc 97.0cu in
In-line 4 fuel injection
Compression ratio: 9.1:1
Bore: 75.0mm 2.9in
Stroke: 90.0mm 3.5in
Valve type/No: Overhead 16
Transmission: Manual
No. of forward speeds: 5
Wheels driven: Front
Springs F/R: Coil/Coil
Brake system: PA
Brakes F/R: Disc/Drum
Steering: Rack & pinion
Wheelbase: 230.1cm 90.6in
Track F: 145.0cm 57.1in
Track R: 145.0cm 57.1in
Length: 375.4cm 147.8in
Width: 166.9cm 65.7in
Height: 127.0cm 50.0in
Kerb weight: 987.4kg 2175.0lb
Fuel: 45.0L 9.9gal
11.9galUS

362 Honda

Legend Saloon Automatic
1988 Japan
128.0mph 206.0kmh
0-50mph 80.5kmh: 6.5secs
0-60mph 96.5kmh: 8.6secs
0-1/4 mile: 17.1secs
0-1km: 30.9secs
177.0bhp 132.0kW 179.4PS
@ 6000rpm
165.3lbft 224.0Nm @ 4500rpm
127.2bhp/ton 125.1bhp/tonne
66.2bhp/L 49.3kW/L 67.1PS/L
49.2ft/sec 15.0m/sec
21.1mph 33.9kmh/1000rpm
18.4mpg 15.3mpgUS
15.4L/100km
Petrol 4-stroke piston
2675cc 163.0cu in
Vee 6 fuel injection
Compression ratio: 9.0:1
Bore: 87.0mm 3.4in
Stroke: 75.0mm 2.9in
Valve type/No: Overhead 24
Transmission: Automatic
No. of forward speeds: 4
Wheels driven: Front
Springs F/R: Coil/Coil
Brake system: PA ABS
Brakes F/R: Disc/Disc
Steering: Rack & pinion PA
Wheelbase: 276.0cm 108.7in
Track F: 149.0cm 58.7in
Track R: 145.0cm 57.1in
Length: 481.0cm 189.4in
Width: 173.5cm 68.3in
Height: 139.0cm 54.7in
Ground clearance: 15.0cm
.9in
Kerb weight: 1415.0kg

3116.7lb
Fuel: 68.2L 15.0gal 18.0galUS

363 Honda

Shuttle 1.6i RT4 4WD
1988 Japan
107.0mph 172.2kmh
0-50mph 80.5kmh: 6.8secs
0-60mph 96.5kmh: 10.0secs
0-1/4 mile: 17.1secs
0-1km: 32.3secs
116.0bhp 86.5kW 117.6PS
@ 6300rpm
104.1lbft 141.0Nm @ 5300rpm
107.7bhp/ton 105.9bhp/tonne
73.0bhp/L 54.4kW/L 74.0PS/L
61.9ft/sec 18.9m/sec
19.2mph 30.9kmh/1000rpm
22.9mpg 19.1mpgUS
12.3L/100km
Petrol 4-stroke piston
1590cc 97.0cu in
In-line 4 fuel injection
Compression ratio: 9.1:1
Bore: 75.0mm 2.9in
Stroke: 90.0mm 3.5in
Valve type/No: Overhead 16
Transmission: Manual
No. of forward speeds: 5
Wheels driven: 4-wheel drive
Springs F/R: Coil/Coil
Brake system: PA
Brakes F/R: Disc/Drum
Steering: Rack & pinion PA
Wheelbase: 250.0cm 98.4in
Track F: 144.0cm 56.7in
Track R: 145.0cm 57.1in
Length: 410.5cm 161.6in
Width: 169.0cm 66.5in
Height: 151.5cm 59.6in
Ground clearance: 19.0cm
7.5in
Kerb weight: 1095.0kg
2411.9lb
Fuel: 45.0L 9.9gal 11.9galUS

364 Honda

Civic Si
1989 Japan
115.0mph 185.0kmh
0-50mph 80.5kmh: 6.6secs
0-60mph 96.5kmh: 9.4secs
0-1/4 mile: 17.0secs
108.0bhp 80.5kW 109.5PS
@ 6000rpm
100.0lbft 135.5Nm @ 5000rpm
107.0bhp/ton 105.3bhp/tonne
67.9bhp/L 50.6kW/L 68.9PS/L
59.0ft/sec 18.0m/sec
34.2mpg 28.5mpgUS
8.3L/100km
Petrol 4-stroke piston
1590cc 97.0cu in
In-line 4 fuel injection

Compression ratio: 9.1:1
Bore: 75.0mm 2.9in
Stroke: 90.0mm 3.5in
Valve type/No: Overhead 16
Transmission: Manual
No. of forward speeds: 5
Wheels driven: Front
Springs F/R: Coil/Coil
Brake system: PA
Brakes F/R: Disc/Drum
Steering: Rack & pinion
Wheelbase: 249.9cm 98.4in
Track F: 145.0cm 57.1in
Track R: 145.5cm 57.3in
Length: 396.5cm 156.1in
Width: 166.6cm 65.6in
Height: 133.4cm 52.5in
Kerb weight: 1026.0kg
2260.0lb
Fuel: 45.0L 9.9gal 11.9galUS

365 Honda

Prelude Si 4WS
1989 Japan
127.0mph 204.3kmh
0-50mph 80.5kmh: 6.8secs
0-60mph 96.5kmh: 9.4secs
0-1/4 mile: 17.0secs
135.0bhp 100.7kW 136.9PS
@ 6200rpm
127.0lbft 172.1Nm @ 4500rpm
111.8bhp/ton 109.9bhp/tonne
68.9bhp/L 51.4kW/L 69.9PS/L
64.4ft/sec 19.6m/sec
24.6mpg 20.5mpgUS
11.5L/100km
Petrol 4-stroke piston
1958cc 119.5cu in
In-line 4 fuel injection
Compression ratio: 9.0:1
Bore: 81.0mm 3.2in
Stroke: 95.0mm 3.7in
Valve type/No: Overhead 16
Transmission: Manual
No. of forward speeds: 5
Wheels driven: Front
Springs F/R: Coil/Coil
Brake system: PA
Brakes F/R: Disc/Disc
Steering: Rack & pinion PA
Wheelbase: 256.5cm 101.0in
Track F: 148.1cm 58.3in
Track R: 147.1cm 57.9in
Length: 446.0cm 175.6in
Width: 169.4cm 66.7in
Height: 129.5cm 51.0in
Kerb weight: 1228.1kg
2705.0lb
Fuel: 60.2L 13.2gal 15.9galUS

366 Honda

Shuttle 1.4 GL Automatic
1989 Japan
100.0mph 160.9kmh

Compression ratio: 9.1:1
Bore: 75.0mm 2.9in
Stroke: 90.0mm 3.5in
Valve type/No: Overhead 16
Transmission: Manual
No. of forward speeds: 5
Wheels driven: Front
Springs F/R: Coil/Coil
Brake system: PA
Brakes F/R: Disc/Drum
Steering: Rack & pinion
Wheelbase: 249.9cm 98.4in
Track F: 145.0cm 57.1in
Track R: 145.5cm 57.3in
Length: 396.5cm 156.1in
Width: 166.6cm 65.6in
Height: 133.4cm 52.5in
Kerb weight: 1026.0kg
2260.0lb
Fuel: 45.0L 9.9gal 11.9galUS

0-50mph 80.5kmh: 8.8secs
0-60mph 96.5kmh: 12.2secs
0-1/4 mile: 19.1secs
0-1km: 35.4secs
90.0bhp 67.1kW 91.2PS
@ 6300rpm
82.7lbft 112.0Nm @ 4500rpm
93.3bhp/ton 91.7bhp/tonne
64.5bhp/L 48.1kW/L 65.4PS/L
54.4ft/sec 16.6m/sec
20.9mph 33.6kmh/1000rpm
23.3mpg 19.4mpgUS
12.1L/100km
Petrol 4-stroke piston
1396cc 85.0cu in
In-line 4 1 Carburettor
Compression ratio: 9.3:1
Bore: 75.0mm 2.9in
Stroke: 79.0mm 3.1in
Valve type/No: Overhead 16
Transmission: Automatic
No. of forward speeds: 4
Wheels driven: Front
Springs F/R: Coil/Coil
Brake system: PA
Brakes F/R: Disc/Drum
Steering: Rack & pinion PA
Wheelbase: 250.0cm 98.4in
Track F: 144.0cm 56.7in
Track R: 145.0cm 57.1in
Length: 410.5cm 161.6in
Width: 169.0cm 66.5in
Height: 151.5cm 59.6in
Ground clearance: 19.0cm
7.5in
Kerb weight: 981.0kg 2160.8lb
Fuel: 45.0L 9.9gal 11.9galUS

367 Honda

Accord 2.0 EXi
1990 Japan
125.0mph 201.1kmh
0-50mph 80.5kmh: 7.1secs
0-60mph 96.5kmh: 9.9secs
0-1/4 mile: 17.5secs
0-1km: 31.9secs
133.0bhp 99.2kW 134.8PS
@ 5300rpm
132.1lbft 179.0Nm @ 5000rpm
106.5bhp/ton 104.7bhp/tonne
66.6bhp/L 49.7kW/L 67.5PS/L
50.9ft/sec 15.5m/sec
22.7mph 36.5kmh/1000rpm
26.1mpg 21.7mpgUS
10.8L/100km
Petrol 4-stroke piston
1997cc 122.0cu in
In-line 4 fuel injection
Compression ratio: 9.5:1
Bore: 85.0mm 3.3in
Stroke: 88.0mm 3.5in
Valve type/No: Overhead 16
Transmission: Manual
No. of forward speeds: 5

372 Honda
Legend

Wheels driven: Front
Springs F/R: Coil/Coil
Brake system: PA ABS
Brakes F/R: Disc/Drum
Steering: Rack & pinion PA
Wheelbase: 272.0cm 107.1in
Track F: 147.5cm 58.1in
Track R: 148.0cm 58.3in
Length: 468.5cm 184.4in
Width: 169.5cm 66.7in
Height: 139.0cm 54.7in
Kerb weight: 1270.0kg 2797.4lb
Fuel: 65.1L 14.3gal 17.2galUS

368 Honda
Concerto 1.6i 16
1990 Japan
121.0mph 194.7kmh
0-50mph 80.5kmh: 6.8secs
0-60mph 96.5kmh: 9.3secs
0-1/4 mile: 17.2secs
0-1km: 31.6secs
130.0bhp 96.9kW 131.8PS @ 6800rpm
105.5lbft 143.0Nm @ 5700rpm
116.0bhp/ton 114.0bhp/tonne
81.8bhp/L 61.0kW/L 82.9PS/L
66.9ft/sec 20.4m/sec
18.1mph 29.1kmh/1000rpm
25.7mpg 21.4mpgUS
11.0L/100km
Petrol 4-stroke piston
1590cc 97.0cu in
In-line 4 fuel injection
Compression ratio: 9.5:1
Bore: 75.0mm 2.9in
Stroke: 90.0mm 3.5in
Valve type/No: Overhead 16
Transmission: Manual
No. of forward speeds: 5
Wheels driven: Front
Springs F/R: Coil/Coil

Brake system: PA ABS
Brakes F/R: Disc/Disc
Steering: Rack & pinion PA
Wheelbase: 255.0cm 100.4in
Track F: 147.5cm 58.1in
Track R: 147.0cm 57.9in
Length: 426.5cm 167.9in
Width: 194.0cm 76.4in
Height: 140.0cm 55.1in
Ground clearance: 15.0cm 5.9in
Kerb weight: 1140.0kg 2511.0lb
Fuel: 55.1L 12.1gal 14.5galUS

369 Honda
CRX 1.6i VT
1990 Japan
132.0mph 212.4kmh
0-50mph 80.5kmh: 6.2secs
0-60mph 96.5kmh: 8.0secs
0-1/4 mile: 16.4secs
0-1km: 29.6secs
150.0bhp 111.9kW 152.1PS @ 7600rpm
106.3lbft 144.0Nm @ 7100rpm
148.8bhp/ton 146.3bhp/tonne
94.0bhp/L 70.1kW/L 95.3PS/L
64.0ft/sec 19.5m/sec
18.7mph 30.1kmh/1000rpm
28.1mpg 23.4mpgUS
10.1L/100km
Petrol 4-stroke piston
1595cc 97.0cu in
In-line 4 fuel injection
Compression ratio: 10.2:1
Bore: 81.0mm 3.2in
Stroke: 77.0mm 3.0in
Valve type/No: Overhead 16
Transmission: Manual
No. of forward speeds: 5
Wheels driven: Front
Springs F/R: Coil/Coil

Brake system: PA ABS
Brakes F/R: Disc/Disc
Steering: Rack & pinion
Wheelbase: 229.9cm 90.5in
Track F: 145.3cm 57.2in
Track R: 145.3cm 57.2in
Length: 377.4cm 148.6in
Width: 167.4cm 65.9in
Height: 127.0cm 50.0in
Kerb weight: 1025.0kg 2257.7lb
Fuel: 45.0L 9.9gal 11.9galUS

370 Honda
Civic 1.5 VEi
1991 Japan
112.0mph 180.2kmh
0-50mph 80.5kmh: 7.6secs
0-60mph 96.5kmh: 11.2secs
0-1/4 mile: 17.8secs
0-1km: 33.3secs
90.0bhp 67.1kW 91.2PS @ 5500rpm
95.0lbft 128.7Nm @ 4500rpm
94.8bhp/ton 93.2bhp/tonne
60.3bhp/L 44.9kW/L 61.1PS/L
50.9ft/sec 15.5m/sec
23.6mph 38.0kmh/1000rpm
31.9mpg 26.6mpgUS
8.9L/100km
Petrol 4-stroke piston
1493cc 91.1cu in
In-line 4 fuel injection
Compression ratio: 9.3:1
Bore: 75.0mm 2.9in
Stroke: 84.5mm 3.3in
Valve type/No: Overhead 16
Transmission: Manual
No. of forward speeds: 5
Wheels driven: Front
Springs F/R: Coil/Coil
Brake system: PA
Brakes F/R: Disc/Drum

Steering: Rack & pinion PA
Wheelbase: 257.0cm 101.2in
Track F: 147.3cm 58.0in
Track R: 146.6cm 57.7in
Length: 407.9cm 160.6in
Width: 169.4cm 66.7in
Height: 134.6cm 53.0in
Kerb weight: 965.2kg 2126.0lb
Fuel: 45.0L 9.9gal 11.9galUS

371 Honda
Civic 1.6i VT
1991 Japan
128.0mph 206.0kmh
0-50mph 80.5kmh: 5.9secs
0-60mph 96.5kmh: 7.6secs
0-1/4 mile: 16.1secs
0-1km: 29.3secs
150.0bhp 111.9kW 152.1PS @ 7600rpm
106.3lbft 144.0Nm @ 7100rpm
146.7bhp/ton 144.2bhp/tonne
94.0bhp/L 70.1kW/L 95.3PS/L
64.4ft/sec 19.6m/sec
18.7mph 30.1kmh/1000rpm
30.0mpg 25.0mpgUS
9.4L/100km
Petrol 4-stroke piston
1595cc 97.0cu in
In-line 4 fuel injection
Compression ratio: 10.2:1
Bore: 81.0mm 3.2in
Stroke: 77.4mm 3.0in
Valve type/No: Overhead 16
Transmission: Manual
No. of forward speeds: 5
Wheels driven: Front
Springs F/R: Coil/Coil
Brake system: PA
Brakes F/R: Disc/Disc
Steering: Rack & pinion
Wheelbase: 249.9cm 98.4in
Track F: 145.0cm 57.1in

Track R: 145.5cm 57.3in
Length: 396.5cm 156.1in
Width: 167.9cm 66.1in
Height: 133.1cm 52.4in
Ground clearance: 16.0cm
6.3in
Kerb weight: 1040.0kg
2290.7lb
Fuel: 45.0L 9.9gal 11.9galUS

372 Honda
Legend
1991 Japan
142.0mph 228.5kmh
0-50mph 80.5kmh: 6.3secs
0-60mph 96.5kmh: 8.2secs
0-1/4 mile: 16.5secs
0-1km: 29.6secs
205.0bhp 152.9kW 207.8PS
@ 5500rpm
216.2lbft 293.0Nm @ 4400rpm
129.5bhp/ton 127.3bhp/tonne
63.9bhp/L 47.7kW/L 64.8PS/L
50.6ft/sec 15.4m/sec
25.6mph 41.2kmh/1000rpm
19.1mpg 15.9mpgUS
14.8L/100km
Petrol 4-stroke piston
3206cc 196.0cu in
Vee 6 fuel injection
Compression ratio: 9.6:1
Bore: 90.0mm 3.5in
Stroke: 84.0mm 3.3in
Valve type/No: Overhead 24
Transmission: Automatic
No. of forward speeds: 4
Wheels driven: Front
Springs F/R: Coil/Coil
Brake system: PA ABS
Brakes F/R: Disc/Disc
Steering: Rack & pinion PA
Wheelbase: 290.8cm 114.5in

Track F: 154.9cm 61.0in
Track R: 153.9cm 60.6in
Length: 494.8cm 194.8in
Width: 180.8cm 71.2in
Height: 141.0cm 55.5in
Ground clearance: 15.0cm
5.9in
Kerb weight: 1610.0kg
3546.3lb
Fuel: 68.0L 14.9gal 18.0galUS

373 Honda
NSX
1991 Japan
162.0mph 260.7kmh
0-50mph 80.5kmh: 4.4secs
0-60mph 96.5kmh: 5.8secs
0-1/4 mile: 14.2secs
0-1km: 25.5secs
270.0bhp 201.3kW 273.7PS
@ 7100rpm
210.3lbft 285.0Nm @ 3500rpm
204.9bhp/ton 201.5bhp/tonne
90.7bhp/L 67.6kW/L 91.9PS/L
60.5ft/sec 18.5m/sec
23.0mph 37.0kmh/1000rpm
19.6mpg 16.3mpgUS
14.4L/100km
Petrol 4-stroke piston
2977cc 182.0cu in
Vee 6 fuel injection
Compression ratio: 10.2:1
Bore: 90.0mm 3.5in
Stroke: 78.0mm 3.1in
Valve type/No: Overhead 24
Transmission: Manual
No. of forward speeds: 5
Wheels driven: Rear
Springs F/R: Coil/Coil
Brake system: PA ABS
Brakes F/R: Disc/Disc
Steering: Rack & pinion

Wheelbase: 253.0cm 99.6in
Track F: 150.9cm 59.4in
Track R: 152.9cm 60.2in
Length: 440.4cm 173.4in
Width: 181.1cm 71.3in
Height: 117.1cm 46.1in
Ground clearance: 13.5cm
5.3in
Kerb weight: 1340.0kg
2951.5lb
Fuel: 70.1L 15.4gal 18.5galUS

374 Honda
Prelude Si
1991 Japan
130.0mph 209.2kmh
0-50mph 80.5kmh: 6.3secs
0-60mph 96.5kmh: 8.8secs
0-1/4 mile: 16.4secs
140.0bhp 104.4kW 141.9PS
@ 5800rpm
135.0lbft 182.9Nm @ 5000rpm
115.9bhp/ton 114.0bhp/tonne
68.1bhp/L 50.8kW/L 69.0PS/L
60.3ft/sec 18.4m/sec
28.8mpg 24.0mpgUS
9.8L/100km
Petrol 4-stroke piston
2056cc 125.4cu in
In-line 4 fuel injection
Compression ratio: 9.4:1
Bore: 83.0mm 3.3in
Stroke: 95.0mm 3.7in
Valve type/No: Overhead 16
Transmission: Manual
No. of forward speeds: 5
Wheels driven: Front
Springs F/R: Coil/Coil
Brake system: PA ABS
Brakes F/R: Disc/Disc
Steering: Rack & pinion PA
Wheelbase: 256.5cm 101.0in

Track F: 148.1cm 58.3in
Track R: 147.1cm 57.9in
Length: 451.1cm 177.6in
Width: 170.9cm 67.3in
Height: 129.5cm 51.0in
Kerb weight: 1228.1kg
2705.0lb
Fuel: 60.2L 13.2gal 15.9galUS

375 Honda
Legend Coupe
1992 Japan
143.0mph 230.1kmh
0-50mph 80.5kmh: 6.2secs
0-60mph 96.5kmh: 8.1secs
0-1/4 mile: 16.3secs
0-1km: 29.3secs
205.0bhp 152.9kW 207.8PS
@ 5500rpm
216.0lbft 292.7Nm @ 4400rpm
133.6bhp/ton 131.4bhp/tonne
63.9bhp/L 47.7kW/L 64.8PS/L
50.6ft/sec 15.4m/sec
25.6mph 41.2kmh/1000rpm
21.5mpg 17.9mpgUS
13.1L/100km
Petrol 4-stroke piston
3206cc 195.6cu in
Vee 6 fuel injection
Compression ratio: 9.6:1
Bore: 90.0mm 3.5in
Stroke: 84.0mm 3.3in
Valve type/No: Overhead 24
Transmission: Automatic
No. of forward speeds: 4
Wheels driven: Front
Springs F/R: Coil/Coil
Brake system: PA
Brakes F/R: Disc/Disc
Steering: Rack & pinion PA
Wheelbase: 283.0cm 111.4in
Track F: 154.9cm 61.0in

373 Honda
NSX

Track R: 153.9cm 60.6in
Length: 488.4cm 192.3in
Width: 180.3cm 71.2in
Height: 136.9cm 53.9in
Ground clearance: 15.0cm
5.9in
Kerb weight: 1559.9kg
3436.0lb
Fuel: 68.2L 15.0gal 18.0galUS

376 Honda
NSX Auto
1992 Japan
158.0mph 254.2kmh
0-50mph 80.5kmh: 5.1secs
0-60mph 96.5kmh: 6.8secs
0-1/4 mile: 14.9secs
0-1km: 29.6secs
255.0bhp 190.1kW 258.5PS
@ 6800rpm
209.6lbft 284.0Nm @ 5400rpm
189.3bhp/ton 186.1bhp/tonne
85.7bhp/L 63.9kW/L 86.8PS/L
58.0ft/sec 17.7m/sec
22.8mph 36.7kmh/1000rpm
18.3mpg 15.2mpgUS
15.4L/100km
Petrol 4-stroke piston
2977cc 182.0cu in
Vee 6 fuel injection
Compression ratio: 10.2:1
Bore: 90.0mm 3.5in
Stroke: 78.0mm 3.1in
Valve type/No: Overhead 24
Transmission: Automatic
No. of forward speeds: 4
Wheels driven: Rear
Springs F/R: Coil/Coil
Brake system: PA ABS
Brakes F/R: Disc/Disc
Steering: Rack & pinion PA
Wheelbase: 253.0cm 99.6in
Track F: 151.0cm 59.4in
Track R: 153.0cm 60.2in
Length: 440.5cm 173.4in
Width: 181.0cm 71.3in
Height: 117.0cm 46.1in
Ground clearance: 13.4cm
5.3in
Kerb weight: 1370.0kg
3017.6lb
Fuel: 70.0L 15.4gal 18.5galUS

377 Honda
Prelude Si 4WS
1992 Japan
126.0mph 202.7kmh
0-50mph 80.5kmh: 5.8secs
0-60mph 96.5kmh: 7.9secs
0-1/4 mile: 16.0secs
160.0bhp 119.3kW 162.2PS
@ 5800rpm
156.0lbft 211.4Nm @ 4500rpm
123.6bhp/ton 121.5bhp/tonne

70.8bhp/L 52.8kW/L 71.8PS/L
60.3ft/sec 18.4m/sec
28.8mpg 24.0mpgUS
9.8L/100km
Petrol 4-stroke piston
2259cc 137.8cu in
In-line 4 fuel injection
Compression ratio: 9.8:1
Bore: 87.0mm 3.4in
Stroke: 95.0mm 3.7in
Valve type/No: Overhead 16
Transmission: Manual
No. of forward speeds: 5
Wheels driven: Front
Springs F/R: Coil/Coil
Brake system: PA ABS
Brakes F/R: Disc/Disc
Steering: Rack & pinion PA
Wheelbase: 255.0cm 100.4in
Track F: 152.4cm 60.0in
Track R: 151.4cm 59.6in
Length: 444.0cm 174.8in
Width: 176.5cm 69.5in
Height: 129.0cm 50.8in
Ground clearance: 15.0cm
5.9in
Kerb weight: 1316.6kg
2900.0lb
Fuel: 60.2L 13.2gal 15.9galUS

378 Hyundai
Excel GLS
1986 Korea
92.0mph 148.0kmh
0-60mph 96.5kmh: 12.9secs
0-1/4 mile: 19.0secs
68.0bhp 50.7kW 68.9PS
@ 5500rpm
82.0lbft 111.1Nm @ 3500rpm
68.3bhp/ton 67.2bhp/tonne
46.3bhp/L 34.5kW/L 47.0PS/L
49.3ft/sec 15.0m/sec
37.8mpg 31.5mpgUS
7.5L/100km
Petrol 4-stroke piston
1468cc 89.6cu in
In-line 4 1 Carburettor
Compression ratio: 9.5:1
Bore: 75.5mm 3.0in
Stroke: 82.0mm 3.2in
Valve type/No: Overhead 8
Transmission: Manual
No. of forward speeds: 5
Wheels driven: Front
Springs F/R: Coil/Coil
Brake system: PA
Brakes F/R: Disc/Drum
Steering: Rack & pinion
Wheelbase: 238.0cm 93.7in
Track F: 137.4cm 54.1in
Track R: 134.1cm 52.8in
Length: 408.7cm 160.9in
Width: 160.3cm 63.1in
Height: 137.4cm 54.1in

Kerb weight: 1012.4kg
2230.0lb
Fuel: 40.1L 8.8gal 10.6galUS

379 Hyundai
Sonata 2.0i GLS
1989 Korea
108.0mph 173.8kmh
0-50mph 80.5kmh: 8.5secs
0-60mph 96.5kmh: 11.7secs
0-1/4 mile: 18.9secs
0-1km: 34.5secs
100.0bhp 74.6kW 101.4PS
@ 5000rpm
121.0lbft 164.0Nm @ 3500rpm
83.2bhp/ton 81.8bhp/tonne
50.1bhp/L 37.3kW/L 50.8PS/L
48.1ft/sec 14.7m/sec
22.3mph 35.9kmh/1000rpm
23.0mpg 19.2mpgUS
12.3L/100km
Petrol 4-stroke piston
1997cc 122.0cu in
In-line 4 fuel injection
Compression ratio: 8.6:1
Bore: 85.0mm 3.3in
Stroke: 88.0mm 3.5in
Valve type/No: Overhead 8
Transmission: Manual
No. of forward speeds: 5
Wheels driven: Front
Springs F/R: Coil/Coil
Brake system: PA
Brakes F/R: Disc/Drum
Steering: Rack & pinion PA
Wheelbase: 265.0cm 104.3in
Track F: 145.5cm 57.3in
Track R: 144.0cm 56.7in
Length: 468.0cm 184.3in
Width: 175.1cm 68.9in
Height: 141.1cm 55.6in
Ground clearance: 16.0cm
6.3in
Kerb weight: 1222.0kg
2691.6lb
Fuel: 60.1L 13.2gal 15.9galUS

380 Hyundai
Sonata 2.4i GLS
1990 Korea
115.0mph 185.0kmh
0-50mph 80.5kmh: 9.1secs
0-60mph 96.5kmh: 12.3secs
0-1/4 mile: 19.0secs
0-1km: 34.5secs
115.0bhp 85.8kW 116.6PS
@ 4500rpm
141.7lbft 192.0Nm @ 3500rpm
90.0bhp/ton 88.5bhp/tonne
48.9bhp/L 36.5kW/L 49.6PS/L
49.2ft/sec 15.0m/sec
26.1mph 42.0kmh/1000rpm
22.7mpg 18.9mpgUS
12.4L/100km

Petrol 4-stroke piston
2351cc 143.0cu in
In-line 4 fuel injection
Compression ratio: 8.6:1
Bore: 86.5mm 3.4in
Stroke: 100.0mm 3.9in
Valve type/No: Overhead 8
Transmission: Automatic
No. of forward speeds: 4
Wheels driven: Front
Springs F/R: Coil/Coil
Brake system: PA
Brakes F/R: Disc/Drum
Steering: Rack & pinion PA
Wheelbase: 265.0cm 104.3in
Track F: 145.5cm 57.3in
Track R: 144.0cm 56.7in
Length: 468.0cm 184.3in
Width: 175.1cm 68.9in
Height: 141.1cm 55.6in
Ground clearance: 16.0cm
6.3in
Kerb weight: 1300.0kg
2863.4lb
Fuel: 60.1L 13.2gal 15.9galUS

381 Hyundai
Lantra 1.6 Cdi
1991 Korea
114.0mph 183.4kmh
0-50mph 80.5kmh: 7.4secs
0-60mph 96.5kmh: 10.9secs
0-1/4 mile: 18.1secs
0-1km: 33.4secs
112.0bhp 83.5kW 113.5PS
@ 6200rpm
103.3lbft 140.0Nm @ 4500rpm
102.6bhp/ton 100.9bhp/tonne
70.2bhp/L 52.3kW/L 71.1PS/L
50.8ft/sec 15.5m/sec
18.9mph 30.4kmh/1000rpm
27.1mpg 22.6mpgUS
10.4L/100km
Petrol 4-stroke piston
1596cc 97.0cu in
In-line 4 fuel injection
Compression ratio: 9.2:1
Bore: 82.0mm 3.2in
Stroke: 75.0mm 2.9in
Valve type/No: Overhead 16
Transmission: Manual
No. of forward speeds: 5
Wheels driven: Front
Springs F/R: Coil/Coil
Brake system: PA
Brakes F/R: Disc/Drum
Steering: Rack & pinion PA
Wheelbase: 250.2cm 98.5in
Track F: 143.0cm 56.3in
Track R: 143.0cm 56.3in
Length: 435.9cm 171.6in
Width: 166.9cm 65.7in
Height: 138.4cm 54.5in
Ground clearance: 16.0cm

6.3in
Kerb weight: 1110.0kg
2444.9lb
Fuel: 52.0L 11.4gal 13.7galUS

382 Hyundai
S Coupe GSi
1991 Korea
105.0mph 168.9kmh
0-50mph 80.5kmh: 8.9secs
0-60mph 96.5kmh: 12.5secs
0-1/4 mile: 18.5secs
0-1km: 34.7secs
82.0bhp 61.1kW 83.1PS
@ 5500rpm
87.8lbft 119.0Nm @ 4000rpm
85.1bhp/ton 83.7bhp/tonne
55.9bhp/L 41.7kW/L 56.6PS/L
49.3ft/sec 15.0m/sec
19.4mph 31.2kmh/1000rpm
30.8mpg 25.6mpgUS
9.2L/100km
Petrol 4-stroke piston
1468cc 90.0cu in
In-line 4 fuel injection
Compression ratio: 9.4:1
Bore: 75.5mm 3.0in
Stroke: 82.0mm 3.2in
Valve type/No: Overhead 8
Transmission: Manual
No. of forward speeds: 5
Wheels driven: Front
Springs F/R: Coil/Coil
Brake system: PA
Brakes F/R: Disc/Drum
Steering: Rack & pinion PA
Wheelbase: 238.3cm 93.8in
Track F: 139.0cm 54.7in
Track R: 134.0cm 52.8in
Length: 421.3cm 165.9in
Width: 162.6cm 64.0in
Height: 132.8cm 52.3in
Ground clearance: 16.6cm
6.5in
Kerb weight: 980.0kg 2158.6lb
Fuel: 45.0L 9.9gal 11.9galUS

383 Hyundai
X2 1.5 GSi
1991 Korea
100.0mph 160.9kmh
0-50mph 80.5kmh: 9.4secs
0-60mph 96.5kmh: 13.7secs
0-1/4 mile: 19.1secs
0-1km: 36.4secs
83.0bhp 61.9kW 84.1PS
@ 5500rpm
88.6lbft 120.0Nm @ 4000rpm
83.2bhp/ton 81.8bhp/tonne
56.5bhp/L 42.2kW/L 57.3PS/L
49.3ft/sec 15.0m/sec
21.0mph 33.8kmh/1000rpm
28.0mpg 23.3mpgUS
10.1L/100km

Petrol 4-stroke piston
1468cc 90.0cu in
In-line 4 fuel injection
Compression ratio: 9.4:1
Bore: 76.0mm 3.0in
Stroke: 82.0mm 3.2in
Valve type/No: Overhead 8
Transmission: Manual
No. of forward speeds: 5
Wheels driven: Front
Springs F/R: Coil/Coil
Brake system: PA
Brakes F/R: Disc/Drum
Steering: Rack & pinion PA
Wheelbase: 238.3cm 93.8in
Track F: 138.9cm 54.7in
Track R: 133.9cm 52.7in
Length: 410.0cm 161.4in
Width: 160.5cm 63.2in
Height: 137.7cm 54.2in
Kerb weight: 1015.0kg
2235.7lb
Fuel: 45.0L 9.9gal 11.9galUS

384 IAD
Venus
1990 UK
165.0mph 265.5kmh
0-60mph 96.5kmh: 5.0secs
172.0bhp 128.3kW 174.4PS
@ 6000rpm
163.0lbft 220.9Nm @ 5000rpm
79.1bhp/L 59.0kW/L 80.2PS/L
50.0ft/sec 15.2m/sec
Petrol 4-stroke piston
2174cc 132.6cu in
In-line 4 2 Carburettor
Compression ratio: 10.9:1
Bore: 95.3mm 3.8in
Stroke: 76.2mm 3.0in
Valve type/No: Overhead 16
Transmission: Manual
No. of forward speeds: 5
Wheels driven: Rear
Springs F/R: Coil/Coil
Brakes F/R: Disc/Disc
Steering: Rack & pinion PA
Wheelbase: 269.5cm 106.1in
Track F: 167.9cm 66.1in
Track R: 167.9cm 66.1in
Length: 398.8cm 157.0in
Width: 195.1cm 76.8in
Height: 111.0cm 43.7in

385 Infiniti
M30
1990 Japan
128.0mph 206.0kmh
0-50mph 80.5kmh: 7.5secs
0-60mph 96.5kmh: 10.4secs
0-1/4 mile: 17.5secs
162.0bhp 120.8kW 164.2PS
@ 5200rpm
180.0lbft 243.9Nm @ 3600rpm

108.8bhp/ton 107.0bhp/tonne
54.5bhp/L 40.7kW/L 55.3PS/L
47.2ft/sec 14.4m/sec
24.0mpg 20.0mpgUS
11.8L/100km
Petrol 4-stroke piston
2971cc 181.3cu in
Vee 6 fuel injection
Compression ratio: 9.0:1
Bore: 87.1mm 3.4in
Stroke: 83.1mm 3.3in
Valve type/No: Overhead 12
Transmission: Automatic
No. of forward speeds: 4
Wheels driven: Rear
Springs F/R: Coil/Coil
Brake system: PA ABS
Brakes F/R: Disc/Disc
Steering: Rack & pinion PA
Wheelbase: 261.6cm 103.0in
Track F: 143.5cm 56.5in
Track R: 143.5cm 56.5in
Length: 479.6cm 188.8in
Width: 168.9cm 66.5in
Height: 137.9cm 54.3in
Kerb weight: 1514.1kg
3335.0lb
Fuel: 65.1L 14.3gal 17.2galUS

386 Infiniti
Q45
1990 Japan
150.0mph 241.4kmh
0-60mph 96.5kmh: 6.9secs
0-1/4 mile: 15.4secs
278.0bhp 207.3kW 281.9PS
@ 6000rpm
292.0lbft 395.7Nm @ 4000rpm
155.1bhp/ton 152.5bhp/tonne
61.9bhp/L 46.1kW/L 62.7PS/L
54.3ft/sec 16.5m/sec
20.7mpg 17.2mpgUS
13.7L/100km
Petrol 4-stroke piston
4494cc 274.2cu in
Vee 8 fuel injection
Compression ratio: 10.2:1
Bore: 93.0mm 3.7in
Stroke: 82.7mm 3.3in
Valve type/No: Overhead 32
Transmission: Automatic
No. of forward speeds: 4
Wheels driven: Rear
Springs F/R: Coil/Coil
Brake system: PA ABS
Brakes F/R: Disc/Disc
Steering: Rack & pinion PA
Wheelbase: 287.5cm 113.2in
Track F: 157.0cm 61.8in
Track R: 157.0cm 61.8in
Length: 507.5cm 199.8in
Width: 182.6cm 71.9in
Height: 143.0cm 56.3in
Kerb weight: 1822.8kg

4015.0lb
Fuel: 85.2L 18.7gal 22.5galUS

387 Infiniti
G20
1991 Japan
130.0mph 209.2kmh
0-50mph 80.5kmh: 7.4secs
0-60mph 96.5kmh: 10.0secs
0-1/4 mile: 17.6secs
140.0bhp 104.4kW 141.9PS
@ 6400rpm
132.0lbft 178.9Nm @ 4800rpm
104.5bhp/ton 102.8bhp/tonne
70.1bhp/L 52.2kW/L 71.0PS/L
60.3ft/sec 18.3m/sec
28.8mpg 24.0mpgUS
9.8L/100km
Petrol 4-stroke piston
1998cc 121.9cu in
In-line 4 fuel injection
Compression ratio: 9.5:1
Bore: 86.0mm 3.4in
Stroke: 86.0mm 3.4in
Valve type/No: Overhead 16
Transmission: Manual
No. of forward speeds: 5
Wheels driven: Front
Springs F/R: Coil/Coil
Brake system: PA ABS
Brakes F/R: Disc/Disc
Steering: Rack & pinion PA
Wheelbase: 255.0cm 100.4in
Track F: 147.1cm 57.9in
Track R: 146.1cm 57.5in
Length: 444.0cm 174.8in
Width: 169.4cm 66.7in
Height: 138.9cm 54.7in
Ground clearance: 15.7cm
6.2in
Kerb weight: 1362.0kg
3000.0lb
Fuel: 60.2L 13.2gal 15.9galUS

388 Irmsher
GT
1989 Germany
150.0mph 241.4kmh
0-60mph 96.5kmh: 7.9secs
200.0bhp 149.1kW 202.8PS
@ 5200rpm
210.0lbft 284.6Nm @ 4200rpm
153.1bhp/ton 150.6bhp/tonne
55.7bhp/L 41.5kW/L 56.5PS/L
48.4ft/sec 14.7m/sec
Petrol 4-stroke piston
3590cc 219.0cu in
In-line 6 fuel injection
Compression ratio: 9.4:1
Bore: 95.0mm 3.7in
Stroke: 85.0mm 3.3in
Valve type/No: Overhead 12
Transmission: Manual
No. of forward speeds: 5

Wheels driven: Rear
Springs F/R: Coil/Coil
Brake system: PA ABS
Brakes F/R: Disc/Disc
Steering: Rack & pinion PA
Wheelbase: 273.1cm 107.5in
Track F: 145.0cm 57.1in
Track R: 146.8cm 57.8in
Length: 459.0cm 180.7in
Width: 177.8cm 70.0in
Height: 133.9cm 52.7in
Kerb weight: 1328.4kg
2926.0lb
Fuel: 74.9L 16.5gal 19.8galUS

389 Isdera
Imperator
1988 Germany
176.0mph 283.2kmh
0-60mph 96.5kmh: 5.0secs
0-1/4 mile: 13.3secs
390.0bhp 290.8kW 395.4PS
@ 5500rpm
387.0lbft 524.4Nm @ 4100rpm
317.1bhp/ton 311.8bhp/tonne
70.3bhp/L 52.4kW/L 71.3PS/L
57.0ft/sec 17.4m/sec
Petrol 4-stroke piston
5547cc 338.4cu in
Vee 8 fuel injection
Compression ratio: 9.8:1
Bore: 96.5mm 3.8in
Stroke: 94.8mm 3.7in
Valve type/No: Overhead 32
Transmission: Manual
No. of forward speeds: 5
Wheels driven: Rear
Springs F/R: Coil/Coil
Brake system: PA
Brakes F/R: Disc/Disc
Steering: Rack & pinion PA
Wheelbase: 240.0cm 94.5in

Track F: 147.1cm 57.9in
Track R: 139.7cm 55.0in
Length: 421.6cm 166.0in
Width: 182.9cm 72.0in
Height: 113.3cm 44.6in
Kerb weight: 1250.8kg
2755.0lb
Fuel: 113.5L 25.0gal
30.0galUS

390 Isuzu
Impulse Sports Coupe
1986 Japan
100.0mph 160.9kmh
0-60mph 96.5kmh: 12.9secs
0-1/4 mile: 19.0secs
90.0bhp 67.1kW 91.2PS
@ 5000rpm
108.0lbft 146.3Nm @ 3000rpm
73.8bhp/ton 72.6bhp/tonne
46.2bhp/L 34.4kW/L 46.8PS/L
44.9ft/sec 13.7m/sec
34.8mpg 29.0mpgUS
8.1L/100km
Petrol 4-stroke piston
1949cc 118.9cu in
In-line 4 fuel injection
Compression ratio: 9.2:1
Bore: 87.0mm 3.4in
Stroke: 82.0mm 3.2in
Valve type/No: Overhead 8
Transmission: Manual
No. of forward speeds: 5
Wheels driven: Rear
Springs F/R: Coil/Coil
Brake system: PA
Brakes F/R: Disc/Disc
Steering: Rack & pinion PA
Wheelbase: 243.8cm 96.0in
Track F: 135.4cm 53.3in
Track R: 137.9cm 54.3in
Length: 438.4cm 172.6in

Width: 165.6cm 65.2in
Height: 130.6cm 51.4in
Kerb weight: 1239.4kg
2730.0lb
Fuel: 57.1L 12.6gal 15.1galUS

391 Isuzu
Impulse Turbo
1986 Japan
130.0mph 209.2kmh
0-60mph 96.5kmh: 8.5secs
0-1/4 mile: 16.4secs
140.0bhp 104.4kW 141.9PS
@ 5400rpm
166.0lbft 224.9Nm @ 3000rpm
108.9bhp/ton 107.1bhp/tonne
70.2bhp/L 52.4kW/L 71.2PS/L
48.4ft/sec 14.8m/sec
31.2mpg 26.0mpgUS
9.0L/100km
Petrol 4-stroke piston
1994cc 121.7cu in
turbocharged
In-line 4 fuel injection
Compression ratio: 7.9:1
Bore: 88.0mm 3.5in
Stroke: 82.0mm 3.2in
Valve type/No: Overhead 8
Transmission: Manual
No. of forward speeds: 5
Wheels driven: Rear
Springs F/R: Coil/Coil
Brake system: PA
Brakes F/R: Disc/Disc
Steering: Rack & pinion PA
Wheelbase: 243.8cm 96.0in
Track F: 135.4cm 53.3in
Track R: 137.9cm 54.3in
Length: 438.4cm 172.6in
Width: 165.6cm 65.2in
Height: 130.6cm 51.4in
Kerb weight: 1307.5kg

2880.0lb
Fuel: 57.1L 12.6gal 15.1galUS

392 Isuzu
Trooper 3-door TD
1987 Japan
80.0mph 128.7kmh
0-50mph 80.5kmh: 16.6secs
0-60mph 96.5kmh: 26.1secs
0-1/4 mile: Japan
0-1km: 43.2secs
74.0bhp 55.2kW 75.0PS
@ 4000rpm
114.2lbft 154.7Nm @ 2500rpm
45.5bhp/ton 44.7bhp/tonne
33.1bhp/L 24.7kW/L 33.5PS/L
40.2ft/sec 12.3m/sec
20.6mph 33.1kmh/1000rpm
22.3mpg 18.6mpgUS
12.7L/100km
Diesel 4-stroke piston
2238cc 137.0cu in
turbocharged
In-line 4 fuel injection
Compression ratio: 21.0:1
Bore: 88.0mm 3.5in
Stroke: 92.0mm 3.6in
Valve type/No: Overhead 8
Transmission: Manual
No. of forward speeds: 5
Wheels driven: 4-wheel
engageable
Springs F/R: Torsion bar/Leaf
Brake system: PA
Brakes F/R: Disc/Drum
Steering: Recirculating ball PA
Wheelbase: 230.1cm 90.6in
Track F: 139.4cm 54.9in
Track R: 140.0cm 55.1in
Length: 426.7cm 168.0in
Width: 167.6cm 66.0in
Height: 152.4cm 60.0in

389 Isdera
Imperator

392 Isuzu
Trooper 3-door TD

Ground clearance: 22.6cm
8.9in
Kerb weight: 1655.0kg
3645.4lb
Fuel: 83.3L 18.3gal 22.0galUS

393 Isuzu
Impulse Turbo
1988 Japan
125.0mph 201.1kmh
0-50mph 80.5kmh: 6.3secs
0-60mph 96.5kmh: 8.8secs
0-1/4 mile: 16.6secs
140.0bhp 104.4kW 141.9PS
@ 5400rpm
166.0lbft 224.9Nm @ 3000rpm
105.6bhp/ton 103.8bhp/tonne
70.2bhp/L 52.4kW/L 71.2PS/L
48.4ft/sec 14.8m/sec
30.6mpg 25.5mpgUS
9.2L/100km
Petrol 4-stroke piston
1994cc 121.7cu in
turbocharged
In-line 4 fuel injection
Compression ratio: 7.9:1
Bore: 88.0mm 3.5in
Stroke: 82.0mm 3.2in
Valve type/No: Overhead 8
Transmission: Manual
No. of forward speeds: 5
Wheels driven: Rear
Springs F/R: Coil/Coil
Brake system: PA
Brakes F/R: Disc/Disc
Steering: Rack & pinion PA
Wheelbase: 244.1cm 96.1in
Track F: 135.4cm 53.3in
Track R: 137.9cm 54.3in
Length: 438.4cm 172.6in
Width: 166.9cm 65.7in
Height: 130.6cm 51.4in
Kerb weight: 1348.4kg

2970.0lb
Fuel: 57.1L 12.6gal 15.1galUS

394 Isuzu
Impulse XS
1990 Japan
120.0mph 193.1kmh
0-60mph 96.5kmh: 9.3secs
0-1/4 mile: 16.9secs
130.0bhp 96.9kW 131.8PS
@ 6800rpm
102.0lbft 138.2Nm @ 4600rpm
118.9bhp/ton 116.9bhp/tonne
81.9bhp/L 61.0kW/L 83.0PS/L
58.7ft/sec 17.9m/sec
28.8mpg 24.0mpgUS
9.8L/100km
Petrol 4-stroke piston
1588cc 96.9cu in
In-line 4 fuel injection
Compression ratio: 9.8:1
Bore: 80.0mm 3.1in
Stroke: 79.0mm 3.1in
Valve type/No: Overhead 16
Transmission: Manual
No. of forward speeds: 5
Wheels driven: Front
Springs F/R: Coil/Coil
Brake system: PA
Brakes F/R: Disc/Disc
Steering: Rack & pinion PA
Wheelbase: 245.1cm 96.5in
Track F: 143.0cm 56.3in
Track R: 140.7cm 55.4in
Length: 421.6cm 166.0in
Width: 169.4cm 66.7in
Height: 129.8cm 51.1in
Kerb weight: 1112.3kg
2450.0lb
Fuel: 46.9L 10.3gal 12.4galUS

395 Isuzu
Impulse RS

1991 Japan
120.0mph 193.1kmh
0-60mph 96.5kmh: 8.6secs
0-1/4 mile: 16.6secs
160.0bhp 119.3kW 162.2PS
@ 6600rpm
150.0lbft 203.3Nm @ 4800rpm
131.3bhp/ton 129.1bhp/tonne
100.8bhp/L 75.1kW/L
102.1PS/L
57.0ft/sec 17.4m/sec
29.9mpg 24.9mpgUS
9.4L/100km
Petrol 4-stroke piston
1588cc 96.9cu in
turbocharged
In-line 4 fuel injection
Compression ratio: 8.5:1
Bore: 80.0mm 3.1in
Stroke: 79.0mm 3.1in
Valve type/No: Overhead 16
Transmission: Manual
No. of forward speeds: 5
Wheels driven: 4-wheel drive
Springs F/R: Coil/Coil
Brake system: PA ABS
Brakes F/R: Disc/Disc
Steering: Rack & pinion PA
Wheelbase: 245.1cm 96.5in
Track F: 143.0cm 56.3in
Track R: 140.5cm 55.3in
Length: 421.9cm 166.1in
Width: 169.4cm 66.7in
Height: 131.6cm 51.8in
Kerb weight: 1239.4kg
2730.0lb
Fuel: 46.9L 10.3gal 12.4galUS

396 Jaguar
XJ6 3.6
1986 UK
140.0mph 225.3kmh
0-50mph 80.5kmh: 7.4secs

0-60mph 96.5kmh: 9.8secs
0-1/4 mile: 15.8secs
0-1km: 28.9secs
221.0bhp 164.8kW 224.1PS
@ 5250rpm
248.0lbft 336.0Nm @ 4000rpm
133.2bhp/ton 131.0bhp/tonne
61.6bhp/L 45.9kW/L 62.4PS/L
52.8ft/sec 16.1m/sec
28.1mph 45.2kmh/1000rpm
20.7mpg 17.2mpgUS
13.6L/100km
Petrol 4-stroke piston
3590cc 219.0cu in
In-line 6 fuel injection
Compression ratio: 9.6:1
Bore: 91.0mm 3.6in
Stroke: 92.0mm 3.6in
Valve type/No: Overhead 24
Transmission: Manual
No. of forward speeds: 5
Wheels driven: Rear
Springs F/R: Coil/Coil
Brake system: PA
Brakes F/R: Disc/Disc
Steering: Rack & pinion PA
Wheelbase: 287.0cm 113.0in
Track F: 150.0cm 59.1in
Track R: 149.8cm 59.0in
Length: 498.8cm 196.4in
Width: 200.5cm 78.9in
Height: 138.0cm 54.3in
Ground clearance: 12.0cm
4.7in
Kerb weight: 1687.0kg
3715.9lb
Fuel: 89.2L 19.6gal 23.6galUS

397 Jaguar
XJS
1986 UK
132.0mph 212.4kmh
0-60mph 96.5kmh: 8.7secs

0-1/4 mile: 16.5secs
262.0bhp 195.4kW 265.6PS
@ 5000rpm
290.0lbft 393.0Nm @ 3000rpm
148.0bhp/ton 145.5bhp/tonne
49.0bhp/L 36.6kW/L 49.7PS/L
38.3ft/sec 11.7m/sec
15.9mpg 13.2mpgUS
17.8L/100km
Petrol 4-stroke piston
5343cc 326.0cu in
Vee 12 fuel injection
Bore: 90.0mm 3.5in
Stroke: 70.0mm 2.8in
Valve type/No: Overhead 24
Transmission: Automatic
No. of forward speeds: 3
Wheels driven: Rear
Springs F/R: Coil/Coil
Brake system: PA
Brakes F/R: Disc/Disc
Steering: Rack & pinion PA
Wheelbase: 259.1cm 102.0in
Track F: 148.8cm 58.6in
Track R: 150.4cm 59.2in
Length: 486.9cm 191.7in
Width: 179.3cm 70.6in
Height: 120.9cm 47.6in
Kerb weight: 1800.1kg
3965.0lb
Fuel: 90.8L 20.0gal 24.0galUS

398 Jaguar
XJSC
1986 UK
141.0mph 226.9kmh
0-50mph 80.5kmh: 6.8secs
0-60mph 96.5kmh: 8.8secs
0-1/4 mile: 16.7secs
262.0bhp 195.4kW 265.6PS
@ 5000rpm
290.0lbft 393.0Nm @ 3000rpm
146.3bhp/ton 143.9bhp/tonne
49.0bhp/L 36.6kW/L 49.7PS/L
38.3ft/sec 11.7m/sec
15.6mpg 13.0mpgUS
18.1L/100km
Petrol 4-stroke piston
5343cc 326.0cu in
Vee 12 fuel injection
Compression ratio: 11.5:1
Bore: 90.0mm 3.5in
Stroke: 70.0mm 2.8in
Valve type/No: Overhead 24
Transmission: Automatic
No. of forward speeds: 3
Wheels driven: Rear
Springs F/R: Coil/Coil
Brake system: PA
Brakes F/R: Disc/Disc
Steering: Rack & pinion PA
Wheelbase: 259.1cm 102.0in
Track F: 148.8cm 58.6in
Track R: 150.4cm 59.2in

Length: 486.9cm 191.7in
Width: 179.3cm 70.6in
Height: 121.4cm 47.8in
Kerb weight: 1820.5kg
4010.0lb
Fuel: 90.8L 20.0gal 24.0galUS

399 Jaguar
XJSC HE
1986 UK
150.0mph 241.4kmh
0-60mph 96.5kmh: 7.2secs
0-1/4 mile: 15.7secs
295.0bhp 220.0kW 299.1PS
@ 5500rpm
320.0lbft 433.6Nm @ 3250rpm
170.7bhp/ton 167.9bhp/tonne
55.2bhp/L 41.2kW/L 56.0PS/L
42.2ft/sec 12.8m/sec
26.3mph 42.3kmh/1000rpm
16.2mpg 13.5mpgUS
17.4L/100km
Petrol 4-stroke piston
5343cc 326.0cu in
Vee 12 fuel injection
Compression ratio: 12.5:1
Bore: 90.0mm 3.5in
Stroke: 70.0mm 2.8in
Valve type/No: Overhead 24
Transmission: Automatic
No. of forward speeds: 3
Wheels driven: Rear
Springs F/R: Coil/Coil
Brakes F/R: Disc/Disc
Steering: Rack & pinion PA
Wheelbase: 259.1cm 102.0in
Track F: 148.8cm 58.6in
Track R: 150.4cm 59.2in
Length: 476.5cm 187.6in
Width: 179.3cm 70.6in
Height: 126.2cm 49.7in
Kerb weight: 1757.0kg
3870.0lb

400 Jaguar
XJ6
1987 UK
130.0mph 209.2kmh
0-50mph 80.5kmh: 8.3secs
0-60mph 96.5kmh: 11.5secs
0-1/4 mile: 18.3secs
181.0bhp 135.0kW 183.5PS
@ 4750rpm
221.0lbft 299.5Nm @ 3750rpm
103.6bhp/ton 101.8bhp/tonne
50.4bhp/L 37.6kW/L 51.1PS/L
47.8ft/sec 14.6m/sec
21.0mpg 17.5mpgUS
13.4L/100km
Petrol 4-stroke piston
3590cc 219.0cu in
In-line 6 fuel injection
Compression ratio: 8.2:1
Bore: 91.0mm 3.6in

Stroke: 92.0mm 3.6in
Valve type/No: Overhead 24
Transmission: Automatic
No. of forward speeds: 4
Wheels driven: Rear
Springs F/R: Coil/Coil
Brake system: PA ABS
Brakes F/R: Disc/Disc
Steering: Rack & pinion PA
Wheelbase: 287.0cm 113.0in
Track F: 150.1cm 59.1in
Track R: 149.9cm 59.0in
Length: 498.9cm 196.4in
Width: 200.4cm 78.9in
Height: 137.9cm 54.3in
Ground clearance: 11.9cm
4.7in
Kerb weight: 1777.4kg
3915.0lb
Fuel: 87.8L 19.3gal 23.2galUS

401 Jaguar
XJ6 2.9
1987 UK
118.0mph 189.9kmh
0-50mph 80.5kmh: 7.0secs
0-60mph 96.5kmh: 9.9secs
0-1/4 mile: 17.5secs
0-1km: 32.0secs
165.0bhp 123.0kW 167.3PS
@ 5600rpm
176.0lbft 238.5Nm @ 4000rpm
101.2bhp/ton 99.5bhp/tonne
56.5bhp/L 42.1kW/L 57.3PS/L
45.7ft/sec 14.0m/sec
26.6mph 42.8kmh/1000rpm
19.3mpg 16.1mpgUS
14.6L/100km
Petrol 4-stroke piston
2919cc 178.0cu in
In-line 6 fuel injection
Compression ratio: 12.6:1
Bore: 91.0mm 3.6in
Stroke: 74.8mm 2.9in
Valve type/No: Overhead 12
Transmission: Manual
No. of forward speeds: 5
Wheels driven: Rear
Springs F/R: Coil/Coil
Brake system: PA
Brakes F/R: Disc/Drum
Steering: Rack & pinion PA
Wheelbase: 287.0cm 113.0in
Track F: 150.0cm 59.1in
Track R: 149.8cm 59.0in
Length: 498.8cm 196.4in
Width: 200.5cm 78.9in
Height: 138.0cm 54.3in
Ground clearance: 12.0cm
4.7in
Kerb weight: 1658.0kg
3652.0lb
Fuel: 88.7L 19.5gal
23.4galUS

402 Jaguar
XJS 3.6 Automatic
1987 UK
134.0mph 215.6kmh
0-50mph 80.5kmh: 5.8secs
0-60mph 96.5kmh: 7.8secs
0-1/4 mile: 16.0secs
0-1km: 29.4secs
221.0bhp 164.8kW 224.1PS
@ 5250rpm
248.0lbft 336.0Nm @ 4000rpm
139.6bhp/ton 137.3bhp/tonne
61.6bhp/L 45.9kW/L 62.4PS/L
52.8ft/sec 16.1m/sec
29.1mph 46.8kmh/1000rpm
18.0mpg 15.0mpgUS
15.7L/100km
Petrol 4-stroke piston
3590cc 219.0cu in
In-line 6 fuel injection
Compression ratio: 9.6:1
Bore: 91.0mm 3.6in
Stroke: 92.0mm 3.6in
Valve type/No: Overhead 24
Transmission: Automatic
No. of forward speeds: 4
Wheels driven: Rear
Springs F/R: Coil/Coil
Brake system: PA
Brakes F/R: Disc/Disc
Steering: Rack & pinion PA
Wheelbase: 259.1cm 102.0in
Track F: 148.1cm 58.3in
Track R: 149.6cm 58.9in
Length: 476.4cm 187.6in
Width: 179.3cm 70.6in
Height: 126.4cm 49.8in
Ground clearance: 14.0cm
5.5in
Kerb weight: 1610.0kg
3546.3lb
Fuel: 91.0L 20.0gal 24.0galUS

403 Jaguar
XJS
1988 UK
140.0mph 225.3kmh
0-50mph 80.5kmh: 6.7secs
0-60mph 96.5kmh: 8.6secs
0-1/4 mile: 16.5secs
262.0bhp 195.4kW 265.6PS
@ 5000rpm
290.0lbft 393.0Nm @ 5000rpm
145.3bhp/ton 142.8bhp/tonne
49.0bhp/L 36.6kW/L 49.7PS/L
38.3ft/sec 11.7m/sec
16.8mpg 14.0mpgUS
16.8L/100km
Petrol 4-stroke piston
5344cc 326.0cu in
Vee 12 fuel injection
Compression ratio: 11.5:1
Bore: 90.0mm 3.5in
Stroke: 70.0mm 2.8in

406 Jaguar
4.0 Sovereign

Valve type/No: Overhead 24
Transmission: Automatic
No. of forward speeds: 3
Wheels driven: Rear
Springs F/R: Coil/Coil
Brake system: PA
Brakes F/R: Disc/Disc
Steering: Rack & pinion PA
Wheelbase: 259.1cm 102.0in
Track F: 148.8cm 58.6in
Track R: 150.4cm 59.2in
Length: 486.9cm 191.7in
Width: 179.3cm 70.6in
Height: 121.4cm 47.8in
Kerb weight: 1834.2kg
4040.0lb
Fuel: 90.8L 20.0gal 24.0galUS

404 Jaguar
XJS Koenig
1988 UK
150.0mph 241.4kmh
0-60mph 96.5kmh: 7.5secs
0-1/4 mile: 15.5secs
280.0bhp 208.8kW 283.9PS
@ 5000rpm
300.0lbft 406.5Nm @ 3000rpm
155.6bhp/ton 153.0bhp/tonne
52.4bhp/L 39.1kW/L 53.1PS/L
38.3ft/sec 11.7m/sec
Petrol 4-stroke piston
5343cc 326.0cu in
Vee 12 fuel injection
Compression ratio: 11.5:1
Bore: 90.0mm 3.5in
Stroke: 70.0mm 2.8in
Valve type/No: Overhead 24
Transmission: Automatic
No. of forward speeds: 3
Wheels driven: Rear
Springs F/R: Coil/Coil
Brake system: PA
Brakes F/R: Disc/Disc

Steering: Rack & pinion PA
Wheelbase: 259.1cm 102.0in
Length: 486.9cm 191.7in
Width: 214.6cm 84.5in
Height: 121.4cm 47.8in
Kerb weight: 1829.6kg
4030.0lb
Fuel: 90.8L 20.0gal 24.0galUS

405 Jaguar
XJS V12 Convertible
1988 UK
146.0mph 234.9kmh
0-50mph 80.5kmh: 6.3secs
0-60mph 96.5kmh: 8.0secs
0-1/4 mile: 16.3secs
0-1km: 29.2secs
291.0bhp 217.0kW 295.0PS
@ 5500rpm
317.3lbft 430.0Nm @ 3000rpm
161.3bhp/ton 158.6bhp/tonne
54.4bhp/L 40.6kW/L 55.2PS/L
42.2ft/sec 12.8m/sec
26.2mph 42.2kmh/1000rpm
13.8mpg 11.5mpgUS
20.5L/100km
Petrol 4-stroke piston
5345cc 326.0cu in
Vee 12 fuel injection
Compression ratio: 12.5:1
Bore: 90.0mm 3.5in
Stroke: 70.0mm 2.8in
Valve type/No: Overhead 24
Transmission: Automatic
No. of forward speeds: 3
Wheels driven: Rear
Springs F/R: Coil/Coil
Brake system: PA ABS
Brakes F/R: Disc/Disc
Steering: Rack & pinion PA
Wheelbase: 259.0cm 102.0in
Track F: 148.1cm 58.3in
Track R: 149.6cm 58.9in

Length: 476.5cm 187.6in
Width: 179.3cm 70.6in
Height: 126.2cm 49.7in
Ground clearance: 14.0cm
5.5in
Kerb weight: 1835.0kg
4041.8lb
Fuel: 81.9L 18.0gal 21.6galUS

406 Jaguar
4.0 Sovereign
1989 UK
141.0mph 226.9kmh
0-50mph 80.5kmh: 6.3secs
0-60mph 96.5kmh: 8.3secs
0-1/4 mile: 16.0secs
0-1km: 29.2secs
235.0bhp 175.2kW 238.3PS
@ 4750rpm
285.6lbft 387.0Nm @ 3750rpm
136.6bhp/ton 134.3bhp/tonne
59.0bhp/L 44.0kW/L 59.9PS/L
53.0ft/sec 16.1m/sec
29.0mph 46.7kmh/1000rpm
18.5mpg 15.4mpgUS
15.3L/100km
Petrol 4-stroke piston
3980cc 243.0cu in
In-line 6 fuel injection
Compression ratio: 9.5:1
Bore: 91.0mm 3.6in
Stroke: 102.0mm 4.0in
Valve type/No: Overhead 24
Transmission: Automatic
No. of forward speeds: 4
Wheels driven: Rear
Springs F/R: Coil/Coil
Brake system: PA ABS
Brakes F/R: Disc/Disc
Steering: Rack & pinion PA
Wheelbase: 287.0cm 113.0in
Track F: 150.0cm 59.1in
Track R: 149.8cm 59.0in

Length: 498.8cm 196.4in
Width: 201.5cm 79.3in
Height: 138.0cm 54.3in
Ground clearance: 12.0cm
4.7in
Kerb weight: 1750.0kg
3854.6lb
Fuel: 89.2L 19.6gal 23.6galUS

407 Jaguar
XJ220
1989 UK
200.0mph 321.8kmh
0-60mph 96.5kmh: 3.5secs
530.0bhp 395.2kW 537.3PS
@ 7000rpm
400.0lbft 542.0Nm @ 5000rpm
345.1bhp/ton 339.4bhp/tonne
85.2bhp/L 63.5kW/L 86.4PS/L
59.7ft/sec 18.2m/sec
Petrol 4-stroke piston
6222cc 379.6cu in
Vee 12
Compression ratio: 10.0:1
Bore: 92.0mm 3.6in
Stroke: 78.0mm 3.1in
Valve type/No: Overhead 48
Transmission: Manual
No. of forward speeds: 5
Wheels driven: 4-wheel drive
Springs F/R: Coil/Coil
Brake system: ABS
Brakes F/R: Disc/Disc
Steering: Rack & pinion
Wheelbase: 284.5cm 112.0in
Length: 513.1cm 202.0in
Width: 199.9cm 78.7in
Height: 115.1cm 45.3in
Kerb weight: 1561.8kg
3440.0lb

408 Jaguar
XJS Convertible

407 Jaguar
XJ220

1989 UK
150.0mph 241.4kmh
0-60mph 96.5kmh: 8.7secs
0-1/4 mile: 16.2secs
262.0bhp 195.4kW 265.6PS
@ 5000rpm
290.0lbft 393.0Nm @ 5000rpm
140.1bhp/ton 137.7bhp/tonne
49.0bhp/L 36.6kW/L 49.7PS/L
38.3ft/sec 11.7m/sec
Petrol 4-stroke piston
5344cc 326.0cu in
Vee 12 fuel injection
Compression ratio: 11.5:1
Bore: 90.0mm 3.5in
Stroke: 70.0mm 2.8in
Valve type/No: Overhead 24
Transmission: Automatic
No. of forward speeds: 3
Wheels driven: Rear
Springs F/R: Coil/Coil
Brake system: PA ABS
Brakes F/R: Disc/Disc
Steering: Rack & pinion PA
Wheelbase: 259.1cm 102.0in
Track F: 148.8cm 58.6in
Track R: 150.4cm 59.2in
Length: 486.9cm 191.7in
Width: 179.3cm 70.6in
Height: 121.4cm 47.8in
Kerb weight: 1902.3kg
4190.0lb
Fuel: 90.8L 20.0gal 24.0galUS

409 Jaguar
XJS V12 Convertible
1989 UK
141.0mph 226.9kmh
0-60mph 96.5kmh: 9.9secs
0-1/4 mile: 17.5secs
262.0bhp 195.4kW 265.6PS
@ 5000rpm
290.0lbft 393.0Nm @ 5000rpm

141.2bhp/ton 138.9bhp/tonne
49.0bhp/L 36.5kW/L 49.7PS/L
38.3ft/sec 11.7m/sec
15.6mpg 13.0mpgUS
18.1L/100km
Petrol 4-stroke piston
5345cc 326.1cu in
Vee 12 fuel injection
Compression ratio: 11.5:1
Bore: 90.0mm 3.5in
Stroke: 70.0mm 2.8in
Valve type/No: Overhead 24
Transmission: Automatic
No. of forward speeds: 3
Wheels driven: Rear
Springs F/R: Coil/Coil
Brake system: PA ABS
Brakes F/R: Disc/Disc
Steering: Rack & pinion PA
Wheelbase: 259.1cm 102.0in
Track F: 148.8cm 58.6in
Track R: 150.4cm 59.2in
Length: 486.9cm 191.7in
Width: 179.3cm 70.6in
Height: 125.5cm 49.4in
Kerb weight: 1886.4kg
4155.0lb
Fuel: 81.8L 18.0gal 21.6galUS

410 Jaguar
XJ6 Vanden Plas
1990 UK
136.0mph 218.8kmh
0-60mph 96.5kmh: 9.2secs
0-1/4 mile: 16.9secs
223.0bhp 166.3kW 226.1PS
@ 4750rpm
278.0lbft 376.7Nm @ 3650rpm
120.2bhp/ton 118.2bhp/tonne
56.0bhp/L 41.8kW/L 56.8PS/L
53.0ft/sec 16.1m/sec
21.6mpg 18.0mpgUS
13.1L/100km

Petrol 4-stroke piston
3980cc 242.8cu in
In-line 6 fuel injection
Compression ratio: 9.6:1
Bore: 91.0mm 3.6in
Stroke: 102.0mm 4.0in
Valve type/No: Overhead 24
Transmission: Automatic
No. of forward speeds: 4
Wheels driven: Rear
Springs F/R: Coil/Coil
Brake system: PA ABS
Brakes F/R: Disc/Disc
Steering: Rack & pinion PA
Wheelbase: 287.0cm 113.0in
Track F: 150.1cm 59.1in
Track R: 149.9cm 59.0in
Length: 498.9cm 196.4in
Width: 200.4cm 78.9in
Height: 137.9cm 54.3in
Kerb weight: 1886.4kg
4155.0lb
Fuel: 87.8L 19.3gal 23.2galUS

411 Jaguar
XJR 4.0
1990 UK
143.0mph 230.1kmh
0-50mph 80.5kmh: 6.3secs
0-60mph 96.5kmh: 8.3secs
0-1/4 mile: 16.6secs
0-1km: 29.1secs
251.0bhp 187.2kW 254.5PS
@ 5250rpm
278.2lbft 377.0Nm @ 4000rpm
142.6bhp/ton 140.2bhp/tonne
63.1bhp/L 47.0kW/L 63.9PS/L
58.6ft/sec 17.8m/sec
29.0mph 46.7kmh/1000rpm
17.2mpg 14.3mpgUS
16.4L/100km
Petrol 4-stroke piston
3980cc 243.0cu in

In-line 6 fuel injection
Compression ratio: 9.7:1
Bore: 91.0mm 3.6in
Stroke: 102.0mm 4.0in
Valve type/No: Overhead 24
Transmission: Automatic
No. of forward speeds: 4
Wheels driven: Rear
Springs F/R: Coil/Coil
Brake system: PA ABS
Brakes F/R: Disc/Disc
Steering: Rack & pinion PA
Wheelbase: 287.0cm 113.0in
Track F: 150.1cm 59.1in
Track R: 149.9cm 59.0in
Length: 498.9cm 196.4in
Width: 201.4cm 79.3in
Height: 137.9cm 54.3in
Ground clearance: 11.9cm
4.7in
Kerb weight: 1790.0kg
3942.7lb
Fuel: 89.2L 19.6gal 23.6galUS

412 Jaguar
XJRS
1990 UK
150.0mph 241.4kmh
0-60mph 96.5kmh: 7.7secs
0-1/4 mile: 17.5secs
286.0bhp 213.3kW 290.0PS
@ 5150rpm
310.0lbft 420.1Nm @ 2800rpm
161.4bhp/ton 158.7bhp/tonne
53.5bhp/L 39.9kW/L 54.3PS/L
39.5ft/sec 12.0m/sec
Petrol 4-stroke piston
5344cc 326.0cu in
Vee 12 fuel injection
Compression ratio: 11.5:1
Bore: 90.0mm 3.5in
Stroke: 70.0mm 2.8in
Valve type/No: Overhead 24

Transmission: Automatic
No. of forward speeds: 3
Wheels driven: Rear
Springs F/R: Coil/Coil
Brake system: PA ABS
Brakes F/R: Disc/Disc
Steering: Rack & pinion PA
Wheelbase: 259.1cm 102.0in
Track F: 150.1cm 59.1in
Track R: 151.6cm 59.7in
Length: 476.5cm 187.6in
Width: 188.2cm 74.1in
Height: 126.2cm 49.7in
Kerb weight: 1802.4kg
3970.0lb
Fuel: 90.8L 20.0gal 24.0galUS

413 Jaguar

XJS
1990 UK
135.0mph 217.2kmh
0-50mph 80.5kmh: 7.7secs
0-60mph 96.5kmh: 9.9secs
0-1/4 mile: 17.5secs
262.0bhp 195.4kW 265.6PS
@ 5000rpm
290.0lbft 393.0Nm @ 3000rpm
146.2bhp/ton 143.7bhp/tonne
49.0bhp/L 36.6kW/L 49.7PS/L
38.3ft/sec 11.7m/sec
22.5mpg 18.7mpgUS
12.6L/100km
Petrol 4-stroke piston
5344cc 326.0cu in
Vee 12 fuel injection
Compression ratio: 11.5:1
Bore: 90.0mm 3.5in
Stroke: 70.0mm 2.8in
Valve type/No: Overhead 24
Transmission: Automatic
No. of forward speeds: 3
Wheels driven: Rear
Springs F/R: Coil/Coil
Brake system: PA ABS
Brakes F/R: Disc/Disc
Steering: Rack & pinion PA
Wheelbase: 259.1cm 102.0in
Track F: 148.8cm 58.6in
Track R: 150.4cm 59.2in
Length: 486.9cm 191.7in
Width: 179.3cm 70.6in
Height: 121.4cm 47.8in
Kerb weight: 1822.8kg
4015.0lb
Fuel: 90.8L 20.0gal 24.0galUS

414 Jaguar

XJS Lister Le Mans
1990 UK
200.0mph 321.8kmh
0-60mph 96.5kmh: 4.4secs
0-1/4 mile: 12.6secs
496.0bhp 369.9kW 502.9PS
@ 6200rpm

500.0lbft 677.5Nm @ 3850rpm
279.9bhp/ton 275.2bhp/tonne
70.9bhp/L 52.9kW/L 71.9PS/L
57.0ft/sec 17.4m/sec
Petrol 4-stroke piston
6996cc 426.8cu in
Vee 12 fuel injection
Compression ratio: 11.2:1
Bore: 94.0mm 3.7in
Stroke: 84.0mm 3.3in
Valve type/No: Overhead 24
Transmission: Manual
No. of forward speeds: 5
Wheels driven: Rear
Springs F/R: Coil/Coil
Brake system: PA ABS
Brakes F/R: Disc/Disc
Steering: Rack & pinion PA
Wheelbase: 259.1cm 102.0in
Track F: 186.7cm 73.5in
Track R: 193.0cm 76.0in
Length: 466.6cm 183.7in
Width: 196.9cm 77.5in
Height: 120.7cm 47.5in
Kerb weight: 1802.4kg
3970.0lb
Fuel: 128.3L 28.2gal
33.9galUS

415 Jaguar

XJS Railton
1990 UK
135.0mph 217.2kmh
0-60mph 96.5kmh: 9.9secs
0-1/4 mile: 17.5secs
286.0bhp 213.3kW 290.0PS
@ 5000rpm
310.0lbft 420.1Nm @ 5000rpm
153.4bhp/ton 150.9bhp/tonne
53.5bhp/L 39.9kW/L 54.3PS/L
38.3ft/sec 11.7m/sec
Petrol 4-stroke piston
5344cc 326.0cu in
Vee 12 fuel injection
Compression ratio: 11.5:1
Bore: 90.0mm 3.5in
Stroke: 70.0mm 2.8in
Valve type/No: Overhead 24
Transmission: Automatic
No. of forward speeds: 3
Wheels driven: Rear
Springs F/R: Coil/Coil
Brake system: PA ABS
Brakes F/R: Disc/Disc
Steering: Rack & pinion PA
Wheelbase: 259.1cm 102.0in
Track F: 148.8cm 58.6in
Track R: 150.4cm 59.2in
Length: 480.1cm 189.0in
Width: 181.9cm 71.6in
Height: 121.4cm 47.8in
Kerb weight: 1895.4kg
4175.0lb
Fuel: 90.8L 20.0gal 24.0galUS

416 Jaguar

XJ6 3.2
1991 UK
136.0mph 218.8kmh
0-50mph 80.5kmh: 6.1secs
0-60mph 96.5kmh: 8.3secs
0-1/4 mile: 16.1secs
0-1km: 29.7secs
200.0bhp 149.1kW 202.8PS
@ 5250rpm
219.9lbft 298.0Nm @ 4000rpm
121.8bhp/ton 119.8bhp/tonne
61.7bhp/L 46.0kW/L 62.6PS/L
47.7ft/sec 14.5m/sec
26.6mph 42.8kmh/1000rpm
19.9mpg 16.6mpgUS
14.2L/100km
Petrol 4-stroke piston
3239cc 198.0cu in
In-line 6 fuel injection
Compression ratio: 9.7:1
Bore: 91.0mm 3.6in
Stroke: 83.0mm 3.3in
Valve type/No: Overhead 24
Transmission: Manual
No. of forward speeds: 5
Wheels driven: Rear
Springs F/R: Coil/Coil
Brake system: PA ABS
Brakes F/R: Disc/Disc
Steering: Rack & pinion PA
Wheelbase: 287.0cm 113.0in
Track F: 150.0cm 59.1in
Track R: 149.8cm 59.0in
Length: 498.8cm 196.4in
Width: 201.5cm 79.3in
Height: 138.0cm 54.3in
Ground clearance: 12.0cm
4.7in
Kerb weight: 1670.0kg
3678.4lb
Fuel: 86.4L 19.0gal
22.8galUS

417 Jaguar

XJS
1991 UK
147.0mph 236.5kmh
0-60mph 96.5kmh: 7.8secs
280.0bhp 208.8kW 283.9PS
@ 5550rpm
306.0lbft 414.6Nm @ 2800rpm
155.8bhp/ton 153.2bhp/tonne
52.4bhp/L 39.1kW/L 53.1PS/L
42.5ft/sec 12.9m/sec
Petrol 4-stroke piston
5345cc 326.1cu in
Vee 12 fuel injection
Compression ratio: 11.5:1
Bore: 90.0mm 3.5in
Stroke: 70.0mm 2.8in
Valve type/No: Overhead 24
Transmission: Automatic
No. of forward speeds: 3

Wheels driven: Rear
Springs F/R: Coil/Coil
Brake system: PA ABS
Brakes F/R: Disc/Disc
Steering: Rack & pinion PA
Wheelbase: 259.1cm 102.0in
Track F: 148.8cm 58.6in
Track R: 150.4cm 59.2in
Length: 476.5cm 187.6in
Width: 188.2cm 74.1in
Height: 125.5cm 49.4in
Kerb weight: 1827.3kg
4025.0lb
Fuel: 88.9L 19.5gal
23.5galUS

418 Jaguar

XJS 4.0 Auto
1991 UK
138.0mph 222.0kmh
0-50mph 80.5kmh: 6.6secs
0-60mph 96.5kmh: 8.7secs
0-1/4 mile: 16.7secs
0-1km: 30.0secs
223.0bhp 166.3kW 226.1PS
@ 4750rpm
278.2lbft 377.0Nm @ 3650rpm
140.7bhp/ton 138.3bhp/tonne
56.0bhp/L 41.8kW/L 56.8PS/L
53.0ft/sec 16.1m/sec
29.1mph 46.8kmh/1000rpm
18.8mpg 15.7mpgUS
15.0L/100km
Petrol 4-stroke piston
3980cc 243.0cu in
In-line 6 fuel injection
Compression ratio: 9.5:1
Bore: 91.0mm 3.6in
Stroke: 102.0mm 4.0in
Valve type/No: Overhead 24
Transmission: Automatic
No. of forward speeds: 4
Wheels driven: Rear
Springs F/R: Coil/Coil
Brake system: PA ABS
Brakes F/R: Disc/Disc
Steering: Rack & pinion PA
Wheelbase: 259.1cm 102.0in
Track F: 148.9cm 58.6in
Track R: 150.4cm 59.2in
Length: 476.4cm 187.6in
Width: 179.3cm 70.6in
Height: 124.7cm 49.1in
Ground clearance: 12.8cm
5.0in
Kerb weight: 1612.0kg
3550.7lb
Fuel: 55.1L 12.1gal
14.5galUS

419 Jaguar

XJS Convertible
1991 UK
135.0mph 217.2kmh

0-50mph 80.5kmh: 7.7secs
0-60mph 96.5kmh: 9.9secs
0-1/4 mile: 17.5secs
262.0bhp 195.4kW 265.6PS
@ 5000rpm
290.0lbft 393.0Nm @ 3000rpm
136.5bhp/ton 134.2bhp/tonne
49.0bhp/L 36.6kW/L 49.7PS/L
38.3ft/sec 11.7m/sec
22.5mpg 18.7mpgUS
12.6L/100km
Petrol 4-stroke piston
5344cc 326.0cu in
Vee 12 fuel injection
Compression ratio: 11.5:1
Bore: 90.0mm 3.5in
Stroke: 70.0mm 2.8in
Valve type/No: Overhead 24
Transmission: Automatic
No. of forward speeds: 3
Wheels driven: Rear
Springs F/R: Coil/Coil
Brake system: PA ABS
Brakes F/R: Disc/Disc
Steering: Rack & pinion PA
Wheelbase: 259.1cm 102.0in
Track F: 128.5cm 50.6in
Track R: 150.4cm 59.2in
Length: 486.9cm 191.7in
Width: 179.3cm 70.6in
Height: 121.4cm 47.8in
Kerb weight: 1952.2kg
4300.0lb
Fuel: 90.8L 20.0gal
24.0galUS

420 Koenig
Competition
1989 Germany
217.0mph 349.2kmh
0-60mph 96.5kmh: 4.0secs
800.0bhp 596.6kW 811.1PS

@ 6250rpm
664.0lbft 899.7Nm @ 5000rpm
512.0bhp/ton 503.5bhp/tonne
161.9bhp/L 120.7kW/L
164.1PS/L
53.3ft/sec 16.2m/sec
Petrol 4-stroke piston
4942cc 301.5cu in
turbochargedsupercharged
Flat 12 fuel injection
Compression ratio: 7.5:1
Bore: 82.0mm 3.2in
Stroke: 78.0mm 3.1in
Valve type/No: Overhead 48
Transmission: Manual
No. of forward speeds: 5
Wheels driven: Rear
Springs F/R: Coil/Coil
Brake system: PA
Brakes F/R: Disc/Disc
Steering: Rack & pinion
Wheelbase: 255.0cm 100.4in
Track F: 185.9cm 73.2in
Track R: 215.9cm 85.0in
Length: 442.0cm 174.0in
Width: 219.5cm 86.4in
Height: 114.0cm 44.9in
Kerb weight: 1589.0kg
3500.0lb
Fuel: 104.8L 23.0gal
27.7galUS

421 Koenig
Competition Evolution
1990 Germany
230.0mph 370.1kmh
0-60mph 96.5kmh: 3.5secs
1000.0bhp 745.7kW 1013.9PS
@ 6700rpm
738.0lbft 1000.0Nm @
4500rpm
640.0bhp/ton 629.3bhp/tonne

202.3bhp/L 150.9kW/L
205.1PS/L
57.5ft/sec 17.5m/sec
Petrol 4-stroke piston
4942cc 301.5cu in
turbocharged
Flat 12 fuel injection
Compression ratio: 7.5:1
Bore: 82.0mm 3.2in
Stroke: 78.4mm 3.1in
Valve type/No: Overhead 48
Transmission: Manual
No. of forward speeds: 5
Wheels driven: Rear
Springs F/R: Coil/Coil
Brake system: PA ABS
Brakes F/R: Disc/Disc
Steering: Rack & pinion
Wheelbase: 255.0cm 100.4in
Track F: 185.9cm 73.2in
Track R: 215.9cm 85.0in
Length: 460.5cm 181.3in
Width: 219.5cm 86.4in
Height: 115.8cm 45.6in
Kerb weight: 1589.0kg
3500.0lb
Fuel: 100.3L 22.0gal
26.5galUS

422 Koenig
C62
1991 Germany
235.0mph 378.1kmh
0-60mph 96.5kmh: 3.5secs
588.0bhp 438.5kW 596.2PS
@ 6300rpm
533.0lbft 722.2Nm @ 4500rpm
543.1bhp/ton 534.1bhp/tonne
174.6bhp/L 130.2kW/L
177.0PS/L
Petrol 4-stroke piston
3368cc 205.5cu in

turbocharged
Flat 6 fuel injection
Transmission: Manual
No. of forward speeds: 5
Wheels driven: Rear
Springs F/R: Coil/Coil
Brakes F/R: Disc/Disc
Steering: Rack & pinion
Length: 495.8cm 195.2in
Width: 205.2cm 80.8in
Height: 111.8cm 44.0in
Kerb weight: 1100.9kg
2425.0lb

423 Lada
Riva 1500 Estate
1986 USSR
89.0mph 143.2kmh
0-50mph 80.5kmh: 11.0secs
0-60mph 96.5kmh: 16.4secs
0-1/4 mile: 20.3secs
0-1km: 38.3secs
71.0bhp 52.9kW 72.0PS
@ 5600rpm
78.0lbft 105.7Nm @ 3400rpm
68.1bhp/ton 67.0bhp/tonne
48.9bhp/L 36.5kW/L 49.6PS/L
49.0ft/sec 14.9m/sec
20.0mph 32.2kmh/1000rpm
30.1mpg 25.1mpgUS
9.4L/100km
Petrol 4-stroke piston
1452cc 89.0cu in
In-line 4 1 Carburettor
Compression ratio: 8.5:1
Bore: 76.0mm 3.0in
Stroke: 80.0mm 3.1in
Valve type/No: Overhead 8
Transmission: Manual
No. of forward speeds: 5
Wheels driven: Rear
Springs F/R: Coil/Coil

423 Lada
Riva 1500 Estate

Brake system: PA
Brakes F/R: Disc/Drum
Steering: Worm & roller
Wheelbase: 242.4cm 95.4in
Track F: 136.5cm 53.7in
Track R: 132.1cm 52.0in
Length: 411.5cm 162.0in
Width: 162.0cm 63.8in
Height: 145.8cm 57.4in
Ground clearance: 17.9cm
7.0in
Kerb weight: 1060.0kg
2334.8lb
Fuel: 45.5L 10.0gal 12.0galUS

424 Lada

Niva Cossack Cabrio
1987 USSR
81.0mph 130.3kmh
0-50mph 80.5kmh: 12.5secs
0-60mph 96.5kmh: 18.9secs
0-1/4 mile: 21.1secs
0-1km: 40.6secs
78.0bhp 58.2kW 79.1PS
@ 5400rpm
88.0lbft 119.2Nm @ 3400rpm
73.8bhp/ton 72.6bhp/tonne
49.7bhp/L 37.1kW/L 50.4PS/L
47.2ft/sec 14.4m/sec
19.2mph 30.9kmh/1000rpm
18.8mpg 15.7mpgUS
15.0L/100km
Petrol 4-stroke piston
1569cc 96.0cu in
In-line 4 1 Carburettor
Compression ratio: 8.5:1
Bore: 79.0mm 3.1in
Stroke: 80.0mm 3.1in
Valve type/No: Overhead 8
Transmission: Manual
No. of forward speeds: 5
Wheels driven: 4-wheel drive
Springs F/R: Coil/Coil
Brake system: PA
Brakes F/R: Disc/Drum
Steering: Worm & roller
Wheelbase: 219.7cm 86.5in
Track F: 142.2cm 56.0in
Track R: 139.7cm 55.0in
Length: 370.8cm 146.0in
Width: 167.6cm 66.0in
Height: 163.8cm 64.5in
Ground clearance: 23.5cm
9.3in
Kerb weight: 1075.0kg
2367.8lb
Fuel: 45.5L 10.0gal 12.0galUS

425 Lada

Samara 1300 SL
1987 USSR
96.0mph 154.5kmh
0-50mph 80.5kmh: 9.5secs
0-60mph 96.5kmh: 14.0secs

0-1/4 mile: 19.6secs
0-1km: 36.5secs
65.0bhp 48.5kW 65.9PS
@ 5600rpm
70.8lbft 95.9Nm @ 3300rpm
71.6bhp/ton 70.4bhp/tonne
50.5bhp/L 37.6kW/L 51.2PS/L
43.6ft/sec 13.2m/sec
20.9mph 33.6kmh/1000rpm
31.1mpg 25.9mpgUS
9.1L/100km
Petrol 4-stroke piston
1288cc 79.0cu in
In-line 4 1 Carburettor
Compression ratio: 9.9:1
Bore: 76.0mm 3.0in
Stroke: 71.0mm 2.8in
Valve type/No: Overhead 8
Transmission: Manual
No. of forward speeds: 5
Wheels driven: Front
Springs F/R: Coil/Coil
Brake system: PA
Brakes F/R: Disc/Drum
Steering: Rack & pinion
Wheelbase: 246.0cm 96.9in
Track F: 139.0cm 54.7in
Track R: 136.0cm 53.5in
Length: 400.6cm 157.7in
Width: 162.0cm 63.8in
Height: 133.5cm 52.6in
Ground clearance: 15.2cm
6.0in
Kerb weight: 923.0kg 2033.0lb
Fuel: 43.2L 9.5gal 11.4galUS

426 Lada

Samara 1300L 5-door
1989 USSR
94.0mph 151.2kmh
0-50mph 80.5kmh: 9.4secs
0-60mph 96.5kmh: 13.4secs
0-1/4 mile: 19.5secs
0-1km: 36.6secs
65.0bhp 48.5kW 65.9PS
@ 5600rpm
71.6lbft 97.0Nm @ 3500rpm
70.8bhp/ton 69.7bhp/tonne
50.5bhp/L 37.6kW/L 51.2PS/L
43.6ft/sec 13.2m/sec
20.9mph 33.6kmh/1000rpm
30.5mpg 25.4mpgUS
9.3L/100km
Petrol 4-stroke piston
1288cc 79.0cu in
In-line 4 1 Carburettor
Compression ratio: 9.9:1
Bore: 76.0mm 3.0in
Stroke: 71.0mm 2.8in
Valve type/No: Overhead 8
Transmission: Manual
No. of forward speeds: 5
Wheels driven: Front
Springs F/R: Coil/Coil

Brake system: PA
Brakes F/R: Disc/Drum
Steering: Rack & pinion
Wheelbase: 246.0cm 96.9in
Track F: 139.0cm 54.7in
Track R: 136.0cm 53.5in
Length: 400.6cm 157.7in
Width: 162.0cm 63.8in
Height: 133.5cm 52.6in
Ground clearance: 15.2cm
6.0in
Kerb weight: 933.0kg 2055.1lb
Fuel: 43.2L 9.5gal
11.4galUS

427 Lamborghini

Countach 5000S
1986 Italy
173.0mph 278.4kmh
0-60mph 96.5kmh: 5.2secs
0-1/4 mile: 13.7secs
420.0bhp 313.2kW 425.8PS
@ 7000rpm
341.0lbft 462.1Nm @ 5000rpm
286.4bhp/ton 281.6bhp/tonne
81.3bhp/L 60.6kW/L 82.4PS/L
57.4ft/sec 17.5m/sec
23.4mph 37.7kmh/1000rpm
12.0mpg 10.0mpgUS
23.5L/100km
Petrol 4-stroke piston
5167cc 315.2cu in
Vee 12 fuel injection
Compression ratio: 9.5:1
Bore: 85.5mm 3.4in
Stroke: 75.0mm 2.9in
Valve type/No: Overhead 48
Transmission: Manual
No. of forward speeds: 5
Wheels driven: Rear
Springs F/R: Coil/Coil
Brake system: PA
Brakes F/R: Disc/Disc
Steering: Rack & pinion
Wheelbase: 249.9cm 98.4in
Track F: 153.7cm 60.5in
Track R: 160.5cm 63.2in
Length: 420.1cm 165.4in
Width: 199.9cm 78.7in
Height: 106.9cm 42.1in
Ground clearance: 12.4cm
4.9in
Kerb weight: 1491.4kg
3285.0lb
Fuel: 120.0L 26.4gal
31.7galUS

428 Lamborghini

Jalpa 3500
1986 Italy
145.0mph 233.3kmh
0-50mph 80.5kmh: 4.9secs
0-60mph 96.5kmh: 6.2secs
0-1/4 mile: 14.7secs

0-1km: 26.8secs
255.0bhp 190.1kW 258.5PS
@ 7000rpm
231.0lbft 313.0Nm @ 3500rpm
184.3bhp/ton 181.2bhp/tonne
73.2bhp/L 54.6kW/L 74.2PS/L
57.4ft/sec 17.5m/sec
20.6mph 33.1kmh/1000rpm
15.9mpg 13.2mpgUS
17.8L/100km
Petrol 4-stroke piston
3485cc 213.0cu in
Vee 8 4 Carburettor
Compression ratio: 9.2:1
Bore: 86.0mm 3.4in
Stroke: 75.0mm 2.9in
Valve type/No: Overhead 16
Transmission: Manual
No. of forward speeds: 5
Wheels driven: Rear
Springs F/R: Coil/Coil
Brake system: PA
Brakes F/R: Disc/Disc
Steering: Rack & pinion
Wheelbase: 245.0cm 96.5in
Track F: 150.0cm 59.1in
Track R: 154.4cm 60.8in
Length: 433.0cm 170.5in
Width: 188.0cm 74.0in
Height: 114.0cm 44.9in
Ground clearance: 10.2cm
4.0in
Kerb weight: 1407.0kg
3099.1lb
Fuel: 80.1L 17.6gal 21.1galUS

429 Lamborghini

Jalpa
1988 Italy
154.0mph 247.8kmh
0-60mph 96.5kmh: 6.8secs
0-1/4 mile: 15.5secs
247.0bhp 184.2kW 250.4PS
@ 7000rpm
260.0lbft 352.3Nm @ 3400rpm
167.4bhp/ton 164.6bhp/tonne
70.9bhp/L 52.8kW/L 71.9PS/L
57.4ft/sec 17.5m/sec
16.2mpg 13.5mpgUS
17.4L/100km
Petrol 4-stroke piston
3485cc 212.6cu in
Vee 8 4 Carburettor
Compression ratio: 9.0:1
Bore: 86.0mm 3.4in
Stroke: 75.0mm 2.9in
Valve type/No: Overhead 16
Transmission: Manual
No. of forward speeds: 5
Wheels driven: Rear
Springs F/R: Coil/Coil
Brake system: PA
Brakes F/R: Disc/Disc
Steering: Rack & pinion

429
Lamborghini
Jalpa

Wheelbase: 245.1cm 96.5in
Track F: 148.1cm 58.3in
Track R: 153.7cm 60.5in
Length: 421.6cm 166.0in
Width: 165.4cm 65.1in
Height: 111.5cm 43.9in
Ground clearance: 12.7cm
5.0in
Kerb weight: 1500.5kg
3305.0lb
Fuel: 79.9L 17.6gal 21.1galUS

430 Lamborghini
LM129
1988 Italy
124.0mph 199.5kmh
0-50mph 80.5kmh: 6.1secs
0-60mph 96.5kmh: 7.8secs
0-1/4 mile: 15.8secs
415.0bhp 309.5kW 420.8PS
@ 6900rpm
363.0lbft 491.9Nm @ 4600rpm
139.3bhp/ton 136.9bhp/tonne
80.3bhp/L 59.9kW/L 81.4PS/L
56.5ft/sec 17.2m/sec
9.0mpg 7.5mpgUS
31.4L/100km
Petrol 4-stroke piston
5167cc 315.2cu in
Vee 12 6 Carburettor
Compression ratio: 9.5:1
Bore: 85.5mm 3.4in
Stroke: 75.0mm 2.9in
Valve type/No: Overhead 48
Transmission: Manual
No. of forward speeds: 5
Wheels driven: 4-wheel drive
Springs F/R: Coil/Coil
Brake system: PA
Brakes F/R: Disc/Drum
Steering: Recirculating ball PA
Wheelbase: 300.0cm 118.1in
Track F: 161.5cm 63.6in

Track R: 161.5cm 63.6in
Length: 495.0cm 194.9in
Width: 204.0cm 80.3in
Height: 183.9cm 72.4in
Kerb weight: 3030.4kg
6675.0lb
Fuel: 230.1L 50.6gal
60.8galUS

431 Lamborghini
Countach 25th Anniversary
1989 Italy
179.0mph 288.0kmh
0-50mph 80.5kmh: 3.8secs
0-60mph 96.5kmh: 4.7secs
0-1/4 mile: 12.9secs
455.0bhp 339.3kW 461.3PS
@ 7000rpm
368.0lbft 498.6Nm @ 5000rpm
310.7bhp/ton 305.5bhp/tonne
88.1bhp/L 65.7kW/L 89.3PS/L
57.4ft/sec 17.5m/sec
12.0mpg 10.0mpgUS
23.5L/100km
Petrol 4-stroke piston
5167cc 315.2cu in
Vee 12 6 Carburettor
Compression ratio: 9.5:1
Bore: 85.5mm 3.4in
Stroke: 75.0mm 2.9in
Valve type/No: Overhead 48
Transmission: Manual
No. of forward speeds: 5
Wheels driven: Rear
Springs F/R: Coil/Coil
Brake system: PA
Brakes F/R: Disc/Disc
Steering: Rack & pinion
Wheelbase: 245.1cm 96.5in
Track F: 153.7cm 60.5in
Track R: 160.5cm 63.2in
Length: 420.1cm 165.4in
Width: 199.9cm 78.7in

Height: 106.9cm 42.1in
Kerb weight: 1489.1kg
3280.0lb
Fuel: 120.0L 26.4gal
31.7galUS

432 Lamborghini
Diablo
1991 Italy
202.0mph 325.0kmh
0-50mph 80.5kmh: 3.8secs
0-60mph 96.5kmh: 4.5secs
0-1/4 mile: 13.3secs
492.0bhp 366.9kW 498.8PS
@ 6800rpm
428.0lbft 579.9Nm @ 5200rpm
304.4bhp/ton 299.4bhp/tonne
86.2bhp/L 64.3kW/L 87.4PS/L
59.5ft/sec 18.1m/sec
13.2mpg 11.0mpgUS
21.4L/100km
Petrol 4-stroke piston
5707cc 348.2cu in
Vee 12 fuel injection
Compression ratio: 10.0:1
Bore: 87.0mm 3.4in
Stroke: 80.0mm 3.1in
Valve type/No: Overhead 48
Transmission: Manual
No. of forward speeds: 5
Wheels driven: Rear
Springs F/R: Coil/Coil
Brake system: PA
Brakes F/R: Disc/Disc
Steering: Rack & pinion
Wheelbase: 264.9cm 104.3in
Track F: 150.9cm 59.4in
Track R: 164.1cm 64.6in
Length: 446.0cm 175.6in
Width: 204.0cm 80.3in
Height: 110.5cm 43.5in
Ground clearance: 14.0cm
5.5in

Kerb weight: 1643.5kg
3620.0lb
Fuel: 99.9L 22.0gal 26.4galUS

433 Lancia
Delta HF 4WD
1986 Italy
130.0mph 209.2kmh
0-50mph 80.5kmh: 4.9secs
0-60mph 96.5kmh: 6.6secs
0-1/4 mile: 15.2secs
0-1km: 29.0secs
165.0bhp 123.0kW 167.3PS
@ 5250rpm
188.0lbft 254.7Nm @ 2500rpm
135.3bhp/ton 133.1bhp/tonne
82.7bhp/L 61.7kW/L 83.8PS/L
51.6ft/sec 15.7m/sec
23.9mph 38.5kmh/1000rpm
19.8mpg 16.5mpgUS
14.3L/100km
Petrol 4-stroke piston
1995cc 122.0cu in
In-line 4 fuel injection
Compression ratio: 8.0:1
Bore: 84.0mm 3.3in
Stroke: 90.0mm 3.5in
Valve type/No: Overhead 8
Transmission: Manual
No. of forward speeds: 5
Wheels driven: 4-wheel drive
Springs F/R: Coil/Coil
Brake system: PA
Brakes F/R: Disc/Disc
Steering: Rack & pinion PA
Wheelbase: 247.5cm 97.4in
Track F: 140.2cm 55.2in
Track R: 140.0cm 55.1in
Length: 389.5cm 153.3in
Width: 162.0cm 63.8in
Height: 138.0cm 54.3in
Ground clearance: 15.2cm
6.0in

Kerb weight: 1240.0kg
2731.3lb
Fuel: 56.9L 12.5gal 15.0galUS

434 Lancia
Delta HF Turbo ie
1986 Italy
122.0mph 196.3kmh
0-50mph 80.5kmh: 6.0secs
0-60mph 96.5kmh: 8.5secs
0-1/4 mile: 16.1secs
0-1km: 30.1secs
140.0bhp 104.4kW 141.9PS
@ 5500rpm
141.0lbft 191.1Nm @ 3500rpm
133.8bhp/ton 131.6bhp/tonne
88.3bhp/L 65.9kW/L 89.5PS/L
42.9ft/sec 13.1m/sec
20.9mpg 33.6kmh/1000rpm
24.5mpg 20.4mpgUS
11.5L/100km
Petrol 4-stroke piston
1585cc 97.0cu in
turbocharged
In-line 4 fuel injection
Compression ratio: 8.0:1
Bore: 84.0mm 3.3in
Stroke: 71.5mm 2.8in
Valve type/No: Overhead 8
Transmission: Manual
No. of forward speeds: 5
Wheels driven: Front
Springs F/R: Coil/Coil
Brake system: PA
Brakes F/R: Disc/Disc
Steering: Rack & pinion
Wheelbase: 247.5cm 97.4in
Track F: 140.2cm 55.2in
Track R: 140.0cm 55.1in
Length: 389.5cm 153.3in
Width: 162.0cm 63.8in
Height: 138.0cm 54.3in
Ground clearance: 15.2cm
6.0in
Kerb weight: 1064.0kg
2343.6lb
Fuel: 57.3L 12.6gal
15.1galUS

435 Lancia
Prisma 1600ie LX
1986 Italy
119.0mph 191.5kmh
0-50mph 80.5kmh: 6.9secs
0-60mph 96.5kmh: 9.5secs
0-1/4 mile: 17.3secs
0-1km: 31.7secs
108.0bhp 80.5kW 109.5PS
@ 5900rpm
135.4lbft 183.5Nm @ 3500rpm
103.8bhp/ton 102.1bhp/tonne
68.1bhp/L 50.8kW/L 69.1PS/L
46.0ft/sec 14.1m/sec
18.6mph 29.9kmh/1000rpm

27.1mpg 22.6mpgUS
10.4L/100km
Petrol 4-stroke piston
1585cc 97.0cu in
In-line 4 fuel injection
Compression ratio: 9.7:1
Bore: 84.0mm 3.3in
Stroke: 71.5mm 2.8in
Valve type/No: Overhead 8
Transmission: Manual
No. of forward speeds: 5
Wheels driven: Front
Springs F/R: Coil/Coil
Brake system: PA
Brakes F/R: Disc/Disc
Steering: Rack & pinion
Wheelbase: 247.5cm 97.4in
Track F: 140.0cm 55.1in
Track R: 140.0cm 55.1in
Length: 418.0cm 164.6in
Width: 162.0cm 63.8in
Height: 138.5cm 54.5in
Ground clearance: 15.2cm
6.0in
Kerb weight: 1058.0kg
2330.4lb
Fuel: 56.9L 12.5gal 15.0galUS

436 Lancia
Thema 2.0ie
1986 Italy
118.0mph 189.9kmh
0-50mph 80.5kmh: 7.1secs
0-60mph 96.5kmh: 10.3secs
0-1/4 mile: 17.3secs
0-1km: 31.9secs
120.0bhp 89.5kW 121.7PS
@ 5250rpm
123.0lbft 166.7Nm @ 3300rpm
103.2bhp/ton 101.5bhp/tonne
60.1bhp/L 44.8kW/L 61.0PS/L
51.6ft/sec 15.7m/sec
21.6mph 34.8kmh/1000rpm
25.9mpg 21.6mpgUS
10.9L/100km
Petrol 4-stroke piston
1995cc 121.7cu in
In-line 4 fuel injection
Compression ratio: 9.8:1
Bore: 84.0mm 3.3in
Stroke: 90.0mm 3.5in
Valve type/No: Overhead 8
Transmission: Manual
No. of forward speeds: 5
Wheels driven: Front
Springs F/R: Coil/Coil
Brake system: PA
Brakes F/R: Disc/Disc
Steering: Rack & pinion PA
Wheelbase: 265.9cm 104.7in
Track F: 148.8cm 58.6in
Track R: 148.3cm 58.4in
Length: 459.2cm 180.8in
Width: 177.5cm 69.9in

Height: 143.3cm 56.4in
Ground clearance: 17.8cm
7.0in
Kerb weight: 1182.2kg
2604.0lb
Fuel: 70.1L 15.4gal 18.5galUS

437 Lancia
Y10 Fire
1986 Italy
94.0mph 151.2kmh
0-50mph 80.5kmh: 11.7secs
0-60mph 96.5kmh: 16.7secs
0-1/4 mile: 20.6secs
0-1km: 38.6secs
45.0bhp 33.6kW 45.6PS
@ 5000rpm
59.0lbft 79.9Nm @ 2750rpm
58.7bhp/ton 57.7bhp/tonne
45.0bhp/L 33.6kW/L 45.7PS/L
35.6ft/sec 10.8m/sec
21.3mph 34.3kmh/1000rpm
40.6mpg 33.8mpgUS
7.0L/100km
Petrol 4-stroke piston
999cc 60.9cu in
In-line 4 1 Carburettor
Compression ratio: 9.8:1
Bore: 70.0mm 2.8in
Stroke: 64.9mm 2.6in
Valve type/No: Overhead 8
Transmission: Manual
No. of forward speeds: 5
Wheels driven: Front
Springs F/R: Coil/Coil
Brake system: PA
Brakes F/R: Disc/Drum
Steering: Rack & pinion
Wheelbase: 215.9cm 85.0in
Track F: 128.0cm 50.4in
Track R: 127.8cm 50.3in
Length: 339.1cm 133.5in
Width: 150.6cm 59.3in
Height: 124.5cm 49.0in
Ground clearance: 15.2cm
6.0in
Kerb weight: 779.5kg 1717.0lb
Fuel: 46.9L 10.3gal 12.4galUS

438 Lancia
Thema 8.32
1987 Italy
149.0mph 239.7kmh
0-60mph 96.5kmh: 6.8secs
215.0bhp 160.3kW 218.0PS
@ 6750rpm
209.0lbft 283.2Nm @ 4500rpm
156.4bhp/ton 153.8bhp/tonne
73.4bhp/L 54.8kW/L 74.5PS/L
52.5ft/sec 16.0m/sec
Petrol 4-stroke piston
2927cc 178.6cu in
Vee 8 fuel injection
Compression ratio: 10.5:1

Bore: 81.0mm 3.2in
Stroke: 71.0mm 2.8in
Valve type/No: Overhead 32
Transmission: Manual
No. of forward speeds: 5
Wheels driven: Front
Springs F/R: Coil/Coil
Brake system: PA ABS
Brakes F/R: Disc/Disc
Steering: Rack & pinion PA
Wheelbase: 267.0cm 105.1in
Track F: 149.9cm 59.0in
Track R: 148.8cm 58.6in
Length: 491.0cm 193.3in
Width: 173.7cm 68.4in
Height: 143.8cm 56.6in
Kerb weight: 1398.3kg
3080.0lb
Fuel: 70.4L 15.5gal 18.6galUS

439 Lancia
Delta HF Integrale
1988 Italy
130.0mph 209.2kmh
0-50mph 80.5kmh: 4.7secs
0-60mph 96.5kmh: 6.4secs
0-1/4 mile: 14.8secs
0-1km: 27.9secs
185.0bhp 137.9kW 187.6PS
@ 5300rpm
223.6lbft 303.0Nm @ 3500rpm
148.5bhp/ton 146.0bhp/tonne
92.7bhp/L 69.1kW/L 94.0PS/L
52.1ft/sec 15.9m/sec
23.4mph 37.7kmh/1000rpm
17.6mpg 14.7mpgUS
16.1L/100km
Petrol 4-stroke piston
1995cc 122.0cu in
turbocharged
In-line 4 fuel injection
Compression ratio: 8.0:1
Bore: 87.0mm 3.4in
Stroke: 90.0mm 3.5in
Valve type/No: Side 8
Transmission: Manual
No. of forward speeds: 5
Wheels driven: 4-wheel drive
Springs F/R: Coil/Coil
Brake system: PA
Brakes F/R: Disc/Disc
Steering: Rack & pinion PA
Wheelbase: 248.0cm 97.6in
Track F: 142.6cm 56.1in
Track R: 140.6cm 55.4in
Length: 390.0cm 153.5in
Width: 170.0cm 66.9in
Height: 138.0cm 54.3in
Ground clearance: 15.2cm
6.0in
Kerb weight: 1267.0kg
2790.7lb
Fuel: 56.9L 12.5gal
15.0galUS

441 Lancia
Delta Integrale 16v

440 Lancia
Thema 8.32
1988 Italy
140.0mph 225.3kmh
0-50mph 80.5kmh: 5.5secs
0-60mph 96.5kmh: 7.2secs
0-1/4 mile: 15.2secs
0-1km: 28.4secs
215.0bhp 160.3kW 218.0PS @ 6750rpm
209.6lbft 284.0Nm @ 4500rpm
153.6bhp/ton 151.1bhp/tonne
73.4bhp/L 54.8kW/L 74.5PS/L
52.5ft/sec 16.0m/sec
21.8mpg 35.1kmh/1000rpm
15.6mpg 13.0mpgUS
18.1L/100km
Petrol 4-stroke piston
2927cc 179.0cu in
Vee 8 fuel injection
Compression ratio: 10.5:1
Bore: 81.0mm 3.2in
Stroke: 71.0mm 2.8in
Valve type/No: Overhead 32
Transmission: Manual
No. of forward speeds: 5
Wheels driven: Front
Springs F/R: Coil/Coil
Brake system: PA ABS
Brakes F/R: Disc/Disc
Steering: Rack & pinion PA
Wheelbase: 266.0cm 104.7in
Track F: 149.4cm 58.8in
Track R: 148.4cm 58.4in
Length: 459.0cm 180.7in
Width: 173.3cm 68.2in
Height: 143.3cm 56.4in
Ground clearance: 17.9cm 7.0in
Kerb weight: 1423.0kg 3134.4lb
Fuel: 70.1L 15.4gal 18.5galUS

441 Lancia
Delta Integrale 16v
1989 Italy
129.0mph 207.6kmh
0-50mph 80.5kmh: 4.7secs
0-60mph 96.5kmh: 6.3secs
0-1/4 mile: 14.9secs
0-1km: 28.2secs
200.0bhp 149.1kW 202.8PS @ 5500rpm
219.9lbft 298.0Nm @ 3000rpm
157.4bhp/ton 154.8bhp/tonne
100.2bhp/L 74.8kW/L
101.6PS/L
54.1ft/sec 16.5m/sec
23.1mph 37.2kmh/1000rpm
19.1mpg 15.9mpgUS
14.8L/100km
Petrol 4-stroke piston
1995cc 122.0cu in
turbocharged
In-line 4 fuel injection
Compression ratio: 8.0:1
Bore: 84.0mm 3.3in
Stroke: 90.0mm 3.5in
Valve type/No: Overhead 16
Transmission: Manual
No. of forward speeds: 5
Wheels driven: 4-wheel drive
Springs F/R: Coil/Coil
Brake system: PA ABS
Brakes F/R: Disc/Disc
Steering: Rack & pinion PA
Wheelbase: 248.0cm 97.6in
Track F: 144.8cm 57.0in
Track R: 144.0cm 56.7in
Length: 389.5cm 153.3in
Width: 168.6cm 66.4in
Height: 136.5cm 53.7in
Ground clearance: 15.2cm 6.0in
Kerb weight: 1292.0kg 2845.8lb
Fuel: 56.9L 12.5gal 15.0galUS

442 Lancia
Thema 2.0ie 16v
1989 Italy
126.0mph 202.7kmh
0-50mph 80.5kmh: 6.6secs
0-60mph 96.5kmh: 8.8secs
0-1/4 mile: 16.7secs
0-1km: 30.3secs
150.0bhp 111.9kW 152.1PS @ 6000rpm
138.7lbft 188.0Nm @ 4000rpm
117.3bhp/ton 115.4bhp/tonne
75.2bhp/L 56.1kW/L 76.2PS/L
59.0ft/sec 18.0m/sec
21.8mph 35.1kmh/1000rpm
24.4mpg 20.3mpgUS
11.6L/100km
Petrol 4-stroke piston
1995cc 122.0cu in
In-line 4 fuel injection
Compression ratio: 9.8:1
Bore: 84.0mm 3.3in
Stroke: 90.0mm 3.5in
Valve type/No: Overhead 16
Transmission: Manual
No. of forward speeds: 5
Wheels driven: Front
Springs F/R: Coil/Coil
Brake system: PA
Brakes F/R: Disc/Disc
Steering: Rack & pinion PA
Wheelbase: 266.0cm 104.7in
Track F: 148.2cm 58.3in
Track R: 147.2cm 58.0in
Length: 459.0cm 180.7in
Width: 175.2cm 69.0in
Height: 143.3cm 56.4in
Ground clearance: 17.9cm 7.0in
Kerb weight: 1300.0kg 2863.4lb
Fuel: 70.1L 15.4gal 18.5galUS

443 Lancia
Dedra 1.8i
1990 Italy
118.0mph 189.9kmh
0-50mph 80.5kmh: 8.3secs
0-60mph 96.5kmh: 11.5secs
0-1/4 mile: 18.1secs
0-1km: 33.6secs
110.0bhp 82.0kW 111.5PS @ 6000rpm
104.8lbft 142.0Nm @ 3000rpm
90.2bhp/ton 88.7bhp/tonne
62.6bhp/L 46.7kW/L 63.5PS/L
52.0ft/sec 15.8m/sec
19.4mph 31.2kmh/1000rpm
25.2mpg 21.0mpgUS
11.2L/100km
Petrol 4-stroke piston
1756cc 107.0cu in
In-line 4 fuel injection
Compression ratio: 9.5:1
Bore: 84.0mm 3.3in
Stroke: 79.2mm 3.1in
Valve type/No: Overhead 8
Transmission: Manual
No. of forward speeds: 5
Wheels driven: Front
Springs F/R: Coil/Coil
Brake system: PA ABS
Brakes F/R: Disc/Disc
Steering: Rack & pinion PA
Wheelbase: 254.0cm 100.0in
Track F: 143.5cm 56.5in
Track R: 141.5cm 55.7in
Length: 434.1cm 170.9in
Width: 169.9cm 66.9in
Height: 142.7cm 56.2in
Kerb weight: 1240.0kg 2731.3lb
Fuel: 63.2L 13.9gal 16.7galUS

444 Lancia
Dedra 2.0ie
1990 Italy

126.0mph 202.7kmh
0-50mph 80.5kmh: 7.1secs
0-60mph 96.5kmh: 10.0secs
0-1/4 mile: 17.6secs
0-1km: 31.5secs
120.0bhp 89.5kW 121.7PS
@ 5750rpm
119.6lbft 162.0Nm @ 3300rpm
96.8bhp/ton 95.2bhp/tonne
60.1bhp/L 44.8kW/L 61.0PS/L
56.5ft/sec 17.2m/sec
21.9mph 35.2kmh/1000rpm
23.7mpg 19.7mpgUS
11.9L/100km
Petrol 4-stroke piston
1995cc 122.0cu in
In-line 4 fuel injection
Compression ratio: 9.5:1
Bore: 84.0mm 3.3in
Stroke: 90.0mm 3.5in
Valve type/No: Overhead 8
Transmission: Manual
No. of forward speeds: 5
Wheels driven: Front
Springs F/R: Coil/Coil
Brake system: PA ABS
Brakes F/R: Disc/Drum
Steering: Rack & pinion PA
Wheelbase: 254.0cm 100.0in
Track F: 143.5cm 56.5in
Track R: 141.5cm 55.7in
Length: 434.1cm 170.9in
Width: 169.9cm 66.9in
Height: 142.7cm 56.2in
Kerb weight: 1260.0kg
2775.3lb
Fuel: 63.2L 13.9gal 16.7galUS

445 Lancia
Thema 2.0ie 16v SE Turbo
1990 Italy
143.0mph 230.1kmh
0-50mph 80.5kmh: 5.3secs
0-60mph 96.5kmh: 6.8secs
0-1/4 mile: 15.2secs
0-1km: 27.7secs
185.0bhp 137.9kW 187.6PS
@ 5500rpm
236.2lbft 320.0Nm @ 3500rpm
136.8bhp/ton 134.5bhp/tonne
92.7bhp/L 69.1kW/L 94.0PS/L
54.1ft/sec 16.5m/sec
24.7mph 39.7kmh/1000rpm
21.1mpg 17.6mpgUS
13.4L/100km
Petrol 4-stroke piston
1995cc 122.0cu in
turbocharged
In-line 4 fuel injection
Compression ratio: 8.0:1
Bore: 84.0mm 3.3in
Stroke: 90.0mm 3.5in
Valve type/No: Overhead 16
Transmission: Manual

No. of forward speeds: 5
Wheels driven: Front
Springs F/R: Coil/Coil
Brake system: PA ABS
Brakes F/R: Disc/Disc
Steering: Rack & pinion PA
Wheelbase: 266.0cm 104.7in
Track F: 148.2cm 58.3in
Track R: 147.2cm 58.0in
Length: 459.0cm 180.7in
Width: 175.2cm 69.0in
Height: 143.3cm 56.4in
Ground clearance: 17.8cm
7.0in
Kerb weight: 1375.0kg
3028.6lb
Fuel: 70.1L 15.4gal 18.5galUS

446 Lancia
Y10 GTie
1990 Italy
107.0mph 172.2kmh
0-50mph 80.5kmh: 8.0secs
0-60mph 96.5kmh: 11.5secs
0-1/4 mile: 18.1secs
0-1km: 33.8secs
78.0bhp 58.2kW 79.1PS
@ 5750rpm
75.3lbft 102.0Nm @ 3250rpm
91.2bhp/ton 89.7bhp/tonne
59.9bhp/L 44.7kW/L 60.8PS/L
45.2ft/sec 13.8m/sec
19.2mph 30.9kmh/1000rpm
29.8mpg 24.8mpgUS
9.5L/100km
Petrol 4-stroke piston
1301cc 79.0cu in
In-line 4 fuel injection
Compression ratio: 9.5:1
Bore: 76.0mm 3.0in
Stroke: 72.0mm 2.8in
Valve type/No: Overhead 8
Transmission: Manual
No. of forward speeds: 5
Wheels driven: Front
Springs F/R: Coil/Coil
Brake system: PA
Brakes F/R: Disc/Drum
Steering: Rack & pinion
Wheelbase: 215.9cm 85.0in

Track F: 128.2cm 50.5in
Track R: 127.8cm 50.3in
Length: 339.2cm 133.5in
Width: 150.7cm 59.3in
Height: 124.5cm 49.0in
Ground clearance: 14.0cm
5.5in
Kerb weight: 870.0kg 1916.3lb
Fuel: 46.9L 10.3gal 12.4galUS

447 Lancia
Dedra 2000 Turbo
1991 Italy
133.0mph 214.0kmh
0-50mph 80.5kmh: 5.5secs
0-60mph 96.5kmh: 7.4secs
0-1/4 mile: 15.9secs
0-1km: 28.8secs
165.0bhp 123.0kW 167.3PS
@ 5500rpm
182.3lbft 247.0Nm @ 3000rpm
126.2bhp/ton 124.1bhp/tonne
82.7bhp/L 61.7kW/L 83.8PS/L
54.1ft/sec 16.5m/sec
22.5mph 36.2kmh/1000rpm
22.1mpg 18.4mpgUS
12.8L/100km
Petrol 4-stroke piston
1995cc 122.0cu in
turbocharged
In-line 4 fuel injection
Compression ratio: 7.5:1
Bore: 84.0mm 3.3in
Stroke: 90.0mm 3.5in
Valve type/No: Overhead 8
Transmission: Manual
No. of forward speeds: 5
Wheels driven: Front
Springs F/R: Coil/Coil
Brake system: PA ABS
Brakes F/R: Disc/Disc
Steering: Rack & pinion PA
Wheelbase: 254.0cm 100.0in
Track F: 143.6cm 56.5in
Track R: 141.5cm 55.7in
Length: 434.3cm 171.0in
Width: 170.0cm 66.9in
Height: 143.0cm 56.3in
Kerb weight: 1330.0kg

2929.5lb
Fuel: 63.0L 13.8gal 16.6galUS

448 Land Rover
Ninety County Turbo Diesel
1987 UK
76.0mph 122.3kmh
0-50mph 80.5kmh: 14.2secs
0-60mph 96.5kmh: 22.3secs
0-1/4 mile: 22.1secs
0-1km: 42.3secs
85.0bhp 63.4kW 86.2PS
@ 4000rpm
150.0lbft 203.3Nm @ 1800rpm
50.5bhp/ton 49.7bhp/tonne
34.1bhp/L 25.4kW/L 34.5PS/L
42.4ft/sec 12.9m/sec
20.2mph 32.5kmh/1000rpm
18.0mpg 15.0mpgUS
15.7L/100km
Diesel 4-stroke piston
2494cc 152.0cu in
turbocharged
In-line 4 fuel injection
Compression ratio: 21.0:1
Bore: 90.5mm 3.6in
Stroke: 97.0mm 3.8in
Valve type/No: Overhead 8
Transmission: Manual
No. of forward speeds: 5
Wheels driven: 4-wheel drive
Springs F/R: Coil/Coil
Brake system: PA
Brakes F/R: Disc/Drum
Steering: Worm & roller PA
Wheelbase: 236.0cm 92.9in
Track F: 148.6cm 58.5in
Track R: 148.6cm 58.5in
Length: 388.3cm 152.9in
Width: 179.0cm 70.5in
Height: 198.9cm 78.3in
Ground clearance: 25.4cm
10.0in
Kerb weight: 1710.0kg
3766.5lb
Fuel: 54.6L 12.0gal 14.4galUS

449 Land Rover
One Ten County V8

**449 Land
Rover**
One Ten County
V8

1987 UK
88.0mph 141.6kmh
0-50mph 80.5kmh: 10.4secs
0-60mph 96.5kmh: 15.1secs
0-1/4 mile: 19.8secs
0-1km: 37.9secs
134.0bhp 99.9kW 135.9PS
@ 5000rpm
187.0lbft 253.4Nm @ 2500rpm
68.1bhp/ton 66.9bhp/tonne
38.0bhp/L 28.3kW/L 38.5PS/L
38.9ft/sec 11.8m/sec
23.0mph 37.0kmh/1000rpm
13.4mpg 11.2mpgUS
21.1L/100km
Petrol 4-stroke piston
3528cc 215.0cu in
Vee 8 2 Carburettor
Compression ratio: 8.0:1
Bore: 88.9mm 3.5in
Stroke: 71.1mm 2.8in
Valve type/No: Overhead 16
Transmission: Manual
No. of forward speeds: 5
Wheels driven: 4-wheel drive
Springs F/R: Coil/Coil
Brake system: PA
Brakes F/R: Disc/Drum
Steering: Worm & roller PA
Wheelbase: 279.4cm 110.0in
Track F: 148.6cm 58.5in
Track R: 148.6cm 58.5in
Length: 459.9cm 181.1in
Width: 179.0cm 70.5in
Height: 203.5cm 80.1in
Ground clearance: 38.1cm
15.0in
Kerb weight: 2002.0kg
4409.7lb
Fuel: 79.6L 17.5gal
 21.0galUS

450 Land Rover

Ninety County V8
1989 UK
92.0mph 148.0kmh
0-50mph 80.5kmh: 8.9secs
0-60mph 96.5kmh: 13.6secs
0-1/4 mile: 18.8secs
0-1km: 35.5secs
134.0bhp 99.9kW 135.9PS
@ 5000rpm
186.7lbft 253.0Nm @ 2500rpm
81.5bhp/ton 80.1bhp/tonne
38.0bhp/L 28.3kW/L 38.5PS/L
38.9ft/sec 11.8m/sec
21.1mph 33.9kmh/1000rpm
12.1mpg 10.1mpgUS
23.3L/100km
Petrol 4-stroke piston
3528cc 215.0cu in
Vee 8 2 Carburettor
Compression ratio: 8.0:1
Bore: 88.9mm 3.5in
Stroke: 71.1mm 2.8in
Valve type/No: Overhead 16
Transmission: Manual
No. of forward speeds: 5
Wheels driven: 4-wheel
engageable
Springs F/R: Coil/Coil
Brake system: PA
Brakes F/R: Disc/Drum
Steering: Worm & roller
Wheelbase: 236.0cm 92.9in
Track F: 148.6cm 58.5in
Track R: 148.6cm 58.5in
Length: 388.3cm 152.9in
Width: 179.0cm 70.5in
Height: 198.9cm 78.3in
Ground clearance: 25.5cm
10.0in
Kerb weight: 1672.0kg
3682.8lb
Fuel: 54.6L 12.0gal 14.4galUS

451 Land Rover

Discovery TDi
1990 UK
92.0mph 148.0kmh
0-50mph 80.5kmh: 11.5secs
0-60mph 96.5kmh: 17.1secs
0-1/4 mile: 20.5secs
0-1km: 38.8secs
111.0bhp 82.8kW 112.5PS
@ 4000rpm
195.6lbft 265.0Nm @ 1800rpm
56.1bhp/ton 55.2bhp/tonne
44.5bhp/L 33.2kW/L 45.1PS/L
42.4ft/sec 12.9m/sec
25.1mph 40.4kmh/1000rpm
23.9mpg 19.9mpgUS
11.8L/100km
Diesel 4-stroke piston
2495cc 152.0cu in
turbocharged
In-line 4 fuel injection
Compression ratio: 19.5:1
Bore: 90.5mm 3.6in
Stroke: 97.0mm 3.8in
Valve type/No: Overhead 8
Transmission: Manual
No. of forward speeds: 5
Wheels driven: 4-wheel drive
Springs F/R: Coil/Coil
Brake system: PA
Brakes F/R: Disc/Disc
Steering: Recirculating ball PA
Wheelbase: 254.0cm 100.0in
Track F: 148.6cm 58.5in
Track R: 148.6cm 58.5in
Length: 452.1cm 178.0in
Width: 179.3cm 70.6in
Height: 191.9cm 75.6in
Ground clearance: 19.1cm
7.5in
Kerb weight: 2012.0kg
4431.7lb
Fuel: 88.7L 19.5gal 23.4galUS

452 Land Rover

Discovery V8
1990 UK
99.0mph 159.3kmh
0-50mph 80.5kmh: 8.7secs
0-60mph 96.5kmh: 12.8secs
0-1/4 mile: 18.8secs
0-1km: 35.7secs
145.0bhp 108.1kW 147.0PS
@ 5000rpm
191.9lbft 260.0Nm @ 2800rpm
78.3bhp/ton 77.0bhp/tonne
41.1bhp/L 30.6kW/L 41.7PS/L
38.9ft/sec 11.8m/sec
25.1mph 40.4kmh/1000rpm
14.0mpg 11.7mpgUS
20.2L/100km
Petrol 4-stroke piston
3528cc 215.0cu in
Vee 8 2 Carburettor
Compression ratio: 8.0:1
Bore: 88.9mm 3.5in
Stroke: 71.1mm 2.8in
Valve type/No: Overhead 16
Transmission: Manual
No. of forward speeds: 5
Wheels driven: 4-wheel drive
Springs F/R: Coil/Coil
Brake system: PA
Brakes F/R: Disc/Disc
Steering: Recirculating ball PA
Wheelbase: 254.0cm 100.0in
Track F: 148.6cm 58.5in
Track R: 148.6cm 58.5in
Length: 452.1cm 178.0in
Width: 179.3cm 70.6in
Height: 191.9cm 75.6in
Ground clearance: 19.1cm
7.5in
Kerb weight: 1882.0kg
4145.4lb
Fuel: 88.7L 19.5gal
23.4galUS

454 Lexus
LS400

453 Land Rover

Discovery V8i 5DR
1991 UK
107.0mph 172.2kmh
0-50mph 80.5kmh: 8.2secs
0-60mph 96.5kmh: 11.7secs
0-1/4 mile: 17.7secs
0-1km: 33.6secs
164.0bhp 122.3kW 166.3PS
@ 4750rpm
211.8lbft 287.0Nm @ 2600rpm
88.5bhp/ton 87.0bhp/tonne
46.5bhp/L 34.7kW/L 47.1PS/L
36.9ft/sec 11.2m/sec
25.1mph 40.4kmh/1000rpm
16.5mpg 13.7mpgUS
17.1L/100km
Petrol 4-stroke piston
3528cc 215.0cu in
Vee 8 fuel injection
Compression ratio: 9.4:1
Bore: 89.0mm 3.5in
Stroke: 71.0mm 2.8in
Valve type/No: Overhead 16
Transmission: Manual
No. of forward speeds: 5
Wheels driven: 4-wheel drive
Springs F/R: Coil/Coil
Brake system: PA
Brakes F/R: Disc/Disc
Steering: Recirculating ball PA
Wheelbase: 254.0cm 100.0in
Track F: 148.6cm 58.5in
Track R: 148.6cm 58.5in
Length: 452.1cm 178.0in
Width: 179.3cm 70.6in
Height: 191.9cm 75.6in
Ground clearance: 19.0cm
7.5in
Kerb weight: 1885.0kg
4152.0lb
Fuel: 88.7L 19.5gal 23.
4galUS

454 Lexus

LS400
1990 Japan
148.0mph 238.1kmh
0-50mph 80.5kmh: 6.3secs
0-60mph 96.5kmh: 8.3secs
0-1/4 mile: 16.3secs
0-1km: 29.1secs
241.0bhp 179.7kW 244.3PS
@ 5400rpm
258.3lbft 350.0Nm @ 4400rpm
138.9bhp/ton 136.5bhp/tonne
60.7bhp/L 45.3kW/L 61.6PS/L
48.7ft/sec 14.8m/sec
27.1mph 43.6kmh/1000rpm
19.7mpg 16.4mpgUS
14.3L/100km
Petrol 4-stroke piston
3969cc 242.0cu in
Vee 8 fuel injection

Compression ratio: 10.0:1
Bore: 87.5mm 3.4in
Stroke: 82.5mm 3.2in
Valve type/No: Overhead 32
Transmission: Automatic
No. of forward speeds: 4
Wheels driven: Rear
Springs F/R: Coil/Coil
Brake system: PA ABS
Brakes F/R: Disc/Disc
Steering: Rack & pinion PA
Wheelbase: 281.4cm 110.8in
Track F: 156.5cm 61.6in
Track R: 155.4cm 61.2in
Length: 499.4cm 196.6in
Width: 181.9cm 71.6in
Height: 142.5cm 56.1in
Ground clearance: 15.7cm
6.2in
Kerb weight: 1765.0kg
3887.7lb
Fuel: 85.1L 18.7gal 22.5galUS

455 Lexus

SC400
1991 Japan
150.0mph 241.4kmh
0-50mph 80.5kmh: 5.3secs
0-60mph 96.5kmh: 6.9secs
0-1/4 mile: 15.2secs
250.0bhp 186.4kW 253.5PS
@ 5600rpm
260.0lbft 352.3Nm @ 4400rpm
153.8bhp/ton 151.3bhp/tonne
63.0bhp/L 47.0kW/L 63.9PS/L
50.6ft/sec 15.4m/sec
24.0mpg 20.0mpgUS
11.8L/100km
Petrol 4-stroke piston
3969cc 242.2cu in
Vee 8 fuel injection
Compression ratio: 10.0:1
Bore: 87.5mm 3.4in
Stroke: 82.5mm 3.2in
Valve type/No: Overhead 32
Transmission: Automatic
No. of forward speeds: 4
Wheels driven: Rear
Springs F/R: Coil/Coil
Brake system: PA ABS
Brakes F/R: Disc/Disc
Steering: Rack & pinion PA
Wheelbase: 269.0cm 105.9in
Track F: 152.1cm 59.9in
Track R: 152.4cm 60.0in
Length: 485.4cm 191.1in
Width: 179.1cm 70.5in
Height: 133.6cm 52.6in
Ground clearance: 14.0cm
5.5in
Kerb weight: 1652.6kg
3640.0lb
Fuel: 78.0L 17.1gal 20.6galUS

456 Lexus

ES300
1992 Japan
135.0mph 217.2kmh
0-50mph 80.5kmh: 6.8secs
0-60mph 96.5kmh: 9.0secs
0-1/4 mile: 17.0secs
185.0bhp 137.9kW 187.6PS
@ 5200rpm
195.0lbft 264.2Nm @ 4400rpm
121.5bhp/ton 119.5bhp/tonne
62.5bhp/L 46.6kW/L 63.4PS/L
46.7ft/sec 14.2m/sec
25.2mpg 21.0mpgUS
11.2L/100km
Petrol 4-stroke piston
2959cc 180.5cu in
Vee 6 fuel injection
Compression ratio: 9.6:1
Bore: 87.5mm 3.4in
Stroke: 82.0mm 3.2in
Valve type/No: Overhead 24
Transmission: Automatic
No. of forward speeds: 4
Wheels driven: Front
Springs F/R: Coil/Coil
Brake system: PA ABS
Brakes F/R: Disc/Disc
Steering: Rack & pinion PA
Wheelbase: 261.9cm 103.1in
Track F: 154.9cm 61.0in
Track R: 149.9cm 59.0in
Length: 477.0cm 187.8in
Width: 177.8cm 70.0in
Height: 136.9cm 53.9in
Ground clearance: 13.0cm
5.1in
Kerb weight: 1548.1kg
3410.0lb
Fuel: 70.0L 15.4gal 18.5galUS

457 Lincoln

Continental Mk VII
1989 USA
125.0mph 201.1kmh
0-50mph 80.5kmh: 6.0secs
0-60mph 96.5kmh: 8.0secs
0-1/4 mile: 16.3secs
225.0bhp 167.8kW 228.1PS
@ 4200rpm
300.0lbft 406.5Nm @ 3200rpm
133.9bhp/ton 131.6bhp/tonne
45.5bhp/L 33.9kW/L 46.2PS/L
35.0ft/sec 10.7m/sec
19.2mpg 16.0mpgUS
14.7L/100km
Petrol 4-stroke piston
4942cc 301.5cu in
Vee 8 fuel injection
Compression ratio: 9.0:1
Bore: 101.6mm 4.0in
Stroke: 76.2mm 3.0in
Valve type/No: Overhead 16
Transmission: Automatic

No. of forward speeds: 4
Wheels driven: Rear
Springs F/R: Gas/Gas
Brake system: PA ABS
Brakes F/R: Disc/Disc
Steering: Rack & pinion PA
Wheelbase: 275.6cm 108.5in
Track F: 148.3cm 58.4in
Track R: 149.9cm 59.0in
Length: 515.1cm 202.8in
Width: 180.1cm 70.9in
Height: 137.7cm 54.2in
Kerb weight: 1709.3kg
3765.0lb
Fuel: 83.6L 18.4gal 22.1galUS

458 Lincoln

Continental Mk VII LSC
1990 USA
125.0mph 201.1kmh
0-50mph 80.5kmh: 6.1secs
0-60mph 96.5kmh: 8.3secs
0-1/4 mile: 16.3secs
225.0bhp 167.8kW 228.1PS
@ 4200rpm
300.0lbft 406.5Nm @ 4200rpm
133.9bhp/ton 131.6bhp/tonne
45.5bhp/L 33.9kW/L 46.2PS/L
35.0ft/sec 10.7m/sec
19.2mpg 16.0mpgUS
14.7L/100km
Petrol 4-stroke piston
4942cc 301.5cu in
Vee 8 fuel injection
Compression ratio: 9.0:1
Bore: 101.6mm 4.0in
Stroke: 76.2mm 3.0in
Valve type/No: Overhead 16
Transmission: Automatic
No. of forward speeds: 4
Wheels driven: Rear
Springs F/R: Gas/Gas
Brake system: PA ABS
Brakes F/R: Disc/Disc
Steering: Rack & pinion PA
Wheelbase: 275.6cm 108.5in
Track F: 148.3cm 58.4in
Track R: 149.9cm 59.0in
Length: 515.1cm 202.8in
Width: 180.1cm 70.9in
Height: 137.7cm 54.2in
Kerb weight: 1709.3kg
3765.0lb
Fuel: 83.6L 18.4gal 22.1galUS

459 Lotus

Esprit Turbo
1986 UK
152.0mph 244.6kmh
0-50mph 80.5kmh: 4.4secs
0-60mph 96.5kmh: 5.6secs
0-1/4 mile: 14.3secs
215.0bhp 160.3kW 218.0PS
@ 6250rpm

194.0lbft 262.9Nm @ 5000rpm
175.1bhp/ton 172.2bhp/tonne
98.9bhp/L 73.7kW/L 100.3PS/L
52.1ft/sec 15.9m/sec
18.6mpg 15.5mpgUS
15.2L/100km
Petrol 4-stroke piston
2174cc 132.6cu in
turbocharged
In-line 4 fuel injection
Compression ratio: 8.5:1
Bore: 95.3mm 3.8in
Stroke: 76.2mm 3.0in
Valve type/No: Overhead 16
Transmission: Manual
No. of forward speeds: 5
Wheels driven: Rear
Springs F/R: Coil/Coil
Brake system: PA
Brakes F/R: Disc/Disc
Steering: Rack & pinion
Wheelbase: 243.8cm 96.0in
Track F: 153.7cm 60.5in
Track R: 155.4cm 61.2in
Length: 429.3cm 169.0in
Width: 185.4cm 73.0in
Height: 113.0cm 44.5in
Ground clearance: 14.7cm
5.8in
Kerb weight: 1248.5kg
2750.0lb
Fuel: 85.9L 18.9gal 22.7galUS

460 Lotus
Excel SE
1986 UK
134.0mph 215.6kmh
0-50mph 80.5kmh: 5.2secs
0-60mph 96.5kmh: 6.8secs
0-1/4 mile: 15.3secs
0-1km: 28.1secs
180.0bhp 134.2kW 182.5PS

@ 6500rpm
165.0lbft 223.6Nm @ 5000rpm
156.2bhp/ton 153.6bhp/tonne
82.8bhp/L 61.7kW/L 83.9PS/L
54.2ft/sec 16.5m/sec
21.0mph 33.8kmh/1000rpm
19.6mpg 16.3mpgUS
14 4L/100km
Petrol 4-stroke piston
2174cc 132.6cu in
In-line 4 2 Carburettor
Compression ratio: 10.9:1
Bore: 95.3mm 3.8in
Stroke: 76.2mm 3.0in
Valve type/No: Overhead 16
Transmission: Manual
No. of forward speeds: 5
Wheels driven: Rear
Springs F/R: Coil/Coil
Brake system: PA
Brakes F/R: Disc/Disc
Steering: Rack & pinion PA
Wheelbase: 248.2cm 97.7in
Track F: 147.8cm 58.2in
Track R: 147.8cm 58.2in
Length: 439.7cm 173.1in
Width: 181.6cm 71.5in
Height: 120.7cm 47.5in
Ground clearance: 12.2cm
4.8in
Kerb weight: 1171.8kg
2581.0lb
Fuel: 66.9L 14.7gal 17.7galUS

461 Lotus
Elan Autocrosser
1987 UK
115.0mph 185.0kmh
0-50mph 80.5kmh: 4.2secs
0-60mph 96.5kmh: 5.3secs
0-1/4 mile: 13.9secs
163.0bhp 121.5kW 165.3PS

@ 8000rpm
118.0lbft 159.9Nm @ 7000rpm
259.9bhp/ton 255.5bhp/tonne
102.3bhp/L 76.2kW/L
103.7PS/L
63.8ft/sec 19.4m/sec
7.2mpg 6.0mpgUS
39.2L/100km
Petrol 4-stroke piston
1594cc 97.2cu in
In-line 4 2 Carburettor
Compression ratio: 14.0:1
Bore: 83.5mm 3.3in
Stroke: 72.8mm 2.9in
Valve type/No: Overhead 16
Transmission: Manual
No. of forward speeds: 4
Wheels driven: Rear
Springs F/R: Coil/Coil
Brakes F/R: Disc/Disc
Steering: Rack & pinion
Wheelbase: 213.4cm 84.0in
Track F: 128.3cm 50.5in
Track R: 132.1cm 52.0in
Length: 368.3cm 145.0in
Width: 161.3cm 63.5in
Height: 109.2cm 43.0in
Ground clearance: 6.4cm
2.5in
Kerb weight: 637.9kg 1405.0lb
Fuel: 38.6L 8.5gal 10.2galUS

462 Lotus
Esprit Turbo HC
1987 UK
146.0mph 234.9kmh
0-50mph 80.5kmh: 4.4secs
0-60mph 96.5kmh: 5.6secs
0-1/4 mile: 14.4secs
0-1km: 26.2secs
215.0bhp 160.3kW 218.0PS
@ 6000rpm

220.0lbft 298.1Nm @ 4250rpm
190.6bhp/ton 187.4bhp/tonne
98.9bhp/L 73.7kW/L 100.3PS/L
50.0ft/sec 15.2m/sec
22.5mph 36.2kmh/1000rpm
20.9mpg 17.4mpgUS
13.5L/100km
Petrol 4-stroke piston
2174cc 133.0cu in
turbocharged
In-line 4 1 Carburettor
Compression ratio: 8.0:1
Bore: 95.3mm 3.8in
Stroke: 76.2mm 3.0in
Valve type/No: Overhead 16
Transmission: Manual
No. of forward speeds: 5
Wheels driven: Rear
Springs F/R: Coil/Coil
Brake system: PA
Brakes F/R: Disc/Disc
Steering: Rack & pinion
Wheelbase: 243.8cm 96.0in
Track F: 151.1cm 59.5in
Track R: 151.1cm 59.5in
Length: 422.4cm 166.3in
Width: 185.4cm 73.0in
Height: 101.7cm 40.0in
Kerb weight: 1147.0kg
2526.4lb
Fuel: 86.0L 18.9gal 22.7galUS

463 Lotus
Excel SA
1987 UK
128.0mph 206.0kmh
0-50mph 80.5kmh: 6.2secs
0-60mph 96.5kmh: 8.2secs
0-1/4 mile: 16.9secs
0-1km: 31.0secs
180.0bhp 134.2kW 182.5PS
@ 6500rpm

464 Lotus
Esprit Turbo

165.0lbft 223.6Nm @ 5000rpm
155.5bhp/ton 152.9bhp/tonne
82.8bhp/L 61.7kW/L 83.9PS/L
54.2ft/sec 16.5m/sec
25.0mph 40.2kmh/1000rpm
22.2mpg 18.5mpgUS
12.7L/100km
Petrol 4-stroke piston
2174cc 133.0cu in
In-line 4 2 Carburettor
Compression ratio: 10.9:1
Bore: 95.3mm 3.8in
Stroke: 76.2mm 3.0in
Valve type/No: Overhead 16
Transmission: Automatic
No. of forward speeds: 4
Wheels driven: Rear
Springs F/R: Coil/Coil
Brake system: PA
Brakes F/R: Disc/Disc
Steering: Rack & pinion PA
Wheelbase: 248.3cm 97.8in
Track F: 147.9cm 58.2in
Track R: 147.9cm 58.2in
Length: 439.8cm 173.1in
Width: 181.6cm 71.5in
Height: 120.7cm 47.5in
Ground clearance: 12.2cm
4.8in
Kerb weight: 1177.0kg
2592.5lb
Fuel: 66.9L 14.7gal 17.7galUS

464 Lotus
Esprit Turbo
1988 UK
153.0mph 246.2kmh
0-50mph 80.5kmh: 4.3secs
0-60mph 96.5kmh: 5.4secs
0-1/4 mile: 13.7secs
0-1km: 25.0secs
215.0bhp 160.3kW 218.0PS
@ 6000rpm
220.0lbft 298.1Nm @ 4250rpm
157.8bhp/ton 155.2bhp/tonne
98.9bhp/L 73.7kW/L 100.3PS/L
50.0ft/sec 15.2m/sec
23.7mph 38.1kmh/1000rpm
20.4mpg 17.0mpgUS
13.8L/100km
Petrol 4-stroke piston
2174cc 132.6cu in
turbocharged
In-line 4 2 Carburettor
Compression ratio: 8.0:1
Bore: 95.3mm 3.8in
Stroke: 76.2mm 3.0in
Valve type/No: Overhead 16
Transmission: Manual
No. of forward speeds: 5
Wheels driven: Rear
Springs F/R: Coil/Coil
Brake system: PA
Brakes F/R: Disc/Disc

Steering: Rack & pinion
Wheelbase: 243.8cm 96.0in
Track F: 152.4cm 60.0in
Track R: 155.4cm 61.2in
Length: 433.1cm 170.5in
Width: 185.9cm 73.2in
Height: 115.1cm 45.3in
Ground clearance: 14.7cm
5.8in
Kerb weight: 1385.6kg
3052.0lb
Fuel: 78.7L 17.3gal 20.8galUS

465 Lotus
Esprit Turbo
1989 UK
156.0mph 251.0kmh
0-60mph 96.5kmh: 5.2secs
0-1/4 mile: 13.6secs
228.0bhp 170.0kW 231.2PS
@ 6500rpm
218.0lbft 295.4Nm @ 4000rpm
179.5bhp/ton 176.5bhp/tonne
104.9bhp/L 78.2kW/L
106.3PS/L
54.2ft/sec 16.5m/sec
20.4mpg 17.0mpgUS
13.8L/100km
Petrol 4-stroke piston
2174cc 132.6cu in
turbocharged
In-line 4 fuel injection
Compression ratio: 8.0:1
Bore: 95.3mm 3.8in
Stroke: 76.2mm 3.0in
Valve type/No: Overhead 16
Transmission: Manual
No. of forward speeds: 5
Wheels driven: Rear
Springs F/R: Coil/Coil
Brake system: PA
Brakes F/R: Disc/Disc
Steering: Rack & pinion PA
Wheelbase: 243.8cm 96.0in
Track F: 152.4cm 60.0in
Track R: 155.4cm 61.2in
Length: 433.1cm 170.5in
Width: 185.9cm 73.2in
Height: 115.1cm 45.3in
Kerb weight: 1291.6kg
2845.0lb
Fuel: 70.0L 15.4gal 18.5galUS

466 Lotus
Esprit Turbo SE
1989 UK
161.0mph 259.0kmh
0-50mph 80.5kmh: 3.8secs
0-60mph 96.5kmh: 4.9secs
0-1/4 mile: 13.5secs
0-1km: 25.3secs
264.0bhp 196.9kW 267.7PS
@ 6500rpm
261.3lbft 354.0Nm @ 3900rpm

201.9bhp/ton 198.5bhp/tonne
121.4bhp/L 90.6kW/L
123.1PS/L
54.2ft/sec 16.5m/sec
23.1mph 37.2kmh/1000rpm
23.5mpg 19.6mpgUS
12.0L/100km
Petrol 4-stroke piston
2174cc 133.0cu in
turbocharged
In-line 4 fuel injection
Compression ratio: 8.1:1
Bore: 93.5mm 3.7in
Stroke: 76.2mm 3.0in
Valve type/No: Overhead 16
Transmission: Manual
No. of forward speeds: 5
Wheels driven: Rear
Springs F/R: Coil/Coil
Brake system: PA
Brakes F/R: Disc/Disc
Steering: Rack & pinion
Wheelbase: 243.8cm 96.0in
Track F: 152.4cm 60.0in
Track R: 155.4cm 61.2in
Length: 433.0cm 170.5in
Width: 186.0cm 73.2in
Height: 115.0cm 45.3in
Ground clearance: 14.6cm
5.7in
Kerb weight: 1330.0kg
2929.5lb
Fuel: 70.1L 15.4gal 18.5galUS

467 Lotus
Elan
1990 UK
137.0mph 220.4kmh
0-60mph 96.5kmh: 6.7secs
0-1/4 mile: 15.4secs
165.0bhp 123.0kW 167.3PS
@ 6600rpm
148.0lbft 200.5Nm @ 4200rpm
164.3bhp/ton 161.5bhp/tonne
103.9bhp/L 77.5kW/L
105.3PS/L
57.0ft/sec 17.4m/sec
Petrol 4-stroke piston
1588cc 96.9cu in
turbocharged
In-line 4 fuel injection
Compression ratio: 8.2:1
Bore: 80.0mm 3.1in
Stroke: 79.0mm 3.1in
Valve type/No: Overhead 16
Transmission: Manual
No. of forward speeds: 5
Wheels driven: Front
Springs F/R: Coil/Coil
Brakes F/R: Disc/Disc
Steering: Rack & pinion PA
Wheelbase: 225.0cm 88.6in
Track F: 148.6cm 58.5in
Track R: 148.6cm 58.5in

Length: 380.2cm 149.7in
Width: 173.5cm 68.3in
Height: 122.9cm 48.4in
Kerb weight: 1021.5kg
2250.0lb
Fuel: 38.6L 8.5gal 10.2galUS

468 Lotus
Elan SE
1990 UK
137.0mph 220.4kmh
0-50mph 80.5kmh: 5.0secs
0-60mph 96.5kmh: 6.5secs
0-1/4 mile: 15.0secs
0-1km: 27.4secs
165.0bhp 123.0kW 167.3PS
@ 6600rpm
147.6lbft 200.0Nm @ 4200rpm
164.0bhp/ton 161.3bhp/tonne
103.9bhp/L 77.5kW/L
105.3PS/L
57.0ft/sec 17.4m/sec
20.8mph 33.5kmh/1000rpm
20.1mpg 16.7mpgUS
14.1L/100km
Petrol 4-stroke piston
1588cc 97.0cu in
turbocharged
In-line 4 fuel injection
Compression ratio: 8.2:1
Bore: 80.0mm 3.1in
Stroke: 79.0mm 3.1in
Valve type/No: Overhead 16
Transmission: Manual
No. of forward speeds: 5
Wheels driven: Front
Springs F/R: Coil/Coil
Brake system: PA
Brakes F/R: Disc/Disc
Steering: Rack & pinion PA
Wheelbase: 225.0cm 88.6in
Track F: 148.6cm 58.5in
Track R: 148.6cm 58.5in
Length: 380.2cm 149.7in
Width: 188.7cm 74.3in
Height: 122.9cm 48.4in
Ground clearance: 13.0cm
5.1in
Kerb weight: 1023.0kg
2253.3lb
Fuel: 46.4L 10.2gal 12.3galUS

469 Lotus
Esprit Turbo SE
1990 UK
165.0mph 265.5kmh
0-50mph 80.5kmh: 4.1secs
0-60mph 96.5kmh: 5.1secs
0-1/4 mile: 13.7secs
264.0bhp 196.9kW 267.7PS
@ 6500rpm
261.0lbft 353.7Nm @ 3900rpm
206.4bhp/ton 203.0bhp/tonne
121.4bhp/L 90.6kW/L

471 Lynx
D Type

123.1PS/L
54.2ft/sec 16.5m/sec
19.8mpg 16.5mpgUS
14.3L/100km
Petrol 4-stroke piston
2174cc 132.6cu in
turbocharged
In-line 4 fuel injection
Compression ratio: 8.0:1
Bore: 95.3mm 3.8in
Stroke: 76.2mm 3.0in
Valve type/No: Overhead 16
Transmission: Manual
No. of forward speeds: 5
Wheels driven: Rear
Springs F/R: Coil/Coil
Brake system: PA
Brakes F/R: Disc/Disc
Steering: Rack & pinion PA
Wheelbase: 243.8cm 96.0in
Track F: 152.4cm 60.0in
Track R: 155.4cm 61.2in
Length: 433.1cm 170.5in
Width: 185.9cm 73.2in
Height: 115.1cm 45.3in
Kerb weight: 1300.7kg
2865.0lb
Fuel: 70.0L 15.4gal 18.5galUS

470 Lotus
Elan SE
1991 UK
137.0mph 220.4kmh
0-50mph 80.5kmh: 5.1secs
0-60mph 96.5kmh: 6.6secs
0-1/4 mile: 15.2secs
162.0bhp 120.8kW 164.2PS
@ 6600rpm
148.0lbft 200.5Nm @ 4200rpm
148.1bhp/ton 145.6bhp/tonne
102.0bhp/L 76.1kW/L
103.4PS/L
57.0ft/sec 17.4m/sec

31.2mpg 26.0mpgUS
9.0L/100km
Petrol 4-stroke piston
1588cc 96.9cu in
turbocharged
In-line 4 fuel injection
Compression ratio: 8.2:1
Bore: 80.0mm 3.1in
Stroke: 79.0mm 3.1in
Valve type/No: Overhead 16
Transmission: Manual
No. of forward speeds: 5
Wheels driven: Front
Springs F/R: Coil/Coil
Brake system: PA ABS
Brakes F/R: Disc/Disc
Steering: Rack & pinion PA
Wheelbase: 225.0cm 88.6in
Track F: 148.6cm 58.5in
Track R: 148.6cm 58.5in
Length: 386.6cm 152.2in
Width: 173.2cm 68.2in
Height: 122.9cm 48.4in
Ground clearance: 12.7cm
5.0in
Kerb weight: 1112.3kg
2450.0lb
Fuel: 38.6L 8.5gal 10.2galUS

471 Lynx
D Type
1987 UK
150.0mph 241.4kmh
0-60mph 96.5kmh: 5.3secs
0-1/4 mile: 13.9secs
285.0bhp 212.5kW 288.9PS
@ 5300rpm
299.7bhp/ton 294.7bhp/tonne
75.4bhp/L 56.2kW/L 76.4PS/L
61.4ft/sec 18.7m/sec
Petrol 4-stroke piston
3781cc 230.7cu in
In-line 6 3 Carburettor

Compression ratio: 9.0:1
Bore: 87.0mm 3.4in
Stroke: 106.0mm 4.2in
Valve type/No: Overhead 12
Transmission: Manual
No. of forward speeds: 4
Wheels driven: Rear
Springs F/R: Torsion bar/Coil
Brake system: PA
Brakes F/R: Disc/Disc
Steering: Rack & pinion
Wheelbase: 229.9cm 90.5in
Track F: 127.0cm 50.0in
Track R: 127.0cm 50.0in
Length: 391.2cm 154.0in
Width: 168.9cm 66.5in
Height: 80.0cm 31.5in
Kerb weight: 967.0kg 2130.0lb
Fuel: 100.3L 22.0gal
26.5galUS

472 Lynx
XJS Eventer
1988 UK
151.0mph 243.0kmh
0-60mph 96.5kmh: 7.6secs
0-1/4 mile: 15.6secs
295.0bhp 220.0kW 299.1PS
@ 5500rpm
320.0lbft 433.6Nm @ 3250rpm
166.4bhp/ton 163.7bhp/tonne
55.2bhp/L 41.2kW/L 56.0PS/L
42.2ft/sec 12.8m/sec
Petrol 4-stroke piston
5345cc 326.1cu in
Vee 12 fuel injection
Compression ratio: 11.5:1
Bore: 90.0mm 3.5in
Stroke: 70.0mm 2.8in
Valve type/No: Overhead 24
Transmission: Automatic
No. of forward speeds: 3
Wheels driven: Rear

Springs F/R: Coil/Coil
Brake system: PA
Brakes F/R: Disc/Disc
Steering: Rack & pinion PA
Wheelbase: 259.1cm 102.0in
Track F: 148.8cm 58.6in
Track R: 150.4cm 59.2in
Length: 476.5cm 187.6in
Width: 179.3cm 70.6in
Height: 128.8cm 50.7in
Kerb weight: 1802.4kg
3970.0lb
Fuel: 84.0L 18.5gal 22.2galUS

473 Maserati
Biturbo Spyder
1987 Italy
128.0mph 206.0kmh
0-50mph 80.5kmh: 5.3secs
0-60mph 96.5kmh: 7.2secs
0-1/4 mile: 16.0secs
0-1km: 29.5secs
192.0bhp 143.2kW 194.7PS
@ 5500rpm
220.0lbft 298.1Nm @ 3000rpm
156.8bhp/ton 154.2bhp/tonne
77.1bhp/L 57.5kW/L 78.1PS/L
37.9ft/sec 11.5m/sec
22.9mph 36.8kmh/1000rpm
18.5mpg 15.4mpgUS
15.3L/100km
Petrol 4-stroke piston
2491cc 152.0cu in
turbocharged
Vee 6 1 Carburettor
Compression ratio: 7.4:1
Bore: 91.6mm 3.6in
Stroke: 63.0mm 2.5in
Valve type/No: Overhead 18
Transmission: Manual
No. of forward speeds: 5
Wheels driven: Rear
Springs F/R: Coil/Coil

475 Maserati
Biturbo

Brake system: PA
Brakes F/R: Disc/Disc
Steering: Rack & pinion PA
Wheelbase: 240.0cm 94.5in
Track F: 142.0cm 55.9in
Track R: 143.1cm 56.3in
Length: 404.3cm 159.2in
Width: 171.4cm 67.5in
Height: 131.0cm 51.6in
Ground clearance: 12.5cm
4.9in
Kerb weight: 1245.0kg
2742.3lb
Fuel: 80.1L 17.6gal 21.1galUS

474 Maserati
430
1988 Italy
130.0mph 209.2kmh
0-60mph 96.5kmh: 6.5secs
225.0bhp 167.8kW 228.1PS
@ 5500rpm
266.0lbft 360.4Nm @ 3500rpm
192.4bhp/ton 189.2bhp/tonne
80.6bhp/L 60.1kW/L 81.8PS/L
40.3ft/sec 12.3m/sec
Petrol 4-stroke piston
2790cc 170.2cu in
turbocharged
Vee 6 fuel injection
Compression ratio: 7.4:1
Bore: 94.0mm 3.7in
Stroke: 67.0mm 2.6in
Valve type/No: Overhead 12
Transmission: Manual
No. of forward speeds: 5
Wheels driven: Rear
Springs F/R: Coil/Coil
Brake system: PA
Brakes F/R: Disc/Disc
Steering: Rack & pinion PA
Wheelbase: 260.1cm 102.4in
Track F: 144.3cm 56.8in

Track R: 145.3cm 57.2in
Length: 439.9cm 173.2in
Width: 173.0cm 68.1in
Height: 130.8cm 51.5in
Kerb weight: 1189.5kg
2620.0lb
Fuel: 81.8L 18.0gal 21.6galUS

475 Maserati
Biturbo
1988 Italy
129.0mph 207.6kmh
0-50mph 80.5kmh: 5.3secs
0-60mph 96.5kmh: 6.9secs
0-1/4 mile: 15.3secs
187.0bhp 139.4kW 189.6PS
@ 5500rpm
240.0lbft 325.2Nm @ 3000rpm
157.8bhp/ton 155.1bhp/tonne
75.1bhp/L 56.0kW/L 76.1PS/L
37.9ft/sec 11.5m/sec
16.2mpg 13.5mpgUS
17.4L/100km
Petrol 4-stroke piston
2491cc 152.0cu in
turbocharged
Vee 6 fuel injection
Compression ratio: 7.3:1
Bore: 91.6mm 3.6in
Stroke: 63.0mm 2.5in
Valve type/No: Overhead 12
Transmission: Manual
No. of forward speeds: 5
Wheels driven: Rear
Springs F/R: Coil/Coil
Brake system: PA
Brakes F/R: Disc/Disc
Steering: Rack & pinion PA
Wheelbase: 251.5cm 99.0in
Track F: 142.0cm 55.9in
Track R: 143.3cm 56.4in
Length: 415.3cm 163.5in
Width: 171.5cm 67.5in

Height: 130.6cm 51.4in
Kerb weight: 1205.4kg
2655.0lb
Fuel: 66.6L 14.6gal 17.6galUS

476 Maserati
430
1989 Italy
145.0mph 233.3kmh
0-60mph 96.5kmh: 6.3secs
0-1/4 mile: 15.0secs
225.0bhp 167.8kW 228.1PS
@ 5600rpm
246.0lbft 333.3Nm @ 3500rpm
165.0bhp/ton 162.2bhp/tonne
80.6bhp/L 60.1kW/L 81.8PS/L
41.1ft/sec 12.5m/sec
20.4mpg 17.0mpgUS
13.8L/100km
Petrol 4-stroke piston
2790cc 170.2cu in
turbocharged
Vee 6 fuel injection
Compression ratio: 7.7:1
Bore: 94.0mm 3.7in
Stroke: 67.0mm 2.6in
Valve type/No: Overhead 18
Transmission: Manual
No. of forward speeds: 5
Wheels driven: Rear
Springs F/R: Coil/Coil
Brake system: PA
Brakes F/R: Disc/Disc
Steering: Rack & pinion PA
Wheelbase: 260.1cm 102.4in
Track F: 144.3cm 56.8in
Track R: 145.3cm 57.2in
Length: 439.9cm 173.2in
Width: 173.0cm 68.1in
Height: 135.9cm 53.5in
Kerb weight: 1387.0kg
3055.0lb
Fuel: 75.7L 16.6gal 20.0galUS

477 Maserati
Spyder
1989 Italy
130.0mph 209.2kmh
0-50mph 80.5kmh: 5.2secs
0-60mph 96.5kmh: 6.7secs
0-1/4 mile: 15.2secs
225.0bhp 167.8kW 228.1PS
@ 5600rpm
246.0lbft 333.3Nm @ 3500rpm
168.0bhp/ton 165.2bhp/tonne
80.6bhp/L 60.1kW/L 81.8PS/L
41.1ft/sec 12.5m/sec
18.6mpg 15.5mpgUS
15.2L/100km
Petrol 4-stroke piston
2790cc 170.2cu in
turbocharged
Vee 6 fuel injection
Compression ratio: 7.4:1
Bore: 94.0mm 3.7in
Stroke: 67.0mm 2.6in
Valve type/No: Overhead 18
Transmission: Manual
No. of forward speeds: 5
Wheels driven: Rear
Springs F/R: Coil/Coil
Brake system: PA
Brakes F/R: Disc/Disc
Steering: Rack & pinion PA
Wheelbase: 240.0cm 94.5in
Track F: 150.1cm 59.1in
Track R: 147.6cm 58.1in
Length: 404.4cm 159.2in
Width: 171.5cm 67.5in
Height: 131.1cm 51.6in
Kerb weight: 1362.0kg
3000.0lb
Fuel: 68.1L 15.0gal 18.0galUS

478 Maserati
228
1990 Italy

142.0mph 228.5kmh
0-50mph 80.5kmh: 5.2secs
0-60mph 96.5kmh: 6.7secs
0-1/4 mile: 15.2secs
225.0bhp 167.8kW 228.1PS
@ 5600rpm
246.0lbft 333.3Nm @ 3500rpm
166.9bhp/ton 164.1bhp/tonne
80.6bhp/L 60.1kW/L 81.8PS/L
41.1ft/sec 12.5m/sec
18.6mpg 15.5mpgUS
15.2L/100km
Petrol 4-stroke piston
2790cc 170.2cu in
turbocharged
Vee 6 fuel injection
Compression ratio: 7.4:1
Bore: 94.0mm 3.7in
Stroke: 67.0mm 2.6in
Valve type/No: Overhead 18
Transmission: Manual
No. of forward speeds: 5
Wheels driven: Rear
Springs F/R: Coil/Coil
Brake system: PA
Brakes F/R: Disc/Disc
Steering: Rack & pinion PA
Wheelbase: 260.1cm 102.4in
Track F: 153.9cm 60.6in
Track R: 154.9cm 61.0in
Length: 446.0cm 175.6in
Width: 186.4cm 73.4in
Height: 133.1cm 52.4in
Kerb weight: 1371.1kg
3020.0lb
Fuel: 65.9L 14.5gal 17.4galUS

479 Mazda
323 1.5 GLX Estate
1986 Japan
98.0mph 157.7kmh
0-50mph 80.5kmh: 8.5secs
0-60mph 96.5kmh: 12.3secs
0-1/4 mile: 19.4secs
0-1km: 36.3secs
74.0bhp 55.2kW 75.0PS
@ 5500rpm
84.6lbft 114.6Nm @ 3500rpm
77.7bhp/ton 76.4bhp/tonne
49.7bhp/L 37.0kW/L 50.4PS/L
48.1ft/sec 14.7m/sec
21.8mph 35.1km/h/1000rpm
20.4mpg 17.0mpgUS
13.8L/100km
Petrol 4-stroke piston
1490cc 91.0cu in
In-line 4 1 Carburettor
Compression ratio: 9.0:1
Bore: 77.0mm 3.0in
Stroke: 80.0mm 3.1in
Valve type/No: Overhead 8
Transmission: Manual
No. of forward speeds: 5
Wheels driven: Front

Springs F/R: Coil/Coil
Brake system: PA
Brakes F/R: Disc/Drum
Steering: Rack & pinion
Wheelbase: 240.0cm 94.5in
Track F: 139.0cm 54.7in
Track R: 141.5cm 55.7in
Length: 422.5cm 166.3in
Width: 164.5cm 64.8in
Height: 143.0cm 56.3in
Ground clearance: 15.5cm
6.1in
Kerb weight: 969.0kg 2134.4lb
Fuel: 45.0L 9.9gal 11.9galUS

480 Mazda
323 LX
1986 Japan
104.0mph 167.3kmh
0-60mph 96.5kmh: 11.1secs
0-1/4 mile: 18.2secs
82.0bhp 61.1kW 83.1PS
@ 5000rpm
92.0lbft 124.7Nm @ 2500rpm
79.7bhp/ton 78.4bhp/tonne
51.3bhp/L 38.3kW/L 52.1PS/L
45.7ft/sec 13.9m/sec
29.4mpg 24.5mpgUS
9.6L/100km
Petrol 4-stroke piston
1597cc 97.4cu in
In-line 4 fuel injection
Compression ratio: 9.3:1
Bore: 78.0mm 3.1in
Stroke: 83.6mm 3.3in
Valve type/No: Overhead 8
Transmission: Automatic
No. of forward speeds: 3
Wheels driven: Front
Springs F/R: Coil/Coil
Brake system: PA
Brakes F/R: Disc/Drum
Steering: Rack & pinion PA
Wheelbase: 240.0cm 94.5in
Track F: 138.9cm 54.7in
Track R: 141.5cm 55.7in
Length: 411.0cm 161.8in
Width: 164.6cm 64.8in
Height: 138.9cm 54.7in
Kerb weight: 1046.5kg
2305.0lb
Fuel: 45.0L 9.9gal 11.9galUS

481 Mazda
323 Turbo 4x4 Lux
1986 Japan
125.0mph 201.1kmh
0-50mph 80.5kmh: 5.6secs
0-60mph 96.5kmh: 7.9secs
0-1/4 mile: 16.5secs
0-1km: 29.6secs
148.0bhp 110.4kW 150.0PS
@ 6000rpm
143.8lbft 194.9Nm @ 5000rpm

135.7bhp/ton 133.4bhp/tonne
92.7bhp/L 69.1kW/L 94.0PS/L
54.8ft/sec 16.7m/sec
20.2mph 32.5kmh/1000rpm
21.3mpg 17.7mpgUS
13.3L/100km
Petrol 4-stroke piston
1597cc 97.0cu in
turbocharged
In-line 4 fuel injection
Compression ratio: 7.9:1
Bore: 78.0mm 3.1in
Stroke: 83.6mm 3.3in
Valve type/No: Overhead 16
Transmission: Manual
No. of forward speeds: 5
Wheels driven: 4-wheel drive
Springs F/R: Coil/Coil
Brake system: PA
Brakes F/R: Disc/Disc
Steering: Rack & pinion PA
Wheelbase: 240.0cm 94.5in
Track F: 140.5cm 55.3in
Track R: 142.5cm 56.1in
Length: 399.0cm 157.1in
Width: 164.5cm 64.8in
Height: 135.5cm 53.3in
Ground clearance: 15.3cm
6.0in
Kerb weight: 1109.0kg
2442.7lb
Fuel: 52.8L 11.6gal 13.9galUS

482 Mazda
626 2.0i 5-door
1986 Japan
121.0mph 194.7kmh
0-50mph 80.5kmh: 7.5secs
0-60mph 96.5kmh: 9.8secs
0-1/4 mile: 17.8secs
0-1km: 32.4secs
118.0bhp 88.0kW 119.6PS
@ 5400rpm
125.8lbft 170.5Nm @ 4000rpm
101.9bhp/ton 100.2bhp/tonne
59.1bhp/L 44.0kW/L 59.9PS/L
50.8ft/sec 15.5m/sec
21.8mph 35.1km/h/1000rpm
29.5mpg 24.6mpgUS
9.6L/100km
Petrol 4-stroke piston
1998cc 122.0cu in
In-line 4 fuel injection
Compression ratio: 10.0:1
Bore: 86.0mm 3.4in
Stroke: 86.0mm 3.4in
Valve type/No: Overhead 8
Transmission: Manual
No. of forward speeds: 5
Wheels driven: Front
Springs F/R: Coil/Coil
Brake system: PA
Brakes F/R: Disc/Disc
Steering: Rack & pinion PA

Wheelbase: 251.0cm 98.8in
Track F: 143.0cm 56.3in
Track R: 142.5cm 56.1in
Length: 443.0cm 174.4in
Width: 169.0cm 66.5in
Height: 136.4cm 53.7in
Ground clearance: 16.5cm
6.5in
Kerb weight: 1178.0kg
2594.7lb
Fuel: 59.1L 13.0gal 15.6galUS

483 Mazda
626 GT
1986 Japan
120.0mph 193.1kmh
0-50mph 80.5kmh: 5.6secs
0-60mph 96.5kmh: 7.8secs
0-1/4 mile: 16.0secs
120.0bhp 89.5kW 121.7PS
@ 5000rpm
150.0lbft 203.3Nm @ 3000rpm
100.1bhp/ton 98.4bhp/tonne
60.1bhp/L 44.8kW/L 60.9PS/L
47.1ft/sec 14.3m/sec
28.8mpg 24.0mpgUS
9.8L/100km
Petrol 4-stroke piston
1998cc 121.9cu in
turbocharged
In-line 4 fuel injection
Compression ratio: 7.8:1
Bore: 86.0mm 3.4in
Stroke: 86.0mm 3.4in
Valve type/No: Overhead 8
Transmission: Manual
No. of forward speeds: 5
Wheels driven: Front
Springs F/R: Coil/Coil
Brake system: PA
Brakes F/R: Disc/Disc
Steering: Rack & pinion PA
Wheelbase: 251.0cm 98.8in
Track F: 143.0cm 56.3in
Track R: 142.5cm 56.1in
Length: 451.6cm 177.8in
Width: 168.9cm 66.5in
Height: 136.4cm 53.7in
Kerb weight: 1219.0kg
2685.0lb
Fuel: 59.8L 13.1gal 15.8galUS

484 Mazda
RX-7
1986 Japan
135.0mph 217.2kmh
0-50mph 80.5kmh: 5.7secs
0-60mph 96.5kmh: 8.5secs
0-1/4 mile: 16.2secs
0-1km: 29.8secs
150.0bhp 111.9kW 152.1PS
@ 6500rpm
134.5lbft 182.2Nm @ 3000rpm
124.9bhp/ton 122.8bhp/tonne

114.7bhp/L 85.5kW/L
116.3PS/L
21.4mph 34.4kmh/1000rpm
18.7mpg 15.6mpgUS
15.1L/100km
Petrol Wankel rotary
1308cc 80.0cu in
Rotary 2 fuel injection
Compression ratio: 9.4:1
Valve type/No: Ports
Transmission: Manual
No. of forward speeds: 5
Wheels driven: Rear
Springs F/R: Coil/Coil
Brake system: PA
Brakes F/R: Disc/Disc
Steering: Rack & pinion
Wheelbase: 243.0cm 95.7in
Track F: 145.0cm 57.1in
Track R: 144.0cm 56.7in
Length: 429.0cm 168.9in
Width: 169.0cm 66.5in
Height: 126.5cm 49.8in
Ground clearance: 16.0cm
6.3in
Kerb weight: 1221.0kg
2689.4lb
Fuel: 63.2L 13.9gal 16.7galUS

485 Mazda

RX-7 GXL
1986 Japan
128.0mph 206.0kmh
0-60mph 96.5kmh: 7.9secs
0-1/4 mile: 16.1secs
146.0bhp 108.9kW 148.0PS
@ 6500rpm
138.0lbft 187.0Nm @ 3500rpm
121.1bhp/ton 119.1bhp/tonne
111.6bhp/L 83.2kW/L
113.2PS/L
22.8mpg 19.0mpgUS
12.4L/100km
Petrol Wankel rotary
1308cc 79.8cu in
Rotary 2 fuel injection
Compression ratio: 9.4:1
Transmission: Manual
No. of forward speeds: 5
Wheels driven: Rear
Springs F/R: Coil/Coil
Brake system: PA
Brakes F/R: Disc/Disc
Steering: Rack & pinion PA
Wheelbase: 243.1cm 95.7in
Track F: 145.0cm 57.1in
Track R: 144.0cm 56.7in
Length: 429.0cm 168.9in
Width: 168.9cm 66.5in
Height: 126.5cm 49.8in
Ground clearance: 14.7cm
5.8in
Kerb weight: 1225.8kg
2700.0lb

Fuel: 62.8L 13.8gal 16.6galUS

486 Mazda

RX-7 Turbo
1986 Japan
150.0mph 241.4kmh
0-50mph 80.5kmh: 5.1secs
0-60mph 96.5kmh: 6.6secs
0-1/4 mile: 15.2secs
182.0bhp 135.7kW 184.5PS
@ 6500rpm
183.0lbft 248.0Nm @ 3500rpm
144.1bhp/ton 141.6bhp/tonne
139.1bhp/L 103.8kW/L
141.1PS/L
23.1mpg 19.2mpgUS
12.3L/100km
Petrol Wankel rotary
1308cc 79.8cu in
turbocharged
Rotary 2 fuel injection
Compression ratio: 8.5:1
Transmission: Manual
No. of forward speeds: 5
Wheels driven: Rear
Springs F/R: Coil/Coil
Brake system: PA
Brakes F/R: Disc/Disc
Steering: Rack & pinion PA
Wheelbase: 243.1cm 95.7in
Track F: 145.0cm 57.1in
Track R: 144.0cm 56.7in
Length: 429.0cm 168.9in
Width: 168.9cm 66.5in
Height: 126.5cm 49.8in
Kerb weight: 1284.8kg
2830.0lb
Fuel: 62.8L 13.8gal 16.6galUS

487 Mazda

323 1.5 GLX
1987 Japan
101.0mph 162.5kmh
0-50mph 80.5kmh: 7.7secs
0-60mph 96.5kmh: 11.1secs
0-1/4 mile: 18.2secs
0-1km: 34.0secs
72.0bhp 53.7kW 73.0PS
@ 5700rpm
82.0lbft 111.1Nm @ 2200rpm
77.2bhp/ton 75.9bhp/tonne
48.1bhp/L 35.8kW/L 48.7PS/L
48.6ft/sec 14.8m/sec
22.0mph 35.4kmh/1000rpm
30.1mpg 25.1mpgUS
9.4L/100km
Petrol 4-stroke piston
1498cc 91.0cu in
In-line 4 1 Carburettor
Compression ratio: 9.0:1
Bore: 74.0mm 2.9in
Stroke: 78.0mm 3.1in
Valve type/No: Overhead 8
Transmission: Manual

No. of forward speeds: 5
Wheels driven: Front
Springs F/R: Coil/Coil
Brake system: PA
Brakes F/R: Disc/Drum
Steering: Rack & pinion PA
Wheelbase: 240.0cm 94.5in
Track F: 140.5cm 55.3in
Track R: 142.5cm 56.1in
Length: 419.5cm 165.2in
Width: 164.5cm 64.8in
Height: 139.0cm 54.7in
Ground clearance: 15.3cm
6.0in
Kerb weight: 948.0kg 2088.1lb
Fuel: 47.8L 10.5gal 12.6galUS

488 Mazda

626 2.0 GLS Executive
1987 Japan
113.0mph 181.8kmh
0-50mph 80.5kmh: 7.6secs
0-60mph 96.5kmh: 10.4secs
0-1/4 mile: 17.9secs
0-1km: 32.7secs
108.0bhp 80.5kW 109.5PS
@ 5300rpm
121.0lbft 164.0Nm @ 3300rpm
91.8bhp/ton 90.3bhp/tonne
54.0bhp/L 40.3kW/L 54.8PS/L
49.9ft/sec 15.2m/sec
20.7mph 33.3kmh/1000rpm
24.8mpg 20.7mpgUS
11.4L/100km
Petrol 4-stroke piston
1998cc 122.0cu in
In-line 4 1 Carburettor
Compression ratio: 9.5:1
Bore: 86.0mm 3.4in
Stroke: 86.0mm 3.4in
Valve type/No: Overhead 12
Transmission: Manual
No. of forward speeds: 5
Wheels driven: Front
Springs F/R: Coil/Coil
Brake system: PA
Brakes F/R: Disc/Disc
Steering: Rack & pinion PA
Wheelbase: 257.5cm 101.4in
Track F: 145.5cm 57.3in
Track R: 146.5cm 57.7in
Length: 451.5cm 177.8in
Width: 169.0cm 66.5in
Height: 137.5cm 54.1in
Ground clearance: 16.5cm
6.5in
Kerb weight: 1196.0kg
2634.4lb
Fuel: 60.1L 13.2gal 15.9galUS

489 Mazda

121 1.3 LX Sun Top
1988 Japan
92.0mph 148.0kmh

No. of forward speeds: 5
Wheels driven: Front
Springs F/R: Coil/Coil
Brake system: PA
Brakes F/R: Disc/Drum
Steering: Rack & pinion PA
Wheelbase: 240.0cm 94.5in
Track F: 140.5cm 55.3in
Track R: 142.5cm 56.1in
Length: 419.5cm 165.2in
Width: 164.5cm 64.8in
Height: 139.0cm 54.7in
Ground clearance: 15.3cm
6.0in
Kerb weight: 948.0kg 2088.1lb
Fuel: 47.8L 10.5gal 12.6galUS

0-50mph 80.5kmh: 7.7secs
0-60mph 96.5kmh: 11.0secs
0-1/4 mile: 18.1secs
0-1km: 34.3secs
65.0bhp 48.5kW 65.9PS
@ 5600rpm
76.0lbft 103.0Nm @ 3600rpm
84.5bhp/ton 83.1bhp/tonne
49.1bhp/L 36.6kW/L 49.8PS/L
51.2ft/sec 15.6m/sec
22.0mph 35.4kmh/1000rpm
30.8mpg 25.6mpgUS
9.2L/100km
Petrol 4-stroke piston
1324cc 81.0cu in
In-line 4 1 Carburettor
Compression ratio: 9.4:1
Bore: 71.0mm 2.8in
Stroke: 83.6mm 3.3in
Valve type/No: Overhead 8
Transmission: Manual
No. of forward speeds: 5
Wheels driven: Front
Springs F/R: Coil/Coil
Brake system: PA
Brakes F/R: Disc/Drum
Steering: Rack & pinion
Wheelbase: 229.5cm 90.4in
Track F: 140.0cm 55.1in
Track R: 138.5cm 54.5in
Length: 347.5cm 136.8in
Width: 160.5cm 63.2in
Height: 150.5cm 59.3in
Ground clearance: 17.9cm
7.0in
Kerb weight: 782.0kg 1722.5lb
Fuel: 38.2L 8.4gal 10.1galUS

490 Mazda

323 GTX
1988 Japan
119.0mph 191.5kmh
0-60mph 96.5kmh: 8.7secs
0-1/4 mile: 16.7secs
132.0bhp 98.4kW 133.8PS
@ 6000rpm
136.0lbft 184.3Nm @ 3000rpm
111.8bhp/ton 109.9bhp/tonne
82.6bhp/L 61.6kW/L 83.8PS/L
54.8ft/sec 16.7m/sec
25.8mpg 21.5mpgUS
10.9L/100km
Petrol 4-stroke piston
1597cc 97.4cu in
turbocharged
In-line 4 fuel injection
Compression ratio: 7.9:1
Bore: 78.0mm 3.1in
Stroke: 83.6mm 3.3in
Valve type/No: Overhead 16
Transmission: Manual
No. of forward speeds: 5
Wheels driven: 4-wheel drive
Springs F/R: Coil/Coil

Brake system: PA
Brakes F/R: Disc/Disc
Steering: Rack & pinion PA
Wheelbase: 240.0cm 94.5in
Track F: 140.0cm 55.1in
Track R: 142.5cm 56.1in
Length: 411.0cm 161.8in
Width: 164.6cm 64.8in
Height: 139.4cm 54.9in
Kerb weight: 1200.8kg
2645.0lb
Fuel: 50.0L 11.0gal 13.2galUS

491 Mazda
323 Turbo 4x4
1988 Japan
124.0mph 199.5kmh
0-50mph 80.5kmh: 5.8secs
0-60mph 96.5kmh: 7.8secs
0-1/4 mile: 15.9secs
0-1km: 29.8secs
148.0bhp 110.4kW 150.0PS
@ 6000rpm
144.0lbft 195.1Nm @ 5000rpm
128.6bhp/ton 126.5bhp/tonne
92.7bhp/L 69.1kW/L 94.0PS/L
55.2ft/sec 16.8m/sec
20.2mph 32.5kmh/1000rpm
20.7mpg 17.2mpgUS
13.6L/100km
Petrol 4-stroke piston
1597cc 97.0cu in
turbocharged
In-line 4 fuel injection
Compression ratio: 7.9:1
Bore: 78.0mm 3.1in
Stroke: 84.0mm 3.3in
Valve type/No: Overhead 16
Transmission: Manual
No. of forward speeds: 5
Wheels driven: 4-wheel drive
Springs F/R: Coil/Coil
Brake system: PA
Brakes F/R: Disc/Disc
Steering: Rack & pinion PA
Wheelbase: 240.0cm 94.5in
Track F: 140.5cm 55.3in
Track R: 142.5cm 56.1in
Length: 399.0cm 157.1in
Width: 164.5cm 64.8in
Height: 133.5cm 52.6in
Ground clearance: 15.3cm
6.0in
Kerb weight: 1170.0kg
2577.0lb
Fuel: 52.8L 11.6gal 13.9galUS

492 Mazda
626 2.0 GLX Executive Estate
1988 Japan
110.0mph 177.0kmh
0-50mph 80.5kmh: 8.5secs
0-60mph 96.5kmh: 11.7secs
0-1/4 mile: 18.1secs

0-1km: 33.3secs
108.0bhp 80.5kW 109.5PS
@ 5300rpm
121.0lbft 164.0Nm @ 3300rpm
88.0bhp/ton 86.5bhp/tonne
54.0bhp/L 40.3kW/L 54.8PS/L
49.9ft/sec 15.2m/sec
21.6mph 34.8kmh/1000rpm
22.9mpg 19.1mpgUS
12.3L/100km
Petrol 4-stroke piston
1998cc 122.0cu in
In-line 4 1 Carburettor
Compression ratio: 9.5:1
Bore: 86.0mm 3.4in
Stroke: 86.0mm 3.4in
Valve type/No: Overhead 12
Transmission: Manual
No. of forward speeds: 5
Wheels driven: Front
Springs F/R: Coil/Coil
Brake system: PA
Brakes F/R: Disc/Disc
Steering: Rack & pinion PA
Wheelbase: 257.5cm 101.4in
Track F: 145.5cm 57.3in
Track R: 146.5cm 57.7in
Length: 459.0cm 180.7in
Width: 169.0cm 66.5in
Height: 143.0cm 56.3in
Ground clearance: 16.5cm
6.5in
Kerb weight: 1248.0kg
2748.9lb
Fuel: 60.1L 13.2gal 15.9galUS

493 Mazda
626 2.0i GT 4WS
1988 Japan
124.0mph 199.5kmh
0-50mph 80.5kmh: 6.9secs
0-60mph 96.5kmh: 9.4secs
0-1/4 mile: 17.3secs
0-1km: 31.6secs
146.0bhp 108.9kW 148.0PS
@ 6000rpm
134.3lbft 182.0Nm @ 4000rpm
112.5bhp/ton 110.6bhp/tonne
73.1bhp/L 54.5kW/L 74.1PS/L
56.5ft/sec 17.2m/sec
20.3mph 32.7kmh/1000rpm
24.5mpg 20.4mpgUS
11.5L/100km
Petrol 4-stroke piston
1998cc 122.0cu in
In-line 4 fuel injection
Compression ratio: 10.0:1
Bore: 86.0mm 3.4in
Stroke: 86.0mm 3.4in
Valve type/No: Overhead 16
Transmission: Manual
No. of forward speeds: 5
Wheels driven: Front
Springs F/R: Coil/Coil

Brake system: PA ABS
Brakes F/R: Disc/Disc
Steering: Rack & pinion PA
Wheelbase: 257.5cm 101.4in
Track F: 145.5cm 57.3in
Track R: 146.5cm 57.7in
Length: 451.5cm 177.8in
Width: 169.0cm 66.5in
Height: 137.5cm 54.1in
Ground clearance: 16.5cm
6.5in
Kerb weight: 1320.0kg
2907.5lb
Fuel: 60.1L 13.2gal 15.9galUS

494 Mazda
929
1988 Japan
121.0mph 194.7kmh
0-60mph 96.5kmh: 10.5secs
0-1/4 mile: 17.5secs
158.0bhp 117.8kW 160.2PS
@ 5500rpm
170.0lbft 230.4Nm @ 4000rpm
102.4bhp/ton 100.7bhp/tonne
53.5bhp/L 39.9kW/L 54.2PS/L
46.6ft/sec 14.2m/sec
22.2mpg 18.5mpgUS
12.7L/100km
Petrol 4-stroke piston
2954cc 180.2cu in
Vee 6 fuel injection
Compression ratio: 8.5:1
Bore: 90.0mm 3.5in
Stroke: 77.4mm 3.0in
Valve type/No: Overhead 18
Transmission: Automatic
No. of forward speeds: 4
Wheels driven: Rear
Springs F/R: Coil/Coil
Brake system: PA ABS
Brakes F/R: Disc/Disc
Steering: Rack & pinion PA
Wheelbase: 271.0cm 106.7in
Track F: 144.5cm 56.9in
Track R: 146.1cm 57.5in
Length: 490.5cm 193.1in
Width: 169.9cm 66.9in
Height: 139.4cm 54.9in
Kerb weight: 1568.6kg
3455.0lb
Fuel: 70.0L 15.4gal 18.5galUS

495 Mazda
MX-6
1988 Japan
131.0mph 210.8kmh
0-60mph 96.5kmh: 7.5secs
0-1/4 mile: 15.7secs
145.0bhp 108.1kW 147.0PS
@ 4300rpm
190.0lbft 257.5Nm @ 3500rpm
115.6bhp/ton 113.7bhp/tonne
66.4bhp/L 49.5kW/L 67.3PS/L

44.2ft/sec 13.5m/sec
24.0mpg 20.0mpgUS
11.8L/100km
Petrol 4-stroke piston
2184cc 133.2cu in
turbocharged
In-line 4 fuel injection
Compression ratio: 7.8:1
Bore: 86.0mm 3.4in
Stroke: 94.0mm 3.7in
Valve type/No: Overhead 12
Transmission: Manual
No. of forward speeds: 5
Wheels driven: Front
Springs F/R: Coil/Coil
Brake system: PA
Brakes F/R: Disc/Disc
Steering: Rack & pinion PA
Wheelbase: 251.5cm 99.0in
Track F: 145.5cm 57.3in
Track R: 146.6cm 57.7in
Length: 449.6cm 177.0in
Width: 168.9cm 66.5in
Height: 135.9cm 53.5in
Ground clearance: 14.0cm
5.5in
Kerb weight: 1275.7kg
2810.0lb
Fuel: 60.2L 13.2gal 15.9galUS

496 Mazda
MX-6 GT
1988 Japan
131.0mph 210.8kmh
0-50mph 80.5kmh: 5.2secs
0-60mph 96.5kmh: 7.2secs
0-1/4 mile: 15.5secs
145.0bhp 108.1kW 147.0PS
@ 4300rpm
190.0lbft 257.5Nm @ 3500rpm
120.1bhp/ton 118.1bhp/tonne
66.4bhp/L 49.5kW/L 67.3PS/L
44.2ft/sec 13.5m/sec
28.8mpg 24.0mpgUS
9.8L/100km
Petrol 4-stroke piston
2184cc 133.2cu in
turbocharged
In-line 4 fuel injection
Compression ratio: 7.8:1
Bore: 86.0mm 3.4in
Stroke: 94.0mm 3.7in
Valve type/No: Overhead 8
Transmission: Manual
No. of forward speeds: 5
Wheels driven: Front
Springs F/R: Coil/Coil
Brake system: PA
Brakes F/R: Disc/Disc
Steering: Rack & pinion PA
Wheelbase: 251.5cm 99.0in
Track F: 145.5cm 57.3in
Track R: 146.6cm 57.7in
Length: 449.6cm 177.0in

Width: 168.9cm 66.5in
Height: 135.9cm 53.5in
Kerb weight: 1228.1kg
2705.0lb
Fuel: 60.2L 13.2gal 15.9galUS

497 Mazda

RX-7 Convertible
1988 Japan
125.0mph 201.1kmh
0-50mph 80.5kmh: 7.2secs
0-60mph 96.5kmh: 9.7secs
0-1/4 mile: 17.5secs
146.0bhp 108.9kW 148.0PS
@ 6500rpm
138.0lbft 187.0Nm @ 3500rpm
107.9bhp/ton 106.1bhp/tonne
111.6bhp/L 83.2kW/L
113.2PS/L
19.2mpg 16.0mpgUS
14.7L/100km
Petrol Wankel rotary
1308cc 79.8cu in
Rotary 2 fuel injection
Compression ratio: 9.4:1
Transmission: Manual
No. of forward speeds: 5
Wheels driven: Rear
Springs F/R: Coil/Coil
Brake system: PA
Brakes F/R: Disc/Disc
Steering: Rack & pinion PA
Wheelbase: 243.1cm 95.7in
Track F: 145.0cm 57.1in
Track R: 144.0cm 56.7in
Length: 429.0cm 168.9in
Width: 168.9cm 66.5in
Height: 126.5cm 49.8in
Ground clearance: 12.2cm
4.8in
Kerb weight: 1375.6kg
3030.0lb
Fuel: 62.8L 13.8gal 16.6galUS

498 Mazda

RX-7 GTU
1988 Japan
128.0mph 206.0kmh
0-60mph 96.5kmh: 8.5secs
0-1/4 mile: 16.5secs
146.0bhp 108.9kW 148.0PS
@ 6500rpm
138.0lbft 187.0Nm @ 3500rpm
118.7bhp/ton 116.7bhp/tonne
111.6bhp/L 83.2kW/L
113.2PS/L
25.2mpg 21.0mpgUS
11.2L/100km
Petrol Wankel rotary
1308cc 79.8cu in
Rotary 2 fuel injection
Compression ratio: 9.4:1
Transmission: Manual
No. of forward speeds: 5

Wheels driven: Rear
Springs F/R: Coil/Coil
Brake system: PA
Brakes F/R: Disc/Disc
Steering: Rack & pinion
Wheelbase: 243.1cm 95.7in
Track F: 145.0cm 57.1in
Track RX: 144.0cm 56.7in
Length: 429.0cm 168.9in
Width: 168.9cm 66.5in
Height: 126.5cm 49.8in
Kerb weight: 1250.8kg
2755.0lb
Fuel: 62.8L 13.8gal 16.6galUS

499 Mazda

RX-7 Turbo
1988 Japan
140.0mph 225.3kmh
0-50mph 80.5kmh: 5.1secs
0-60mph 96.5kmh: 6.6secs
0-1/4 mile: 15.1secs
182.0bhp 135.7kW 184.5PS
@ 6500rpm
183.0lbft 248.0Nm @ 3500rpm
143.0bhp/ton 140.7bhp/tonne
139.1bhp/L 103.8kW/L
141.1PS/L
22.8mpg 19.0mpgUS
12.4L/100km
Petrol Wankel rotary
1308cc 79.8cu in
turbocharged
Rotary 2 fuel injection
Compression ratio: 8.5:1
Transmission: Manual
No. of forward speeds: 5
Wheels driven: Rear
Springs F/R: Coil/Coil
Brake system: PA ABS
Brakes F/R: Disc/Disc
Steering: Rack & pinion PA
Wheelbase: 243.1cm 95.7in
Track F: 145.0cm 57.1in
Track R: 144.0cm 56.7in
Length: 429.0cm 168.9in
Width: 168.9cm 66.5in
Height: 126.5cm 49.8in
Kerb weight: 1293.9kg
2850.0lb
Fuel: 62.8L 13.8gal 16.6galUS

500 Mazda

323 1.6i
1989 Japan
116.0mph 186.6kmh
0-50mph 80.5kmh: 6.3secs
0-60mph 96.5kmh: 8.8secs
0-1/4 mile: 16.7secs
0-1km: 33.4secs
105.0bhp 78.3kW 106.5PS
@ 6000rpm
101.8lbft 138.0Nm @ 4200rpm
109.6bhp/ton 107.8bhp/tonne

65.7bhp/L 49.0kW/L 66.7PS/L
54.8ft/sec 16.7m/sec
19.6mph 31.5kmh/1000rpm
27.5mpg 22.9mpgUS
10.3L/100km
Petrol 4-stroke piston
1597cc 97.0cu in
In-line 4 fuel injection
Compression ratio: 10.5:1
Bore: 78.0mm 3.1in
Stroke: 83.6mm 3.3in
Valve type/No: Overhead 8
Transmission: Manual
No. of forward speeds: 5
Wheels driven: Front
Springs F/R: Coil/Coil
Brake system: PA
Brakes F/R: Disc/Disc
Steering: Rack & pinion PA
Wheelbase: 240.0cm 94.5in
Track F: 140.5cm 55.3in
Track R: 142.5cm 56.1in
Length: 399.0cm 157.1in
Width: 164.5cm 64.8in
Height: 135.5cm 53.3in
Ground clearance: 15.3cm
6.0in
Kerb weight: 974.0kg 2145.4lb
Fuel: 48.2L 10.6gal 12.7galUS

501 Mazda

323 1.8 GT
1989 Japan
128.0mph 206.0kmh
0-50mph 80.5kmh: 6.3secs
0-60mph 96.5kmh: 8.6secs
0-1/4 mile: 16.8secs
0-1km: 30.3secs
140.0bhp 104.4kW 141.9PS
@ 6500rpm
120.3lbft 163.0Nm @ 4700rpm
132.1bhp/ton 129.9bhp/tonne
76.1bhp/L 56.7kW/L 77.1PS/L
60.5ft/sec 18.4m/sec
19.3mph 31.1kmh/1000rpm
28.6mpg 23.8mpgUS
9.9L/100km
Petrol 4-stroke piston
1840cc 112.0cu in
In-line 4 fuel injection
Compression ratio: 9.8:1
Bore: 83.0mm 3.3in
Stroke: 85.0mm 3.3in
Valve type/No: Overhead 16
Transmission: Manual
No. of forward speeds: 5
Wheels driven: Front
Springs F/R: Coil/Coil
Brake system: PA
Brakes F/R: Disc/Disc
Steering: Rack & pinion PA
Wheelbase: 250.0cm 98.4in
Track F: 143.0cm 56.3in
Track R: 143.5cm 56.5in

Length: 426.0cm 167.7in
Width: 167.5cm 65.9in
Height: 133.5cm 52.6in
Ground clearance: 19.0cm
7.5in
Kerb weight: 1078.0kg
2374.4lb
Fuel: 55.0L 12.1gal 14.5galUS

502 Mazda

MPV
1989 Japan
110.0mph 177.0kmh
0-50mph 80.5kmh: 8.5secs
0-60mph 96.5kmh: 11.7secs
0-1/4 mile: 18.4secs
150.0bhp 111.9kW 152.1PS
@ 5000rpm
165.0lbft 223.6Nm @ 4000rpm
87.6bhp/ton 86.1bhp/tonne
50.8bhp/L 37.9kW/L 51.5PS/L
42.4ft/sec 12.9m/sec
25.2mpg 21.0mpgUS
11.2L/100km
Petrol 4-stroke piston
2954cc 180.2cu in
Vee 6 fuel injection
Compression ratio: 8.5:1
Bore: 90.0mm 3.5in
Stroke: 77.4mm 3.0in
Valve type/No: Overhead 18
Transmission: Automatic
No. of forward speeds: 4
Wheels driven: Rear
Springs F/R: Coil/Coil
Brake system: PA
Brakes F/R: Disc/Drum
Steering: Rack & pinion PA
Wheelbase: 280.4cm 110.4in
Track F: 152.4cm 60.0in
Track R: 153.9cm 60.6in
Length: 446.5cm 175.8in
Width: 182.6cm 71.9in
Height: 173.0cm 68.1in
Ground clearance: 15.7cm
6.2in
Kerb weight: 1741.1kg
3835.0lb
Fuel: 74.2L 16.3gal 19.6galUS

503 Mazda

MX-5 Miata
1989 Japan
117.0mph 188.3kmh
0-50mph 80.5kmh: 7.0secs
0-60mph 96.5kmh: 9.5secs
0-1/4 mile: 17.0secs
116.0bhp 86.5kW 117.6PS
@ 6500rpm
100.0lbft 135.5Nm @ 5500rpm
117.8bhp/ton 115.9bhp/tonne
72.6bhp/L 54.2kW/L 73.6PS/L
59.4ft/sec 18.1m/sec
30.6mpg 25.5mpgUS

9.2L/100km
Petrol 4-stroke piston
1597cc 97.4cu in
In-line 4 fuel injection
Compression ratio: 9.4:1
Bore: 78.0mm 3.1in
Stroke: 83.6mm 3.3in
Valve type/No: Overhead 16
Transmission: Manual
No. of forward speeds: 5
Wheels driven: Rear
Springs F/R: Coil/Coil
Brake system: PA
Brakes F/R: Disc/Disc
Steering: Rack & pinion PA
Wheelbase: 226.6cm 89.2in
Track F: 141.0cm 55.5in
Track R: 142.7cm 56.2in
Length: 394.7cm 155.4in
Width: 167.4cm 65.9in
Height: 122.4cm 48.2in
Ground clearance: 11.4cm
4.5in
Kerb weight: 1001.1kg
2205.0lb
Fuel: 45.0L 9.9gal 11.9galUS

504 Mazda
MX-6 GT 4WS
1989 Japan
124.0mph 199.5kmh
0-50mph 80.5kmh: 6.0secs
0-60mph 96.5kmh: 8.2secs
0-1/4 mile: 16.0secs
145.0bhp 108.1kW 147.0PS
@ 4300rpm
190.0lbft 257.5Nm @ 3500rpm
112.4bhp/ton 110.5bhp/tonne
66.4bhp/L 49.5kW/L 67.3PS/L
44.2ft/sec 13.5m/sec
22.8mpg 19.0mpgUS
12.4L/100km

Petrol 4-stroke piston
2184cc 133.2cu in
turbocharged
In-line 4 fuel injection
Compression ratio: 7.8:1
Bore: 86.0mm 3.4in
Stroke: 94.0mm 3.7in
Valve type/No: Overhead 12
Transmission: Manual
No. of forward speeds: 5
Wheels driven: Front
Springs F/R: Coil/Coil
Brake system: PA
Brakes F/R: Disc/Disc
Steering: Rack & pinion PA
Wheelbase: 251.5cm 99.0in
Track F: 145.5cm 57.3in
Track R: 146.6cm 57.7in
Length: 449.6cm 177.0in
Width: 168.9cm 66.5in
Height: 135.9cm 53.5in
Kerb weight: 1312.1kg
2890.0lb
Fuel: 60.2L 13.2gal 15.9galUS

505 Mazda
RX-7 GTUs
1989 Japan
130.0mph 209.2kmh
0-60mph 96.5kmh: 8.6secs
0-1/4 mile: 16.7secs
160.0bhp 119.3kW 162.2PS
@ 7000rpm
140.0lbft 189.7Nm @ 4000rpm
129.1bhp/ton 127.0bhp/tonne
122.3bhp/L 91.2kW/L
124.0PS/L
22.8mpg 19.0mpgUS
12.4L/100km
Petrol Wankel rotary
1308cc 79.8cu in
Rotary 2 fuel injection

Compression ratio: 9.7:1
Transmission: Manual
No. of forward speeds: 5
Wheels driven: Rear
Springs F/R: Coil/Coil
Brake system: PA
Brakes F/R: Disc/Disc
Steering: Rack & pinion PA
Wheelbase: 243.1cm 95.7in
Track F: 145.0cm 57.1in
Track R: 144.0cm 56.7in
Length: 431.5cm 169.9in
Width: 168.9cm 66.5in
Height: 126.5cm 49.8in
Kerb weight: 1259.8kg
2775.0lb
Fuel: 70.0L 15.4gal 18.5galUS

506 Mazda
RX-7 Turbo II
1989 Japan
150.0mph 241.4kmh
0-50mph 80.5kmh: 5.0secs
0-60mph 96.5kmh: 6.7secs
0-1/4 mile: 14.9secs
0-1km: 27.0secs
200.0bhp 149.1kW 202.8PS
@ 6500rpm
195.6lbft 265.0Nm @ 3500rpm
152.9bhp/ton 150.4bhp/tonne
152.9bhp/L 114.0kW/L
155.0PS/L
24.3mph 39.1kmh/1000rpm
14.3mpg 11.9mpgUS
19.8L/100km
Petrol Wankel rotary
1308cc 80.0cu in
turbocharged
Rotary 2 fuel injection
Compression ratio: 9.0:1
Valve type/No: Ports
Transmission: Manual

No. of forward speeds: 5
Wheels driven: Rear
Springs F/R: Coil/Coil
Brake system: PA ABS
Brakes F/R: Disc/Disc
Steering: Rack & pinion PA
Wheelbase: 243.0cm 95.7in
Track F: 145.0cm 57.1in
Track R: 144.0cm 56.7in
Length: 431.5cm 169.9in
Width: 169.0cm 66.5in
Height: 126.5cm 49.8in
Ground clearance: 16.0cm
6.3in
Kerb weight: 1330.0kg
2929.5lb
Fuel: 71.9L 15.8gal 19.0galUS

507 Mazda
MX-5 Miata Millen
1990 Japan
126.0mph 202.7kmh
0-60mph 96.5kmh: 6.4secs
0-1/4 mile: 14.7secs
230.0bhp 171.5kW 233.2PS
@ 5500rpm
235.0lbft 318.4Nm @ 5000rpm
230.0bhp/ton 226.2bhp/tonne
144.0bhp/L 107.4kW/L
146.0PS/L
50.3ft/sec 15.3m/sec
18.0mpg 15.0mpgUS
15.7L/100km
Petrol 4-stroke piston
1597cc 97.4cu in
turbocharged
In-line 4 fuel injection
Compression ratio: 7.2:1
Bore: 78.0mm 3.1in
Stroke: 83.6mm 3.3in
Valve type/No: Overhead 16
Transmission: Manual

506 Mazda
RX-7 Turbo II

No. of forward speeds: 5
Wheels driven: Rear
Springs F/R: Coil/Coil
Brake system: PA
Brakes F/R: Disc/Disc
Steering: Rack & pinion
Wheelbase: 226.6cm 89.2in
Track F: 143.3cm 56.4in
Track R: 144.5cm 56.9in
Length: 394.7cm 155.4in
Width: 167.4cm 65.9in
Height: 117.6cm 46.3in
Kerb weight: 1017.0kg
2240.0lb
Fuel: 45.0L 9.9gal 11.9galUS

508 Mazda

MX-6 GT
1990 Japan
124.0mph 199.5kmh
0-50mph 80.5kmh: 6.0secs
0-60mph 96.5kmh: 8.2secs
0-1/4 mile: 16.0secs
145.0bhp 108.1kW 147.0PS
@ 4300rpm
190.0lbft 257.5Nm @ 3500rpm
112.4bhp/ton 110.5bhp/tonne
66.4bhp/L 49.5kW/L 67.3PS/L
44.2ft/sec 13.5m/sec
22.8mpg 19.0mpgUS
12.4L/100km
Petrol 4-stroke piston
2184cc 133.2cu in
turbocharged
In-line 4 fuel injection
Compression ratio: 7.8:1
Bore: 86.0mm 3.4in
Stroke: 94.0mm 3.7in
Valve type/No: Overhead 12
Transmission: Manual
No. of forward speeds: 5
Wheels driven: Front
Springs F/R: Coil/Coil
Brake system: PA
Brakes F/R: Disc/Disc
Steering: Rack & pinion PA
Wheelbase: 251.5cm 99.0in
Track F: 145.5cm 57.3in
Track R: 146.6cm 57.7in
Length: 449.6cm 177.0in
Width: 168.9cm 66.5in
Height: 135.9cm 53.5in
Kerb weight: 1312.1kg
2890.0lb
Fuel: 60.2L 13.2gal 15.9galUS

509 Mazda

Protege
1990 Japan
130.0mph 209.2kmh
0-60mph 96.5kmh: 9.1secs
0-1/4 mile: 16.8secs
125.0bhp 93.2kW 126.7PS
@ 6500rpm

114.0lbft 154.5Nm @ 4500rpm
112.2bhp/ton 110.3bhp/tonne
68.0bhp/L 50.7kW/L 68.9PS/L
60.5ft/sec 18.4m/sec
31.2mpg 26.0mpgUS
9.0L/100km
Petrol 4-stroke piston
1839cc 112.2cu in
In-line 4 fuel injection
Compression ratio: 9.0:1
Bore: 83.0mm 3.3in
Stroke: 85.0mm 3.3in
Valve type/No: Overhead 16
Transmission: Manual
No. of forward speeds: 5
Wheels driven: Front
Springs F/R: Coil/Coil
Brake system: PA
Brakes F/R: Disc/Disc
Steering: Rack & pinion PA
Wheelbase: 249.9cm 98.4in
Track F: 143.0cm 56.3in
Track R: 143.5cm 56.5in
Length: 435.6cm 171.5in
Width: 167.4cm 65.9in
Height: 137.4cm 54.1in
Kerb weight: 1132.7kg
2495.0lb
Fuel: 54.9L 12.1gal 14.5galUS

510 Mazda

RX-7 Turbo
1990 Japan
140.0mph 225.3kmh
0-50mph 80.5kmh: 4.8secs
0-60mph 96.5kmh: 6.4secs
0-1/4 mile: 15.0secs
200.0bhp 149.1kW 202.8PS
@ 6500rpm
196.0lbft 265.6Nm @ 3500rpm
150.1bhp/ton 147.6bhp/tonne
152.9bhp/L 114.0kW/L
155.0PS/L
19.2mpg 16.0mpgUS
14.7L/100km
Petrol Wankel rotary
1308cc 79.8cu in
turbocharged
Rotary 2 fuel injection
Compression ratio: 9.0:1
Transmission: Manual
No. of forward speeds: 5
Wheels driven: Rear
Springs F/R: Coil/Coil
Brake system: PA ABS
Brakes F/R: Disc/Disc
Steering: Rack & pinion PA
Wheelbase: 243.1cm 95.7in
Track F: 145.0cm 57.1in
Track R: 144.0cm 56.7in
Length: 431.5cm 169.9in
Width: 168.9cm 66.5in
Height: 126.5cm 49.8in
Kerb weight: 1355.2kg

2985.0lb
Fuel: 70.0L 15.4gal 18.5galUS

511 Mazda

121 GSX
1991 Japan
91.0mph 146.4kmh
0-50mph 80.5kmh: 10.9secs
0-60mph 96.5kmh: 15.3secs
0-1/4 mile: 20.4secs
0-1km: 38.4secs
72.4bhp 54.0kW 73.4PS
@ 6000rpm
78.2lbft 106.0Nm @ 3700rpm
86.9bhp/ton 85.5bhp/tonne
54.7bhp/L 40.8kW/L 55.4PS/L
54.8ft/sec 16.7m/sec
24.7mph 39.7kmh/1000rpm
30.7mpg 25.6mpgUS
9.2L/100km
Petrol 4-stroke piston
1324cc 81.0cu in
In-line 4 fuel injection
Compression ratio: 9.4:1
Bore: 71.0mm 2.8in
Stroke: 83.6mm 3.3in
Valve type/No: Overhead 16
Transmission: Automatic
No. of forward speeds: 4
Wheels driven: Front
Springs F/R: Coil/Coil
Brake system: PA
Brakes F/R: Disc/Drum
Steering: Rack & pinion
Wheelbase: 239.0cm 94.1in
Track F: 142.0cm 55.9in
Track R: 140.0cm 55.1in
Length: 381.0cm 150.0in
Width: 165.0cm 65.0in
Height: 147.0cm 57.9in
Ground clearance: 14.8cm
5.8in
Kerb weight: 847.0kg 1865.6lb
Fuel: 43.0L 9.4gal 11.4galUS

512 Mazda

323 1.8 GT
1991 Japan
123.0mph 197.9kmh
0-50mph 80.5kmh: 5.9secs
0-60mph 96.5kmh: 7.8secs
0-1/4 mile: 16.3secs
0-1km: 29.9secs
140.0bhp 104.4kW 141.9PS
@ 6500rpm
120.3lbft 163.0Nm @ 4700rpm
132.1bhp/ton 129.9bhp/tonne
76.1bhp/L 56.7kW/L 77.1PS/L
60.5ft/sec 18.4m/sec
19.3mph 31.1kmh/1000rpm
28.5mpg 23.7mpgUS
9.9L/100km
Petrol 4-stroke piston
1840cc 112.0cu in

In-line 4 fuel injection
Compression ratio: 9.8:1
Bore: 83.0mm 3.3in
Stroke: 85.0mm 3.3in
Valve type/No: Overhead 16
Transmission: Manual
No. of forward speeds: 5
Wheels driven: Front
Springs F/R: Coil/Coil
Brake system: PA
Brakes F/R: Disc/Disc
Steering: Rack & pinion PA
Wheelbase: 245.0cm 96.5in
Track F: 143.0cm 56.3in
Track R: 143.5cm 56.5in
Length: 399.5cm 157.3in
Width: 167.5cm 65.9in
Height: 138.0cm 54.3in
Ground clearance: 19.0cm
7.5in
Kerb weight: 1078.0kg
2374.4lb
Fuel: 55.0L 12.1gal 14.5galUS

513 Mazda

MX-3 V6
1991 Japan
125.0mph 201.1kmh
0-50mph 80.5kmh: 6.6secs
0-60mph 96.5kmh: 8.9secs
0-1/4 mile: 16.9secs
0-1km: 31.2secs
134.0bhp 99.9kW 135.9PS
@ 6800rpm
118.0lbft 159.9Nm @ 5300rpm
122.7bhp/ton 120.6bhp/tonne
72.6bhp/L 54.2kW/L 73.6PS/L
51.8ft/sec 15.8m/sec
18.8mph 30.2kmh/1000rpm
25.0mpg 20.8mpgUS
11.3L/100km
Petrol 4-stroke piston
1845cc 112.6cu in
Vee 6 fuel injection
Compression ratio: 9.2:1
Bore: 75.0mm 2.9in
Stroke: 69.6mm 2.7in
Valve type/No: Overhead 24
Transmission: Manual
No. of forward speeds: 5
Wheels driven: Front
Springs F/R: Coil/Coil
Brake system: PA ABS
Brakes F/R: Disc/Disc
Steering: Rack & pinion PA
Wheelbase: 245.5cm 96.7in
Track F: 146.0cm 57.5in
Track R: 146.5cm 57.7in
Length: 422.0cm 166.1in
Width: 169.5cm 66.7in
Height: 131.0cm 51.6in
Ground clearance: 13.0cm
5.1in
Kerb weight: 1110.9kg

2447.0lb
Fuel: 50.0L 11.0gal 13.2galUS

514 Mazda

RX-7 Cartech
1991 Japan
178.0mph 286.4kmh
0-60mph 96.5kmh: 5.5secs
0-1/4 mile: 13.9secs
344.0bhp 256.5kW 348.8PS
@ 6500rpm
285.0lbft 386.2Nm @ 4900rpm
270.4bhp/ton 265.9bhp/tonne
263.0bhp/L 196.1kW/L
266.6PS/L
Petrol Wankel rotary
1308cc 79.8cu in
turbocharged
Rotary 2 fuel injection
Compression ratio: 9.0:1
Transmission: Manual
No. of forward speeds: 5
Wheels driven: Rear
Springs F/R: Coil/Coil
Brake system: PA ABS
Brakes F/R: Disc/Disc
Steering: Rack & pinion PA
Wheelbase: 243.1cm 95.7in
Track F: 145.0cm 57.1in
Track R: 144.0cm 56.7in
Length: 431.5cm 169.9in
Width: 168.9cm 66.5in
Height: 123.2cm 48.5in
Kerb weight: 1293.9kg
2850.0lb

515 Mazda

RX-7 Infini IV
1991 Japan
112.0mph 180.2kmh
0-60mph 96.5kmh: 7.0secs
0-1/4 mile: 14.9secs
215.0bhp 160.3kW 218.0PS
@ 6500rpm
206.0lbft 279.1Nm @ 3500rpm
173.9bhp/ton 171.0bhp/tonne
164.4bhp/L 122.6kW/L
166.6PS/L
17.4mpg 14.5mpgUS
16.2L/100km
Petrol Wankel rotary
1308cc 79.8cu in
Rotary 2 fuel injection
Compression ratio: 9.0:1
Transmission: Manual
No. of forward speeds: 5
Wheels driven: Rear
Springs F/R: Coil/Coil
Brake system: PA
Brakes F/R: Disc/Disc
Steering: Rack & pinion PA
Wheelbase: 243.1cm 95.7in
Track F: 145.0cm 57.1in
Track R: 144.0cm 56.7in

Length: 431.5cm 169.9in
Width: 168.9cm 66.5in
Height: 126.5cm 49.8in
Kerb weight: 1257.6kg
2770.0lb
Fuel: 70.0L 15.4gal 18.5galUS

516 Mazda

RX-7 Mariah Mode Six
1991 Japan
170.0mph 273.5kmh
0-60mph 96.5kmh: 4.8secs
380.0bhp 283.4kW 385.3PS
@ 6800rpm
330.0lbft 447.2Nm @ 6000rpm
279.1bhp/ton 274.4bhp/tonne
290.5bhp/L 216.6kW/L
294.5PS/L
Petrol Wankel rotary
1308cc 79.8cu in
turbocharged
Rotary 2 fuel injection
Compression ratio: 9.0:1
Transmission: Manual
No. of forward speeds: 5
Wheels driven: Rear
Springs F/R: Coil/Coil
Brake system: PA ABS
Brakes F/R: Disc/Disc
Steering: Rack & pinion PA
Wheelbase: 243.1cm 95.7in
Track F: 153.2cm 60.3in
Track R: 153.7cm 60.5in
Length: 431.5cm 169.9in
Width: 168.9cm 66.5in
Height: 123.2cm 48.5in
Kerb weight: 1384.7kg
3050.0lb
Fuel: 70.0L 15.4gal 18.5galUS

517 Mazda

RX-7 Turbo
1991 Japan
149.0mph 239.7kmh
0-50mph 80.5kmh: 4.8secs
0-60mph 96.5kmh: 6.4secs
0-1/4 mile: 15.0secs
200.0bhp 149.1kW 202.8PS
@ 6500rpm
196.0lbft 265.6Nm @ 3500rpm
149.8bhp/ton 147.3bhp/tonne
152.9bhp/L 114.0kW/L
155.0PS/L
19.2mpg 16.0mpgUS
14.7L/100km
Petrol Wankel rotary
1308cc 79.8cu in
turbocharged
Rotary 2 fuel injection
Compression ratio: 9.0:1
Transmission: Manual
No. of forward speeds: 5
Wheels driven: Rear
Springs F/R: Coil/Coil

Brake system: PA ABS
Brakes F/R: Disc/Disc
Steering: Rack & pinion PA
Wheelbase: 243.1cm 95.7in
Track F: 144.0cm 56.7in
Length: 431.5cm 169.9in
Width: 168.9cm 66.5in
Height: 126.5cm 49.8in
Kerb weight: 1357.5kg
2990.0lb
Fuel: 70.0L 15.4gal 18.5galUS

518 Mazda

MX-3 1.6 Auto
1992 Japan
102.0mph 164.1kmh
0-50mph 80.5kmh: 9.7secs
0-60mph 96.5kmh: 13.4secs
0-1/4 mile: 19.6secs
0-1km: 36.0secs
88.0bhp 65.6kW 89.2PS
@ 5300rpm
99.0lbft 134.1Nm @ 4000rpm
82.1bhp/ton 80.7bhp/tonne
55.1bhp/L 41.1kW/L 55.8PS/L
48.4ft/sec 14.8m/sec
25.2mph 40.5kmh/1000rpm
27.9mpg 23.2mpgUS
10.1L/100km
Petrol 4-stroke piston
1598cc 97.5cu in
In-line 4 fuel injection
Compression ratio: 9.0:1
Bore: 78.0mm 3.1in
Stroke: 83.6mm 3.3in
Valve type/No: Overhead 16
Transmission: Automatic
No. of forward speeds: 4
Wheels driven: Front
Springs F/R: Coil/Coil
Brake system: PA
Brakes F/R: Disc/Disc
Steering: Rack & pinion PA
Wheelbase: 245.6cm 96.7in
Track F: 146.1cm 57.5in
Track R: 146.6cm 57.7in
Length: 421.9cm 166.1in
Width: 169.4cm 66.7in
Height: 131.1cm 51.6in
Kerb weight: 1090.0kg
2401.0lb
Fuel: 50.0L 11.0gal 13.2galUS

519 Mazda

MX-3 GS
1992 Japan
120.0mph 193.1kmh
0-50mph 80.5kmh: 6.6secs
0-60mph 96.5kmh: 9.2secs
0-1/4 mile: 16.8secs
130.0bhp 96.9kW 131.8PS
@ 6500rpm
115.0lbft 155.8Nm @ 4500rpm

114.4bhp/ton 112.5bhp/tonne
70.5bhp/L 52.6kW/L 71.5PS/L
49.5ft/sec 15.1m/sec
28.8mpg 24.0mpgUS
9.8L/100km
Petrol 4-stroke piston
1844cc 112.5cu in
Vee 6 fuel injection
Compression ratio: 9.2:1
Bore: 75.0mm 2.9in
Stroke: 69.6mm 2.7in
Valve type/No: Overhead 24
Transmission: Manual
No. of forward speeds: 5
Wheels driven: Front
Springs F/R: Coil/Coil
Brake system: PA
Brakes F/R: Disc/Disc
Steering: Rack & pinion PA
Wheelbase: 244.6cm 96.3in
Track F: 146.1cm 57.5in
Track R: 146.6cm 57.7in
Length: 420.9cm 165.7in
Width: 169.4cm 66.7in
Height: 131.1cm 51.6in
Ground clearance: 13.0cm
5.1in
Kerb weight: 1155.4kg
2545.0lb
Fuel: 50.0L 11.0gal 13.2galUS

520 Mercedes-Benz

190E 2.3-16
1986 Germany
139.0mph 223.7kmh
0-50mph 80.5kmh: 5.7secs
0-60mph 96.5kmh: 7.8secs
0-1/4 mile: 16.2secs
167.0bhp 124.5kW 169.3PS
@ 5800rpm
162.0lbft 219.5Nm @ 4750rpm
124.1bhp/ton 122.0bhp/tonne
72.6bhp/L 54.2kW/L 73.6PS/L
50.9ft/sec 15.5m/sec
25.2mpg 21.0mpgUS
11.2L/100km
Petrol 4-stroke piston
2299cc 140.3cu in
In-line 4 fuel injection
Compression ratio: 8.0:1
Bore: 95.5mm 3.8in
Stroke: 80.3mm 3.2in
Valve type/No: Overhead 16
Transmission: Manual
No. of forward speeds: 5
Wheels driven: Rear
Springs F/R: Coil/Coil
Brake system: PA ABS
Brakes F/R: Disc/Disc
Steering: Recirculating ball PA
Wheelbase: 266.4cm 104.9in
Track F: 144.5cm 56.9in
Track R: 143.0cm 56.3in
Length: 443.0cm 174.4in

Width: 170.7cm 67.2in
Height: 136.1cm 53.6in
Ground clearance: 12.7cm
5.0in
Kerb weight: 1368.8kg
3015.0lb
Fuel: 79.5L 17.5gal 21.0galUS

521 Mercedes-Benz
200
1986 Germany
121.0mph 194.7kmh
0-50mph 80.5kmh: 9.0secs
0-60mph 96.5kmh: 13.0secs
0-1/4 mile: 19.1secs
0-1km: 34.9secs
109.0bhp 81.3kW 110.5PS
@ 5200rpm
125.0lbft 169.4Nm @ 2500rpm
83.9bhp/ton 82.5bhp/tonne
54.6bhp/L 40.7kW/L 55.3PS/L
45.6ft/sec 13.9m/sec
26.4mph 42.5kmh/1000rpm
23.6mpg 19.7mpgUS
12.0L/100km
Petrol 4-stroke piston
1997cc 121.8cu in
In-line 4 1 Carburettor
Compression ratio: 9.0:1
Bore: 89.0mm 3.5in
Stroke: 80.2mm 3.2in
Valve type/No: Overhead 8
Transmission: Manual
No. of forward speeds: 5
Wheels driven: Rear
Springs F/R: Coil/Coil
Brake system: PA
Brakes F/R: Disc/Disc
Steering: Recirculating ball PA
Wheelbase: 279.9cm 110.2in
Track F: 149.6cm 58.9in
Track R: 148.8cm 58.6in
Length: 474.0cm 186.6in
Width: 174.0cm 68.5in
Height: 144.0cm 56.7in
Ground clearance: 15.2cm
6.0in
Kerb weight: 1321.1kg
2910.0lb
Fuel: 70.1L 15.4gal 18.5galUS

522 Mercedes-Benz
300E
1986 Germany
140.0mph 225.3kmh
0-60mph 96.5kmh: 7.5secs
0-1/4 mile: 16.0secs
177.0bhp 132.0kW 179.4PS
@ 5700rpm
137.0lbft 185.6Nm @ 5750rpm
123.1bhp/ton 121.1bhp/tonne
59.8bhp/L 44.6kW/L 60.6PS/L
50.0ft/sec 15.3m/sec
29.6mph 47.6kmh/1000rpm

26.4mpg 22.0mpgUS
10.7L/100km
Petrol 4-stroke piston
2962cc 180.7cu in
In-line 6 fuel injection
Compression ratio: 9.2:1
Bore: 88.5mm 3.5in
Stroke: 80.3mm 3.2in
Valve type/No: Overhead 12
Transmission: Manual
No. of forward speeds: 5
Wheels driven: Rear
Springs F/R: Coil/Coil
Brake system: PA ABS
Brakes F/R: Disc/Disc
Steering: Recirculating ball PA
Wheelbase: 279.9cm 110.2in
Track F: 149.6cm 58.9in
Track R: 148.8cm 58.6in
Length: 475.5cm 187.2in
Width: 174.0cm 68.5in
Height: 144.5cm 56.9in
Kerb weight: 1461.9kg
3220.0lb
Fuel: 70.0L 15.4gal 18.5galUS

523 Mercedes-Benz
300TE
1986 Germany
127.0mph 204.3kmh
0-50mph 80.5kmh: 6.3secs
0-60mph 96.5kmh: 8.7secs
0-1/4 mile: 17.1secs
0-1km: 31.0secs
188.0bhp 140.2kW 190.6PS
@ 5700rpm
191.0lbft 258.8Nm @ 4400rpm
123.1bhp/ton 121.1bhp/tonne
63.5bhp/L 47.3kW/L 64.3PS/L
50.0ft/sec 15.2m/sec
23.3mph 37.5kmh/1000rpm
21.5mpg 17.9mpgUS
13.1L/100km
Petrol 4-stroke piston
2962cc 181.0cu in
In-line 6 fuel injection
Compression ratio: 9.2:1
Bore: 88.5mm 3.5in
Stroke: 80.2mm 3.2in
Valve type/No: Overhead 12
Transmission: Automatic
No. of forward speeds: 4
Wheels driven: Rear
Springs F/R: Coil/Coil
Brake system: PA
Brakes F/R: Disc/Disc
Steering: Recirculating ball PA
Wheelbase: 280.0cm 110.2in
Track F: 149.7cm 58.9in
Track R: 148.8cm 58.6in
Length: 476.5cm 187.6in
Width: 174.0cm 68.5in
Height: 149.0cm 58.7in
Ground clearance: 17.9cm

7.0in
Kerb weight: 1553.0kg
3420.7lb
Fuel: 79.1L 17.4gal 20.9galUS

524 Mercedes-Benz
420SE
1986 Germany
138.0mph 222.0kmh
0-50mph 80.5kmh: 5.8secs
0-60mph 96.5kmh: 7.6secs
0-1/4 mile: 16.1secs
0-1km: 29.0secs
218.0bhp 162.6kW 221.0PS
@ 5200rpm
243.0lbft 329.3Nm @ 3750rpm
135.2bhp/ton 132.9bhp/tonne
51.9bhp/L 38.7kW/L 52.7PS/L
44.9ft/sec 13.7m/sec
29.8mph 47.9kmh/1000rpm
18.4mpg 15.3mpgUS
15.4L/100km
Petrol 4-stroke piston
4196cc 256.0cu in
Vee 8 fuel injection
Compression ratio: 9.0:1
Bore: 92.0mm 3.6in
Stroke: 78.9mm 3.1in
Valve type/No: Overhead 16
Transmission: Automatic
No. of forward speeds: 4
Wheels driven: Rear
Springs F/R: Coil/Coil
Brake system: PA
Brakes F/R: Disc/Disc
Steering: Recirculating ball PA
Wheelbase: 293.5cm 115.6in
Track F: 155.5cm 61.2in
Track R: 152.7cm 60.1in
Length: 502.0cm 197.6in
Width: 182.0cm 71.7in
Height: 143.7cm 56.6in
Ground clearance: 17.9cm
7.0in
Kerb weight: 1640.0kg
3612.3lb
Fuel: 102.8L 22.6gal
27.2galUS

525 Mercedes-Benz
560SEC
1986 Germany
145.0mph 233.3kmh
0-60mph 96.5kmh: 7.0secs
0-1/4 mile: 15.5secs
238.0bhp 177.5kW 241.3PS
@ 5200rpm
287.0lbft 388.9Nm @ 3500rpm
140.7bhp/ton 138.3bhp/tonne
42.9bhp/L 32.0kW/L 43.5PS/L
53.9ft/sec 16.4m/sec
17.4mpg 14.5mpgUS
16.2L/100km
Petrol 4-stroke piston

5547cc 338.4cu in
Vee 8 fuel injection
Bore: 96.5mm 3.8in
Stroke: 94.8mm 3.7in
Valve type/No: Overhead 16
Transmission: Automatic
No. of forward speeds: 4
Wheels driven: Rear
Springs F/R: Coil/Coil
Brake system: PA ABS
Brakes F/R: Disc/Disc
Steering: Recirculating ball PA
Wheelbase: 284.5cm 112.0in
Track F: 155.4cm 61.2in
Track R: 152.7cm 60.1in
Length: 483.1cm 190.2in
Width: 182.9cm 72.0in
Height: 141.2cm 55.6in
Kerb weight: 1720.7kg
3790.0lb
Fuel: 90.1L 19.8gal 23.8galUS

526 Mercedes-Benz
560SEL
1986 Germany
151.0mph 243.0kmh
0-50mph 80.5kmh: 5.5secs
0-60mph 96.5kmh: 7.1secs
0-1/4 mile: 15.8secs
0-1km: 28.4secs
295.0bhp 220.0kW 299.1PS
@ 5000rpm
335.0lbft 453.9Nm @ 3750rpm
168.5bhp/ton 165.7bhp/tonne
53.2bhp/L 39.7kW/L 53.9PS/L
51.8ft/sec 15.8m/sec
28.2mph 45.4kmh/1000rpm
18.6mpg 15.5mpgUS
15.2L/100km
Petrol 4-stroke piston
5547cc 338.0cu in
Vee 8 fuel injection
Compression ratio: 10.0:1
Bore: 96.5mm 3.8in
Stroke: 94.8mm 3.7in
Valve type/No: Overhead 16
Transmission: Automatic
No. of forward speeds: 4
Wheels driven: Rear
Springs F/R: Coil/Coil
Brake system: PA
Brakes F/R: Disc/Disc
Steering: Recirculating ball PA
Wheelbase: 307.0cm 120.9in
Track F: 155.5cm 61.2in
Track R: 152.7cm 60.1in
Length: 516.0cm 203.1in
Width: 200.6cm 79.0in
Height: 144.6cm 56.9in
Ground clearance: 15.1cm
5.9in
Kerb weight: 1780.0kg
3920.7lb
Fuel: 90.1L 19.8gal 23.8galUS

527 Mercedes-Benz

560SL
1986 Germany
130.0mph 209.2kmh
0-60mph 96.5kmh: 7.5secs
0-1/4 mile: 15.8secs
227.0bhp 169.3kW 230.1PS
@ 4750rpm
279.0lbft 378.0Nm @ 3250rpm
138.5bhp/ton 136.2bhp/tonne
40.9bhp/L 30.5kW/L 41.5PS/L
49.2ft/sec 15.0m/sec
18.0mpg 15.0mpgUS
15.7L/100km
Petrol 4-stroke piston
5547cc 338.4cu in
Vee 8 fuel injection
Compression ratio: 9.0:1
Bore: 96.5mm 3.8in
Stroke: 94.8mm 3.7in
Valve type/No: Overhead 16
Transmission: Automatic
No. of forward speeds: 4
Wheels driven: Rear
Springs F/R: Coil/Coil
Brake system: PA ABS
Brakes F/R: Disc/Disc
Steering: Recirculating ball PA
Wheelbase: 245.6cm 96.7in
Track F: 146.3cm 57.6in
Track R: 146.6cm 57.7in
Length: 458.0cm 180.3in
Width: 179.1cm 70.5in
Height: 129.0cm 50.8in
Kerb weight: 1666.2kg
3670.0lb
Fuel: 85.2L 18.7gal 22.5galUS

528 Mercedes-Benz

190E 2.6
1987 Germany
134.0mph 215.6kmh
0-50mph 80.5kmh: 6.2secs
0-60mph 96.5kmh: 8.3secs
0-1/4 mile: 16.5secs
0-1km: 30.1secs
166.0bhp 123.8kW 168.3PS
@ 5800rpm
168.0lbft 227.6Nm @ 4600rpm
134.5bhp/ton 132.3bhp/tonne
63.9bhp/L 47.6kW/L 64.8PS/L
50.9ft/sec 15.5m/sec
23.2mph 37.3kmh/1000rpm
21.0mpg 17.5mpgUS
13.5L/100km
Petrol 4-stroke piston
2599cc 159.0cu in
In-line 6 fuel injection
Compression ratio: 9.2:1
Bore: 82.9mm 3.3in
Stroke: 80.2mm 3.2in
Valve type/No: Overhead 12
Transmission: Automatic
No. of forward speeds: 4

Wheels driven: Rear
Springs F/R: Coil/Coil
Brake system: PA
Brakes F/R: Disc/Disc
Steering: Recirculating ball PA
Wheelbase: 266.5cm 104.9in
Track F: 142.9cm 56.3in
Track R: 142.9cm 56.3in
Length: 443.0cm 174.4in
Width: 170.6cm 67.2in
Height: 136.1cm 53.6in
Ground clearance: 15.5cm
6.1in
Kerb weight: 1255.0kg
2764.3lb
Fuel: 55.1L 12.1gal 14.5galUS

529 Mercedes-Benz

300CE
1987 Germany
141.0mph 226.9kmh
0-50mph 80.5kmh: 5.6secs
0-60mph 96.5kmh: 7.5secs
0-1/4 mile: 16.0secs
0-1km: 29.0secs
188.0bhp 140.2kW 190.6PS
@ 5700rpm
191.6lbft 259.6Nm @ 4400rpm
130.1bhp/ton 127.9bhp/tonne
63.5bhp/L 47.3kW/L 64.3PS/L
50.0ft/sec 15.3m/sec
23.3mph 37.5kmh/1000rpm
21.4mpg 17.8mpgUS
13.2L/100km
Petrol 4-stroke piston
2962cc 181.0cu in
In-line 6 fuel injection
Compression ratio: 9.2:1
Bore: 88.5mm 3.5in
Stroke: 80.3mm 3.2in
Valve type/No: Overhead 12
Transmission: Automatic
No. of forward speeds: 4
Wheels driven: Rear
Springs F/R: Coil/Coil
Brake system: PA ABS
Brakes F/R: Disc/Disc
Steering: Recirculating ball PA
Wheelbase: 271.5cm 106.9in
Track F: 149.7cm 58.9in
Track R: 148.8cm 58.6in
Length: 465.5cm 183.3in
Width: 188.4cm 74.2in
Height: 141.0cm 55.5in
Ground clearance: 19.0cm
7.5in
Kerb weight: 1470.0kg
3237.9lb
Fuel: 70.1L 15.4gal 18.5galUS

530 Mercedes-Benz

300TD
1987 Germany
115.0mph 185.0kmh

0-50mph 80.5kmh: 7.2secs
0-60mph 96.5kmh: 10.3secs
0-1/4 mile: 17.5secs
143.0bhp 106.6kW 145.0PS
@ 4600rpm
195.0lbft 264.2Nm @ 2400rpm
86.2bhp/ton 84.8bhp/tonne
47.7bhp/L 35.6kW/L 48.4PS/L
42.3ft/sec 12.9m/sec
26.4mpg 22.0mpgUS
10.7L/100km
Diesel 4-stroke piston
2996cc 182.8cu in
turbocharged
In-line 6 fuel injection
Compression ratio: 22.0:1
Bore: 87.0mm 3.4in
Stroke: 84.0mm 3.3in
Valve type/No: Overhead 12
Transmission: Automatic
No. of forward speeds: 4
Wheels driven: Rear
Springs F/R: Coil/Coil
Brake system: PA ABS
Brakes F/R: /Disc
Steering: Recirculating ball PA
Wheelbase: 279.9cm 110.2in
Track F: 149.6cm 58.9in
Track R: 148.8cm 58.6in
Length: 478.0cm 188.2in
Width: 174.0cm 68.5in
Height: 151.9cm 59.8in
Kerb weight: 1686.6kg
3715.0lb
Fuel: 71.9L 15.8gal 19.0galUS

531 Mercedes-Benz

420SEL
1987 Germany
131.0mph 210.8kmh
0-50mph 80.5kmh: 6.7secs
0-60mph 96.5kmh: 8.7secs
0-1/4 mile: 16.7secs
201.0bhp 149.9kW 203.8PS
@ 5200rpm
228.0lbft 308.9Nm @ 3600rpm
115.9bhp/ton 114.0bhp/tonne
47.9bhp/L 35.7kW/L 48.6PS/L
44.9ft/sec 13.7m/sec
19.2mpg 16.0mpgUS
14.7L/100km
Petrol 4-stroke piston
4196cc 256.0cu in
Vee 8 fuel injection
Compression ratio: 9.0:1
Bore: 92.0mm 3.6in
Stroke: 78.9mm 3.1in
Valve type/No: Overhead 16
Transmission: Automatic
No. of forward speeds: 4
Wheels driven: Rear
Springs F/R: Coil/Coil
Brake system: PA ABS
Brakes F/R: Disc/Disc

Steering: Recirculating ball PA
Wheelbase: 307.6cm 121.1in
Track F: 155.4cm 61.2in
Track R: 152.7cm 60.1in
Length: 528.6cm 208.1in
Width: 182.1cm 71.7in
Height: 144.0cm 56.7in
Kerb weight: 1763.8kg
3885.0lb
Fuel: 90.1L 19.8gal 23.8galUS

532 Mercedes-Benz

560SEC Cabriolet Straman
1987 Germany
146.0mph 234.9kmh
0-60mph 96.5kmh: 6.9secs
0-1/4 mile: 15.4secs
238.0bhp 177.5kW 241.3PS
@ 4800rpm
287.0lbft 388.9Nm @ 3500rpm
134.1bhp/ton 131.9bhp/tonne
42.9bhp/L 32.0kW/L 43.5PS/L
49.7ft/sec 15.2m/sec
Petrol 4-stroke piston
5547cc 338.4cu in
Vee 8 fuel injection
Compression ratio: 9.0:1
Bore: 96.5mm 3.8in
Stroke: 94.8mm 3.7in
Valve type/No: Overhead 16
Transmission: Automatic
No. of forward speeds: 4
Wheels driven: Rear
Springs F/R: Coil/Coil
Brake system: PA ABS
Brakes F/R: Disc/Disc
Steering: Recirculating ball PA
Wheelbase: 284.5cm 112.0in
Track F: 155.4cm 61.2in
Track R: 152.7cm 60.1in
Length: 506.0cm 199.2in
Width: 182.9cm 72.0in
Height: 139.7cm 55.0in
Kerb weight: 1804.6kg
3975.0lb
Fuel: 90.1L 19.8gal 23.8galUS

533 Mercedes-Benz

260E Auto
1988 Germany
134.0mph 215.6kmh
0-50mph 80.5kmh: 6.7secs
0-60mph 96.5kmh: 8.8secs
0-1/4 mile: 16.9secs
0-1km: 30.5secs
166.0bhp 123.8kW 168.3PS
@ 5800rpm
168.3lbft 228.0Nm @ 4600rpm
118.6bhp/ton 116.6bhp/tonne
63.9bhp/L 47.6kW/L 64.8PS/L
50.9ft/sec 15.5m/sec
21.9mph 35.2kmh/1000rpm
23.1mpg 19.2mpgUS
12.2L/100km

Petrol 4-stroke piston
2599cc 159.0cu in
In-line 6 fuel injection
Compression ratio: 9.2:1
Bore: 82.9mm 3.3in
Stroke: 80.2mm 3.2in
Valve type/No: Overhead 12
Transmission: Automatic
No. of forward speeds: 4
Wheels driven: Rear
Springs F/R: Coil/Coil
Brake system: PA ABS
Brakes F/R: Disc/Disc
Steering: Recirculating ball PA
Wheelbase: 280.0cm 110.2in
Track F: 149.7cm 58.9in
Track R: 148.8cm 58.6in
Length: 474.0cm 186.6in
Width: 188.4cm 74.2in
Height: 144.6cm 56.9in
Ground clearance: 17.9cm
7.0in
Kerb weight: 1423.0kg
3134.4lb
Fuel: 70.1L 15.4gal 18.5galUS

534 Mercedes-Benz

300CE
1988 Germany
137.0mph 220.4kmh
0-60mph 96.5kmh: 8.0secs
177.0bhp 132.0kW 179.4PS
@ 5700rpm
188.0lbft 254.7Nm @ 4400rpm
119.8bhp/ton 117.8bhp/tonne
49.8bhp/L 44.6kW/L 60.6PS/L
50.0ft/sec 15.3m/sec
Petrol 4-stroke piston
2962cc 180.7cu in
In-line 6 fuel injection
Compression ratio: 9.2:1
Bore: 88.5mm 3.5in
Stroke: 80.3mm 3.2in
Valve type/No: Overhead 12
Transmission: Automatic
No. of forward speeds: 4
Wheels driven: Rear
Springs F/R: Coil/Coil
Brake system: PA ABS
Brakes F/R: Disc/Disc
Steering: Recirculating ball PA
Wheelbase: 271.5cm 106.9in
Track F: 149.6cm 58.9in
Track R: 148.8cm 58.6in
Length: 467.1cm 183.9in
Width: 174.0cm 68.5in
Height: 141.0cm 55.5in
Kerb weight: 1502.7kg
3310.0lb
Fuel: 70.0L 15.4gal 18.5galUS

535 Mercedes-Benz

300E 4Matic
1988 Germany

136.0mph 218.8kmh
0-50mph 80.5kmh: 6.3secs
0-60mph 96.5kmh: 8.3secs
0-1/4 mile: 16.3secs
0-1km: 29.6secs
188.0bhp 140.2kW 190.6PS
@ 5700rpm
191.9lbft 260.0Nm @ 4400rpm
122.7bhp/ton 120.7bhp/tonne
63.5bhp/L 47.3kW/L 64.3PS/L
50.0ft/sec 15.2m/sec
23.5mph 37.8kmh/1000rpm
18.6mpg 15.5mpgUS
15.2L/100km
Petrol 4-stroke piston
2962cc 181.0cu in
In-line 6 fuel injection
Compression ratio: 9.2:1
Bore: 88.5mm 3.5in
Stroke: 80.2mm 3.2in
Valve type/No: Overhead 12
Transmission: Automatic
No. of forward speeds: 4
Wheels driven: Rear
Springs F/R: Coil/Coil
Brake system: PA ABS
Brakes F/R: Disc/Disc
Steering: Recirculating ball PA
Wheelbase: 280.0cm 110.2in
Track F: 149.7cm 58.9in
Track R: 148.8cm 58.6in
Length: 474.0cm 186.6in
Width: 188.4cm 74.2in
Height: 144.6cm 56.9in
Ground clearance: 17.9cm
7.0in
Kerb weight: 1558.0kg
3431.7lb
Fuel: 70.1L 15.4gal 18.5galUS

536 Mercedes-Benz

300E AMG Hammer
1988 Germany
183.0mph 294.4kmh
0-60mph 96.5kmh: 5.2secs
0-1/4 mile: 13.6secs
375.0bhp 279.6kW 380.2PS
@ 5500rpm
407.0lbft 551.5Nm @ 4000rpm
231.1bhp/ton 227.2bhp/tonne
63.0bhp/L 47.0kW/L 63.9PS/L
57.0ft/sec 17.4m/sec
Petrol 4-stroke piston
5953cc 363.2cu in
Vee 8 fuel injection
Compression ratio: 9.8:1
Bore: 100.0mm 3.9in
Stroke: 94.8mm 3.7in
Valve type/No: Overhead 32
Transmission: Automatic
No. of forward speeds: 4
Wheels driven: Rear
Springs F/R: Coil/Coil
Brake system: PA ABS

Brakes F/R: Disc/Disc
Steering: Recirculating ball PA
Wheelbase: 279.9cm 110.2in
Track F: 149.6cm 58.9in
Track R: 148.8cm 58.6in
Length: 474.0cm 186.6in
Width: 174.0cm 68.5in
Height: 137.4cm 54.1in
Kerb weight: 1650.3kg
3635.0lb
Fuel: 70.0L 15.4gal 18.5galUS

537 Mercedes-Benz

190
1989 Germany
110.0mph 177.0kmh
0-50mph 80.5kmh: 8.0secs
0-60mph 96.5kmh: 11.2secs
0-1/4 mile: 17.8secs
0-1km: 33.2secs
105.0bhp 78.3kW 106.5PS
@ 5200rpm
125.5lbft 170.0Nm @ 2500rpm
94.5bhp/ton 92.9bhp/tonne
52.6bhp/L 39.2kW/L 53.3PS/L
45.5ft/sec 13.9m/sec
26.1mph 42.0kmh/1000rpm
28.0mpg 23.3mpgUS
10.1L/100km
Petrol 4-stroke piston
1997cc 122.0cu in
In-line 4 1 Carburettor
Compression ratio: 9.1:1
Bore: 89.0mm 3.5in
Stroke: 80.0mm 3.1in
Valve type/No: Overhead 8
Transmission: Manual
No. of forward speeds: 5
Wheels driven: Rear
Springs F/R: Coil/Coil
Brake system: PA ABS
Brakes F/R: Disc/Disc
Steering: Recirculating ball PA
Wheelbase: 266.5cm 104.9in
Track F: 143.7cm 56.6in
Track R: 141.8cm 55.8in
Length: 442.0cm 174.0in
Width: 183.5cm 72.2in
Height: 139.0cm 54.7in
Ground clearance: 15.5cm
6.1in
Kerb weight: 1130.0kg
2489.0lb
Fuel: 55.1L 12.1gal 14.5galUS

538 Mercedes-Benz

190E 2.5-16
1989 Germany
144.0mph 231.7kmh
0-50mph 80.5kmh: 5.4secs
0-60mph 96.5kmh: 7.2secs
0-1/4 mile: 15.1secs
0-1km: 28.3secs
197.0bhp 146.9kW 199.7PS

@ 6200rpm
176.4lbft 239.0Nm @ 4500rpm
147.3bhp/ton 144.8bhp/tonne
78.9bhp/L 58.8kW/L 80.0PS/L
59.2ft/sec 18.0m/sec
22.5mph 36.2kmh/1000rpm
22.0mpg 18.3mpgUS
12.8L/100km
Petrol 4-stroke piston
2498cc 152.0cu in
In-line 4 fuel injection
Compression ratio: 9.7:1
Bore: 95.5mm 3.8in
Stroke: 87.3mm 3.4in
Valve type/No: Overhead 16
Transmission: Manual
No. of forward speeds: 5
Wheels driven: Rear
Springs F/R: Coil/Coil
Brake system: PA ABS
Brakes F/R: Disc/Disc
Steering: Recirculating ball PA
Wheelbase: 226.5cm 89.2in
Track F: 144.6cm 56.9in
Track R: 142.9cm 56.3in
Length: 443.0cm 174.4in
Width: 160.6cm 63.2in
Height: 136.1cm 53.6in
Ground clearance: 15.5cm
6.1in
Kerb weight: 1360.0kg
2995.6lb
Fuel: 70.1L 15.4gal 18.5galUS

539 Mercedes-Benz

190E 2.6
1989 Germany
129.0mph 207.6kmh
0-60mph 96.5kmh: 8.7secs
0-1/4 mile: 16.8secs
158.0bhp 117.8kW 160.2PS
@ 5800rpm
162.0lbft 219.5Nm @ 4600rpm
117.4bhp/ton 115.4bhp/tonne
60.8bhp/L 45.3kW/L 61.6PS/L
50.9ft/sec 15.5m/sec
24.6mpg 20.5mpgUS
11.5L/100km
Petrol 4-stroke piston
2599cc 158.6cu in
In-line 6 fuel injection
Compression ratio: 9.2:1
Bore: 82.9mm 3.3in
Stroke: 80.3mm 3.2in
Valve type/No: Overhead 12
Transmission: Automatic
No. of forward speeds: 4
Wheels driven: Rear
Springs F/R: Coil/Coil
Brake system: PA ABS
Brakes F/R: Disc/Disc
Steering: Recirculating ball PA
Wheelbase: 266.4cm 104.9in
Track F: 143.8cm 56.6in

Track R: 141.7cm 55.8in
Length: 444.8cm 175.1in
Width: 168.9cm 66.5in
Height: 138.9cm 54.7in
Kerb weight: 1368.8kg
3015.0lb
Fuel: 54.9L 12.1gal 14.5galUS

540 Mercedes-Benz

190E AMG Baby Hammer
1989 Germany
140.0mph 225.3kmh
0-60mph 96.5kmh: 7.0secs
0-1/4 mile: 15.5secs
234.0bhp 174.5kW 237.2PS
@ 5750rpm
234.0lbft 317.1Nm @ 4500rpm
184.9bhp/ton 181.8bhp/tonne
73.0bhp/L 54.4kW/L 74.0PS/L
52.9ft/sec 16.1m/sec
Petrol 4-stroke piston
3205cc 195.5cu in
In-line 6 fuel injection
Compression ratio: 10.0:1
Bore: 90.0mm 3.5in
Stroke: 84.0mm 3.3in
Valve type/No: Overhead 12
Transmission: Automatic
No. of forward speeds: 4
Wheels driven: Rear
Springs F/R: Coil/Coil
Brake system: PA ABS
Brakes F/R: Disc/Disc
Steering: Recirculating ball PA
Wheelbase: 266.4cm 104.9in
Track F: 145.5cm 57.3in
Track R: 143.0cm 56.3in
Length: 444.8cm 175.1in
Width: 167.9cm 66.1in
Height: 135.9cm 53.5in
Kerb weight: 1287.1kg
2835.0lb
Fuel: 61.7L 13.6gal 16.3galUS

541 Mercedes-Benz

200E Automatic
1989 Germany
120.0mph 193.1kmh
0-50mph 80.5kmh: 8.9secs
0-60mph 96.5kmh: 12.3secs
0-1/4 mile: 18.5secs
0-1km: 33.6secs
122.0bhp 91.0kW 123.7PS
@ 5100rpm
131.4lbft 178.0Nm @ 3500rpm
89.3bhp/ton 87.8bhp/tonne
61.1bhp/L 45.6kW/L 61.9PS/L
44.8ft/sec 13.6m/sec
20.7mph 33.3kmh/1000rpm
23.3mpg 19.4mpgUS
12.1L/100km
Petrol 4-stroke piston
1997cc 122.0cu in
In-line 4 fuel injection

Compression ratio: 9.1:1
Bore: 89.0mm 3.5in
Stroke: 80.2mm 3.2in
Valve type/No: Overhead 8
Transmission: Automatic
No. of forward speeds: 4
Wheels driven: Rear
Springs F/R: Coil/Coil
Brake system: PA ABS
Brakes F/R: Disc/Disc
Steering: Recirculating ball PA
Wheelbase: 280.0cm 110.2in
Track F: 149.7cm 58.9in
Track R: 148.8cm 58.6in
Length: 447.0cm 176.0in
Width: 188.4cm 74.2in
Height: 144.6cm 56.9in
Ground clearance: 17.9cm
7.0in
Kerb weight: 1390.0kg
3061.7lb
Fuel: 70.1L 15.4gal 18.5galUS

542 Mercedes-Benz

300CE Cabriolet Straman
1989 Germany
137.0mph 220.4kmh
0-60mph 96.5kmh: 8.5secs
177.0bhp 132.0kW 179.4PS
@ 5700rpm
188.0lbft 254.7Nm @ 4400rpm
120.1bhp/ton 118.1bhp/tonne
59.8bhp/L 44.6kW/L 60.6PS/L
50.0ft/sec 15.3m/sec
Petrol 4-stroke piston
2962cc 180.7cu in
In-line 6 fuel injection
Compression ratio: 9.2:1
Bore: 88.5mm 3.5in
Stroke: 80.3mm 3.2in
Valve type/No: Overhead 12
Transmission: Automatic
No. of forward speeds: 4
Wheels driven: Rear
Springs F/R: Coil/Coil
Brake system: PA ABS
Brakes F/R: Disc/Disc
Steering: Recirculating ball PA
Wheelbase: 271.5cm 106.9in
Track F: 149.6cm 58.9in
Track R: 148.8cm 58.6in
Length: 467.1cm 183.9in
Width: 174.0cm 68.5in
Height: 141.0cm 55.5in
Kerb weight: 1498.2kg
3300.0lb
Fuel: 79.1L 17.4gal 20.9galUS

543 Mercedes-Benz

300E
1989 Germany
140.0mph 225.3kmh
0-60mph 96.5kmh: 8.3secs
0-1/4 mile: 16.4secs

177.0bhp 132.0kW 179.4PS
@ 5700rpm
188.0lbft 254.7Nm @ 4400rpm
121.4bhp/ton 119.4bhp/tonne
59.8bhp/L 44.6kW/L 60.6PS/L
50.0ft/sec 15.3m/sec
24.6mpg 20.5mpgUS
11.5L/100km
Petrol 4-stroke piston
2962cc 180.7cu in
In-line 6 fuel injection
Compression ratio: 9.2:1
Bore: 88.5mm 3.5in
Stroke: 80.3mm 3.2in
Valve type/No: Overhead 12
Transmission: Automatic
No. of forward speeds: 4
Wheels driven: Rear
Springs F/R: Coil/Coil
Brake system: PA ABS
Brakes F/R: Disc/Disc
Steering: Recirculating ball PA
Wheelbase: 279.9cm 110.2in
Track F: 149.6cm 58.9in
Track R: 148.8cm 58.6in
Length: 475.5cm 187.2in
Width: 174.0cm 68.5in
Height: 144.5cm 56.9in
Kerb weight: 1482.3kg
3265.0lb
Fuel: 70.0L 15.4gal 18.5galUS

544 Mercedes-Benz

560SL
1989 Germany
137.0mph 220.4kmh
0-60mph 96.5kmh: 6.8secs
0-1/4 mile: 15.2secs
227.0bhp 169.3kW 230.1PS
@ 4750rpm
279.0lbft 378.0Nm @ 3250rpm
142.4bhp/ton 140.1bhp/tonne
40.9bhp/L 30.5kW/L 41.5PS/L
49.2ft/sec 15.0m/sec
19.8mpg 16.5mpgUS
14.3L/100km
Petrol 4-stroke piston
5547cc 338.4cu in
Vee 8 fuel injection
Compression ratio: 9.0:1
Bore: 96.5mm 3.8in
Stroke: 94.8mm 3.7in
Valve type/No: Overhead 16
Transmission: Automatic
No. of forward speeds: 4
Wheels driven: Rear
Springs F/R: Coil/Coil
Brake system: PA ABS
Brakes F/R: Disc/Disc
Steering: Recirculating ball PA
Wheelbase: 245.6cm 96.7in
Track F: 146.6cm 57.7in
Track R: 146.6cm 57.7in
Length: 458.0cm 180.3in

Width: 179.1cm 70.5in
Height: 129.8cm 51.1in
Kerb weight: 1620.8kg
3570.0lb
Fuel: 85.2L 18.7gal 22.5galUS

545 Mercedes-Benz

190E 2.5-16 Evolution II
1990 Germany
155.0mph 249.4kmh
0-60mph 96.5kmh: 7.1secs
235.0bhp 175.2kW 238.3PS
@ 7200rpm
181.0lbft 245.3Nm @ 5000rpm
178.1bhp/ton 175.2bhp/tonne
95.4bhp/L 71.1kW/L 96.7PS/L
65.2ft/sec 19.9m/sec
Petrol 4-stroke piston
2463cc 150.3cu in
In-line 4 fuel injection
Compression ratio: 10.5:1
Bore: 97.3mm 3.8in
Stroke: 82.8mm 3.3in
Valve type/No: Overhead 16
Transmission: Manual
No. of forward speeds: 5
Wheels driven: Rear
Springs F/R: Coil/Coil
Brake system: PA ABS
Brakes F/R: Disc/Disc
Steering: Recirculating ball PA
Wheelbase: 266.4cm 104.9in
Track F: 147.6cm 58.1in
Track R: 147.6cm 58.1in
Length: 452.9cm 178.3in
Width: 172.0cm 67.7in
Height: 134.1cm 52.8in
Kerb weight: 1341.6kg
2955.0lb
Fuel: 70.0L 15.4gal 18.5galUS

546 Mercedes-Benz

300CE
1990 Germany
151.0mph 243.0kmh
0-60mph 96.5kmh: 8.5secs
0-1/4 mile: 16.6secs
217.0bhp 161.8kW 220.0PS
@ 6400rpm
195.0lbft 264.2Nm @ 4600rpm
139.3bhp/ton 137.0bhp/tonne
73.3bhp/L 54.6kW/L 74.3PS/L
56.2ft/sec 17.1m/sec
23.4mpg 19.5mpgUS
12.1L/100km
Petrol 4-stroke piston
2962cc 180.7cu in
In-line 6 fuel injection
Compression ratio: 10.0:1
Bore: 88.5mm 3.5in
Stroke: 80.3mm 3.2in
Valve type/No: Overhead 24
Transmission: Automatic
No. of forward speeds: 4

550
Mercedes-Benz
500SL

Wheels driven: Rear
Springs F/R: Coil/Coil
Brake system: PA ABS
Brakes F/R: Disc/Disc
Steering: Recirculating ball PA
Wheelbase: 271.5cm 106.9in
Track F: 150.1cm 59.1in
Track R: 149.1cm 58.7in
Length: 467.1cm 183.9in
Width: 174.0cm 68.5in
Height: 139.4cm 54.9in
Kerb weight: 1584.5kg
3490.0lb

547 Mercedes-Benz
300CE AMG Hammer
1990 Germany
185.0mph 297.7kmh
0-60mph 96.5kmh: 5.0secs
0-1/4 mile: 13.2secs
375.0bhp 279.6kW 380.2PS
@ 5500rpm
407.0lbft 551.5Nm @ 4000rpm
226.4bhp/ton 222.6bhp/tonne
63.0bhp/L 47.0kW/L 63.9PS/L
57.0ft/sec 17.4m/sec
Petrol 4-stroke piston
5953cc 363.2cu in
Vee 8 fuel injection
Compression ratio: 9.2:1
Bore: 100.1mm 3.9in
Stroke: 94.7mm 3.7in
Valve type/No: Overhead 32
Transmission: Automatic
No. of forward speeds: 4
Wheels driven: Rear
Springs F/R: Coil/Coil
Brake system: PA ABS
Brakes F/R: Disc/Disc
Steering: Recirculating ball PA
Wheelbase: 271.5cm 106.9in
Track F: 151.1cm 59.5in
Track R: 150.1cm 59.1in

Length: 466.1cm 183.5in
Width: 174.0cm 68.5in
Height: 136.9cm 53.9in
Kerb weight: 1684.3kg
3710.0lb
Fuel: 70.0L 15.4gal 18.5galUS

548 Mercedes-Benz
300D
1990 Germany
115.0mph 185.0kmh
0-50mph 80.5kmh: 10.9secs
0-60mph 96.5kmh: 15.5secs
0-1/4 mile: 20.5secs
0-1km: 36.8secs
109.0bhp 81.3kW 110.5PS
@ 4600rpm
136.5lbft 185.0Nm @ 2800rpm
74.1bhp/ton 72.9bhp/tonne
36.8bhp/L 27.4kW/L 37.3PS/L
42.3ft/sec 12.9m/sec
25.0mph 40.2kmh/1000rpm
28.6mpg 23.8mpgUS
9.9L/100km
Diesel 4-stroke piston
2962cc 181.0cu in
In-line 6 fuel injection
Compression ratio: 22.0:1
Bore: 87.0mm 3.4in
Stroke: 84.0mm 3.3in
Valve type/No: Overhead 12
Transmission: Automatic
No. of forward speeds: 4
Wheels driven: Rear
Springs F/R: Coil/Coil
Brake system: PA ABS
Brakes F/R: Disc/Disc
Steering: Recirculating ball PA
Wheelbase: 280.0cm 110.2in
Track F: 149.7cm 58.9in
Track R: 148.8cm 58.6in
Length: 474.0cm 186.6in
Width: 188.4cm 74.2in

Height: 144.6cm 56.9in
Ground clearance: 17.9cm
7.0in
Kerb weight: 1495.0kg
3292.9lb
Fuel: 70.1L 15.4gal 18.5galUS

549 Mercedes-Benz
300E-24
1990 Germany
148.0mph 238.1kmh
0-50mph 80.5kmh: 6.0secs
0-60mph 96.5kmh: 7.8secs
0-1/4 mile: 16.0secs
0-1km: 28.8secs
231.0bhp 172.3kW 234.2PS
@ 6300rpm
200.0lbft 271.0Nm @ 4600rpm
152.5bhp/ton 150.0bhp/tonne
78.0bhp/L 58.2kW/L 79.1PS/L
55.3ft/sec 16.8m/sec
21.9mph 35.2kmh/1000rpm
18.5mpg 15.4mpgUS
15.3L/100km
Petrol 4-stroke piston
2962cc 181.0cu in
In-line 6 fuel injection
Compression ratio: 10.1:1
Bore: 88.5mm 3.5in
Stroke: 80.2mm 3.2in
Valve type/No: Overhead 24
Transmission: Manual
No. of forward speeds: 5
Wheels driven: Rear
Springs F/R: Coil/Coil
Brake system: PA ABS
Brakes F/R: Disc/Disc
Steering: Recirculating ball PA
Wheelbase: 280.0cm 110.2in
Track F: 149.7cm 58.9in
Track R: 148.8cm 58.6in
Length: 474.0cm 186.6in
Width: 188.4cm 74.2in

Height: 144.6cm 56.9in
Ground clearance: 17.9cm
7.0in
Kerb weight: 1540.0kg
3392.1lb
Fuel: 70.1L 15.4gal 18.5galUS

550 Mercedes-Benz
500SL
1990 Germany
160.0mph 257.4kmh
0-50mph 80.5kmh: 4.7secs
0-60mph 96.5kmh: 5.9secs
0-1/4 mile: 14.4secs
0-1km: 25.3secs
326.0bhp 243.1kW 330.5PS
@ 5500rpm
339.5lbft 460.0Nm @ 4000rpm
175.2bhp/ton 172.3bhp/tonne
65.5bhp/L 48.9kW/L 66.5PS/L
51.2ft/sec 15.6m/sec
28.3mph 45.5kmh/1000rpm
16.2mpg 13.5mpgUS
17.4L/100km
Petrol 4-stroke piston
4973cc 303.0cu in
Vee 8 fuel injection
Compression ratio: 10.0:1
Bore: 96.5mm 3.8in
Stroke: 85.0mm 3.3in
Valve type/No: Overhead 32
Transmission: Automatic
No. of forward speeds: 5
Wheels driven: Rear
Springs F/R: Coil/Coil
Brake system: PA ABS
Brakes F/R: Disc/Disc
Steering: Recirculating ball PA
Wheelbase: 251.5cm 99.0in
Track F: 153.5cm 60.4in
Track R: 152.3cm 60.0in
Length: 447.0cm 176.0in
Width: 181.2cm 71.3in

Height: 130.3cm 51.3in
Ground clearance: 15.3cm
6.0in
Kerb weight: 1892.0kg
4167.4lb
Fuel: 90.1L 19.8gal 23.8galUS

551 Mercedes-Benz

190E 1.8 Auto
1991 Germany
120.0mph 193.1kmh
0-50mph 80.5kmh: 8.8secs
0-60mph 96.5kmh: 12.3secs
0-1/4 mile: 19.1secs
0-1km: 34.4secs
109.0bhp 81.3kW 110.5PS
@ 5500rpm
110.7lbft 150.0Nm @ 3700rpm
90.5bhp/ton 89.0bhp/tonne
60.7bhp/L 45.2kW/L 61.5PS/L
43.2ft/sec 13.2m/sec
20.4mph 32.8kmh/1000rpm
26.1mpg 21.7mpgUS
10.8L/100km
Petrol 4-stroke piston
1797cc 110.0cu in
In-line 4 fuel injection
Compression ratio: 9.1:1
Bore: 89.0mm 3.5in
Stroke: 72.0mm 2.8in
Valve type/No: Overhead 8
Transmission: Automatic
No. of forward speeds: 4
Wheels driven: Rear
Springs F/R: Coil/Coil
Brake system: PA
Brakes F/R: Disc/Disc
Steering: Recirculating ball PA
Wheelbase: 266.4cm 104.9in
Track F: 143.8cm 56.6in
Track R: 141.7cm 55.8in
Length: 442.0cm 174.0in

Width: 183.4cm 72.2in
Height: 138.9cm 54.7in
Ground clearance: 15.5cm
6.1in
Kerb weight: 1225.0kg
2698.2lb
Fuel: 55.1L 12.1gal 14.5galUS

552 Mercedes-Benz

300SL-24 5-speed Auto
1991 Germany
135.0mph 217.2kmh
0-50mph 80.5kmh: 6.5secs
0-60mph 96.5kmh: 8.6secs
0-1/4 mile: 16.7secs
0-1km: 29.9secs
231.0bhp 172.3kW 234.2PS
@ 6300rpm
201.5lbft 273.0Nm @ 4600rpm
130.9bhp/ton 128.7bhp/tonne
78.0bhp/L 58.2kW/L 79.1PS/L
55.1ft/sec 16.8m/sec
28.9mph 46.5kmh/1000rpm
19.2mpg 16.0mpgUS
14.7L/100km
Petrol 4-stroke piston
2960cc 181.0cu in
In-line 6 fuel injection
Compression ratio: 10.0:1
Bore: 89.0mm 3.5in
Stroke: 80.0mm 3.1in
Valve type/No: Overhead 24
Transmission: Automatic
No. of forward speeds: 5
Wheels driven: Rear
Brake system: PA ABS
Brakes F/R: Disc/Disc
Steering: Recirculating ball PA
Wheelbase: 251.5cm 99.0in
Track F: 153.4cm 60.4in
Track R: 152.1cm 59.9in

Length: 446.8cm 175.9in
Width: 181.1cm 71.3in
Height: 130.0cm 51.2in
Ground clearance: 15.2cm
6.0in
Kerb weight: 1795.0kg
3953.7lb
Fuel: 80.1L 17.6gal 21.1galUS

553 Mercedes-Benz

400SE
1991 Germany
146.0mph 234.9kmh
0-50mph 80.5kmh: 6.4secs
0-60mph 96.5kmh: 8.4secs
0-1/4 mile: 16.5secs
0-1km: 29.6secs
286.0bhp 213.3kW 290.0PS
@ 5700rpm
302.0lbft 409.2Nm @ 3900rpm
137.2bhp/ton 134.9bhp/tonne
68.2bhp/L 50.8kW/L 69.1PS/L
49.2ft/sec 15.0m/sec
27.7mph 44.6kmh/1000rpm
14.6mpg 12.2mpgUS
19.3L/100km
Petrol 4-stroke piston
4196cc 256.0cu in
Vee 8 fuel injection
Compression ratio: 10.0:1
Bore: 92.0mm 3.6in
Stroke: 79.0mm 3.1in
Valve type/No: Overhead 32
Transmission: Automatic
No. of forward speeds: 4
Wheels driven: Rear
Springs F/R: Coil/Coil
Brake system: PA
Brakes F/R: Disc/Disc
Steering: Recirculating ball PA
Wheelbase: 303.8cm 119.6in
Track F: 160.0cm 63.0in

Track R: 157.2cm 61.9in
Length: 511.0cm 201.2in
Width: 188.5cm 74.2in
Height: 149.6cm 58.9in
Kerb weight: 2120.2kg
4670.0lb
Fuel: 100.1L 22.0gal
26.4galUS

554 Mercedes-Benz

500E
1991 Germany
159.0mph 255.8kmh
0-50mph 80.5kmh: 4.9secs
0-60mph 96.5kmh: 6.3secs
0-1/4 mile: 14.7secs
0-1km: 26.3secs
326.0bhp 243.1kW 330.5PS
@ 5700rpm
354.0lbft 479.7Nm @ 3900rpm
191.7bhp/ton 188.5bhp/tonne
65.5bhp/L 48.9kW/L 66.5PS/L
53.0ft/sec 16.1m/sec
26.4mph 42.5kmh/1000rpm
16.5mpg 13.7mpgUS
17.1L/100km
Petrol 4-stroke piston
4973cc 303.4cu in
Vee 8 fuel injection
Compression ratio: 10.0:1
Bore: 97.0mm 3.8in
Stroke: 85.0mm 3.3in
Valve type/No: Overhead 32
Transmission: Automatic
No. of forward speeds: 4
Wheels driven: Rear
Springs F/R: Coil/Coil
Brake system: PA ABS
Brakes F/R: Disc/Disc
Steering: Rack & pinion PA
Wheelbase: 279.9cm 110.2in
Track F: 149.6cm 58.9in

**556
Mercedes-Benz**
G-Wagen 300 GD
LWB

Track R: 148.8cm 58.6in
Length: 474.0cm 186.6in
Width: 188.5cm 74.2in
Height: 144.5cm 56.9in
Ground clearance: 17.8cm
7.0in
Kerb weight: 1729.7kg
3810.0lb
Fuel: 90.1L 19.8gal 23.8galUS

555 Mercedes-Benz
500SL
1991 Germany
155.0mph 249.4kmh
0-50mph 80.5kmh: 4.9secs
0-60mph 96.5kmh: 6.4secs
0-1/4 mile: 14.9secs
322.0bhp 240.1kW 326.5PS
@ 5500rpm
332.0lbft 449.9Nm @ 4000rpm
179.9bhp/ton 176.9bhp/tonne
64.7bhp/L 48.3kW/L 65.6PS/L
51.2ft/sec 15.6m/sec
19.2mpg 16.0mpgUS
14.7L/100km
Petrol 4-stroke piston
4973cc 303.4cu in
Vee 8 fuel injection
Compression ratio: 10.0:1
Bore: 96.5mm 3.8in
Stroke: 85.0mm 3.3in
Valve type/No: Overhead 32
Transmission: Automatic
No. of forward speeds: 4
Wheels driven: Rear
Springs F/R: Coil/Coil
Brake system: PA ABS
Brakes F/R: Disc/Disc
Steering: Recirculating ball PA
Wheelbase: 251.5cm 99.0in
Track F: 153.4cm 60.4in
Track R: 152.4cm 60.0in
Length: 447.0cm 176.0in
Width: 181.1cm 71.3in
Height: 128.8cm 50.7in
Kerb weight: 1820.5kg
4010.0lb
Fuel: 79.9L 17.6gal 21.1galUS

556 Mercedes-Benz
G-Wagen 300 GD LWB
1991 Germany
84.0mph 135.2kmh
0-50mph 80.5kmh: 16.1secs
0-60mph 96.5kmh: 25.4secs
0-1/4 mile: 23.4secs
0-1km: 43.6secs
113.0bhp 84.3kW 114.6PS
@ 4600rpm
141.0lbft 191.0Nm @ 2900rpm
50.1bhp/ton 49.3bhp/tonne
37.7bhp/L 28.1kW/L 38.2PS/L
42.3ft/sec 12.9m/sec
20.5mph 33.0kmh/1000rpm

17.1mpg 14.2mpgUS
16.5L/100km
Diesel 4-stroke piston
2996cc 183.0cu in
In-line 6 fuel injection
Compression ratio: 22.0:1
Bore: 87.0mm 3.4in
Stroke: 84.0mm 3.3in
Valve type/No: Overhead 12
Transmission: Manual
No. of forward speeds: 5
Wheels driven: 4-wheel drive
Springs F/R: Coil/Coil
Brake system: PA ABS
Brakes F/R: Disc/Drum
Steering: Recirculating ball PA
Wheelbase: 285.0cm 112.2in
Track F: 142.5cm 56.1in
Track R: 142.5cm 56.1in
Length: 439.4cm 173.0in
Width: 169.9cm 66.9in
Height: 197.6cm 77.8in
Ground clearance: 20.8cm
8.2in
Kerb weight: 2293.0kg
5050.7lb
Fuel: 95.1L 20.9gal 25.1galUS

557 Mercedes-Benz
500SEL
1992 Germany
155.0mph 249.4kmh
0-50mph 80.5kmh: 5.6secs
0-60mph 96.5kmh: 7.5secs
0-1/4 mile: 15.8secs
322.0bhp 240.1kW 326.5PS
@ 5700rpm
354.0lbft 479.7Nm @ 3900rpm
148.6bhp/ton 146.1bhp/tonne
64.7bhp/L 48.3kW/L 65.6PS/L
13.8mpg 11.5mpgUS
20.5L/100km
Petrol 4-stroke piston
4973cc 303.4cu in
Vee 8 fuel injection
Compression ratio: 10.0:1
Transmission: Automatic
No. of forward speeds: 4
Wheels driven: Rear
Springs F/R: Coil/Coil
Brake system: PA ABS
Brakes F/R: Disc/Disc
Steering: Recirculating ball PA
Wheelbase: 313.9cm 123.6in
Track F: 160.3cm 63.1in
Track R: 157.5cm 62.0in
Length: 521.2cm 205.2in
Width: 188.7cm 74.3in
Height: 149.6cm 58.9in
Ground clearance: 17.8cm
7.0in
Kerb weight: 2204.2kg
4855.0lb
Fuel: 99.9L 22.0gal 26.4galUS

558 Mercedes-Benz
600SEL
1992 Germany
155.0mph 249.4kmh
0-60mph 96.5kmh: 6.1secs
408.0bhp 304.2kW 413.7PS
@ 5200rpm
428.0lbft 579.9Nm @ 3800rpm
189.2bhp/ton 186.1bhp/tonne
68.1bhp/L 50.8kW/L 69.1PS/L
45.6ft/sec 13.9m/sec
Petrol 4-stroke piston
5987cc 365.3cu in
Vee 12 fuel injection
Compression ratio: 10.0:1
Bore: 89.0mm 3.5in
Stroke: 80.2mm 3.2in
Valve type/No: Overhead 48
Transmission: Automatic
No. of forward speeds: 4
Wheels driven: Rear
Springs F/R: Coil/Coil
Brake system: ABS
Brakes F/R: Disc/Disc
Steering: Recirculating ball PA
Wheelbase: 313.9cm 123.6in
Track F: 160.3cm 63.1in
Track R: 157.5cm 62.0in
Length: 521.2cm 205.2in
Width: 188.7cm 74.3in
Height: 149.1cm 58.7in
Kerb weight: 2192.8kg
4830.0lb
Fuel: 112.4L 24.7gal
29.7galUS

559 Mercury
Sable LS
1986 USA
115.0mph 185.0kmh
0-60mph 96.5kmh: 11.7secs
0-1/4 mile: 18.7secs
140.0bhp 104.4kW 141.9PS
@ 4800rpm
160.0lbft 216.8Nm @ 3000rpm
95.6bhp/ton 94.0bhp/tonne
46.9bhp/L 35.0kW/L 47.5PS/L
42.0ft/sec 12.8m/sec
23.8mpg 19.8mpgUS
11.9L/100km
Petrol 4-stroke piston
2986cc 182.2cu in
Vee 6 fuel injection
Compression ratio: 9.3:1
Bore: 89.0mm 3.5in
Stroke: 80.0mm 3.1in
Valve type/No: Overhead 12
Transmission: Automatic
No. of forward speeds: 4
Wheels driven: Front
Springs F/R: Coil/Coil
Brake system: PA
Brakes F/R: Disc/Drum
Steering: Rack & pinion PA

Wheelbase: 269.2cm 106.0in
Track F: 156.2cm 61.5in
Track R: 153.7cm 60.5in
Length: 484.9cm 190.9in
Width: 180.1cm 70.9in
Height: 138.2cm 54.4in
Kerb weight: 1489.1kg
3280.0lb
Fuel: 60.6L 13.3gal 16.0galUS

560 Mercury
Tracer
1987 Japan
105.0mph 168.9kmh
0-50mph 80.5kmh: 7.4secs
0-60mph 96.5kmh: 10.3secs
0-1/4 mile: 17.6secs
82.0bhp 61.1kW 83.1PS
@ 5000rpm
92.0lbft 124.7Nm @ 2500rpm
80.0bhp/ton 78.7bhp/tonne
51.3bhp/L 38.3kW/L 52.1PS/L
45.7ft/sec 13.9m/sec
39.6mpg 33.0mpgUS
7.1L/100km
Petrol 4-stroke piston
1597cc 97.4cu in
In-line 4 fuel injection
Compression ratio: 9.3:1
Bore: 78.0mm 3.1in
Stroke: 83.6mm 3.3in
Valve type/No: Overhead 8
Transmission: Manual
No. of forward speeds: 5
Wheels driven: Front
Springs F/R: Coil/Coil
Brake system: PA
Brakes F/R: Disc/Drum
Steering: Rack & pinion PA
Wheelbase: 240.5cm 94.7in
Track F: 139.4cm 54.9in
Track R: 142.2cm 56.0in
Length: 411.5cm 162.0in
Width: 165.6cm 65.2in
Height: 134.6cm 53.0in
Ground clearance: 15.2cm
6.0in
Kerb weight: 1041.9kg
2295.0lb
Fuel: 45.0L 9.9gal 11.9galUS

561 Mercury
Cougar XR-7
1989 USA
143.0mph 230.1kmh
0-50mph 80.5kmh: 5.5secs
0-60mph 96.5kmh: 7.4secs
0-1/4 mile: 15.9secs
210.0bhp 156.6kW 212.9PS
@ 4000rpm
315.0lbft 426.8Nm @ 2600rpm
126.8bhp/ton 124.7bhp/tonne
55.3bhp/L 41.2kW/L 56.1PS/L
37.7ft/sec 11.5m/sec

20.4mpg 17.0mpgUS
13.8L/100km
Petrol 4-stroke piston
3797cc 231.7cu in
supercharged
Vee 6 fuel injection
Compression ratio: 8.2:1
Bore: 96.8mm 3.8in
Stroke: 86.0mm 3.4in
Valve type/No: Overhead 12
Transmission: Manual
No. of forward speeds: 5
Wheels driven: Rear
Springs F/R: Coil/Coil
Brake system: PA ABS
Brakes F/R: Disc/Disc
Steering: Rack & pinion PA
Wheelbase: 287.0cm 113.0in
Track F: 156.0cm 61.4in
Track R: 152.9cm 60.2in
Length: 504.7cm 198.7in
Width: 184.7cm 72.7in
Height: 133.9cm 52.7in
Kerb weight: 1684.3kg
3710.0lb
Fuel: 71.9L 15.8gal 19.0galUS

562 Mercury
Capri XR-2
1990 Australia
124.0mph 199.5kmh
0-60mph 96.5kmh: 8.3secs
0-1/4 mile: 16.3secs
132.0bhp 98.4kW 133.8PS
@ 6000rpm
136.0lbft 184.3Nm @ 3000rpm
113.3bhp/ton 111.4bhp/tonne
82.6bhp/L 61.6kW/L 83.7PS/L
54.8ft/sec 16.7m/sec
30.0mpg 25.0mpgUS
9.4L/100km
Petrol 4-stroke piston

1598cc 97.5cu in
turbocharged
In-line 4 fuel injection
Compression ratio: 7.9:1
Bore: 78.0mm 3.1in
Stroke: 83.6mm 3.3in
Valve type/No: Overhead 16
Transmission: Manual
No. of forward speeds: 5
Wheels driven: Front
Springs F/R: Coil/Coil
Brake system: PA
Brakes F/R: Disc/Disc
Steering: Rack & pinion PA
Wheelbase: 240.5cm 94.7in
Track F: 139.4cm 54.9in
Track R: 142.2cm 56.0in
Length: 421.9cm 166.1in
Width: 164.1cm 64.6in
Height: 127.5cm 50.2in
Kerb weight: 1184.9kg
2610.0lb
Fuel: 42.0L 9.2gal 11.1galUS

563 Mercury
Tracer LTS
1991 USA
125.0mph 201.1kmh
0-60mph 96.5kmh: 9.1secs
0-1/4 mile: 17.0secs
127.0bhp 94.7kW 128.8PS
@ 6500rpm
114.0lbft 154.5Nm @ 4500rpm
112.9bhp/ton 111.0bhp/tonne
69.0bhp/L 51.5kW/L 70.0PS/L
60.5ft/sec 18.4m/sec
32.4mpg 27.0mpgUS
8.7L/100km
Petrol 4-stroke piston
1840cc 112.3cu in
In-line 4 fuel injection
Compression ratio: 9.0:1

Bore: 83.0mm 3.3in
Stroke: 85.0mm 3.3in
Valve type/No: Overhead 16
Transmission: Manual
No. of forward speeds: 5
Wheels driven: Front
Springs F/R: Coil/Coil
Brake system: PA
Brakes F/R: Disc/Disc
Steering: Rack & pinion PA
Wheelbase: 249.9cm 98.4in
Track F: 143.5cm 56.5in
Track R: 143.5cm 56.5in
Length: 434.1cm 170.9in
Width: 169.4cm 66.7in
Height: 133.9cm 52.7in
Kerb weight: 1144.1kg
2520.0lb
Fuel: 45.0L 9.9gal 11.9galUS

564 Merkur
Scorpio
1987 Germany
130.0mph 209.2kmh
0-50mph 80.5kmh: 7.4secs
0-60mph 96.5kmh: 10.1secs
0-1/4 mile: 17.5secs
144.0bhp 107.4kW 146.0PS
@ 5500rpm
162.0lbft 219.5Nm @ 3000rpm
97.9bhp/ton 96.3bhp/tonne
49.1bhp/L 36.6kW/L 49.8PS/L
43.2ft/sec 13.2m/sec
25.2mpg 21.0mpgUS
11.2L/100km
Petrol 4-stroke piston
2933cc 178.9cu in
Vee 6 fuel injection
Compression ratio: 9.0:1
Bore: 93.0mm 3.7in
Stroke: 72.0mm 2.8in
Valve type/No: Overhead 12

Transmission: Automatic
No. of forward speeds: 4
Wheels driven: Rear
Springs F/R: Coil/Coil
Brake system: PA ABS
Brakes F/R: Disc/Disc
Steering: Rack & pinion PA
Wheelbase: 276.1cm 108.7in
Track F: 147.6cm 58.1in
Track R: 147.6cm 58.1in
Length: 473.5cm 186.4in
Width: 176.5cm 69.5in
Height: 138.9cm 54.7in
Ground clearance: 15.2cm
6.0in
Kerb weight: 1495.9kg
3295.0lb
Fuel: 64.0L 14.1gal 16.9galUS

565 Merkur
XR4Ti
1988 Germany
120.0mph 193.1kmh
0-50mph 80.5kmh: 5.9secs
0-60mph 96.5kmh: 8.1secs
0-1/4 mile: 16.2secs
170.0bhp 126.8kW 172.4PS
@ 5200rpm
195.0lbft 264.2Nm @ 3800rpm
129.1bhp/ton 126.9bhp/tonne
74.1bhp/L 55.3kW/L 75.2PS/L
45.1ft/sec 13.7m/sec
28.8mpg 24.0mpgUS
9.8L/100km
Petrol 4-stroke piston
2293cc 139.9cu in
turbocharged
In-line 4 fuel injection
Compression ratio: 8.0:1
Bore: 96.0mm 3.8in
Stroke: 79.2mm 3.1in
Valve type/No: Overhead 8

563 Mercury
Tracer LTS

566 MG
Maestro Turbo

Transmission: Manual
No. of forward speeds: 5
Wheels driven: Rear
Springs F/R: Coil/Coil
Brake system: PA
Brakes F/R: Disc/Drum
Steering: Rack & pinion PA
Wheelbase: 260.9cm 102.7in
Track F: 145.3cm 57.2in
Track R: 146.8cm 57.8in
Length: 453.1cm 178.4in
Width: 172.7cm 68.0in
Height: 139.2cm 54.8in
Kerb weight: 1339.3kg
2950.0lb
Fuel: 56.8L 12.5gal 15.0galUS

566 MG
Maestro Turbo
1989 UK
131.0mph 210.8kmh
0-50mph 80.5kmh: 5.2secs
0-60mph 96.5kmh: 6.9secs
0-1/4 mile: 15.4secs
0-1km: 28.4secs
152.0bhp 113.3kW 154.1PS
@ 5100rpm
169.0lbft 229.0Nm @ 3500rpm
143.1bhp/ton 140.7bhp/tonne
76.2bhp/L 56.8kW/L 77.3PS/L
49.6ft/sec 15.1m/sec
25.0mph 40.2kmh/1000rpm
20.9mpg 17.4mpgUS
13.5L/100km
Petrol 4-stroke piston
1994cc 122.0cu in
turbocharged
In-line 4 1 Carburettor
Compression ratio: 8.5:1
Bore: 84.5mm 3.3in
Stroke: 89.0mm 3.5in
Valve type/No: Overhead 8
Transmission: Manual

No. of forward speeds: 5
Wheels driven: Front
Springs F/R: Coil/Coil
Brake system: PA
Brakes F/R: Disc/Drum
Steering: Rack & pinion PA
Wheelbase: 250.6cm 98.7in
Track F: 148.1cm 58.3in
Track R: 145.5cm 57.3in
Length: 400.3cm 157.6in
Width: 193.0cm 76.0in
Height: 142.0cm 55.9in
Ground clearance: 17.9cm
7.0in
Kerb weight: 1080.0kg
2378.8lb
Fuel: 50.0L 11.0gal 13.2galUS

567 MG
B British Motor Heritage
1990 UK
105.0mph 168.9kmh
0-60mph 96.5kmh: 12.8secs
0-1/4 mile: 19.0secs
92.0bhp 68.6kW 93.3PS
@ 5400rpm
110.0lbft 149.1Nm @ 3000rpm
89.4bhp/ton 87.9bhp/tonne
51.2bhp/L 38.1kW/L 51.9PS/L
52.5ft/sec 16.0m/sec
30.0mpg 25.0mpgUS
9.4L/100km
Petrol 4-stroke piston
1798cc 109.7cu in
In-line 4 2 Carburettor
Compression ratio: 8.8:1
Bore: 80.3mm 3.2in
Stroke: 89.0mm 3.5in
Valve type/No: Overhead 8
Transmission: Manual with
overdrive
Wheels driven: Rear
Springs F/R: Coil/Leaf

Brake system: PA
Brakes F/R: Disc/Drum
Steering: Rack & pinion
Wheelbase: 231.1cm 91.0in
Track F: 125.2cm 49.3in
Track R: 125.2cm 49.3in
Length: 387.9cm 152.7in
Width: 152.1cm 59.9in
Height: 126.5cm 49.8in
Kerb weight: 1046.5kg
2305.0lb
Fuel: 45.4L 10.0gal 12.0galUS

568 Mitsubishi
Cordia Turbo
1986 Japan
115.0mph 185.0kmh
0-60mph 96.5kmh: 8.5secs
0-1/4 mile: 16.5secs
116.0bhp 86.5kW 117.6PS
@ 5500rpm
129.0lbft 174.8Nm @ 3000rpm
103.1bhp/ton 101.4bhp/tonne
64.6bhp/L 48.2kW/L 65.5PS/L
52.9ft/sec 16.1m/sec
24.6mpg 20.5mpgUS
11.5L/100km
Petrol 4-stroke piston
1795cc 109.5cu in
turbocharged
In-line 4 fuel injection
Compression ratio: 7.5:1
Bore: 80.6mm 3.2in
Stroke: 88.0mm 3.5in
Valve type/No: Overhead 8
Transmission: Manual
No. of forward speeds: 5
Wheels driven: Front
Springs F/R: Coil/Coil
Brake system: PA
Brakes F/R: Disc/Drum
Steering: Rack & pinion PA
Wheelbase: 244.6cm 96.3in

Track F: 141.0cm 55.5in
Track R: 137.4cm 54.1in
Length: 439.4cm 173.0in
Width: 166.1cm 65.4in
Height: 125.5cm 49.4in
Kerb weight: 1144.1kg
2520.0lb
Fuel: 48.4L 10.6gal 12.8galUS

569 Mitsubishi
Lancer 1.5 GLX Estate
1986 Japan
97.0mph 156.1kmh
0-50mph 80.5kmh: 10.1secs
0-60mph 96.5kmh: 14.0secs
0-1/4 mile: 19.7secs
0-1km: 38.6secs
74.0bhp 55.2kW 75.0PS
@ 5500rpm
87.0lbft 117.9Nm @ 4000rpm
76.6bhp/ton 75.3bhp/tonne
50.4bhp/L 37.6kW/L 51.1PS/L
49.3ft/sec 15.0m/sec
20.8mph 33.5kmh/1000rpm
27.8mpg 23.1mpgUS
10.2L/100km
Petrol 4-stroke piston
1468cc 89.6cu in
In-line 4 1 Carburettor
Compression ratio: 9.5:1
Bore: 75.5mm 3.0in
Stroke: 82.0mm 3.2in
Valve type/No: Overhead 8
Transmission: Manual
No. of forward speeds: 5
Wheels driven: Front
Springs F/R: Coil/Coil
Brake system: PA
Brakes F/R: Disc/Drum
Steering: Rack & pinion PA
Wheelbase: 238.0cm 93.7in
Track F: 138.9cm 54.7in
Track R: 133.9cm 52.7in

571 Mitsubishi
Galant 2000 GLS
Automatic

Length: 413.3cm 162.7in
Width: 163.3cm 64.3in
Height: 142.0cm 55.9in
Ground clearance: 15.0cm
5.9in
Kerb weight: 982.5kg 2164.0lb
Fuel: 46.9L 10.3gal 12.4galUS

570 Mitsubishi
Colt 1500 GLX 5-door
1987 Japan
95.0mph 152.9kmh
0-50mph 80.5kmh: 8.5secs
0-60mph 96.5kmh: 12.2secs
0-1/4 mile: 18.9secs
0-1km: 35.4secs
74.0bhp 55.2kW 75.0PS
@ 5500rpm
87.0lbft 117.9Nm @ 4000rpm
85.5bhp/ton 84.1bhp/tonne
50.4bhp/L 37.6kW/L 51.1PS/L
49.3ft/sec 15.0m/sec
21.1mph 33.9kmh/1000rpm
29.2mpg 24.3mpgUS
9.7L/100km
Petrol 4-stroke piston
1468cc 90.0cu in
In-line 4 1 Carburettor
Compression ratio: 9.5:1
Bore: 75.5mm 3.0in
Stroke: 82.0mm 3.2in
Valve type/No: Overhead 8
Transmission: Manual
No. of forward speeds: 5
Wheels driven: Front
Springs F/R: Coil/Coil
Brake system: PA
Brakes F/R: Disc/Drum
Steering: Rack & pinion PA
Wheelbase: 238.0cm 93.7in
Track F: 141.0cm 55.5in
Track R: 134.0cm 52.8in
Length: 396.5cm 156.1in

Width: 163.5cm 64.4in
Height: 136.0cm 53.5in
Ground clearance: 15.5cm
6.1in
Kerb weight: 880.0kg 1938.3lb
Fuel: 45.0L 9.9gal 11.9galUS

571 Mitsubishi
Galant 2000 GLS Automatic
1987 Japan
105.0mph 168.9kmh
0-50mph 80.5kmh: 9.7secs
0-60mph 96.5kmh: 13.1secs
0-1/4 mile: 19.6secs
0-1km: 36.0secs
101.0bhp 75.3kW 102.4PS
@ 6000rpm
114.0lbft 154.5Nm @ 3500rpm
90.1bhp/ton 88.6bhp/tonne
50.6bhp/L 37.7kW/L 51.3PS/L
57.7ft/sec 17.6m/sec
27.7mph 44.6kmh/1000rpm
25.3mpg 21.1mpgUS
11.2L/100km
Petrol 4-stroke piston
1997cc 122.0cu in
In-line 4 1 Carburettor
Compression ratio: 9.5:1
Bore: 85.0mm 3.3in
Stroke: 88.0mm 3.5in
Valve type/No: Overhead 8
Transmission: Automatic
No. of forward speeds: 4
Wheels driven: Front
Springs F/R: Coil/Coil
Brake system: PA
Brakes F/R: Disc/Disc
Steering: Rack & pinion PA
Wheelbase: 259.0cm 102.0in
Track F: 144.5cm 56.9in
Track R: 140.5cm 55.3in
Length: 456.0cm 179.5in
Width: 195.5cm 77.0in

Height: 139.9cm 55.1in
Ground clearance: 15.6cm
6.1in
Kerb weight: 1140.0kg
2511.0lb
Fuel: 59.6L 13.1gal 15.7galUS

572 Mitsubishi
Galant Sapporo
1987 Japan
115.0mph 185.0kmh
0-50mph 80.5kmh: 7.7secs
0-60mph 96.5kmh: 10.4secs
0-1/4 mile: 18.0secs
0-1km: 33.0secs
127.0bhp 94.7kW 128.8PS
@ 5000rpm
142.0lbft 192.4Nm @ 4000rpm
104.9bhp/ton 103.2bhp/tonne
54.0bhp/L 40.3kW/L 54.8PS/L
54.7ft/sec 16.7m/sec
26.1mph 42.0kmh/1000rpm
24.8mpg 20.7mpgUS
11.4L/100km
Petrol 4-stroke piston
2351cc 143.0cu in
In-line 4 fuel injection
Compression ratio: 9.5:1
Bore: 86.5mm 3.4in
Stroke: 100.0mm 3.9in
Valve type/No: Overhead 8
Transmission: Automatic
No. of forward speeds: 4
Wheels driven: Front
Springs F/R: Coil/Coil
Brake system: PA ABS
Brakes F/R: Disc/Disc
Steering: Rack & pinion PA
Wheelbase: 260.0cm 102.4in
Track F: 144.5cm 56.9in
Track R: 141.5cm 55.7in
Length: 466.0cm 183.5in
Width: 195.5cm 77.0in

Height: 137.0cm 53.9in
Ground clearance: 15.4cm
6.1in
Kerb weight: 1231.0kg
2711.4lb
Fuel: 60.1L 13.2gal 15.9galUS

573 Mitsubishi
Shogun Turbo Diesel 5-door
1987 Japan
83.0mph 133.5kmh
0-50mph 80.5kmh: 11.7secs
0-60mph 96.5kmh: 17.8secs
0-1/4 mile: 20.6secs
0-1km: 40.0secs
84.0bhp 62.6kW 85.2PS
@ 4200rpm
148.0lbft 200.5Nm @ 2000rpm
48.5bhp/ton 47.7bhp/tonne
33.9bhp/L 25.3kW/L 34.4PS/L
43.6ft/sec 13.3m/sec
20.1mph 32.3kmh/1000rpm
22.1mpg 18.4mpgUS
12.8L/100km
Diesel 4-stroke piston
2477cc 151.0cu in
turbocharged
In-line 4 fuel injection
Compression ratio: 21.0:1
Bore: 91.0mm 3.6in
Stroke: 95.0mm 3.7in
Valve type/No: Overhead 8
Transmission: Manual
No. of forward speeds: 5
Wheels driven: 4-wheel
engageable
Springs F/R: Torsion bar/Leaf
Brake system: PA
Brakes F/R: Disc/Drum
Steering: Recirculating ball PA
Wheelbase: 269.5cm 106.1in
Track F: 140.0cm 55.1in
Track R: 137.5cm 54.1in

Length: 460.0cm 181.1in
Width: 168.0cm 66.1in
Height: 188.0cm 74.0in
Ground clearance: 20.5cm
8.1in
Kerb weight: 1760.0kg
3876.6lb
Fuel: 91.9L 20.2gal 24.3galUS

574 Mitsubishi

Starion 2000 Turbo
1987 Japan
136.0mph 218.8kmh
0-50mph 80.5kmh: 5.3secs
0-60mph 96.5kmh: 6.9secs
0-1/4 mile: 16.0secs
0-1km: 28.6secs
177.0bhp 132.0kW 179.4PS
@ 6000rpm
214.0lbft 290.0Nm @ 3500rpm
137.6bhp/ton 135.3bhp/tonne
88.6bhp/L 66.1kW/L 89.9PS/L
57.7ft/sec 17.6m/sec
23.9mph 38.5kmh/1000rpm
19.0mpg 15.8mpgUS
14.9L/100km
Petrol 4-stroke piston
1997cc 122.0cu in
turbocharged
In-line 4 fuel injection
Compression ratio: 7.6:1
Bore: 85.0mm 3.3in
Stroke: 88.0mm 3.5in
Valve type/No: Overhead 8
Transmission: Manual
No. of forward speeds: 5
Wheels driven: Rear
Springs F/R: Coil/Coil
Brake system: PA ABS
Brakes F/R: Disc/Disc
Steering: Recirculating ball PA
Wheelbase: 243.5cm 95.9in
Track F: 146.5cm 57.7in
Track R: 145.5cm 57.3in
Length: 443.0cm 174.4in
Width: 174.5cm 68.7in
Height: 131.5cm 51.8in
Ground clearance: 15.5cm
6.1in
Kerb weight: 1308.0kg
2881.1lb
Fuel: 75.1L 16.5gal 19.8galUS

575 Mitsubishi

Cordia Turbo
1988 Japan
115.0mph 185.0kmh
0-50mph 80.5kmh: 5.8secs
0-60mph 96.5kmh: 8.4secs
0-1/4 mile: 16.4secs
116.0bhp 86.5kW 117.6PS
@ 5500rpm
129.0lbft 174.8Nm @ 3000rpm
105.6bhp/ton 103.9bhp/tonne

64.6bhp/L 48.2kW/L 65.5PS/L
52.9ft/sec 16.1m/sec
25.2mpg 21.0mpgUS
11.2L/100km
Petrol 4-stroke piston
1795cc 109.5cu in
turbocharged
In-line 4 fuel injection
Compression ratio: 7.5:1
Bore: 80.6mm 3.2in
Stroke: 88.0mm 3.5in
Valve type/No: Overhead 8
Transmission: Manual
No. of forward speeds: 5
Wheels driven: Front
Springs F/R: Coil/Coil
Brake system: PA
Brakes F/R: Disc/Drum
Steering: Rack & pinion PA
Wheelbase: 244.6cm 96.3in
Track F: 141.0cm 55.5in
Track R: 137.4cm 54.1in
Length: 439.4cm 173.0in
Width: 166.1cm 65.4in
Height: 125.5cm 49.4in
Kerb weight: 1116.8kg
2460.0lb
Fuel: 48.4L 10.6gal 12.8galUS

576 Mitsubishi

Galant 2000 GLSi
1988 Japan
112.0mph 180.2kmh
0-50mph 80.5kmh: 7.7secs
0-60mph 96.5kmh: 10.9secs
0-1/4 mile: 17.4secs
0-1km: 32.7secs
110.0bhp 82.0kW 111.5PS
@ 5500rpm
118.1lbft 160.0Nm @ 4500rpm
93.8bhp/ton 92.2bhp/tonne
55.1bhp/L 41.1kW/L 55.8PS/L
52.9ft/sec 16.1m/sec
21.6mph 34.8kmh/1000rpm
21.2mpg 17.7mpgUS
13.3L/100km
Petrol 4-stroke piston
1997cc 122.0cu in
In-line 4 fuel injection
Compression ratio: 9.0:1
Bore: 85.0mm 3.3in
Stroke: 88.0mm 3.5in
Valve type/No: Overhead 8
Transmission: Manual
No. of forward speeds: 5
Wheels driven: Front
Springs F/R: Coil/Coil
Brake system: PA ABS
Brakes F/R: Disc/Disc
Steering: Rack & pinion PA
Wheelbase: 260.0cm 102.4in
Track F: 146.0cm 57.5in
Track R: 145.0cm 57.1in
Length: 454.0cm 178.7in

Width: 169.5cm 66.7in
Height: 141.5cm 55.7in
Ground clearance: 17.9cm
7.0in
Kerb weight: 1193.0kg
2627.7lb
Fuel: 60.1L 13.2gal 15.9galUS

577 Mitsubishi

Galant GTi-16v
1988 Japan
122.0mph 196.3kmh
0-50mph 80.5kmh: 6.2secs
0-60mph 96.5kmh: 8.7secs
0-1/4 mile: 16.6secs
0-1km: 30.4secs
144.0bhp 107.4kW 146.0PS
@ 6500rpm
125.5lbft 170.0Nm @ 5000rpm
119.2bhp/ton 117.2bhp/tonne
72.1bhp/L 53.8kW/L 73.1PS/L
62.5ft/sec 19.1m/sec
19.8mph 31.9kmh/1000rpm
24.4mpg 20.3mpgUS
11.6L/100km
Petrol 4-stroke piston
1997cc 122.0cu in
In-line 4 fuel injection
Compression ratio: 9.8:1
Bore: 85.0mm 3.3in
Stroke: 88.0mm 3.5in
Valve type/No: Overhead 16
Transmission: Manual
No. of forward speeds: 5
Wheels driven: Front
Springs F/R: Coil/Coil
Brake system: PA ABS
Brakes F/R: Disc/Disc
Steering: Rack & pinion PA
Wheelbase: 260.0cm 102.4in
Track F: 146.0cm 57.5in
Track R: 145.0cm 57.1in
Length: 454.0cm 178.7in
Width: 169.5cm 66.7in
Height: 141.5cm 55.7in
Ground clearance: 17.9cm
7.0in
Kerb weight: 1229.0kg
2707.0lb
Fuel: 60.1L 13.2gal 15.9galUS

578 Mitsubishi

Lancer GTi 16v
1988 Japan
116.0mph 186.6kmh
0-50mph 80.5kmh: 6.4secs
0-60mph 96.5kmh: 9.0secs
0-1/4 mile: 17.1secs
0-1km: 31.2secs
123.0bhp 91.7kW 124.7PS
@ 6500rpm
104.8lbft 142.0Nm @ 5000rpm
116.4bhp/ton 114.4bhp/tonne
77.1bhp/L 57.5kW/L 78.1PS/L

53.3ft/sec 16.2m/sec
19.3mph 31.1kmh/1000rpm
25.9mpg 21.6mpgUS
10.9L/100km
Petrol 4-stroke piston
1596cc 97.0cu in
In-line 4 fuel injection
Compression ratio: 10.0:1
Bore: 82.3mm 3.2in
Stroke: 75.0mm 2.9in
Valve type/No: Overhead 16
Transmission: Manual
No. of forward speeds: 5
Wheels driven: Front
Springs F/R: Coil/Coil
Brake system: PA
Brakes F/R: Disc/Disc
Steering: Rack & pinion PA
Wheelbase: 245.5cm 96.7in
Track F: 143.0cm 56.3in
Track R: 143.0cm 56.3in
Length: 423.5cm 166.7in
Width: 167.0cm 65.7in
Height: 167.0cm 65.7in
Ground clearance: 15.5cm
6.1in
Kerb weight: 1075.0kg
2367.8lb
Fuel: 50.0L 11.0gal 13.2galUS

579 Mitsubishi

Mirage Turbo
1988 Japan
118.0mph 189.9kmh
0-50mph 80.5kmh: 5.8secs
0-60mph 96.5kmh: 8.2secs
0-1/4 mile: 16.3secs
105.0bhp 78.3kW 106.5PS
@ 5500rpm
122.0lbft 165.3Nm @ 3500rpm
106.7bhp/ton 104.9bhp/tonne
65.7bhp/L 49.0kW/L 66.7PS/L
51.8ft/sec 15.8m/sec
30.0mpg 25.0mpgUS
9.4L/100km
Petrol 4-stroke piston
1597cc 97.4cu in
turbocharged
In-line 4 fuel injection
Compression ratio: 7.6:1
Bore: 76.9mm 3.0in
Stroke: 86.0mm 3.4in
Valve type/No: Overhead 8
Transmission: Manual
No. of forward speeds: 5
Wheels driven: Front
Springs F/R: Coil/Coil
Brake system: PA
Brakes F/R: Disc/Drum
Steering: Rack & pinion PA
Wheelbase: 238.0cm 93.7in
Track F: 138.9cm 54.7in
Track R: 134.1cm 52.8in
Length: 399.5cm 157.3in

Width: 163.6cm 64.4in
Height: 129.0cm 50.8in
Kerb weight: 1001.1kg
2205.0lb
Fuel: 45.0L 9.9gal 11.9galUS

580 Mitsubishi
Eclipse
1989 Japan
143.0mph 230.1kmh
0-50mph 80.5kmh: 5.5secs
0-60mph 96.5kmh: 7.2secs
0-1/4 mile: 15.8secs
190.0bhp 141.7kW 192.6PS
@ 6000rpm
203.0lbft 275.1Nm @ 3000rpm
149.6bhp/ton 147.1bhp/tonne
96.1bhp/L 71.7kW/L 97.4PS/L
57.7ft/sec 17.6m/sec
27.6mpg 23.0mpgUS
10.2L/100km
Petrol 4-stroke piston
1977cc 120.6cu in
turbocharged
In-line 4 fuel injection
Compression ratio: 7.8:1
Bore: 85.0mm 3.3in
Stroke: 88.0mm 3.5in
Valve type/No: Overhead 16
Transmission: Manual
No. of forward speeds: 5
Wheels driven: Front
Springs F/R: Coil/Coil
Brake system: PA
Brakes F/R: Disc/Disc
Steering: Rack & pinion PA
Wheelbase: 246.9cm 97.2in
Track F: 146.6cm 57.7in
Track R: 145.0cm 57.1in
Length: 433.1cm 170.5in
Width: 168.9cm 66.5in
Height: 130.6cm 51.4in
Ground clearance: 16.5cm
6.5in
Kerb weight: 1291.6kg
2845.0lb
Fuel: 60.2L 13.2gal 15.9galUS

581 Mitsubishi
Galant GS
1989 Japan
120.0mph 193.1kmh
0-50mph 80.5kmh: 6.8secs
0-60mph 96.5kmh: 9.7secs
0-1/4 mile: 18.2secs
135.0bhp 100.7kW 136.9PS
@ 6000rpm
125.0lbft 169.4Nm @ 5000rpm
105.0bhp/ton 103.2bhp/tonne
67.6bhp/L 50.4kW/L 68.5PS/L
57.7ft/sec 17.6m/sec
27.5mpg 22.9mpgUS
10.3L/100km
Petrol 4-stroke piston

1997cc 121.8cu in
In-line 4 fuel injection
Compression ratio: 9.0:1
Bore: 85.0mm 3.3in
Stroke: 88.0mm 3.5in
Valve type/No: Overhead 16
Transmission: Manual
No. of forward speeds: 5
Wheels driven: Front
Springs F/R: Coil/Coil
Brake system: PA ABS
Brakes F/R: Disc/Disc
Steering: Rack & pinion PA
Wheelbase: 260.1cm 102.4in
Track F: 146.1cm 57.5in
Track R: 145.0cm 57.1in
Length: 467.1cm 183.9in
Width: 169.4cm 66.7in
Height: 141.0cm 55.5in
Kerb weight: 1307.5kg
2880.0lb
Fuel: 60.2L 13.2gal
15.9galUS

582 Mitsubishi
Shogun V6 5-door
1989 Japan
102.0mph 164.1kmh
0-50mph 80.5kmh: 9.0secs
0-60mph 96.5kmh: 12.8secs
0-1/4 mile: 18.8secs
0-1km: 35.2secs
141.0bhp 105.1kW 143.0PS
@ 5000rpm
166.1lbft 225.0Nm @ 3000rpm
75.5bhp/ton 74.2bhp/tonne
47.4bhp/L 35.4kW/L 48.1PS/L
41.5ft/sec 12.7m/sec
20.9mph 33.6kmh/1000rpm
19.7mpg 16.4mpgUS
14.3L/100km
Petrol 4-stroke piston
2972cc 181.0cu in
Vee 6 fuel injection
Compression ratio: 8.9:1
Bore: 91.1mm 3.6in
Stroke: 76.0mm 3.0in
Valve type/No: Overhead 12
Transmission: Manual
No. of forward speeds: 5
Wheels driven: 4-wheel drive
Springs F/R: Torsion bar/Leaf
Brake system: PA
Brakes F/R: Disc/Drum
Steering: Recirculating ball PA
Wheelbase: 269.5cm 106.1in
Track F: 140.0cm 55.1in
Track R: 137.5cm 54.1in
Length: 460.0cm 181.1in
Width: 168.0cm 66.1in
Height: 188.0cm 74.0in
Ground clearance: 20.5cm
8.1in
Kerb weight: 1900.0kg

4185.0lb
Fuel: 91.9L 20.2gal 24.3galUS

583 Mitsubishi
Starion 2.6 Turbo
1989 Japan
130.0mph 209.2kmh
0-50mph 80.5kmh: 5.6secs
0-60mph 96.5kmh: 7.8secs
0-1/4 mile: 15.9secs
0-1km: 29.4secs
153.0bhp 114.1kW 155.1PS
@ 5000rpm
207.4lbft 281.0Nm @ 2500rpm
113.2bhp/ton 111.3bhp/tonne
59.9bhp/L 44.6kW/L 60.7PS/L
53.6ft/sec 16.3m/sec
23.7mph 38.1kmh/1000rpm
18.8mpg 15.7mpgUS
15.0L/100km
Petrol 4-stroke piston
2555cc 156.0cu in
turbocharged
In-line 4 fuel injection
Compression ratio: 7.0:1
Bore: 91.1mm 3.6in
Stroke: 98.0mm 3.9in
Valve type/No: Overhead 12
Transmission: Manual
No. of forward speeds: 5
Wheels driven: Rear
Springs F/R: Coil/Coil
Brake system: PA
Brakes F/R: Disc/Disc
Steering: Recirculating ball PA
Wheelbase: 243.5cm 95.9in
Track F: 146.5cm 57.7in
Track R: 145.5cm 57.3in
Length: 443.0cm 174.4in
Width: 174.5cm 68.7in
Height: 131.5cm 51.8in
Ground clearance: 15.5cm
6.1in
Kerb weight: 1375.0kg
3028.6lb
Fuel: 75.1L 16.5gal 19.8galUS

584 Mitsubishi
Starion ESI-R
1989 Japan
127.0mph 204.3kmh
0-50mph 80.5kmh: 5.9secs
0-60mph 96.5kmh: 8.0secs
0-1/4 mile: 16.3secs
188.0bhp 140.2kW 190.6PS
@ 5000rpm
234.0lbft 317.1Nm @ 2500rpm
135.8bhp/ton 133.6bhp/tonne
73.6bhp/L 54.9kW/L 74.6PS/L
53.6ft/sec 16.3m/sec
24.0mpg 20.0mpgUS
11.8L/100km
Petrol 4-stroke piston
2555cc 155.9cu in

turbocharged
In-line 4 fuel injection
Compression ratio: 7.0:1
Bore: 91.5mm 3.6in
Stroke: 98.0mm 3.9in
Valve type/No: Overhead 8
Transmission: Manual
No. of forward speeds: 5
Wheels driven: Rear
Springs F/R: Coil/Coil
Brake system: PA ABS
Brakes F/R: Disc/Disc
Steering: Recirculating ball PA
Wheelbase: 243.6cm 95.9in
Track F: 146.6cm 57.7in
Track R: 145.5cm 57.3in
Length: 439.9cm 173.2in
Width: 173.5cm 68.3in
Height: 127.5cm 50.2in
Kerb weight: 1407.4kg
3100.0lb
Fuel: 74.9L 16.5gal
19.8galUS

585 Mitsubishi
Galant 2000 GLSi Coupe
1990 Japan
113.0mph 181.8kmh
0-50mph 80.5kmh: 7.4secs
0-60mph 96.5kmh: 10.4secs
0-1/4 mile: 17.9secs
0-1km: 33.1secs
111.0bhp 82.8kW 112.5PS
@ 5500rpm
118.1lbft 160.0Nm @ 4500rpm
90.4bhp/ton 88.9bhp/tonne
55.6bhp/L 41.5kW/L 56.4PS/L
52.9ft/sec 16.1m/sec
23.4mph 37.7kmh/1000rpm
23.0mpg 19.2mpgUS
12.3L/100km
Petrol 4-stroke piston
1996cc 122.0cu in
In-line 4 fuel injection
Compression ratio: 9.0:1
Bore: 85.0mm 3.3in
Stroke: 88.0mm 3.5in
Valve type/No: Overhead 8
Transmission: Manual
No. of forward speeds: 5
Wheels driven: Front
Springs F/R: Coil/Coil
Brake system: PA ABS
Brakes F/R: Disc/Disc
Steering: Rack & pinion PA
Wheelbase: 259.8cm 102.3in
Track F: 145.8cm 57.4in
Track R: 144.8cm 57.0in
Length: 453.9cm 178.7in
Width: 169.4cm 66.7in
Height: 140.0cm 55.1in
Ground clearance: 16.8cm
6.6in
Kerb weight: 1248.0kg

2748.9lb
Fuel: 60.1L 13.2gal 15.9galUS

586 Mitsubishi

Lancer GLXi 4WD Liftback
1990 Japan
105.0mph 168.9kmh
0-50mph 80.5kmh: 8.2secs
0-60mph 96.5kmh: 11.7secs
0-1/4 mile: 18.3secs
0-1km: 34.4secs
95.0bhp 70.8kW 96.3PS
@ 5500rpm
104.1lbft 141.0Nm @ 4000rpm
79.8bhp/ton 78.5bhp/tonne
54.1bhp/L 40.4kW/L 54.9PS/L
51.8ft/sec 15.8m/sec
19.8mph 31.9kmh/1000rpm
24.6mpg 20.5mpgUS
11.5L/100km
Petrol 4-stroke piston
1755cc 107.0cu in
In-line 4 fuel injection
Compression ratio: 9.5:1
Bore: 81.0mm 3.2in
Stroke: 86.0mm 3.4in
Valve type/No: Overhead 8
Transmission: Manual
No. of forward speeds: 5
Wheels driven: 4-wheel drive
Springs F/R: Coil/Coil
Brake system: PA
Brakes F/R: Disc/Drum
Steering: Rack & pinion PA
Wheelbase: 245.5cm 96.7in
Track F: 142.5cm 56.1in
Track R: 142.0cm 55.9in
Length: 423.5cm 166.7in
Width: 167.0cm 65.7in
Height: 142.5cm 56.1in
Ground clearance: 15.5cm
6.1in
Kerb weight: 1210.0kg
2665.2lb
Fuel: 50.0L 11.0gal 13.2galUS

587 Mitsubishi

Lancer Liftback 1800 GTi-16v
1990 Japan
122.0mph 196.3kmh
0-50mph 80.5kmh: 6.2secs
0-60mph 96.5kmh: 8.5secs
0-1/4 mile: 16.7secs
0-1km: 30.0secs
134.0bhp 99.9kW 135.9PS
@ 6500rpm
119.6lbft 162.0Nm @ 4500rpm
117.1bhp/ton 115.1bhp/tonne
73.0bhp/L 54.4kW/L 74.0PS/L
62.5ft/sec 19.1m/sec
19.1mph 30.7kmh/1000rpm
22.9mpg 19.1mpgUS
12.3L/100km
Petrol 4-stroke piston

1836cc 112.0cu in
In-line 4 fuel injection
Compression ratio: 10.5:1
Bore: 81.5mm 3.2in
Stroke: 88.0mm 3.5in
Valve type/No: Overhead 16
Transmission: Manual
No. of forward speeds: 5
Wheels driven: Front
Springs F/R: Coil/Coil
Brake system: PA
Brakes F/R: Disc/Disc
Steering: Rack & pinion PA
Wheelbase: 245.5cm 96.7in
Track F: 143.0cm 56.3in
Track R: 143.0cm 56.3in
Length: 424.5cm 167.1in
Width: 167.0cm 65.7in
Height: 140.5cm 55.3in
Ground clearance: 15.3cm
6.0in
Kerb weight: 1164.0kg
2563.9lb

588 Mitsubishi

3000GT VR-4
1991 Japan
159.0mph 255.8kmh
0-50mph 80.5kmh: 4.8secs
0-60mph 96.5kmh: 6.3secs
0-1/4 mile: 14.7secs
300.0bhp 223.7kW 304.2PS
@ 6000rpm
307.0lbft 416.0Nm @ 2500rpm
173.4bhp/ton 170.5bhp/tonne
100.9bhp/L 75.3kW/L
102.3PS/L
49.8ft/sec 15.2m/sec
21.6mpg 18.0mpgUS
13.1L/100km
Petrol 4-stroke piston
2972cc 181.3cu in
turbocharged
Vee 6 fuel injection
Compression ratio: 8.0:1
Bore: 91.1mm 3.6in
Stroke: 76.0mm 3.0in
Valve type/No: Overhead 24
Transmission: Manual
No. of forward speeds: 5
Wheels driven: 4-wheel drive
Springs F/R: Coil/Coil
Brake system: PA ABS
Brakes F/R: Disc/Disc
Steering: Rack & pinion PA
Wheelbase: 246.9cm 97.2in
Track F: 156.0cm 61.4in
Track R: 158.0cm 62.2in
Length: 458.5cm 180.5in
Width: 183.9cm 72.4in
Height: 128.5cm 50.6in
Ground clearance: 14.5cm
5.7in

Kerb weight: 1759.2kg
3875.0lb
Fuel: 74.9L 16.5gal 19.8galUS

589 Mitsubishi

Colt 1800 GTI-16v
1991 Japan
121.0mph 194.7kmh
0-50mph 80.5kmh: 6.0secs
0-60mph 96.5kmh: 8.3secs
0-1/4 mile: 16.5secs
0-1km: 30.3secs
134.0bhp 99.9kW 135.9PS
@ 6500rpm
119.6lbft 162.0Nm @ 4500rpm
127.2bhp/ton 125.1bhp/tonne
73.0bhp/L 54.4kW/L 74.0PS/L
62.5ft/sec 19.1m/sec
19.3mph 31.1kmh/1000rpm
26.9mpg 22.4mpgUS
10.5L/100km
Petrol 4-stroke piston
1836cc 112.0cu in
In-line 4 fuel injection
Compression ratio: 10.5:1
Bore: 81.5mm 3.2in
Stroke: 88.0mm 3.5in
Valve type/No: Overhead 16
Transmission: Manual
No. of forward speeds: 5
Wheels driven: Front
Springs F/R: Coil/Coil
Brake system: PA
Brakes F/R: Disc/Disc
Steering: Rack & pinion PA
Wheelbase: 238.5cm 93.9in
Track F: 143.0cm 56.3in
Track R: 143.0cm 56.3in
Length: 396.0cm 155.9in
Width: 166.9cm 65.7in
Height: 137.9cm 54.3in
Ground clearance: 12.4cm
4.9in
Kerb weight: 1071.0kg
2359.0lb
Fuel: 50.0L 11.0gal
13.2galUS

590 Mitsubishi

Diamante LS
1991 Japan
130.0mph 209.2kmh
0-50mph 80.5kmh: 6.6secs
0-60mph 96.5kmh: 9.0secs
0-1/4 mile: 16.7secs
202.0bhp 150.6kW 204.8PS
@ 6000rpm
199.0lbft 269.6Nm @ 3000rpm
120.7bhp/ton 118.6bhp/tonne
68.0bhp/L 50.7kW/L 68.9PS/L
49.8ft/sec 15.2m/sec
22.8mpg 19.0mpgUS
12.4L/100km
Petrol 4-stroke piston

2972cc 181.3cu in
Vee 6 fuel injection
Compression ratio: 10.0:1
Bore: 91.0mm 3.6in
Stroke: 76.0mm 3.0in
Valve type/No: Overhead 24
Transmission: Automatic
No. of forward speeds: 4
Wheels driven: Front
Springs F/R: Coil/Coil
Brake system: PA ABS
Brakes F/R: Disc/Disc
Steering: Rack & pinion PA
Wheelbase: 272.0cm 107.1in
Track F: 153.4cm 60.4in
Track R: 152.9cm 60.2in
Length: 483.1cm 190.2in
Width: 177.5cm 69.9in
Height: 141.0cm 55.5in
Ground clearance: 15.2cm
6.0in
Kerb weight: 1702.5kg
3750.0lb
Fuel: 71.9L 15.8gal
19.0galUS

591 Mitsubishi

Galant VR-4
1991 Japan
130.0mph 209.2kmh
0-60mph 96.5kmh: 7.9secs
0-1/4 mile: 15.8secs
195.0bhp 145.4kW 197.7PS
@ 6000rpm
203.0lbft 275.1Nm @ 3000rpm
127.7bhp/ton 125.6bhp/tonne
97.6bhp/L 72.8kW/L 99.0PS/L
57.7ft/sec 17.6m/sec
25.2mpg 21.0mpgUS
11.2L/100km
Petrol 4-stroke piston
1997cc 121.8cu in
turbocharged
In-line 4 fuel injection
Compression ratio: 7.8:1
Bore: 85.0mm 3.3in
Stroke: 88.0mm 3.5in
Valve type/No: Overhead 16
Transmission: Manual
No. of forward speeds: 5
Wheels driven: 4-wheel drive
Springs F/R: Coil/Coil
Brake system: PA ABS
Brakes F/R: Disc/Disc
Steering: Rack & pinion PA
Wheelbase: 260.1cm 102.4in
Track F: 146.1cm 57.5in
Track R: 145.0cm 57.1in
Length: 467.1cm 183.9in
Width: 169.4cm 66.7in
Height: 141.0cm 55.5in
Kerb weight: 1552.7kg
3420.0lb
Fuel: 60.2L 13.2gal 15.9galUS

592 Mitsubishi
Shogun V6 LWB

592 Mitsubishi
Shogun V6 LWB
1991 Japan
105.0mph 168.9kmh
0-50mph 80.5kmh: 9.4secs
0-60mph 96.5kmh: 13.1secs
0-1/4 mile: 19.7secs
0-1km: 30.6secs
147.0bhp 109.6kW 149.0PS
@ 5000rpm
174.2lbft 236.0Nm @ 4000rpm
72.7bhp/ton 71.5bhp/tonne
49.5bhp/L 36.9kW/L 50.1PS/L
41.5ft/sec 12.7m/sec
20.9mph 33.6kmh/1000rpm
16.1mpg 13.4mpgUS
17.5L/100km
Petrol 4-stroke piston
2972cc 181.0cu in
Vee 6 fuel injection
Compression ratio: 8.9:1
Bore: 91.1mm 3.6in
Stroke: 76.0mm 3.0in
Valve type/No: Overhead 12
Transmission: Manual
No. of forward speeds: 5
Wheels driven: 4-wheel
engageable
Springs F/R: Coil/Torsion bar
Brake system: PA ABS
Brakes F/R: Disc/Disc
Steering: Rack & pinion PA
Wheelbase: 272.5cm 107.3in
Track F: 148.6cm 58.5in
Track R: 148.6cm 58.5in
Length: 472.4cm 186.0in
Width: 178.6cm 70.3in
Height: 186.9cm 73.6in
Ground clearance: 21.6cm
8.5in
Kerb weight: 2055.0kg
4526.4lb
Fuel: 92.0L 20.2gal 24.3galUS

593 Mitsubishi
Sigma
1991 Japan
130.0mph 209.2kmh
0-50mph 80.5kmh: 7.2secs
0-60mph 96.5kmh: 9.3secs
0-1/4 mile: 17.2secs
0-1km: 30.8secs
202.0bhp 150.6kW 204.8PS
@ 6000rpm
200.0lbft 271.0Nm @ 3000rpm
126.3bhp/ton 124.2bhp/tonne
68.0bhp/L 50.7kW/L 68.9PS/L
49.8ft/sec 15.2m/sec
24.6mph 39.6kmh/1000rpm
18.3mpg 15.2mpgUS
15.4L/100km
Petrol 4-stroke piston
2972cc 181.0cu in
Vee 6 fuel injection
Compression ratio: 10.0:1
Bore: 91.0mm 3.6in
Stroke: 76.0mm 3.0in
Valve type/No: Overhead 24
Transmission: Automatic
No. of forward speeds: 4
Wheels driven: Front
Springs F/R: Coil/Coil
Brake system: PA ABS
Brakes F/R: Disc/Disc
Steering: Rack & pinion PA
Wheelbase: 272.0cm 107.1in
Track F: 153.4cm 60.4in
Track R: 152.9cm 60.2in
Length: 475.0cm 187.0in
Width: 177.5cm 69.9in
Height: 143.5cm 56.5in
Ground clearance: 15.5cm
6.1in
Kerb weight: 1626.0kg
3581.5lb
Fuel: 72.8L 16.0gal
19.2galUS

594 Mitsubishi
Space Runner 1800-16v GLXi
1991 Japan
115.0mph 185.0kmh
0-50mph 80.5kmh: 7.5secs
0-60mph 96.5kmh: 10.0secs
0-1/4 mile: 18.1secs
0-1km: 33.4secs
121.0bhp 90.2kW 122.7PS
@ 6000rpm
119.0lbft 161.2Nm @ 4500rpm
97.7bhp/ton 96.0bhp/tonne
66.0bhp/L 49.2kW/L 66.9PS/L
58.3ft/sec 17.8m/sec
20.3mph 32.7kmh/1000rpm
27.2mpg 22.6mpgUS
10.4L/100km
Petrol 4-stroke piston
1834cc 111.9cu in
In-line 4 fuel injection
Compression ratio: 10.0:1
Bore: 81.0mm 3.2in
Stroke: 89.0mm 3.5in
Valve type/No: Overhead 16
Transmission: Manual
No. of forward speeds: 5
Wheels driven: Front
Springs F/R: Coil/Coil
Brake system: PA
Brakes F/R: Disc/Drum
Steering: Rack & pinion PA
Wheelbase: 252.0cm 99.2in
Track F: 146.1cm 57.5in
Track R: 146.1cm 57.5in
Length: 429.0cm 168.9in
Width: 169.4cm 66.7in
Height: 162.6cm 64.0in
Ground clearance: 16.0cm
6.3in
Kerb weight: 1259.8kg
2775.0lb
Fuel: 55.1L 12.1gal
14.5galUS

595 Morgan
Plus 8
1987 UK
126.0mph 202.7kmh
0-50mph 80.5kmh: 4.3secs
0-60mph 96.5kmh: 5.6secs
0-1/4 mile: 14.4secs
0-1km: 26.7secs
190.0bhp 141.7kW 192.6PS
@ 5280rpm
220.0lbft 298.1Nm @ 4000rpm
211.2bhp/ton 207.6bhp/tonne
53.8bhp/L 40.2kW/L 54.6PS/L
41.1ft/sec 12.5m/sec
27.5mph 44.2kmh/1000rpm
20.9mpg 17.4mpgUS
13.5L/100km
Petrol 4-stroke piston
3528cc 215.0cu in
Vee 8 fuel injection
Compression ratio: 9.7:1
Bore: 88.9mm 3.5in
Stroke: 71.1mm 2.8in
Valve type/No: Overhead 16
Transmission: Manual
No. of forward speeds: 5
Wheels driven: Rear
Springs F/R: Coil/Leaf
Brake system: PA
Brakes F/R: Disc/Drum
Steering: Rack & pinion
Wheelbase: 248.9cm 98.0in
Track F: 134.6cm 53.0in
Track R: 137.2cm 54.0in
Length: 396.2cm 156.0in
Width: 160.0cm 63.0in
Height: 132.1cm 52.0in
Ground clearance: 15.2cm
6.0in
Kerb weight: 915.0kg 2015.4lb
Fuel: 63.7L 14.0gal
16.8galUS

595 Morgan
Plus 8

596 Morgan
Plus 8
1988 UK
125.0mph 201.1kmh
0-50mph 80.5kmh: 5.3secs
0-60mph 96.5kmh: 7.4secs
0-1/4 mile: 15.6secs
190.0bhp 141.7kW 192.6PS
@ 5280rpm
220.0lbft 298.1Nm @ 4000rpm
198.9bhp/ton 195.6bhp/tonne
53.8bhp/L 40.2kW/L 54.6PS/L
41.1ft/sec 12.5m/sec
23.4mpg 19.5mpgUS
12.1L/100km
Petrol 4-stroke piston
3528cc 215.2cu in
Vee 8 1 Carburettor
Compression ratio: 9.8:1
Bore: 89.0mm 3.5in
Stroke: 71.0mm 2.8in
Valve type/No: Overhead 16
Transmission: Manual
No. of forward speeds: 5
Wheels driven: Rear
Springs F/R: Coil/Leaf
Brakes F/R: Disc/Drum
Steering: Rack & pinion
Wheelbase: 248.9cm 98.0in
Track F: 132.1cm 52.0in
Track R: 134.6cm 53.0in
Length: 398.8cm 157.0in
Width: 157.5cm 62.0in
Height: 132.1cm 52.0in
Kerb weight: 971.6kg 2140.0lb
Fuel: 66.2L 14.6gal 17.5galUS

597 Morgan
Plus 8
1991 UK
122.0mph 196.3kmh
0-50mph 80.5kmh: 4.7secs
0-60mph 96.5kmh: 6.1secs
0-1/4 mile: 15.1secs
0-1km: 27.8secs
190.0bhp 141.7kW 192.6PS
@ 4750rpm
230.3lbft 312.0Nm @ 2600rpm
206.7bhp/ton 203.2bhp/tonne
48.1bhp/L 35.9kW/L 48.8PS/L
36.9ft/sec 11.3m/sec
27.6mph 44.4kmh/1000rpm
20.1mpg 16.7mpgUS
14.1L/100km
Petrol 4-stroke piston
3946cc 241.0cu in
Vee 8 fuel injection
Compression ratio: 9.3:1
Bore: 94.0mm 3.7in
Stroke: 71.1mm 2.8in
Valve type/No: Overhead 16
Transmission: Manual
No. of forward speeds: 5
Wheels driven: Rear
Springs F/R: Coil/Leaf
Brake system: PA
Brakes F/R: Disc/Drum
Steering: Rack & pinion
Wheelbase: 248.9cm 98.0in
Track F: 137.1cm 54.0in
Track R: 137.1cm 54.0in
Length: 396.2cm 156.0in
Width: 160.0cm 63.0in
Height: 121.9cm 48.0in
Kerb weight: 935.0kg 2059.5lb
Fuel: 54.6L 12.0gal 14.4galUS

598 MVS
Venturi GT
1987 France
0-60mph 96.5kmh: 6.0secs
0-1/4 mile: 14.8secs
200.0bhp 149.1kW 202.8PS

598 MVS
Venturi GT

@ 5750rpm
195.0lbft 264.2Nm @ 2500rpm
172.6bhp/ton 169.8bhp/tonne
81.4bhp/L 60.7kW/L 82.5PS/L
39.6ft/sec 12.1m/sec
Petrol 4-stroke piston
2458cc 150.0cu in
turbocharged
Vee 6 fuel injection
Compression ratio: 8.6:1
Bore: 91.0mm 3.6in
Stroke: 63.0mm 2.5in
Valve type/No: Overhead 12
Transmission: Manual
No. of forward speeds: 5
Wheels driven: Rear
Springs F/R: Coil/Coil
Brake system: PA
Brakes F/R: Disc/Disc
Steering: Rack & pinion PA
Wheelbase: 240.0cm 94.5in
Track F: 146.6cm 57.7in
Track R: 147.1cm 57.9in
Length: 408.9cm 161.0in
Width: 170.2cm 67.0in
Height: 116.8cm 46.0in
Kerb weight: 1178.1kg
2595.0lb
Fuel: 68.9L 15.1gal 18.2galUS

599 Nissan
300ZX
1986 Japan
128.0mph 206.0kmh
0-50mph 80.5kmh: 6.5secs
0-60mph 96.5kmh: 9.1secs
0-1/4 mile: 16.9secs
160.0bhp 119.3kW 162.2PS
@ 5200rpm
174.0lbft 235.8Nm @ 4000rpm
110.6bhp/ton 108.8bhp/tonne
54.0bhp/L 40.3kW/L 54.8PS/L
47.2ft/sec 14.4m/sec
25.9mph 41.7kmh/1000rpm
21.6mpg 18.0mpgUS
13.1L/100km
Petrol 4-stroke piston
2960cc 180.6cu in
Vee 6 fuel injection
Compression ratio: 9.0:1
Bore: 87.0mm 3.4in
Stroke: 83.0mm 3.3in
Valve type/No: Overhead 12
Transmission: Manual
No. of forward speeds: 5
Wheels driven: Rear
Springs F/R: Coil/Coil
Brake system: PA
Brakes F/R: Disc/Disc
Steering: Rack & pinion PA
Wheelbase: 231.9cm 91.3in
Track F: 145.5cm 57.3in
Track R: 147.6cm 58.1in
Length: 433.6cm 170.7in

Width: 172.5cm 67.9in
Height: 129.5cm 51.0in
Kerb weight: 1471.0kg
3240.0lb
Fuel: 71.9L 15.8gal 19.0galUS

600 Nissan
300ZX Turbo
1986 Japan
133.0mph 214.0kmh
0-50mph 80.5kmh: 5.3secs
0-60mph 96.5kmh: 7.2secs
0-1/4 mile: 15.7secs
200.0bhp 149.1kW 202.8PS
@ 5200rpm
227.0lbft 307.6Nm @ 3600rpm
134.7bhp/ton 132.5bhp/tonne
67.6bhp/L 50.4kW/L 68.5PS/L
47.2ft/sec 14.4m/sec
26.9mph 43.3kmh/1000rpm
20.4mpg 17.0mpgUS
13.8L/100km
Petrol 4-stroke piston
2960cc 180.6cu in
turbocharged
Vee 6 fuel injection
Compression ratio: 9.0:1
Bore: 87.0mm 3.4in
Stroke: 83.0mm 3.3in
Valve type/No: Overhead 12
Transmission: Manual
No. of forward speeds: 5
Wheels driven: Rear
Springs F/R: Coil/Coil
Brake system: PA
Brakes F/R: Disc/Disc
Steering: Rack & pinion PA
Wheelbase: 231.9cm 91.3in
Track F: 143.5cm 56.5in
Track R: 145.5cm 57.3in
Length: 433.6cm 170.7in
Width: 172.5cm 67.9in
Height: 129.5cm 51.0in
Kerb weight: 1509.5kg
3325.0lb
Fuel: 71.9L 15.8gal 19.0galUS

601 Nissan
Bluebird 1.8 ZX
1986 Japan
118.0mph 189.9kmh
0-50mph 80.5kmh: 6.3secs
0-60mph 96.5kmh: 8.6secs
0-1/4 mile: 16.5secs
0-1km: 30.6secs
135.0bhp 100.7kW 136.9PS
@ 6000rpm
141.0lbft 191.1Nm @ 4000rpm
108.5bhp/ton 106.7bhp/tonne
74.6bhp/L 55.6kW/L 75.7PS/L
54.8ft/sec 16.7m/sec
24.2mph 38.9kmh/1000rpm
23.8mpg 19.8mpgUS
11.9L/100km

Petrol 4-stroke piston
1809cc 110.0cu in
turbocharged
In-line 4 fuel injection
Compression ratio: 8.0:1
Bore: 83.0mm 3.3in
Stroke: 83.6mm 3.3in
Valve type/No: Overhead 8
Transmission: Manual
No. of forward speeds: 5
Wheels driven: Front
Springs F/R: Coil/Coil
Brake system: PA
Brakes F/R: Disc/Disc
Steering: Rack & pinion PA
Wheelbase: 255.0cm 100.4in
Track F: 146.0cm 57.5in
Track R: 146.0cm 57.5in
Length: 440.5cm 173.4in
Width: 169.0cm 66.5in
Height: 139.5cm 54.9in
Ground clearance: 15.0cm
5.9in
Kerb weight: 1265.0kg
2786.3lb
Fuel: 60.1L 13.2gal 15.9galUS

602 Nissan
Laurel 2.4 SGL Automatic
1986 Japan
105.0mph 168.9kmh
0-50mph 80.5kmh: 8.6secs
0-60mph 96.5kmh: 11.6secs
0-1/4 mile: 18.4secs
0-1km: 33.6secs
128.0bhp 95.4kW 129.8PS
@ 5600rpm
135.0lbft 182.9Nm @ 4400rpm
98.5bhp/ton 96.8bhp/tonne
53.5bhp/L 39.9kW/L 54.2PS/L
45.1ft/sec 13.8m/sec
25.1mph 40.4kmh/1000rpm
20.0mpg 16.7mpgUS
14.1L/100km
Petrol 4-stroke piston
2393cc 146.0cu in
In-line 6 fuel injection
Compression ratio: 8.9:1
Bore: 83.0mm 3.3in
Stroke: 73.7mm 2.9in
Valve type/No: Overhead 12
Transmission: Automatic
No. of forward speeds: 4
Wheels driven: Rear
Springs F/R: Coil/Coil
Brake system: PA
Brakes F/R: Disc/Drum
Steering: Rack & pinion PA
Wheelbase: 267.0cm 105.1in
Track F: 142.0cm 55.9in
Track R: 140.7cm 55.4in
Length: 464.8cm 183.0in
Width: 168.9cm 66.5in
Height: 142.5cm 56.1in

Ground clearance: 17.8cm
7.0in
Kerb weight: 1321.6kg
2911.0lb
Fuel: 65.1L 14.3gal 17.2galUS

603 Nissan
Sentra Sport Coupe SE
1986 Japan
100.0mph 160.9kmh
0-50mph 80.5kmh: 9.6secs
0-60mph 96.5kmh: 13.3secs
0-1/4 mile: 19.0secs
70.0bhp 52.2kW 71.0PS
@ 5000rpm
94.0lbft 127.4Nm @ 2800rpm
65.7bhp/ton 64.6bhp/tonne
43.8bhp/L 32.7kW/L 44.4PS/L
48.1ft/sec 14.7m/sec
32.4mpg 27.0mpgUS
8.7L/100km
Petrol 4-stroke piston
1597cc 97.4cu in
In-line 4 1 Carburettor
Compression ratio: 9.4:1
Bore: 76.0mm 3.0in
Stroke: 88.0mm 3.5in
Valve type/No: Overhead 8
Transmission: Manual
No. of forward speeds: 5
Wheels driven: Front
Springs F/R: Coil/Coil
Brake system: PA
Brakes F/R: Disc/Drum
Steering: Rack & pinion PA
Wheelbase: 243.1cm 95.7in
Track F: 143.5cm 56.5in
Track R: 143.5cm 56.5in
Length: 422.9cm 166.5in
Width: 166.6cm 65.6in
Height: 132.6cm 52.2in
Ground clearance: 14.0cm
5.5in
Kerb weight: 1082.8kg
2385.0lb
Fuel: 50.0L 11.0gal 13.2galUS

604 Nissan
Sunny 1.3 LX 5-door
1986 Japan
93.0mph 149.6kmh
0-50mph 80.5kmh: 9.7secs
0-60mph 96.5kmh: 13.7secs
0-1/4 mile: 19.7secs
0-1km: 37.1secs
60.0bhp 44.7kW 60.8PS
@ 5600rpm
74.0lbft 100.3Nm @ 3600rpm
64.9bhp/ton 63.8bhp/tonne
47.2bhp/L 35.2kW/L 47.9PS/L
42.9ft/sec 13.1m/sec
18.4mph 29.6kmh/1000rpm
31.0mpg 25.8mpgUS
9.1L/100km

606 Nissan
300ZX Turbo

Petrol 4-stroke piston
1270cc 77.0cu in
In-line 4 1 Carburettor
Compression ratio: 9.0:1
Bore: 76.0mm 3.0in
Stroke: 70.0mm 2.8in
Valve type/No: Overhead 8
Transmission: Manual
No. of forward speeds: 5
Wheels driven: Front
Springs F/R: Coil/Coil
Brake system: PA
Brakes F/R: Disc/Drum
Steering: Rack & pinion
Wheelbase: 243.0cm 95.7in
Track F: 143.5cm 56.5in
Track R: 143.0cm 56.3in
Length: 403.0cm 158.7in
Width: 164.0cm 64.6in
Height: 138.0cm 54.3in
Ground clearance: 14.9cm
5.9in
Kerb weight: 940.0kg 2070.5lb
Fuel: 50.0L 11.0gal 13.2galUS

605 Nissan
300ZX
1987 Japan
125.0mph 201.1kmh
0-60mph 96.5kmh: 8.8secs
0-1/4 mile: 16.5secs
160.0bhp 119.3kW 162.2PS
@ 5200rpm
174.0lbft 235.8Nm @ 4000rpm
108.3bhp/ton 106.5bhp/tonne
54.0bhp/L 40.3kW/L 54.8PS/L
47.2ft/sec 14.4m/sec
23.8mpg 19.8mpgUS
11.9L/100km
Petrol 4-stroke piston
2960cc 180.6cu in
Vee 6 fuel injection
Compression ratio: 9.0:1

Bore: 87.0mm 3.4in
Stroke: 83.0mm 3.3in
Valve type/No: Overhead 24
Transmission: Manual
No. of forward speeds: 5
Wheels driven: Rear
Springs F/R: Coil/Coil
Brakes F/R: Disc/Disc
Steering: Rack & pinion PA
Wheelbase: 252.0cm 99.2in
Track F: 141.5cm 55.7in
Track R: 143.5cm 56.5in
Length: 453.4cm 178.5in
Width: 172.5cm 67.9in
Height: 126.5cm 49.8in
Kerb weight: 1502.7kg
3310.0lb
Fuel: 71.9L 15.8gal 19.0galUS

606 Nissan
300ZX Turbo
1987 Japan
144.0mph 231.7kmh
0-50mph 80.5kmh: 5.2secs
0-60mph 96.5kmh: 7.0secs
0-1/4 mile: 15.5secs
0-1km: 28.1secs
225.0bhp 167.8kW 228.1PS
@ 5400rpm
240.0lbft 325.2Nm @ 4400rpm
154.8bhp/ton 152.2bhp/tonne
76.0bhp/L 56.7kW/L 77.1PS/L
49.0ft/sec 14.9m/sec
28.1mph 45.2kmh/1000rpm
19.2mpg 16.0mpgUS
14.7L/100km
Petrol 4-stroke piston
2960cc 181.0cu in
turbocharged
Vee 6 fuel injection
Compression ratio: 7.8:1
Bore: 87.0mm 3.4in
Stroke: 83.0mm 3.3in

Valve type/No: Overhead 12
Transmission: Manual
No. of forward speeds: 5
Wheels driven: Rear
Springs F/R: Coil/Coil
Brake system: PA
Brakes F/R: Disc/Disc
Steering: Rack & pinion PA
Wheelbase: 252.0cm 99.2in
Track F: 145.5cm 57.3in
Track R: 147.3cm 58.0in
Length: 460.5cm 181.3in
Width: 172.5cm 67.9in
Height: 131.0cm 51.6in
Ground clearance: 15.0cm
5.9in
Kerb weight: 1478.0kg
3255.5lb
Fuel: 77.3L 17.0gal 20.4galUS

607 Nissan
Micra SGL 5-DR
1987 Japan
89.0mph 143.2kmh
0-50mph 80.5kmh: 9.8secs
0-60mph 96.5kmh: 14.5secs
0-1/4 mile: 19.6secs
0-1km: 37.4secs
55.0bhp 41.0kW 55.8PS
@ 6000rpm
56.0lbft 75.9Nm @ 3600rpm
80.5bhp/ton 79.1bhp/tonne
55.7bhp/L 41.5kW/L 56.4PS/L
44.7ft/sec 13.6m/sec
17.8mph 28.6kmh/1000rpm
40.9mpg 34.1mpgUS
6.9L/100km
Petrol 4-stroke piston
988cc 60.0cu in
In-line 4 1 Carburettor
Compression ratio: 10.3:1
Bore: 68.0mm 2.7in
Stroke: 68.0mm 2.7in

Valve type/No: Overhead 8
Transmission: Manual
No. of forward speeds: 5
Wheels driven: Front
Springs F/R: Coil/Coil
Brake system: PA
Brakes F/R: Disc/Drum
Steering: Rack & pinion
Wheelbase: 230.0cm 90.6in
Track F: 134.6cm 53.0in
Track R: 133.1cm 52.4in
Length: 364.5cm 143.5in
Width: 156.0cm 61.4in
Height: 139.5cm 54.9in
Ground clearance: 15.2cm
6.0in
Kerb weight: 695.0kg 1530.8lb
Fuel: 40.0L 8.8gal 10.6galUS

608 Nissan
Pulsar NX SE
1987 Japan
115.0mph 185.0kmh
0-50mph 80.5kmh: 7.2secs
0-60mph 96.5kmh: 10.3secs
0-1/4 mile: 17.8secs
113.0bhp 84.3kW 114.6PS
@ 6400rpm
99.0lbft 134.1Nm @ 4800rpm
97.3bhp/ton 95.7bhp/tonne
70.8bhp/L 52.8kW/L 71.7PS/L
58.5ft/sec 17.8m/sec
28.8mpg 24.0mpgUS
9.8L/100km
Petrol 4-stroke piston
1597cc 97.4cu in
In-line 4 fuel injection
Compression ratio: 9.0:1
Bore: 78.0mm 3.1in
Stroke: 83.6mm 3.3in
Valve type/No: Overhead 16
Transmission: Manual
No. of forward speeds: 5

Wheels driven: Front
Springs F/R: Coil/Coil
Brake system: PA
Brakes F/R: Disc/Drum
Steering: Rack & pinion PA
Wheelbase: 243.1cm 95.7in
Track F: 144.0cm 56.7in
Track R: 144.0cm 56.7in
Length: 422.9cm 166.5in
Width: 166.9cm 65.7in
Height: 129.0cm 50.8in
Ground clearance: 14.7cm
5.8in
Kerb weight: 1180.4kg
2600.0lb
Fuel: 50.0L 11.0gal 13.2galUS

609 Nissan
Sunny ZX Coupe
1987 Japan
118.0mph 189.9kmh
0-50mph 80.5kmh: 6.6secs
0-60mph 96.5kmh: 9.2secs
0-1/4 mile: 17.2secs
0-1km: 31.5secs
120.0bhp 89.5kW 121.7PS
@ 6600rpm
102.0lbft 138.2Nm @ 5200rpm
112.3bhp/ton 110.4bhp/tonne
75.1bhp/L 56.0kW/L 76.1PS/L
60.3ft/sec 18.4m/sec
18.1mph 29.1kmh/1000rpm
27.2mpg 22.6mpgUS
10.4L/100km
Petrol 4-stroke piston
1598cc 97.0cu in
In-line 4 fuel injection
Compression ratio: 10.0:1
Bore: 78.0mm 3.1in
Stroke: 83.6mm 3.3in
Valve type/No: Overhead 16
Transmission: Manual
No. of forward speeds: 5
Wheels driven: Front
Springs F/R: Coil/Coil
Brake system: PA
Brakes F/R: Disc/Disc
Steering: Rack & pinion PA
Wheelbase: 243.0cm 95.7in
Track F: 143.5cm 56.5in
Track R: 143.5cm 56.5in
Length: 423.5cm 166.7in
Width: 166.5cm 65.6in
Height: 132.5cm 52.2in
Ground clearance: 14.0cm
5.5in
Kerb weight: 1087.0kg
2394.3lb
Fuel: 50.0L 11.0gal 13.2galUS

610 Nissan
Sunny 1.6 SLX Coupe
1987 Japan
105.0mph 168.9kmh

0-50mph 80.5kmh: 8.1secs
0-60mph 96.5kmh: 11.6secs
0-1/4 mile: 18.2secs
0-1km: 34.0secs
84.0bhp 62.6kW 85.2PS
@ 5600rpm
97.0lbft 131.4Nm @ 3200rpm
86.7bhp/ton 85.3bhp/tonne
52.6bhp/L 39.2kW/L 53.3PS/L
50.2ft/sec 15.3m/sec
20.9mph 33.6kmh/1000rpm
28.7mpg 23.9mpgUS
9.8L/100km
Petrol 4-stroke piston
1597cc 97.0cu in
In-line 4 1 Carburettor
Compression ratio: 9.6:1
Bore: 76.0mm 3.0in
Stroke: 82.0mm 3.2in
Valve type/No: Overhead 8
Transmission: Manual
No. of forward speeds: 5
Wheels driven: Front
Springs F/R: Coil/Coil
Brake system: PA
Brakes F/R: Disc/Drum
Steering: Rack & pinion PA
Wheelbase: 243.0cm 95.7in
Track F: 143.5cm 56.5in
Track R: 143.5cm 56.5in
Length: 423.5cm 166.7in
Width: 166.5cm 65.6in
Height: 132.5cm 52.2in
Ground clearance: 14.0cm
5.5in
Kerb weight: 985.0kg 2169.6lb
Fuel: 50.0L 11.0gal 13.2galUS

611 Nissan
200SX SE V6
1988 Japan
119.0mph 191.5kmh
0-50mph 80.5kmh: 6.1secs
0-60mph 96.5kmh: 8.4secs
0-1/4 mile: 16.4secs
165.0bhp 123.0kW 167.3PS
@ 5200rpm
168.0lbft 227.6Nm @ 3200rpm
123.0bhp/ton 120.9bhp/tonne
55.7bhp/L 41.6kW/L 56.5PS/L
47.2ft/sec 14.4m/sec
23.4mpg 19.5mpgUS
12.1L/100km
Petrol 4-stroke piston
2960cc 180.6cu in
Vee 6 fuel injection
Compression ratio: 9.0:1
Bore: 87.0mm 3.4in
Stroke: 83.0mm 3.3in
Valve type/No: Overhead 12
Transmission: Manual
No. of forward speeds: 5
Wheels driven: Rear
Springs F/R: Coil/Coil

Brake system: PA
Brakes F/R: Disc/Disc
Steering: Rack & pinion PA
Wheelbase: 242.6cm 95.5in
Track F: 138.9cm 54.7in
Track R: 142.5cm 56.1in
Length: 446.0cm 175.6in
Width: 166.9cm 65.7in
Height: 128.0cm 50.4in
Kerb weight: 1364.3kg
3005.0lb
Fuel: 53.0L 11.6gal 14.0galUS

612 Nissan
240SX
1988 Japan
125.0mph 201.1kmh
0-50mph 80.5kmh: 6.3secs
0-60mph 96.5kmh: 8.8secs
0-1/4 mile: 16.5secs
140.0bhp 104.4kW 141.9PS
@ 5600rpm
152.0lbft 206.0Nm @ 4400rpm
112.0bhp/ton 110.1bhp/tonne
58.6bhp/L 43.7kW/L 59.4PS/L
58.8ft/sec 17.9m/sec
30.0mpg 25.0mpgUS
9.4L/100km
Petrol 4-stroke piston
2389cc 145.8cu in
In-line 4 fuel injection
Compression ratio: 8.7:1
Bore: 89.0mm 3.5in
Stroke: 96.0mm 3.8in
Valve type/No: Overhead 12
Transmission: Manual
No. of forward speeds: 5
Wheels driven: Rear
Springs F/R: Coil/Coil
Brake system: PA
Brakes F/R: Disc/Disc
Steering: Rack & pinion PA
Wheelbase: 247.4cm 97.4in
Track F: 146.1cm 57.5in
Track R: 146.1cm 57.5in
Length: 452.1cm 178.0in
Width: 168.9cm 66.5in
Height: 129.0cm 50.8in
Ground clearance: 14.5cm
5.7in
Kerb weight: 1271.2kg
2800.0lb
Fuel: 60.2L 13.2gal 15.9galUS

613 Nissan
300ZX
1988 Japan
125.0mph 201.1kmh
0-50mph 80.5kmh: 5.2secs
0-60mph 96.5kmh: 7.4secs
0-1/4 mile: 16.4secs
205.0bhp 152.9kW 207.8PS
@ 5200rpm
227.0lbft 307.6Nm @ 3600rpm

140.6bhp/ton 138.3bhp/tonne
69.3bhp/L 51.6kW/L 70.2PS/L
47.2ft/sec 14.4m/sec
25.2mpg 21.0mpgUS
11.2L/100km
Petrol 4-stroke piston
2960cc 180.6cu in
turbocharged
Vee 6 fuel injection
Compression ratio: 8.3:1
Bore: 87.0mm 3.4in
Stroke: 83.0mm 3.3in
Valve type/No: Overhead 12
Transmission: Manual
No. of forward speeds: 5
Wheels driven: Rear
Springs F/R: Coil/Coil
Brake system: PA
Brakes F/R: Disc/Disc
Steering: Rack & pinion PA
Wheelbase: 231.9cm 91.3in
Track F: 143.5cm 56.5in
Track R: 145.5cm 57.3in
Length: 440.4cm 173.4in
Width: 172.5cm 67.9in
Height: 126.2cm 49.7in
Kerb weight: 1482.3kg
3265.0lb
Fuel: 71.9L 15.8gal 19.0galUS

614 Nissan
200SX
1989 Japan
141.0mph 226.9kmh
0-50mph 80.5kmh: 5.2secs
0-60mph 96.5kmh: 7.2secs
0-1/4 mile: 15.3secs
0-1km: 28.0secs
171.0bhp 127.5kW 173.4PS
@ 6400rpm
168.3lbft 228.0Nm @ 4000rpm
141.7bhp/ton 139.4bhp/tonne
94.5bhp/L 70.5kW/L 95.8PS/L
58.5ft/sec 17.8m/sec
21.4mph 34.4kmh/1000rpm
19.5mpg 16.2mpgUS
14.5L/100km
Petrol 4-stroke piston
1809cc 110.0cu in
turbocharged
In-line 4 fuel injection
Compression ratio: 8.5:1
Bore: 83.0mm 3.3in
Stroke: 83.6mm 3.3in
Valve type/No: Overhead 16
Transmission: Manual
No. of forward speeds: 5
Wheels driven: Rear
Springs F/R: Coil/Coil
Brake system: PA ABS
Brakes F/R: Disc/Disc
Steering: Rack & pinion PA
Wheelbase: 247.5cm 97.4in
Track F: 146.5cm 57.7in

Track R: 146.5cm 57.7in
Length: 453.5cm 178.5in
Width: 169.0cm 66.5in
Height: 129.0cm 50.8in
Ground clearance: 14.5cm
5.7in
Kerb weight: 1227.0kg
2702.6lb
Fuel: 60.1L 13.2gal 15.9galUS

615 Nissan

300ZX
1989 Japan
148.0mph 238.1kmh
0-50mph 80.5kmh: 5.4secs
0-60mph 96.5kmh: 7.1secs
0-1/4 mile: 15.5secs
222.0bhp 165.5kW 225.1PS
@ 6400rpm
198.0lbft 268.3Nm @ 4800rpm
154.4bhp/ton 151.9bhp/tonne
75.0bhp/L 55.9kW/L 76.0PS/L
58.1ft/sec 17.7m/sec
28.2mpg 23.5mpgUS
10.0L/100km
Petrol 4-stroke piston
2960cc 180.6cu in
Vee 6 fuel injection
Compression ratio: 10.1:1
Bore: 87.0mm 3.4in
Stroke: 83.0mm 3.3in
Valve type/No: Overhead 24
Transmission: Manual
No. of forward speeds: 5
Wheels driven: Rear
Springs F/R: Coil/Coil
Brake system: PA ABS
Brakes F/R: Disc/Disc
Steering: Rack & pinion PA
Wheelbase: 245.1cm 96.5in
Track F: 149.6cm 58.9in
Track R: 153.4cm 60.4in

Length: 430.5cm 169.5in
Width: 179.1cm 70.5in
Height: 125.5cm 49.4in
Ground clearance: 12.7cm
5.0in
Kerb weight: 1461.9kg
3220.0lb
Fuel: 70.8L 15.6gal 18.7galUS

616 Nissan

Maxima
1989 Japan
122.0mph 196.3kmh
0-50mph 80.5kmh: 7.0secs
0-60mph 96.5kmh: 9.4secs
0-1/4 mile: 16.8secs
0-1km: 31.1secs
172.0bhp 128.3kW 174.4PS
@ 5600rpm
134.3lbft 182.0Nm @ 2800rpm
123.0bhp/ton 121.0bhp/tonne
58.1bhp/L 43.3kW/L 58.9PS/L
50.9ft/sec 15.5m/sec
29.2mph 47.0kmh/1000rpm
16.8mpg 14.0mpgUS
16.8L/100km
Petrol 4-stroke piston
2960cc 181.0cu in
Vee 6 fuel injection
Compression ratio: 10.0:1
Bore: 87.0mm 3.4in
Stroke: 83.0mm 3.3in
Valve type/No: Overhead 12
Transmission: Automatic
No. of forward speeds: 4
Wheels driven: Front
Springs F/R: Coil/Coil
Brake system: PA
Brakes F/R: Disc/Disc
Steering: Rack & pinion PA
Wheelbase: 264.9cm 104.3in
Track F: 150.9cm 59.4in

Track R: 149.1cm 58.7in
Length: 478.0cm 188.2in
Width: 162.3cm 63.9in
Height: 140.5cm 55.3in
Kerb weight: 1421.5kg
3131.0lb
Fuel: 70.1L 15.4gal 18.5galUS

617 Nissan

Maxima SE
1989 Japan
120.0mph 193.1kmh
0-50mph 80.5kmh: 6.3secs
0-60mph 96.5kmh: 8.5secs
0-1/4 mile: 16.6secs
160.0bhp 119.3kW 162.2PS
@ 5200rpm
182.0lbft 246.6Nm @ 2800rpm
114.0bhp/ton 112.1bhp/tonne
54.0bhp/L 40.3kW/L 54.8PS/L
47.2ft/sec 14.4m/sec
24.6mpg 20.5mpgUS
11.5L/100km
Petrol 4-stroke piston
2960cc 180.6cu in
Vee 6 fuel injection
Compression ratio: 9.0:1
Bore: 87.0mm 3.4in
Stroke: 83.0mm 3.3in
Valve type/No: Overhead 12
Transmission: Manual
No. of forward speeds: 5
Wheels driven: Front
Springs F/R: Coil/Coil
Brake system: PA ABS
Brakes F/R: Disc/Disc
Steering: Rack & pinion PA
Wheelbase: 264.9cm 104.3in
Track F: 150.9cm 59.4in
Track R: 149.1cm 58.7in
Length: 476.5cm 187.6in
Width: 176.0cm 69.3in

Height: 140.0cm 55.1in
Ground clearance: 12.7cm
5.0in
Kerb weight: 1427.8kg
3145.0lb
Fuel: 70.0L 15.4gal 18.5galUS

618 Nissan

Micra 1.2 GSX
1989 Japan
93.0mph 149.6kmh
0-50mph 80.5kmh: 9.6secs
0-60mph 96.5kmh: 14.0secs
0-1/4 mile: 19.6secs
0-1km: 36.5secs
60.0bhp 44.7kW 60.8PS
@ 5600rpm
69.4lbft 94.0Nm @ 3200rpm
76.7bhp/ton 75.4bhp/tonne
48.6bhp/L 36.2kW/L 49.3PS/L
47.8ft/sec 14.6m/sec
22.8mph 36.7kmh/1000rpm
32.3mpg 26.9mpgUS
8.7L/100km
Petrol 4-stroke piston
1235cc 75.0cu in
In-line 4 1 Carburettor
Compression ratio: 9.0:1
Bore: 71.0mm 2.8in
Stroke: 78.0mm 3.1in
Valve type/No: Overhead 8
Transmission: Manual
No. of forward speeds: 5
Wheels driven: Front
Springs F/R: Coil/Coil
Brake system: PA
Brakes F/R: Disc/Drum
Steering: Rack & pinion
Wheelbase: 230.0cm 90.6in
Track F: 134.6cm 53.0in
Track R: 133.1cm 52.4in
Length: 364.5cm 143.5in
Width: 156.0cm 61.4in
Height: 139.5cm 54.9in
Ground clearance: 15.2cm
6.0in
Kerb weight: 796.0kg 1753.3lb
Fuel: 40.0L 8.8gal 10.6galUS

619 Nissan

Prairie
1989 Japan
98.0mph 157.7kmh
0-50mph 80.5kmh: 10.3secs
0-60mph 96.5kmh: 14.4secs
0-1/4 mile: 19.4secs
0-1km: 36.1secs
98.0bhp 73.1kW 99.4PS
@ 5600rpm
118.8lbft 161.0Nm @ 3200rpm
76.0bhp/ton 74.7bhp/tonne
49.6bhp/L 37.0kW/L 50.3PS/L
53.8ft/sec 16.4m/sec
25.7mph 41.4kmh/1000rpm

619 Nissan
Prairie

20.6mpg 17.2mpgUS
13.7L/100km
Petrol 4-stroke piston
1974cc 120.0cu in
In-line 4 1 Carburettor
Compression ratio: 9.4:1
Bore: 84.5mm 3.3in
Stroke: 88.0mm 3.5in
Valve type/No: Overhead 8
Transmission: Automatic
No. of forward speeds: 4
Wheels driven: Front
Springs F/R: Coil/Coil
Brake system: PA ABS
Brakes F/R: Disc/Drum
Steering: Rack & pinion PA
Wheelbase: 261.0cm 102.8in
Track F: 146.0cm 57.5in
Track R: 143.0cm 56.3in
Length: 436.0cm 171.7in
Width: 169.0cm 66.5in
Height: 163.0cm 64.2in
Ground clearance: 19.5cm
7.7in
Kerb weight: 1312.0kg
2889.9lb
Fuel: 65.1L 14.3gal 17.2galUS

620 Nissan
Sunny 1.4 LS
1989 Japan
105.0mph 168.9kmh
0-50mph 80.5kmh: 8.0secs
0-60mph 96.5kmh: 12.0secs
0-1/4 mile: 18.5secs
0-1km: 34.4secs
83.0bhp 61.9kW 84.1PS
@ 6200rpm
81.9lbft 111.0Nm @ 4000rpm
87.2bhp/ton 85.7bhp/tonne
59.6bhp/L 44.5kW/L 60.4PS/L
55.5ft/sec 16.9m/sec
21.5mph 34.6kmh/1000rpm
28.9mpg 24.1mpgUS
9.8L/100km
Petrol 4-stroke piston
1392cc 85.0cu in
In-line 4 1 Carburettor
Compression ratio: 9.4:1
Bore: 73.6mm 2.9in
Stroke: 81.8mm 3.2in
Valve type/No: Overhead 12
Transmission: Manual
No. of forward speeds: 5
Wheels driven: Front
Springs F/R: Coil/Coil
Brake system: PA
Brakes F/R: Disc/Drum
Steering: Rack & pinion
Wheelbase: 243.0cm 95.7in
Track F: 143.5cm 56.5in
Track R: 143.0cm 56.3in
Length: 403.0cm 158.7in
Width: 164.0cm 64.6in

Height: 138.0cm 54.3in
Ground clearance: 14.9cm
5.9in
Kerb weight: 968.0kg 2132.2lb
Fuel: 50.0L 11.0gal 13.2galUS

621 Nissan
300ZX
1990 Japan
158.0mph 254.2kmh
0-50mph 80.5kmh: 4.3secs
0-60mph 96.5kmh: 5.6secs
0-1/4 mile: 14.4secs
0-1km: 25.6secs
280.0bhp 208.8kW 283.9PS
@ 6400rpm
274.5lbft 372.0Nm @ 3600rpm
180.2bhp/ton 177.2bhp/tonne
94.6bhp/L 70.5kW/L 95.9PS/L
58.1ft/sec 17.7m/sec
25.7mph 41.4kmh/1000rpm
17.0mpg 14.2mpgUS
16.6L/100km
Petrol 4-stroke piston
2960cc 181.0cu in
turbocharged
Vee 6 fuel injection
Compression ratio: 8.5:1
Bore: 87.0mm 3.4in
Stroke: 83.0mm 3.3in
Valve type/No: Overhead 24
Transmission: Manual
No. of forward speeds: 5
Wheels driven: Rear
Springs F/R: Coil/Coil
Brake system: PA ABS
Brakes F/R: Disc/Disc
Steering: Rack & pinion PA
Wheelbase: 257.0cm 101.2in
Track F: 149.5cm 58.9in
Track R: 153.5cm 60.4in
Length: 452.5cm 178.1in
Width: 180.0cm 70.9in
Height: 125.5cm 49.4in
Kerb weight: 1580.0kg
3480.2lb
Fuel: 71.9L 15.8gal 19.0galUS

622 Nissan
300ZX Turbo
1990 Japan
155.0mph 249.4kmh
0-50mph 80.5kmh: 5.0secs
0-60mph 96.5kmh: 6.5secs
0-1/4 mile: 15.0secs
300.0bhp 223.7kW 304.2PS
@ 6400rpm
283.0lbft 383.5Nm @ 3600rpm
193.1bhp/ton 189.9bhp/tonne
101.3bhp/L 75.6kW/L
102.8PS/L
58.1ft/sec 17.7m/sec
Petrol 4-stroke piston
2960cc 180.6cu in

turbocharged
Vee 6 fuel injection
Compression ratio: 8.5:1
Bore: 87.0mm 3.4in
Stroke: 83.0mm 3.3in
Valve type/No: Overhead 24
Transmission: Manual
No. of forward speeds: 5
Wheels driven: Rear
Springs F/R: Coil/Coil
Brake system: PA ABS
Brakes F/R: Disc/Disc
Steering: Rack & pinion PA
Wheelbase: 245.1cm 96.5in
Track F: 149.6cm 58.9in
Track R: 155.4cm 61.2in
Length: 430.5cm 169.5in
Width: 179.1cm 70.5in
Height: 125.0cm 49.2in
Kerb weight: 1579.9kg
3480.0lb
Fuel: 70.8L 15.6gal 18.7galUS

623 Nissan
Primera 1.6 LS
1990 UK
112.8mph 181.5kmh
0-50mph 80.5kmh: 8.7secs
0-60mph 96.5kmh: 12.4secs
0-1/4 mile: 18.3secs
0-1km: 34.2secs
95.0bhp 70.8kW 96.3PS
@ 6000rpm
98.9lbft 134.0Nm @ 4000rpm
84.7bhp/ton 83.3bhp/tonne
59.5bhp/L 44.4kW/L 60.3PS/L
56.5ft/sec 17.2m/sec
22.7mph 36.5kmh/1000rpm
29.2mpg 24.3mpgUS
9.7L/100km
Petrol 4-stroke piston
1597cc 97.0cu in
In-line 4 1 Carburettor
Compression ratio: 9.8:1
Bore: 76.0mm 3.0in
Stroke: 86.0mm 3.4in
Valve type/No: Overhead 16
Transmission: Manual
No. of forward speeds: 5
Wheels driven: Front
Springs F/R: Coil/Coil
Brake system: PA
Brakes F/R: Disc/Drum
Steering: Rack & pinion PA
Wheelbase: 255.0cm 100.4in
Track F: 147.0cm 57.9in
Track R: 146.0cm 57.5in
Length: 440.0cm 173.2in
Width: 170.0cm 66.9in
Height: 139.0cm 54.7in
Ground clearance: 17.9cm
7.0in
Kerb weight: 1140.0kg
2511.0lb

Fuel: 60.1L 13.2gal 15.9galUS

624 Nissan
Skyline GT-R
1990 Japan
156.0mph 251.0kmh
0-60mph 96.5kmh: 5.6secs
0-1/4 mile: 13.6secs
276.0bhp 205.8kW 279.8PS
@ 6800rpm
260.0lbft 352.3Nm @ 4400rpm
196.3bhp/ton 193.0bhp/tonne
107.5bhp/L 80.1kW/L
109.0PS/L
54.8ft/sec 16.7m/sec
Petrol 4-stroke piston
2568cc 156.7cu in
turbocharged
In-line 6 fuel injection
Compression ratio: 8.5:1
Bore: 86.0mm 3.4in
Stroke: 73.7mm 2.9in
Valve type/No: Overhead 24
Transmission: Manual
No. of forward speeds: 5
Wheels driven: 4-wheel drive
Springs F/R: Coil/Coil
Brake system: PA ABS
Brakes F/R: Disc/Disc
Steering: Rack & pinion PA
Wheelbase: 261.6cm 103.0in
Track F: 148.1cm 58.3in
Track R: 148.1cm 58.3in
Length: 454.7cm 179.0in
Width: 175.5cm 69.1in
Height: 134.1cm 52.8in
Kerb weight: 1430.1kg
3150.0lb
Fuel: 71.9L 15.8gal 19.0galUS

625 Nissan
100NX
1991 Japan
116.0mph 186.6kmh
0-50mph 80.5kmh: 7.3secs
0-60mph 96.5kmh: 10.1secs
0-1/4 mile: 17.7secs
0-1km: 32.6secs
95.0bhp 70.8kW 96.3PS
@ 6000rpm
98.9lbft 134.0Nm @ 4000rpm
93.8bhp/ton 92.2bhp/tonne
59.5bhp/L 44.4kW/L 60.3PS/L
57.7ft/sec 17.6m/sec
21.1mph 33.9kmh/1000rpm
27.2mpg 22.6mpgUS
10.4L/100km
Petrol 4-stroke piston
1597cc 97.0cu in
In-line 4 1 Carburettor
Compression ratio: 9.5:1
Bore: 76.0mm 3.0in
Stroke: 88.0mm 3.5in
Valve type/No: Overhead 16

Transmission: Manual
No. of forward speeds: 5
Wheels driven: Front
Springs F/R: Coil/Coil
Brake system: PA
Brakes F/R: Disc/Drum
Steering: Rack & pinion PA
Wheelbase: 243.1cm 95.7in
Track F: 143.5cm 56.5in
Track R: 143.5cm 56.5in
Length: 414.0cm 163.0in
Width: 167.9cm 66.1in
Height: 131.1cm 51.6in
Kerb weight: 1030.0kg
2268.7lb
Fuel: 50.0L 11.0gal 13.2galUS

626 Nissan

240SX
1991 Japan
115.0mph 185.0kmh
0-50mph 80.5kmh: 6.4secs
0-60mph 96.5kmh: 8.8secs
0-1/4 mile: 16.7secs
155.0bhp 115.6kW 157.1PS
@ 5600rpm
160.0lbft 216.8Nm @ 4400rpm
124.0bhp/ton 121.9bhp/tonne
64.9bhp/L 48.4kW/L 65.8PS/L
58.8ft/sec 17.9m/sec
28.6mpg 23.8mpgUS
9.9L/100km
Petrol 4-stroke piston
2389cc 145.8cu in
In-line 4 fuel injection
Compression ratio: 8.6:1
Bore: 89.0mm 3.5in
Stroke: 96.0mm 3.8in
Valve type/No: Overhead 16
Transmission: Manual
No. of forward speeds: 5
Wheels driven: Rear
Springs F/R: Coil/Coil
Brake system: PA ABS
Brakes F/R: Disc/Disc
Steering: Rack & pinion PA
Wheelbase: 247.4cm 97.4in
Track F: 146.6cm 57.7in
Track R: 146.1cm 57.5in
Length: 452.1cm 178.0in
Width: 168.9cm 66.5in
Height: 129.0cm 50.8in
Kerb weight: 1271.2kg
2800.0lb
Fuel: 60.2L 13.2gal 15.9galUS

627 Nissan

300ZX Turbo Millen Super GTZ
1991 Japan
166.0mph 267.1kmh
0-50mph 80.5kmh: 4.0secs
0-60mph 96.5kmh: 5.0secs
0-1/4 mile: 13.5secs
460.0bhp 343.0kW 466.4PS

@ 6500rpm
430.0lbft 582.7Nm @ 4750rpm
295.2bhp/ton 290.3bhp/tonne
155.4bhp/L 115.9kW/L
157.6PS/L
59.0ft/sec 18.0m/sec
19.2mpg 16.0mpgUS
14.7L/100km
Petrol 4-stroke piston
2960cc 180.6cu in
turbocharged
Vee 6 fuel injection
Compression ratio: 10.1:1
Bore: 87.0mm 3.4in
Stroke: 83.0mm 3.3in
Valve type/No: Overhead 24
Transmission: Manual
No. of forward speeds: 5
Wheels driven: Rear
Springs F/R: Coil/Coil
Brake system: PA ABS
Brakes F/R: Disc/Disc
Steering: Rack & pinion PA
Wheelbase: 245.1cm 96.5in
Track F: 149.6cm 58.9in
Track R: 153.4cm 60.4in
Length: 430.5cm 169.5in
Width: 179.1cm 70.5in
Height: 125.5cm 49.4in
Ground clearance: 12.7cm
5.0in
Kerb weight: 1584.5kg
3490.0lb
Fuel: 70.8L 15.6gal 18.7galUS

628 Nissan

300ZX Twin Turbo
1991 Japan
155.0mph 249.4kmh
0-50mph 80.5kmh: 5.0secs
0-60mph 96.5kmh: 6.5secs
0-1/4 mile: 15.0secs
300.0bhp 223.7kW 304.2PS
@ 6400rpm
283.0lbft 383.5Nm @ 3600rpm
193.4bhp/ton 190.2bhp/tonne
101.3bhp/L 75.6kW/L
102.8PS/L
58.1ft/sec 17.7m/sec
23.5mpg 19.6mpgUS
12.0L/100km
Petrol 4-stroke piston
2960cc 180.6cu in
turbocharged
Vee 6 fuel injection
Compression ratio: 8.5:1
Bore: 87.0mm 3.4in
Stroke: 83.0mm 3.3in
Valve type/No: Overhead 24
Transmission: Manual
No. of forward speeds: 5
Wheels driven: Rear
Springs F/R: Coil/Coil
Brake system: PA ABS

Brakes F/R: Disc/Disc
Steering: Rack & pinion PA
Wheelbase: 245.1cm 96.5in
Track F: 149.6cm 58.9in
Track R: 155.4cm 61.2in
Length: 430.5cm 169.5in
Width: 179.1cm 70.5in
Height: 125.0cm 49.2in
Kerb weight: 1577.6kg
3475.0lb
Fuel: 70.8L 15.6gal 18.7galUS

629 Nissan

NX2000
1991 Japan
130.0mph 209.2kmh
0-60mph 96.5kmh: 8.1secs
0-1/4 mile: 16.6secs
140.0bhp 104.4kW 141.9PS
@ 6400rpm
130.0lbft 176.2Nm @ 4800rpm
119.5bhp/ton 117.5bhp/tonne
70.1bhp/L 52.2kW/L 71.0PS/L
60.3ft/sec 18.3m/sec
36.0mpg 30.0mpgUS
7.8L/100km
Petrol 4-stroke piston
1998cc 121.9cu in
In-line 4 fuel injection
Compression ratio: 9.5:1
Bore: 86.0mm 3.4in
Stroke: 86.0mm 3.4in
Valve type/No: Overhead 16
Transmission: Manual
No. of forward speeds: 5
Wheels driven: Front
Springs F/R: Coil/Coil
Brake system: PA ABS
Brakes F/R: Disc/Disc
Steering: Rack & pinion PA
Wheelbase: 243.1cm 95.7in
Track F: 144.5cm 56.9in
Track R: 142.5cm 56.1in
Length: 412.5cm 162.4in
Width: 167.9cm 66.1in
Height: 131.6cm 51.8in
Kerb weight: 1191.7kg
2625.0lb
Fuel: 50.0L 11.0gal 13.2galUS

630 Nissan

Primera 2.0 GSX
1991 UK
125.0mph 201.1kmh
0-50mph 80.5kmh: 6.4secs
0-60mph 96.5kmh: 8.9secs
0-1/4 mile: 16.9secs
0-1km: 31.1secs
121.0bhp 90.2kW 122.7PS
@ 6000rpm
124.7lbft 169.0Nm @ 6000rpm
100.4bhp/ton 98.8bhp/tonne
60.6bhp/L 45.2kW/L 61.4PS/L
56.5ft/sec 17.2m/sec

22.6mph 36.4kmh/1000rpm
26.0mpg 21.6mpgUS
10.9L/100km
Petrol 4-stroke piston
1998cc 122.0cu in
In-line 4 fuel injection
Compression ratio: 9.5:1
Bore: 86.0mm 3.4in
Stroke: 86.0mm 3.4in
Valve type/No: Overhead 16
Transmission: Manual
No. of forward speeds: 5
Wheels driven: Front
Springs F/R: Coil/Coil
Brake system: PA
Brakes F/R: Disc/Disc
Steering: Rack & pinion PA
Wheelbase: 254.8cm 100.3in
Track F: 146.8cm 57.8in
Track R: 145.8cm 57.4in
Length: 439.9cm 173.2in
Width: 169.9cm 66.9in
Height: 139.4cm 54.9in
Ground clearance: 14.0cm
5.5in
Kerb weight: 1225.0kg
2698.2lb
Fuel: 60.1L 13.2gal 15.9galUS

631 Nissan

Primera 2.0E ZX
1991 UK
132.0mph 212.4kmh
0-50mph 80.5kmh: 6.5secs
0-60mph 96.5kmh: 8.7secs
0-1/4 mile: 17.9secs
0-1km: 30.7secs
150.0bhp 111.9kW 152.1PS
@ 6400rpm
133.6lbft 181.0Nm @ 4800rpm
120.1bhp/ton 118.1bhp/tonne
75.1bhp/L 56.0kW/L 76.1PS/L
60.3ft/sec 18.3m/sec
21.1mph 33.9kmh/1000rpm
26.8mpg 22.3mpgUS
10.5L/100km
Petrol 4-stroke piston
1998cc 122.0cu in
In-line 4 fuel injection
Compression ratio: 10.0:1
Bore: 86.0mm 3.4in
Stroke: 86.0mm 3.4in
Valve type/No: Overhead 16
Transmission: Manual
No. of forward speeds: 5
Wheels driven: Front
Springs F/R: Coil/Coil
Brake system: PA ABS
Brakes F/R: Disc/Disc
Steering: Rack & pin on PA
Wheelbase: 255.0cm 100.4in
Track F: 147.0cm 57.9in
Track R: 146.0cm 57.5in
Length: 440.0cm 173.2in

Width: 170.0cm 66.9in
Height: 139.5cm 54.9in
Ground clearance: 14.0cm
5.5in
Kerb weight: 1270.0kg
2797.4lb
Fuel: 60.1L 13.2gal 15.9galUS

632 Nissan

Sentra SE-R
1991 Japan
125.0mph 201.1kmh
0-50mph 80.5kmh: 5.9secs
0-60mph 96.5kmh: 8.1secs
0-1/4 mile: 16.2secs
140.0bhp 104.4kW 141.9PS
@ 6400rpm
132.0lbft 178.9Nm @ 4800rpm
120.6bhp/ton 118.6bhp/tonne
70.1bhp/L 52.2kW/L 71.0PS/L
60.3ft/sec 18.3m/sec
19.2mpg 16.0mpgUS
14.7L/100km
Petrol 4-stroke piston
1998cc 121.9cu in
In-line 4 fuel injection
Compression ratio: 9.5:1
Bore: 86.0mm 3.4in
Stroke: 86.0mm 3.4in
Valve type/No: Overhead 16
Transmission: Manual
No. of forward speeds: 5
Wheels driven: Front
Springs F/R: Coil/Coil
Brake system: PA ABS
Brakes F/R: Disc/Disc
Steering: Rack & pinion PA
Wheelbase: 243.1cm 95.7in
Track F: 144.5cm 56.9in
Track R: 143.0cm 56.3in
Length: 432.6cm 170.3in
Width: 166.6cm 65.6in
Height: 136.9cm 53.9in
Ground clearance: 15.2cm
6.0in
Kerb weight: 1180.4kg
2600.0lb
Fuel: 50.0L 11.0gal 13.2galUS

633 Nissan

Sunny 1.6 GS 5-door
1991 Japan
112.0mph 180.2kmh
0-50mph 80.5kmh: 7.5secs
0-60mph 96.5kmh: 10.5secs
0-1/4 mile: 17.9secs
0-1km: 32.9secs
95.0bhp 70.8kW 96.3PS
@ 6000rpm
98.9lbft 134.0Nm @ 4000rpm
91.1bhp/ton 89.6bhp/tonne
59.5bhp/L 44.4kW/L 60.3PS/L
53.8ft/sec 16.4m/sec
23.6mph 38.0kmh/1000rpm

31.6mpg 26.3mpgUS
8.9L/100km
Petrol 4-stroke piston
1597cc 97.0cu in
In-line 4 1 Carburettor
Compression ratio: 9.5:1
Bore: 74.0mm 2.9in
Stroke: 82.0mm 3.2in
Valve type/No: Overhead 16
Transmission: Manual
No. of forward speeds: 5
Wheels driven: Front
Springs F/R: Coil/Coil
Brake system: PA
Brakes F/R: Disc/Drum
Steering: Rack & pinion PA
Wheelbase: 242.8cm 95.6in
Length: 414.3cm 163.1in
Width: 166.9cm 65.7in
Height: 139.4cm 54.9in
Kerb weight: 1060.0kg
2334.8lb
Fuel: 50.0L 11.0gal 13.2galUS

634 Nissan

200SX
1992 Japan
140.0mph 225.3kmh
0-50mph 80.5kmh: 5.1secs
0-60mph 96.5kmh: 6.8secs
0-1/4 mile: 15.3secs
0-1km: 27.8secs
164.0bhp 122.3kW 166.3PS
@ 6400rpm
163.0lbft 220.9Nm @ 4000rpm
133.6bhp/ton 131.4bhp/tonne
90.7bhp/L 67.6kW/L 91.9PS/L
58.8ft/sec 17.9m/sec
21.4mph 34.4kmh/1000rpm
19.7mpg 16.4mpgUS
14.3L/100km
Petrol 4-stroke piston
1809cc 110.4cu in
turbocharged
In-line 4 fuel injection
Compression ratio: 8.5:1
Bore: 83.0mm 3.3in
Stroke: 84.0mm 3.3in
Valve type/No: Overhead 16
Transmission: Manual
No. of forward speeds: 5
Wheels driven: Rear
Springs F/R: Coil/Coil
Brake system: PA ABS
Brakes F/R: Disc/Disc
Steering: Rack & pinion PA
Wheelbase: 247.5cm 97.4in
Track F: 146.5cm 57.7in
Track R: 146.5cm 57.7in
Length: 453.5cm 178.5in
Width: 169.0cm 66.5in
Height: 129.0cm 50.8in
Kerb weight: 1248.0kg
2748.9lb

Fuel: 60.1L 13.2gal 15.9galUS

635 Nissan

Sunny 2.0 GTi
1992 Japan
132.0mph 212.4kmh
0-50mph 80.5kmh: 5.7secs
0-60mph 96.5kmh: 7.5secs
0-1/4 mile: 16.0secs
0-1km: 29.1secs
143.0bhp 106.6kW 145.0PS
@ 6400rpm
131.0lbft 177.5Nm @ 4800rpm
130.9bhp/ton 128.7bhp/tonne
71.6bhp/L 53.4kW/L 72.6PS/L
60.3ft/sec 18.3m/sec
19.8mph 31.9kmh/1000rpm
27.8mpg 23.1mpgUS
10.2L/100km
Petrol 4-stroke piston
1998cc 121.9cu in
In-line 4 fuel injection
Compression ratio: 10.0:1
Bore: 86.0mm 3.4in
Stroke: 86.0mm 3.4in
Valve type/No: Overhead 16
Transmission: Manual
No. of forward speeds: 5
Wheels driven: Front
Springs F/R: Coil/Coil
Brake system: PA ABS
Brakes F/R: Disc/Disc
Steering: Rack & pinion PA
Wheelbase: 243.1cm 95.7in
Track F: 144.5cm 56.9in
Track R: 142.0cm 55.9in
Length: 397.5cm 156.5in
Width: 168.9cm 66.5in
Height: 139.4cm 54.9in
Ground clearance: 12.4cm
4.9in
Kerb weight: 1110.9kg
2447.0lb
Fuel: 50.0L 11.0gal 13.2galUS

636 Oldsmobile

Cutlass Calais International
Series
1988 USA
120.0mph 193.1kmh
0-50mph 80.5kmh: 5.7secs
0-60mph 96.5kmh: 8.2secs
0-1/4 mile: 16.2secs
150.0bhp 111.9kW 152.1PS
@ 5200rpm
160.0lbft 216.8Nm @ 4000rpm
131.8bhp/ton 129.6bhp/tonne
66.4bhp/L 49.5kW/L 67.3PS/L
48.4ft/sec 14.7m/sec
27.6mpg 23.0mpgUS
10.2L/100km
Petrol 4-stroke piston
2260cc 137.9cu in
In-line 4 fuel injection

Compression ratio: 9.5:1
Bore: 92.0mm 3.6in
Stroke: 85.0mm 3.3in
Valve type/No: Overhead 16
Transmission: Manual
No. of forward speeds: 5
Wheels driven: Front
Springs F/R: Coil/Coil
Brake system: PA
Brakes F/R: Disc/Drum
Steering: Rack & pinion PA
Wheelbase: 262.6cm 103.4in
Track F: 141.0cm 55.5in
Track R: 140.2cm 55.2in
Length: 454.2cm 178.8in
Width: 169.9cm 66.9in
Height: 133.1cm 52.4in
Kerb weight: 1157.7kg
2550.0lb
Fuel: 51.5L 11.3gal 13.6galUS

637 Oldsmobile

Calais HO Quad 4
1989 USA
125.0mph 201.1kmh
0-50mph 80.5kmh: 5.9secs
0-60mph 96.5kmh: 7.6secs
0-1/4 mile: 16.5secs
180.0bhp 134.2kW 182.5PS
@ 6200rpm
160.0lbft 216.8Nm @ 5200rpm
147.1bhp/ton 144.7bhp/tonne
79.6bhp/L 59.4kW/L 80.7PS/L
57.7ft/sec 17.6m/sec
32.4mpg 27.0mpgUS
8.7L/100km
Petrol 4-stroke piston
2260cc 137.9cu in
In-line 4 fuel injection
Compression ratio: 10.0:1
Bore: 92.0mm 3.6in
Stroke: 85.0mm 3.3in
Valve type/No: Overhead 16
Transmission: Manual
No. of forward speeds: 5
Wheels driven: Front
Springs F/R: Coil/Coil
Brake system: PA
Brakes F/R: Disc/Drum
Steering: Rack & pinion PA
Wheelbase: 262.6cm 103.4in
Track F: 141.2cm 55.6in
Track R: 140.2cm 55.2in
Length: 454.2cm 178.8in
Width: 169.4cm 66.7in
Height: 133.1cm 52.4in
Ground clearance: 14.7cm
5.8in
Kerb weight: 1244.0kg
2740.0lb
Fuel: 51.5L 11.3gal 13.6galUS

638 Oldsmobile

Cutlass Calais International

637 Oldsmobile
Calais HO Quad 4

Series HO
1989 USA
125.0mph 201.1kmh
0-50mph 80.5kmh: 5.9secs
0-60mph 96.5kmh: 7.6secs
0-1/4 mile: 16.5secs
180.0bhp 134.2kW 182.5PS
@ 6200rpm
160.0lbft 216.8Nm @ 5200rpm
147.1bhp/ton 144.7bhp/tonne
79.6bhp/L 59.4kW/L 80.7PS/L
57.7ft/sec 17.6m/sec
32.4mpg 27.0mpgUS
8.7L/100km
Petrol 4-stroke piston
2260cc 137.9cu in
In-line 4 fuel injection
Compression ratio: 10.0:1
Bore: 92.0mm 3.6in
Stroke: 85.0mm 3.3in
Valve type/No: Overhead 16

Transmission: Manual
No. of forward speeds: 5
Wheels driven: Front
Springs F/R: Coil/Coil
Brake system: PA
Brakes F/R: Disc/Drum
Steering: Rack & pinion PA
Wheelbase: 262.6cm 103.4in
Track F: 141.2cm 55.6in
Track R: 140.2cm 55.2in
Length: 454.2cm 178.8in
Width: 169.4cm 66.7in
Height: 133.1cm 52.4in
Kerb weight: 1244.0kg
2740.0lb
Fuel: 51.5L 11.3gal 13.6galUS

639 Oldsmobile
Toronado Trofeo
1990 USA
125.0mph 201.1kmh

0-50mph 80.5kmh: 7.3secs
0-60mph 96.5kmh: 10.4secs
0-1/4 mile: 17.6secs
170.0bhp 126.8kW 172.4PS
@ 4800rpm
220.0lbft 298.1Nm @ 3200rpm
107.1bhp/ton 105.3bhp/tonne
44.8bhp/L 33.4kW/L 45.5PS/L
45.3ft/sec 13.8m/sec
21.4mpg 17.8mpgUS
13.2L/100km
Petrol 4-stroke piston
3791cc 231.3cu in
Vee 6 fuel injection
Compression ratio: 8.5:1
Bore: 96.5mm 3.8in
Stroke: 86.4mm 3.4in
Valve type/No: Overhead 12
Transmission: Automatic
No. of forward speeds: 4
Wheels driven: Front

Springs F/R: Coil/Leaf
Brake system: PA ABS
Brakes F/R: Disc/Disc
Steering: Rack & pinion PA
Wheelbase: 274.3cm 108.0in
Track F: 152.1cm 59.9in
Track R: 152.1cm 59.9in
Length: 508.8cm 200.3in
Width: 184.9cm 72.8in
Height: 134.6cm 53.0in
Kerb weight: 1614.0kg
3555.0lb
Fuel: 69.3L 15.2gal 18.3galUS

640 Opel
Calibra 2.0i 16v
1991 Germany
137.0mph 220.4kmh
0-50mph 80.5kmh: 5.8secs
0-60mph 96.5kmh: 7.6secs
0-1/4 mile: 15.9secs

640 Opel
Calibra 2.0i 16v

642 Panther
Solo

150.0bhp 111.9kW 152.1PS
@ 6000rpm
145.0lbft 196.5Nm @ 4800rpm
125.4bhp/ton 123.3bhp/tonne
75.1bhp/L 56.0kW/L 76.1PS/L
56.5ft/sec 17.2m/sec
30.0mpg 25.0mpgUS
9.4L/100km
Petrol 4-stroke piston
1998cc 121.9cu in
In-line 4 fuel injection
Compression ratio: 10.5:1
Bore: 86.0mm 3.4in
Stroke: 86.0mm 3.4in
Valve type/No: Overhead 16
Transmission: Manual
No. of forward speeds: 5
Wheels driven: Front
Springs F/R: Coil/Coil
Brake system: PA ABS
Brakes F/R: Disc/Disc
Steering: Rack & pinion PA
Wheelbase: 260.1cm 102.4in
Track F: 142.5cm 56.1in
Track R: 144.5cm 56.9in
Length: 449.3cm 176.9in
Width: 167.6cm 66.0in
Height: 128.8cm 50.7in
Kerb weight: 1216.7kg
2680.0lb
Fuel: 65.1L 14.3gal 17.2galUS

641 Opel
Omega Lotus
1991 Germany
174.0mph 280.0kmh
0-60mph 96.5kmh: 4.9secs
0-1/4 mile: 13.5secs
377.0bhp 281.1kW 382.2PS
@ 5200rpm
409.0lbft 554.2Nm @ 4200rpm
227.0bhp/ton 223.2bhp/tonne
103.7bhp/L 77.3kW/L

105.1PS/L
48.4ft/sec 14.7m/sec
Petrol 4-stroke piston
3637cc 221.9cu in
turbocharged
In-line 6 fuel injection
Compression ratio: 8.2:1
Bore: 95.3mm 3.8in
Stroke: 85.0mm 3.3in
Valve type/No: Overhead 24
Transmission: Manual
No. of forward speeds: 6
Wheels driven: Rear
Springs F/R: Coil/Coil
Brake system: PA ABS
Brakes F/R: Disc/Disc
Steering: Rack & pinion
Wheelbase: 273.1cm 107.5in
Track F: 148.8cm 58.6in
Track R: 153.4cm 60.4in
Length: 476.8cm 187.7in
Width: 181.1cm 71.3in
Height: 143.3cm 56.4in
Kerb weight: 1688.9kg
3720.0lb

642 Panther
Solo
1990 UK
146.0mph 234.9kmh
0-50mph 80.5kmh: 5.3secs
0-60mph 96.5kmh: 6.8secs
0-1/4 mile: 15.3secs
0-1km: 27.4secs
204.0bhp 152.1kW 206.8PS
@ 6000rpm
203.7lbft 276.0Nm @ 4500rpm
154.8bhp/ton 152.2bhp/tonne
102.4bhp/L 76.3kW/L
103.8PS/L
50.5ft/sec 15.4m/sec
23.5mph 37.8kmh/1000rpm
19.4mpg 16.2mpgUS

14.6L/100km
Petrol 4-stroke piston
1993cc 122.0cu in
turbocharged
In-line 4 fuel injection
Compression ratio: 8.0:1
Bore: 91.0mm 3.6in
Stroke: 77.0mm 3.0in
Valve type/No: Overhead 16
Transmission: Manual
No. of forward speeds: 5
Wheels driven: 4-wheel drive
Springs F/R: Coil/Coil
Brake system: PA ABS
Brakes F/R: Disc/Disc
Steering: Rack & pinion
Wheelbase: 254.0cm 100.0in
Track F: 152.9cm 60.2in
Track R: 151.4cm 59.6in
Length: 433.4cm 170.6in
Width: 179.0cm 70.5in
Height: 118.0cm 46.5in
Ground clearance: 12.7cm
5.0in
Kerb weight: 1340.0kg
2951.5lb
Fuel: 54.6L 12.0gal 14.4galUS

643 Parradine
V-12
1991 UK
170.0mph 273.5kmh
0-60mph 96.5kmh: 5.4secs
277.0bhp 206.6kW 280.8PS
@ 5150rpm
397.0lbft 537.9Nm @ 2800rpm
225.6bhp/ton 221.9bhp/tonne
51.8bhp/L 38.7kW/L 52.6PS/L
39.5ft/sec 12.0m/sec
Petrol 4-stroke piston
5343cc 326.0cu in
Vee 12 fuel injection
Compression ratio: 11.5:1

Bore: 90.0mm 3.5in
Stroke: 70.0mm 2.8in
Valve type/No: Overhead 24
Transmission: Manual
No. of forward speeds: 5
Wheels driven: Rear
Springs F/R: Coil/Coil
Brake system: ABS
Brakes F/R: Disc/Disc
Steering: Rack & pinion PA
Wheelbase: 236.2cm 93.0in
Length: 418.6cm 164.8in
Width: 187.2cm 73.7in
Height: 134.4cm 52.9in
Kerb weight: 1248.5kg
2750.0lb

644 Peugeot
205 GTi
1986 France
122.0mph 196.3kmh
0-50mph 80.5kmh: 6.5secs
0-60mph 96.5kmh: 8.7secs
0-1/4 mile: 17.4secs
0-1km: 31.1secs
115.0bhp 85.8kW 116.6PS
@ 6250rpm
98.4lbft 133.3Nm @ 4000rpm
132.3bhp/ton 130.1bhp/tonne
72.8bhp/L 54.3kW/L 73.8PS/L
49.8ft/sec 15.2m/sec
18.6mph 29.9kmh/1000rpm
29.9mpg 24.9mpgUS
9.4L/100km
Petrol 4-stroke piston
1580cc 96.0cu in
In-line 4 fuel injection
Compression ratio: 9.8:1
Bore: 83.0mm 3.3in
Stroke: 73.0mm 2.9in
Valve type/No: Overhead 8
Transmission: Manual
No. of forward speeds: 5

Wheels driven: Front
Springs F/R: Coil/Torsion bar
Brake system: PA
Brakes F/R: Disc/Drum
Steering: Rack & pinion
Wheelbase: 242.0cm 95.3in
Track F: 139.3cm 54.8in
Track R: 133.2cm 52.4in
Length: 370.5cm 145.9in
Width: 157.2cm 61.9in
Height: 135.5cm 53.3in
Ground clearance: 12.0cm
4.7in
Kerb weight: 883.7kg 1946.5lb
Fuel: 50.0L 11.0gal 13.2galUS

645 Peugeot
309 1.6 GR
1986 France
99.0mph 159.3kmh
0-50mph 80.5kmh: 8.9secs
0-60mph 96.5kmh: 12.9secs
0-1/4 mile: 19.0secs
0-1km: 35.7secs
80.0bhp 59.7kW 81.1PS
@ 5600rpm
97.7lbft 132.4Nm @ 2800rpm
90.5bhp/ton 89.0bhp/tonne
50.6bhp/L 37.8kW/L 51.3PS/L
44.6ft/sec 13.6m/sec
22.2mph 35.7kmh/1000rpm
29.8mpg 24.8mpgUS
9.5L/100km
Petrol 4-stroke piston
1580cc 96.4cu in
In-line 4 1 Carburettor
Compression ratio: 9.4:1
Bore: 83.0mm 3.3in
Stroke: 73.0mm 2.9in
Valve type/No: Overhead 8
Transmission: Manual
No. of forward speeds: 5
Wheels driven: Front

Springs F/R: Coil/Torsion bar
Brake system: PA
Brakes F/R: Disc/Drum
Steering: Rack & pinion
Wheelbase: 246.9cm 97.2in
Track F: 140.7cm 55.4in
Track R: 137.4cm 54.1in
Length: 404.9cm 159.4in
Width: 162.6cm 64.0in
Height: 137.9cm 54.3in
Ground clearance: 16.5cm
6.5in
Kerb weight: 898.9kg 1980.0lb
Fuel: 55.1L 12.1gal 14.5galUS

646 Peugeot
309 SR Injection
1986 France
122.0mph 196.3kmh
0-50mph 80.5kmh: 7.1secs
0-60mph 96.5kmh: 9.7secs
0-1/4 mile: 17.9secs
0-1km: 32.1secs
115.0bhp 85.8kW 116.6PS
@ 6250rpm
98.4lbft 133.3Nm @ 4000rpm
127.5bhp/ton 125.4bhp/tonne
72.8bhp/L 54.3kW/L 73.8PS/L
49.8ft/sec 15.2m/sec
18.7mph 30.1kmh/1000rpm
30.9mpg 25.7mpgUS
9.1L/100km
Petrol 4-stroke piston
1580cc 96.0cu in
In-line 4 fuel injection
Compression ratio: 9.8:1
Bore: 83.0mm 3.3in
Stroke: 73.0mm 2.9in
Valve type/No: Overhead 8
Transmission: Manual
No. of forward speeds: 5
Wheels driven: Front
Springs F/R: Coil/Coil

Brake system: PA
Brakes F/R: Disc/Drum
Steering: Rack & pinion
Wheelbase: 246.9cm 97.2in
Track F: 140.8cm 55.4in
Track R: 137.6cm 54.2in
Length: 405.1cm 159.5in
Width: 162.8cm 64.1in
Height: 138.0cm 54.3in
Ground clearance: 16.5cm
6.5in
Kerb weight: 917.0kg 2019.8lb
Fuel: 55.1L 12.1gal 14.5galUS

647 Peugeot
505 GTi Family Estate
1986 France
109.0mph 175.4kmh
0-50mph 80.5kmh: 6.3secs
0-60mph 96.5kmh: 10.5secs
0-1/4 mile: 17.6secs
0-1km: 33.0secs
130.0bhp 96.9kW 131.8PS
@ 5750rpm
139.0lbft 188.3Nm @ 4250rpm
94.2bhp/ton 92.7bhp/tonne
60.0bhp/L 44.8kW/L 60.9PS/L
55.9ft/sec 17.1m/sec
21.2mph 34.1kmh/1000rpm
20.6mpg 17.2mpgUS
13.7L/100km
Petrol 4-stroke piston
2165cc 132.0cu in
In-line 4 fuel injection
Compression ratio: 9.8:1
Bore: 88.0mm 3.5in
Stroke: 89.0mm 3.5in
Valve type/No: Overhead 8
Transmission: Manual
No. of forward speeds: 5
Wheels driven: Rear
Springs F/R: Coil/Coil
Brake system: PA

Brakes F/R: Disc/Drum
Steering: Rack & pinion PA
Wheelbase: 290.0cm 114.2in
Track F: 149.6cm 58.9in
Track R: 145.0cm 57.1in
Length: 490.1cm 193.0in
Width: 173.7cm 68.4in
Height: 154.4cm 60.8in
Ground clearance: 13.0cm
5.1in
Kerb weight: 1403.0kg
3090.3lb
Fuel: 70.1L 15.4gal 18.5galUS

648 Peugeot
205 GTi 1.9
1987 France
123.0mph 197.9kmh
0-50mph 80.5kmh: 5.9secs
0-60mph 96.5kmh: 7.8secs
0-1/4 mile: 16.3secs
0-1km: 30.2secs
130.0bhp 96.9kW 131.8PS
@ 6000rpm
118.7lbft 160.8Nm @ 4750rpm
145.3bhp/ton 142.9bhp/tonne
68.2bhp/L 50.9kW/L 69.2PS/L
57.7ft/sec 17.6m/sec
20.8mph 33.5kmh/1000rpm
28.1mpg 23.4mpgUS
10.1L/100km
Petrol 4-stroke piston
1905cc 116.0cu in
In-line 4 fuel injection
Compression ratio: 9.6:1
Bore: 83.0mm 3.3in
Stroke: 88.0mm 3.5in
Valve type/No: Overhead 8
Transmission: Manual
No. of forward speeds: 5
Wheels driven: Front
Springs F/R: Coil/Coil
Brake system: PA

644 Peugeot
205 GTi

Brakes F/R: Disc/Disc
Steering: Rack & pinion
Wheelbase: 242.0cm 95.3in
Track F: 139.3cm 54.8in
Track R: 133.2cm 52.4in
Length: 370.5cm 145.9in
Width: 157.2cm 61.9in
Height: 135.5cm 53.3in
Ground clearance: 12.0cm
4.7in
Kerb weight: 910.0kg 2004.4lb
Fuel: 50.0L 11.0gal 13.2galUS

649 Peugeot

309 GTi
1987 France
122.0mph 196.3kmh
0-50mph 80.5kmh: 6.4secs
0-60mph 96.5kmh: 8.7secs
0-1/4 mile: 17.0secs
0-1km: 31.2secs
130.0bhp 96.9kW 131.8PS
@ 6000rpm
119.0lbft 161.2Nm @ 4750rpm
134.6bhp/ton 132.4bhp/tonne
68.2bhp/L 50.9kW/L 69.2PS/L
57.7ft/sec 17.6m/sec
20.9mph 33.6kmh/1000rpm
28.3mpg 23.6mpgUS
10.0L/100km
Petrol 4-stroke piston
1905cc 116.0cu in
In-line 4 fuel injection
Compression ratio: 9.6:1
Bore: 83.0mm 3.3in
Stroke: 88.0mm 3.5in
Valve type/No: Overhead 8
Transmission: Manual
No. of forward speeds: 5
Wheels driven: Front
Springs F/R: Coil/Torsion bar
Brake system: PA
Brakes F/R: Disc/Disc
Steering: Rack & pinion PA
Wheelbase: 246.9cm 97.2in
Track F: 140.8cm 55.4in
Track R: 137.6cm 54.2in
Length: 405.1cm 159.5in
Width: 162.8cm 64.1in
Height: 138.0cm 54.3in
Ground clearance: 16.5cm
6.5in
Kerb weight: 982.0kg 2163.0lb
Fuel: 55.1L 12.1gal 14.5galUS

650 Peugeot

505 STX
1987 France
125.0mph 201.1kmh
0-50mph 80.5kmh: 6.8secs
0-60mph 96.5kmh: 9.8secs
0-1/4 mile: 17.2secs
145.0bhp 108.1kW 147.0PS
@ 5000rpm

173.0lbft 234.4Nm @ 3750rpm
102.5bhp/ton 100.7bhp/tonne
50.9bhp/L 37.9kW/L 51.6PS/L
39.9ft/sec 12.1m/sec
28.8mpg 24.0mpgUS
9.8L/100km
Petrol 4-stroke piston
2849cc 173.8cu in
Vee 6 fuel injection
Compression ratio: 9.5:1
Bore: 91.0mm 3.6in
Stroke: 72.9mm 2.9in
Valve type/No: Overhead 12
Transmission: Manual
No. of forward speeds: 5
Wheels driven: Rear
Springs F/R: Coil/Coil
Brake system: PA ABS
Brakes F/R: Disc/Disc
Steering: Rack & pinion PA
Wheelbase: 274.3cm 108.0in
Track F: 150.1cm 59.1in
Track R: 145.5cm 57.3in
Length: 474.2cm 186.7in
Width: 173.5cm 68.3in
Height: 141.2cm 55.6in
Kerb weight: 1439.2kg
3170.0lb
Fuel: 68.1L 15.0gal 18.0galUS

651 Peugeot

405 1.6 GL
1988 France
108.0mph 173.8kmh
0-50mph 80.5kmh: 7.9secs
0-60mph 96.5kmh: 10.9secs
0-1/4 mile: 18.1secs
0-1km: 33.7secs
92.0bhp 68.6kW 93.3PS
@ 6000rpm
99.0lbft 134.1Nm @ 2600rpm
91.0bhp/ton 89.5bhp/tonne
58.2bhp/L 43.4kW/L 59.0PS/L
47.8ft/sec 14.6m/sec
21.2mph 34.1kmh/1000rpm
27.9mpg 23.2mpgUS
10.1L/100km
Petrol 4-stroke piston
1580cc 96.0cu in
In-line 4 1 Carburettor
Compression ratio: 8.9:1
Bore: 83.0mm 3.3in
Stroke: 73.0mm 2.9in
Valve type/No: Overhead 8
Transmission: Manual
No. of forward speeds: 5
Wheels driven: Front
Springs F/R: Coil/Torsion bar
Brake system: PA
Brakes F/R: Disc/Drum
Steering: Rack & pinion
Wheelbase: 266.9cm 105.1in
Track F: 145.0cm 57.1in
Track R: 143.6cm 56.5in

Length: 440.8cm 173.5in
Width: 169.4cm 66.7in
Height: 140.6cm 55.4in
Ground clearance: 15.3cm
6.0in
Kerb weight: 1028.0kg
2264.3lb
Fuel: 70.1L 15.4gal 18.5galUS

652 Peugeot

405 GRD
1988 France
102.0mph 164.1kmh
0-50mph 80.5kmh: 10.9secs
0-60mph 96.5kmh: 15.4secs
0-1/4 mile: 20.8secs
0-1km: 38.7secs
70.0bhp 52.2kW 71.0PS
@ 4600rpm
88.6lbft 120.0Nm @ 2000rpm
63.3bhp/ton 62.2bhp/tonne
36.7bhp/L 27.4kW/L 37.2PS/L
44.2ft/sec 13.5m/sec
22.2mph 35.7kmh/1000rpm
38.2mpg 31.8mpgUS
7.4L/100km
Diesel 4-stroke piston
1905cc 116.0cu in
In-line 4 fuel injection
Compression ratio: 22.5:1
Bore: 83.0mm 3.3in
Stroke: 88.0mm 3.5in
Valve type/No: Overhead 8
Transmission: Manual
No. of forward speeds: 5
Wheels driven: Front
Springs F/R: Coil/Torsion bar
Brake system: PA
Brakes F/R: Disc/Drum
Steering: Rack & pinion PA
Wheelbase: 266.9cm 105.1in
Track F: 145.0cm 57.1in
Track R: 143.6cm 56.5in
Length: 440.8cm 173.5in
Width: 169.4cm 66.7in
Height: 140.6cm 55.4in
Ground clearance: 15.3cm
6.0in
Kerb weight: 1125.0kg
2478.0lb
Fuel: 70.1L 15.4gal 18.5galUS

653 Peugeot

405 Mi16
1988 France
134.0mph 215.6kmh
0-50mph 80.5kmh: 5.9secs
0-60mph 96.5kmh: 8.0secs
0-1/4 mile: 16.3secs
0-1km: 29.7secs
160.0bhp 119.3kW 162.2PS
@ 6500rpm
133.6lbft 181.0Nm @ 5000rpm
142.0bhp/ton 139.6bhp/tonne

84.0bhp/L 62.6kW/L 85.1PS/L
62.5ft/sec 19.1m/sec
19.9mph 32.0kmh/1000rpm
26.6mpg 22.1mpgUS
10.6L/100km
Petrol 4-stroke piston
1905cc 116.0cu in
In-line 4 fuel injection
Compression ratio: 10.4:1
Bore: 83.0mm 3.3in
Stroke: 88.0mm 3.5in
Valve type/No: Overhead 16
Transmission: Manual
No. of forward speeds: 5
Wheels driven: Front
Springs F/R: Coil/Torsion bar
Brake system: PA ABS
Brakes F/R: Disc/Disc
Steering: Rack & pinion PA
Wheelbase: 266.9cm 105.1in
Track F: 145.0cm 57.1in
Track R: 143.6cm 56.5in
Length: 440.8cm 173.5in
Width: 169.4cm 66.7in
Height: 140.6cm 55.4in
Ground clearance: 15.3cm
6.0in
Kerb weight: 1146.0kg
2524.2lb
Fuel: 70.1L 15.4gal 18.5galUS

654 Peugeot

405 SRi
1988 France
118.0mph 189.9kmh
0-50mph 80.5kmh: 7.4secs
0-60mph 96.5kmh: 10.3secs
0-1/4 mile: 17.9secs
0-1km: 33.2secs
125.0bhp 93.2kW 126.7PS
@ 5500rpm
130.6lbft 177.0Nm @ 4500rpm
122.2bhp/ton 120.2bhp/tonne
65.6bhp/L 48.9kW/L 66.5PS/L
52.9ft/sec 16.1m/sec
22.0mph 35.4kmh/1000rpm
23.8mpg 19.8mpgUS
11.9L/100km
Petrol 4-stroke piston
1905cc 116.0cu in
In-line 4 fuel injection
Compression ratio: 9.3:1
Bore: 83.0mm 3.3in
Stroke: 88.0mm 3.5in
Valve type/No: Overhead 8
Transmission: Manual
No. of forward speeds: 5
Wheels driven: Front
Springs F/R: Coil/Torsion bar
Brake system: PA
Brakes F/R: Disc/Disc
Steering: Rack & pinion PA
Wheelbase: 266.9cm 105.1in
Track F: 145.0cm 57.1in

Track R: 143.6cm 56.5in
Length: 440.8cm 173.5in
Width: 169.4cm 66.7in
Height: 140.6cm 55.4in
Ground clearance: 15.3cm
6.0in
Kerb weight: 1040.0kg
2290.7lb
Fuel: 70.1L 15.4gal 18.5galUS

655 Peugeot
205 1.1GL
1989 France
99.0mph 159.3kmh
0-50mph 80.5kmh: 9.3secs
0-60mph 96.5kmh: 13.3secs
0-1/4 mile: 19.4secs
0-1km: 32.6secs
55.0bhp 41.0kW 55.8PS
@ 5800rpm
63.5lbft 86.0Nm @ 3200rpm
70.8bhp/ton 69.6bhp/tonne
48.9bhp/L 36.5kW/L 49.6PS/L
43.8ft/sec 13.3m/sec
21.0mph 33.8kmh/1000rpm
39.5mpg 32.9mpgUS
7.2L/100km
Petrol 4-stroke piston
1124cc 69.0cu in
In-line 4 1 Carburettor
Compression ratio: 9.4:1
Bore: 72.0mm 2.8in
Stroke: 69.0mm 2.7in
Valve type/No: Overhead 8
Transmission: Manual
No. of forward speeds: 4
Wheels driven: Front
Springs F/R: Coil/Torsion bar
Brake system: PA
Brakes F/R: Disc/Drum
Steering: Rack & pinion
Wheelbase: 242.0cm 95.3in
Track F: 139.3cm 54.8in
Track R: 133.2cm 52.4in
Length: 370.5cm 145.9in
Width: 157.2cm 61.9in
Height: 137.0cm 53.9in
Ground clearance: 12.0cm
4.7in
Kerb weight: 790.0kg 1740.1lb
Fuel: 50.0L 11.0gal 13.2galUS

656 Peugeot
205 CJ
1989 France
101.0mph 162.5kmh
0-50mph 80.5kmh: 7.3secs
0-60mph 96.5kmh: 11.7secs
0-1/4 mile: 18.4secs
0-1km: 45.4secs
65.0bhp 48.5kW 65.9PS
@ 5400rpm
83.4lbft 113.0Nm @ 3000rpm
76.0bhp/ton 74.7bhp/tonne

47.8bhp/L 35.6kW/L 48.5PS/L
45.4ft/sec 13.9m/sec
17.6mph 28.3kmh/1000rpm
37.2mpg 31.0mpgUS
7.6L/100km
Petrol 4-stroke piston
1360cc 83.0cu in
In-line 4 1 Carburettor
Compression ratio: 9.3:1
Bore: 75.0mm 2.9in
Stroke: 77.0mm 3.0in
Valve type/No: Overhead 8
Transmission: Manual
No. of forward speeds: 5
Wheels driven: Front
Springs F/R: Coil/Torsion bar
Brake system: PA
Brakes F/R: Disc/Drum
Steering: Rack & pinion
Wheelbase: 242.0cm 95.3in
Track F: 139.3cm 54.8in
Track R: 133.2cm 52.4in
Length: 370.5cm 145.9in
Width: 157.2cm 61.9in
Height: 135.5cm 53.3in
Ground clearance: 12.0cm
4.7in
Kerb weight: 870.0kg 1916.3lb
Fuel: 50.0L 11.0gal 13.2galUS

657 Peugeot
309 GLX
1989 France
95.0mph 152.9kmh
0-50mph 80.5kmh: 11.3secs
0-60mph 96.5kmh: 16.5secs
0-1/4 mile: 20.6secs
0-1km: 38.7secs
65.0bhp 48.5kW 65.9PS
@ 5600rpm
76.0lbft 103.0Nm @ 2800rpm
75.1bhp/ton 73.9bhp/tonne
50.2bhp/L 37.5kW/L 50.9PS/L
42.9ft/sec 13.1m/sec
20.2mph 32.5kmh/1000rpm
28.2mpg 23.5mpgUS
10.0L/100km
Petrol 4-stroke piston
1294cc 79.0cu in
In-line 4 1 Carburettor
Compression ratio: 9.4:1
Bore: 76.7mm 3.0in
Stroke: 70.0mm 2.8in
Valve type/No: Overhead 8
Transmission: Manual
No. of forward speeds: 5
Wheels driven: Front
Springs F/R: Coil/Torsion bar
Brake system: PA
Brakes F/R: Disc/Drum
Steering: Rack & pinion
Wheelbase: 246.9cm 97.2in
Track F: 140.8cm 55.4in
Track R: 137.6cm 54.2in

Length: 405.1cm 159.5in
Width: 162.8cm 64.1in
Height: 138.0cm 54.3in
Ground clearance: 16.5cm
6.5in
Kerb weight: 880.0kg 1938.3lb
Fuel: 55.1L 12.1gal 14.5galUS

658 Peugeot
309 GTi
1989 France
125.0mph 201.1kmh
0-50mph 80.5kmh: 6.6secs
0-60mph 96.5kmh: 8.8secs
0-1/4 mile: 16.5secs
0-1km: 29.7secs
130.0bhp 96.9kW 131.8PS
@ 6000rpm
118.8lbft 161.0Nm @ 4750rpm
133.8bhp/ton 131.6bhp/tonne
68.2bhp/L 50.9kW/L 69.2PS/L
57.7ft/sec 17.6m/sec
20.9mph 33.6kmh/1000rpm
25.2mpg 21.0mpgUS
11.2L/100km
Petrol 4-stroke piston
1905cc 116.0cu in
In-line 4 fuel injection
Compression ratio: 9.6:1
Bore: 83.0mm 3.3in
Stroke: 88.0mm 3.5in
Valve type/No: Overhead 8
Transmission: Manual
No. of forward speeds: 5
Wheels driven: Front
Springs F/R: Coil/Torsion bar
Brake system: PA
Brakes F/R: Disc/Disc
Steering: Rack & pinion PA
Wheelbase: 246.9cm 97.2in
Track F: 140.8cm 55.4in
Track R: 136.6cm 53.8in
Length: 405.1cm 159.5in
Width: 162.8cm 64.1in
Height: 138.0cm 54.3in
Ground clearance: 16.5cm
6.5in
Kerb weight: 988.0kg 2176.2lb
Fuel: 55.1L 12.1gal 14.5galUS

659 Peugeot
405 GTD Turbo
1989 France
109.0mph 175.4kmh
0-50mph 80.5kmh: 8.8secs
0-60mph 96.5kmh: 12.2secs
0-1/4 mile: 16.8secs
0-1km: 33.9secs
92.0bhp 68.6kW 93.3PS
@ 4300rpm
132.8lbft 180.0Nm @ 2200rpm
82.7bhp/ton 81.3bhp/tonne
52.0bhp/L 38.8kW/L 52.7PS/L
41.3ft/sec 12.6m/sec

25.3mph 40.7kmh/1000rpm
31.3mpg 26.1mpgUS
9.0L/100km
Diesel 4-stroke piston
1769cc 108.0cu in
turbocharged
In-line 4 fuel injection
Compression ratio: 22.5:1
Bore: 80.0mm 3.1in
Stroke: 88.0mm 3.5in
Valve type/No: Overhead 8
Transmission: Manual
No. of forward speeds: 5
Wheels driven: Front
Springs F/R: Coil/Torsion bar
Brake system: PA
Brakes F/R: Disc/Drum
Steering: Rack & pinion PA
Wheelbase: 266.9cm 105.1in
Track F: 145.0cm 57.1in
Track R: 143.6cm 56.5in
Length: 440.8cm 173.5in
Width: 169.4cm 66.7in
Height: 140.6cm 55.4in
Ground clearance: 15.3cm
6.0in
Kerb weight: 1131.0kg
2491.2lb
Fuel: 70.1L 15.4gal 18.5galUS

660 Peugeot
405 Mi16
1989 France
130.0mph 209.2kmh
0-50mph 80.5kmh: 6.6secs
0-60mph 96.5kmh: 9.0secs
0-1/4 mile: 16.8secs
150.0bhp 111.9kW 152.1PS
@ 6400rpm
128.0lbft 173.4Nm @ 5000rpm
124.7bhp/ton 122.6bhp/tonne
78.7bhp/L 58.7kW/L 79.8PS/L
61.5ft/sec 18.8m/sec
26.4mpg 22.0mpgUS
10.7L/100km
Petrol 4-stroke piston
1905cc 116.2cu in
In-line 4 fuel injection
Compression ratio: 9.5:1
Bore: 83.0mm 3.3in
Stroke: 88.0mm 3.5in
Valve type/No: Overhead 16
Transmission: Manual
No. of forward speeds: 5
Wheels driven: Front
Springs F/R: Coil/Torsion bar
Brake system: PA
Brakes F/R: Disc/Disc
Steering: Rack & pinion PA
Wheelbase: 267.0cm 105.1in
Track F: 144.0cm 56.7in
Track R: 143.3cm 56.4in
Length: 451.4cm 177.7in
Width: 171.7cm 67.6in

664 Peugeot
405 Mi16x4

Height: 140.7cm 55.4in
Ground clearance: 14.5cm
5.7in
Kerb weight: 1223.5kg
2695.0lb
Fuel: 65.1L 14.3gal 17.2galUS

661 Peugeot
309 Style 5-door
1990 France
96.0mph 154.5kmh
0-50mph 80.5kmh: 10.1secs
0-60mph 96.5kmh: 14.7secs
0-1/4 mile: 20.0secs
0-1km: 37.3secs
65.0bhp 48.5kW 65.9PS
@ 5600rpm
76.0lbft 103.0Nm @ 2800rpm
76.4bhp/ton 75.1bhp/tonne
50.2bhp/L 37.5kW/L 50.9PS/L
42.9ft/sec 13.1m/sec
21.0mph 33.8kmh/1000rpm
30.4mpg 25.3mpgUS
9.3L/100km
Petrol 4-stroke piston
1294cc 79.0cu in
In-line 4 1 Carburettor
Compression ratio: 9.4:1
Bore: 77.0mm 3.0in
Stroke: 70.0mm 2.8in
Valve type/No: Overhead 8
Transmission: Manual
No. of forward speeds: 5
Wheels driven: Front
Springs F/R: Coil/Coil
Brake system: PA
Brakes F/R: Disc/Drum
Steering: Rack & pinion
Wheelbase: 246.9cm 97.2in
Track F: 140.8cm 55.4in
Track R: 137.6cm 54.2in
Length: 405.1cm 159.5in
Width: 162.8cm 64.1in

Height: 138.0cm 54.3in
Ground clearance: 16.5cm
6.5in
Kerb weight: 865.0kg 1905.3lb
Fuel: 55.1L 12.1gal 14.5galUS

662 Peugeot
405 GLx4
1990 France
117.0mph 188.3kmh
0-50mph 80.5kmh: 7.7secs
0-60mph 96.5kmh: 10.9secs
0-1/4 mile: 18.0secs
0-1km: 33.5secs
110.0bhp 82.0kW 111.5PS
@ 6000rpm
118.1lbft 160.0Nm @ 3000rpm
88.6bhp/ton 87.2bhp/tonne
57.7bhp/L 43.1kW/L 58.5PS/L
57.7ft/sec 17.6m/sec
20.3mph 32.7kmh/1000rpm
24.4mpg 20.3mpgUS
11.6L/100km
Petrol 4-stroke piston
1905cc 116.0cu in
In-line 4 1 Carburettor
Compression ratio: 9.3:1
Bore: 83.0mm 3.3in
Stroke: 88.0mm 3.5in
Valve type/No: Overhead 8
Transmission: Manual
No. of forward speeds: 5
Wheels driven: 4-wheel drive
Springs F/R: Coil/Coil
Brake system: PA ABS
Brakes F/R: Disc/Disc
Steering: Rack & pinion PA
Wheelbase: 266.9cm 105.1in
Track F: 145.0cm 57.1in
Track R: 143.6cm 56.5in
Length: 440.8cm 173.5in
Width: 169.4cm 66.7in
Height: 140.6cm 55.4in

Ground clearance: 15.3cm
6.0in
Kerb weight: 1262.0kg
2779.7lb
Fuel: 70.5L 15.5gal 18.6galUS

663 Peugeot
405 GR Injection Auto
1990 France
121.0mph 194.7kmh
0-50mph 80.5kmh: 9.0secs
0-60mph 96.5kmh: 11.9secs
0-1/4 mile: 18.8secs
0-1km: 34.0secs
125.0bhp 93.2kW 126.7PS
@ 5500rpm
129.2lbft 175.0Nm @ 4500rpm
110.5bhp/ton 108.7bhp/tonne
65.6bhp/L 48.9kW/L 66.5PS/L
52.9ft/sec 16.1m/sec
23.0mph 37.0kmh/1000rpm
27.3mpg 22.7mpgUS
10.3L/100km
Petrol 4-stroke piston
1905cc 116.0cu in
In-line 4 fuel injection
Compression ratio: 9.3:1
Bore: 83.0mm 3.3in
Stroke: 88.0mm 3.5in
Valve type/No: Overhead 8
Transmission: Automatic
No. of forward speeds: 4
Wheels driven: Front
Springs F/R: Coil/Coil
Brake system: PA ABS
Brakes F/R: Disc/Disc
Steering: Rack & pinion PA
Wheelbase: 266.9cm 105.1in
Track F: 145.0cm 57.1in
Track R: 143.6cm 56.5in
Length: 440.8cm 173.5in
Width: 169.4cm 66.7in
Height: 140.6cm 55.4in

Ground clearance: 15.3cm
6.0in
Kerb weight: 1150.0kg
2533.0lb
Fuel: 70.1L 15.4gal 18.5galUS

664 Peugeot
405 Mi16x4
1990 France
127.0mph 204.3kmh
0-50mph 80.5kmh: 6.8secs
0-60mph 96.5kmh: 9.5secs
0-1/4 mile: 17.1secs
0-1km: 30.6secs
160.0bhp 119.3kW 162.2PS
@ 6500rpm
132.8lbft 180.0Nm @ 5000rpm
126.1bhp/ton 124.0bhp/tonne
84.0bhp/L 62.6kW/L 85.1PS/L
62.5ft/sec 19.1m/sec
19.1mph 30.7kmh/1000rpm
26.0mpg 21.6mpgUS
10.9L/100km
Petrol 4-stroke piston
1905cc 116.0cu in
In-line 4 fuel injection
Compression ratio: 10.4:1
Bore: 83.0mm 3.3in
Stroke: 88.0mm 3.5in
Valve type/No: Overhead 16
Transmission: Manual
No. of forward speeds: 5
Wheels driven: 4-wheel drive
Springs F/R: Coil/Gas
Brake system: PA ABS
Brakes F/R: Disc/Disc
Steering: Rack & pinion PA
Wheelbase: 266.9cm 105.1in
Track F: 145.0cm 57.1in
Track R: 143.6cm 56.5in
Length: 440.8cm 173.5in
Width: 169.4cm 66.7in
Height: 140.8cm 55.4in

Ground clearance: 15.3cm
6.0in
Kerb weight: 1290.0kg
2841.4lb
Fuel: 70.1L 15.4gal 18.5galUS

665 Peugeot
405 Turbo 16 Pike's Peak
1990 France
125.0mph 201.1kmh
0-60mph 96.5kmh: 6.2secs
640.0bhp 477.2kW 648.9PS
@ 7000rpm
542.0lbft 734.4Nm @ 6000rpm
764.6bhp/ton 751.8bhp/tonne
336.0bhp/L 250.5kW/L
340.6PS/L
67.3ft/sec 20.5m/sec
Petrol 4-stroke piston
1905cc 116.2cu in
turbocharged
In-line 4 fuel injection
Compression ratio: 7.0:1
Bore: 83.0mm 3.3in
Stroke: 88.0mm 3.5in
Valve type/No: Overhead 16
Transmission: Manual
No. of forward speeds: 6
Wheels driven: 4-wheel drive
Springs F/R: Coil/Coil
Brakes F/R: Disc/Disc
Steering: Rack & pinion PA
Wheelbase: 288.8cm 113.7in
Track F: 151.9cm 59.8in
Track R: 151.9cm 59.8in
Length: 424.9cm 167.3in
Width: 176.0cm 69.3in
Height: 119.9cm 47.2in
Kerb weight: 851.2kg 1875.0lb
Fuel: 60.2L 13.2gal 15.9galUS

666 Peugeot
605 SRi
1990 France
123.0mph 197.9kmh
0-50mph 80.5kmh: 8.5secs
0-60mph 96.5kmh: 11.8secs
0-1/4 mile: 18.2secs
0-1km: 33.4secs
122.0bhp 91.0kW 123.7PS
@ 5600rpm
126.0lbft 170.7Nm @ 4000rpm
88.8bhp/ton 87.3bhp/tonne
61.1bhp/L 45.5kW/L 61.9PS/L
52.7ft/sec 16.0m/sec
22.2mph 35.7kmh/1000rpm
24.7mpg 20.6mpgUS
11.4L/100km
Petrol 4-stroke piston
1998cc 121.9cu in
In-line 4 fuel injection
Compression ratio: 8.8:1
Bore: 86.0mm 3.4in
Stroke: 86.0mm 3.4in

Valve type/No: Overhead 8
Transmission: Manual
No. of forward speeds: 5
Wheels driven: Front
Springs F/R: Coil/Coil
Brake system: PA ABS
Brakes F/R: Disc/Disc
Steering: Rack & pinion PA
Wheelbase: 279.9cm 110.2in
Track F: 152.7cm 60.1in
Track R: 152.7cm 60.1in
Length: 472.2cm 185.9in
Width: 179.8cm 70.8in
Height: 141.5cm 55.7in
Kerb weight: 1397.0kg
3077.0lb
Fuel: 79.6L 17.5gal 21.0galUS

667 Peugeot
106 XR
1991 France
97.0mph 156.1kmh
0-50mph 80.5kmh: 9.5secs
0-60mph 96.5kmh: 13.8secs
0-1/4 mile: 19.5secs
0-1km: 36.4secs
60.0bhp 44.7kW 60.8PS
@ 5800rpm
66.4lbft 90.0Nm @ 3200rpm
77.2bhp/ton 75.9bhp/tonne
53.4bhp/L 39.8kW/L 54.1PS/L
43.8ft/sec 13.3m/sec
20.2mph 32.5kmh/1000rpm
32.1mpg 26.7mpgUS
8.8L/100km
Petrol 4-stroke piston
1124cc 69.0cu in
In-line 4 fuel injection
Compression ratio: 9.4:1
Bore: 72.0mm 2.8in
Stroke: 69.0mm 2.7in
Valve type/No: Overhead 8
Transmission: Manual
No. of forward speeds: 5
Wheels driven: Front
Springs F/R: Coil/Torsion bar
Brake system: PA
Brakes F/R: Disc/Drum
Steering: Rack & pinion
Wheelbase: 238.3cm 93.8in
Track F: 135.6cm 53.4in
Track R: 132.3cm 52.1in
Length: 356.4cm 140.3in
Width: 158.8cm 62.5in
Height: 136.7cm 53.8in
Kerb weight: 790.0kg 1740.1lb
Fuel: 45.0L 9.9gal 11.9galUS

668 Peugeot
106 XSi
1991 France
118.0mph 189.9kmh
0-50mph 80.5kmh: 6.9secs
0-60mph 96.5kmh: 9.7secs

0-1/4 mile: 17.5secs
0-1km: 31.8secs
100.0bhp 74.6kW 101.4PS
@ 6800rpm
89.0lbft 120.6Nm @ 4200rpm
113.8bhp/ton 111.9bhp/tonne
73.5bhp/L 54.8kW/L 74.5PS/L
57.2ft/sec 17.4m/sec
17.7mph 28.5kmh/1000rpm
26.9mpg 22.4mpgUS
10.5L/100km
Petrol 4-stroke piston
1360cc 83.0cu in
In-line 4 fuel injection
Compression ratio: 9.6:1
Bore: 75.0mm 2.9in
Stroke: 77.0mm 3.0in
Valve type/No: Overhead 8
Transmission: Manual
No. of forward speeds: 5
Wheels driven: Front
Springs F/R: Coil/Torsion bar
Brake system: PA
Brakes F/R: Disc/Drum
Steering: Rack & pinion
Wheelbase: 238.3cm 93.8in
Track F: 135.6cm 53.4in
Track R: 132.3cm 52.1in
Length: 356.4cm 140.3in
Width: 158.8cm 62.5in
Height: 136.7cm 53.8in
Kerb weight: 893.9kg 1969.0lb
Fuel: 45.0L 9.9gal 11.9galUS

669 Peugeot
205 D Turbo
1991 France
103.0mph 165.7kmh
0-50mph 80.5kmh: 8.9secs
0-60mph 96.5kmh: 12.3secs
0-1/4 mile: 18.9secs
0-1km: 35.1secs
78.0bhp 58.2kW 79.1PS
@ 4300rpm
115.9lbft 157.0Nm @ 2100rpm
83.5bhp/ton 82.1bhp/tonne
44.1bhp/L 32.9kW/L 44.7PS/L
41.3ft/sec 12.6m/sec
24.3mph 39.1kmh/1000rpm
34.4mpg 28.6mpgUS
8.2L/100km
Diesel 4-stroke piston
1769cc 108.0cu in
turbocharged
In-line 4 fuel injection
Compression ratio: 22.0:1
Bore: 80.0mm 3.1in
Stroke: 88.0mm 3.5in
Valve type/No: Overhead 8
Transmission: Manual
No. of forward speeds: 5
Wheels driven: Front
Springs F/R: Coil/Torsion bar
Brake system: PA

Brakes F/R: Disc/Drum
Steering: Rack & pinion PA
Wheelbase: 241.3cm 95.0in
Track F: 136.4cm 53.7in
Track R: 131.3cm 51.7in
Length: 370.6cm 145.9in
Width: 154.9cm 61.0in
Height: 137.2cm 54.0in
Ground clearance: 11.9cm
4.7in
Kerb weight: 950.0kg 2092.5lb
Fuel: 50.0L 11.0gal 13.2galUS

670 Peugeot
605 SR TD
1991 France
116.0mph 186.6kmh
0-50mph 80.5kmh: 9.5secs
0-60mph 96.5kmh: 12.8secs
0-1/4 mile: 19.3secs
0-1km: 35.1secs
110.0bhp 82.0kW 111.5PS
@ 4300rpm
183.0lbft 248.0Nm @ 2000rpm
74.8bhp/ton 73.6bhp/tonne
52.7bhp/L 39.3kW/L 53.4PS/L
43.2ft/sec 13.2m/sec
27.4mph 44.1kmh/1000rpm
33.3mpg 27.7mpgUS
8.5L/100km
Diesel 4-stroke piston
2088cc 127.0cu in
turbocharged
In-line 4 fuel injection
Compression ratio: 21.5:1
Bore: 85.0mm 3.3in
Stroke: 92.0mm 3.6in
Valve type/No: Overhead 8
Transmission: Manual
No. of forward speeds: 5
Wheels driven: Front
Springs F/R: Coil/Coil
Brake system: PA ABS
Brakes F/R: Disc/Disc
Steering: Rack & pinion PA
Wheelbase: 279.9cm 110.2in
Track F: 152.7cm 60.1in
Track R: 152.7cm 60.1in
Length: 472.2cm 185.9in
Width: 179.8cm 70.8in
Height: 141.5cm 55.7in
Kerb weight: 1495.0kg
3292.9lb
Fuel: 79.6L 17.5gal 21.0galUS

671 Peugeot
605 SV 3.0 Auto
1991 France
133.0mph 214.0kmh
0-50mph 80.5kmh: 7.8secs
0-60mph 96.5kmh: 10.6secs
0-1/4 mile: 17.5secs
0-1km: 31.8secs
170.0bhp 126.8kW 172.4PS

@ 6000rpm
177.1lbft 240.0Nm @ 6400rpm
112.3bhp/ton 110.4bhp/tonne
57.1bhp/L 42.6kW/L 57.9PS/L
47.8ft/sec 14.6m/sec
25.1mph 40.4kmh/1000rpm
19.8mpg 16.5mpgUS
14.3L/100km
Petrol 4-stroke piston
2975cc 182.0cu in
Vee 6 fuel injection
Compression ratio: 9.5:1
Bore: 93.0mm 3.7in
Stroke: 73.0mm 2.9in
Valve type/No: Overhead 12
Transmission: Automatic
No. of forward speeds: 4
Wheels driven: Front
Springs F/R: Coil/Coil
Brake system: PA ABS
Brakes F/R: Disc/Disc
Steering: Rack & pinion PA
Wheelbase: 279.9cm 110.2in
Track F: 152.7cm 60.1in
Track R: 152.7cm 60.1in
Length: 472.2cm 185.9in
Width: 179.8cm 70.8in
Height: 141.5cm 55.7in
Kerb weight: 1540.0kg
3392.1lb
Fuel: 80.1L 17.6gal 21.1galUS

672 Peugeot
605 SVE 24
1991 France
144.0mph 231.7kmh
0-50mph 80.5kmh: 6.1secs
0-60mph 96.5kmh: 7.9secs
0-1/4 mile: 16.3secs
0-1km: 29.5secs
200.0bhp 149.1kW 202.8PS
@ 6000rpm

191.9lbft 260.0Nm @ 3600rpm
128.3bhp/ton 126.2bhp/tonne
67.2bhp/L 50.1kW/L 68.2PS/L
47.8ft/sec 14.6m/sec
23.1mph 37.2kmh/1000rpm
20.4mpg 17.0mpgUS
13.8L/100km
Petrol 4-stroke piston
2975cc 182.0cu in
Vee 6 fuel injection
Compression ratio: 9.5:1
Bore: 93.0mm 3.7in
Stroke: 73.0mm 2.9in
Valve type/No: Overhead 24
Transmission: Manual
No. of forward speeds: 5
Wheels driven: Front
Springs F/R: Coil/Coil
Brake system: PA ABS
Brakes F/R: Disc/Disc
Steering: Rack & pinion PA
Wheelbase: 280.0cm 110.2in
Track F: 152.7cm 60.1in
Track R: 153.6cm 60.5in
Length: 472.3cm 185.9in
Width: 179.9cm 70.8in
Height: 141.7cm 55.8in
Kerb weight: 1585.0kg
3491.2lb
Fuel: 79.5L 17.5gal 21.0galUS

673 Peugeot
106 XT
1992 France
109.0mph 175.4kmh
0-50mph 80.5kmh: 8.1secs
0-60mph 96.5kmh: 11.0secs
0-1/4 mile: 18.5secs
0-1km: 31.4secs
75.0bhp 55.9kW 76.0PS
@ 5800rpm
84.0lbft 113.8Nm @ 3800rpm

93.2bhp/ton 91.7bhp/tonne
55.1bhp/L 41.1kW/L 55.9PS/L
48.8ft/sec 14.9m/sec
21.3mph 34.3kmh/1000rpm
32.7mpg 27.2mpgUS
8.6L/100km
Petrol 4-stroke piston
1360cc 83.0cu in
In-line 4 1 Carburettor
Compression ratio: 9.3:1
Bore: 75.0mm 2.9in
Stroke: 77.0mm 3.0in
Valve type/No: Overhead 8
Transmission: Manual
No. of forward speeds: 5
Wheels driven: Front
Springs F/R: Coil/Torsion bar
Brake system: PA
Brakes F/R: Disc/Drum
Steering: Rack & pinion
Wheelbase: 238.5cm 93.9in
Track F: 135.6cm 53.4in
Track R: 132.3cm 52.1in
Length: 356.4cm 140.3in
Width: 159.0cm 62.6in
Height: 136.7cm 53.8in
Kerb weight: 818.0kg 1801.8lb
Fuel: 45.0L 9.9gal 11.9galUS

674 Plymouth
Laser RS
1989 Japan
143.0mph 230.1kmh
0-50mph 80.5kmh: 5.5secs
0-60mph 96.5kmh: 7.2secs
0-1/4 mile: 15.8secs
190.0bhp 141.7kW 192.6PS
@ 6000rpm
203.0lbft 275.1Nm @ 3000rpm
149.6bhp/ton 147.1bhp/tonne
96.1bhp/L 71.7kW/L 97.4PS/L
57.7ft/sec 17.6m/sec

27.6mpg 23.0mpgUS
10.2L/100km
Petrol 4-stroke piston
1977cc 120.6cu in
turbocharged
In-line 4 fuel injection
Compression ratio: 7.8:1
Bore: 85.0mm 3.3in
Stroke: 88.0mm 3.5in
Valve type/No: Overhead 16
Transmission: Manual
No. of forward speeds: 5
Wheels driven: Front
Springs F/R: Coil/Coil
Brake system: PA
Brakes F/R: Disc/Disc
Steering: Rack & pinion PA
Wheelbase: 246.9cm 97.2in
Track F: 146.6cm 57.7in
Track R: 145.0cm 57.1in
Length: 433.1cm 170.5in
Width: 168.9cm 66.5in
Height: 130.6cm 51.4in
Kerb weight: 1291.6kg
2845.0lb
Fuel: 60.2L 13.2gal 15.9galUS

675 Plymouth
Laser RS Turbo
1990 USA
143.0mph 230.1kmh
0-60mph 96.5kmh: 6.9secs
0-1/4 mile: 15.4secs
190.0bhp 141.7kW 192.6PS
@ 6000rpm
203.0lbft 275.1Nm @ 4400rpm
149.6bhp/ton 147.1bhp/tonne
95.1bhp/L 70.9kW/L 96.5PS/L
57.7ft/sec 17.6m/sec
29.5mpg 24.6mpgUS
9.6L/100km
Petrol 4-stroke piston

672 Peugeot
605 SVE 24

1997cc 121.8cu in
turbocharged
In-line 4 fuel injection
Compression ratio: 7.8:1
Bore: 85.0mm 3.3in
Stroke: 88.0mm 3.5in
Valve type/No: Overhead 16
Transmission: Manual
No. of forward speeds: 5
Wheels driven: Front
Springs F/R: Coil/Coil
Brake system: PA
Brakes F/R: Disc/Disc
Steering: Rack & pinion PA
Wheelbase: 246.9cm 97.2in
Track F: 146.6cm 57.7in
Track R: 145.0cm 57.1in
Length: 433.1cm 170.5in
Width: 168.9cm 66.5in
Height: 126.5cm 49.8in
Kerb weight: 1291.6kg 2845.0lb
Fuel: 60.2L 13.2gal 15.9galUS

676 Plymouth
Laser RS Turbo
1991 USA
125.0mph 201.1kmh
0-60mph 96.5kmh: 8.0secs
0-1/4 mile: 16.2secs
190.0bhp 141.7kW 192.6PS
@ 6000rpm
203.0lbft 275.1Nm @ 3000rpm
149.3bhp/ton 146.8bhp/tonne
95.1bhp/L 70.9kW/L 96.5PS/L
57.7ft/sec 17.6m/sec
24.0mpg 20.0mpgUS
11.8L/100km
Petrol 4-stroke piston
1997cc 121.8cu in
turbocharged
In-line 4 fuel injection
Compression ratio: 7.8:1
Bore: 85.0mm 3.3in
Stroke: 88.0mm 3.5in
Valve type/No: Overhead 16
Transmission: Automatic
No. of forward speeds: 4
Wheels driven: Front
Springs F/R: Coil/Coil
Brake system: PA ABS
Brakes F/R: Disc/Disc
Steering: Rack & pinion PA
Wheelbase: 246.9cm 97.2in
Track F: 146.6cm 57.7in
Track R: 145.0cm 57.1in
Length: 433.1cm 170.5in
Width: 168.9cm 66.5in
Height: 126.5cm 49.8in
Kerb weight: 1293.9kg 2850.0lb
Fuel: 60.2L 13.2gal 15.9galUS

677 Pontiac
Fiero GT
1986 USA

125.0mph 201.1kmh
0-50mph 80.5kmh: 5.5secs
0-60mph 96.5kmh: 7.7secs
0-1/4 mile: 15.9secs
140.0bhp 104.4kW 141.9PS
@ 5200rpm
170.0lbft 230.4Nm @ 3600rpm
109.6bhp/ton 107.8bhp/tonne
49.3bhp/L 36.8kW/L 50.0PS/L
43.2ft/sec 13.2m/sec
27.0mpg 22.5mpgUS
10.5L/100km
Petrol 4-stroke piston
2837cc 173.1cu in
Vee 6 fuel injection
Compression ratio: 8.9:1
Bore: 89.0mm 3.5in
Stroke: 76.0mm 3.0in
Valve type/No: Overhead 12
Transmission: Manual
No. of forward speeds: 4
Wheels driven: Rear
Springs F/R: Coil/Coil
Brake system: PA
Brakes F/R: Disc/Disc
Steering: Rack & pinion
Wheelbase: 237.2cm 93.4in
Track F: 146.8cm 57.8in
Track R: 149.1cm 58.7in
Length: 419.4cm 165.1in
Width: 175.0cm 68.9in
Height: 119.1cm 46.9in
Kerb weight: 1298.4kg
2860.0lb
Fuel: 38.6L 8.5gal
10.2galUS

678 Pontiac
Bonneville SE
1987 USA
120.0mph 193.1kmh
0-50mph 80.5kmh: 6.8secs
0-60mph 96.5kmh: 9.7secs
0-1/4 mile: 17.3secs
150.0bhp 111.9kW 152.1PS
@ 4400rpm
200.0lbft 271.0Nm @ 2000rpm
98.1bhp/ton 96.5bhp/tonne
39.6bhp/L 29.5kW/L 40.1PS/L
41.6ft/sec 12.7m/sec
22.2mpg 18.5mpgUS
12.7L/100km
Petrol 4-stroke piston
3791cc 231.3cu in
Vee 6 fuel injection
Compression ratio: 8.5:1
Bore: 96.5mm 3.8in
Stroke: 86.4mm 3.4in
Valve type/No: Overhead 12
Transmission: Automatic
No. of forward speeds: 4
Wheels driven: Front
Springs F/R: Coil/Coil
Brake system: PA

Brakes F/R: Disc/Drum
Steering: Rack & pinion PA
Wheelbase: 281.4cm 110.8in
Track F: 153.2cm 60.3in
Track R: 151.9cm 59.8in
Length: 504.7cm 198.7in
Width: 183.1cm 72.1in
Height: 141.0cm 55.5in
Ground clearance: 16.5cm
6.5in
Kerb weight: 1554.9kg
3425.0lb
Fuel: 68.1L 15.0gal 18.0galUS

679 Pontiac
Fiero Formula
1987 USA
125.0mph 201.1kmh
0-50mph 80.5kmh: 5.5secs
0-60mph 96.5kmh: 8.0secs
0-1/4 mile: 16.0secs
135.0bhp 100.7kW 136.9PS
@ 4500rpm
165.0lbft 223.6Nm @ 3600rpm
109.0bhp/ton 107.2bhp/tonne
47.6bhp/L 35.5kW/L 48.2PS/L
37.4ft/sec 11.4m/sec
20.7mpg 17.2mpgUS
13.7L/100km
Petrol 4-stroke piston
2837cc 173.1cu in
Vee 6 fuel injection
Compression ratio: 8.5:1
Bore: 89.0mm 3.5in
Stroke: 76.0mm 3.0in
Valve type/No: Overhead 12
Transmission: Manual
No. of forward speeds: 5
Wheels driven: Rear
Springs F/R: Coil/Coil
Brakes F/R: Disc/Disc
Steering: Rack & pinion
Wheelbase: 237.2cm 93.4in
Track F: 151.6cm 59.7in
Track R: 153.4cm 60.4in
Length: 414.3cm 163.1in
Width: 175.0cm 68.9in
Height: 119.1cm 46.9in
Kerb weight: 1259.8kg
2775.0lb
Fuel: 45.0L 9.9gal 11.9galUS

680 Pontiac
Le Mans
1987 Korea
100.0mph 160.9kmh
0-60mph 96.5kmh: 12.1secs
0-1/4 mile: 18.7secs
74.0bhp 55.2kW 75.0PS
@ 5200rpm
88.0lbft 119.2Nm @ 3400rpm
72.1bhp/ton 70.9bhp/tonne
46.3bhp/L 34.5kW/L 46.9PS/L
46.4ft/sec 14.1m/sec

36.6mpg 30.5mpgUS
7.7L/100km
Petrol 4-stroke piston
1598cc 97.5cu in
In-line 4 fuel injection
Compression ratio: 8.6:1
Bore: 79.0mm 3.1in
Stroke: 81.5mm 3.2in
Valve type/No: Overhead 8
Transmission: Manual
No. of forward speeds: 5
Wheels driven: Front
Springs F/R: Coil/Coil
Brake system: PA
Brakes F/R: Disc/Drum
Steering: Rack & pinion PA
Wheelbase: 252.0cm 99.2in
Track F: 140.7cm 55.4in
Track R: 140.7cm 55.4in
Length: 436.6cm 171.9in
Width: 166.4cm 65.5in
Height: 138.9cm 54.7in
Kerb weight: 1044.2kg
2300.0lb
Fuel: 49.2L 10.8gal
13.0galUS

681 Pontiac
Bonneville SE
1988 USA
115.0mph 185.0kmh
0-60mph 96.5kmh: 9.5secs
0-1/4 mile: 17.2secs
165.0bhp 123.0kW 167.3PS
@ 5200rpm
210.0lbft 284.6Nm @ 2000rpm
105.0bhp/ton 103.2bhp/tonne
43.5bhp/L 32.5kW/L 44.1PS/L
49.1ft/sec 15.0m/sec
21.6mpg 18.0mpgUS
13.1L/100km
Petrol 4-stroke piston
3791cc 231.3cu in
Vee 6 fuel injection
Compression ratio: 8.5:1
Bore: 96.5mm 3.8in
Stroke: 86.4mm 3.4in
Valve type/No: Overhead 12
Transmission: Automatic
No. of forward speeds: 4
Wheels driven: Front
Springs F/R: Coil/Coil
Brake system: PA
Brakes F/R: Disc/Drum
Steering: Rack & pinion PA
Wheelbase: 281.4cm 110.8in
Track F: 153.2cm 60.3in
Track R: 151.9cm 59.8in
Length: 503.7cm 198.3in
Width: 183.9cm 72.4in
Height: 141.0cm 55.5in
Kerb weight: 1598.1kg
3520.0lb
Fuel: 68.1L 15.0gal 18.0galUS

682 Pontiac
Sunbird GT
1988 USA
125.0mph 201.1kmh
0-50mph 80.5kmh: 5.6secs
0-60mph 96.5kmh: 7.8secs
0-1/4 mile: 15.9secs
165.0bhp 123.0kW 167.3PS
@ 5600rpm
175.0lbft 237.1Nm @ 4000rpm
146.1bhp/ton 143.6bhp/tonne
82.3bhp/L 61.4kW/L 83.4PS/L
52.7ft/sec 16.0m/sec
31.2mpg 26.0mpgUS
9.0L/100km
Petrol 4-stroke piston
2005cc 122.3cu in
turbocharged
In-line 4 fuel injection
Compression ratio: 8.0:1
Bore: 86.0mm 3.4in
Stroke: 86.0mm 3.4in
Valve type/No: Overhead 8
Transmission: Manual
No. of forward speeds: 5
Wheels driven: Front
Springs F/R: Coil/Coil
Brake system: PA
Brakes F/R: Disc/Drum
Steering: Rack & pinion PA
Wheelbase: 257.0cm 101.2in
Track F: 141.2cm 55.6in
Track R: 140.2cm 55.2in
Length: 452.6cm 178.2in
Width: 165.1cm 65.0in
Height: 128.0cm 50.4in
Kerb weight: 1148.6kg
2530.0lb
Fuel: 51.5L 11.3gal 13.6galUS

683 Pontiac
Firebird T/A Turbo Pontiac
Engineering
1989 USA
178.0mph 286.4kmh
0-50mph 80.5kmh: 5.1secs
0-60mph 96.5kmh: 6.5secs
0-1/4 mile: 14.8secs
380.0bhp 283.4kW 385.3PS
@ 5000rpm
425.0lbft 575.9Nm @ 4500rpm
238.1bhp/ton 234.1bhp/tonne
100.2bhp/L 74.7kW/L
101.6PS/L
47.2ft/sec 14.4m/sec
Petrol 4-stroke piston
3791cc 231.3cu in
turbocharged
Vee 6 fuel injection
Compression ratio: 8.6:1
Bore: 96.5mm 3.8in
Stroke: 86.4mm 3.4in
Valve type/No: Overhead 12
Transmission: Manual

No. of forward speeds: 6
Wheels driven: Rear
Springs F/R: Coil/Coil
Brake system: PA
Brakes F/R: Disc/Disc
Steering: Recirculating ball PA
Wheelbase: 256.5cm 101.0in
Track F: 154.2cm 60.7in
Track R: 156.5cm 61.6in
Length: 486.7cm 191.6in
Width: 183.9cm 72.4in
Height: 127.0cm 50.0in
Kerb weight: 1623.0kg
3575.0lb

684 Pontiac
Firebird TransAm 20th
Anniversary
1989 USA
155.0mph 249.4kmh
0-50mph 80.5kmh: 3.8secs
0-60mph 96.5kmh: 5.1secs
0-1/4 mile: 13.7secs
250.0bhp 186.4kW 253.5PS
@ 4000rpm
340.0lbft 460.7Nm @ 2800rpm
164.7bhp/ton 162.0bhp/tonne
65.9bhp/L 49.2kW/L 66.9PS/L
37.8ft/sec 11.5m/sec
Petrol 4-stroke piston
3791cc 231.3cu in
turbocharged
Vee 6 fuel injection
Compression ratio: 8.0:1
Bore: 96.5mm 3.8in
Stroke: 86.4mm 3.4in
Valve type/No: Overhead 12
Transmission: Automatic
No. of forward speeds: 4
Wheels driven: Rear
Springs F/R: Coil/Coil
Brake system: PA
Brakes F/R: Disc/Disc
Steering: Recirculating ball PA
Wheelbase: 256.5cm 101.0in
Track F: 154.2cm 60.7in
Track R: 156.5cm 61.6in
Length: 483.4cm 190.3in
Width: 183.9cm 72.4in
Height: 126.5cm 49.8in
Kerb weight: 1543.6kg
3400.0lb
Fuel: 58.7L 12.9gal 15.5galUS

685 Pontiac
Firebird TransAm Turbo
1989 USA
150.0mph 241.4kmh
0-60mph 96.5kmh: 5.3secs
0-1/4 mile: 13.9secs
250.0bhp 186.4kW 253.5PS
@ 4000rpm
340.0lbft 460.7Nm @ 2800rpm
163.0bhp/ton 160.3bhp/tonne

65.9bhp/L 49.2kW/L 66.9PS/L
37.8ft/sec 11.5m/sec
16.8mpg 14.0mpgUS
16.8L/100km
Petrol 4-stroke piston
3791cc 231.3cu in
turbocharged
Vee 6 fuel injection
Compression ratio: 8.0:1
Bore: 96.5mm 3.8in
Stroke: 86.4mm 3.4in
Valve type/No: Overhead 12
Transmission: Automatic
No. of forward speeds: 4
Wheels driven: Rear
Springs F/R: Coil/Coil
Brake system: PA
Brakes F/R: Disc/Disc
Steering: Recirculating ball PA
Wheelbase: 256.5cm 101.0in
Track F: 154.2cm 60.7in
Track R: 156.5cm 61.6in
Length: 483.4cm 190.3in
Width: 183.9cm 72.4in
Height: 126.5cm 49.8in
Kerb weight: 1559.5kg
3435.0lb
Fuel: 58.7L 12.9gal 15.5galUS

686 Pontiac
Grand Am Turbo
1989 USA
120.0mph 193.1kmh
0-50mph 80.5kmh: 5.7secs
0-60mph 96.5kmh: 7.9secs
0-1/4 mile: 16.1secs
165.0bhp 123.0kW 167.3PS
@ 5600rpm
175.0lbft 237.1Nm @ 4000rpm
123.4bhp/ton 121.3bhp/tonne
82.6bhp/L 61.6kW/L 83.7PS/L
52.7ft/sec 16.0m/sec
Petrol 4-stroke piston
1998cc 121.9cu in
turbocharged
In-line 4 fuel injection
Compression ratio: 8.0:1
Bore: 86.0mm 3.4in
Stroke: 86.0mm 3.4in
Valve type/No: Overhead 8
Transmission: Automatic
No. of forward speeds: 3
Wheels driven: Front
Springs F/R: Coil/Coil
Brake system: PA
Brakes F/R: Disc/Drum
Steering: Rack & pinion PA
Wheelbase: 262.6cm 103.4in
Track F: 141.2cm 55.6in
Track R: 140.2cm 55.2in
Length: 457.5cm 180.1in
Width: 168.9cm 66.5in
Height: 133.4cm 52.5in
Kerb weight: 1359.7kg

2995.0lb
Fuel: 51.5L 11.3gal 13.6galUS

687 Pontiac
Grand Prix McLaren Turbo
1989 USA
128.0mph 206.0kmh
0-50mph 80.5kmh: 5.2secs
0-60mph 96.5kmh: 7.0secs
0-1/4 mile: 15.3secs
205.0bhp 152.9kW 207.8PS
@ 5200rpm
225.0lbft 304.9Nm @ 2400rpm
133.1bhp/ton 130.9bhp/tonne
65.5bhp/L 48.9kW/L 66.4PS/L
47.8ft/sec 14.6m/sec
25.2mpg 21.0mpgUS
11.2L/100km
Petrol 4-stroke piston
3128cc 190.8cu in
turbocharged
Vee 6 fuel injection
Compression ratio: 8.7:1
Bore: 88.9mm 3.5in
Stroke: 84.0mm 3.3in
Valve type/No: Overhead 12
Transmission: Automatic
No. of forward speeds: 4
Wheels driven: Front
Springs F/R: Coil/Coil
Brake system: PA ABS
Brakes F/R: Disc/Disc
Steering: Rack & pinion PA
Wheelbase: 273.1cm 107.5in
Track F: 151.1cm 59.5in
Track R: 147.3cm 58.0in
Length: 492.5cm 193.9in
Width: 182.6cm 71.9in
Height: 134.1cm 52.8in
Kerb weight: 1566.3kg
3450.0lb
Fuel: 60.6L 13.3gal 16.0galUS

688 Pontiac
Firebird Formula
1990 USA
135.0mph 217.2kmh
0-50mph 80.5kmh: 5.0secs
0-60mph 96.5kmh: 6.6secs
0-1/4 mile: 15.2secs
225.0bhp 167.8kW 228.1PS
@ 4400rpm
300.0lbft 406.5Nm @ 3200rpm
151.8bhp/ton 149.3bhp/tonne
44.9bhp/L 33.5kW/L 45.5PS/L
42.5ft/sec 13.0m/sec
Petrol 4-stroke piston
5012cc 305.8cu in
Vee 8 fuel injection
Compression ratio: 9.3:1
Bore: 95.0mm 3.7in
Stroke: 88.4mm 3.5in
Valve type/No: Overhead 16
Transmission: Manual

No. of forward speeds: 5
Wheels driven: Rear
Springs F/R: Coil/Coil
Brake system: PA
Brakes F/R: Disc/Disc
Steering: Recirculating ball PA
Wheelbase: 256.5cm 101.0in
Track F: 154.2cm 60.7in
Track R: 156.5cm 61.6in
Length: 486.7cm 191.6in
Width: 182.9cm 72.0in
Height: 126.2cm 49.7in
Kerb weight: 1507.3kg
3320.0lb
Fuel: 58.7L 12.9gal 15.5galUS

689 Pontiac
Firebird TransAm GTA
1990 USA
136.0mph 218.8kmh
0-50mph 80.5kmh: 5.3secs
0-60mph 96.5kmh: 7.1secs
0-1/4 mile: 15.5secs
235.0bhp 175.2kW 238.3PS
@ 4400rpm
340.0lbft 460.7Nm @ 3200rpm
149.1bhp/ton 146.6bhp/tonne
41.0bhp/L 30.6kW/L 41.5PS/L
42.5ft/sec 13.0m/sec
21.3mpg 17.7mpgUS
13.3L/100km
Petrol 4-stroke piston
5735cc 349.9cu in
Vee 8 fuel injection
Compression ratio: 9.3:1
Bore: 101.6mm 4.0in
Stroke: 88.4mm 3.5in
Valve type/No: Overhead 16
Transmission: Automatic
No. of forward speeds: 4
Wheels driven: Rear
Springs F/R: Coil/Coil
Brake system: PA
Brakes F/R: Disc/Disc
Steering: Recirculating ball PA
Wheelbase: 256.5cm 101.0in
Track F: 154.2cm 60.7in
Track R: 156.5cm 61.6in
Length: 486.7cm 191.6in
Width: 182.9cm 72.0in
Height: 126.2cm 49.7in
Kerb weight: 1602.6kg
3530.0lb
Fuel: 58.7L 12.9gal 15.5galUS

690 Pontiac
Grand Prix STE Turbo
1990 USA
124.0mph 199.5kmh
0-50mph 80.5kmh: 5.6secs
0-60mph 96.5kmh: 7.7secs
0-1/4 mile: 15.8secs
205.0bhp 152.9kW 207.8PS
@ 4800rpm

220.0lbft 298.1Nm @ 3200rpm
132.3bhp/ton 130.1bhp/tonne
65.5bhp/L 48.9kW/L 66.4PS/L
44.1ft/sec 13.4m/sec
24.5mpg 20.4mpgUS
11.5L/100km
Petrol 4-stroke piston
3128cc 190.8cu in
turbocharged
Vee 6 fuel injection
Compression ratio: 8.8:1
Bore: 88.9mm 3.5in
Stroke: 84.0mm 3.3in
Valve type/No: Overhead 12
Transmission: Automatic
No. of forward speeds: 4
Wheels driven: Front
Springs F/R: Coil/Leaf
Brake system: PA ABS
Brakes F/R: Disc/Disc
Steering: Rack & pinion PA
Wheelbase: 273.1cm 107.5in
Track F: 151.1cm 59.5in
Track R: 147.3cm 58.0in
Length: 494.8cm 194.8in
Width: 180.1cm 70.9in
Height: 137.9cm 54.3in
Kerb weight: 1575.4kg
3470.0lb
Fuel: 60.6L 13.3gal 16.0galUS

691 Pontiac
Firebird GTA
1991 USA
150.0mph 241.4kmh
0-50mph 80.5kmh: 5.1secs
0-60mph 96.5kmh: 6.6secs
0-1/4 mile: 15.2secs
240.0bhp 179.0kW 243.3PS
@ 4400rpm
340.0lbft 460.7Nm @ 3200rpm
155.6bhp/ton 153.0bhp/tonne
41.9bhp/L 31.2kW/L 42.4PS/L
42.5ft/sec 13.0m/sec
21.3mpg 17.7mpgUS
13.3L/100km
Petrol 4-stroke piston
5733cc 349.8cu in
Vee 8 fuel injection
Compression ratio: 9.3:1
Bore: 101.6mm 4.0in
Stroke: 88.4mm 3.5in
Valve type/No: Overhead 16

Transmission: Automatic
No. of forward speeds: 4
Wheels driven: Rear
Springs F/R: Coil/Coil
Brake system: PA
Brakes F/R: Disc/Disc
Steering: Recirculating ball PA
Wheelbase: 256.5cm 101.0in
Track F: 154.2cm 60.7in
Track R: 156.5cm 61.6in
Length: 495.6cm 195.1in
Width: 182.9cm 72.0in
Height: 125.7cm 49.5in
Kerb weight: 1568.6kg
3455.0lb
Fuel: 58.7L 12.9gal 15.5galUS

692 Pontiac
Firebird TDM Technologies
1991 USA
192.0mph 308.9kmh
0-60mph 96.5kmh: 4.1secs
0-1/4 mile: 12.7secs
685.0bhp 510.8kW 694.5PS
@ 4500rpm
785.0lbft 1063.7Nm @
4500rpm
426.2bhp/ton 419.1bhp/tonne
119.5bhp/L 89.1kW/L
121.1PS/L
43.5ft/sec 13.3m/sec
Petrol 4-stroke piston
5733cc 349.8cu in
turbocharged
Vee 8
Compression ratio: 9.0:1
Bore: 101.6mm 4.0in
Stroke: 88.4mm 3.5in
Valve type/No: Overhead 16
Transmission: Manual
No. of forward speeds: 6
Wheels driven: Rear
Springs F/R: Coil/Leaf
Brakes F/R: Disc/Disc
Steering: Recirculating ball PA
Wheelbase: 256.5cm 101.0in
Track F: 154.2cm 60.7in
Track R: 157.2cm 61.9in
Length: 486.7cm 191.6in
Width: 182.9cm 72.0in
Height: 126.2cm 49.7in
Kerb weight: 1634.4kg
3600.0lb

693 Pontiac
Firebird TransAm Convertible
1991 USA
130.0mph 209.2kmh
0-50mph 80.5kmh: 6.0secs
0-60mph 96.5kmh: 7.8secs
0-1/4 mile: 16.2secs
205.0bhp 152.9kW 207.8PS
@ 4200rpm
285.0lbft 386.2Nm @ 3200rpm
133.5bhp/ton 131.3bhp/tonne
41.0bhp/L 30.6kW/L 41.5PS/L
40.6ft/sec 12.4m/sec
22.8mpg 19.0mpgUS
12.4L/100km
Petrol 4-stroke piston
5002cc 305.2cu in
Vee 8 fuel injection
Compression ratio: 9.3:1
Bore: 94.9mm 3.7in
Stroke: 88.4mm 3.5in
Valve type/No: Overhead 16
Transmission: Manual
No. of forward speeds: 5
Wheels driven: Rear
Springs F/R: Coil/Coil
Brake system: PA
Brakes F/R: Disc/Drum
Steering: Recirculating ball PA
Wheelbase: 256.5cm 101.0in
Track F: 154.2cm 60.7in
Track R: 156.5cm 61.6in
Length: 495.6cm 195.1in
Width: 183.9cm 72.4in
Height: 126.2cm 49.7in
Kerb weight: 1561.8kg
3440.0lb
Fuel: 58.7L 12.9gal 15.5galUS

694 Pontiac
Grand Prix GTP
1991 USA
125.0mph 201.1kmh
0-60mph 96.5kmh: 8.4secs
0-1/4 mile: 16.3secs
200.0bhp 149.1kW 202.8PS
@ 5000rpm
215.0lbft 291.3Nm @ 4000rpm
129.1bhp/ton 126.9bhp/tonne
59.7bhp/L 44.5kW/L 60.5PS/L
46.0ft/sec 14.0m/sec
25.2mpg 21.0mpgUS
11.2L/100km

694 Pontiac
Grand Prix GTP

Petrol 4-stroke piston
3352cc 204.5cu in
Vee 6 fuel injection
Compression ratio: 9.3:1
Bore: 92.0mm 3.6in
Stroke: 84.0mm 3.3in
Valve type/No: Overhead 24
Transmission: Automatic
No. of forward speeds: 4
Wheels driven: Front
Springs F/R: Coil/Leaf
Brake system: PA ABS
Brakes F/R: Disc/Disc
Steering: Rack & pinion PA
Wheelbase: 273.1cm 107.5in
Track F: 151.1cm 59.5in
Track R: 147.3cm 58.0in
Length: 494.8cm 194.8in
Width: 180.1cm 70.9in
Height: 137.9cm 54.3in
Kerb weight: 1575.4kg
3470.0lb
Fuel: 62.4L 13.7gal
16.5galUS

695 Pontiac
Bonneville SSEi
1992 USA
128.0mph 206.0kmh
0-50mph 80.5kmh: 6.3secs
0-60mph 96.5kmh: 9.0secs
0-1/4 mile: 16.7secs
205.0bhp 152.9kW 207.8PS
@ 4600rpm
260.0lbft 352.3Nm @ 2800rpm
127.7bhp/ton 125.6bhp/tonne
54.1bhp/L 40.3kW/L 54.8PS/L
43.4ft/sec 13.2m/sec
24.0mpg 20.0mpgUS
11.8L/100km
Petrol 4-stroke piston
3791cc 231.3cu in
supercharged
Vee 6 fuel injection
Compression ratio: 8.5:1
Bore: 96.5mm 3.8in
Stroke: 86.4mm 3.4in
Valve type/No: Overhead 12
Transmission: Automatic
No. of forward speeds: 4
Wheels driven: Front
Springs F/R: Coil/Coil
Brake system: PA ABS
Brakes F/R: Disc/Drum
Steering: Rack & pinion PA
Wheelbase: 281.4cm 110.8in
Track F: 153.7cm 60.5in
Track R: 152.9cm 60.2in
Length: 511.0cm 201.2in
Width: 186.9cm 73.6in
Height: 141.0cm 55.5in
Ground clearance: 13.5cm
5.3in
Kerb weight: 1632.1kg

3595.0lb
Fuel: 68.1L 15.0gal 18.0galUS

696 Porsche
911 Cabriolet
1986 Germany
130.0mph 209.2kmh
0-60mph 96.5kmh: 5.7secs
0-1/4 mile: 14.3secs
200.0bhp 149.1kW 202.8PS
@ 5900rpm
185.0lbft 250.7Nm @ 4800rpm
162.9bhp/ton 160.2bhp/tonne
63.2bhp/L 47.1kW/L 64.1PS/L
48.0ft/sec 14.6m/sec
23.2mph 37.3kmh/1000rpm
22.3mpg 18.6mpgUS
12.6L/100km
Petrol 4-stroke piston
3164cc 193.0cu in
Flat 6 fuel injection
Compression ratio: 9.5:1
Bore: 95.0mm 3.7in
Stroke: 74.4mm 2.9in
Valve type/No: Overhead 12
Transmission: Manual
No. of forward speeds: 5
Wheels driven: Rear
Springs F/R: Torsion
bar/Torsion bar
Brake system: PA
Brakes F/R: Disc/Disc
Steering: Rack & pinion
Wheelbase: 227.3cm 89.5in
Track F: 136.9cm 53.9in
Track R: 137.9cm 54.3in
Length: 429.0cm 168.9in
Width: 165.1cm 65.0in
Height: 131.1cm 51.6in
Kerb weight: 1248.5kg
2750.0lb

697 Porsche
911 Carrera SE
1986 Germany
149.0mph 239.7kmh
0-50mph 80.5kmh: 4.3secs
0-60mph 96.5kmh: 5.6secs
0-1/4 mile: 14.4secs
0-1km: 26.4secs
231.0bhp 172.3kW 234.2PS
@ 5900rpm
209.6lbft 284.0Nm @ 4800rpm
186.1bhp/ton 183.0bhp/tonne
73.0bhp/L 54.4kW/L 74.0PS/L
48.0ft/sec 14.6m/sec
24.0mph 38.6kmh/1000rpm
19.2mpg 16.0mpgUS
14.7L/100km
Petrol 4-stroke piston
3164cc 193.0cu in
Flat 6 fuel injection
Compression ratio: 10.3:1
Bore: 95.0mm 3.7in

Stroke: 74.4mm 2.9in
Valve type/No: Overhead 12
Transmission: Manual
No. of forward speeds: 5
Wheels driven: Rear
Springs F/R: Torsion
bar/Torsion bar
Brake system: PA
Brakes F/R: Disc/Disc
Steering: Rack & pinion
Wheelbase: 227.3cm 89.5in
Track F: 143.3cm 56.4in
Track R: 150.1cm 59.1in
Length: 429.0cm 168.9in
Width: 177.5cm 69.9in
Height: 131.1cm 51.6in
Ground clearance: 11.9cm
4.7in
Kerb weight: 1262.1kg
2780.0lb
Fuel: 85.1L 18.7gal 22.5galUS

698 Porsche
911 Turbo
1986 Germany
153.0mph 246.2kmh
0-60mph 96.5kmh: 5.0secs
0-1/4 mile: 13.4secs
282.0bhp 210.3kW 285.9PS
@ 5500rpm
278.0lbft 376.7Nm @ 4000rpm
206.4bhp/ton 203.0bhp/tonne
85.5bhp/L 63.7kW/L 86.7PS/L
44.8ft/sec 13.6m/sec
28.6mph 46.0kmh/1000rpm
20.4mpg 17.0mpgUS
13.8L/100km
Petrol 4-stroke piston
3299cc 201.3cu in
turbocharged
Flat 6 fuel injection
Compression ratio: 7.0:1
Bore: 97.0mm 3.8in
Stroke: 74.4mm 2.9in
Valve type/No: Overhead 12
Transmission: Manual
No. of forward speeds: 4
Wheels driven: Rear
Springs F/R: Torsion
bar/Torsion bar
Brake system: PA
Brakes F/R: Disc/Disc
Steering: Rack & pinion
Wheelbase: 227.3cm 89.5in
Track F: 143.3cm 56.4in
Track R: 149.1cm 58.7in
Length: 429.0cm 168.9in
Width: 177.5cm 69.9in
Height: 131.1cm 51.6in
Kerb weight: 1389.2kg 3060.0lb

699 Porsche
924 S
1986 Germany

132.0mph 212.4kmh
0-60mph 96.5kmh: 7.8secs
0-1/4 mile: 15.8secs
147.0bhp 109.6kW 149.0PS
@ 5800rpm
140.0lbft 189.7Nm @ 3000rpm
119.1bhp/ton 117.1bhp/tonne
59.3bhp/L 44.2kW/L 60.1PS/L
50.1ft/sec 15.2m/sec
26.4mpg 22.0mpgUS
10.7L/100km
Petrol 4-stroke piston
2479cc 151.2cu in
In-line 4 fuel injection
Compression ratio: 9.7:1
Bore: 100.0mm 3.9in
Stroke: 78.9mm 3.1in
Valve type/No: Overhead 8
Transmission: Manual
No. of forward speeds: 5
Wheels driven: Rear
Springs F/R: Coil/Torsion bar
Brake system: PA
Brakes F/R: Disc/Disc
Steering: Rack & pinion PA
Wheelbase: 240.0cm 94.5in
Track F: 142.0cm 55.9in
Track R: 139.2cm 54.8in
Length: 429.0cm 168.9in
Width: 168.4cm 66.3in
Height: 127.5cm 50.2in
Kerb weight: 1255.3kg
2765.0lb
Fuel: 65.9L 14.5gal 17.4galUS

700 Porsche
928 S
1986 Germany
152.0mph 244.6kmh
0-60mph 96.5kmh: 6.3secs
0-1/4 mile: 14.7secs
288.0bhp 214.8kW 292.0PS
@ 5750rpm
302.0lbft 409.2Nm @ 2700rpm
182.2bhp/ton 179.2bhp/tonne
58.1bhp/L 43.3kW/L 58.9PS/L
49.7ft/sec 15.1m/sec
32.9mph 52.9kmh/1000rpm
19.6mpg 16.3mpgUS
14.4L/100km
Petrol 4-stroke piston
4957cc 302.4cu in
Vee 8 fuel injection
Compression ratio: 10.0:1
Bore: 100.0mm 3.9in
Stroke: 78.9mm 3.1in
Valve type/No: Overhead 16
Transmission: Automatic
No. of forward speeds: 4
Wheels driven: Rear
Springs F/R: Coil/Coil
Brake system: PA ABS
Brakes F/R: Disc/Disc
Steering: Rack & pinion PA

704 Porsche
944 Turbo

Wheelbase: 249.9cm 98.4in
Track F: 154.9cm 61.0in
Track R: 152.1cm 59.9in
Length: 446.3cm 175.7in
Width: 183.6cm 72.3in
Height: 128.3cm 50.5in
Kerb weight: 1607.2kg
3540.0lb

701 Porsche
928 S4 Automatic
1986 Germany
161.0mph 259.0kmh
0-50mph 80.5kmh: 4.7secs
0-60mph 96.5kmh: 6.2secs
0-1/4 mile: 14.5secs
0-1km: 26.2secs
320.0bhp 238.6kW 324.4PS
@ 6000rpm
317.0lbft 429.5Nm @ 3000rpm
205.2bhp/ton 201.8bhp/tonne
64.6bhp/L 48.1kW/L 65.4PS/L
51.8ft/sec 15.8m/sec
78.0mph 45.1kmh/1000rpm
17.0mpg 14.2mpgUS
16.6L/100km
Petrol 4-stroke piston
4957cc 302.0cu in
Vee 8 fuel injection
Compression ratio: 10.0:1
Bore: 100.0mm 3.9in
Stroke: 78.9mm 3.1in
Valve type/No: Overhead 32
Transmission: Automatic
No. of forward speeds: 4
Wheels driven: Rear
Springs F/R: Coil/Coil
Brake system: PA ABS
Brakes F/R: Disc/Disc
Steering: Rack & pinion PA
Wheelbase: 250.0cm 98.4in
Track F: 155.1cm 61.1in
Track R: 154.6cm 60.9in

Length: 452.0cm 178.0in
Width: 183.6cm 72.3in
Height: 128.2cm 50.5in
Ground clearance: 16.5cm
6.5in
Kerb weight: 1586.0kg
3493.4lb
Fuel: 86.4L 19.0gal 22.8galUS

702 Porsche
944
1986 Germany
123.0mph 197.9kmh
0-60mph 96.5kmh: 8.9secs
0-1/4 mile: 16.6secs
147.0bhp 109.6kW 149.0PS
@ 5800rpm
144.0lbft 195.1Nm @ 3000rpm
113.5bhp/ton 111.6bhp/tonne
59.3bhp/L 44.2kW/L 60.1PS/L
50.1ft/sec 15.2m/sec
26.7mph 43.0kmh/1000rpm
26.5mpg 22.1mpgUS
10.6L/100km
Petrol 4-stroke piston
2479cc 151.2cu in
In-line 4 fuel injection
Compression ratio: 9.7:1
Bore: 100.0mm 3.9in
Stroke: 78.9mm 3.1in
Valve type/No: Overhead 8
Transmission: Manual
No. of forward speeds: 5
Wheels driven: Rear
Springs F/R: Coil/Torsion bar
Brake system: PA
Brakes F/R: Disc/Disc
Steering: Rack & pinion PA
Track F: 147.8cm 58.2in
Track R: 145.0cm 57.1in
Length: 429.0cm 168.9in
Width: 173.5cm 68.3in

Height: 127.5cm 50.2in
Kerb weight: 1316.6kg
2900.0lb

703 Porsche
944 S
1986 Germany
141.0mph 226.9kmh
0-50mph 80.5kmh: 5.1secs
0-60mph 96.5kmh: 6.7secs
0-1/4 mile: 15.1secs
0-1km: 27.8secs
190.0bhp 141.7kW 192.6PS
@ 6000rpm
170.0lbft 230.4Nm @ 4300rpm
151.0bhp/ton 148.4bhp/tonne
76.6bhp/L 57.1kW/L 77.7PS/L
51.8ft/sec 15.8m/sec
22.3mph 35.9kmh/1000rpm
23.5mpg 19.6mpgUS
12.0L/100km
Petrol 4-stroke piston
2479cc 151.0cu in
In-line 4 fuel injection
Compression ratio: 10.9:1
Bore: 100.0mm 3.9in
Stroke: 78.9mm 3.1in
Valve type/No: Overhead 16
Transmission: Manual
No. of forward speeds: 5
Wheels driven: Rear
Springs F/R: Coil/Torsion bar
Brake system: PA
Brakes F/R: Disc/Disc
Steering: Rack & pinion PA
Wheelbase: 240.0cm 94.5in
Track F: 147.7cm 58.1in
Track R: 145.1cm 57.1in
Length: 420.0cm 165.4in
Width: 173.5cm 68.3in
Height: 127.5cm 50.2in
Ground clearance: 13.8cm
5.4in

Kerb weight: 1280.0kg
2819.4lb
Fuel: 80.1L 17.6gal 21.1galUS

704 Porsche
944 Turbo
1986 Germany
158.0mph 254.2kmh
0-50mph 80.5kmh: 4.5secs
0-60mph 96.5kmh: 6.0secs
0-1/4 mile: 14.8secs
0-1km: 27.0secs
220.0bhp 164.0kW 223.0PS
@ 5800rpm
243.0lbft 329.3Nm @ 3500rpm
165.6bhp/ton 162.9bhp/tonne
88.7bhp/L 66.2kW/L 90.0PS/L
50.1ft/sec 15.2m/sec
25.6mph 41.2kmh/1000rpm
22.9mpg 19.1mpgUS
12.3L/100km
Petrol 4-stroke piston
2479cc 151.2cu in
In-line 4 fuel injection
turbocharged
Compression ratio: 8.0:1
Bore: 100.0mm 3.9in
Stroke: 78.9mm 3.1in
Valve type/No: Overhead 8
Transmission: Manual
No. of forward speeds: 5
Wheels driven: Rear
Springs F/R: Coil/Torsion bar
Brake system: PA
Brakes F/R: Disc/Disc
Steering: Rack & pinion PA
Wheelbase: 240.0cm 94.5in
Track F: 147.8cm 58.2in
Track R: 145.0cm 57.1in
Length: 419.9cm 165.3in
Width: 173.5cm 68.3in
Height: 127.5cm 50.2in
Ground clearance: 11.9cm

197

4.7in
Kerb weight: 1350.6kg
2975.0lb
Fuel: 80.1L 17.6gal 21.1galUS

705 Porsche
911 Turbo Gemballa
Avalanche
1987 Germany
165.0mph 265.5kmh
0-60mph 96.5kmh: 4.9secs
0-1/4 mile: 13.0secs
282.0bhp 210.3kW 285.9PS
@ 5500rpm
278.0lbft 376.7Nm @ 4000rpm
191.1bhp/ton 187.9bhp/tonne
85.5bhp/L 63.7kW/L 86.7PS/L
44.8ft/sec 13.6m/sec
Petrol 4-stroke piston
3299cc 201.3cu in
turbocharged
Flat 6 fuel injection
Compression ratio: 7.0:1
Bore: 97.0mm 3.8in
Stroke: 74.4mm 2.9in
Valve type/No: Overhead 12
Transmission: Manual
No. of forward speeds: 4
Wheels driven: Rear
Springs F/R: Torsion
bar/Torsion bar
Brake system: PA
Brakes F/R: Disc/Disc
Steering: Rack & pinion
Wheelbase: 227.3cm 89.5in
Track F: 144.8cm 57.0in
Track R: 151.9cm 59.8in
Length: 428.0cm 168.5in
Width: 192.8cm 75.9in
Height: 131.1cm 51.6in
Kerb weight: 1500.5kg
3305.0lb
Fuel: 85.2L 18.7gal 22.5galUS

706 Porsche
911 Turbo Ruf 3.4
1987 Germany
158.0mph 254.2kmh
0-60mph 96.5kmh: 4.5secs
0-1/4 mile: 13.0secs
374.0bhp 278.9kW 379.2PS
@ 6000rpm
354.0lbft 479.7Nm @ 4800rpm
299.2bhp/ton 294.2bhp/tonne
111.1bhp/L 82.9kW/L
112.6PS/L
28.4mph 45.7kmh/1000rpm
Petrol 4-stroke piston
3366cc 205.4cu in
turbocharged
Flat 6 fuel injection
Compression ratio: 7.0:1
Valve type/No: Overhead 12
Transmission: Manual

No. of forward speeds: 5
Wheels driven: Rear
Springs F/R: Torsion
bar/Torsion bar
Brake system: PA
Brakes F/R: Disc/Disc
Steering: Rack & pinion
Wheelbase: 227.3cm 89.5in
Track F: 146.6cm 57.7in
Track R: 150.1cm 59.1in
Length: 429.0cm 168.9in
Width: 177.5cm 69.9in
Height: 128.0cm 50.4in
Kerb weight: 1271.2kg
2800.0lb

707 Porsche
911 Turbo Slant-Nose
1987 Germany
157.0mph 252.6kmh
0-60mph 96.5kmh: 5.0secs
0-1/4 mile: 13.4secs
282.0bhp 210.3kW 285.9PS
@ 5500rpm
278.0lbft 376.7Nm @ 4000rpm
212.3bhp/ton 208.8bhp/tonne
85.5bhp/L 63.7kW/L 86.7PS/L
44.8ft/sec 13.6m/sec
Petrol 4-stroke piston
3299cc 201.3cu in
turbocharged
Flat 6 fuel injection
Compression ratio: 7.0:1
Bore: 97.0mm 3.8in
Stroke: 74.4mm 2.9in
Valve type/No: Overhead 12
Transmission: Manual
No. of forward speeds: 4
Wheels driven: Rear
Springs F/R: Torsion
bar/Torsion bar
Brake system: PA
Brakes F/R: Disc/Disc
Steering: Rack & pinion
Wheelbase: 227.3cm 89.5in
Track F: 143.3cm 56.4in
Track R: 149.1cm 58.7in
Length: 429.0cm 168.9in
Width: 177.5cm 69.9in
Height: 131.1cm 51.6in
Kerb weight: 1350.6kg
2975.0lb
Fuel: 85.2L 18.7gal 22.5galUS

708 Porsche
924 S
1987 Germany
136.0mph 218.8kmh
0-50mph 80.5kmh: 5.6secs
0-60mph 96.5kmh: 8.0secs
0-1/4 mile: 16.3secs
0-1km: 29.2secs
160.0bhp 119.3kW 162.2PS
@ 5900rpm

155.0lbft 210.0Nm @ 4500rpm
136.2bhp/ton 133.9bhp/tonne
64.5bhp/L 48.1kW/L 65.4PS/L
51.0ft/sec 15.5m/sec
22.2mph 35.7kmh/1000rpm
22.6mpg 18.8mpgUS
12.5L/100km
Petrol 4-stroke piston
2479cc 151.0cu in
In-line 4 fuel injection
Compression ratio: 10.2:1
Bore: 100.0mm 3.9in
Stroke: 78.9mm 3.1in
Valve type/No: Overhead 8
Transmission: Manual
No. of forward speeds: 5
Wheels driven: Rear
Springs F/R: Coil/Coil
Brake system: PA
Brakes F/R: Disc/Disc
Steering: Rack & pinion PA
Wheelbase: 235.5cm 92.7in
Track F: 142.0cm 55.9in
Track R: 137.2cm 54.0in
Length: 419.9cm 165.3in
Width: 165.6cm 65.2in
Height: 127.0cm 50.0in
Ground clearance: 15.2cm
6.0in
Kerb weight: 1195.0kg
2632.2lb
Fuel: 66.0L 14.5gal 17.4galUS

709 Porsche
959 S
1987 Germany
201.0mph 323.4kmh
0-60mph 96.5kmh: 4.7secs
0-1/4 mile: 11.9secs
600.0bhp 447.4kW 608.3PS
@ 6500rpm
369.0lbft 500.0Nm @ 5500rpm
451.0bhp/ton 443.5bhp/tonne
210.6bhp/L 157.0kW/L
213.5PS/L
47.7ft/sec 14.5m/sec
Petrol 4-stroke piston
2849cc 173.8cu in
turbocharged
Flat 6 fuel injection
Compression ratio: 8.0:1
Bore: 95.0mm 3.7in
Stroke: 67.0mm 2.6in
Valve type/No: Overhead 24
Transmission: Manual
No. of forward speeds: 6
Wheels driven: 4-wheel drive
Springs F/R: Coil/Coil
Brake system: PA ABS
Brakes F/R: Disc/Disc
Steering: Rack & pinion PA
Wheelbase: 229.9cm 90.5in
Track F: 150.4cm 59.2in
Track R: 154.9cm 61.0in

Length: 426.0cm 167.7in
Width: 183.9cm 72.4in
Height: 128.0cm 50.4in
Kerb weight: 1352.9kg
2980.0lb

710 Porsche
911 Cabriolet
1988 Germany
149.0mph 239.7kmh
0-60mph 96.5kmh: 6.5secs
0-1/4 mile: 15.0secs
214.0bhp 159.6kW 217.0PS
@ 5900rpm
195.0lbft 264.2Nm @ 4800rpm
163.0bhp/ton 160.3bhp/tonne
67.6bhp/L 50.4kW/L 68.6PS/L
48.0ft/sec 14.6m/sec
27.0mpg 22.5mpgUS
10.5L/100km
Petrol 4-stroke piston
3164cc 193.0cu in
Flat 6 fuel injection
Compression ratio: 9.5:1
Bore: 95.0mm 3.7in
Stroke: 74.4mm 2.9in
Valve type/No: Overhead 12
Transmission: Manual
No. of forward speeds: 5
Wheels driven: Rear
Springs F/R: Coil/Torsion bar
Brake system: PA
Brakes F/R: Disc/Disc
Steering: Rack & pinion
Wheelbase: 227.3cm 89.5in
Track F: 136.9cm 53.9in
Track R: 137.9cm 54.3in
Length: 429.0cm 168.9in
Width: 165.1cm 65.0in
Height: 131.1cm 51.6in
Kerb weight: 1334.8kg
2940.0lb
Fuel: 85.2L 18.7gal 22.5galUS

711 Porsche
911 Carrera
1988 Germany
149.0mph 239.7kmh
0-50mph 80.5kmh: 5.4secs
0-60mph 96.5kmh: 7.0secs
0-1/4 mile: 15.1secs
214.0bhp 159.6kW 217.0PS
@ 5900rpm
195.0lbft 264.2Nm @ 4800rpm
173.0bhp/ton 170.2bhp/tonne
67.6bhp/L 50.4kW/L 68.6PS/L
48.0ft/sec 14.6m/sec
24.6mph 39.6kmh/1000rpm
21.6mpg 18.0mpgUS
13.1L/100km
Petrol 4-stroke piston
3164cc 193.0cu in
Flat 6 fuel injection
Compression ratio: 9.5:1

Bore: 95.0mm 3.7in
Stroke: 74.4mm 2.9in
Valve type/No: Overhead 12
Transmission: Manual
No. of forward speeds: 5
Wheels driven: Rear
Springs F/R: Torsion
bar/Torsion bar
Brake system: PA
Brakes F/R: Disc/Disc
Steering: Rack & pinion
Wheelbase: 227.3cm 89.5in
Track F: 136.9cm 53.9in
Track R: 137.9cm 54.3in
Length: 429.0cm 168.9in
Width: 165.1cm 65.0in
Height: 132.1cm 52.0in
Ground clearance: 13.0cm
5.1in
Kerb weight: 1257.6kg
2770.0lb
Fuel: 85.2L 18.7gal 22.5galUS

712 Porsche
911 Carrera Club Sport
1988 Germany
152.0mph 244.6kmh
0-50mph 80.5kmh: 4.0secs
0-60mph 96.5kmh: 5.2secs
0-1/4 mile: 13.9secs
0-1km: 25.3secs
231.0bhp 172.3kW 234.2PS
@ 5900rpm
210.0lbft 284.6Nm @ 4800rpm
198.4bhp/ton 195.1bhp/tonne
73.0bhp/L 54.4kW/L 74.0PS/L
48.0ft/sec 14.6m/sec
24.3mph 39.1kmh/1000rpm
20.6mpg 17.2mpgUS
13.7L/100km
Petrol 4-stroke piston
3164cc 193.0cu in
Flat 6 fuel injection
Compression ratio: 10.3:1
Bore: 95.0mm 3.7in
Stroke: 74.4mm 2.9in
Valve type/No: Overhead 12
Transmission: Manual
No. of forward speeds: 5
Wheels driven: Rear
Springs F/R: Torsion
bar/Torsion bar
Brake system: PA
Brakes F/R: Disc/Disc
Steering: Rack & pinion
Wheelbase: 227.2cm 89.4in
Track F: 143.2cm 56.4in
Track R: 150.1cm 59.1in
Length: 429.0cm 168.9in
Width: 177.5cm 69.9in
Height: 131.0cm 51.6in
Ground clearance: 12.0cm
4.7in
Kerb weight: 1184.0kg

2608.0lb
Fuel: 85.1L 18.7gal 22.5galUS

713 Porsche
911 Club Sport
1988 Germany
149.0mph 239.7kmh
0-50mph 80.5kmh: 4.5secs
0-60mph 96.5kmh: 5.7secs
0-1/4 mile: 14.3secs
214.0bhp 159.6kW 217.0PS
@ 5900rpm
195.0lbft 264.2Nm @ 4800rpm
184.4bhp/ton 181.3bhp/tonne
67.6bhp/L 50.4kW/L 68.6PS/L
48.0ft/sec 14.6m/sec
21.6mpg 18.0mpgUS
13.1L/100km
Petrol 4-stroke piston
3164cc 193.0cu in
Flat 6 fuel injection
Compression ratio: 9.5:1
Bore: 95.0mm 3.7in
Stroke: 74.4mm 2.9in
Valve type/No: Overhead 12
Transmission: Manual
No. of forward speeds: 5
Wheels driven: Rear
Springs F/R: Torsion
bar/Torsion bar
Brake system: PA
Brakes F/R: Disc/Disc
Steering: Rack & pinion
Wheelbase: 227.3cm 89.5in
Track F: 140.0cm 55.1in
Track R: 140.7cm 55.4in
Length: 429.0cm 168.9in
Width: 165.1cm 65.0in
Height: 132.1cm 52.0in
Ground clearance: 13.0cm
5.1in
Kerb weight: 1180.4kg
2600.0lb
Fuel: 85.2L 18.7gal 22.5galUS

714 Porsche
911 Speedster
1988 Germany
155.0mph 249.4kmh
0-60mph 96.5kmh: 6.5secs
0-1/4 mile: 14.5secs
231.0bhp 172.3kW 234.2PS
@ 5900rpm
209.0lbft 283.2Nm @ 4800rpm
200.6bhp/ton 197.2bhp/tonne
73.0bhp/L 54.4kW/L 74.0PS/L
48.0ft/sec 14.6m/sec
Petrol 4-stroke piston
3164cc 193.0cu in
Flat 6 fuel injection
Compression ratio: 9.5:1
Bore: 95.0mm 3.7in
Stroke: 74.4mm 2.9in
Valve type/No: Overhead 12

Transmission: Manual
No. of forward speeds: 5
Wheels driven: Rear
Springs F/R: Torsion
bar/Torsion bar
Brake system: PA
Brakes F/R: Disc/Disc
Steering: Rack & pinion
Wheelbase: 227.3cm 89.5in
Track F: 139.7cm 55.0in
Track R: 140.5cm 55.3in
Length: 429.0cm 168.9in
Width: 165.1cm 65.0in
Height: 130.0cm 51.2in
Kerb weight: 1171.3kg
2580.0lb
Fuel: 79.5L 17.5gal 21.0galUS

715 Porsche
911 Turbo
1988 Germany
155.0mph 249.4kmh
0-50mph 80.5kmh: 3.4secs
0-60mph 96.5kmh: 5.0secs
0-1/4 mile: 13.4secs
282.0bhp 210.3kW 285.9PS
@ 5500rpm
278.0lbft 376.7Nm @ 4000rpm
206.8bhp/ton 203.3bhp/tonne
85.5bhp/L 63.7kW/L 86.7PS/L
44.8ft/sec 13.6m/sec
28.6mph 46.0kmh/1000rpm
19.2mpg 16.0mpgUS
14.7L/100km
Petrol 4-stroke piston
3299cc 201.3cu in
turbocharged
Flat 6 fuel injection
Compression ratio: 7.0:1
Bore: 97.0mm 3.8in
Stroke: 74.4mm 2.9in
Valve type/No: Overhead 12
Transmission: Manual
No. of forward speeds: 4
Wheels driven: Rear
Springs F/R: Torsion
bar/Torsion bar
Brake system: PA
Brakes F/R: Disc/Disc
Steering: Rack & pinion
Wheelbase: 227.3cm 89.5in
Track F: 143.3cm 56.4in
Track R: 150.1cm 59.1in
Length: 429.0cm 168.9in
Width: 177.5cm 69.9in
Height: 131.1cm 51.6in
Ground clearance: 11.9cm
4.7in
Kerb weight: 1387.0kg
3055.0lb
Fuel: 85.2L 18.7gal 22.5galUS

716 Porsche
911 Turbo Koenig RS

1988 Germany
201.0mph 323.4kmh
0-60mph 96.5kmh: 4.0secs
0-1/4 mile: 11.6secs
520.0bhp 387.8kW 527.2PS
@ 6300rpm
490.0lbft 664.0Nm @ 5500rpm
416.0bhp/ton 409.1bhp/tonne
154.5bhp/L 115.2kW/L
156.6PS/L
51.3ft/sec 15.6m/sec
Petrol 4-stroke piston
3366cc 205.4cu in
turbocharged
Flat 6 fuel injection
Compression ratio: 7.5:1
Bore: 98.0mm 3.9in
Stroke: 74.4mm 2.9in
Valve type/No: Overhead 12
Transmission: Manual
No. of forward speeds: 5
Wheels driven: Rear
Springs F/R: Torsion
bar/Torsion bar
Brake system: PA
Brakes F/R: Disc/Disc
Steering: Rack & pinion
Wheelbase: 227.3cm 89.5in
Track F: 143.3cm 56.4in
Track R: 150.1cm 59.1in
Length: 429.0cm 168.9in
Width: 177.5cm 69.9in
Height: 131.1cm 51.6in
Kerb weight: 1271.2kg
2800.0lb
Fuel: 85.2L 18.7gal 22.5galUS

717 Porsche
911 Turbo Ruf Twin Turbo
1988 Germany
211.0mph 339.5kmh
0-60mph 96.5kmh: 4.0secs
0-1/4 mile: 11.7secs
469.0bhp 349.7kW 475.5PS
@ 5950rpm
457.0lbft 619.2Nm @ 5100rpm
415.2bhp/ton 408.3bhp/tonne
139.3bhp/L 103.9kW/L
141.3PS/L
48.4ft/sec 14.8m/sec
Petrol 4-stroke piston
3366cc 205.4cu in
turbocharged
Flat 6 fuel injection
Compression ratio: 7.5:1
Bore: 98.0mm 3.9in
Stroke: 74.4mm 2.9in
Valve type/No: Overhead 12
Transmission: Manual
No. of forward speeds: 5
Wheels driven: Rear
Springs F/R: Torsion
bar/Torsion bar
Brake system: PA

Brakes F/R: Disc/Disc
Steering: Rack & pinion
Wheelbase: 227.3cm 89.5in
Track F: 143.3cm 56.4in
Track R: 150.1cm 59.1in
Length: 429.0cm 168.9in
Width: 177.5cm 69.9in
Height: 131.1cm 51.6in
Kerb weight: 1148.6kg
2530.0lb
Fuel: 85.2L 18.7gal 22.5galUS

718 Porsche

924 S
1988 Germany
138.0mph 222.0kmh
0-50mph 80.5kmh: 5.3secs
0-60mph 96.5kmh: 7.5secs
0-1/4 mile: 15.8secs
147.0bhp 109.6kW 149.0PS
@ 5800rpm
140.0lbft 189.7Nm @ 3000rpm
117.0bhp/ton 115.0bhp/tonne
59.3bhp/L 44.2kW/L 60.1PS/L
50.1ft/sec 15.2m/sec
26.7mph 43.0kmh/1000rpm
26.4mpg 22.0mpgUS
10.7L/100km
Petrol 4-stroke piston
2479cc 151.2cu in
In-line 4 fuel injection
Compression ratio: 9.7:1
Bore: 100.0mm 3.9in
Stroke: 78.9mm 3.1in
Valve type/No: Overhead 8
Transmission: Manual
No. of forward speeds: 5
Wheels driven: Rear
Springs F/R: Coil/Torsion bar
Brake system: PA
Brakes F/R: Disc/Disc
Steering: Rack & pinion PA
Wheelbase: 240.0cm 94.5in
Track F: 142.0cm 55.9in
Track R: 139.2cm 54.8in
Length: 429.0cm 168.9in
Width: 168.4cm 66.3in
Height: 127.5cm 50.2in
Ground clearance: 11.9cm
4.7in
Kerb weight: 1278.0kg
2815.0lb
Fuel: 65.9L 14.5gal 17.4galUS

719 Porsche

928 S4
1988 Germany
165.0mph 265.5kmh
0-50mph 80.5kmh: 4.4secs
0-60mph 96.5kmh: 5.5secs
0-1/4 mile: 13.9secs
316.0bhp 235.6kW 320.4PS
@ 6000rpm
317.0lbft 429.5Nm @ 3000rpm

200.8bhp/ton 197.5bhp/tonne
63.7bhp/L 47.5kW/L 64.6PS/L
51.8ft/sec 15.8m/sec
33.0mph 53.1kmh/1000rpm
19.8mpg 16.5mpgUS
14.3L/100km
Petrol 4-stroke piston
4957cc 302.4cu in
Vee 8 fuel injection
Compression ratio: 10.0:1
Bore: 100.0mm 3.9in
Stroke: 78.9mm 3.1in
Valve type/No: Overhead 32
Transmission: Manual
No. of forward speeds: 5
Wheels driven: Rear
Springs F/R: Coil/Coil
Brake system: PA ABS
Brakes F/R: Disc/Disc
Steering: Rack & pinion PA
Wheelbase: 249.9cm 98.4in
Track F: 154.9cm 61.0in
Track R: 154.7cm 60.9in
Length: 452.4cm 178.1in
Width: 183.6cm 72.3in
Height: 128.3cm 50.5in
Ground clearance: 11.9cm
4.7in
Kerb weight: 1600.3kg
3525.0lb
Fuel: 85.9L 18.9gal 22.7galUS

720 Porsche

944
1988 Germany
132.0mph 212.4kmh
0-50mph 80.5kmh: 6.4secs
0-60mph 96.5kmh: 8.7secs
0-1/4 mile: 16.4secs
147.0bhp 109.6kW 149.0PS
@ 5800rpm
140.0lbft 189.7Nm @ 3000rpm
111.1bhp/ton 109.2bhp/tonne
59.3bhp/L 44.2kW/L 60.1PS/L
50.1ft/sec 15.2m/sec
26.7mph 43.0kmh/1000rpm
25.2mpg 21.0mpgUS
11.2L/100km
Petrol 4-stroke piston
2479cc 151.2cu in
In-line 4 fuel injection
Compression ratio: 9.7:1
Bore: 100.0mm 3.9in
Stroke: 78.9mm 3.1in
Valve type/No: Overhead 8
Transmission: Manual
No. of forward speeds: 5
Wheels driven: Rear
Springs F/R: Coil/Torsion bar
Brake system: PA
Brakes F/R: Disc/Disc
Steering: Rack & pinion PA
Wheelbase: 240.0cm 94.5in
Track F: 147.8cm 58.2in

Track R: 145.0cm 57.1in
Length: 429.0cm 168.9in
Width: 173.5cm 68.3in
Height: 127.5cm 50.2in
Ground clearance: 11.9cm
4.7in
Kerb weight: 1346.1kg
2965.0lb
Fuel: 79.9L 17.6gal 21.1galUS

721 Porsche

944 S
1988 Germany
140.0mph 225.3kmh
0-50mph 80.5kmh: 5.0secs
0-60mph 96.5kmh: 7.5secs
0-1/4 mile: 15.8secs
188.0bhp 140.2kW 190.6PS
@ 6000rpm
170.0lbft 230.4Nm @ 4300rpm
141.5bhp/ton 139.2bhp/tonne
75.8bhp/L 56.5kW/L 76.9PS/L
51.8ft/sec 15.8m/sec
20.7mph 33.3kmh/1000rpm
23.4mpg 19.5mpgUS
12.1L/100km
Petrol 4-stroke piston
2479cc 151.2cu in
In-line 4 fuel injection
Compression ratio: 10.9:1
Bore: 100.0mm 3.9in
Stroke: 78.9mm 3.1in
Valve type/No: Overhead 16
Transmission: Manual
No. of forward speeds: 5
Wheels driven: Rear
Springs F/R: Coil/Torsion bar
Brake system: PA
Brakes F/R: Disc/Disc
Steering: Rack & pinion PA
Wheelbase: 240.0cm 94.5in
Track F: 147.8cm 58.2in
Track R: 145.0cm 57.1in
Length: 429.0cm 168.9in
Width: 173.5cm 68.3in
Height: 127.5cm 50.2in
Ground clearance: 11.9cm
4.7in
Kerb weight: 1350.6kg
2975.0lb
Fuel: 79.9L 17.6gal 21.1galUS

722 Porsche

944 Turbo
1988 Germany
155.0mph 249.4kmh
0-50mph 80.5kmh: 4.7secs
0-60mph 96.5kmh: 6.0secs
0-1/4 mile: 14.6secs
217.0bhp 161.8kW 220.0PS
@ 5800rpm
243.0lbft 329.3Nm @ 3500rpm
156.0bhp/ton 153.4bhp/tonne
87.5bhp/L 65.3kW/L 88.7PS/L

50.1ft/sec 15.2m/sec
26.7mph 43.0kmh/1000rpm
22.8mpg 19.0mpgUS
12.4L/100km
Petrol 4-stroke piston
2479cc 151.2cu in
turbocharged
In-line 4 fuel injection
Compression ratio: 8.0:1
Bore: 100.0mm 3.9in
Stroke: 78.9mm 3.1in
Valve type/No: Overhead 8
Transmission: Manual
No. of forward speeds: 5
Wheels driven: Rear
Springs F/R: Coil/Torsion bar
Brake system: PA
Brakes F/R: Disc/Disc
Steering: Rack & pinion PA
Wheelbase: 240.0cm 94.5in
Track F: 147.8cm 58.2in
Track R: 145.0cm 57.1in
Length: 429.0cm 168.9in
Width: 173.5cm 68.3in
Height: 127.5cm 50.2in
Ground clearance: 11.9cm
4.7in
Kerb weight: 1414.2kg
3115.0lb
Fuel: 79.9L 17.6gal 21.1galUS

723 Porsche

944 Turbo SE
1988 Germany
154.0mph 247.8kmh
0-50mph 80.5kmh: 4.6secs
0-60mph 96.5kmh: 5.7secs
0-1/4 mile: 14.2secs
0-1km: 25.8secs
250.0bhp 186.4kW 253.5PS
@ 6000rpm
258.3lbft 350.0Nm @ 4000rpm
208.2bhp/ton 204.7bhp/tonne
100.8bhp/L 75.2kW/L
102.2PS/L
51.8ft/sec 15.8m/sec
25.3mph 40.7kmh/1000rpm
19.1mpg 15.9mpgUS
14.8L/100km
Petrol 4-stroke piston
2479cc 151.0cu in
turbocharged
In-line 4 fuel injection
Compression ratio: 8.0:1
Bore: 100.0mm 3.9in
Stroke: 78.9mm 3.1in
Valve type/No: Overhead 8
Transmission: Manual
No. of forward speeds: 5
Wheels driven: Rear
Springs F/R: Coil/Torsion bar
Brake system: PA ABS
Brakes F/R: Disc/Disc
Steering: Rack & pinion PA

Wheelbase: 240.0cm 94.5in
Track F: 147.7cm 58.1in
Track R: 145.1cm 57.1in
Length: 420.0cm 165.4in
Width: 173.5cm 68.3in
Height: 127.5cm 50.2in
Ground clearance: 12.0cm
4.7in
Kerb weight: 1221.0kg
2689.4lb
Fuel: 80.1L 17.6gal 21.1galUS

724 Porsche

959 Comfort
1988 Germany
198.0mph 318.6kmh
0-60mph 96.5kmh: 4.0secs
0-1/4 mile: 12.4secs
450.0bhp 335.6kW 456.2PS
@ 6500rpm
370.0lbft 501.4Nm @ 5500rpm
316.0bhp/ton 310.7bhp/tonne
157.9bhp/L 117.8kW/L
160.1PS/L
47.7ft/sec 14.5m/sec
Petrol 4-stroke piston
2849cc 173.8cu in
turbocharged
Flat 6 fuel injection
Compression ratio: 8.3:1
Bore: 95.0mm 3.7in
Stroke: 67.0mm 2.6in
Valve type/No: Overhead 24
Transmission: Manual
No. of forward speeds: 6
Wheels driven: 4-wheel drive
Springs F/R: Coil/Coil
Brake system: ABS
Brakes F/R: Disc/Disc
Steering: Rack & pinion PA
Wheelbase: 227.3cm 89.5in
Track F: 150.4cm 59.2in
Track R: 154.9cm 61.0in
Length: 426.0cm 167.7in
Width: 183.9cm 72.4in
Height: 128.0cm 50.4in
Kerb weight: 1448.3kg
3190.0lb
Fuel: 90.1L 19.8gal 23.8galUS

725 Porsche

959 Sport
1988 Germany
198.0mph 318.6kmh
0-60mph 96.5kmh: 3.6secs
0-1/4 mile: 11.9secs
450.0bhp 335.6kW 456.2PS
@ 6500rpm
370.0lbft 501.4Nm @ 5500rpm
339.4bhp/ton 333.7bhp/tonne
157.9bhp/L 117.8kW/L
160.1PS/L
47.7ft/sec 14.5m/sec
Petrol 4-stroke piston

2849cc 173.8cu in
turbocharged
Flat 6 fuel injection
Compression ratio: 8.3:1
Bore: 95.0mm 3.7in
Stroke: 67.0mm 2.6in
Valve type/No: Overhead 24
Transmission: Manual
No. of forward speeds: 6
Wheels driven: 4-wheel drive
Springs F/R: Coil/Coil
Brake system: ABS
Brakes F/R: Disc/Disc
Steering: Rack & pinion PA
Wheelbase: 227.3cm 89.5in
Track F: 150.4cm 59.2in
Track R: 154.9cm 61.0in
Length: 426.0cm 167.7in
Width: 183.9cm 72.4in
Height: 128.0cm 50.4in
Kerb weight: 1348.4kg
2970.0lb
Fuel: 90.1L 19.8gal 23.8galUS

726 Porsche

911 3.3 Turbo
1989 Germany
158.0mph 254.2kmh
0-50mph 80.5kmh: 3.8secs
0-60mph 96.5kmh: 4.9secs
0-1/4 mile: 13.1secs
0-1km: 24.1secs
300.0bhp 223.7kW 304.2PS
@ 5500rpm
317.3lbft 430.0Nm @ 4000rpm
220.3bhp/ton 216.6bhp/tonne
90.9bhp/L 67.8kW/L 92.2PS/L
44.8ft/sec 13.6m/sec
27.4mph 44.1kmh/1000rpm
16.6mpg 13.8mpgUS
17.0L/100km
Petrol 4-stroke piston
3299cc 201.0cu in
turbocharged
Flat 6 fuel injection
Compression ratio: 7.0:1
Bore: 97.0mm 3.8in
Stroke: 74.4mm 2.9in
Valve type/No: Overhead 12
Transmission: Manual
No. of forward speeds: 5
Wheels driven: Rear
Springs F/R: Torsion
bar/Torsion bar
Brake system: PA
Brakes F/R: Disc/Disc
Steering: Rack & pinion
Wheelbase: 227.2cm 89.4in
Track F: 143.2cm 56.4in
Track R: 150.1cm 59.1in
Length: 429.0cm 168.9in
Width: 177.5cm 69.9in
Height: 131.0cm 51.6in
Ground clearance: 12.0cm

4.7in
Kerb weight: 1385.0kg
3050.7lb
Fuel: 85.1L 18.7gal 22.5galUS

727 Porsche

911 Carrera 4
1989 Germany
161.0mph 259.0kmh
0-50mph 80.5kmh: 3.8secs
0-60mph 96.5kmh: 4.9secs
0-1/4 mile: 13.5secs
250.0bhp 186.4kW 253.5PS
@ 6100rpm
229.0lbft 310.3Nm @ 4800rpm
168.7bhp/ton 165.9bhp/tonne
69.4bhp/L 51.8kW/L 70.4PS/L
51.0ft/sec 15.5m/sec
20.4mpg 17.0mpgUS
13.8L/100km
Petrol 4-stroke piston
3600cc 219.6cu in
Flat 6 fuel injection
Compression ratio: 11.3:1
Bore: 100.0mm 3.9in
Stroke: 76.4mm 3.0in
Valve type/No: Overhead 12
Transmission: Manual
No. of forward speeds: 5
Wheels driven: 4-wheel drive
Springs F/R: Coil/Coil
Brake system: PA ABS
Brakes F/R: Disc/Disc
Steering: Rack & pinion PA
Wheelbase: 227.1cm 89.4in
Track F: 137.9cm 54.3in
Track R: 137.4cm 54.1in
Length: 424.9cm 167.3in
Width: 165.1cm 65.0in
Height: 132.1cm 52.0in
Ground clearance: 16.0cm
6.3in
Kerb weight: 1507.3kg
3320.0lb
Fuel: 76.8L 16.9gal 20.3galUS

728 Porsche

911 Carrera Cabriolet
1989 Germany
149.0mph 239.7kmh
0-50mph 80.5kmh: 4.7secs
0-60mph 96.5kmh: 6.1secs
0-1/4 mile: 14.8secs
214.0bhp 159.6kW 217.0PS
@ 5900rpm
195.0lbft 264.2Nm @ 4800rpm
173.0bhp/ton 170.2bhp/tonne
67.6bhp/L 50.4kW/L 68.6PS/L
48.0ft/sec 14.6m/sec
24.6mph 39.6kmh/1000rpm
24.0mpg 20.0mpgUS
11.8L/100km
Petrol 4-stroke piston
3164cc 193.0cu in

Flat 6 fuel injection
Compression ratio: 9.5:1
Bore: 95.0mm 3.7in
Stroke: 74.4mm 2.9in
Valve type/No: Overhead 12
Transmission: Manual
No. of forward speeds: 5
Wheels driven: Rear
Springs F/R: Torsion
bar/Torsion bar
Brake system: PA
Brakes F/R: Disc/Disc
Steering: Rack & pinion
Wheelbase: 227.3cm 89.5in
Track F: 136.9cm 53.9in
Track R: 137.9cm 54.3in
Length: 429.0cm 168.9in
Width: 165.1cm 65.0in
Height: 132.1cm 52.0in
Ground clearance: 13.0cm
5.1in
Kerb weight: 1257.6kg
2770.0lb
Fuel: 85.2L 18.7gal 22.5galUS

729 Porsche

911 Club Sport
1989 Germany
154.0mph 247.8kmh
0-60mph 96.5kmh: 5.3secs
0-1/4 mile: 13.9secs
214.0bhp 159.6kW 217.0PS
@ 5900rpm
195.0lbft 264.2Nm @ 4800rpm
184.4bhp/ton 181.3bhp/tonne
67.6bhp/L 50.4kW/L 68.6PS/L
48.0ft/sec 14.6m/sec
Petrol 4-stroke piston
3164cc 193.0cu in
Flat 6 fuel injection
Compression ratio: 9.5:1
Bore: 95.0mm 3.7in
Stroke: 74.4mm 2.9in
Valve type/No: Overhead 12
Transmission: Manual
No. of forward speeds: 5
Wheels driven: Rear
Springs F/R: Torsion
bar/Torsion bar
Brake system: PA
Brakes F/R: Disc/Disc
Steering: Rack & pinion
Wheelbase: 227.3cm 89.5in
Track F: 140.0cm 55.1in
Track R: 140.7cm 55.4in
Length: 429.0cm 168.9in
Width: 165.1cm 65.0in
Height: 132.1cm 52.0in
Kerb weight: 1180.4kg
2600.0lb
Fuel: 85.2L 18.7gal 22.5galUS

730 Porsche

911 Turbo

1989 Germany
159.0mph 255.8kmh
0-50mph 80.5kmh: 3.8secs
0-60mph 96.5kmh: 4.8secs
0-1/4 mile: 13.6secs
282.0bhp 210.3kW 285.9PS
@ 5500rpm
288.0lbft 390.2Nm @ 4000rpm
206.8bhp/ton 203.3bhp/tonne
85.5bhp/L 63.7kW/L 86.7PS/L
44.8ft/sec 13.6m/sec
27.6mph 44.4kmh/1000rpm
19.2mpg 16.0mpgUS
14.7L/100km
Petrol 4-stroke piston
3299cc 201.3cu in
turbocharged
Flat 6 fuel injection
Compression ratio: 7.0:1
Bore: 97.0mm 3.8in
Stroke: 74.4mm 2.9in
Valve type/No: Overhead 12
Transmission: Manual
No. of forward speeds: 5
Wheels driven: Rear
Springs F/R: Torsion
bar/Torsion bar
Brake system: PA
Brakes F/R: Disc/Disc
Steering: Rack & pinion
Wheelbase: 227.3cm 89.5in
Track F: 143.3cm 56.4in
Track R: 150.1cm 59.1in
Length: 429.0cm 168.9in
Width: 177.5cm 69.9in
Height: 131.1cm 51.6in
Ground clearance: 11.9cm
4.7in
Kerb weight: 1387.0kg
3055.0lb
Fuel: 85.2L 18.7gal 22.5galUS

731 Porsche
911 Turbo Motorsport Design
1989 Germany
183.0mph 294.4kmh
0-50mph 80.5kmh: 3.1secs
0-60mph 96.5kmh: 4.0secs
0-1/4 mile: 12.4secs
488.0bhp 363.9kW 494.8PS
@ 6250rpm
490.0lbft 664.0Nm @ 5000rpm
366.8bhp/ton 360.7bhp/tonne
147.9bhp/L 110.3kW/L
150.0PS/L
50.9ft/sec 15.5m/sec
Petrol 4-stroke piston
3299cc 201.3cu in
turbocharged
Flat 6 fuel injection
Compression ratio: 7.0:1
Bore: 97.0mm 3.8in
Stroke: 74.4mm 2.9in
Valve type/No: Overhead 12

Transmission: Manual
No. of forward speeds: 4
Wheels driven: Rear
Springs F/R: Torsion
bar/Torsion bar
Brake system: PA
Brakes F/R: Disc/Disc
Steering: Rack & pinion PA
Wheelbase: 227.1cm 89.4in
Track F: 143.3cm 56.4in
Track R: 150.1cm 59.1in
Length: 429.0cm 168.9in
Width: 177.5cm 69.9in
Height: 129.8cm 51.1in
Kerb weight: 1352.9kg
2980.0lb

732 Porsche
928 Cabrio Strosek
1989 Germany
155.0mph 249.4kmh
0-60mph 96.5kmh: 5.5secs
0-1/4 mile: 13.9secs
325.0bhp 242.3kW 329.5PS
@ 6000rpm
317.0lbft 429.5Nm @ 3000rpm
206.5bhp/ton 203.1bhp/tonne
65.6bhp/L 48.9kW/L 66.5PS/L
51.8ft/sec 15.8m/sec
Petrol 4-stroke piston
4957cc 302.4cu in
Vee 8 fuel injection
Compression ratio: 10.0:1
Bore: 100.0mm 3.9in
Stroke: 78.9mm 3.1in
Valve type/No: Overhead 32
Transmission: Manual
No. of forward speeds: 5
Wheels driven: Rear
Springs F/R: Coil/Coil
Brake system: PA ABS
Brakes F/R: Disc/Disc
Steering: Rack & pinion PA
Wheelbase: 249.9cm 98.4in
Track F: 160.0cm 63.0in
Track R: 159.8cm 62.9in
Length: 452.4cm 178.1in
Width: 195.6cm 77.0in
Height: 125.7cm 49.5in
Kerb weight: 1600.3kg
3525.0lb
Fuel: 85.9L 18.9gal 22.7galUS

733 Porsche
928 S4 Koenig
1989 Germany
186.0mph 299.3kmh
0-60mph 96.5kmh: 4.8secs
430.0bhp 320.6kW 436.0PS
@ 6000rpm
470.0lbft 636.9Nm @ 3000rpm
253.5bhp/ton 249.2bhp/tonne
86.7bhp/L 64.7kW/L 87.9PS/L
51.8ft/sec 15.8m/sec

Petrol 4-stroke piston
4957cc 302.4cu in
supercharged
Vee 8 fuel injection
Compression ratio: 10.1:1
Bore: 100.0mm 3.9in
Stroke: 78.9mm 3.1in
Valve type/No: Overhead 32
Transmission: Manual
No. of forward speeds: 5
Wheels driven: Rear
Springs F/R: Coil/Coil
Brake system: PA ABS
Brakes F/R: Disc/Disc
Steering: Rack & pinion PA
Wheelbase: 249.9cm 98.4in
Track F: 154.9cm 61.0in
Track R: 154.7cm 60.9in
Length: 452.4cm 178.1in
Width: 183.6cm 72.3in
Height: 127.0cm 50.0in
Kerb weight: 1725.2kg
3800.0lb
Fuel: 85.9L 18.9gal 22.7galUS

734 Porsche
928 S4 SE
1989 Germany
162.0mph 260.7kmh
0-50mph 80.5kmh: 4.3secs
0-60mph 96.5kmh: 5.5secs
0-1/4 mile: 13.8secs
0-1km: 25.1secs
320.0bhp 238.6kW 324.4PS
@ 6000rpm
317.3lbft 430.0Nm @ 3000rpm
206.0bhp/ton 202.5bhp/tonne
64.6bhp/L 48.1kW/L 65.4PS/L
51.8ft/sec 15.8m/sec
26.1mph 42.0kmh/1000rpm
18.6mpg 15.5mpgUS
15.2L/100km
Petrol 4-stroke piston
4957cc 302.0cu in
Vee 8 fuel injection
Compression ratio: 10.0:1
Bore: 100.0mm 3.9in
Stroke: 78.9mm 3.1in
Valve type/No: Overhead 32
Transmission: Manual
No. of forward speeds: 5
Wheels driven: Rear
Springs F/R: Coil/Coil
Brake system: PA ABS
Brakes F/R: Disc/Disc
Steering: Rack & pinion PA
Wheelbase: 250.0cm 98.4in
Track F: 151.1cm 59.5in
Track R: 154.6cm 60.9in
Length: 452.0cm 178.0in
Width: 183.6cm 72.3in
Height: 128.2cm 50.5in
Ground clearance: 16.5cm
6.5in

Kerb weight: 1580.0kg
3480.2lb
Fuel: 93.7L 20.6gal 24.8galUS

735 Porsche
944
1989 Germany
139.0mph 223.7kmh
0-50mph 80.5kmh: 5.1secs
0-60mph 96.5kmh: 7.0secs
0-1/4 mile: 15.7secs
0-1km: 28.5secs
165.0bhp 123.0kW 167.3PS
@ 5800rpm
166.1lbft 225.0Nm @ 4200rpm
127.1bhp/ton 125.0bhp/tonne
61.5bhp/L 45.9kW/L 62.4PS/L
50.1ft/sec 15.3m/sec
22.3mph 35.9kmh/1000rpm
20.7mpg 17.2mpgUS
13.6L/100km
Petrol 4-stroke piston
2681cc 164.0cu in
In-line 4 fuel injection
Compression ratio: 10.9:1
Bore: 104.0mm 4.1in
Stroke: 79.0mm 3.1in
Valve type/No: Overhead 8
Transmission: Manual
No. of forward speeds: 5
Wheels driven: Rear
Springs F/R: Coil/Torsion bar
Brake system: PA ABS
Brakes F/R: Disc/Disc
Steering: Rack & pinion PA
Wheelbase: 240.0cm 94.5in
Track F: 147.7cm 58.1in
Track R: 145.1cm 57.1in
Length: 420.0cm 165.4in
Width: 173.5cm 68.3in
Height: 127.5cm 50.2in
Ground clearance: 12.0cm
4.7in
Kerb weight: 1320.0kg
2907.5lb
Fuel: 80.1L 17.6gal 21.1galUS

736 Porsche
944 S2
1989 Germany
147.0mph 236.5kmh
0-50mph 80.5kmh: 4.5secs
0-60mph 96.5kmh: 6.0secs
0-1/4 mile: 14.4secs
0-1km: 26.1secs
211.0bhp 157.3kW 213.9PS
@ 5800rpm
206.6lbft 280.0Nm @ 4000rpm
158.9bhp/ton 156.3bhp/tonne
70.6bhp/L 52.6kW/L 71.5PS/L
55.7ft/sec 17.0m/sec
23.7mph 38.1kmh/1000rpm
20.7mpg 17.2mpgUS
13.6L/100km

Petrol 4-stroke piston
2990cc 182.0cu in
In-line 4 fuel injection
Compression ratio: 10.9:1
Bore: 104.0mm 4.1in
Stroke: 88.0mm 3.5in
Valve type/No: Overhead 16
Transmission: Manual
No. of forward speeds: 5
Wheels driven: Rear
Springs F/R: Coil/Torsion bar
Brake system: PA ABS
Brakes F/R: Disc/Disc
Steering: Rack & pinion PA
Wheelbase: 240.0cm 94.5in
Track F: 147.7cm 58.1in
Track R: 145.1cm 57.1in
Length: 420.0cm 165.4in
Width: 173.5cm 68.3in
Height: 127.5cm 50.2in
Ground clearance: 12.0cm
4.7in
Kerb weight: 1350.0kg
2973.6lb
Fuel: 80.1L 17.6gal 21.1galUS

737 Porsche

944 Turbo
1989 Germany
162.0mph 260.7kmh
0-50mph 80.5kmh: 4.3secs
0-60mph 96.5kmh: 5.5secs
0-1/4 mile: 14.2secs
250.0bhp 186.4kW 253.5PS
@ 6000rpm
258.0lbft 349.6Nm @ 4000rpm
175.3bhp/ton 172.3bhp/tonne
100.8bhp/L 75.2kW/L
102.2PS/L
51.8ft/sec 15.8m/sec
26.7mph 43.0kmh/1000rpm
22.2mpg 18.5mpgUS
12.7L/100km
Petrol 4-stroke piston
2479cc 151.2cu in
turbocharged
In-line 4 fuel injection
Compression ratio: 8.0:1
Bore: 100.0mm 3.9in
Stroke: 78.9mm 3.1in
Valve type/No: Overhead 8
Transmission: Manual
No. of forward speeds: 5
Wheels driven: Rear
Springs F/R: Coil/Coil
Brake system: PA ABS
Brakes F/R: Disc/Disc
Steering: Rack & pinion PA
Wheelbase: 240.0cm 94.5in
Track F: 147.8cm 58.2in
Track R: 144.3cm 56.8in
Length: 429.0cm 168.9in
Width: 173.5cm 68.3in
Height: 127.5cm 50.2in

Ground clearance: 11.9cm
4.7in
Kerb weight: 1450.5kg
3195.0lb
Fuel: 79.9L 17.6gal 21.1galUS

738 Porsche

959
1989 Germany
198.0mph 318.6kmh
0-50mph 80.5kmh: 2.8secs
0-60mph 96.5kmh: 3.6secs
0-1/4 mile: 11.9secs
450.0bhp 335.6kW 456.2PS
@ 6500rpm
370.0lbft 501.4Nm @ 5500rpm
316.0bhp/ton 310.7bhp/tonne
157.9bhp/L 117.8kW/L
160.1PS/L
47.7ft/sec 14.5m/sec
Petrol 4-stroke piston
2849cc 173.8cu in
turbocharged
Flat 6 fuel injection
Compression ratio: 8.3:1
Bore: 95.0mm 3.7in
Stroke: 67.0mm 2.6in
Valve type/No: Overhead 24
Transmission: Manual
No. of forward speeds: 6
Wheels driven: 4-wheel drive
Springs F/R: Coil/Coil
Brake system: ABS
Brakes F/R: Disc/Disc
Steering: Rack & pinion PA
Wheelbase: 227.1cm 89.4in
Track F: 150.4cm 59.2in
Track R: 154.9cm 61.0in
Length: 426.0cm 167.7in
Width: 183.9cm 72.4in
Height: 128.0cm 50.4in
Kerb weight: 1448.3kg
3190.0lb
Fuel: 90.1L 19.8gal 23.8galUS

739 Porsche

911 Carrera 2
1990 Germany
160.0mph 257.4kmh
0-50mph 80.5kmh: 4.0secs
0-60mph 96.5kmh: 5.1secs
0-1/4 mile: 13.6secs
0-1km: 24.6secs
250.0bhp 186.4kW 253.5PS
@ 6100rpm
228.8lbft 310.0Nm @ 4800rpm
184.2bhp/ton 181.2bhp/tonne
69.4bhp/L 51.8kW/L 70.4PS/L
51.0ft/sec 15.5m/sec
24.2mph 38.9kmh/1000rpm
20.4mpg 17.0mpgUS
13.8L/100km
Petrol 4-stroke piston
3600cc 220.0cu in

Flat 6 fuel injection
Compression ratio: 11.3:1
Bore: 100.0mm 3.9in
Stroke: 76.4mm 3.0in
Valve type/No: Overhead 12
Transmission: Manual
No. of forward speeds: 5
Wheels driven: Rear
Springs F/R: Coil/Coil
Brake system: PA ABS
Brakes F/R: Disc/Disc
Steering: Rack & pinion PA
Wheelbase: 227.2cm 89.4in
Track F: 138.0cm 54.3in
Track R: 137.4cm 54.1in
Length: 425.0cm 167.3in
Width: 165.2cm 65.0in
Height: 132.0cm 52.0in
Ground clearance: 12.0cm
4.7in
Kerb weight: 1380.0kg
3039.6lb
Fuel: 76.9L 16.9gal 20.3galUS

740 Porsche

911 Carrera 2 Tiptronic
1990 Germany
159.0mph 255.8kmh
0-60mph 96.5kmh: 6.9secs
0-1/4 mile: 15.0secs
247.0bhp 184.2kW 250.4PS
@ 6100rpm
228.0lbft 308.9Nm @ 4800rpm
174.3bhp/ton 171.4bhp/tonne
68.5bhp/L 51.1kW/L 69.5PS/L
51.0ft/sec 15.6m/sec
21.0mpg 17.5mpgUS
13.4L/100km
Petrol 4-stroke piston
3605cc 219.9cu in
Flat 6 fuel injection
Compression ratio: 11.3:1
Bore: 100.0mm 3.9in
Stroke: 76.5mm 3.0in
Valve type/No: Overhead 12
Transmission: Automatic
No. of forward speeds: 4
Wheels driven: Rear
Springs F/R: Coil/Coil
Brake system: PA ABS
Brakes F/R: Disc/Disc
Steering: Rack & pinion PA
Wheelbase: 227.1cm 89.4in
Track F: 137.9cm 54.3in
Track R: 137.4cm 54.1in
Length: 427.5cm 168.3in
Width: 165.1cm 65.0in
Height: 132.1cm 52.0in
Kerb weight: 1441.4kg
3175.0lb
Fuel: 76.8L 16.9gal 20.3galUS

741 Porsche

911 Carrera 4

1990 Germany
161.0mph 259.0kmh
0-50mph 80.5kmh: 4.5secs
0-60mph 96.5kmh: 5.8secs
0-1/4 mile: 14.4secs
250.0bhp 186.4kW 253.5PS
@ 6100rpm
229.0lbft 310.3Nm @ 4800rpm
168.7bhp/ton 165.9bhp/tonne
69.4bhp/L 51.8kW/L 70.4PS/L
51.0ft/sec 15.5m/sec
20.4mpg 17.0mpgUS
13.8L/100km
Petrol 4-stroke piston
3600cc 219.6cu in
Flat 6 fuel injection
Compression ratio: 11.3:1
Bore: 100.0mm 3.9in
Stroke: 76.4mm 3.0in
Valve type/No: Overhead 12
Transmission: Manual
No. of forward speeds: 5
Wheels driven: 4-wheel drive
Springs F/R: Coil/Coil
Brake system: PA ABS
Brakes F/R: Disc/Disc
Steering: Rack & pinion PA
Wheelbase: 227.3cm 89.5in
Track F: 137.9cm 54.3in
Track R: 137.4cm 54.1in
Length: 424.9cm 167.3in
Width: 165.1cm 65.0in
Height: 132.1cm 52.0in
Kerb weight: 1507.3kg
3320.0lb
Fuel: 76.8L 16.9gal 20.3galUS

742 Porsche

911 Speedster
1990 Germany
149.0mph 239.7kmh
0-60mph 96.5kmh: 6.0secs
0-1/4 mile: 14.5secs
214.0bhp 159.6kW 217.0PS
@ 5900rpm
195.0lbft 264.2Nm @ 4800rpm
163.9bhp/ton 161.1bhp/tonne
67.6bhp/L 50.4kW/L 68.6PS/L
48.3ft/sec 14.7m/sec
24.0mpg 20.0mpgUS
11.8L/100km
Petrol 4-stroke piston
3164cc 193.0cu in
Flat 6 fuel injection
Compression ratio: 9.5:1
Bore: 87.0mm 3.4in
Stroke: 75.0mm 2.9in
Valve type/No: Overhead 12
Transmission: Manual
No. of forward speeds: 5
Wheels driven: Rear
Springs F/R: Torsion
bar/Torsion bar
Brake system: PA

Brakes F/R: Disc/Disc
Steering: Rack & pinion
Wheelbase: 227.3cm 89.5in
Track F: 143.3cm 56.4in
Track R: 149.1cm 58.7in
Length: 429.0cm 168.9in
Width: 177.5cm 69.9in
Height: 128.0cm 50.4in
Kerb weight: 1327.9kg
2925.0lb
Fuel: 85.2L 18.7gal 22.5galUS

743 Porsche
911 Turbo
1990 Germany
159.0mph 255.8kmh
0-60mph 96.5kmh: 5.1secs
0-1/4 mile: 13.6secs
300.0bhp 223.7kW 304.2PS
@ 5500rpm
317.0lbft 429.5Nm @ 4000rpm
220.1bhp/ton 216.3bhp/tonne
90.9bhp/L 67.8kW/L 92.2PS/L
44.8ft/sec 13.6m/sec
19.8mpg 16.5mpgUS
14.3L/100km
Petrol 4-stroke piston
3299cc 201.3cu in
turbocharged
Flat 6 fuel injection
Compression ratio: 7.0:1
Bore: 97.0mm 3.8in
Stroke: 74.4mm 2.9in
Valve type/No: Overhead 12
Transmission: Manual
No. of forward speeds: 5
Wheels driven: Rear
Springs F/R: Torsion
bar/Torsion bar
Brake system: PA
Brakes F/R: Disc/Disc
Steering: Rack & pinion
Wheelbase: 227.3cm 89.5in
Track F: 143.3cm 56.4in
Track R: 150.1cm 59.1in
Length: 429.0cm 168.9in
Width: 177.5cm 69.9in
Height: 131.1cm 51.6in
Kerb weight: 1387.0kg
3055.0lb
Fuel: 85.2L 18.7gal 22.5galUS

744 Porsche
911 Turbo RS Tuning
1990 Germany
216.0mph 347.5kmh
0-60mph 96.5kmh: 4.6secs
580.0bhp 432.5kW 588.0PS
@ 6900rpm
465.0lbft 630.1Nm @ 5500rpm
172.2bhp/L 128.4kW/L
174.6PS/L
56.2ft/sec 17.1m/sec
Petrol 4-stroke piston

3368cc 205.5cu in
turbocharged
Flat 6 fuel injection
Bore: 98.0mm 3.9in
Stroke: 74.4mm 2.9in
Valve type/No: Overhead 12
Transmission: Manual
No. of forward speeds: 6
Wheels driven: Rear
Springs F/R: Coil/Coil
Brake system: PA
Brakes F/R: Disc/Disc
Steering: Rack & pinion
Wheelbase: 227.1cm 89.4in
Track F: 143.3cm 56.4in
Track R: 149.1cm 58.7in
Length: 429.0cm 168.9in
Width: 177.5cm 69.9in
Height: 131.1cm 51.6in
Fuel: 85.2L 18.7gal 22.5galUS

745 Porsche
944 S2
1990 Germany
149.0mph 239.7kmh
0-60mph 96.5kmh: 6.7secs
0-1/4 mile: 14.7secs
208.0bhp 155.1kW 210.9PS
@ 5800rpm
207.0lbft 280.5Nm @ 4100rpm
156.1bhp/ton 153.5bhp/tonne
69.6bhp/L 51.9kW/L 70.5PS/L
55.7ft/sec 17.0m/sec
22.2mpg 18.5mpgUS
12.7L/100km
Petrol 4-stroke piston
2990cc 182.4cu in
In-line 4 fuel injection
Compression ratio: 10.9:1
Bore: 104.0mm 4.1in
Stroke: 88.0mm 3.5in
Valve type/No: Overhead 16
Transmission: Manual
No. of forward speeds: 5
Wheels driven: Rear
Springs F/R: Coil/Torsion bar
Brake system: PA ABS
Brakes F/R: Disc/Disc
Steering: Rack & pinion PA
Wheelbase: 240.0cm 94.5in
Track F: 147.8cm 58.2in
Track R: 145.0cm 57.1in
Length: 429.0cm 168.9in
Width: 173.5cm 68.3in
Height: 127.5cm 50.2in
Kerb weight: 1355.2kg
2985.0lb
Fuel: 79.9L 17.6gal 21.1galUS

746 Porsche
944 S2 Cabriolet
1990 Germany
149.0mph 239.7kmh
0-50mph 80.5kmh: 5.2secs

0-60mph 96.5kmh: 7.1secs
0-1/4 mile: 15.3secs
208.0bhp 155.1kW 210.9PS
@ 5800rpm
207.0lbft 280.5Nm @ 4100rpm
150.5bhp/ton 148.0bhp/tonne
69.6bhp/L 51.9kW/L 70.5PS/L
55.7ft/sec 17.0m/sec
22.2mpg 18.5mpgUS
12.7L/100km
Petrol 4-stroke piston
2990cc 182.4cu in
In-line 4 fuel injection
Compression ratio: 10.9:1
Bore: 104.0mm 4.1in
Stroke: 88.0mm 3.5in
Valve type/No: Overhead 16
Transmission: Manual
No. of forward speeds: 5
Wheels driven: Rear
Springs F/R: Coil/Torsion bar
Brake system: PA ABS
Brakes F/R: Disc/Disc
Steering: Rack & pinion PA
Wheelbase: 240.0cm 94.5in
Track F: 147.8cm 58.2in
Track R: 145.0cm 57.1in
Length: 429.0cm 168.9in
Width: 173.5cm 68.3in
Height: 127.5cm 50.2in
Kerb weight: 1405.1kg
3095.0lb
Fuel: 79.9L 17.6gal 21.1galUS

747 Porsche
Carrera 2 Cabriolet Tiptronic
1990 Germany
159.0mph 255.8kmh
0-50mph 80.5kmh: 4.8secs
0-60mph 96.5kmh: 6.2secs
0-1/4 mile: 14.6secs
0-1km: 26.1secs
250.0bhp 186.4kW 253.5PS
@ 6100rpm
228.8lbft 310.0Nm @ 4800rpm
211.9bhp/ton 208.3bhp/tonne
69.4bhp/L 51.8kW/L 70.4PS/L
51.0ft/sec 15.5m/sec
26.8mpg 43.1kmh/1000rpm
17.4mpg 14.5mpgUS
16.2L/100km
Petrol 4-stroke piston
3600cc 220.0cu in
Flat 6 fuel injection
Compression ratio: 11.3:1
Bore: 100.0mm 3.9in
Stroke: 76.4mm 3.0in
Valve type/No: Overhead 24
Transmission: Automatic
No. of forward speeds: 4
Wheels driven: Rear
Springs F/R: Coil/Coil
Brake system: PA ABS
Brakes F/R: Disc/Disc

Steering: Rack & pinion PA
Wheelbase: 227.2cm 89.4in
Track F: 138.0cm 54.3in
Track R: 137.4cm 54.1in
Length: 425.0cm 167.3in
Width: 165.2cm 65.0in
Height: 132.0cm 52.0in
Ground clearance: 12.0cm
4.7in
Kerb weight: 1200.0kg
2643.2lb
Fuel: 76.9L 16.9gal 20.3galUS

748 Porsche
911 Carrera Turbo
1991 Germany
168.0mph 270.3kmh
0-60mph 96.5kmh: 4.6secs
0-1/4 mile: 12.9secs
320.0bhp 238.6kW 324.4PS
@ 5750rpm
332.0lbft 449.9Nm @ 4500rpm
218.9bhp/ton 215.2bhp/tonne
97.0bhp/L 72.3kW/L 98.3PS/L
46.8ft/sec 14.3m/sec
18.0mpg 15.0mpgUS
15.7L/100km
Petrol 4-stroke piston
3299cc 201.3cu in
turbocharged
Flat 6 fuel injection
Compression ratio: 7.0:1
Bore: 97.0mm 3.8in
Stroke: 74.4mm 2.9in
Valve type/No: Overhead 12
Transmission: Manual
No. of forward speeds: 5
Wheels driven: Rear
Springs F/R: Coil/Coil
Brake system: PA ABS
Brakes F/R: Disc/Disc
Steering: Rack & pinion PA
Wheelbase: 227.1cm 89.4in
Track F: 143.5cm 56.5in
Track R: 149.4cm 58.8in
Length: 427.5cm 168.3in
Width: 177.8cm 70.0in
Height: 131.1cm 51.6in
Kerb weight: 1486.8kg
3275.0lb
Fuel: 76.8L 16.9gal 20.3galUS

749 Porsche
911 Ruf
1991 Germany
196.0mph 315.4kmh
0-60mph 96.5kmh: 3.8secs
0-1/4 mile: 12.0secs
455.0bhp 339.3kW 461.3PS
@ 5800rpm
482.0lbft 653.1Nm @ 4500rpm
335.8bhp/ton 330.2bhp/tonne
129.8bhp/L 96.8kW/L
131.6PS/L

752 Porsche
911 Turbo

47.2ft/sec 14.4m/sec
Petrol 4-stroke piston
3504cc 213.8cu in
turbocharged
Flat 6 fuel injection
Compression ratio: 7.6:1
Bore: 100.0mm 3.9in
Stroke: 74.4mm 2.9in
Valve type/No: Overhead 12
Transmission: Manual
No. of forward speeds: 6
Wheels driven: Rear
Springs F/R: Coil/Coil
Brake system: ABS
Brakes F/R: Disc/Disc
Steering: Rack & pinion
Wheelbase: 227.1cm 89.4in
Track F: 136.9cm 53.9in
Track R: 137.9cm 54.3in
Length: 427.5cm 168.3in
Width: 165.1cm 65.0in
Height: 127.3cm 50.1in
Kerb weight: 1377.9kg
3035.0lb

750 Porsche

911 Ruf CTR
1991 Germany
208.0mph 334.7kmh
0-60mph 96.5kmh: 3.9secs
0-1/4 mile: 11.9secs
469.0bhp 349.7kW 475.5PS
@ 5950rpm
408.0lbft 552.8Nm @ 5100rpm
345.6bhp/ton 339.8bhp/tonne
139.3bhp/L 103.9kW/L
141.2PS/L
48.4ft/sec 14.8m/sec
16.8mpg 14.0mpgUS
16.8L/100km
Petrol 4-stroke piston
3367cc 205.4cu in
turbocharged

Flat 6 fuel injection
Compression ratio: 7.5:1
Bore: 98.0mm 3.9in
Stroke: 74.4mm 2.9in
Valve type/No: Overhead 12
Transmission: Manual
No. of forward speeds: 6
Wheels driven: Rear
Springs F/R: Coil/Coil
Brake system: PA
Brakes F/R: Disc/Disc
Steering: Rack & pinion
Wheelbase: 227.3cm 89.5in
Track F: 143.3cm 56.4in
Track R: 149.1cm 58.7in
Length: 429.0cm 168.9in
Width: 177.5cm 69.9in
Height: 131.1cm 51.6in
Kerb weight: 1380.2kg
3040.0lb
Fuel: 85.2L 18.7gal 22.5galUS
751 Porsche
911 Ruf TR2
1991 Germany
196.1mph 315.5kmh
0-60mph 96.5kmh: 3.8secs
0-1/4 mile: 12.0secs
455.0bhp 339.3kW 461.3PS
@ 5800rpm
483.0lbft 654.5Nm @ 4500rpm
335.3bhp/ton 329.7bhp/tonne
129.8bhp/L 96.8kW/L
131.6PS/L
Petrol 4-stroke piston
3504cc 213.8cu in
Flat 6
Valve type/No: Overhead 12
Transmission: Manual
No. of forward speeds: 6
Wheels driven: Rear
Wheelbase: 227.1cm 89.4in
Length: 427.5cm 168.3in
Width: 177.8cm 70.0in

Height: 129.5cm 51.0in
Kerb weight: 1380.2kg
3040.0lb

752 Porsche
911 Turbo
1991 Germany
171.0mph 275.1kmh
0-50mph 80.5kmh: 3.7secs
0-60mph 96.5kmh: 4.7secs
0-1/4 mile: 13.3secs
0-1km: 24.0secs
320.0bhp 238.6kW 324.4PS
@ 5750rpm
332.1lbft 450.0Nm @ 4500rpm
222.9bhp/ton 219.2bhp/tonne
97.0bhp/L 72.3kW/L 98.3PS/L
46.5ft/sec 14.2m/sec
27.8mph 44.7kmh/1000rpm
15.1mpg 12.6mpgUS
18.7L/100km
Petrol 4-stroke piston
3299cc 201.0cu in
turbocharged
Flat 6 fuel injection
Compression ratio: 7.0:1
Bore: 97.0mm 3.8in
Stroke: 74.0mm 2.9in
Valve type/No: Overhead 12
Transmission: Manual
No. of forward speeds: 5
Wheels driven: Rear
Springs F/R: Coil/Coil
Brake system: PA ABS
Brakes F/R: Disc/Disc
Steering: Rack & pinion PA
Wheelbase: 227.1cm 89.4in
Track F: 143.3cm 56.4in
Track R: 149.1cm 58.7in
Length: 424.9cm 167.3in
Width: 177.3cm 69.8in
Height: 130.8cm 51.5in
Ground clearance: 11.9cm

4.7in
Kerb weight: 1460.0kg
3215.9lb
Fuel: 77.3L 17.0gal
20.4galUS

753 Porsche

911 Turbo Gemballa Mirage
1991 Germany
205.0mph 329.8kmh
0-60mph 96.5kmh: 4.1secs
0-1/4 mile: 12.3secs
490.0bhp 365.4kW 496.8PS
@ 5900rpm
420.0lbft 569.1Nm
368.3bhp/ton 362.2bhp/tonne
145.5bhp/L 108.5kW/L
147.5PS/L
48.0ft/sec 14.6m/sec
Petrol 4-stroke piston
3367cc 205.4cu in
turbocharged
Flat 6 fuel injection
Compression ratio: 7.5:1
Bore: 98.0mm 3.9in
Stroke: 74.4mm 2.9in
Valve type/No: Overhead 12
Transmission: Manual
No. of forward speeds: 6
Wheels driven: Rear
Springs F/R: Torsion
bar/Torsion bar
Brakes F/R: Disc/Disc
Steering: Rack & pinion
Wheelbase: 227.3cm 89.5in
Track F: 143.3cm 56.4in
Track R: 150.1cm 59.1in
Length: 435.1cm 171.3in
Width: 210.1cm 82.7in
Height: 115.1cm 45.3in
Kerb weight: 1352.9kg
2980.0lb
Fuel: 85.2L 18.7gal 22.5galUS

754 Porsche

911 Carrera RS
1992 Germany
161.0mph 259.0kmh
0-50mph 80.5kmh: 3.8secs
0-60mph 96.5kmh: 4.9secs
0-1/4 mile: 13.4secs
0-1km: 24.3secs
260.0bhp 193.9kW 263.6PS
@ 6100rpm
239.9lbft 325.0Nm @ 4800rpm
221.3bhp/ton 217.6bhp/tonne
72.2bhp/L 53.9kW/L 73.2PS/L
51.0ft/sec 15.5m/sec
24.2mph 38.9kmh/1000rpm
Petrol 4-stroke piston
3600cc 220.0cu in
Flat 6 fuel injection
Compression ratio: 11.3:1
Bore: 100.0mm 3.9in
Stroke: 76.4mm 3.0in
Valve type/No: Overhead 12
Transmission: Manual
No. of forward speeds: 5
Wheels driven: Rear
Springs F/R: Coil/Coil
Brake system: PA ABS
Brakes F/R: Disc/Disc
Steering: Rack & pinion PA
Wheelbase: 227.2cm 89.4in
Track F: 143.4cm 56.5in
Track R: 149.3cm 58.8in
Length: 425.0cm 167.3in
Width: 177.5cm 69.9in
Height: 131.0cm 51.6in
Ground clearance: 12.0cm 4.7in
Kerb weight: 1195.0kg 2632.2lb
Fuel: 77.0L 16.9gal 20.3galUS

755 Proton

1.5 SE Aeroback
1989 Malaysia
99.0mph 159.3kmh
0-50mph 80.5kmh: 10.1secs
0-60mph 96.5kmh: 14.4secs
0-1/4 mile: 19.8secs
0-1km: 37.0secs
75.0bhp 55.9kW 76.0PS
@ 5500rpm
87.1lbft 118.0Nm @ 3500rpm

82.2bhp/ton 80.8bhp/tonne
51.1bhp/L 38.1kW/L 51.8PS/L
49.3ft/sec 15.0m/sec
23.2mph 37.3kmh/1000rpm
28.0mpg 23.3mpgUS
10.1L/100km
Petrol 4-stroke piston
1468cc 90.0cu in
In-line 4 1 Carburettor
Compression ratio: 9.5:1
Bore: 75.5mm 3.0in
Stroke: 82.0mm 3.2in
Valve type/No: Overhead 8
Transmission: Manual
No. of forward speeds: 5
Wheels driven: Front
Springs F/R: Coil/Coil
Brakes F/R: Disc/Drum
Steering: Rack & pinion PA
Wheelbase: 238.0cm 93.7in
Track F: 139.0cm 54.7in
Track R: 134.2cm 52.8in
Length: 413.5cm 162.8in
Width: 163.0cm 64.2in
Height: 136.0cm 53.5in
Ground clearance: 18.5cm 7.3in
Kerb weight: 928.0kg 2044.0lb
Fuel: 45.5L 10.0gal 12.0galUS

756 Proton

1.5 SE Triple Valve Aeroback
1991 Malaysia
100.0mph 160.9kmh
0-50mph 80.5kmh: 8.7secs
0-60mph 96.5kmh: 12.4secs
0-1/4 mile: 17.9secs
0-1km: 33.0secs
86.0bhp 64.1kW 87.2PS
@ 6000rpm
87.1lbft 118.0Nm @ 4000rpm
88.3bhp/ton 86.9bhp/tonne
58.6bhp/L 43.7kW/L 59.4PS/L
53.8ft/sec 16.4m/sec
20.9mph 33.6kmh/1000rpm
28.5mpg 23.7mpgUS
9.9L/100km
Petrol 4-stroke piston
1468cc 90.0cu in
In-line 4 1 Carburettor
Compression ratio: 9.2:1

Bore: 76.0mm 3.0in
Stroke: 82.0mm 3.2in
Valve type/No: Overhead 12
Transmission: Manual
No. of forward speeds: 5
Wheels driven: Front
Springs F/R: Coil/Coil
Brake system: PA
Brakes F/R: Disc/Drum
Steering: Rack & pinion PA
Wheelbase: 238.0cm 93.7in
Track F: 138.9cm 54.7in
Track R: 134.1cm 52.8in
Length: 414.0cm 163.0in
Width: 163.6cm 64.4in
Height: 135.9cm 53.5in
Ground clearance: 15.0cm 5.9in
Kerb weight: 990.0kg 2180.6lb
Fuel: 45.5L 10.0gal 12.0galUS

757 Range Rover

Vogue
1986 UK
106.0mph 170.6kmh
0-50mph 80.5kmh: 8.2secs
0-60mph 96.5kmh: 11.9secs
0-1/4 mile: 18.4secs
0-1km: 34.3secs
165.0bhp 123.0kW 167.3PS
@ 4750rpm
207.4lbft 281.0Nm @ 3200rpm
85.1bhp/ton 83.7bhp/tonne
46.8bhp/L 34.9kW/L 47.4PS/L
36.9ft/sec 11.3m/sec
25.8mph 41.5kmh/1000rpm
15.0mpg 12.5mpgUS
18.8L/100km
Petrol 4-stroke piston
3528cc 215.2cu in
Vee 8 fuel injection
Compression ratio: 9.3:1
Bore: 88.9mm 3.5in
Stroke: 71.1mm 2.8in
Valve type/No: Overhead 16
Transmission: Manual
No. of forward speeds: 5
Wheels driven: 4-wheel drive
Springs F/R: Coil/Coil
Brake system: PA
Brakes F/R: Disc/Disc

Steering: Recirculating ball PA
Wheelbase: 259.1cm 102.0in
Track F: 148.6cm 58.5in
Track R: 148.6cm 58.5in
Length: 447.0cm 176.0in
Width: 177.8cm 70.0in
Height: 177.8cm 70.0in
Ground clearance: 19.1cm 7.5in
Kerb weight: 1972.2kg
4344.0lb
Fuel: 79.6L 17.5gal 21.0galUS

758 Range Rover

V8
1987 UK
90.0mph 144.8kmh
0-50mph 80.5kmh: 9.2secs
0-60mph 96.5kmh: 12.6secs
0-1/4 mile: 19.0secs
150.0bhp 111.9kW 152.1PS
@ 4750rpm
195.0lbft 264.2Nm @ 3000rpm
77.9bhp/ton 76.6bhp/tonne
42.5bhp/L 31.7kW/L 43.1PS/L
36.9ft/sec 11.3m/sec
16.2mpg 13.5mpgUS
17.4L/100km
Petrol 4-stroke piston
3532cc 215.5cu in
Vee 8 fuel injection
Compression ratio: 8.1:1
Bore: 88.9mm 3.5in
Stroke: 71.1mm 2.8in
Valve type/No: Overhead 16
Transmission: Automatic
No. of forward speeds: 4
Wheels driven: 4-wheel drive
Springs F/R: Coil/Coil
Brake system: PA
Brakes F/R: Disc/Disc
Steering: Recirculating ball PA
Wheelbase: 254.0cm 100.0in
Track F: 148.6cm 58.5in
Track R: 148.6cm 58.5in
Length: 444.5cm 175.0in
Width: 181.9cm 71.6in
Height: 179.8cm 70.8in
Kerb weight: 1959.0kg
4315.0lb
Fuel: 75.7L 16.6gal 20.0galUS

759 Range Rover

Vogue Turbo D
1988 UK
93.0mph 149.6kmh
0-50mph 80.5kmh: 11.6secs
0-60mph 96.5kmh: 16.5secs
0-1/4 mile: 20.6secs
0-1km: 38.6secs
112.0bhp 83.5kW 113.5PS
@ 4200rpm
183.0lbft 248.0Nm @ 2400rpm
56.5bhp/ton 55.6bhp/tonne
46.8bhp/L 34.9kW/L 47.5PS/L

756 Proton
1.5 SE Triple Valve
Aeroback

Steering: Worm & roller PA
Wheelbase: 254.0cm 100.0in
Track F: 148.6cm 58.5in
Track R: 148.6cm 58.5in
Length: 444.5cm 175.0in
Width: 181.4cm 71.4in
Height: 179.8cm 70.8in
Kerb weight: 2036.2kg
4485.0lb
Fuel: 75.7L 16.6gal 20.0galUS

41.3ft/sec 12.6m/sec
25.2mph 40.5kmh/1000rpm
21.4mpg 17.8mpgUS
13.2L/100km
Diesel 4-stroke piston
2392cc 146.0cu in
turbocharged
In-line 4 fuel injection
Compression ratio: 22.0:1
Bore: 92.0mm 3.6in
Stroke: 90.0mm 3.5in
Valve type/No: Overhead 8
Transmission: Manual
No. of forward speeds: 5
Wheels driven: 4-wheel drive
Springs F/R: Coil/Coil
Brake system: PA
Brakes F/R: Disc/Disc
Steering: Recirculating ball PA
Wheelbase: 259.1cm 102.0in
Track F: 148.6cm 58.5in
Track R: 148.6cm 58.5in
Length: 447.0cm 176.0in
Width: 177.8cm 70.0in
Height: 177.8cm 70.0in
Ground clearance: 19.1cm
7.5in
Kerb weight: 2016.0kg
4440.5lb
Fuel: 81.9L 18.0gal 21.6galUS

760 Range Rover
Vogue
1989 UK
106.0mph 170.6kmh
0-50mph 80.5kmh: 8.4secs
0-60mph 96.5kmh: 12.1secs
0-1/4 mile: 18.3secs
0-1km: 34.3secs
168.0bhp 125.3kW 170.3PS
@ 4750rpm
207.4lbft 281.0Nm @ 3200rpm
90.4bhp/ton 88.9bhp/tonne
47.6bhp/L 35.5kW/L 48.3PS/L
36.9ft/sec 11.3m/sec
25.4mph 40.9kmh/1000rpm
18.7mpg 15.6mpgUS
15.1L/100km

Petrol 4-stroke piston
3528cc 215.0cu in
Vee 8 fuel injection
Compression ratio: 9.3:1
Bore: 88.9mm 3.5in
Stroke: 71.1mm 2.8in
Valve type/No: Overhead 16
Transmission: Manual
No. of forward speeds: 5
Wheels driven: 4-wheel drive
Springs F/R: Coil/Coil
Brake system: PA
Brakes F/R: Disc/Disc
Steering: Recirculating ball PA
Wheelbase: 254.0cm 100.0in
Track F: 148.6cm 58.5in
Track R: 148.6cm 58.5in
Length: 447.0cm 176.0in
Width: 177.8cm 70.0in
Height: 177.8cm 70.0in
Ground clearance: 19.1cm
7.5in
Kerb weight: 1890.0kg
4163.0lb
Fuel: 76.4L 16.8gal 20.2galUS

761 Range Rover
Vogue SE
1989 UK
102.0mph 164.1kmh
0-50mph 80.5kmh: 8.4secs
0-60mph 96.5kmh: 11.7secs
0-1/4 mile: 18.9secs
0-1km: 35.0secs
165.0bhp 123.0kW 167.3PS
@ 4750rpm
207.4lbft 281.0Nm @ 3200rpm
88.2bhp/ton 86.7bhp/tonne
46.1bhp/L 34.3kW/L 46.7PS/L
36.9ft/sec 11.3m/sec
26.6mph 42.8kmh/1000rpm
15.8mpg 13.2mpgUS
17.9L/100km
Petrol 4-stroke piston
3582cc 219.0cu in
Vee 8 fuel injection
Compression ratio: 9.3:1
Bore: 88.9mm 3.5in

757 Range Rover
Vogue

Stroke: 71.1mm 2.8in
Valve type/No: Overhead 16
Transmission: Automatic
No. of forward speeds: 4
Wheels driven: 4-wheel drive
Springs F/R: Coil/Coil
Brake system: PA
Brakes F/R: Disc/Disc
Steering: Recirculating ball PA
Wheelbase: 259.1cm 102.0in
Track F: 148.6cm 58.5in
Track R: 148.6cm 58.5in
Length: 447.0cm 176.0in
Width: 177.8cm 70.0in
Height: 177.8cm 70.0in
Ground clearance: 19.1cm
7.5in
Kerb weight: 1903.0kg
4191.6lb
Fuel: 79.6L 17.5gal 21.0galUS

762 Range Rover
County
1990 UK
109.0mph 175.4kmh
0-60mph 96.5kmh: 12.3secs
0-1/4 mile: 18.7secs
178.0bhp 132.7kW 180.5PS
@ 4750rpm
220.0lbft 298.1Nm @ 3250rpm
88.9bhp/ton 87.4bhp/tonne
45.1bhp/L 33.6kW/L 45.7PS/L
36.9ft/sec 11.3m/sec
16.2mpg 13.5mpgUS
17.4L/100km
Petrol 4-stroke piston
3947cc 240.8cu in
Vee 8 fuel injection
Compression ratio: 8.1:1
Bore: 94.0mm 3.7in
Stroke: 71.1mm 2.8in
Valve type/No: Overhead 16
Transmission: Automatic
No. of forward speeds: 4
Wheels driven: 4-wheel drive
Springs F/R: Coil/Coil
Brake system: PA ABS
Brakes F/R: Disc/Disc

763 Range Rover
Vogue SE
1990 UK
110.0mph 177.0kmh
0-50mph 80.5kmh: 8.2secs
0-60mph 96.5kmh: 11.3secs
0-1/4 mile: 18.5secs
0-1km: 34.7secs
185.0bhp 137.9kW 187.6PS
@ 4650rpm
235.4lbft 319.0Nm @ 2600rpm
94.6bhp/ton 93.1bhp/tonne
46.9bhp/L 34.9kW/L 47.5PS/L
36.2ft/sec 11.0m/sec
26.9mph 43.3kmh/1000rpm
14.6mpg 12.2mpgUS
19.3L/100km
Petrol 4-stroke piston
3947cc 241.0cu in
Vee 8 fuel injection
Compression ratio: 9.3:1
Bore: 94.0mm 3.7in
Stroke: 71.1mm 2.8in
Valve type/No: Overhead 16
Transmission: Automatic
No. of forward speeds: 4
Wheels driven: 4-wheel drive
Springs F/R: Coil/Coil
Brake system: PA ABS
Brakes F/R: Disc/Disc
Steering: Recirculating ball PA
Wheelbase: 254.0cm 100.0in
Track F: 148.6cm 58.5in
Track R: 148.6cm 58.5in
Length: 447.0cm 176.0in
Width: 177.8cm 70.0in
Height: 177.8cm 70.0in
Ground clearance: 19.1cm
7.5in
Kerb weight: 1988.0kg
4378.8lb
Fuel: 79.6L 17.5gal
21.0galUS

764 Range Rover
Vogue SE
1992 UK
108.0mph 173.8kmh
0-50mph 80.5kmh: 7.9secs
0-60mph 96.5kmh: 10.8secs
0-1/4 mile: 18.2secs
0-1km: 33.6secs
178.0bhp 132.7kW 180.5PS

@ 4750rpm
220.0lbft 298.1Nm @ 3250rpm
90.1bhp/ton 88.6bhp/tonne
45.1bhp/L 33.6kW/L 45.7PS/L
36.9ft/sec 11.2m/sec
26.9mph 43.3kmh/1000rpm
14.2mpg 11.8mpgUS
19.9L/100km
Petrol 4-stroke piston
3947cc 240.8cu in
Vee 8 fuel injection
Compression ratio: 9.3:1
Bore: 94.0mm 3.7in
Stroke: 71.0mm 2.8in
Valve type/No: Overhead 16
Transmission: Automatic
No. of forward speeds: 4
Wheels driven: 4-wheel drive
Springs F/R: Coil/Coil
Brake system: PA
Brakes F/R: Disc/Disc
Steering: Recirculating ball PA
Wheelbase: 254.0cm 100.0in
Track F: 148.6cm 58.5in
Track R: 148.6cm 58.5in
Length: 447.0cm 176.0in
Width: 177.8cm 70.0in
Height: 177.8cm 70.0in
Ground clearance: 19.1cm
7.5in
Kerb weight: 2009.9kg
4427.0lb
Fuel: 81.9L 18.0gal 21.6galUS

765 Reliant
Scimitar Ti
1986 UK
124.0mph 199.5kmh
0-50mph 80.5kmh: 5.4secs
0-60mph 96.5kmh: 7.2secs
0-1/4 mile: 15.5secs
0-1km: 28.4secs

135.0bhp 100.7kW 136.9PS
@ 6000rpm
143.0lbft 193.8Nm @ 4000rpm
154.4bhp/ton 151.9bhp/tonne
74.6bhp/L 55.6kW/L 75.7PS/L
54.8ft/sec 16.7m/sec
20.6mph 33.1kmh/1000rpm
21.7mpg 18.1mpgUS
13.0L/100km
Petrol 4-stroke piston
1809cc 110.0cu in
turbocharged
In-line 4 fuel injection
Compression ratio: 8.0:1
Bore: 83.0mm 3.3in
Stroke: 83.6mm 3.3in
Valve type/No: Overhead 8
Transmission: Manual
No. of forward speeds: 5
Wheels driven: Rear
Springs F/R: Coil/Coil
Brake system: PA
Brakes F/R: Disc/Drum
Steering: Rack & pinion
Wheelbase: 213.3cm 84.0in
Track F: 130.1cm 51.2in
Track R: 132.0cm 52.0in
Length: 388.6cm 153.0in
Width: 158.2cm 62.3in
Height: 123.9cm 48.8in
Ground clearance: 15.2cm
6.0in
Kerb weight: 889.0kg 1958.1lb
Fuel: 45.5L 10.0gal 12.0galUS

766 Reliant
Scimitar SST 1800Ti
1990 UK
132.0mph 212.4kmh
0-50mph 80.5kmh: 5.2secs
0-60mph 96.5kmh: 7.0secs
0-1/4 mile: 15.5secs

0-1km: 28.3secs
135.0bhp 100.7kW 136.9PS
@ 6000rpm
141.0lbft 191.0Nm @ 4000rpm
147.6bhp/ton 145.2bhp/tonne
74.6bhp/L 55.6kW/L 75.7PS/L
54.8ft/sec 16.7m/sec
20.7mph 33.3kmh/1000rpm
21.7mpg 18.1mpgUS
13.0L/100km
Petrol 4-stroke piston
1809cc 110.0cu in
turbocharged
In-line 4 fuel injection
Compression ratio: 8.1:1
Bore: 83.0mm 3.3in
Stroke: 83.6mm 3.3in
Valve type/No: Overhead 8
Transmission: Manual
No. of forward speeds: 5
Wheels driven: Rear
Springs F/R: Coil/Coil
Brake system: PA
Brakes F/R: Disc/Drum
Steering: Rack & pinion
Wheelbase: 213.4cm 84.0in
Length: 388.6cm 153.0in
Width: 177.8cm 70.0in
Height: 127.0cm 50.0in
Kerb weight: 930.0kg 2048.5lb
Fuel: 45.5L 10.0gal 12.0galUS

767 Renault
21 GTS
1986 France
114.0mph 183.4kmh
0-50mph 80.5kmh: 7.6secs
0-60mph 96.5kmh: 10.9secs
0-1/4 mile: 17.8secs
0-1km: 33.1secs
90.0bhp 67.1kW 91.2PS
@ 5500rpm

102.0lbft 138.2Nm @ 3500rpm
92.3bhp/ton 90.7bhp/tonne
52.3bhp/L 39.0kW/L 53.0PS/L
50.3ft/sec 15.3m/sec
23.9mph 38.5kmh/1000rpm
29.7mpg 24.7mpgUS
9.5L/100km
Petrol 4-stroke piston
1721cc 105.0cu in
In-line 4 1 Carburettor
Compression ratio: 10.0:1
Bore: 81.0mm 3.2in
Stroke: 83.5mm 3.3in
Valve type/No: Overhead 8
Transmission: Manual
No. of forward speeds: 5
Wheels driven: Front
Springs F/R: Coil/Torsion bar
Brake system: PA
Brakes F/R: Disc/Drum
Steering: Rack & pinion
Wheelbase: 265.9cm 104.7in
Track F: 143.0cm 56.3in
Track R: 140.2cm 55.2in
Length: 446.2cm 175.7in
Width: 171.4cm 67.5in
Height: 141.4cm 55.7in
Ground clearance: 15.3cm
6.0in
Kerb weight: 992.0kg 2185.0lb
Fuel: 66.0L 14.5gal 17.4galUS

768 Renault
21 Savanna GTX
1986 France
122.0mph 196.3kmh
0-50mph 80.5kmh: 6.8secs
0-60mph 96.5kmh: 9.8secs
0-1/4 mile: 17.6secs
0-1km: 32.1secs
120.0bhp 89.5kW 121.7PS
@ 5500rpm

766 Reliant
Scimitar SST
1800Ti

124.0lbft 168.0Nm @ 4500rpm
107.0bhp/ton 105.3bhp/tonne
60.1bhp/L 44.8kW/L 61.0PS/L
49.3ft/sec 15.0m/sec
22.8mph 36.7kmh/1000rpm
27.8mpg 23.1mpgUS
10.2L/100km
Petrol 4-stroke piston
1995cc 122.0cu in
In-line 4 fuel injection
Compression ratio: 10.0:1
Bore: 88.0mm 3.5in
Stroke: 82.0mm 3.2in
Valve type/No: Overhead 8
Transmission: Manual
No. of forward speeds: 5
Wheels driven: Front
Springs F/R: Coil/Torsion bar
Brake system: PA
Brakes F/R: Disc/Drum
Steering: Rack & pinion
Wheelbase: 280.9cm 110.6in
Track F: 142.9cm 56.3in
Track R: 140.6cm 55.4in
Length: 464.4cm 182.8in
Width: 171.5cm 67.5in
Height: 142.7cm 56.2in
Ground clearance: 15.3cm
6.0in
Kerb weight: 1140.0kg
2511.0lb
Fuel: 66.0L 14.5gal 17.4galUS

769 Renault

5 GT Turbo
1986 France
126.0mph 202.7kmh
0-50mph 80.5kmh: 5.3secs
0-60mph 96.5kmh: 7.1secs
0-1/4 mile: 15.7secs
0-1km: 29.1secs
115.0bhp 85.8kW 116.6PS

@ 5750rpm
121.0lbft 164.0Nm @ 3000rpm
147.3bhp/ton 144.8bhp/tonne
82.3bhp/L 61.4kW/L 83.5PS/L
48.4ft/sec 14.8m/sec
21.7mph 34.9kmh/1000rpm
22.0mpg 18.3mpgUS
12.8L/100km
Petrol 4-stroke piston
1397cc 85.2cu in
turbocharged
In-line 4 1 Carburettor
Compression ratio: 7.9:1
Bore: 76.0mm 3.0in
Stroke: 77.0mm 3.0in
Valve type/No: Overhead 8
Transmission: Manual
No. of forward speeds: 5
Wheels driven: Front
Springs F/R: Coil/Torsion bar
Brake system: PA
Brakes F/R: Disc/Disc
Steering: Rack & pinion
Wheelbase: 240.8cm 94.8in
Track F: 132.3cm 52.1in
Track R: 131.1cm 51.6in
Length: 358.9cm 141.3in
Width: 159.5cm 62.8in
Height: 136.7cm 53.8in
Ground clearance: 20.3cm
8.0in
Kerb weight: 794.0kg 1749.0lb
Fuel: 50.0L 11.0gal 13.2galUS

770 Renault

5 GTL 5-door
1986 France
99.0mph 159.3kmh
0-50mph 80.5kmh: 9.6secs
0-60mph 96.5kmh: 14.8secs
0-1/4 mile: 19.5secs
0-1km: 37.1secs

60.0bhp 44.7kW 60.8PS
@ 5250rpm
77.0lbft 104.3Nm @ 2500rpm
79.5bhp/ton 78.2bhp/tonne
42.9bhp/L 32.0kW/L 43.5PS/L
44.2ft/sec 13.5m/sec
23.1mph 37.2kmh/1000rpm
35.5mpg 29.6mpgUS
8.0L/100km
Petrol 4-stroke piston
1397cc 85.0cu in
In-line 4 1 Carburettor
Compression ratio: 9.2:1
Bore: 76.0mm 3.0in
Stroke: 77.0mm 3.0in
Valve type/No: Overhead 8
Transmission: Manual
No. of forward speeds: 5
Wheels driven: Front
Springs F/R: Coil/Torsion bar
Brake system: PA
Brakes F/R: Disc/Drum
Steering: Rack & pinion
Wheelbase: 240.7cm 94.8in
Track F: 132.3cm 52.1in
Track R: 128.0cm 50.4in
Length: 359.1cm 141.4in
Width: 158.4cm 62.4in
Height: 139.7cm 55.0in
Ground clearance: 12.0cm
4.7in
Kerb weight: 767.0kg 1689.4lb
Fuel: 43.2L 9.5gal 11.4galUS

771 Renault

9 Turbo
1986 France
116.0mph 186.6kmh
0-50mph 80.5kmh: 6.4secs
0-60mph 96.5kmh: 9.0secs
0-1/4 mile: 16.6secs
0-1km: 31.7secs

105.0bhp 78.3kW 106.5PS
@ 5500rpm
119.0lbft 161.2Nm @ 2500rpm
115.3bhp/ton 113.4bhp/tonne
75.2bhp/L 56.0kW/L 76.2PS/L
46.3ft/sec 14.1m/sec
21.3mph 34.3kmh/1000rpm
29.4mpg 24.5mpgUS
9.6L/100km
Petrol 4-stroke piston
1397cc 85.0cu in
turbocharged
In-line 4 1 Carburettor
Compression ratio: 8.0:1
Bore: 76.0mm 3.0in
Stroke: 77.0mm 3.0in
Valve type/No: Overhead 8
Transmission: Manual
No. of forward speeds: 5
Wheels driven: Front
Springs F/R: Coil/Torsion bar
Brake system: PA
Brakes F/R: Disc/Drum
Steering: Rack & pinion
Wheelbase: 247.7cm 97.5in
Track F: 138.5cm 54.5in
Track R: 134.7cm 53.0in
Length: 406.3cm 160.0in
Width: 166.8cm 65.7in
Height: 140.5cm 55.3in
Ground clearance: 12.0cm
4.7in
Kerb weight: 926.0kg 2039.6lb
Fuel: 46.9L 10.3gal 12.4galUS

772 Renault

GTA V6 Turbo
1986 France
152.0mph 244.6kmh
0-50mph 80.5kmh: 4.8secs
0-60mph 96.5kmh: 6.3secs
0-1/4 mile: 15.3secs

769 Renault
5 GT Turbo

777 Renault
21 Ti

0-1km: 27.4secs	0-1/4 mile: 16.4secs	0-1/4 mile: 16.0secs	0-1/4 mile: 17.0secs
200.0bhp 149.1kW 202.8PS	0-1km: 30.4secs	0-1km: 29.6secs	0-1km: 31.6secs
@ 5750rpm	115.0bhp 85.8kW 116.6PS	120.0bhp 89.5kW 121.7PS	90.0bhp 67.1kW 91.2PS
214.0lbft 290.0Nm @ 2500rpm	@ 5750rpm	@ 5750rpm	@ 5500rpm
171.3bhp/ton 168.5bhp/tonne	121.0lbft 164.0Nm @ 3000rpm	121.0lbft 164.0Nm @ 3750rpm	102.0lbft 138.2Nm @ 3500rpm
81.4bhp/L 60.7kW/L 82.5PS/L	124.7bhp/ton 122.6bhp/tonne	142.7bhp/ton 140.3bhp/tonne	110.9bhp/ton 109.1bhp/tonne
39.6ft/sec 12.1m/sec	82.3bhp/L 61.4kW/L 83.5PS/L	85.9bhp/L 64.0kW/L 87.1PS/L	52.3bhp/L 39.0kW/L 53.0PS/L
26.4mph 42.5kmh/1000rpm	48.4ft/sec 14.8m/sec	48.4ft/sec 14.8m/sec	50.3ft/sec 15.3m/sec
20.5mpg 17.1mpgUS	20.3mph 32.7kmh/1000rpm	21.9mph 35.2kmh/1000rpm	21.8mph 35.1kmh/1000rpm
13.8L/100km	27.3mpg 22.7mpgUS	28.4mpg 23.6mpgUS	30.0mpg 25.0mpgUS
Petrol 4-stroke piston	10.3L/100km	9.9L/100km	9.4L/100km
2458cc 150.0cu in	Petrol 4-stroke piston	Petrol 4-stroke piston	Petrol 4-stroke piston
turbocharged	1397cc 85.0cu in	1397cc 85.0cu in	1721cc 105.0cu in
Vee 6 fuel injection	turbocharged	turbocharged	In-line 4 1 Carburettor
Compression ratio: 8.6:1	In-line 4 1 Carburettor	In-line 4 1 Carburettor	Compression ratio: 10.0:1
Bore: 91.0mm 3.6in	Compression ratio: 7.9:1	Compression ratio: 7.9:1	Bore: 81.0mm 3.2in
Stroke: 63.0mm 2.5in	Bore: 76.0mm 3.0in	Bore: 76.0mm 3.0in	Stroke: 83.5mm 3.3in
Valve type/No: Overhead 12	Stroke: 77.0mm 3.0in	Stroke: 77.0mm 3.0in	Valve type/No: Overhead 8
Transmission: Manual	Valve type/No: Overhead 8	Valve type/No: Overhead 8	Transmission: Manual
No. of forward speeds: 5	Transmission: Manual	Transmission: Manual	No. of forward speeds: 5
Wheels driven: Rear	No. of forward speeds: 5	No. of forward speeds: 5	Wheels driven: Front
Springs F/R: Coil/Coil	Wheels driven: Front	Wheels driven: Front	Springs F/R: Coil/Torsion bar
Brake system: PA	Springs F/R: Coil/Torsion bar	Springs F/R: Coil/Torsion bar	Brake system: PA
Brakes F/R: Disc/Disc	Brake system: PA	Brake system: PA	Brakes F/R: Disc/Drum
Steering: Rack & pinion	Brakes F/R: Disc/Disc	Brakes F/R: Disc/Disc	Steering: Rack & pinion
Wheelbase: 233.9cm 92.1in	Steering: Rack & pinion	Steering: Rack & pinion	Wheelbase: 240.7cm 94.8in
Track F: 149.3cm 58.8in	Wheelbase: 248.4cm 97.8in	Wheelbase: 240.7cm 94.8in	Track F: 132.3cm 52.1in
Track R: 146.2cm 57.6in	Track F: 149.0cm 58.7in	Track F: 132.3cm 52.1in	Track R: 131.0cm 51.6in
Length: 433.0cm 170.5in	Track R: 135.0cm 53.1in	Track R: 131.0cm 51.6in	Length: 358.9cm 141.3in
Width: 175.5cm 69.1in	Length: 397.3cm 156.4in	Length: 358.9cm 141.3in	Width: 159.5cm 62.8in
Height: 119.7cm 47.1in	Width: 163.1cm 64.2in	Width: 159.5cm 62.8in	Height: 136.6cm 53.8in
Ground clearance: 14.0cm	Height: 141.0cm 55.5in	Height: 136.6cm 53.8in	Ground clearance: 20.3cm
5.5in	Ground clearance: 12.0cm	Ground clearance: 20.3cm	8.0in
Kerb weight: 1187.0kg	4.7in	8.0in	Kerb weight: 825.0kg 1817.2lb
2614.5lb	Kerb weight: 938.0kg 2066.1lb	Kerb weight: 855.0kg 1883.3lb	Fuel: 43.2L 9.5gal 11.4galUS
Fuel: 71.9L 15.8gal 19.0galUS	Fuel: 46.9L 10.3gal 12.4galUS	Fuel: 50.0L 11.0gal 13.2galUS	

773 Renault
11 Turbo
1987 France
123.0mph 197.9kmh
0-50mph 80.5kmh: 5.8secs
0-60mph 96.5kmh: 7.9secs

774 Renault
5 GT Turbo
1987 France
122.0mph 196.3kmh
0-50mph 80.5kmh: 5.5secs
0-60mph 96.5kmh: 7.3secs

775 Renault
5 GTX 3-DR
1987 France
112.0mph 180.2kmh
0-50mph 80.5kmh: 6.8secs
0-60mph 96.5kmh: 9.5secs

776 Renault
Alliance GTA
1987 France
110.0mph 177.0kmh
0-50mph 80.5kmh: 8.6secs
0-60mph 96.5kmh: 11.1secs
0-1/4 mile: 18.0secs

95.0bhp 70.8kW 96.3PS
@ 5250rpm
114.0lbft 154.5Nm @ 2750rpm
88.5bhp/ton 87.0bhp/tonne
48.3bhp/L 36.0kW/L 49.0PS/L
53.4ft/sec 16.3m/sec
30.6mpg 25.5mpgUS
9.2L/100km
Petrol 4-stroke piston
1965cc 119.9cu in
In-line 4 fuel injection
Compression ratio: 9.5:1
Bore: 82.0mm 3.2in
Stroke: 93.0mm 3.7in
Valve type/No: Overhead 8
Transmission: Manual
No. of forward speeds: 5
Wheels driven: Front
Springs F/R: Coil/Torsion bar
Brake system: PA
Brakes F/R: Disc/Drum
Steering: Rack & pinion PA
Wheelbase: 248.4cm 97.8in
Track F: 140.2cm 55.2in
Track R: 134.1cm 52.8in
Length: 416.1cm 163.8in
Width: 165.1cm 65.0in
Height: 134.9cm 53.1in
Kerb weight: 1091.9kg
2405.0lb
Fuel: 47.3L 10.4gal 12.5galUS

777 Renault
21 Ti
1988 France
122.0mph 196.3kmh
0-50mph 80.5kmh: 7.3secs
0-60mph 96.5kmh: 10.3secs
0-1/4 mile: 17.7secs
0-1km: 32.3secs
120.0bhp 89.5kW 121.7PS
@ 5500rpm

121.0lbft 164.0Nm @ 4500rpm
110.4bhp/ton 108.6bhp/tonne
60.1bhp/L 44.8kW/L 61.0PS/L
55.3ft/sec 16.9m/sec
22.8mph 36.7kmh/1000rpm
25.7mpg 21.4mpgUS
11.0L/100km
Petrol 4-stroke piston
1995cc 122.0cu in
In-line 4 fuel injection
Compression ratio: 10.0:1
Bore: 88.0mm 3.5in
Stroke: 92.0mm 3.6in
Valve type/No: Overhead 8
Transmission: Manual
No. of forward speeds: 5
Wheels driven: Front
Springs F/R: Coil/Torsion bar
Brake system: PA
Brakes F/R: Disc/Drum
Steering: Rack & pinion PA
Wheelbase: 265.9cm 104.7in
Track F: 143.0cm 56.3in
Track R: 140.2cm 55.2in
Length: 446.2cm 175.7in
Width: 171.4cm 67.5in
Height: 141.4cm 55.7in
Ground clearance: 15.3cm
6.0in
Kerb weight: 1105.0kg
2433.9lb
Fuel: 66.0L 14.5gal 17.4galUS

778 Renault
21 Turbo
1988 France
139.0mph 223.7kmh
0-50mph 80.5kmh: 5.8secs
0-60mph 96.5kmh: 7.9secs
0-1/4 mile: 15.5secs
0-1km: 28.1secs
175.0bhp 130.5kW 177.4PS

@ 5200rpm
199.3lbft 270.0Nm @ 3000rpm
145.2bhp/ton 142.7bhp/tonne
87.7bhp/L 65.4kW/L 88.9PS/L
46.7ft/sec 14.2m/sec
23.9mph 38.5kmh/1000rpm
18.3mpg 15.2mpgUS
15.4L/100km
Petrol 4-stroke piston
1995cc 122.0cu in
turbocharged
In-line 4 fuel injection
Compression ratio: 8.0:1
Bore: 88.0mm 3.5in
Stroke: 82.0mm 3.2in
Valve type/No: Overhead 8
Transmission: Manual
No. of forward speeds: 5
Wheels driven: Front
Springs F/R: Coil/Coil
Brake system: PA ABS
Brakes F/R: Disc/Disc
Steering: Rack & pinion PA
Wheelbase: 259.8cm 102.3in
Track F: 145.3cm 57.2in
Track R: 140.5cm 55.3in
Length: 446.3cm 175.7in
Width: 169.4cm 66.7in
Height: 141.5cm 55.7in
Ground clearance: 15.2cm
6.0in
Kerb weight: 1226.0kg
2700.4lb
Fuel: 66.0L 14.5gal 17.4galUS

779 Renault
GTA V6
1988 France
141.0mph 226.9kmh
0-50mph 80.5kmh: 5.6secs
0-60mph 96.5kmh: 7.5secs
0-1/4 mile: 16.0secs

0-1km: 29.4secs
160.0bhp 119.3kW 162.2PS
@ 5750rpm
163.1lbft 221.0Nm @ 3500rpm
141.9bhp/ton 139.5bhp/tonne
56.2bhp/L 41.9kW/L 56.9PS/L
45.8ft/sec 14.0m/sec
24.4mph 39.3kmh/1000rpm
19.1mpg 15.9mpgUS
14.8L/100km
Petrol 4-stroke piston
2849cc 174.0cu in
Vee 6 2 Carburettor
Compression ratio: 9.5:1
Bore: 91.0mm 3.6in
Stroke: 73.0mm 2.9in
Valve type/No: Overhead 12
Transmission: Manual
No. of forward speeds: 5
Wheels driven: Rear
Springs F/R: Coil/Coil
Brake system: PA
Brakes F/R: Disc/Disc
Steering: Rack & pinion
Wheelbase: 233.0cm 91.7in
Track F: 149.3cm 58.8in
Track R: 146.2cm 57.6in
Length: 433.0cm 170.5in
Width: 175.5cm 69.1in
Height: 119.7cm 47.1in
Ground clearance: 14.0cm
5.5in
Kerb weight: 1147.0kg
2526.4lb
Fuel: 71.9L 15.8gal 19.0galUS

780 Renault
19 TSE
1989 France
108.0mph 173.8kmh
0-50mph 80.5kmh: 9.0secs
0-60mph 96.5kmh: 12.8secs

779 Renault
GTA V6

0-1/4 mile: 18.7secs
0-1km: 34.0secs
80.0bhp 59.7kW 81.1PS
@ 5750rpm
81.2lbft 110.0Nm @ 2750rpm
82.6bhp/ton 81.2bhp/tonne
57.5bhp/L 42.9kW/L 58.3PS/L
48.4ft/sec 14.8m/sec
19.9mpg 32.0kmh/1000rpm
34.9mpg 29.1mpgUS
8.1L/100km
Petrol 4-stroke piston
1390cc 85.0cu in
In-line 4 1 Carburettor
Compression ratio: 9.5:1
Bore: 75.8mm 3.0in
Stroke: 77.0mm 3.0in
Valve type/No: Overhead 8
Transmission: Manual
No. of forward speeds: 5
Wheels driven: Front
Springs F/R: Coil/Torsion bar
Brake system: PA
Brakes F/R: Disc/Drum
Steering: Rack & pinion
Wheelbase: 254.5cm 100.2in
Track F: 141.8cm 55.8in
Track R: 141.7cm 55.8in
Length: 415.5cm 163.6in
Width: 167.6cm 66.0in
Height: 141.2cm 55.6in
Ground clearance: 12.0cm
4.7in
Kerb weight: 985.0kg 2169.6lb
Fuel: 55.1L 12.1gal 14.5galUS

781 Renault
19 TXE
1989 France
112.0mph 180.2kmh
0-50mph 80.5kmh: 8.2secs
0-60mph 96.5kmh: 11.7secs
0-1/4 mile: 18.4secs
0-1km: 33.7secs
92.0bhp 68.6kW 93.3PS
@ 5750rpm
98.9lbft 134.0Nm @ 3000rpm
90.0bhp/ton 88.5bhp/tonne
53.5bhp/L 39.9kW/L 54.2PS/L
52.5ft/sec 16.0m/sec
21.3mph 34.3kmh/1000rpm
26.1mpg 21.7mpgUS
10.8L/100km
Petrol 4-stroke piston
1721cc 105.0cu in
In-line 4 1 Carburettor
Compression ratio: 9.5:1
Bore: 81.0mm 3.2in
Stroke: 83.5mm 3.3in
Valve type/No: Overhead 8
Transmission: Manual
No. of forward speeds: 5
Wheels driven: Front
Springs F/R: Coil/Torsion bar

Brake system: PA
Brakes F/R: Disc/Drum
Steering: Rack & pinion PA
Wheelbase: 254.5cm 100.2in
Track F: 141.8cm 55.8in
Track R: 141.7cm 55.8in
Length: 415.5cm 163.6in
Width: 167.6cm 66.0in
Height: 141.2cm 55.6in
Ground clearance: 12.0cm
4.7in
Kerb weight: 1040.0kg
2290.7lb
Fuel: 55.0L 12.1gal 14.5galUS

782 Renault
21 GTS Hatchback
1989 France
112.0mph 180.2kmh
0-50mph 80.5kmh: 8.3secs
0-60mph 96.5kmh: 11.6secs
0-1/4 mile: 18.3secs
0-1km: 34.0secs
92.0bhp 68.6kW 93.3PS
@ 5750rpm
99.6lbft 135.0Nm @ 3000rpm
86.3bhp/ton 84.9bhp/tonne
53.5bhp/L 39.9kW/L 54.2PS/L
52.5ft/sec 16.0m/sec
23.1mph 37.2kmh/1000rpm
28.1mpg 23.4mpgUS
10.1L/100km
Petrol 4-stroke piston
1721cc 105.0cu in
In-line 4 1 Carburettor
Compression ratio: 9.5:1
Bore: 81.0mm 3.2in
Stroke: 83.5mm 3.3in
Valve type/No: Overhead 8
Transmission: Manual
No. of forward speeds: 5
Wheels driven: Front
Springs F/R: Coil/Torsion bar
Brake system: PA
Brakes F/R: Disc/Drum
Steering: Rack & pinion
Wheelbase: 265.9cm 104.7in
Track F: 143.5cm 56.5in
Track R: 139.9cm 55.1in
Length: 446.0cm 175.6in
Width: 172.4cm 67.9in
Height: 139.9cm 55.1in
Ground clearance: 17.9cm
7.0in
Kerb weight: 1084.0kg
2387.7lb
Fuel: 66.0L 14.5gal 17.4galUS

783 Renault
25 TXi
1989 France
128.0mph 206.0kmh
0-50mph 80.5kmh: 7.4secs
0-60mph 96.5kmh: 10.0secs

0-1/4 mile: 17.3secs
0-1km: 31.8secs
140.0bhp 104.4kW 141.9PS
@ 6000rpm
129.9lbft 176.0Nm @ 4300rpm
103.6bhp/ton 101.9bhp/tonne
70.2bhp/L 52.3kW/L 71.1PS/L
53.8ft/sec 16.4m/sec
21.1mph 33.9kmh/1000rpm
22.5mpg 18.7mpgUS
12.6L/100km
Petrol 4-stroke piston
1995cc 122.0cu in
In-line 4 fuel injection
Compression ratio: 9.3:1
Bore: 88.0mm 3.5in
Stroke: 82.0mm 3.2in
Valve type/No: Overhead 12
Transmission: Manual
No. of forward speeds: 5
Wheels driven: Front
Springs F/R: Coil/Coil
Brake system: PA
Brakes F/R: Disc/Disc
Steering: Rack & pinion PA
Wheelbase: 272.3cm 107.2in
Track F: 150.2cm 59.1in
Track R: 147.9cm 58.2in
Length: 471.3cm 185.6in
Width: 180.6cm 71.1in
Height: 141.5cm 55.7in
Ground clearance: 12.0cm
4.7in
Kerb weight: 1374.0kg
3026.4lb
Fuel: 71.9L 15.8gal 19.0galUS

784 Renault
5 Campus
1989 France
92.0mph 148.0kmh
0-50mph 80.5kmh: 10.8secs
0-60mph 96.5kmh: 16.3secs
0-1/4 mile: 19.5secs
0-1km: 38.0secs
47.0bhp 35.0kW 47.6PS
@ 5250rpm
58.3lbft 79.0Nm @ 2500rpm
63.8bhp/ton 62.7bhp/tonne
42.4bhp/L 31.6kW/L 43.0PS/L
41.3ft/sec 12.6m/sec
18.3mph 29.4kmh/1000rpm
41.0mpg 34.1mpgUS
6.9L/100km
Petrol 4-stroke piston
1108cc 68.0cu in
In-line 4 1 Carburettor
Compression ratio: 9.5:1
Bore: 70.0mm 2.8in
Stroke: 72.0mm 2.8in
Valve type/No: Overhead 8
Transmission: Manual
No. of forward speeds: 4
Wheels driven: Front

Springs F/R: Coil/Coil
Brake system: PA
Brakes F/R: Disc/Drum
Steering: Rack & pinion
Wheelbase: 246.9cm 97.2in
Track F: 132.3cm 52.1in
Track R: 129.0cm 50.8in
Length: 365.0cm 143.7in
Width: 159.0cm 62.6in
Height: 139.0cm 54.7in
Ground clearance: 12.0cm
4.7in
Kerb weight: 749.0kg 1649.8lb
Fuel: 43.2L 9.5gal 11.4galUS

785 Renault
Espace TXE
1989 France
109.0mph 175.4kmh
0-50mph 80.5kmh: 7.7secs
0-60mph 96.5kmh: 10.8secs
0-1/4 mile: 18.0secs
0-1km: 33.5secs
120.0bhp 89.5kW 121.7PS
@ 5500rpm
124.0lbft 168.0Nm @ 4500rpm
97.8bhp/ton 96.1bhp/tonne
60.1bhp/L 44.8kW/L 61.0PS/L
49.3ft/sec 15.0m/sec
21.2mph 34.1kmh/1000rpm
26.7mpg 22.2mpgUS
10.6L/100km
Petrol 4-stroke piston
1995cc 122.0cu in
In-line 4 fuel injection
Compression ratio: 10.0:1
Bore: 88.0mm 3.5in
Stroke: 82.0mm 3.2in
Valve type/No: Overhead 8
Transmission: Manual
No. of forward speeds: 5
Wheels driven: Front
Springs F/R: Coil/Torsion bar
Brake system: PA
Brakes F/R: Disc/Disc
Steering: Rack & pinion PA
Wheelbase: 258.0cm 101.6in
Track F: 146.7cm 57.8in
Track R: 148.9cm 58.6in
Length: 436.5cm 171.9in
Width: 177.7cm 70.0in
Height: 166.0cm 65.4in
Ground clearance: 15.0cm
5.9in
Kerb weight: 1248.0kg
2748.9lb
Fuel: 63.2L 13.9gal 16.7galUS

786 Renault
19 TXE Chamade
1990 France
111.0mph 178.6kmh
0-50mph 80.5kmh: 8.2secs
0-60mph 96.5kmh: 11.4secs

785 Renault
Espace TXE

0-1/4 mile: 18.1secs
0-1km: 30.3secs
92.0bhp 68.6kW 93.3PS
@ 5750rpm
98.9lbft 134.0Nm @ 3000rpm
91.7bhp/ton 90.2bhp/tonne
53.5bhp/L 39.9kW/L 54.2PS/L
52.9ft/sec 16.1m/sec
21.3mph 34.3kmh/1000rpm
30.9mpg 25.7mpgUS
9.1L/100km
Petrol 4-stroke piston
1721cc 105.0cu in
In-line 4 1 Carburettor
Compression ratio: 9.5:1
Bore: 81.0mm 3.2in
Stroke: 84.0mm 3.3in
Valve type/No: Overhead 8
Transmission: Manual
No. of forward speeds: 5
Wheels driven: Front
Springs F/R: Coil/Torsion bar
Brake system: PA
Brakes F/R: Disc/Drum
Steering: Rack & pinion
Wheelbase: 254.0cm 100.0in
Track F: 141.8cm 55.8in
Track R: 140.6cm 55.4in
Length: 426.2cm 167.8in
Width: 168.4cm 66.3in
Height: 141.2cm 55.6in
Ground clearance: 12.0cm
4.7in
Kerb weight: 1020.0kg
2246.7lb
Fuel: 55.1L 12.1gal 14.5galUS

787 Renault
21 Turbo Quadra
1990 France
140.0mph 225.3kmh
0-50mph 80.5kmh: 5.5secs
0-60mph 96.5kmh: 7.8secs

0-1/4 mile: 15.4secs
0-1km: 28.9secs
175.0bhp 130.5kW 177.4PS
@ 5200rpm
199.0lbft 269.6Nm @ 3000rpm
124.0bhp/ton 122.0bhp/tonne
87.7bhp/L 65.4kW/L 88.9PS/L
46.7ft/sec 14.2m/sec
23.9mph 38.5kmh/1000rpm
22.1mpg 18.4mpgUS
12.8L/100km
Petrol 4-stroke piston
1995cc 122.0cu in
turbocharged
In-line 4 fuel injection
Compression ratio: 8.0:1
Bore: 88.0mm 3.5in
Stroke: 82.0mm 3.2in
Valve type/No: Overhead 8
Transmission: Manual
No. of forward speeds: 5
Wheels driven: Front
Springs F/R: Coil/Coil
Brake system: PA ABS
Brakes F/R: Disc/Disc
Steering: Rack & pinion PA
Wheelbase: 259.8cm 102.3in
Track F: 145.3cm 57.2in
Track R: 140.0cm 55.1in
Length: 452.6cm 178.2in
Width: 176.0cm 69.3in
Height: 140.0cm 55.1in
Ground clearance: 15.2cm
6.0in
Kerb weight: 1434.6kg
3160.0lb
Fuel: 61.9L 13.6gal 16.3galUS

788 Renault
21 TXi
1990 France
126.0mph 202.7kmh
0-50mph 80.5kmh: 6.9secs

0-60mph 96.5kmh: 9.8secs
0-1/4 mile: 17.1secs
0-1km: 31.5secs
140.0bhp 104.4kW 141.9PS
@ 6000rpm
129.9lbft 176.0Nm @ 4300rpm
114.8bhp/ton 112.9bhp/tonne
70.2bhp/L 52.3kW/L 71.1PS/L
53.8ft/sec 16.4m/sec
21.1mph 33.9kmh/1000rpm
25.5mpg 21.2mpgUS
11.1L/100km
Petrol 4-stroke piston
1995cc 122.0cu in
In-line 4 fuel injection
Compression ratio: 9.3:1
Bore: 88.0mm 3.5in
Stroke: 82.0mm 3.2in
Valve type/No: Overhead 12
Transmission: Manual
No. of forward speeds: 5
Wheels driven: Front
Springs F/R: Coil/Coil
Brake system: PA ABS
Brakes F/R: Disc/Disc
Steering: Rack & pinion PA
Wheelbase: 265.9cm 104.7in
Track F: 143.5cm 56.5in
Track R: 139.9cm 55.1in
Length: 446.0cm 175.6in
Width: 172.4cm 67.9in
Height: 139.9cm 55.1in
Ground clearance: 17.9cm
7.0in
Kerb weight: 1240.0kg
2731.3lb
Fuel: 66.0L 14.5gal 17.4galUS

789 Renault
25 V6 2.9 Auto
1990 France
129.0mph 207.6kmh
0-50mph 80.5kmh: 7.2secs

0-60mph 96.5kmh: 9.5secs
0-1/4 mile: 17.4secs
0-1km: 31.4secs
160.0bhp 119.3kW 162.2PS
@ 5400rpm
169.7lbft 230.0Nm @ 2500rpm
115.8bhp/ton 113.9bhp/tonne
56.2bhp/L 41.9kW/L 56.9PS/L
43.0ft/sec 13.1m/sec
21.2mph 34.1kmh/1000rpm
21.2mpg 17.7mpgUS
13.3L/100km
Petrol 4-stroke piston
2849cc 174.0cu in
Vee 6 fuel injection
Compression ratio: 9.5:1
Bore: 91.0mm 3.6in
Stroke: 73.0mm 2.9in
Valve type/No: Overhead 24
Transmission: Automatic
No. of forward speeds: 4
Wheels driven: Front
Springs F/R: Coil/Coil
Brake system: PA
Brakes F/R: Disc/Disc
Steering: Rack & pinion PA
Wheelbase: 272.3cm 107.2in
Track F: 150.2cm 59.1in
Track R: 147.9cm 58.2in
Length: 471.3cm 185.6in
Width: 180.6cm 71.1in
Height: 141.5cm 55.7in
Ground clearance: 12.0cm
4.7in
Kerb weight: 1405.0kg
3094.7lb
Fuel: 71.9L 15.8gal 19.0galUS

790 Renault
19 16v
1991 France
130.0mph 209.2kmh
0-50mph 80.5kmh: 7.1secs

0-60mph 96.5kmh: 9.5secs
0-1/4 mile: 17.3secs
0-1km: 31.4secs
137.0bhp 102.2kW 138.9PS
@ 6500rpm
118.8lbft 161.0Nm @ 4250rpm
123.3bhp/ton 121.2bhp/tonne
77.7bhp/L 57.9kW/L 78.7PS/L
59.8ft/sec 18.2m/sec
20.3mph 32.7kmh/1000rpm
28.2mpg 23.5mpgUS
10.0L/100km
Petrol 4-stroke piston
1764cc 108.0cu in
In-line 4 fuel injection
Compression ratio: 10.0:1
Bore: 82.0mm 3.2in
Stroke: 84.0mm 3.3in
Valve type/No: Overhead 16
Transmission: Manual
No. of forward speeds: 5
Wheels driven: Front
Springs F/R: Coil/Torsion bar
Brake system: PA ABS
Brakes F/R: Disc/Disc
Steering: Rack & pinion PA
Wheelbase: 254.0cm 100.0in
Track F: 141.7cm 55.8in
Track R: 141.5cm 55.7in
Length: 415.3cm 163.5in
Width: 167.4cm 65.9in
Height: 141.5cm 55.7in
Ground clearance: 11.9cm
4.7in
Kerb weight: 1130.0kg
2489.0lb
Fuel: 55.1L 12.1gal 14.5galUS

791 Renault
Alpine A610 Turbo
1991 France
165.0mph 265.5kmh
0-60mph 96.5kmh: 5.7secs
250.0bhp 186.4kW 253.5PS
@ 5750rpm
258.0lbft 349.6Nm @ 2900rpm
179.2bhp/ton 176.2bhp/tonne
84.0bhp/L 62.7kW/L 85.2PS/L
45.8ft/sec 14.0m/sec
Petrol 4-stroke piston
2975cc 181.5cu in
turbocharged
Vee 6 fuel injection
Compression ratio: 7.6:1
Bore: 91.0mm 3.6in
Stroke: 73.0mm 2.9in
Valve type/No: Overhead 12
Transmission: Manual
No. of forward speeds: 5
Wheels driven: Rear
Brake system: PA ABS
Brakes F/R: Disc/Disc
Steering: Rack & pinion
Wheelbase: 233.9cm 92.1in

Length: 444.5cm 175.0in
Width: 175.5cm 69.1in
Height: 119.6cm 47.1in
Kerb weight: 1418.7kg
3125.0lb
Fuel: 75.7L 16.6gal 20.0galUS

792 Renault
Clio 1.2 RN
1991 France
100.1mph 161.0kmh
0-50mph 80.5kmh: 9.8secs
0-60mph 96.5kmh: 14.0secs
0-1/4 mile: 19.7secs
0-1km: 36.5secs
60.0bhp 44.7kW 60.8PS
@ 6000rpm
62.7lbft 85.0Nm @ 3500rpm
72.0bhp/ton 70.8bhp/tonne
51.2bhp/L 38.2kW/L 51.9PS/L
42.7ft/sec 13.0m/sec
18.8mph 30.2kmh/1000rpm
36.8mpg 30.6mpgUS
7.7L/100km
Petrol 4-stroke piston
1171cc 71.0cu in
In-line 4 fuel injection
Compression ratio: 9.2:1
Bore: 75.8mm 3.0in
Stroke: 64.9mm 2.6in
Valve type/No: Overhead 8
Transmission: Manual
No. of forward speeds: 5
Wheels driven: Front
Springs F/R: Coil/Torsion bar
Brakes F/R: Disc/Drum
Steering: Rack & pinion
Wheelbase: 247.2cm 97.3in
Track F: 135.8cm 53.5in
Track R: 132.4cm 52.1in
Length: 370.9cm 146.0in
Width: 162.5cm 64.0in
Height: 139.5cm 54.9in
Kerb weight: 847.0kg 1865.6lb
Fuel: 43.0L 9.4gal 11.4galUS

793 Renault
Clio 1.4RT
1991 France
107.0mph 172.2kmh
0-50mph 80.5kmh: 8.2secs
0-60mph 96.5kmh: 12.0secs
0-1/4 mile: 18.7secs
0-1km: 34.5secs
80.0bhp 59.7kW 81.1PS
@ 5750rpm
79.0lbft 107.0Nm @ 3500rpm
91.9bhp/ton 90.4bhp/tonne
57.5bhp/L 42.9kW/L 58.3PS/L
48.4ft/sec 14.8m/sec
22.0mph 35.4kmh/1000rpm
38.3mpg 31.9mpgUS
7.4L/100km
Petrol 4-stroke piston

1390cc 85.0cu in
In-line 4 fuel injection
Compression ratio: 9.5:1
Bore: 76.0mm 3.0in
Stroke: 77.0mm 3.0in
Valve type/No: Overhead 8
Transmission: Manual
No. of forward speeds: 5
Wheels driven: Front
Springs F/R: Coil/Coil
Brake system: PA
Brakes F/R: Disc/Drum
Steering: Rack & pinion PA
Wheelbase: 247.1cm 97.3in
Track F: 135.6cm 53.4in
Track R: 132.3cm 52.1in
Length: 370.8cm 146.0in
Width: 187.7cm 73.9in
Height: 139.4cm 54.9in
Kerb weight: 885.0kg 1949.3lb
Fuel: 43.2L 9.5gal 11.4galUS

794 Renault
Espace V6
1991 France
118.0mph 189.9kmh
0-50mph 80.5kmh: 6.6secs
0-60mph 96.5kmh: 9.4secs
0-1/4 mile: 17.2secs
0-1km: 31.7secs
153.0bhp 114.1kW 155.1PS
@ 5400rpm
166.1lbft 225.0Nm @ 2500rpm
105.8bhp/ton 104.1bhp/tonne
53.7bhp/L 40.0kW/L 54.4PS/L
43.0ft/sec 13.1m/sec
22.6mph 36.4kmh/1000rpm
19.9mpg 16.6mpgUS
14.2L/100km
Petrol 4-stroke piston
2849cc 174.0cu in
Vee 6 fuel injection
Compression ratio: 9.5:1
Bore: 91.0mm 3.6in
Stroke: 73.0mm 2.9in
Valve type/No: Overhead 12
Transmission: Manual
No. of forward speeds: 5
Wheels driven: Front
Springs F/R: Coil/Torsion bar
Brake system: PA ABS
Brakes F/R: Disc/Disc
Steering: Rack & pinion PA
Wheelbase: 258.0cm 101.6in
Track F: 151.7cm 59.7in
Track R: 150.8cm 59.4in
Length: 442.9cm 174.4in
Width: 179.5cm 70.7in
Height: 180.5cm 71.1in
Ground clearance: 15.0cm
5.9in
Kerb weight: 1470.0kg
3237.9lb
Fuel: 77.0L 16.9gal 20.3galUS

795 Renault
19 16v Cabriolet
1992 France
129.0mph 207.6kmh
0-50mph 80.5kmh: 6.6secs
0-60mph 96.5kmh: 8.8secs
0-1/4 mile: 16.9secs
0-1km: 30.8secs
137.0bhp 102.2kW 138.9PS
@ 6500rpm
119.0lbft 161.2Nm @ 4250rpm
118.1bhp/ton 116.1bhp/tonne
77.7bhp/L 57.9kW/L 78.7PS/L
59.8ft/sec 18.2m/sec
20.3mph 32.7kmh/1000rpm
24.9mpg 20.7mpgUS
11.3L/100km
Petrol 4-stroke piston
1764cc 107.6cu in
In-line 4 fuel injection
Compression ratio: 10.0:1
Bore: 82.0mm 3.2in
Stroke: 84.0mm 3.3in
Valve type/No: Overhead 16
Transmission: Manual
No. of forward speeds: 5
Wheels driven: Front
Springs F/R: Coil/Torsion bar
Brake system: PA
Brakes F/R: Disc/Disc
Steering: Rack & pinion
Wheelbase: 254.0cm 100.0in
Track F: 142.7cm 56.2in
Track R: 140.5cm 55.3in
Length: 415.0cm 163.4in
Width: 168.1cm 66.2in
Height: 136.4cm 53.7in
Ground clearance: 11.9cm
4.7in
Kerb weight: 1179.9kg
2599.0lb
Fuel: 55.1L 12.1gal 14.5galUS

796 Renault
Cleo 16v
1992 France
126.8mph 204.0kmh
0-50mph 80.5kmh: 6.3secs
0-60mph 96.5kmh: 8.6secs
0-1/4 mile: 16.7secs
0-1km: 30.4secs
137.0bhp 102.2kW 138.9PS
@ 6500rpm
118.8lbft 161.0Nm @ 4250rpm
136.6bhp/ton 134.3bhp/tonne
77.7bhp/L 57.9kW/L 78.7PS/L
59.4ft/sec 18.1m/sec
19.4mph 31.2kmh/1000rpm
26.0mpg 21.6mpgUS
10.9L/100km
Petrol 4-stroke piston
1764cc 108.0cu in
In-line 4 fuel injection
Compression ratio: 10.0:1

798 Rolls-Royce
Silver Spirit

Bore: 82.0mm 3.2in
Stroke: 83.5mm 3.3in
Valve type/No: Overhead 16
Transmission: Manual
No. of forward speeds: 5
Wheels driven: Front
Springs F/R: Coil/Torsion bar
Brake system: PA ABS
Brakes F/R: Disc/Disc
Steering: Rack & pinion PA
Wheelbase: 247.2cm 97.3in
Track F: 135.8cm 53.5in
Track R: 132.4cm 52.1in
Length: 370.9cm 146.0in
Width: 162.5cm 64.0in
Height: 139.5cm 54.9in
Kerb weight: 1020.0kg
2246.7lb
Fuel: 50.0L 11.0gal 13.2galUS

797 Rolls-Royce
Silver Shadow II
1980 UK
114.0mph 183.4kmh
0-50mph 80.5kmh: 8.1secs
0-60mph 96.5kmh: 11.3secs
0-1/4 mile: 18.4secs
190.0bhp 141.7kW 192.6PS
@ 4000rpm
290.0lbft 393.0Nm @ 2500rpm
85.6bhp/ton 84.2bhp/tonne
28.1bhp/L 21.0kW/L 28.5PS/L
43.3ft/sec 13.2m/sec
25.5mph 41.0kmh/1000rpm
11.4mpg 9.5mpgUS
24.8L/100km
Petrol 4-stroke piston
6750cc 411.8cu in
Vee 8 2 Carburettor
Compression ratio: 7.3:1
Bore: 104.0mm 4.1in
Stroke: 99.1mm 3.9in
Valve type/No: Overhead 16

Transmission: Automatic
No. of forward speeds: 3
Wheels driven: Rear
Springs F/R: Coil/Coil
Brakes F/R: Disc/Disc
Steering: Rack & pinion PA
Wheelbase: 315.2cm 124.1in
Track F: 146.1cm 57.5in
Track R: 146.1cm 57.5in
Length: 537.2cm 211.5in
Width: 182.4cm 71.8in
Height: 151.9cm 59.8in
Kerb weight: 2256.4kg
4970.0lb
Fuel: 85.2L 18.7gal 22.5galUS

798 Rolls-Royce
Silver Spirit
1981 UK
119.0mph 191.5kmh
0-50mph 80.5kmh: 7.3secs
0-60mph 96.5kmh: 10.0secs
0-1/4 mile: 17.1secs
0-1km: 31.9secs
26.2mph 42.2kmh/1000rpm
14.0mpg 11.7mpgUS
20.2L/100km
Petrol 4-stroke piston
6750cc 411.8cu in
Vee 8 2 Carburettor
Compression ratio: 9.0:1
Bore: 104.1mm 4.1in
Stroke: 99.1mm 3.9in
Valve type/No: Overhead 16
Transmission: Automatic
No. of forward speeds: 3
Wheels driven: Rear
Springs F/R: Coil/Coil
Brake system: PA
Brakes F/R: Disc/Disc
Steering: Rack & pinion PA
Wheelbase: 306.1cm 120.5in
Track F: 153.7cm 60.5in

Track R: 153.7cm 60.5in
Length: 526.8cm 207.4in
Width: 188.7cm 74.3in
Height: 148.6cm 58.5in
Ground clearance: 16.5cm
6.5in
Kerb weight: 2227.8kg
4907.0lb
Fuel: 106.9L 23.5gal
28.2galUS

799 Rolls-Royce
Corniche
1982 UK
129.0mph 207.6kmh
0-50mph 80.5kmh: 7.0secs
0-60mph 96.5kmh: 9.7secs
0-1/4 mile: 17.1secs
0-1km: 31.5secs
26.2mph 42.2kmh/1000rpm
12.3mpg 10.2mpgUS
23.0L/100km
Petrol 4-stroke piston
6750cc 411.8cu in
Vee 8 1 Carburettor
Compression ratio: 9.0:1
Bore: 104.1mm 4.1in
Stroke: 99.1mm 3.9in
Valve type/No: Overhead 16
Transmission: Automatic
No. of forward speeds: 3
Wheels driven: Rear
Springs F/R: Coil/Coil
Brake system: PA
Brakes F/R: Disc/Disc
Steering: Rack & pinion PA
Wheelbase: 304.8cm 120.0in
Track F: 152.4cm 60.0in
Track R: 153.9cm 60.6in
Length: 519.4cm 204.5in
Width: 182.9cm 72.0in
Height: 151.6cm 59.7in
Ground clearance: 15.2cm

6.0in
Kerb weight: 2300.0kg
5066.0lb
Fuel: 106.9L 23.5gal
28.2galUS

800 Rolls-Royce
Silver Spur
1982 UK
108.0mph 173.8kmh
0-50mph 80.5kmh: 9.0secs
0-60mph 96.5kmh: 12.6secs
0-1/4 mile: 18.8secs
260.0bhp 193.9kW 263.6PS
@ 4000rpm
380.0lbft 514.9Nm @ 2500rpm
117.1bhp/ton 115.1bhp/tonne
38.5bhp/L 28.7kW/L 39.0PS/L
43.3ft/sec 13.2m/sec
25.0mph 40.2kmh/1000rpm
12.0mpg 10.0mpgUS
23.5L/100km
Petrol 4-stroke piston
6750cc 411.8cu in
Vee 8 fuel injection
Compression ratio: 8.0:1
Bore: 104.0mm 4.1in
Stroke: 99.1mm 3.9in
Valve type/No: Overhead 16
Transmission: Automatic
No. of forward speeds: 3
Wheels driven: Rear
Springs F/R: Coil/Coil
Brakes F/R: Disc/Disc
Steering: Rack & pinion PA
Wheelbase: 316.2cm 124.5in
Track F: 153.9cm 60.6in
Track R: 153.9cm 60.6in
Length: 538.0cm 211.8in
Width: 189.0cm 74.4in
Height: 148.6cm 58.5in
Ground clearance: 14.0cm
5.5in

Kerb weight: 2258.6kg
4975.0lb
Fuel: 107.9L 23.7gal
28.5galUS

801 Rolls-Royce
Silver Spirit II
1991 UK
128.0mph 206.0kmh
0-50mph 80.5kmh: 7.6secs
0-60mph 96.5kmh: 10.4secs
0-1/4 mile: 17.4secs
0-1km: 32.0secs
226.0bhp 168.5kW 229.1PS
@ 4300rpm
339.5lbft 460.0Nm @ 1500rpm
100.4bhp/ton 98.7bhp/tonne
33.5bhp/L 25.0kW/L 33.9PS/L
46.6ft/sec 14.2m/sec
30.0mph 48.3kmh/1000rpm
13.8mpg 11.5mpgUS
20.5L/100km
Petrol 4-stroke piston
6750cc 412.0cu in
Vee 8 fuel injection
Compression ratio: 8.0:1
Bore: 104.0mm 4.1in
Stroke: 99.0mm 3.9in
Valve type/No: Overhead 16
Transmission: Automatic
No. of forward speeds: 3
Wheels driven: Rear
Springs F/R: Coil/Coil
Brake system: PA ABS
Brakes F/R: Disc/Disc
Steering: Rack & pinion PA
Wheelbase: 306.1cm 120.5in
Track F: 153.7cm 60.5in
Track R: 153.7cm 60.5in
Length: 526.8cm 207.4in
Width: 188.7cm 74.3in
Height: 148.6cm 58.5in
Ground clearance: 13.5cm
5.3in
Kerb weight: 2290.0kg
5044.0lb
Fuel: 107.8L 23.7gal
28.5galUS

802 Rover
825i
1986 UK
130.0mph 209.2kmh
0-50mph 80.5kmh: 6.1secs
0-60mph 96.5kmh: 8.0secs
0-1/4 mile: 16.3secs
0-1km: 29.8secs
173.0bhp 129.0kW 175.4PS
@ 6000rpm
160.0lbft 216.8Nm @ 5000rpm
132.9bhp/ton 130.7bhp/tonne
69.4bhp/L 51.7kW/L 70.3PS/L
49.2ft/sec 15.0m/sec
24.5mph 39.4kmh/1000rpm

26.3mpg 21.9mpgUS
10.7L/100km
Petrol 4-stroke piston
2494cc 152.0cu in
Vee 6 fuel injection
Compression ratio: 9.6:1
Bore: 84.0mm 3.3in
Stroke: 75.0mm 2.9in
Valve type/No: Overhead 24
Transmission: Manual
No. of forward speeds: 5
Wheels driven: Front
Springs F/R: Coil/Coil
Brake system: PA
Brakes F/R: Disc/Disc
Steering: Rack & pinion PA
Wheelbase: 275.9cm 108.6in
Track F: 149.0cm 58.7in
Track R: 145.0cm 57.1in
Length: 469.4cm 184.8in
Width: 194.6cm 76.6in
Height: 139.8cm 55.0in
Ground clearance: 14.5cm
5.7in
Kerb weight: 1324.0kg
2916.3lb
Fuel: 68.2L 15.0gal 18.0galUS

803 Rover
820SE
1987 UK
118.0mph 189.9kmh
0-50mph 80.5kmh: 7.5secs
0-60mph 96.5kmh: 10.2secs
0-1/4 mile: 18.5secs
0-1km: 32.1secs
118.0bhp 88.0kW 119.6PS
@ 5600rpm
119.0lbft 161.2Nm @ 3500rpm
94.3bhp/ton 92.7bhp/tonne
59.2bhp/L 44.1kW/L 60.0PS/L
54.4ft/sec 16.6m/sec
21.7mph 34.9kmh/1000rpm
26.6mpg 22.1mpgUS
10.6L/100km
Petrol 4-stroke piston
1994cc 122.0cu in
In-line 4 fuel injection
Compression ratio: 10.0:1
Bore: 84.2mm 3.3in
Stroke: 89.0mm 3.5in
Valve type/No: Overhead 16
Transmission: Manual
No. of forward speeds: 5
Wheels driven: Front
Springs F/R: Coil/Coil
Brake system: PA
Brakes F/R: Disc/Disc
Steering: Rack & pinion PA
Wheelbase: 275.9cm 108.6in
Track F: 149.0cm 58.7in
Track R: 145.0cm 57.1in
Length: 469.4cm 184.8in
Width: 194.6cm 76.6in

Height: 139.8cm 55.0in
Ground clearance: 14.5cm
5.7in
Kerb weight: 1273.0kg
2804.0lb
Fuel: 68.2L 15.0gal 18.0galUS

804 Rover
Montego 1.6L
1987 UK
106.0mph 170.6kmh
0-50mph 80.5kmh: 8.0secs
0-60mph 96.5kmh: 11.5secs
0-1/4 mile: 18.2secs
0-1km: 33.9secs
86.0bhp 64.1kW 87.2PS
@ 5600rpm
97.0lbft 131.4Nm @ 3500rpm
85.7bhp/ton 84.3bhp/tonne
53.8bhp/L 40.1kW/L 54.6PS/L
53.4ft/sec 16.3m/sec
22.2mph 35.7kmh/1000rpm
29.5mpg 24.6mpgUS
9.6L/100km
Petrol 4-stroke piston
1598cc 97.0cu in
In-line 4 1 Carburettor
Compression ratio: 9.7:1
Bore: 76.2mm 3.0in
Stroke: 87.2mm 3.4in
Valve type/No: Overhead 8
Transmission: Manual
No. of forward speeds: 5
Wheels driven: Front
Springs F/R: Coil/Coil
Brake system: PA
Brakes F/R: Disc/Drum
Steering: Rack & pinion
Wheelbase: 256.5cm 101.0in
Track F: 144.0cm 56.7in
Track R: 145.8cm 57.4in
Length: 446.8cm 175.9in
Width: 171.0cm 67.3in
Height: 142.0cm 55.9in
Ground clearance: 15.7cm
6.2in
Kerb weight: 1020.0kg
2246.7lb
Fuel: 51.0L 11.2gal 13.5galUS

805 Rover
Sterling Automatic
1987 UK
127.0mph 204.3kmh
0-50mph 80.5kmh: 7.6secs
0-60mph 96.5kmh: 10.1secs
0-1/4 mile: 17.7secs
0-1km: 32.0secs
165.0bhp 123.0kW 167.3PS
@ 6000rpm
163.0lbft 220.9Nm @ 4000rpm
124.6bhp/ton 122.5bhp/tonne
66.2bhp/L 49.3kW/L 67.1PS/L
49.2ft/sec 15.0m/sec

25.0mph 40.2kmh/1000rpm
21.0mpg 17.5mpgUS
13.5L/100km
Petrol 4-stroke piston
2494cc 152.0cu in
Vee 6 fuel injection
Compression ratio: 9.6:1
Bore: 84.0mm 3.3in
Stroke: 75.0mm 2.9in
Valve type/No: Overhead 24
Transmission: Automatic
No. of forward speeds: 4
Wheels driven: Front
Springs F/R: Coil/Coil
Brake system: PA
Brakes F/R: Disc/Disc
Steering: Rack & pinion PA
Wheelbase: 275.9cm 108.6in
Track F: 149.0cm 58.7in
Track R: 145.0cm 57.1in
Length: 469.4cm 184.8in
Width: 194.6cm 76.6in
Height: 139.8cm 55.0in
Ground clearance: 14.5cm
5.7in
Kerb weight: 1347.0kg
2967.0lb
Fuel: 68.2L 15.0gal 18.0galUS

806 Rover
820 Fastback
1988 UK
115.0mph 185.0kmh
0-50mph 80.5kmh: 8.3secs
0-60mph 96.5kmh: 11.6secs
0-1/4 mile: 18.4secs
0-1km: 33.8secs
100.0bhp 74.6kW 101.4PS
@ 5400rpm
120.3lbft 163.0Nm @ 3000rpm
76.6bhp/ton 75.4bhp/tonne
50.1bhp/L 37.4kW/L 50.8PS/L
52.5ft/sec 16.0m/sec
22.6mph 36.4kmh/1000rpm
24.4mpg 20.3mpgUS
11.6L/100km
Petrol 4-stroke piston
1994cc 122.0cu in
In-line 4 1 Carburettor
Compression ratio: 10.0:1
Bore: 84.5mm 3.3in
Stroke: 89.0mm 3.5in
Valve type/No: Overhead 8
Transmission: Manual
No. of forward speeds: 5
Wheels driven: Front
Springs F/R: Coil/Coil
Brake system: PA
Brakes F/R: Disc/Disc
Steering: Rack & pinion PA
Wheelbase: 275.9cm 108.6in
Track F: 149.0cm 58.7in
Track R: 145.0cm 57.1in
Length: 469.4cm 184.8in

808 Rover
Vitesse

Width: 194.6cm 76.6in
Height: 139.8cm 55.0in
Ground clearance: 14.5cm
5.7in
Kerb weight: 1327.0kg
2922.9lb
Fuel: 68.2L 15.0gal 18.0galUS

Width: 194.6cm 76.6in
Height: 139.8cm 55.0in
Ground clearance: 14.5cm
5.7in
Kerb weight: 1408.0kg
3101.3lb
Fuel: 68.2L 15.0gal 18.0galUS

Width: 194.6cm 76.6in
Height: 139.8cm 55.0in
Ground clearance: 14.5cm
5.7in
Kerb weight: 1433.0kg
3156.4lb
Fuel: 68.2L 15.0gal 18.0galUS

Length: 446.6cm 175.8in
Width: 171.0cm 67.3in
Height: 142.0cm 55.9in
Ground clearance: 15.3cm
6.0in
Kerb weight: 1100.0kg
2422.9lb
Fuel: 50.0L 11.0gal 13.2galUS

807 Rover
827 SLi Auto
1988 UK
132.0mph 212.4kmh
0-50mph 80.5kmh: 6.4secs
0-60mph 96.5kmh: 8.4secs
0-1/4 mile: 16.5secs
0-1km: 29.8secs
177.0bhp 132.0kW 179.4PS
@ 6000rpm
168.3lbft 228.0Nm @ 4500rpm
127.8bhp/ton 125.7bhp/tonne
66.2bhp/L 49.3kW/L 67.1PS/L
49.2ft/sec 15.0m/sec
21.3mph 34.3kmh/1000rpm
17.9mpg 14.9mpgUS
15.8L/100km
Petrol 4-stroke piston
2675cc 163.0cu in
Vee 6 fuel injection
Compression ratio: 9.4:1
Bore: 87.0mm 3.4in
Stroke: 75.0mm 2.9in
Valve type/No: Overhead 24
Transmission: Automatic
No. of forward speeds: 4
Wheels driven: Front
Springs F/R: Coil/Coil
Brake system: PA ABS
Brakes F/R: Disc/Disc
Steering: Rack & pinion PA
Wheelbase: 275.9cm 108.6in
Track F: 149.0cm 58.7in
Track R: 145.0cm 57.1in
Length: 469.4cm 184.8in

808 Rover
Vitesse
1988 UK
134.0mph 215.6kmh
0-50mph 80.5kmh: 7.3secs
0-60mph 96.5kmh: 9.7secs
0-1/4 mile: 17.0secs
0-1km: 30.7secs
177.0bhp 132.0kW 179.4PS
@ 6000rpm
168.3lbft 228.0Nm @ 4500rpm
125.6bhp/ton 123.5bhp/tonne
66.2bhp/L 49.3kW/L 67.1PS/L
49.2ft/sec 15.0m/sec
22.3mph 35.9kmh/1000rpm
19.9mpg 16.6mpgUS
14.2L/100km
Petrol 4-stroke piston
2675cc 163.0cu in
Vee 6 fuel injection
Compression ratio: 9.4:1
Bore: 87.0mm 3.4in
Stroke: 75.0mm 2.9in
Valve type/No: Overhead 24
Transmission: Automatic
No. of forward speeds: 4
Wheels driven: Front
Springs F/R: Coil/Coil
Brake system: PA ABS
Brakes F/R: Disc/Disc
Steering: Rack & pinion PA
Wheelbase: 275.9cm 108.6in
Track F: 149.0cm 58.7in
Track R: 145.0cm 57.1in
Length: 469.4cm 184.8in

809 Rover
Montego 2.0 DSL Turbo
1989 UK
102.0mph 164.1kmh
0-50mph 80.5kmh: 9.2secs
0-60mph 96.5kmh: 13.2secs
0-1/4 mile: 19.1secs
0-1km: 35.1secs
81.0bhp 60.4kW 82.1PS
@ 4500rpm
116.6lbft 158.0Nm @ 2500rpm
74.9bhp/ton 73.6bhp/tonne
40.6bhp/L 30.3kW/L 41.2PS/L
43.8ft/sec 13.3m/sec
28.7mph 46.2kmh/1000rpm
39.7mpg 33.1mpgUS
7.1L/100km
Diesel 4-stroke piston
1994cc 122.0cu in
turbocharged
In-line 4 fuel injection
Compression ratio: 17.2:1
Bore: 84.5mm 3.3in
Stroke: 89.0mm 3.5in
Valve type/No: Overhead 8
Transmission: Manual
No. of forward speeds: 5
Wheels driven: Front
Springs F/R: Coil/Coil
Brake system: PA
Brakes F/R: Disc/Drum
Steering: Rack & pinion PA
Wheelbase: 257.1cm 101.2in
Track F: 144.0cm 56.7in
Track R: 145.8cm 57.4in

810 Rover
Sterling Catalyst
1989 UK
132.0mph 212.4kmh
0-50mph 80.5kmh: 7.0secs
0-60mph 96.5kmh: 9.3secs
0-1/4 mile: 16.9secs
0-1km: 30.7secs
171.0bhp 127.5kW 173.4PS
@ 6000rpm
162.4lbft 220.0Nm @ 4500rpm
118.5bhp/ton 116.6bhp/tonne
63.9bhp/L 47.7kW/L 64.8PS/L
49.2ft/sec 15.0m/sec
21.3mph 34.3kmh/1000rpm
21.3mpg 17.7mpgUS
13.3L/100km
Petrol 4-stroke piston
2675cc 163.0cu in
Vee 6 fuel injection
Compression ratio: 9.0:1
Bore: 87.0mm 3.4in
Stroke: 75.0mm 2.9in
Valve type/No: Overhead 24
Transmission: Automatic
No. of forward speeds: 4
Wheels driven: Front
Springs F/R: Coil/Coil
Brake system: PA ABS
Brakes F/R: Disc/Disc
Steering: Rack & pinion PA
Wheelbase: 275.9cm 108.6in
Track F: 149.0cm 58.7in
Track R: 145.0cm 57.1in

Length: 469.4cm 184.8in
Width: 194.6cm 76.6in
Height: 139.8cm 55.0in
Ground clearance: 14.5cm
5.7in
Kerb weight: 1467.0kg
3231.3lb
Fuel: 68.2L 15.0gal 18.0galUS

811 Rover
Vitesse
1989 UK
138.0mph 222.0kmh
0-50mph 80.5kmh: 6.2secs
0-60mph 96.5kmh: 8.0secs
0-1/4 mile: 16.6secs
0-1km: 29.5secs
177.0bhp 132.0kW 179.4PS
@ 6000rpm
168.3lbft 228.0Nm @ 4500rpm
126.3bhp/ton 124.2bhp/tonne
66.2bhp/L 49.3kW/L 67.1PS/L
49.2ft/sec 15.0m/sec
22.6mph 36.4kmh/1000rpm
22.1mpg 18.4mpgUS
12.8L/100km
Petrol 4-stroke piston
2675cc 163.0cu in
Vee 6 fuel injection
Compression ratio: 9.4:1
Bore: 87.0mm 3.4in
Stroke: 75.0mm 2.9in
Valve type/No: Overhead 24
Transmission: Manual
No. of forward speeds: 5
Wheels driven: Front
Springs F/R: Coil/Coil
Brake system: PA ABS
Brakes F/R: Disc/Disc
Steering: Rack & pinion PA
Wheelbase: 275.9cm 108.6in
Track F: 149.0cm 58.7in
Track R: 145.0cm 57.1in
Length: 469.4cm 184.8in
Width: 194.6cm 76.6in
Height: 139.8cm 55.0in
Ground clearance: 14.5cm
5.7in
Kerb weight: 1425.0kg
3138.8lb
Fuel: 68.2L 15.0gal 18.0galUS

812 Rover
214 GSi
1990 UK
108.0mph 173.8kmh
0-50mph 80.5kmh: 8.3secs
0-60mph 96.5kmh: 11.7secs
0-1/4 mile: 18.3secs
0-1km: 33.9secs
95.0bhp 70.8kW 96.3PS
@ 6250rpm
91.5lbft 124.0Nm @ 4000rpm
87.8bhp/ton 86.4bhp/tonne

68.0bhp/L 50.7kW/L 69.0PS/L
54.0ft/sec 16.5m/sec
19.8mph 31.9kmh/1000rpm
28.9mpg 24.1mpgUS
9.8L/100km
Petrol 4-stroke piston
1396cc 85.0cu in
In-line 4 fuel injection
Compression ratio: 9.5:1
Bore: 75.0mm 2.9in
Stroke: 79.0mm 3.1in
Valve type/No: Overhead 16
Transmission: Manual
No. of forward speeds: 5
Wheels driven: Front
Springs F/R: Coil/Coil
Brake system: PA ABS
Brakes F/R: Disc/Disc
Steering: Rack & pinion PA
Wheelbase: 255.0cm 100.4in
Length: 422.0cm 166.1in
Width: 194.0cm 76.4in
Height: 140.0cm 55.1in
Ground clearance: 15.0cm
5.9in
Kerb weight: 1100.0kg
2422.9lb
Fuel: 55.1L 12.1gal 14.5galUS

813 Rover
216 GSi
1990 UK
117.0mph 188.3kmh
0-50mph 80.5kmh: 7.6secs
0-60mph 96.5kmh: 10.9secs
0-1/4 mile: 18.2secs
0-1km: 33.4secs
114.0bhp 85.0kW 115.6PS
@ 6300rpm
104.1lbft 141.0Nm @ 5200rpm
99.9bhp/ton 98.3bhp/tonne
71.7bhp/L 53.5kW/L 72.7PS/L
61.9ft/sec 18.9m/sec
19.4mph 31.2kmh/1000rpm
26.0mpg 21.6mpgUS
10.9L/100km
Petrol 4-stroke piston
1590cc 97.0cu in
In-line 4 fuel injection
Compression ratio: 9.1:1
Bore: 75.0mm 2.9in
Stroke: 90.0mm 3.5in
Valve type/No: Overhead 16
Transmission: Manual
No. of forward speeds: 5
Wheels driven: Front
Springs F/R: Coil/Coil
Brake system: PA ABS
Brakes F/R: Disc/Disc
Steering: Rack & pinion PA
Wheelbase: 255.0cm 100.4in
Length: 421.6cm 166.0in
Width: 194.3cm 76.5in
Height: 139.7cm 55.0in

Ground clearance: 15.2cm
6.0in
Kerb weight: 1160.0kg
2555.1lb
Fuel: 55.1L 12.1gal 14.5galUS

814 Rover
414 Si
1990 UK
105.0mph 168.9kmh
0-50mph 80.5kmh: 8.2secs
0-60mph 96.5kmh: 11.5secs
0-1/4 mile: 18.5secs
0-1km: 34.2secs
95.0bhp 70.8kW 96.3PS
@ 6250rpm
91.5lbft 124.0Nm @ 4000rpm
92.4bhp/ton 90.9bhp/tonne
68.0bhp/L 50.7kW/L 69.0PS/L
54.0ft/sec 16.5m/sec
19.8mph 31.9kmh/1000rpm
32.7mpg 27.2mpgUS
8.6L/100km
Petrol 4-stroke piston
1396cc 85.0cu in
In-line 4 fuel injection
Compression ratio: 9.5:1
Bore: 75.0mm 2.9in
Stroke: 79.0mm 3.1in
Valve type/No: Overhead 16
Transmission: Manual
No. of forward speeds: 5
Wheels driven: Front
Springs F/R: Coil/Coil
Brake system: PA ABS
Brakes F/R: Disc/Drum
Steering: Rack & pinion
Wheelbase: 254.0cm 100.0in
Length: 421.6cm 166.0in
Width: 194.1cm 76.4in
Height: 140.0cm 55.1in
Ground clearance: 15.2cm
6.0in
Kerb weight: 1045.0kg
2301.8lb
Fuel: 55.1L 12.1gal 14.5galUS

815 Rover
416 GTi
1990 UK
121.0mph 194.7kmh
0-50mph 80.5kmh: 7.2secs
0-60mph 96.5kmh: 10.0secs
0-1/4 mile: 17.2secs
0-1km: 31.6secs
130.0bhp 96.9kW 131.8PS
@ 6800rpm
105.5lbft 143.0Nm @ 5700rpm
117.0bhp/ton 115.0bhp/tonne
81.8bhp/L 61.0kW/L 82.9PS/L
66.9ft/sec 20.4m/sec
18.2mph 29.3kmh/1000rpm
25.1mpg 20.9mpgUS
11.3L/100km

Petrol 4-stroke piston
1590cc 97.0cu in
In-line 4 fuel injection
Compression ratio: 9.5:1
Bore: 75.0mm 2.9in
Stroke: 90.0mm 3.5in
Valve type/No: Overhead 16
Transmission: Manual
No. of forward speeds: 5
Wheels driven: Front
Springs F/R: Coil/Coil
Brake system: PA ABS
Brakes F/R: Disc/Disc
Steering: Rack & pinion PA
Wheelbase: 254.0cm 100.0in
Length: 421.6cm 166.0in
Width: 194.3cm 76.5in
Height: 139.7cm 55.0in
Ground clearance: 15.2cm 6.0in
Kerb weight: 1130.0kg
2489.0lb
Fuel: 55.1L 12.1gal 14.5galUS

816 Rover
Metro 1.4SL
1990 UK
106.0mph 170.6kmh
0-50mph 80.5kmh: 8.5secs
0-60mph 96.5kmh: 12.0secs
0-1/4 mile: 18.2secs
0-1km: 31.4secs
75.0bhp 55.9kW 76.0PS
@ 5500rpm
86.3lbft 117.0Nm @ 3500rpm
89.7bhp/ton 88.2bhp/tonne
53.7bhp/L 40.1kW/L 54.5PS/L
47.5ft/sec 14.5m/sec
19.9mph 32.0kmh/1000rpm
33.7mpg 28.1mpgUS
8.4L/100km
Petrol 4-stroke piston
1396cc 85.0cu in
In-line 4 1 Carburettor
Compression ratio: 9.7:1
Bore: 75.0mm 2.9in
Stroke: 79.0mm 3.1in
Valve type/No: Overhead 8
Transmission: Manual
No. of forward speeds: 5
Wheels driven: Front
Springs F/R: Gas/Gas
Brake system: PA
Brakes F/R: Disc/Disc
Steering: Rack & pinion
Wheelbase: 226.8cm 89.3in
Track F: 134.1cm 52.8in
Track R: 129.3cm 50.9in
Length: 352.0cm 138.6in
Width: 177.3cm 69.8in
Height: 137.7cm 54.2in
Ground clearance: 15.0cm
5.9in
Kerb weight: 850.0kg 1872.2lb
Fuel: 35.5L 7.8gal 9.4galUS

817 Rover
Metro GTi 16v
1990 UK
115.0mph 185.0kmh
0-50mph 80.5kmh: 7.1secs
0-60mph 96.5kmh: 9.8secs
0-1/4 mile: 17.4secs
0-1km: 31.7secs
95.0bhp 70.8kW 96.3PS
@ 6250rpm
91.5lbft 124.0Nm @ 4000rpm
109.2bhp/ton 107.3bhp/tonne
68.0bhp/L 50.7kW/L 69.0PS/L
54.0ft/sec 16.5m/sec
18.9mph 30.4kmh/1000rpm
31.6mpg 26.3mpgUS
8.9L/100km
Petrol 4-stroke piston
1396cc 85.0cu in
In-line 4 fuel injection
Compression ratio: 9.5:1
Bore: 75.0mm 2.9in
Stroke: 79.0mm 3.1in
Valve type/No: Overhead 16
Transmission: Manual
No. of forward speeds: 5
Wheels driven: Front
Springs F/R: Gas/Gas
Brake system: PA
Brakes F/R: Disc/Drum
Steering: Rack & pinion
Wheelbase: 226.9cm 89.3in
Track F: 134.0cm 52.8in
Track R: 129.4cm 50.9in
Length: 352.1cm 138.6in
Width: 177.5cm 69.9in
Height: 137.7cm 54.2in
Ground clearance: 14.9cm
5.9in
Kerb weight: 885.0kg 1949.3lb
Fuel: 35.5L 7.8gal 9.4galUS

818 Rover
214S
1991 UK
102.0mph 164.1kmh
0-50mph 80.5kmh: 9.2secs
0-60mph 96.5kmh: 13.3secs
0-1/4 mile: 19.3secs
0-1km: 35.9secs
75.0bhp 55.9kW 76.0PS
@ 5700rpm
84.1lbft 114.0Nm @ 3500rpm
74.4bhp/ton 73.2bhp/tonne
53.8bhp/L 40.1kW/L 54.5PS/L
49.2ft/sec 15.0m/sec
19.6mph 31.5kmh/1000rpm
30.0mpg 25.0mpgUS
9.4L/100km
Petrol 4-stroke piston
1394cc 85.0cu in
In-line 4 1 Carburettor
Compression ratio: 9.8:1
Bore: 75.0mm 2.9in

Stroke: 79.0mm 3.1in
Valve type/No: Overhead 8
Transmission: Manual
No. of forward speeds: 5
Wheels driven: Front
Springs F/R: Coil/Coil
Brake system: PA
Brakes F/R: Disc/Drum
Steering: Rack & pinion
Wheelbase: 255.0cm 100.4in
Length: 421.6cm 166.0in
Width: 194.3cm 76.5in
Height: 139.7cm 55.0in
Ground clearance: 15.2cm
6.0in
Kerb weight: 1025.0kg
2257.7lb
Fuel: 55.1L 12.1gal 14.5galUS

819 Rover
216 GTi
1991 UK
126.0mph 202.7kmh
0-50mph 80.5kmh: 6.5secs
0-60mph 96.5kmh: 8.8secs
0-1/4 mile: 16.8secs
0-1km: 30.9secs
130.0bhp 96.9kW 131.8PS
@ 6800rpm
105.5lbft 143.0Nm @ 5700rpm
115.0bhp/ton 113.0bhp/tonne
81.8bhp/L 61.0kW/L 82.9PS/L
66.9ft/sec 20.4m/sec
18.4mph 29.6kmh/1000rpm
24.5mpg 20.4mpgUS
11.5L/100km
Petrol 4-stroke piston
1590cc 97.0cu in
In-line 4 fuel injection
Compression ratio: 9.5:1
Bore: 75.0mm 2.9in
Stroke: 90.0mm 3.5in
Valve type/No: Overhead 16
Transmission: Manual
No. of forward speeds: 5
Wheels driven: Front
Springs F/R: Coil/Coil
Brake system: PA ABS
Brakes F/R: Disc/Disc
Steering: Rack & pinion PA
Wheelbase: 255.0cm 100.4in
Length: 421.9cm 166.1in
Width: 194.1cm 76.4in
Height: 140.0cm 55.1in
Kerb weight: 1150.0kg
2533.0lb
Fuel: 55.1L 12.1gal 14.5galUS

820 Rover
218 SD
1991 UK
97.0mph 156.1kmh
0-50mph 80.5kmh: 12.0secs
0-60mph 96.5kmh: 16.9secs

0-1/4 mile: 20.9secs
0-1km: 38.9secs
66.0bhp 49.2kW 66.9PS
@ 4600rpm
89.3lbft 121.0Nm @ 2500rpm
55.9bhp/ton 55.0bhp/tonne
34.6bhp/L 25.8kW/L 35.1PS/L
44.2ft/sec 13.5m/sec
22.8mph 36.7kmh/1000rpm
35.3mpg 29.4mpgUS
8.0L/100km
Diesel 4-stroke piston
1905cc 116.0cu in
In-line 4 fuel injection
Compression ratio: 23.0:1
Bore: 83.0mm 3.3in
Stroke: 88.0mm 3.5in
Valve type/No: Overhead 8
Transmission: Manual
No. of forward speeds: 5
Wheels driven: Front
Springs F/R: Coil/Coil
Brake system: PA ABS
Brakes F/R: Disc/Drum
Steering: Rack & pinion PA
Wheelbase: 255.0cm 100.4in
Length: 422.0cm 166.1in
Width: 194.0cm 76.4in
Height: 140.0cm 55.1in
Kerb weight: 1200.0kg
2643.2lb
Fuel: 55.1L 12.1gal 14.5galUS

821 Rover
220 GTi
1991 UK
128.0mph 206.0kmh
0-50mph 80.5kmh: 6.5secs
0-60mph 96.5kmh: 8.8secs
0-1/4 mile: 16.9secs
0-1km: 30.8secs
140.0bhp 104 4kW 141.9PS
@ 6000rpm
132.8lbft 180.0Nm @ 4500rpm
121.7bhp/ton 119.7bhp/tonne
70.2bhp/L 52.4kW/L 71.2PS/L
58.3ft/sec 17.8m/sec
20.7mph 33.3kmh/1000rpm
27.1mpg 22.6mpgUS
10.4L/100km
Petrol 4-stroke piston
1994cc 122.0cu in
In-line 4 fuel injection
Compression ratio: 10.0:1
Bore: 84.5mm 3.3in
Stroke: 89.0mm 3.5in
Valve type/No: Overhead 16
Transmission: Manual
No. of forward speeds: 5
Wheels driven: Front
Springs F/R: Coil/Coil
Brake system: PA ABS
Brakes F/R: Disc/Disc
Steering: Rack & pinion PA

Wheelbase: 255.0cm 100.4in
Length: 422.0cm 166.1in
Width: 194.0cm 76.4in
Height: 140.0cm 55.1in
Kerb weight: 1170.0kg
2577.1lb
Fuel: 55.1L 12.1gal 14.5galUS

822 Rover
825 TD
1991 UK
117.0mph 188.3kmh
0-50mph 80.5kmh: 8.4secs
0-60mph 96.5kmh: 11.7secs
0-1/4 mile: 18.5secs
0-1km: 34.1secs
118.0bhp 88.0kW 119.6PS
@ 4200rpm
197.8lbft 268.0Nm @ 2100rpm
82.5bhp/ton 81.1bhp/tonne
47.2bhp/L 35.2kW/L 47.9PS/L
43.2ft/sec 13.2m/sec
28.8mph 46.3kmh/1000rpm
36.2mpg 30.1mpgUS
7.8L/100km
Diesel 4-stroke piston
2500cc 153.0cu in
turbocharged
In-line 4 fuel injection
Compression ratio: 22.1:1
Bore: 92.0mm 3.6in
Stroke: 94.0mm 3.7in
Valve type/No: Overhead 8
Transmission: Manual
No. of forward speeds: 5
Wheels driven: Front
Springs F/R: Coil/Coil
Brake system: PA
Brakes F/R: Disc/Disc
Steering: Rack & pinion PA
Wheelbase: 275.8cm 108.6in
Track F: 149.1cm 58.7in
Track R: 145.0cm 57.1in
Length: 481.1cm 189.4in
Width: 189.0cm 74.4in
Height: 139.7cm 55.0in
Ground clearance: 14.5cm
5.7in
Kerb weight: 1455.0kg
3204.8lb
Fuel: 68.2L 15.0gal 18.0galUS

823 Rover
Mini Cooper S
1991 UK
98.0mph 157.7kmh
0-50mph 80.5kmh: 7.6secs
0-60mph 96.5kmh: 11.0secs
0-1/4 mile: 18.1secs
0-1km: 29.3secs
78.0bhp 58.2kW 79.1PS
@ 6000rpm
78.2lbft 106.0Nm @ 3250rpm
112.4bhp/ton 110.5bhp/tonne

61.2bhp/L 45.6kW/L 62.0PS/L
53.3ft/sec 16.3m/sec
18.6mph 29.9kmh/1000rpm
26.5mpg 22.1mpgUS
10.7L/100km
Petrol 4-stroke piston
1275cc 78.0cu in
In-line 4 2 Carburettor
Compression ratio: 10.2:1
Bore: 70.6mm 2.8in
Stroke: 81.3mm 3.2in
Valve type/No: Overhead 8
Transmission: Manual
No. of forward speeds: 4
Wheels driven: Front
Springs F/R: Coil/Coil
Brake system: PA
Brakes F/R: Disc/Drum
Steering: Rack & pinion
Wheelbase: 203.5cm 80.1in
Track F: 120.7cm 47.5in
Track R: 116.8cm 46.0in
Length: 310.4cm 122.2in
Width: 141.0cm 55.5in
Height: 135.1cm 53.2in
Ground clearance: 15.2cm
6.0in
Kerb weight: 706.0kg 1555.1lb
Fuel: 34.1L 7.5gal 9.0galUS

824 Rover
820i
1992 UK
126.0mph 202.7kmh
0-50mph 80.5kmh: 6.9secs
0-60mph 96.5kmh: 9.2secs
0-1/4 mile: 17.2secs
0-1km: 31.2secs
136.0bhp 101.4kW 137.9PS
@ 6000rpm
136.0lbft 184.3Nm @ 2500rpm
102.1bhp/ton 100.4bhp/tonne
68.2bhp/L 50.9kW/L 69.1PS/L
58.3ft/sec 17.8m/sec
22.5mph 36.2kmh/1000rpm
27.4mpg 22.8mpgUS
10.3L/100km

Petrol 4-stroke piston
1994cc 122.0cu in
In-line 4 fuel injection
Compression ratio: 10.0:1
Bore: 84.5mm 3.3in
Stroke: 89.0mm 3.5in
Valve type/No: Overhead 16
Transmission: Manual
No. of forward speeds: 5
Wheels driven: Front
Springs F/R: Coil/Coil
Brake system: PA ABS
Brakes F/R: Disc/Disc
Steering: Rack & pinion PA
Wheelbase: 276.6cm 108.9in
Track F: 148.3cm 58.4in
Track R: 145.0cm 57.1in
Length: 488.2cm 192.2in
Width: 195.5cm 77.0in
Height: 139.3cm 54.8in
Ground clearance: 14.2cm
5.6in
Kerb weight: 1355.0kg
2984.6lb
Fuel: 68.0L 14.9gal 18.0galUS

825 Rover
Sterling Automatic
1992 UK
131.0mph 210.8kmh
0-50mph 80.5kmh: 7.2secs
0-60mph 96.5kmh: 9.4secs
0-1/4 mile: 17.4secs
0-1km: 31.4secs
169.0bhp 126.0kW 171.3PS
@ 5900rpm
165.0lbft 223.6Nm @ 4500rpm
118.9bhp/ton 117.0bhp/tonne
63.2bhp/L 47.1kW/L 64.0PS/L
48.3ft/sec 14.7m/sec
21.3mph 34.3kmh/1000rpm
19.6mpg 16.3mpgUS
14.4L/100km
Petrol 4-stroke piston
2675cc 163.2cu in
Vee 6 fuel injection
Compression ratio: 9.0:1

Bore: 87.0mm 3.4in
Stroke: 75.0mm 2.9in
Valve type/No: Overhead 24
Transmission: Automatic
No. of forward speeds: 4
Wheels driven: Front
Springs F/R: Coil/Coil
Brake system: PA ABS
Brakes F/R: Disc/Disc
Steering: Rack & pinion PA
Wheelbase: 277.0cm 109.1in
Track F: 160.0cm 63.0in
Track R: 157.2cm 61.9in
Length: 488.0cm 192.1in
Width: 188.5cm 74.2in
Height: 139.0cm 54.7in
Kerb weight: 1445.0kg
3182.8lb
Fuel: 68.2L 15.0gal 18.0galUS

826 Saab
9000
1986 Sweden
124.0mph 199.5kmh
0-60mph 96.5kmh: 8.6secs
0-1/4 mile: 15.6secs
160.0bhp 119.3kW 162.2PS
@ 5500rpm
188.0lbft 254.7Nm @ 3000rpm
119.3bhp/ton 117.3bhp/tonne
80.6bhp/L 60.1kW/L 81.7PS/L
46.9ft/sec 14.3m/sec
25.2mpg 21.0mpgUS
11.2L/100km
Petrol 4-stroke piston
1985cc 121.1cu in
In-line 4 fuel injection
Compression ratio: 9.0:1
Bore: 90.0mm 3.5in
Stroke: 78.0mm 3.1in
Valve type/No: Overhead 16
Transmission: Manual
No. of forward speeds: 5
Wheels driven: Front
Springs F/R: Coil/Coil
Brake system: PA
Brakes F/R: Disc/Disc

Steering: Rack & pinion PA
Wheelbase: 267.2cm 105.2in
Track F: 152.1cm 59.9in
Track R: 149.1cm 58.7in
Length: 462.0cm 181.9in
Width: 176.3cm 69.4in
Height: 142.0cm 55.9in
Ground clearance: 16.5cm
6.5in
Kerb weight: 1364.3kg
3005.0lb
Fuel: 67.7L 14.9gal 17.9galUS

827 Saab
9000i
1986 Sweden
121.0mph 194.7kmh
0-50mph 80.5kmh: 7.0secs
0-60mph 96.5kmh: 9.6secs
0-1/4 mile: 17.9secs
0-1km: 32.7secs
130.0bhp 96.9kW 131.8PS
@ 5500rpm
127.0lbft 172.1Nm @ 3000rpm
100.8bhp/ton 99.2bhp/tonne
65.5bhp/L 48.8kW/L 66.4PS/L
46.9ft/sec 14.3m/sec
23.0mph 37.0kmh/1000rpm
25.7mpg 21.4mpgUS
11.0L/100km
Petrol 4-stroke piston
1985cc 121.0cu in
In-line 4 fuel injection
Compression ratio: 10.0:1
Bore: 90.0mm 3.5in
Stroke: 78.0mm 3.1in
Valve type/No: Overhead 16
Transmission: Manual
No. of forward speeds: 5
Wheels driven: Front
Springs F/R: Coil/Coil
Brake system: PA
Brakes F/R: Disc/Disc
Steering: Rack & pinion PA
Wheelbase: 267.2cm 105.2in
Track F: 152.2cm 59.9in
Track R: 149.2cm 58.7in
Length: 462.0cm 181.9in
Width: 176.4cm 69.4in
Height: 142.0cm 55.9in
Ground clearance: 15.0cm
5.9in
Kerb weight: 1311.0kg
2887.7lb
Fuel: 68.2L 15.0gal 18.0galUS

828 Saab
9000CD
1988 Sweden
140.0mph 225.3kmh
0-50mph 80.5kmh: 5.5secs
0-60mph 96.5kmh: 6.9secs
0-1/4 mile: 15.8secs
0-1km: 28.7secs

825 Rover
Sterling Automatic

175.0bhp 130.5kW 177.4PS
@ 5300rpm
201.5lbft 273.0Nm @ 3000rpm
133.5bhp/ton 131.3bhp/tonne
87.7bhp/L 65.4kW/L 88.9PS/L
45.2ft/sec 13.8m/sec
25.3mph 40.7kmh/1000rpm
20.3mpg 16.9mpgUS
13.9L/100km
Petrol 4-stroke piston
1995cc 122.0cu in
turbocharged
In-line 4 fuel injection
Compression ratio: 9.0:1
Bore: 90.0mm 3.5in
Stroke: 78.0mm 3.1in
Valve type/No: Overhead 16
Transmission: Manual
No. of forward speeds: 5
Wheels driven: Front
Springs F/R: Coil/Coil
Brake system: PA ABS
Brakes F/R: Disc/Disc
Steering: Rack & pinion PA
Wheelbase: 267.2cm 105.2in
Track F: 152.2cm 59.9in
Track R: 149.2cm 58.7in
Length: 478.0cm 188.2in
Width: 176.4cm 69.4in
Height: 142.0cm 55.9in
Ground clearance: 15.0cm
5.9in
Kerb weight: 1333.0kg
2936.1lb
Fuel: 67.8L 14.9gal 17.9galUS

829 Saab
900 Turbo S
1989 Sweden
124.0mph 199.5kmh
0-50mph 80.5kmh: 6.1secs
0-60mph 96.5kmh: 8.9secs
155.0bhp 115.6kW 157.1PS
@ 5000rpm
177.1lbft 240.0Nm @ 3000rpm
125.8bhp/ton 123.7bhp/tonne
78.1bhp/L 58.2kW/L 79.2PS/L
42.6ft/sec 13.0m/sec
22.6mph 36.4kmh/1000rpm
19.1mpg 15.9mpgUS
14.8L/100km
Petrol 4-stroke piston
1985cc 121.0cu in
turbocharged
In-line 4 fuel injection
Compression ratio: 8.5:1
Bore: 90.0mm 3.5in
Stroke: 78.0mm 3.1in
Valve type/No: Overhead 8
Transmission: Manual
No. of forward speeds: 5
Wheels driven: Front
Springs F/R: Coil/Coil
Brake system: PA

829 Saab
900 Turbo S

Brakes F/R: Disc/Disc
Steering: Rack & pinion PA
Wheelbase: 251.7cm 99.1in
Track F: 142.0cm 55.9in
Track R: 143.0cm 56.3in
Length: 468.7cm 184.5in
Width: 169.0cm 66.5in
Height: 142.0cm 55.9in
Ground clearance: 15.0cm
5.9in
Kerb weight: 1253.0kg
2759.9lb
Fuel: 63.2L 13.9gal 16.7galUS

830 Saab
9000 Turbo
1989 Sweden
135.0mph 217.2kmh
0-50mph 80.5kmh: 6.0secs
0-60mph 96.5kmh: 7.7secs
0-1/4 mile: 16.2secs
160.0bhp 119.3kW 162.2PS
@ 5500rpm
188.0lbft 254.7Nm @ 3000rpm
114.9bhp/ton 113.0bhp/tonne
80.6bhp/L 60.1kW/L 81.7PS/L
46.9ft/sec 14.3m/sec
25.2mph 21.0mpgUS
11.2L/100km
Petrol 4-stroke piston
1985cc 121.1cu in
turbocharged
In-line 4 fuel injection
Compression ratio: 9.0:1
Bore: 90.0mm 3.5in
Stroke: 78.0mm 3.1in
Valve type/No: Overhead 16
Transmission: Manual
No. of forward speeds: 5
Wheels driven: Front
Springs F/R: Coil/Coil
Brake system: PA ABS
Brakes F/R: Disc/Disc
Steering: Rack & pinion PA
Wheelbase: 267.0cm 105.1in
Track F: 152.1cm 59.9in
Track R: 149.1cm 58.7in
Length: 462.0cm 181.9in

Width: 176.3cm 69.4in
Height: 142.7cm 56.2in
Kerb weight: 1416.5kg
3120.0lb
Fuel: 67.7L 14.9gal 17.9galUS

831 Saab
Carlsson
1989 Sweden
148.0mph 238.1kmh
0-50mph 80.5kmh: 5.5secs
0-60mph 96.5kmh: 7.3secs
0-1/4 mile: 15.8secs
0-1km: 28.1secs
204.0bhp 152.1kW 206.8PS
@ 6000rpm
214.0lbft 290.0Nm @ 3000rpm
152.5bhp/ton 150.0bhp/tonne
102.8bhp/L 76.6kW/L
104.2PS/L
51.2ft/sec 15.6m/sec
24.2mph 38.9kmh/1000rpm
21.6mpg 18.0mpgUS
13.1L/100km
Petrol 4-stroke piston
1985cc 121.0cu in
turbocharged
In-line 4 fuel injection
Compression ratio: 9.0:1
Bore: 90.0mm 3.5in
Stroke: 78.0mm 3.1in
Valve type/No: Overhead 16
Transmission: Manual
No. of forward speeds: 5
Wheels driven: Front
Springs F/R: Coil/Coil
Brake system: PA ABS
Brakes F/R: Disc/Disc
Steering: Rack & pinion PA
Wheelbase: 267.2cm 105.2in
Track F: 152.2cm 59.9in
Track R: 149.2cm 58.7in
Length: 462.0cm 181.9in
Width: 176.4cm 69.4in
Height: 142.0cm 55.9in
Ground clearance: 15.0cm
5.9in
Kerb weight: 1360.0kg

2995.6lb
Fuel: 67.8L 14.9gal 17.9galUS

832 Saab
9000CD
1990 Sweden
125.0mph 201.1kmh
0-60mph 96.5kmh: 9.7secs
0-1/4 mile: 17.1secs
150.0bhp 111.9kW 152.1PS
@ 5500rpm
156.0lbft 211.4Nm @ 3800rpm
108.9bhp/ton 107.1bhp/tonne
65.5bhp/L 48.8kW/L 66.4PS/L
54.1ft/sec 16.5m/sec
25.2mpg 21.0mpgUS
11.2L/100km
Petrol 4-stroke piston
2290cc 139.7cu in
In-line 4 fuel injection
Compression ratio: 10.0:1
Bore: 90.0mm 3.5in
Stroke: 90.0mm 3.5in
Valve type/No: Overhead 16
Transmission: Manual
No. of forward speeds: 5
Wheels driven: Front
Springs F/R: Coil/Coil
Brake system: PA ABS
Brakes F/R: Disc/Disc
Steering: Rack & pinion PA
Wheelbase: 267.2cm 105.2in
Track F: 152.1cm 59.9in
Track R: 149.1cm 58.7in
Length: 478.0cm 188.2in
Width: 176.3cm 69.4in
Height: 142.0cm 55.9in
Kerb weight: 1400.6kg
3085.0lb
Fuel: 65.9L 14.5gal 17.4galUS

833 Saab
900SPG
1990 Sweden
125.0mph 201.1kmh
0-50mph 80.5kmh: 6.4secs
0-60mph 96.5kmh: 9.1secs
0-1/4 mile: 16.7secs

175.0bhp 130.5kW 177.4PS
@ 5500rpm
195.0lbft 264.2Nm @ 3000rpm
132.9bhp/ton 130.7bhp/tonne
88.2bhp/L 65.7kW/L 89.4PS/L
46.9ft/sec 14.3m/sec
25.7mpg 21.4mpgUS
11.0L/100km
Petrol 4-stroke piston
1985cc 121.1cu in
turbocharged
In-line 4 fuel injection
Compression ratio: 9.0:1
Bore: 89.9mm 3.5in
Stroke: 77.9mm 3.1in
Valve type/No: Overhead 16
Transmission: Manual
No. of forward speeds: 5
Wheels driven: Front
Springs F/R: Coil/Coil
Brake system: PA ABS
Brakes F/R: Disc/Disc
Steering: Rack & pinion PA
Wheelbase: 251.7cm 99.1in
Track F: 145.5cm 57.3in
Track R: 146.6cm 57.7in
Length: 468.6cm 184.5in
Width: 169.4cm 66.7in
Height: 140.5cm 55.3in
Kerb weight: 1339.3kg
2950.0lb
Fuel: 68.1L 15.0gal 18.0galUS

834 Saab

CDS 2.3
1990 Sweden
127.0mph 204.3kmh
0-50mph 80.5kmh: 6.8secs
0-60mph 96.5kmh: 9.3secs
0-1/4 mile: 16.9secs
0-1km: 30.9secs
150.0bhp 111.9kW 152.1PS
@ 5500rpm
156.5lbft 212.0Nm @ 3800rpm
109.7bhp/ton 107.9bhp/tonne
65.5bhp/L 48.8kW/L 66.4PS/L
54.1ft/sec 16.5m/sec
22.8mph 36.7kmh/1000rpm
22.0mpg 18.3mpgUS
12.8L/100km
Petrol 4-stroke piston
2290cc 140.0cu in
In-line 4 fuel injection
Compression ratio: 10.0:1
Bore: 90.0mm 3.5in
Stroke: 90.0mm 3.5in
Valve type/No: Overhead 16
Transmission: Manual
No. of forward speeds: 5
Wheels driven: Front
Springs F/R: Coil/Coil
Brake system: PA ABS
Brakes F/R: Disc/Disc
Steering: Rack & pinion PA

Wheelbase: 267.2cm 105.2in
Track F: 152.2cm 59.9in
Track R: 149.2cm 58.7in
Length: 478.0cm 188.2in
Width: 176.4cm 69.4in
Height: 142.0cm 55.9in
Ground clearance: 15.0cm
5.9in
Kerb weight: 1390.0kg
3061.7lb
Fuel: 66.0L 14.5gal 17.4galUS

835 Saab

9000 2.3 Turbo TCS
1991 Sweden
142.0mph 228.5kmh
0-50mph 80.5kmh: 5.5secs
0-60mph 96.5kmh: 7.5secs
0-1/4 mile: 15.5secs
0-1km: 27.7secs
200.0bhp 149.1kW 202.8PS
@ 5000rpm
243.5lbft 330.0Nm @ 2000rpm
146.8bhp/ton 144.4bhp/tonne
87.3bhp/L 65.1kW/L 88.5PS/L
49.2ft/sec 15.0m/sec
24.0mph 38.6kmh/1000rpm
19.7mpg 16.4mpgUS
14.3L/100km
Petrol 4-stroke piston
2290cc 140.0cu in
turbocharged
In-line 4 fuel injection
Compression ratio: 8.5:1
Bore: 90.0mm 3.5in
Stroke: 90.0mm 3.5in
Valve type/No: Overhead 16
Transmission: Manual
No. of forward speeds: 5
Wheels driven: Front
Springs F/R: Coil/Coil
Brake system: PA ABS
Brakes F/R: Disc/Disc
Steering: Rack & pinion PA
Wheelbase: 266.7cm 105.0in
Track F: 152.1cm 59.9in
Track R: 149.1cm 58.7in
Length: 477.8cm 188.1in
Width: 176.3cm 69.4in
Height: 142.0cm 55.9in
Ground clearance: 15.2cm
6.0in
Kerb weight: 1385.0kg
3050.7lb
Fuel: 62.3L 13.7gal 16.5galUS

836 Saab

9000 Turbo
1991 Sweden
135.0mph 217.2kmh
0-60mph 96.5kmh: 6.8secs
0-1/4 mile: 15.4secs
200.0bhp 149.1kW 202.8PS
@ 5000rpm

244.0lbft 330.6Nm @ 2000rpm
141.1bhp/ton 138.7bhp/tonne
87.3bhp/L 65.1kW/L 88.5PS/L
49.2ft/sec 15.0m/sec
27.6mpg 23.0mpgUS
10.2L/100km
Petrol 4-stroke piston
2290cc 139.7cu in
turbocharged
In-line 4 fuel injection
Compression ratio: 8.5:1
Bore: 90.0mm 3.5in
Stroke: 90.0mm 3.5in
Valve type/No: Overhead 16
Transmission: Manual
No. of forward speeds: 5
Wheels driven: Front
Springs F/R: Coil/Coil
Brake system: PA ABS
Brakes F/R: Disc/Disc
Steering: Rack & pinion PA
Wheelbase: 267.2cm 105.2in
Track F: 149.6cm 58.9in
Track R: 149.1cm 58.7in
Length: 466.6cm 183.7in
Width: 176.3cm 69.4in
Height: 142.0cm 55.9in
Kerb weight: 1441.4kg
3175.0lb
Fuel: 62.1L 13.6gal 16.4galUS

837 Saab

9000CS Turbo S
1991 Sweden
138.6mph 223.0kmh
0-50mph 80.5kmh: 6.2secs
0-60mph 96.5kmh: 8.0secs
0-1/4 mile: 16.3secs
0-1km: 29.2secs
200.0bhp 149.1kW 202.8PS
@ 5000rpm
245.0lbft 332.0Nm @ 2000rpm
143.2bhp/ton 140.8bhp/tonne
87.3bhp/L 65.1kW/L 88.5PS/L
49.2ft/sec 15.0m/sec
21.6mph 34.8kmh/1000rpm
22.2mpg 18.5mpgUS
12.7L/100km
Petrol 4-stroke piston
2290cc 140.0cu in
turbocharged
In-line 4 fuel injection
Compression ratio: 8.5:1
Bore: 90.0mm 3.5in
Stroke: 90.0mm 3.5in
Valve type/No: Overhead 16
Transmission: Manual
No. of forward speeds: 5
Wheels driven: Front
Springs F/R: Coil/Coil
Brake system: PA ABS
Brakes F/R: Disc/Disc
Steering: Rack & pinion PA
Wheelbase: 266.2cm 104.8in

Track F: 152.2cm 59.9in
Track R: 149.2cm 58.7in
Length: 476.1cm 187.4in
Width: 176.4cm 69.4in
Height: 142.0cm 55.9in
Ground clearance: 15.0cm
5.9in
Kerb weight: 1420.0kg
3127.7lb
Fuel: 66.0L 14.5gal 17.4galUS

838 Saab

9000 2.3CS
1992 Sweden
129.0mph 207.6kmh
0-50mph 80.5kmh: 7.0secs
0-60mph 96.5kmh: 9.5secs
0-1/4 mile: 17.4secs
0-1km: 31.4secs
150.0bhp 111.9kW 152.1PS
@ 5500rpm
156.0lbft 211.4Nm @ 3800rpm
109.7bhp/ton 107.9bhp/tonne
65.5bhp/L 48.8kW/L 66.4PS/L
54.1ft/sec 16.5m/sec
23.1mph 37.2kmh/1000rpm
23.6mpg 19.7mpgUS
12.0L/100km
Petrol 4-stroke piston
2290cc 139.7cu in
In-line 4 fuel injection
Compression ratio: 10.0:1
Bore: 90.0mm 3.5in
Stroke: 90.0mm 3.5in
Valve type/No: Overhead 16
Transmission: Manual
No. of forward speeds: 5
Wheels driven: Front
Springs F/R: Coil/Coil
Brake system: PA
Brakes F/R: Disc/Disc
Steering: Rack & pinion PA
Wheelbase: 266.7cm 105.0in
Track F: 152.1cm 59.9in
Track R: 149.1cm 58.7in
Length: 476.3cm 187.5in
Width: 177.8cm 70.0in
Height: 142.0cm 55.9in
Ground clearance: 15.2cm
6.0in
Kerb weight: 1390.1kg
3062.0lb
Fuel: 66.0L 14.5gal 17.4galUS

839 Safir

GT40
1991 UK
193.0mph 310.5kmh
0-60mph 96.5kmh: 4.0secs
0-1/4 mile: 11.9secs
600.0bhp 447.4kW 608.3PS
@ 6000rpm
603.0lbft 817.1Nm @ 4200rpm
564.7bhp/ton 555.3bhp/tonne

74.3bhp/L 55.4kW/L 75.4PS/L
72.8ft/sec 22.2m/sec
Petrol 4-stroke piston
8071cc 492.4cu in
Vee 8 1 Carburettor
Compression ratio: 12.0:1
Bore: 107.5mm 4.2in
Stroke: 111.1mm 4.4in
Valve type/No: Overhead 16
Transmission: Manual
No. of forward speeds: 5
Wheels driven: Rear
Springs F/R: Coil/Coil
Brakes F/R: Disc/Disc
Steering: Rack & pinion
Wheelbase 241.3cm 95.0in
Track F: 144.8cm 57.0in
Track R: 142.2cm 56.0in
Length: 420.4cm 165.5in
Width: 188.0cm 74.0in
Height: 104.1cm 41.0in
Kerb weight: 1080.5kg
2380.0lb

840 Saturn
Sports Coupe
1991 USA
124.0mph 199.5kmh
0-50mph 80.5kmh: 6.2secs
0-60mph 96.5kmh: 8.6secs
0-1/4 mile: 16.6secs
123.0bhp 91.7kW 124.7PS
@ 6000rpm
121.0lbft 164.0Nm @ 4800rpm
116.0bhp/ton 114.1bhp/tonne
64.7bhp/L 48.2kW/L 65.6PS/L
59.0ft/sec 18.0m/sec
31.2mpg 26.0mpgUS
9.0L/100km
Petrol 4-stroke piston
1901cc 116.0cu in
In-line 4 fuel injection
Compression ratio: 9.5:1
Bore: 82.0mm 3.2in
Stroke: 90.0mm 3.5in
Valve type/No: Overhead 16

Transmission: Manual
No. of forward speeds: 5
Wheels driven: Front
Springs F/R: Coil/Coil
Brake system: PA ABS
Brakes F/R: Disc/Disc
Steering: Rack & pinion PA
Wheelbase: 252.0cm 99.2in
Track F: 144.3cm 56.8in
Track R: 141.7cm 55.8in
Length: 446.5cm 175.8in
Width: 171.5cm 67.5in
Height: 128.5cm 50.6in
Ground clearance: 14.2cm
5.6in
Kerb weight: 1078.2kg
2375.0lb
Fuel: 50.0L 11.0gal 13.2galUS

841 Sbarro
Chrono 3.5
1990 Switzerland
124.0mph 199.5kmh
0-60mph 96.5kmh: 3.5secs
500.0bhp 372.8kW 506.9PS
@ 9200rpm
347.0lbft 470.2Nm @ 6500rpm
780.5bhp/ton 767.5bhp/tonne
142.9bhp/L 106.6kW/L
144.9PS/L
84.6ft/sec 25.8m/sec
Petrol 4-stroke piston
3498cc 213.4cu in
In-line 6 fuel injection
Bore: 94.0mm 3.7in
Stroke: 84.0mm 3.3in
Valve type/No: Overhead 24
Transmission: Manual
No. of forward speeds: 5
Wheels driven: Rear
Brakes F/R: Disc/Disc
Wheelbase: 225.0cm 88.6in
Track F: 134.9cm 53.1in
Track R: 150.1cm 59.1in
Length: 305.1cm 120.1in
Width: 186.9cm 73.6in
Height: 100.1cm 39.4in

Kerb weight: 651.5kg 1435.0lb
Fuel: 59.8L 13.1gal 15.8galUS

842 Seat
Malaga 1.2L
1986 Spain
94.0mph 151.2kmh
0-50mph 80.5kmh: 10.9secs
0-60mph 96.5kmh: 15.7secs
0-1/4 mile: 20.4secs
0-1km: 37.9secs
63.0bhp 47.0kW 63.9PS
@ 5800rpm
65.0lbft 88.1Nm @ 3500rpm
68.5bhp/ton 67.4bhp/tonne
52.8bhp/L 39.4kW/L 53.5PS/L
42.9ft/sec 13.0m/sec
19.9mph 32.0kmh/1000rpm
31.3mpg 26.1mpgUS
9.0L/100km
Petrol 4-stroke piston
1193cc 73.0cu in
In-line 4 1 Carburettor
Compression ratio: 9.5:1
Bore: 73.0mm 2.9in
Stroke: 67.5mm 2.7in
Valve type/No: Overhead 8
Transmission: Manual
No. of forward speeds: 5
Wheels driven: Front
Springs F/R: Coil/Leaf
Brake system: PA
Brakes F/R: Disc/Drum
Steering: Rack & pinion
Wheelbase: 244.7cm 96.3in
Track F: 140.0cm 55.1in
Track R: 141.0cm 55.5in
Length: 408.9cm 161.0in
Width: 165.0cm 65.0in
Height: 140.0cm 55.1in
Ground clearance: 15.3cm
6.0in
Kerb weight: 935.0kg 2059.5lb
Fuel: 50.0L 11.0gal
13.2galUS

843 Seat
Ibiza SXi
1989 Spain
108.0mph 173.8kmh
0-50mph 80.5kmh: 7.4secs
0-60mph 96.5kmh: 10.3secs
0-1/4 mile: 17.4secs
0-1km: 33.3secs
100.0bhp 74.6kW 101.4PS
@ 5900rpm
94.5lbft 128.0Nm @ 4700rpm
109.8bhp/ton 108.0bhp/tonne
68.4bhp/L 51.0kW/L 69.4PS/L
42.9ft/sec 13.1m/sec
19.9mph 32.0kmh/1000rpm
34.9mpg 29.1mpgUS
8.1L/100km
Petrol 4-stroke piston
1461cc 89.0cu in
In-line 4 fuel injection
Compression ratio: 11.0:1
Bore: 83.0mm 3.3in
Stroke: 66.5mm 2.6in
Valve type/No: Overhead 8
Transmission: Manual
No. of forward speeds: 5
Wheels driven: Front
Springs F/R: Coil/Coil
Brake system: PA
Brakes F/R: Disc/Drum
Steering: Rack & pinion
Wheelbase: 244.3cm 96.2in
Track F: 142.1cm 55.9in
Track R: 138.7cm 54.6in
Length: 363.8cm 143.2in
Width: 161.0cm 63.4in
Height: 134.0cm 52.8in
Ground clearance: 15.3cm
6.0in
Kerb weight: 926.0kg 2039.6lb
Fuel: 50.0L 11.0gal
13.2galUS

844 Seat
Toledo 1.9 CL Diesel
1992 Spain

843 Seat
Ibiza SXi

103.0mph 165.7kmh
0-50mph 80.5kmh: 11.0secs
0-60mph 96.5kmh: 15.6secs
0-1/4 mile: 20.4secs
0-1km: 37.7secs
68.0bhp 50.7kW 68.9PS
@ 4400rpm
94.0lbft 127.4Nm @ 2400rpm
65.4bhp/ton 64.3bhp/tonne
35.9bhp/L 26.7kW/L 36.4PS/L
46.0ft/sec 14.0m/sec
24.1mph 38.8kmh/1000rpm
37.6mpg 31.3mpgUS
7.5L/100km
Diesel 4-stroke piston
1896cc 115.7cu in
In-line 4 fuel injection
Compression ratio: 23.0:1
Bore: 79.5mm 3.1in
Stroke: 95.5mm 3.8in
Valve type/No: Overhead 8
Transmission: Manual
No. of forward speeds: 5
Wheels driven: Front
Springs F/R: Coil/Coil
Brake system: PA
Brakes F/R: Disc/Drum
Steering: Rack & pinion PA
Wheelbase: 246.9cm 97.2in
Track F: 142.7cm 56.2in
Track R: 142.0cm 55.9in
Length: 432.1cm 170.1in
Width: 166.1cm 65.4in
Height: 142.2cm 56.0in
Kerb weight: 1057.8kg
2330.0lb
Fuel: 60.1L 13.2gal 15.9galUS

845 Seat
Toledo 2.0 GTi
1992 Spain
124.0mph 199.5kmh

0-50mph 80.5kmh: 7.6secs
0-60mph 96.5kmh: 10.6secs
0-1/4 mile: 17.9secs
0-1km: 32.5secs
115.0bhp 85.8kW 116.6PS
@ 5400rpm
122.5lbft 166.0Nm @ 3200rpm
101.3bhp/ton 99.6bhp/tonne
58.0bhp/L 43.2kW/L 58.8PS/L
54.7ft/sec 16.7m/sec
22.2mph 35.7kmh/1000rpm
31.8mpg 26.5mpgUS
8.9L/100km
Petrol 4-stroke piston
1984cc 121.0cu in
In-line 4 fuel injection
Compression ratio: 10.0:1
Bore: 82.5mm 3.2in
Stroke: 92.8mm 3.6in
Valve type/No: Overhead 8
Transmission: Manual
No. of forward speeds: 5
Wheels driven: Front
Springs F/R: Coil/Coil
Brake system: PA ABS
Brakes F/R: Disc/Disc
Steering: Rack & pinion PA
Wheelbase: 247.1cm 97.3in
Track F: 142.9cm 56.3in
Track R: 142.2cm 56.0in
Length: 432.1cm 170.1in
Width: 166.2cm 65.4in
Height: 142.4cm 56.1in
Kerb weight: 1155.0kg
2544.0lb
Fuel: 50.0L 11.0gal 13.2galUS

846 Skoda
Estelle 120 LSE
1986 Czechoslovakia
88.0mph 141.6kmh
0-50mph 80.5kmh: 14.0secs

0-60mph 96.5kmh: 19.9secs
0-1/4 mile: 21.9secs
0-1km: 41.3secs
54.3bhp 40.5kW 55.0PS
@ 5200rpm
66.3lbft 89.8Nm @ 3250rpm
62.4bhp/ton 61.4bhp/tonne
46.2bhp/L 34.5kW/L 46.9PS/L
40.9ft/sec 12.5m/sec
19.4mph 31.2kmh/1000rpm
30.1mpg 25.1mpgUS
9.4L/100km
Petrol 4-stroke piston
1174cc 72.0cu in
In-line 4 1 Carburettor
Compression ratio: 9.5:1
Bore: 72.0mm 2.8in
Stroke: 72.0mm 2.8in
Valve type/No: Overhead 8
Transmission: Manual
No. of forward speeds: 5
Wheels driven: Rear
Springs F/R: Coil/Coil
Brake system: PA
Brakes F/R: Disc/Drum
Steering: Rack & pinion
Wheelbase: 240.0cm 94.5in
Track F: 139.0cm 54.7in
Track R: 135.0cm 53.1in
Length: 420.0cm 165.4in
Width: 161.0cm 63.4in
Height: 140.0cm 55.1in
Ground clearance: 14.5cm
5.7in
Kerb weight: 885.0kg 1949.3lb
Fuel: 35.9L 7.9gal 9.5galUS

847 Skoda
136 Rapide Coupe
1989 Czechoslovakia
92.0mph 148.0kmh
0-50mph 80.5kmh: 10.3secs

0-60mph 96.5kmh: 14.9secs
0-1/4 mile: 19.5secs
0-1km: 36.5secs
62.0bhp 46.2kW 62.9PS
@ 5000rpm
73.8lbft 100.0Nm @ 3000rpm
68.5bhp/ton 67.4bhp/tonne
47.7bhp/L 35.6kW/L 48.4PS/L
39.3ft/sec 12.0m/sec
20.9mph 33.6kmh/1000rpm
30.6mpg 25.5mpgUS
9.2L/100km
Petrol 4-stroke piston
1299cc 79.0cu in
In-line 4 1 Carburettor
Compression ratio: 9.7:1
Bore: 75.5mm 3.0in
Stroke: 72.0mm 2.8in
Valve type/No: Overhead 8
Transmission: Manual
No. of forward speeds: 5
Wheels driven: Rear
Springs F/R: Coil/Coil
Brake system: PA
Brakes F/R: Disc/Drum
Steering: Rack & pinion
Wheelbase: 240.0cm 94.5in
Track F: 139.5cm 54.9in
Track R: 135.5cm 53.3in
Length: 420.0cm 165.4in
Width: 161.0cm 63.4in
Height: 138.0cm 54.3in
Ground clearance: 12.0cm
4.7in
Kerb weight: 920.0kg 2026.4lb
Fuel: 35.9L 7.9gal 9.5galUS

848 Subaru
4WD 3-door Turbo Coupe
1986 Japan
114.0mph 183.4kmh
0-50mph 80.5kmh: 7.4secs

846 Skoda
Estelle 120 LSE

0-60mph 96.5kmh: 10.1secs
0-1/4 mile: 17.4secs
0-1km: 32.6secs
134.0bhp 99.9kW 135.9PS
@ 5600rpm
144.6lbft 195.9Nm @ 2800rpm
118.8bhp/ton 116.9bhp/tonne
75.2bhp/L 56.1kW/L 76.3PS/L
41.1ft/sec 12.5m/sec
19.3mph 31.1kmh/1000rpm
22.6mpg 18.8mpgUS
12.5L/100km
Petrol 4-stroke piston
1781cc 109.0cu in
turbocharged
Flat 4 fuel injection
Compression ratio: 7.7:1
Bore: 92.0mm 3.6in
Stroke: 67.0mm 2.6in
Valve type/No: Overhead 8
Transmission: Automatic
No. of forward speeds: 3
Wheels driven: 4-wheel
engageable
Springs F/R: Coil/Coil
Brake system: PA
Brakes F/R: Disc/Disc
Steering: Rack & pinion PA
Wheelbase: 246.5cm 97.0in
Length: 437.0cm 172.0in
Width: 166.0cm 65.4in
Height: 140.5cm 55.3in
Ground clearance: 17.8cm
7.0in
Kerb weight: 1146.6kg
2525.5lb
Fuel: 60.1L 13.2gal 15.9galUS

849 Subaru

Justy
1987 Japan
91.0mph 146.4kmh
0-50mph 80.5kmh: 9.2secs
0-60mph 96.5kmh: 13.0secs
0-1/4 mile: 19.1secs
0-1km: 34.0secs
67.0bhp 50.0kW 67.9PS
@ 5600rpm
70.8lbft 95.9Nm @ 3600rpm
87.0bhp/ton 85.6bhp/tonne
56.3bhp/L 42.0kW/L 57.1PS/L
50.9ft/sec 15.5m/sec
18.5mph 29.8kmh/1000rpm
32.9mpg 27.4mpgUS
8.6L/100km
Petrol 4-stroke piston
1189cc 73.0cu in
In-line 3 1 Carburettor
Compression ratio: 9.1:1
Bore: 78.0mm 3.1in
Stroke: 83.0mm 3.3in
Valve type/No: Overhead 9
Transmission: Manual
No. of forward speeds: 5

Wheels driven: 4-wheel
engageable
Springs F/R: Coil/Coil
Brake system: PA
Brakes F/R: Disc/Drum
Steering: Rack & pinion
Wheelbase: 228.5cm 90.0in
Track F: 133.0cm 52.4in
Track R: 129.0cm 50.8in
Length: 353.5cm 139.2in
Width: 153.5cm 60.4in
Height: 139.0cm 54.7in
Ground clearance: 15.0cm
5.9in
Kerb weight: 783.0kg 1724.7lb
Fuel: 35.0L 7.7gal 9.2galUS

850 Subaru

XT Turbo 4WD
1987 Japan
121.0mph 194.7kmh
0-50mph 80.5kmh: 6.2secs
0-60mph 96.5kmh: 8.7secs
0-1/4 mile: 16.9secs
0-1km: 31.0secs
134.0bhp 99.9kW 135.9PS
@ 5600rpm
145.0lbft 196.5Nm @ 2800rpm
119.6bhp/ton 117.6bhp/tonne
75.2bhp/L 56.1kW/L 76.3PS/L
41.1ft/sec 12.5m/sec
25.1mph 40.4kmh/1000rpm
25.8mpg 21.5mpgUS
10.9L/100km
Petrol 4-stroke piston
1781cc 109.0cu in
turbocharged
Flat 4 fuel injection
Compression ratio: 7.7:1
Bore: 92.0mm 3.6in
Stroke: 67.0mm 2.6in
Valve type/No: Overhead 8
Transmission: Manual
No. of forward speeds: 5
Wheels driven: 4-wheel drive
Springs F/R: Gas/Gas
Brake system: PA
Brakes F/R: Disc/Disc
Steering: Rack & pinion PA
Wheelbase: 246.5cm 97.0in
Track F: 142.0cm 55.9in
Track R: 142.0cm 55.9in
Length: 449.0cm 176.8in
Width: 168.9cm 66.5in
Height: 133.0cm 52.4in
Ground clearance: 15.5cm
6.1in
Kerb weight: 1139.0kg
2508.8lb
Fuel: 60.1L 13.2gal 15.9galUS

851 Subaru

4WD Turbo Estate
1988 Japan

116.0mph 186.6kmh
0-50mph 80.5kmh: 6.2secs
0-60mph 96.5kmh: 8.5secs
0-1/4 mile: 16.5secs
0-1km: 31.1secs
134.0bhp 99.9kW 135.9PS
@ 5600rpm
134.3lbft 182.0Nm @ 2800rpm
117.1bhp/ton 115.1bhp/tonne
75.2bhp/L 56.1kW/L 76.3PS/L
41.1ft/sec 12.5m/sec
23.6mph 38.0kmh/1000rpm
19.9mpg 16.6mpgUS
14.2L/100km
Petrol 4-stroke piston
1781cc 109.0cu in
turbocharged
Flat 4 fuel injection
Compression ratio: 7.7:1
Bore: 92.0mm 3.6in
Stroke: 67.0mm 2.6in
Valve type/No: Overhead 8
Transmission: Manual
No. of forward speeds: 5
Wheels driven: 4-wheel drive
Springs F/R: Gas/Gas
Brake system: PA
Brakes F/R: Disc/Disc
Steering: Rack & pinion PA
Wheelbase: 246.0cm 96.9in
Track F: 141.0cm 55.5in
Track R: 142.5cm 56.1in
Length: 441.0cm 173.6in
Width: 166.0cm 65.4in
Height: 149.0cm 58.7in
Ground clearance: 16.5cm
6.5in
Kerb weight: 1164.0kg
2563.9lb
Fuel: 60.1L 13.2gal 15.9galUS

852 Subaru

Justy GL II 5-door 4WD
1989 Japan
93.0mph 149.6kmh
0-50mph 80.5kmh: 10.1secs
0-60mph 96.5kmh: 14.4secs
0-1/4 mile: 16.9secs
0-1km: 37.1secs
67.0bhp 50.0kW 67.9PS
@ 5600rpm
71.6lbft 97.0Nm @ 3600rpm
78.7bhp/ton 77.4bhp/tonne
56.3bhp/L 42.0kW/L 57.1PS/L
50.9ft/sec 15.5m/sec
17.6mph 28.3kmh/1000rpm
31.0mpg 25.8mpgUS
9.1L/100km
Petrol 4-stroke piston
1189cc 73.0cu in
In-line 3 1 Carburettor
Compression ratio: 9.1:1
Bore: 78.0mm 3.1in
Stroke: 83.0mm 3.3in

Valve type/No: Overhead 6
Transmission: Manual
No. of forward speeds: 5
Wheels driven: 4-wheel
engageable
Springs F/R: Coil/Coil
Brake system: PA
Brakes F/R: Disc/Drum
Steering: Rack & pinion
Wheelbase: 228.5cm 90.0in
Track F: 133.0cm 52.4in
Track R: 129.0cm 50.8in
Length: 369.5cm 145.5in
Width: 153.5cm 60.4in
Height: 142.0cm 55.9in
Ground clearance: 17.5cm
6.9in
Kerb weight: 866.0kg 1907.5lb
Fuel: 35.0L 7.7gal 9.2galUS

853 Subaru

Legacy FIA Record
1989 Japan
161.0mph 259.0kmh
0-50mph 80.5kmh: 4.5secs
0-60mph 96.5kmh: 6.2secs
0-1/4 mile: 14.6secs
235.0bhp 175.2kW 238.3PS
@ 6400rpm
190.4bhp/ton 187.2bhp/tonne
117.8bhp/L 87.9kW/L
119.5PS/L
52.4ft/sec 16.0m/sec
Petrol 4-stroke piston
1994cc 121.7cu in
turbocharged
Flat 4 fuel injection
Compression ratio: 8.5:1
Bore: 92.0mm 3.6in
Stroke: 75.0mm 2.9in
Valve type/No: Overhead 16
Transmission: Manual
No. of forward speeds: 5
Wheels driven: 4-wheel drive
Springs F/R: Coil/Coil
Brake system: PA
Brakes F/R: Disc/Disc
Steering: Rack & pinion PA
Wheelbase: 258.1cm 101.6in
Track F: 146.1cm 57.5in
Track R: 145.0cm 57.1in
Length: 451.1cm 177.6in
Width: 168.9cm 66.5in
Height: 132.3cm 52.1in
Kerb weight: 1255.3kg
2765.0lb

854 Subaru

XT6
1989 Japan
115.0mph 185.0kmh
0-50mph 80.5kmh: 7.0secs
0-60mph 96.5kmh: 9.7secs
0-1/4 mile: 17.2secs

145.0bhp 108.1kW 147.0PS @ 5200rpm
156.0lbft 211.4Nm @ 4000rpm
112.2bhp/ton 110.3bhp/tonne
54.3bhp/L 40.5kW/L 55.0PS/L
38.1ft/sec 11.6m/sec
29.3mpg 24.4mpgUS
9.6L/100km
Petrol 4-stroke piston
2672cc 163.0cu in
Flat 6 fuel injection
Compression ratio: 9.5:1
Bore: 92.0mm 3.6in
Stroke: 67.0mm 2.6in
Valve type/No: Overhead 12
Transmission: Automatic
No. of forward speeds: 4
Wheels driven: 4-wheel drive
Springs F/R: Gas/Gas
Brake system: PA
Brakes F/R: Disc/Disc
Steering: Rack & pinion PA
Wheelbase: 246.4cm 97.0in
Track F: 143.5cm 56.5in
Track R: 144.0cm 56.7in
Length: 451 1cm 177.6in
Width: 168.9cm 66.5in
Height: 133.6cm 52.6in
Kerb weight: 1314.3kg
2895.0lb
Fuel: 60.2L 13.2gal 15.9galUS

855 Subaru

Legacy 2.2 GX 4WD
1990 Japan
119.0mph 191.5kmh
0-50mph 80.5kmh: 7.8secs
0-60mph 96.5kmh: 10.8secs
0-1/4 mile: 18.0secs
0-1km: 33.3secs
134.0bhp 99.9kW 135.9PS
@ 6000rpm

139.5lbft 189.0Nm @ 4800rpm
106.5bhp/ton 104.7bhp/tonne
60.6bhp/L 45.2kW/L 61.4PS/L
49.2ft/sec 15.0m/sec
20.7mph 33.3kmh/1000rpm
25.0mpg 20.8mpgUS
11.3L/100km
Petrol 4-stroke piston
2212cc 135.0cu in
Flat 4 fuel injection
Compression ratio: 9.5:1
Bore: 96.9mm 3.8in
Stroke: 75.0mm 2.9in
Valve type/No: Overhead 16
Transmission: Manual
No. of forward speeds: 5
Wheels driven: 4-wheel drive
Springs F/R: Coil/Coil
Brakes F/R: Disc/Disc
Steering: Rack & pinion PA
Wheelbase: 258.0cm 101.6in
Track F: 146.5cm 57.7in
Track R: 146.0cm 57.5in
Length: 451.0cm 177.6in
Width: 169.0cm 66.5in
Height: 140.0cm 55.1in
Ground clearance: 17.4cm
6.9in
Kerb weight: 1280.0kg
2819.4lb
Fuel: 60.1L 13.2gal 15.9galUS

856 Subaru

Legacy 2.0 4Cam Turbo
Estate
1991 Japan
134.0mph 215.6kmh
0-50mph 80.5kmh: 5.3secs
0-60mph 96.5kmh: 7.0secs
0-1/4 mile: 15.6secs
0-1km: 28.6secs

197.0bhp 146.9kW 199.7PS @ 6000rpm
193.4lbft 262.0Nm @ 3600rpm
138.2bhp/ton 135.9bhp/tonne
98.8bhp/L 73.7kW/L 100.2PS/L
49.2ft/sec 15.0m/sec
23.4mph 37.7kmh/1000rpm
21.8mpg 18.2mpgUS
13.0L/100km
Petrol 4-stroke piston
1994cc 122.0cu in
turbocharged
Flat 4 fuel injection
Compression ratio: 8.0:1
Bore: 92.0mm 3.6in
Stroke: 75.0mm 2.9in
Valve type/No: Overhead 32
Transmission: Manual
No. of forward speeds: 5
Wheels driven: 4-wheel drive
Springs F/R: Coil/Coil
Brake system: PA ABS
Brakes F/R: Disc/Disc
Steering: Rack & pinion PA
Wheelbase: 258.1cm 101.6in
Track F: 146.1cm 57.5in
Track R: 145.0cm 57.1in
Length: 462.0cm 181.9in
Width: 168.9cm 66.5in
Height: 138.9cm 54.7in
Ground clearance: 17.5cm
6.9in
Kerb weight: 1450.0kg
3193.8lb
Fuel: 60.1L 13.2gal 15.9galUS

857 Subaru

Legacy Sports Sedan
1991 Japan
129.0mph 207.6kmh
0-50mph 80.5kmh: 6.0secs
0-60mph 96.5kmh: 8.1secs

0-1/4 mile: 16.2secs
160.0bhp 119.3kW 162.2PS
@ 5600rpm
181.0lbft 245.3Nm @ 2800rpm
113.8bhp/ton 111.9bhp/tonne
72.3bhp/L 53.9kW/L 73.3PS/L
45.9ft/sec 14.0m/sec
24.0mpg 20.0mpgUS
11.8L/100km
Petrol 4-stroke piston
2212cc 135.0cu in
Flat 4 fuel injection
Compression ratio: 8.0:1
Bore: 96.9mm 3.8in
Stroke: 75.0mm 2.9in
Valve type/No: Overhead 16
Transmission: Manual
No. of forward speeds: 5
Wheels driven: 4-wheel drive
Springs F/R: Coil/Coil
Brake system: PA ABS
Brakes F/R: Disc/Disc
Steering: Rack & pinion PA
Wheelbase: 258.1cm 101.6in
Track F: 146.6cm 57.7in
Track R: 146.6cm 57.7in
Length: 451.1cm 177.6in
Width: 168.9cm 66.5in
Height: 135.9cm 53.5in
Ground clearance: 11.4cm
4.5in
Kerb weight: 1430.1kg
3150.0lb
Fuel: 60.2L 13.2gal 15.9galUS

858 Subaru

SVX
1992 Japan
143.0mph 230.1kmh
0-50mph 80.5kmh: 5.3secs
0-60mph 96.5kmh: 7.3secs
0-1/4 mile: 15.4secs

856 Subaru
Legacy 2.0 4Cam
Turbo Estate

230.0bhp 171.5kW 233.2PS
@ 5400rpm
224.0lbft 303.5Nm @ 4400rpm
143.9bhp/ton 141.5bhp/tonne
69.3bhp/L 51.7kW/L 70.3PS/L
44.2ft/sec 13.5m/sec
21.0mpg 17.5mpgUS
13.4L/100km
Petrol 4-stroke piston
3318cc 202.4cu in
Flat 6 fuel injection
Compression ratio: 10.0:1
Bore: 96.9mm 3.8in
Stroke: 75.0mm 2.9in
Valve type/No: Overhead 24
Transmission: Automatic
No. of forward speeds: 4
Wheels driven: 4-wheel drive
Springs F/R: Coil/Coil
Brake system: PA ABS
Brakes F/R: Disc/Disc
Steering: Rack & pinion PA
Wheelbase: 261.1cm 102.8in
Track F: 149.9cm 59.0in
Track R: 148.1cm 58.3in
Length: 462.5cm 182.1in
Width: 177.0cm 69.7in
Height: 130.0cm 51.2in
Ground clearance: 13.5cm
5.3in
Kerb weight: 1625.3kg
3580.0lb
Fuel: 70.0L 15.4gal 18.5galUS

859 Suzuki
Alto GLA
1986 Japan
80.0mph 128.7kmh
0-50mph 80.5kmh: 16.2secs
0-60mph 96.5kmh: 24.4secs
0-1/4 mile: 22.9secs
0-1km: 43.3secs

40.0bhp 29.8kW 40.5PS
@ 5500rpm
43.0lbft 58.3Nm @ 3000rpm
63.1bhp/ton 62.0bhp/tonne
50.2bhp/L 37.5kW/L 50.9PS/L
43.2ft/sec 13.2m/sec
13.6mph 21.9kmh/1000rpm
35.0mpg 29.1mpgUS
8.1L/100km
Petrol 4-stroke piston
796cc 49.0cu in
In-line 3 1 Carburettor
Compression ratio: 8.7:1
Bore: 68.5mm 2.7in
Stroke: 72.0mm 2.8in
Valve type/No: Overhead 6
Transmission: Automatic
No. of forward speeds: 2
Wheels driven: Front
Springs F/R: Coil/Leaf
Brakes F/R: Disc/Drum
Steering: Rack & pinion
Wheelbase: 215.0cm 84.6in
Track F: 121.5cm 47.8in
Track R: 120.0cm 47.2in
Length: 330.0cm 129.9in
Width: 140.5cm 55.3in
Height: 141.0cm 55.5in
Ground clearance: 17.0cm
6.7in
Kerb weight: 645.0kg 1420.7lb
Fuel: 30.0L 6.6gal 7.9galUS

860 Suzuki
Swift 1.3 GTi
1986 Japan
112.0mph 180.2kmh
0-50mph 80.5kmh: 6.5secs
0-60mph 96.5kmh: 9.1secs
0-1/4 mile: 16.8secs
0-1km: 31.8secs
101.0bhp 75.3kW 102.4PS

@ 6600rpm
79.7lbft 108.0Nm @ 5500rpm
134.3bhp/ton 132.0bhp/tonne
77.8bhp/L 58.0kW/L 78.9PS/L
54.4ft/sec 16.6m/sec
17.2mph 27.7kmh/1000rpm
30.9mpg 25.7mpgUS
9.1L/100km
Petrol 4-stroke piston
1298cc 79.2cu in
In-line 4 fuel injection
Compression ratio: 10.0:1
Bore: 74.0mm 2.9in
Stroke: 75.5mm 3.0in
Valve type/No: Overhead 16
Transmission: Manual
No. of forward speeds: 5
Wheels driven: Front
Springs F/R: Coil/Coil
Brake system: PA
Brakes F/R: Disc/Drum
Steering: Rack & pinion
Wheelbase: 224.5cm 88.4in
Track F: 133.1cm 52.4in
Track R: 130.0cm 51.2in
Length: 358.4cm 141.1in
Width: 152.9cm 60.2in
Height: 134.9cm 53.1in
Ground clearance: 18.0cm
7.1in
Kerb weight: 765.0kg 1685.0lb
Fuel: 33.2L 7.3gal 8.8galUS

861 Suzuki
SJ413V JX
1987 Japan
76.0mph 122.3kmh
0-50mph 80.5kmh: 12.2secs
0-60mph 96.5kmh: 19.4secs
0-1/4 mile: 20.5secs
0-1km: 40.9secs
63.0bhp 47.0kW 63.9PS

@ 6000rpm
73.3lbft 99.3Nm @ 3500rpm
75.9bhp/ton 74.6bhp/tonne
47.6bhp/L 35.5kW/L 48.2PS/L
50.5ft/sec 15.4m/sec
17.6mph 28.3kmh/1000rpm
24.8mpg 20.7mpgUS
11.4L/100km
Petrol 4-stroke piston
1324cc 81.0cu in
In-line 4 1 Carburettor
Compression ratio: 8.9:1
Bore: 74.0mm 2.9in
Stroke: 77.0mm 3.0in
Valve type/No: Overhead 8
Transmission: Manual
No. of forward speeds: 5
Wheels driven: 4-wheel
engageable
Springs F/R: Leaf/Leaf
Brake system: PA
Brakes F/R: Disc/Drum
Steering: Recirculating ball
Wheelbase: 203.0cm 79.9in
Track F: 121.0cm 47.6in
Track R: 122.0cm 48.0in
Length: 343.0cm 135.0in
Width: 146.0cm 57.5in
Height: 168.0cm 66.1in
Ground clearance: 22.5cm
8.9in
Kerb weight: 844.0kg 1859.0lb
Fuel: 40.0L 8.8gal 10.6galUS

862 Suzuki
Swift 1.3 GLX
1987 Japan
102.0mph 164.1kmh
0-50mph 80.5kmh: 7.8secs
0-60mph 96.5kmh: 11.3secs
0-1/4 mile: 18.3secs
0-1km: 34.1secs

860 Suzuki
Swift 1.3 GTi

67.0bhp 50.0kW 67.9PS @ 5300rpm	67.0bhp 50.0kW 67.9PS @ 5300rpm	67.0bhp 50.0kW 67.9PS @ 6000rpm	101.0bhp 75.3kW 102.4PS @ 6450rpm
76.3lbft 103.4Nm @ 3700rpm	76.0lbft 103.0Nm @ 3700rpm	74.5lbft 101.0Nm @ 4950rpm	83.4lbft 113.0Nm @ 4950rpm
93.0bhp/ton 91.4bhp/tonne	93.3bhp/ton 91.8bhp/tonne	89.2bhp/ton 87.7bhp/tonne	126.8bhp/ton 124.7bhp/tonne
50.6bhp/L 37.7kW/L 51.3PS/L	50.6bhp/L 37.7kW/L 51.3PS/L	51.6bhp/L 38.5kW/L 52.3PS/L	77.8bhp/L 58.0kW/L 78.9PS/L
44.6ft/sec 13.6m/sec	44.6ft/sec 13.6m/sec	49.5ft/sec 15.1m/sec	53.2ft/sec 16.2m/sec
22.8mph 36.7kmh/1000rpm	17.3mph 27.8kmh/1000rpm	23.3mph 37.5kmh/1000rpm	17.9mph 28.8kmh/1000rpm
37.0mpg 30.8mpgUS	36.0mpg 30.0mpgUS	35.7mpg 29.7mpgUS	29.2mpg 24.3mpgUS
7.6L/100km	7.8L/100km	7.9L/100km	9.7L/100km
Petrol 4-stroke piston	Petrol 4-stroke piston	Petrol 4-stroke piston	Petrol 4-stroke piston
1324cc 81.0cu in	1324cc 81.0cu in	1298cc 79.0cu in	1298cc 79.0cu in
In-line 4 1 Carburettor	In-line 4 1 Carburettor	In-line 4 1 Carburettor	In-line 4 fuel injection
Compression ratio: 8.9:1	Compression ratio: 8.9:1	Compression ratio: 9.5:1	Compression ratio: 10.0:1
Bore: 74.0mm 2.9in	Bore: 74.0mm 2.9in	Bore: 74.0mm 2.9in	Bore: 74.0mm 2.9in
Stroke: 77.0mm 3.0in	Stroke: 77.0mm 3.0in	Stroke: 75.5mm 3.0in	Stroke: 75.5mm 3.0in
Valve type/No: Overhead 8	Valve type/No: Overhead 8	Valve type/No: Overhead 8	Valve type/No: Overhead 16
Transmission: Manual	Transmission: Automatic	Transmission: Manual	Transmission: Manual
No. of forward speeds: 5	No. of forward speeds: 3	No. of forward speeds: 5	No. of forward speeds: 5
Wheels driven: Front	Wheels driven: Front	Wheels driven: Front	Wheels driven: Front
Springs F/R: Coil/Coil	Springs F/R: Coil/Coil	Springs F/R: Coil/Coil	Springs F/R: Coil/Coil
Brake system: PA	Brake system: PA	Brake system: PA	Brake system: PA
Brakes F/R: Disc/Drum	Brakes F/R: Disc/Drum	Brakes F/R: Disc/Drum	Brakes F/R: Disc/Disc
Steering: Rack & pinion	Steering: Rack & pinion	Steering: Rack & pinion	Steering: Rack & pinion
Wheelbase: 234.5cm 92.3in	Wheelbase: 234.5cm 92.3in	Wheelbase: 245.0cm 96.5in	Wheelbase: 226.5cm 89.2in
Track F: 133.5cm 52.6in	Track F: 133.5cm 52.6in	Track F: 136.5cm 53.7in	Track F: 136.5cm 53.7in
Track R: 130.0cm 51.2in	Track R: 130.0cm 51.2in	Track R: 134.0cm 52.8in	Track R: 134.0cm 52.8in
Length: 377.0cm 148.4in	Length: 377.0cm 148.4in	Length: 381.0cm 150.0in	Length: 371.0cm 146.1in
Width: 154.5cm 60.8in	Width: 154.5cm 60.8in	Width: 159.0cm 62.6in	Width: 159.0cm 62.6in
Height: 135.0cm 53.1in	Height: 135.0cm 53.1in	Height: 138.0cm 54.3in	Height: 135.0cm 53.1in
Ground clearance: 18.0cm 7.1in	Ground clearance: 18.0cm 7.1in	Ground clearance: 17.0cm 6.7in	Ground clearance: 17.0cm 6.7in
Kerb weight: 733.0kg 1614.5lb	Kerb weight: 730.0kg 1607.9lb	Kerb weight: 764.0kg 1682.8lb	Kerb weight: 810.0kg 1784.1lb
Fuel: 33.2L 7.3gal 8.8galUS	Fuel: 33.2L 7.3gal 8.8galUS	Fuel: 40.0L 8.8gal 10.6galUS	Fuel: 40.0L 8.8gal 10.6galUS

863 Suzuki	**864 Suzuki**	**865 Suzuki**	**866 Suzuki**
Swift 1.3 GLX Executive	Swift 1.3 GLX	Swift 1.3 GTi	Vitara JLX
1987 Japan	1989 Japan	1989 Japan	1989 Japan
97.0mph 156.1kmh	103.0mph 165.7kmh	116.0mph 186.6kmh	90.0mph 144.8kmh
0-50mph 80.5kmh: 10.1secs	0-50mph 80.5kmh: 8.2secs	0-50mph 80.5kmh: 6.4secs	0-50mph 80.5kmh: 10.0secs
0-60mph 96.5kmh: 14.5secs	0-60mph 96.5kmh: 11.6secs	0-60mph 96.5kmh: 8.7secs	0-60mph 96.5kmh: 14.5secs
0-1/4 mile: 19.9secs	0-1/4 mile: 18.5secs	0-1/4 mile: 16.8secs	0-1/4 mile: 19.5secs
0-1km: 37.1secs	0-1km: 34.8secs	0-1km: 30.1secs	0-1km: 37.1secs

866 Suzuki
Vitara JLX

75.0bhp 55.9kW 76.0PS
@ 5250rpm
90.8lbft 123.0Nm @ 3100rpm
74.6bhp/ton 73.3bhp/tonne
47.2bhp/L 35.2kW/L 47.8PS/L
51.6ft/sec 15.7m/sec
17.8mph 28.6kmh/1000rpm
24.2mpg 20.2mpgUS
11.7L/100km
Petrol 4-stroke piston
1590cc 97.0cu in
In-line 4 1 Carburettor
Compression ratio: 8.9:1
Bore: 75.0mm 2.9in
Stroke: 90.0mm 3.5in
Valve type/No: Overhead 8
Transmission: Manual
No. of forward speeds: 5
Wheels driven: 4-wheel
engageable
Springs F/R: Coil/Coil
Brake system: PA
Brakes F/R: Disc/Drum
Steering: Recirculating ball
Wheelbase: 220.0cm 86.6in
Track F: 139.5cm 54.9in
Track R: 140.0cm 55.1in
Length: 362.0cm 142.5in
Width: 163.0cm 64.2in
Height: 166.5cm 65.6in
Ground clearance: 20.0cm
7.9in
Kerb weight: 1023.0kg
2253.3lb
Fuel: 43.2L 9.5gal 11.4galUS

867 Suzuki
Swift GT
1990 Japan
105.0mph 168.9kmh
0-60mph 96.5kmh: 9.5secs
0-1/4 mile: 17.0secs
100.0bhp 74.6kW 101.4PS
@ 6500rpm
83.0lbft 112.5Nm @ 5000rpm
119.5bhp/ton 117.5bhp/tonne
77.0bhp/L 57.4kW/L 78.0PS/L
53.6ft/sec 16.4m/sec
37.8mpg 31.5mpgUS
7.5L/100km
Petrol 4-stroke piston
1299cc 79.3cu in
In-line 4 fuel injection
Compression ratio: 10.0:1
Bore: 74.0mm 2.9in
Stroke: 75.5mm 3.0in
Valve type/No: Overhead 16
Transmission: Manual
No. of forward speeds: 5
Wheels driven: Front
Springs F/R: Coil/Coil
Brake system: PA
Brakes F/R: Disc/Disc
Steering: Rack & pinion

Wheelbase: 226.6cm 89.2in
Track F: 136.4cm 53.7in
Track R: 134.1cm 52.8in
Length: 371.1cm 146.1in
Width: 158.5cm 62.4in
Height: 134.9cm 53.1in
Kerb weight: 851.2kg 1875.0lb
Fuel: 40.1L 8.8gal 10.6galUS

868 Toyota
Celica 2.0GT
1986 Japan
133.0mph 214.0kmh
0-50mph 80.5kmh: 6.1secs
0-60mph 96.5kmh: 8.2secs
0-1/4 mile: 16.7secs
0-1km: 30.4secs
147.0bhp 109.6kW 149.0PS
@ 6400rpm
130.0lbft 176.2Nm @ 4800rpm
117.4bhp/ton 115.4bhp/tonne
73.6bhp/L 54.9kW/L 74.6PS/L
60.3ft/sec 18.3m/sec
21.0mph 33.8kmh/1000rpm
25.0mpg 20.8mpgUS
11.3L/100km
Petrol 4-stroke piston
1998cc 121.9cu in
In-line 4 fuel injection
Compression ratio: 9.8:1
Bore: 86.0mm 3.4in
Stroke: 86.0mm 3.4in
Valve type/No: Overhead 16
Transmission: Manual
No. of forward speeds: 5
Wheels driven: Front
Springs F/R: Coil/Coil
Brake system: PA
Brakes F/R: Disc/Disc
Steering: Rack & pinion PA
Wheelbase: 252.5cm 99.4in
Track F: 146.3cm 57.6in
Track R: 142.7cm 56.2in
Length: 436.4cm 171.8in
Width: 170.9cm 67.3in
Ground clearance: 17.8cm
7.0in
Kerb weight: 1273.5kg
2805.0lb
Fuel: 60.1L 13.2gal 15.9galUS

869 Toyota
Celica GTS
1986 Japan
123.0mph 197.9kmh
0-60mph 96.5kmh: 8.6secs
0-1/4 mile: 16.5secs
135.0bhp 100.7kW 136.9PS
@ 6000rpm
125.0lbft 169.4Nm @ 4800rpm
111.2bhp/ton 109.3bhp/tonne
67.6bhp/L 50.4kW/L 68.5PS/L
56.5ft/sec 17.2m/sec
24.3mpg 20.2mpgUS

11.6L/100km
Petrol 4-stroke piston
1998cc 121.9cu in
In-line 4 fuel injection
Compression ratio: 9.2:1
Bore: 86.0mm 3.4in
Stroke: 86.0mm 3.4in
Valve type/No: Overhead 16
Transmission: Manual
No. of forward speeds: 5
Wheels driven: Front
Springs F/R: Coil/Coil
Brake system: PA
Brakes F/R: Disc/Disc
Steering: Rack & pinion PA
Wheelbase: 252.5cm 99.4in
Track F: 147.1cm 57.9in
Track R: 143.5cm 56.5in
Length: 440.9cm 173.6in
Width: 170.9cm 67.3in
Height: 126.5cm 49.8in
Ground clearance: 14.0cm
5.5in
Kerb weight: 1234.9kg
2720.0lb
Fuel: 60.2L 13.2gal 15.9galUS

870 Toyota
MR2
1986 Japan
121.0mph 194.7kmh
0-60mph 96.5kmh: 8.4secs
0-1/4 mile: 16.5secs
112.0bhp 83.5kW 113.5PS
@ 6600rpm
97.0lbft 131.4Nm @ 4800rpm
105.0bhp/ton 103.2bhp/tonne
70.6bhp/L 52.6kW/L 71.5PS/L
55.5ft/sec 16.9m/sec
30.0mpg 25.0mpgUS
9.4L/100km
Petrol 4-stroke piston
1587cc 96.8cu in
In-line 4 fuel injection
Compression ratio: 9.4:1
Bore: 81.0mm 3.2in
Stroke: 77.0mm 3.0in
Valve type/No: Overhead 16
Transmission: Manual
No. of forward speeds: 5
Wheels driven: Rear
Springs F/R: Coil/Coil
Brake system: PA
Brakes F/R: Disc/Disc
Steering: Rack & pinion
Wheelbase: 231.9cm 91.3in
Track F: 144.0cm 56.7in
Track R: 144.5cm 56.9in
Length: 392.4cm 154.5in
Width: 166.6cm 65.6in
Height: 125.0cm 49.2in
Kerb weight: 1085.1kg
2390.0lb
Fuel: 40.9L 9.0gal 10.8galUS

871 Toyota
MR2 T-Bar
1986 Japan
122.0mph 196.3kmh
0-50mph 80.5kmh: 5.7secs
0-60mph 96.5kmh: 7.7secs
0-1/4 mile: 16.5secs
0-1km: 30.7secs
122.0bhp 91.0kW 123.7PS
@ 6600rpm
105.0lbft 142.3Nm @ 5000rpm
116.4bhp/ton 114.4bhp/tonne
76.9bhp/L 57.3kW/L 77.9PS/L
55.5ft/sec 16.9m/sec
18.6mph 29.9kmh/1000rpm
28.6mpg 23.8mpgUS
9.9L/100km
Petrol 4-stroke piston
1587cc 97.0cu in
In-line 4 fuel injection
Compression ratio: 10.0:1
Bore: 81.0mm 3.2in
Stroke: 77.0mm 3.0in
Valve type/No: Overhead 16
Transmission: Manual
No. of forward speeds: 5
Wheels driven: Rear
Springs F/R: Coil/Coil
Brake system: PA
Brakes F/R: Disc/Disc
Steering: Rack & pinion
Wheelbase: 232.0cm 91.3in
Track F: 144.0cm 56.7in
Track R: 144.0cm 56.7in
Length: 392.5cm 154.5in
Width: 166.6cm 65.6in
Height: 124.9cm 49.2in
Ground clearance: 17.9cm
7.0in
Kerb weight: 1066.0kg
2348.0lb
Fuel: 40.9L 9.0gal 10.8galUS

872 Toyota
Supra
1986 Japan
133.0mph 214.0kmh
0-50mph 80.5kmh: 5.1secs
0-60mph 96.5kmh: 7.0secs
0-1/4 mile: 15.4secs
200.0bhp 149.1kW 202.8PS
@ 6000rpm
185.0lbft 250.7Nm @ 4800rpm
131.4bhp/ton 129.2bhp/tonne
67.7bhp/L 50.5kW/L 68.6PS/L
59.7ft/sec 18.2m/sec
21.6mpg 18.0mpgUS
13.1L/100km
Petrol 4-stroke piston
2954cc 180.2cu in
In-line 6 fuel injection
Compression ratio: 9.2:1
Bore: 83.0mm 3.3in
Stroke: 91.0mm 3.6in

Valve type/No: Overhead 24
Transmission: Manual
No. of forward speeds: 5
Wheels driven: Rear
Springs F/R: Coil/Coil
Brake system: PA
Brakes F/R: Disc/Disc
Steering: Rack & pinion PA
Wheelbase: 259.6cm 102.2in
Track F: 148.6cm 58.5in
Track R: 148.6cm 58.5in
Length: 462.0cm 181.9in
Width: 174.5cm 68.7in
Height: 131.1cm 51.6in
Ground clearance: 14.0cm
5.5in
Kerb weight: 1548.1kg
3410.0lb
Fuel: 70.0L 15.4gal 18.5galUS

873 Toyota

Supra 3.0i
1986 Japan
138.0mph 222.0kmh
0-50mph 80.5kmh: 5.6secs
0-60mph 96.5kmh: 7.7secs
0-1/4 mile: 16.0secs
0-1km: 29.4secs
201.0bhp 149.9kW 203.8PS
@ 6000rpm
187.4lbft 253.9Nm @ 4000rpm
131.9bhp/ton 129.7bhp/tonne
68.0bhp/L 50.7kW/L 69.0PS/L
59.7ft/sec 18.2m/sec
23.8mph 38.3kmh/1000rpm
19.4mpg 16.2mpgUS
14.6L/100km
Petrol 4-stroke piston
2954cc 180.0cu in
In-line 6 fuel injection
Compression ratio: 9.2:1
Bore: 83.0mm 3.3in
Stroke: 91.0mm 3.6in
Valve type/No: Overhead 24
Transmission: Manual
No. of forward speeds: 5
Wheels driven: Rear
Springs F/R: Coil/Coil
Brake system: PA ABS
Brakes F/R: Disc/Disc
Steering: Rack & pinion PA
Wheelbase: 259.5cm 102.2in
Track F: 148.5cm 58.5in
Track R: 148.0cm 58.3in
Length: 462.0cm 181.9in
Width: 174.5cm 68.7in
Height: 131.0cm 51.6in
Kerb weight: 1550.0kg 3414.1lb
Fuel: 70.1L 15.4gal 18.5galUS

874 Toyota

Camry 2.0 GLi
1987 Japan
118.0mph 189.9kmh

0-50mph 80.5kmh: 6.4secs
0-60mph 96.5kmh: 9.0secs
0-1/4 mile: 16.9secs
0-1km: 31.2secs
126.0bhp 94.0kW 127.7PS
@ 5600rpm
132.0lbft 178.9Nm @ 4400rpm
107.4bhp/ton 105.6bhp/tonne
63.1bhp/L 47.0kW/L 63.9PS/L
52.7ft/sec 16.0m/sec
23.1mph 37.2kmh/1000rpm
30.5mpg 25.4mpgUS
9.3L/100km
Petrol 4-stroke piston
1998cc 122.0cu in
In-line 4 fuel injection
Compression ratio: 9.8:1
Bore: 86.0mm 3.4in
Stroke: 86.0mm 3.4in
Valve type/No: Overhead 16
Transmission: Manual
No. of forward speeds: 5
Wheels driven: Front
Springs F/R: Coil/Coil
Brake system: PA
Brakes F/R: Disc/Drum
Steering: Rack & pinion PA
Wheelbase: 260.0cm 102.4in
Track F: 147.5cm 58.1in
Track R: 144.5cm 56.9in
Length: 452.0cm 178.0in
Width: 171.0cm 67.3in
Height: 140.0cm 55.1in
Ground clearance: 17.0cm
6.7in
Kerb weight: 1193.0kg
2627.7lb
Fuel: 60.1L 13.2gal 15.9galUS

875 Toyota

Camry 2.0 GLi Estate
1987 Japan
112.0mph 180.2kmh
0-50mph 80.5kmh: 7.1secs
0-60mph 96.5kmh: 10.1secs
0-1/4 mile: 17.7secs
0-1km: 32.5secs
126.0bhp 94.0kW 127.7PS
@ 5600rpm
132.0lbft 178.9Nm @ 4400rpm
99.2bhp/ton 97.6bhp/tonne
63.1bhp/L 47.0kW/L 63.9PS/L
52.7ft/sec 16.0m/sec
23.1mph 37.2kmh/1000rpm
28.7mpg 23.9mpgUS
9.8L/100km
Petrol 4-stroke piston
1998cc 122.0cu in
In-line 4 fuel injection
Compression ratio: 9.8:1
Bore: 86.0mm 3.4in
Stroke: 86.0mm 3.4in
Valve type/No: Overhead 16
Transmission: Manual

No. of forward speeds: 5
Wheels driven: Front
Springs F/R: Coil/Coil
Brake system: PA
Brakes F/R: Disc/Drum
Steering: Rack & pinion PA
Wheelbase: 260.0cm 102.4in
Track F: 147.5cm 58.1in
Track R: 144.5cm 56.9in
Length: 461.0cm 181.5in
Width: 171.0cm 67.3in
Height: 140.0cm 55.1in
Ground clearance: 17.0cm
6.7in
Kerb weight: 1291.0kg
2843.6lb
Fuel: 60.1L 13.2gal 15.9galUS

876 Toyota

Celica GT Cabriolet
1987 Japan
127.0mph 204.3kmh
0-50mph 80.5kmh: 6.3secs
0-60mph 96.5kmh: 8.6secs
0-1/4 mile: 16.9secs
0-1km: 31.2secs
147.0bhp 109.6kW 149.0PS
@ 6400rpm
133.0lbft 180.2Nm @ 4800rpm
125.1bhp/ton 123.0bhp/tonne
73.6bhp/L 54.9kW/L 74.6PS/L
60.3ft/sec 18.3m/sec
21.4mph 34.4kmh/1000rpm
28.7mpg 23.9mpgUS
9.8L/100km
Petrol 4-stroke piston
1998cc 122.0cu in
In-line 4 fuel injection
Compression ratio: 9.8:1
Bore: 86.0mm 3.4in
Stroke: 86.0mm 3.4in
Valve type/No: Overhead 16
Transmission: Manual
No. of forward speeds: 5
Wheels driven: Front
Springs F/R: Coil/Coil
Brake system: PA
Brakes F/R: Disc/Disc
Steering: Rack & pinion PA
Wheelbase: 252.5cm 99.4in
Track F: 146.5cm 57.7in
Track R: 143.0cm 56.3in
Length: 436.5cm 171.9in
Width: 171.0cm 67.3in
Ground clearance: 17.9cm
7.0in
Kerb weight: 1195.0kg
2632.2lb
Fuel: 60.1L 13.2gal 15.9galUS

877 Toyota

Corolla Executive
1987 Japan
104.0mph 167.3kmh
0-50mph 80.5kmh: 7.9secs

No. of forward speeds: 5
Wheels driven: Front
Springs F/R: Coil/Coil
Brake system: PA
Brakes F/R: Disc/Drum
Steering: Rack & pinion PA
Wheelbase: 260.0cm 102.4in
Track F: 147.5cm 58.1in
Track R: 144.5cm 56.9in
Length: 461.0cm 181.5in
Width: 171.0cm 67.3in
Height: 140.0cm 55.1in
Ground clearance: 17.0cm
6.7in
Kerb weight: 1291.0kg
2843.6lb
Fuel: 60.1L 13.2gal 15.9galUS

0-60mph 96.5kmh: 11.1secs
0-1/4 mile: 18.0secs
0-1km: 33.4secs
94.0bhp 70.1kW 95.3PS
@ 6000rpm
100.0lbft 135.5Nm @ 3600rpm
93.7bhp/ton 92.2bhp/tonne
59.2bhp/L 44.2kW/L 60.0PS/L
50.5ft/sec 15.4m/sec
19.9mph 32.0kmh/1000rpm
29.6mpg 24.6mpgUS
9.5L/100km
Petrol 4-stroke piston
1587cc 97.0cu in
In-line 4 1 Carburettor
Compression ratio: 9.5:1
Bore: 81.0mm 3.2in
Stroke: 77.0mm 3.0in
Valve type/No: Overhead 16
Transmission: Manual
No. of forward speeds: 5
Wheels driven: Front
Springs F/R: Coil/Coil
Brake system: PA
Brakes F/R: Disc/Drum
Steering: Rack & pinion PA
Wheelbase: 243.0cm 95.7in
Track F: 143.0cm 56.3in
Track R: 141.0cm 55.5in
Length: 421.5cm 165.9in
Width: 165.5cm 65.2in
Height: 136.5cm 53.7in
Ground clearance: 15.5cm
6.1in
Kerb weight: 1020.0kg
2246.7lb
Fuel: 50.0L 11.0gal 13.2galUS

878 Toyota

Corolla FX-16 GTS
1987 Japan
105.0mph 168.9kmh
0-50mph 80.5kmh: 6.7secs
0-60mph 96.5kmh: 9.2secs
0-1/4 mile: 17.0secs
108.0bhp 80.5kW 109.5PS
@ 6600rpm
96.0lbft 130.1Nm @ 4800rpm
98.1bhp/ton 96.5bhp/tonne
68.0bhp/L 50.7kW/L 69.0PS/L
55.5ft/sec 16.9m/sec
30.6mpg 25.5mpgUS
9.2L/100km
Petrol 4-stroke piston
1587cc 96.8cu in
In-line 4 fuel injection
Compression ratio: 9.4:1
Bore: 81.0mm 3.2in
Stroke: 77.0mm 3.0in
Valve type/No: Overhead 16
Transmission: Manual
No. of forward speeds: 5
Wheels driven: Front
Springs F/R: Coil/Coil

Brake system: PA
Brakes F/R: Disc/Disc
Steering: Rack & pinion PA
Wheelbase: 243.1cm 95.7in
Track F: 143.5cm 56.5in
Track R: 141.5cm 55.7in
Length: 406.4cm 160.0in
Width: 165.6cm 65.2in
Height: 134.1cm 52.8in
Ground clearance: 13.5cm 5.3in
Kerb weight: 1119.1kg 2465.0lb
Fuel: 50.0L 11.0gal 13.2galUS

879 Toyota

Landcruiser
1987 Japan
87.0mph 140.0kmh
0-50mph 80.5kmh: 14.2secs
0-60mph 96.5kmh: 21.6secs
0-1/4 mile: 22.0secs
0-1km: 45.9secs
102.0bhp 76.1kW 103.4PS @ 3500rpm
178.0lbft 241.2Nm @ 1800rpm
47.8bhp/ton 47.0bhp/tonne
25.6bhp/L 19.1kW/L 26.0PS/L
39.1ft/sec 11.9m/sec
26.8mph 43.1kmh/1000rpm
17.9mpg 14.9mpgUS
15.8L/100km
Diesel 4-stroke piston
3980cc 243.0cu in
In-line 6 fuel injection
Compression ratio: 22.7:1
Bore: 91.0mm 3.6in
Stroke: 102.0mm 4.0in
Valve type/No: Overhead 12
Transmission: Manual
No. of forward speeds: 5
Wheels driven: 4-wheel

engageable
Springs F/R: Leaf/Leaf
Brake system: PA
Brakes F/R: Disc/Drum
Steering: Recirculating ball PA
Wheelbase: 273.0cm 107.5in
Track F: 147.5cm 58.1in
Track R: 146.0cm 57.5in
Length: 467.5cm 184.1in
Width: 180.0cm 70.9in
Height: 180.0cm 70.9in
Ground clearance: 18.0cm 7.1in
Kerb weight: 2170.0kg 4779.7lb
Fuel: 90.1L 19.8gal 23.8galUS

880 Toyota

MR2 Supercharged
1987 Japan
135.0mph 217.2kmh
0-50mph 80.5kmh: 5.0secs
0-60mph 96.5kmh: 7.0secs
0-1/4 mile: 15.3secs
145.0bhp 108.1kW 147.0PS @ 6400rpm
140.0lbft 189.7Nm @ 4000rpm
124.0bhp/ton 121.9bhp/tonne
91.4bhp/L 68.1kW/L 92.6PS/L
53.9ft/sec 16.4m/sec
25.3mpg 21.1mpgUS
11.1L/100km
Petrol 4-stroke piston
1587cc 96.8cu in supercharged
In-line 4 fuel injection
Compression ratio: 8.0:1
Bore: 81.0mm 3.2in
Stroke: 77.0mm 3.0in
Valve type/No: Overhead 16
Transmission: Manual
No. of forward speeds: 5

Wheels driven: Rear
Springs F/R: Coil/Coil
Brakes F/R: Disc/Disc
Steering: Rack & pinion
Wheelbase: 231.9cm 91.3in
Track F: 144.0cm 56.7in
Track R: 144.0cm 56.7in
Length: 395.0cm 155.5in
Width: 166.6cm 65.6in
Height: 123.4cm 48.6in
Kerb weight: 1189.5kg 2620.0lb
Fuel: 40.9L 9.0gal 10.8galUS

881 Toyota

Supra
1987 Japan
130.0mph 209.2kmh
0-60mph 96.5kmh: 7.9secs
0-1/4 mile: 16.0secs
200.0bhp 149.1kW 202.8PS @ 6000rpm
185.0lbft 250.7Nm @ 4800rpm
126.9bhp/ton 124.8bhp/tonne
67.7bhp/L 50.5kW/L 68.6PS/L
59.7ft/sec 18.2m/sec
22.8mpg 19.0mpgUS
12.4L/100km
Petrol 4-stroke piston
2954cc 180.2cu in
In-line 6 fuel injection
Compression ratio: 9.2:1
Bore: 83.0mm 3.3in
Stroke: 91.0mm 3.6in
Valve type/No: Overhead 24
Transmission: Manual
No. of forward speeds: 5
Wheels driven: Rear
Springs F/R: Coil/Coil
Brake system: ABS
Brakes F/R: Disc/Disc
Steering: Rack & pinion PA

Wheelbase: 259.6cm 102.2in
Track F: 148.6cm 58.5in
Track R: 148.6cm 58.5in
Length: 462.0cm 181.9in
Width: 174.5cm 68.7in
Height: 131.1cm 51.6in
Kerb weight: 1602.6kg 3530.0lb
Fuel: 70.0L 15.4gal 18.5galUS

882 Toyota

Carina 1.6 GL
1988 Japan
109.0mph 175.4kmh
0-50mph 80.5kmh: 7.6secs
0-60mph 96.5kmh: 10.5secs
0-1/4 mile: 17.9secs
0-1km: 33.3secs
94.0bhp 70.1kW 95.3PS @ 6000rpm
100.4lbft 136.0Nm @ 3600rpm
88.5bhp/ton 87.0bhp/tonne
59.2bhp/L 44.2kW/L 60.0PS/L
51.0ft/sec 15.6m/sec
20.5mph 33.0kmh/1000rpm
28.5mpg 23.7mpgUS
9.9L/100km
Petrol 4-stroke piston
1587cc 97.0cu in
In-line 4 1 Carburettor
Compression ratio: 9.5:1
Bore: 81.0mm 3.2in
Stroke: 77.8mm 3.1in
Valve type/No: Overhead 16
Transmission: Manual
No. of forward speeds: 5
Wheels driven: Front
Springs F/R: Coil/Coil
Brakes F/R: Disc/Drum
Steering: Rack & pinion PA
Wheelbase: 252.5cm 99.4in
Track F: 145.5cm 57.3in

879 Toyota
Landcruiser

884 Toyota
Celica GT4

Track R: 142.5cm 56.1in
Length: 440.0cm 173.2in
Width: 169.0cm 66.5in
Height: 137.0cm 53.9in
Ground clearance: 16.0cm
6.3in
Kerb weight: 1080.0kg
2378.8lb
Fuel: 60.1L 13.2gal 15.9galUS

883 Toyota
Celica All-Trac Turbo
1988 Japan
134.0mph 215.6kmh
0-50mph 80.5kmh: 5.5secs
0-60mph 96.5kmh: 7.7secs
0-1/4 mile: 15.6secs
190.0bhp 141.7kW 192.6PS
@ 6000rpm
190.0lbft 257.5Nm @ 3200rpm
130.1bhp/ton 128.0bhp/tonne
95.1bhp/L 70.9kW/L 96.4PS/L
56.5ft/sec 17.2m/sec
24.0mpg 20.0mpgUS
11.8L/100km
Petrol 4-stroke piston
1998cc 121.9cu in
turbocharged
In-line 4 fuel injection
Compression ratio: 8.5:1
Bore: 86.0mm 3.4in
Stroke: 86.0mm 3.4in
Valve type/No: Overhead 16
Transmission: Manual
No. of forward speeds: 5
Wheels driven: 4-wheel drive
Springs F/R: Coil/Coil
Brake system: PA ABS
Brakes F/R: Disc/Disc
Steering: Rack & pinion PA
Wheelbase: 252.5cm 99.4in
Track F: 147.1cm 57.9in
Track R: 144.5cm 56.9in

Length: 436.6cm 171.9in
Width: 170.9cm 67.3in
Height: 126.5cm 49.8in
Kerb weight: 1484.6kg
3270.0lb
Fuel: 60.2L 13.2gal 15.9galUS

884 Toyota
Celica GT4
1988 Japan
137.0mph 220.4kmh
0-50mph 80.5kmh: 5.6secs
0-60mph 96.5kmh: 7.6secs
0-1/4 mile: 15.7secs
0-1km: 28.9secs
182.0bhp 135.7kW 184.5PS
@ 6000rpm
184.5lbft 250.0Nm @ 3600rpm
125.5bhp/ton 123.4bhp/tonne
91.1bhp/L 67.9kW/L 92.3PS/L
56.5ft/sec 17.2m/sec
23.6mph 38.0kmh/1000rpm
19.2mpg 16.0mpgUS
14.7L/100km
Petrol 4-stroke piston
1998cc 122.0cu in
turbocharged
In-line 4 fuel injection
Compression ratio: 8.5:1
Bore: 86.0mm 3.4in
Stroke: 86.0mm 3.4in
Valve type/No: Overhead 16
Transmission: Manual
No. of forward speeds: 5
Wheels driven: 4-wheel drive
Springs F/R: Coil/Coil
Brake system: PA ABS
Brakes F/R: Disc/Disc
Steering: Rack & pinion PA
Wheelbase: 252.5cm 99.4in
Track F: 146.5cm 57.7in
Track R: 143.0cm 56.3in
Length: 436.5cm 171.9in

Width: 171.0cm 67.3in
Height: 127.0cm 50.0in
Ground clearance: 17.9cm
7.0in
Kerb weight: 1475.0kg
3248.9lb
Fuel: 62.6L 13.7gal 16.5galUS

885 Toyota
Corolla 1.3 GL
1988 Japan
97.0mph 156.1kmh
0-50mph 80.5kmh: 8.7secs
0-60mph 96.5kmh: 12.8secs
0-1/4 mile: 19.0secs
0-1km: 35.9secs
74.0bhp 55.2kW 75.0PS
@ 6200rpm
76.0lbft 103.0Nm @ 4200rpm
76.8bhp/ton 75.5bhp/tonne
57.1bhp/L 42.6kW/L 57.9PS/L
52.5ft/sec 16.0m/sec
19.8mph 31.9kmh/1000rpm
31.2mpg 26.0mpgUS
9.1L/100km
Petrol 4-stroke piston
1295cc 79.0cu in
In-line 4 1 Carburettor
Compression ratio: 9.5:1
Bore: 73.0mm 2.9in
Stroke: 77.4mm 3.0in
Valve type/No: Overhead 8
Transmission: Manual
No. of forward speeds: 5
Wheels driven: Front
Springs F/R: Coil/Coil
Brake system: PA
Brakes F/R: Disc/Drum
Steering: Rack & pinion
Wheelbase: 243.0cm 95.7in
Track F: 143.0cm 56.3in
Track R: 141.0cm 55.5in
Length: 399.5cm 157.3in

Width: 165.5cm 65.2in
Height: 136.5cm 53.7in
Ground clearance: 19.0cm
7.5in
Kerb weight: 980.0kg 2158.6lb
Fuel: 50.0L 11.0gal 13.2galUS

886 Toyota
Corolla 4WD Estate
1988 Japan
97.0mph 156.1kmh
0-50mph 80.5kmh: 8.4secs
0-60mph 96.5kmh: 12.1secs
0-1/4 mile: 18.6secs
0-1km: 35.0secs
94.0bhp 70.1kW 95.3PS
@ 6000rpm
100.4lbft 136.0Nm @ 3600rpm
82.1bhp/ton 80.7bhp/tonne
59.2bhp/L 44.2kW/L 60.0PS/L
51.0ft/sec 15.6m/sec
19.8mph 31.9kmh/1000rpm
21.3mpg 17.7mpgUS
13.3L/100km
Petrol 4-stroke piston
1587cc 97.0cu in
In-line 4 1 Carburettor
Compression ratio: 9.5:1
Bore: 81.0mm 3.2in
Stroke: 77.8mm 3.1in
Valve type/No: Overhead
Transmission: Manual
No. of forward speeds: 5
Wheels driven: 4-wheel drive
Springs F/R: Coil/Coil
Brake system: PA
Brakes F/R: Disc/Drum
Steering: Rack & pinion PA
Wheelbase: 243.0cm 95.7in
Track F: 144.0cm 56.7in
Track R: 138.0cm 54.3in
Length: 425.0cm 167.3in
Width: 165.5cm 65.2in

Height: 145.0cm 57.1in
Ground clearance: 21.5cm
8.5in
Kerb weight: 1165.0kg
2566.1lb
Fuel: 50.0L 11.0gal 13.2galUS

887 Toyota
Landcruiser II TD
1988 Japan
81.0mph 130.3kmh
0-50mph 80.5kmh: 14.9secs
0-60mph 96.5kmh: 23.5secs
0-1/4 mile: 22.3secs
0-1km: 42.4secs
84.0bhp 62.6kW 85.2PS
@ 4000rpm
138.0lbft 187.0Nm @ 2400rpm
47.9bhp/ton 47.1bhp/tonne
34.3bhp/L 25.6kW/L 34.8PS/L
40.2ft/sec 12.3m/sec
20.4mph 32.8kmh/1000rpm
20.2mpg 16.8mpgUS
14.0L/100km
Diesel 4-stroke piston
2446cc 149.0cu in
turbocharged
In-line 4 fuel injection
Compression ratio: 20.0:1
Bore: 92.0mm 3.6in
Stroke: 92.0mm 3.6in
Valve type/No: Overhead 8
Transmission: Manual
No. of forward speeds: 5
Wheels driven: 4-wheel
engageable
Springs F/R: Coil/Coil
Brake system: PA
Brakes F/R: Disc/Drum
Steering: Recirculating ball PA
Wheelbase: 231.0cm 90.9in
Track F: 141.5cm 55.7in
Track R: 146.0cm 57.5in
Length: 406.0cm 159.8in
Width: 179.0cm 70.5in
Height: 187.5cm 73.8in
Ground clearance: 18.0cm
7.1in
Kerb weight: 1784.0kg
3929.5lb
Fuel: 90.1L 19.8gal 23.8galUS

888 Toyota
Supra
1988 Japan
133.0mph 214.0kmh
0-50mph 80.5kmh: 5.0secs
0-60mph 96.5kmh: 7.5secs
0-1/4 mile: 15.9secs
200.0bhp 149.1kW 202.8PS
@ 6000rpm
188.0lbft 254.7Nm @ 3600rpm
129.9bhp/ton 127.7bhp/tonne
67.7bhp/L 50.5kW/L 68.6PS/L

59.7ft/sec 18.2m/sec
21.6mpg 18.0mpgUS
13.1L/100km
Petrol 4-stroke piston
2954cc 180.2cu in
In-line 6 fuel injection
Compression ratio: 9.2:1
Bore: 83.0mm 3.3in
Stroke: 91.0mm 3.6in
Valve type/No: Overhead 24
Transmission: Manual
No. of forward speeds: 5
Wheels driven: Rear
Springs F/R: Coil/Coil
Brake system: PA
Brakes F/R: Disc/Disc
Steering: Rack & pinion PA
Wheelbase: 259.6cm 102.2in
Track F: 148.6cm 58.5in
Track R: 148.6cm 58.5in
Length: 462.0cm 181.9in
Width: 174.5cm 68.7in
Height: 131.1cm 51.6in
Kerb weight: 1566.3kg
3450.0lb
Fuel: 70.0L 15.4gal 18.5galUS

889 Toyota
Supra Turbo
1988 Japan
140.0mph 225.3kmh
0-50mph 80.5kmh: 5.1secs
0-60mph 96.5kmh: 7.0secs
0-1/4 mile: 15.2secs
230.0bhp 171.5kW 233.2PS
@ 5600rpm
246.0lbft 333.3Nm @ 4000rpm
142.5bhp/ton 140.1bhp/tonne
77.9bhp/L 58.1kW/L 78.9PS/L
55.7ft/sec 17.0m/sec
24.0mpg 20.0mpgUS
11.8L/100km
Petrol 4-stroke piston
2954cc 180.2cu in
turbocharged
In-line 6 fuel injection
Compression ratio: 8.4:1
Bore: 83.0mm 3.3in
Stroke: 90.9mm 3.6in
Valve type/No: Overhead 24
Transmission: Manual
No. of forward speeds: 5
Wheels driven: Rear
Springs F/R: Coil/Coil
Brake system: PA ABS
Brakes F/R: Disc/Disc
Steering: Rack & pinion PA
Wheelbase: 259.6cm 102.2in
Track F: 148.6cm 58.5in
Track R: 148.6cm 58.5in
Length: 462.0cm 181.9in
Width: 174.5cm 68.7in
Height: 131.1cm 51.6in
Kerb weight: 1641.2kg

3615.0lb
Fuel: 70.0L 15.4gal 18.5galUS

890 Toyota
Camry V6 GXi
1989 Japan
119.0mph 191.5kmh
0-50mph 80.5kmh: 8.2secs
0-60mph 96.5kmh: 11.1secs
0-1/4 mile: 25.0secs
0-1km: 34.2secs
158.0bhp 117.8kW 160.2PS
@ 5800rpm
152.0lbft 206.0Nm @ 4600rpm
112.8bhp/ton 110.9bhp/tonne
63.0bhp/L 47.0kW/L 63.9PS/L
44.1ft/sec 13.4m/sec
24.2mph 38.9kmh/1000rpm
20.6mpg 17.2mpgUS
13.7L/100km
Petrol 4-stroke piston
2507cc 153.0cu in
Vee 6 fuel injection
Compression ratio: 9.0:1
Bore: 87.5mm 3.4in
Stroke: 69.5mm 2.7in
Valve type/No: Overhead 24
Transmission: Automatic
No. of forward speeds: 4
Wheels driven: Front
Springs F/R: Coil/Coil
Brake system: PA ABS
Brakes F/R: Disc/Disc
Steering: Rack & pinion PA
Wheelbase: 260.0cm 102.4in
Track F: 147.5cm 58.1in
Track R: 144.5cm 56.9in
Length: 452.0cm 178.0in
Width: 171.0cm 67.3in
Height: 140.0cm 55.1in
Ground clearance: 17.0cm
6.7in
Kerb weight: 1425.0kg
3138.8lb
Fuel: 60.1L 13.2gal 15.9galUS

891 Toyota
Carina II 1.6 GL Liftback
1989 Japan
110.0mph 177.0kmh
0-50mph 80.5kmh: 8.7secs
0-60mph 96.5kmh: 12.2secs
0-1/4 mile: 18.6secs
0-1km: 34.0secs
94.0bhp 70.1kW 95.3PS
@ 6000rpm
100.4lbft 136.0Nm @ 3600rpm
87.3bhp/ton 85.8bhp/tonne
59.2bhp/L 44.2kW/L 60.0PS/L
51.0ft/sec 15.6m/sec
20.5mph 33.0kmh/1000rpm
30.0mpg 25.0mpgUS
9.4L/100km
Petrol 4-stroke piston

1587cc 97.0cu in
In-line 4 1 Carburettor
Compression ratio: 9.5:1
Bore: 81.0mm 3.2in
Stroke: 77.8mm 3.1in
Valve type/No: Overhead 16
Transmission: Manual
No. of forward speeds: 5
Wheels driven: Front
Springs F/R: Coil/Coil
Brake system: PA
Brakes F/R: Disc/Disc
Steering: Rack & pinion PA
Wheelbase: 252.5cm 99.4in
Track F: 145.5cm 57.3in
Track R: 142.5cm 56.1in
Length: 440.0cm 173.2in
Width: 169.0cm 66.5in
Height: 137.0cm 53.9in
Ground clearance: 16.0cm
6.3in
Kerb weight: 1095.0kg
2411.9lb
Fuel: 60.1L 13.2gal
15.9galUS

892 Toyota
Corolla GTS
1989 Japan
115.0mph 185.0kmh
0-50mph 80.5kmh: 7.0secs
0-60mph 96.5kmh: 9.3secs
0-1/4 mile: 17.1secs
115.0bhp 85.8kW 116.6PS
@ 6600rpm
100.0lbft 135.5Nm @ 4800rpm
103.2bhp/ton 101.5bhp/tonne
72.5bhp/L 54.0kW/L 73.5PS/L
55.5ft/sec 16.9m/sec
33.6mpg 28.0mpgUS
8.4L/100km
Petrol 4-stroke piston
1587cc 96.8cu in
In-line 4 fuel injection
Compression ratio: 9.4:1
Bore: 81.0mm 3.2in
Stroke: 77.0mm 3.0in
Valve type/No: Overhead 16
Transmission: Manual
No. of forward speeds: 5
Wheels driven: Front
Springs F/R: Coil/Coil
Brake system: PA
Brakes F/R: Disc/Disc
Steering: Rack & pinion PA
Wheelbase: 243.1cm 95.7in
Track F: 144.3cm 56.8in
Track R: 142.5cm 56.1in
Length: 437.6cm 172.3in
Width: 166.6cm 65.6in
Height: 126.0cm 49.6in
Kerb weight: 1132.7kg
2495.0lb
Fuel: 50.0L 11.0gal 13.2galUS

893 Toyota
MR2 Supercharged
1989 Japan
135.0mph 217.2kmh
0-50mph 80.5kmh: 5.3secs
0-60mph 96.5kmh: 7.7secs
0-1/4 mile: 16.0secs
145.0bhp 108.1kW 147.0PS
@ 6400rpm
140.0lbft 189.7Nm @ 4000rpm
124.0bhp/ton 121.9bhp/tonne
91.4bhp/L 68.1kW/L 92.6PS/L
53.9ft/sec 16.4m/sec
27.6mpg 23.0mpgUS
10.2L/100km
Petrol 4-stroke piston
1587cc 96.8cu in
supercharged
In-line 4 fuel injection
Compression ratio: 8.0:1
Bore: 81.0mm 3.2in
Stroke: 77.0mm 3.0in
Valve type/No: Overhead 16
Transmission: Manual
No. of forward speeds: 5
Wheels driven: Rear
Springs F/R: Coil/Coil
Brake system: PA
Brakes F/R: Disc/Disc
Steering: Rack & pinion
Wheelbase: 231.9cm 91.3in
Track F: 144.0cm 56.7in
Track R: 144.0cm 56.7in
Length: 395.0cm 155.5in
Width: 166.6cm 65.6in
Height: 123.4cm 48.6in
Kerb weight: 1189.5kg 2620.0lb
Fuel: 40.9L 9.0gal 10.8galUS

894 Toyota
Supra 3.0i Turbo
1989 Japan

146.0mph 234.9kmh
0-50mph 80.5kmh: 5.3secs
0-60mph 96.5kmh: 6.9secs
0-1/4 mile: 15.9secs
0-1km: 28.0secs
232.0bhp 173.0kW 235.2PS
@ 5600rpm
254.6lbft 345.0Nm @ 3200rpm
149.8bhp/ton 147.3bhp/tonne
78.5bhp/L 58.6kW/L 79.6PS/L
55.7ft/sec 17.0m/sec
25.7mph 41.4kmh/1000rpm
18.8mpg 15.7mpgUS
15.0L/100km
Petrol 4-stroke piston
2954cc 180.0cu in
turbocharged
In-line 6 fuel injection
Compression ratio: 8.4:1
Bore: 83.0mm 3.3in
Stroke: 91.0mm 3.6in
Valve type/No: Overhead 24
Transmission: Manual
No. of forward speeds: 5
Wheels driven: Rear
Springs F/R: Coil/Coil
Brake system: PA ABS
Brakes F/R: Disc/Disc
Steering: Rack & pinion PA
Wheelbase: 259.5cm 102.2in
Track F: 148.5cm 58.5in
Track R: 148.0cm 58.3in
Length: 462.0cm 181.9in
Width: 174.5cm 68.7in
Height: 131.0cm 51.6in
Kerb weight: 1575.0kg
3469.2lb
Fuel: 70.1L 15.4gal 18.5galUS

895 Toyota
Supra Turbo
1989 Japan

140.0mph 225.3kmh
0-50mph 80.5kmh: 5.0secs
0-60mph 96.5kmh: 6.6secs
0-1/4 mile: 15.2secs
233.0bhp 173.7kW 236.2PS
@ 5600rpm
254.0lbft 344.2Nm @ 3200rpm
144.2bhp/ton 141.8bhp/tonne
78.9bhp/L 58.8kW/L 80.0PS/L
55.7ft/sec 17.0m/sec
26.4mpg 22.0mpgUS
10.7L/100km
Petrol 4-stroke piston
2954cc 180.2cu in
turbocharged
In-line 6 fuel injection
Compression ratio: 8.4:1
Bore: 83.0mm 3.3in
Stroke: 90.9mm 3.6in
Valve type/No: Overhead 24
Transmission: Manual
No. of forward speeds: 5
Wheels driven: Rear
Springs F/R: Coil/Coil
Brake system: PA ABS
Brakes F/R: Disc/Disc
Steering: Rack & pinion PA
Wheelbase: 259.6cm 102.2in
Track F: 148.6cm 58.5in
Track R: 148.6cm 58.5in
Length: 462.0cm 181.9in
Width: 174.5cm 68.7in
Height: 131.1cm 51.6in
Kerb weight: 1643.5kg
3620.0lb
Fuel: 70.0L 15.4gal 18.5galUS

896 Toyota
Celica 2.0 GT
1990 Japan
134.0mph 215.6kmh
0-50mph 80.5kmh: 6.2secs

0-60mph 96.5kmh: 8.1secs
0-1/4 mile: 16.5secs
0-1km: 29.7secs
158.0bhp 117.8kW 160.2PS
@ 6600rpm
141.0lbft 191.0Nm @ 4800rpm
131.5bhp/ton 129.3bhp/tonne
79.1bhp/L 59.0kW/L 80.2PS/L
62.1ft/sec 18.9m/sec
21.2mph 34.1kmh/1000rpm
22.5mpg 18.7mpgUS
12.6L/100km
Petrol 4-stroke piston
1998cc 122.0cu in
In-line 4 fuel injection
Compression ratio: 10.0:1
Bore: 86.0mm 3.4in
Stroke: 86.0mm 3.4in
Valve type/No: Overhead 16
Transmission: Manual
No. of forward speeds: 5
Wheels driven: Front
Springs F/R: Coil/Coil
Brake system: PA ABS
Brakes F/R: Disc/Disc
Steering: Rack & pinion PA
Wheelbase: 252.5cm 99.4in
Track F: 146.5cm 57.7in
Track R: 143.0cm 56.3in
Length: 442.0cm 174.0in
Width: 170.5cm 67.1in
Height: 130.0cm 51.2in
Ground clearance: 17.4cm
6.9in
Kerb weight: 1222.0kg
2691.6lb
Fuel: 60.1L 13.2gal 15.9galUS

897 Toyota
Celica All-Trac Turbo
1990 Japan
125.0mph 201.1kmh

894 Toyota
Supra 3.0i Turbo

0-50mph 80.5kmh: 6.6secs
0-60mph 96.5kmh: 9.5secs
0-1/4 mile: 16.8secs
200.0bhp 149.1kW 202.8PS
@ 6000rpm
200.0lbft 271.0Nm @ 3200rpm
137.0bhp/ton 134.7bhp/tonne
100.1bhp/L 74.6kW/L
101.5PS/L
56.5ft/sec 17.2m/sec
Petrol 4-stroke piston
1998cc 121.9cu in
turbocharged
In-line 4 fuel injection
Compression ratio: 8.8:1
Bore: 86.1mm 3.4in
Stroke: 86.1mm 3.4in
Valve type/No: Overhead 16
Transmission: Manual
No. of forward speeds: 5
Wheels driven: 4-wheel drive
Springs F/R: Coil/Coil
Brake system: PA ABS
Brakes F/R: Disc/Disc
Steering: Rack & pinion PA
Wheelbase: 252.5cm 99.4in
Track F: 148.1cm 58.3in
Track R: 144.5cm 56.9in
Length: 442.0cm 174.0in
Width: 174.5cm 68.7in
Height: 128.0cm 50.4in
Kerb weight: 1484.6kg
3270.0lb
Fuel: 60.2L 13.2gal 15.9galUS

898 Toyota

Celica GTS
1990 Japan
125.0mph 201.1kmh
0-60mph 96.5kmh: 10.5secs
0-1/4 mile: 17.6secs
130.0bhp 96.9kW 131.8PS
@ 5400rpm
140.0lbft 189.7Nm @ 4400rpm
97.9bhp/ton 96.2bhp/tonne
60.1bhp/L 44.8kW/L 60.9PS/L
53.7ft/sec 16.4m/sec
29.8mpg 24.8mpgUS
9.5L/100km
Petrol 4-stroke piston
2164cc 132.0cu in
In-line 4 fuel injection
Compression ratio: 9.5:1
Bore: 87.1mm 3.4in
Stroke: 90.9mm 3.6in
Valve type/No: Overhead 16
Transmission: Manual
No. of forward speeds: 5
Wheels driven: Front
Springs F/R: Coil/Coil
Brake system: PA ABS
Brakes F/R: Disc/Disc
Steering: Rack & pinion PA
Wheelbase: 252.5cm 99.4in

Track F: 148.1cm 58.3in
Track R: 144.5cm 56.9in
Length: 442.0cm 174.0in
Width: 174.5cm 68.7in
Height: 128.0cm 50.4in
Kerb weight: 1350.6kg
2975.0lb
Fuel: 60.2L 13.2gal 15.9galUS

899 Toyota

MR2 GT
1990 Japan
139.0mph 223.7kmh
0-50mph 80.5kmh: 5.1secs
0-60mph 96.5kmh: 6.7secs
0-1/4 mile: 15.2secs
0-1km: 27.5secs
158.0bhp 117.8kW 160.2PS
@ 6600rpm
140.2lbft 190.0Nm @ 4800rpm
132.2bhp/ton 130.0bhp/tonne
79.1bhp/L 59.0kW/L 80.2PS/L
62.1ft/sec 18.9m/sec
21.2mph 34.1kmh/1000rpm
27.2mpg 22.6mpgUS
10.4L/100km
Petrol 4-stroke piston
1998cc 122.0cu in
In-line 4 fuel injection
Compression ratio: 10.0:1
Bore: 86.0mm 3.4in
Stroke: 86.0mm 3.4in
Valve type/No: Overhead 16
Transmission: Manual
No. of forward speeds: 5
Wheels driven: Rear
Springs F/R: Coil/Coil
Brake system: PA
Brakes F/R: Disc/Disc
Steering: Rack & pinion
Wheelbase: 240.0cm 94.5in
Track F: 147.0cm 57.9in
Track R: 145.0cm 57.1in
Length: 418.0cm 164.6in
Width: 170.0cm 66.9in
Height: 124.0cm 48.8in
Kerb weight: 1215.0kg
2676.2lb
Fuel: 54.6L 12.0gal 14.4galUS

900 Toyota

MR2 Turbo
1990 Japan
145.0mph 233.3kmh
0-50mph 80.5kmh: 4.6secs
0-60mph 96.5kmh: 6.2secs
0-1/4 mile: 14.7secs
200.0bhp 149.1kW 202.8PS
@ 6000rpm
200.0lbft 271.0Nm @ 3200rpm
158.6bhp/ton 155.9bhp/tonne
100.1bhp/L 74.6kW/L
101.5PS/L
56.5ft/sec 17.2m/sec

22.2mpg 18.5mpgUS
12.7L/100km
Petrol 4-stroke piston
1998cc 121.9cu in
turbocharged
In-line 4 fuel injection
Compression ratio: 8.8:1
Bore: 86.0mm 3.4in
Stroke: 86.0mm 3.4in
Valve type/No: Overhead 16
Transmission: Manual
No. of forward speeds: 5
Wheels driven: Rear
Springs F/R: Coil/Coil
Brake system: PA ABS
Brakes F/R: Disc/Disc
Steering: Rack & pinion
Wheelbase: 240.0cm 94.5in
Track F: 147.1cm 57.9in
Track R: 145.0cm 57.1in
Length: 417.1cm 164.2in
Width: 169.9cm 66.9in
Height: 124.0cm 48.8in
Ground clearance: 14.0cm
5.5in
Kerb weight: 1282.5kg
2825.0lb
Fuel: 54.1L 11.9gal 14.3galUS

901 Toyota

Previa LE
1990 Japan
107.0mph 172.2kmh
0-60mph 96.5kmh: 12.5secs
0-1/4 mile: 18.8secs
138.0bhp 102.9kW 139.9PS
@ 5000rpm
154.0lbft 208.7Nm @ 4000rpm
80.3bhp/ton 78.9bhp/tonne
56.6bhp/L 42.2kW/L 57.4PS/L
47.1ft/sec 14.3m/sec
22.8mpg 19.0mpgUS
12.4L/100km
Petrol 4-stroke piston
2438cc 148.7cu in
In-line 4 fuel injection
Compression ratio: 9.3:1
Bore: 95.0mm 3.7in
Stroke: 86.0mm 3.4in
Valve type/No: Overhead 16
Transmission: Automatic
No. of forward speeds: 4
Wheels driven: Rear
Springs F/R: Coil/Coil
Brake system: PA
Brakes F/R: Disc/Disc
Steering: Rack & pinion PA
Wheelbase: 286.5cm 112.8in
Track F: 156.5cm 61.6in
Track R: 155.4cm 61.2in
Length: 475.0cm 187.0in
Width: 180.1cm 70.9in
Height: 174.5cm 68.7in
Kerb weight: 1747.9kg

3850.0lb
Fuel: 74.9L 16.5gal 19.8galUS

902 Toyota

Starlet GL
1990 Japan
91.0mph 146.4kmh
0-50mph 80.5kmh: 10.3secs
0-60mph 96.5kmh: 14.7secs
0-1/4 mile: 20.1secs
0-1km: 37.4secs
54.0bhp 40.3kW 54.7PS
@ 6000rpm
55.4lbft 75.0Nm @ 3800rpm
70.4bhp/ton 69.2bhp/tonne
54.0bhp/L 40.3kW/L 54.8PS/L
42.0ft/sec 12.8m/sec
19.5mph 31.4kmh/1000rpm
38.1mpg 31.7mpgUS
7.4L/100km
Petrol 4-stroke piston
999cc 61.0cu in
In-line 4 1 Carburettor
Compression ratio: 9.0:1
Bore: 70.5mm 2.8in
Stroke: 64.0mm 2.5in
Valve type/No: Overhead 12
Transmission: Manual
No. of forward speeds: 5
Wheels driven: Front
Springs F/R: Coil/Coil
Brake system: PA
Brakes F/R: Disc/Drum
Steering: Rack & pinion
Wheelbase: 230.1cm 90.6in
Track F: 138.9cm 54.7in
Track R: 136.9cm 53.9in
Length: 372.1cm 146.5in
Width: 160.0cm 63.0in
Height: 138.4cm 54.5in
Ground clearance: 15.2cm
6.0in
Kerb weight: 780.0kg 1718.1lb
Fuel: 40.0L 8.8gal 10.6galUS

903 Toyota

Camry 2.2 GL
1991 Japan
126.0mph 202.7kmh
0-50mph 80.5kmh: 7.0secs
0-60mph 96.5kmh: 9.7secs
0-1/4 mile: 17.4secs
0-1km: 31.7secs
134.0bhp 99.9kW 135.9PS
@ 5400rpm
145.0lbft 196.5Nm @ 4000rpm
103.2bhp/ton 101.5bhp/tonne
61.9bhp/L 46.2kW/L 62.8PS/L
53.7ft/sec 16.4m/sec
23.3mph 37.5kmh/1000rpm
24.5mpg 20.4mpgUS
11.5L/100km
Petrol 4-stroke piston
2164cc 132.0cu in

In-line 4 fuel injection
Compression ratio: 9.8:1
Bore: 87.0mm 3.4in
Stroke: 91.0mm 3.6in
Valve type/No: Overhead 16
Transmission: Manual
No. of forward speeds: 5
Wheels driven: Front
Springs F/R: Coil/Coil
Brake system: PA ABS
Brakes F/R: Disc/Disc
Steering: Rack & pinion PA
Wheelbase: 262.0cm 103.1in
Track F: 154.5cm 60.8in
Track R: 150.0cm 59.1in
Length: 472.5cm 186.0in
Width: 177.0cm 69.7in
Height: 140.0cm 55.1in
Ground clearance: 15.0cm
5.9in
Kerb weight: 1320.0kg
2907.5lb
Fuel: 70.0L 15.4gal 18.5galUS

904 Toyota
Celica GT Convertible
1991 Japan
120.0mph 193.1kmh
0-50mph 80.5kmh: 7.4secs
0-60mph 96.5kmh: 10.8secs
0-1/4 mile: 17.8secs
130.0bhp 96.9kW 131.8PS
@ 5400rpm
140.0lbft 189.7Nm @ 4400rpm
97.2bhp/ton 95.6bhp/tonne
60.1bhp/L 44.8kW/L 60.9PS/L
53.7ft/sec 16.4m/sec
27.6mpg 23.0mpgUS
10.2L/100km
Petrol 4-stroke piston
2164cc 132.0cu in
In-line 4 fuel injection
Compression ratio: 9.5:1
Bore: 87.0mm 3.4in
Stroke: 91.0mm 3.6in
Valve type/No: Overhead 16
Transmission: Manual
No. of forward speeds: 5
Wheels driven: Front
Springs F/R: Coil/Coil
Brake system: PA
Brakes F/R: Disc/Drum
Steering: Rack & pinion PA
Wheelbase: 252.5cm 99.4in
Track F: 147.1cm 57.9in
Track R: 143.5cm 56.5in
Length: 447.0cm 176.0in
Width: 170.4cm 67.1in
Height: 128.5cm 50.6in
Ground clearance: 13.0cm
5.1in
Kerb weight: 1359.7kg
2995.0lb
Fuel: 60.2L 13.2gal 15.9galUS

905 Toyota
Landcruiser VX
1991 Japan
104.0mph 167.3kmh
0-50mph 80.5kmh: 9.2secs
0-60mph 96.5kmh: 12.6secs
0-1/4 mile: 19.2secs
0-1km: 35.0secs
165.0bhp 123.0kW 167.3PS
@ 3600rpm
143.9lbft 195.0Nm @ 1800rpm
74.3bhp/ton 73.1bhp/tonne
39.6bhp/L 29.5kW/L 40.2PS/L
39.4ft/sec 12.0m/sec
26.6mph 42.8kmh/1000rpm
19.8mpg 16.5mpgUS
14.3L/100km
Diesel 4-stroke piston
4164cc 254.0cu in
turbocharged
In-line 6 fuel injection
Compression ratio: 18.6:1
Bore: 94.0mm 3.7in
Stroke: 100.0mm 3.9in
Valve type/No: Overhead 12
Transmission: Manual
No. of forward speeds: 5
Wheels driven: 4-wheel drive
Springs F/R: Coil/Coil
Brake system: PA
Brakes F/R: Disc/Disc
Steering: Recirculating ball PA
Wheelbase: 285.0cm 112.2in
Track F: 157.5cm 62.0in
Track R: 157.5cm 62.0in
Length: 462.8cm 182.2in
Width: 190.0cm 74.8in
Height: 190.0cm 74.8in
Ground clearance: 21.8cm
8.6in
Kerb weight: 2257.0kg
4971.4lb
Fuel: 95.5L 21.0gal 25.2galUS

906 Toyota
MR2
1991 Japan
123.0mph 197.9kmh
0-50mph 80.5kmh: 6.6secs
0-60mph 96.5kmh: 9.3secs
0-1/4 mile: 17.0secs
0-1km: 31.4secs
119.0bhp 88.7kW 120.6PS
@ 5600rpm
93.0lbft 126.0Nm @ 4000rpm
102.6bhp/ton 100.8bhp/tonne
59.6bhp/L 44.4kW/L 60.4PS/L
52.7ft/sec 16.0m/sec
21.3mph 34.3kmh/1000rpm
26.7mpg 22.2mpgUS
10.6L/100km
Petrol 4-stroke piston
1998cc 122.0cu in
In-line 4 fuel injection

Compression ratio: 9.8:1
Bore: 86.0mm 3.4in
Stroke: 86.0mm 3.4in
Valve type/No: Overhead 16
Transmission: Manual
No. of forward speeds: 5
Wheels driven: Rear
Springs F/R: Coil/Coil
Brake system: PA
Brakes F/R: Disc/Disc
Steering: Rack & pinion
Wheelbase: 239.8cm 94.4in
Track F: 146.8cm 57.8in
Track R: 144.8cm 57.0in
Length: 417.8cm 164.5in
Width: 169.9cm 66.9in
Height: 124.0cm 48.8in
Kerb weight: 1180.0kg
2599.1lb
Fuel: 55.1L 12.1gal 14.5galUS

907 Toyota
Previa
1991 Japan
110.0mph 177.0kmh
0-50mph 80.5kmh: 8.8secs
0-60mph 96.5kmh: 12.9secs
0-1/4 mile: 19.1secs
0-1km: 34.7secs
133.0bhp 99.2kW 134.8PS
@ 5000rpm
152.0lbft 206.0Nm @ 4000rpm
78.6bhp/ton 77.3bhp/tonne
54.5bhp/L 40.7kW/L 55.3PS/L
47.1ft/sec 14.3m/sec
22.1mph 35.6kmh/1000rpm
21.6mpg 18.0mpgUS
13.1L/100km
Petrol 4-stroke piston
2438cc 149.0cu in
In-line 4 fuel injection
Compression ratio: 9.3:1
Bore: 95.0mm 3.7in
Stroke: 86.0mm 3.4in
Valve type/No: Overhead 16
Transmission: Manual
No. of forward speeds: 5
Wheels driven: Rear
Springs F/R: Coil/Coil
Brake system: PA
Brakes F/R: Disc/Disc
Steering: Rack & pinion PA
Wheelbase: 286.0cm 112.6in
Track F: 156.0cm 61.4in
Track R: 155.0cm 61.0in
Length: 475.0cm 187.0in
Width: 180.0cm 70.9in
Height: 179.0cm 70.5in
Kerb weight: 1720.0kg 3788.5lb
Fuel: 75.1L 16.5gal 19.8galUS

908 Toyota
Tercel LE
1991 Japan

105.0mph 168.9kmh
0-50mph 80.5kmh: 8.3secs
0-60mph 96.5kmh: 12.2secs
0-1/4 mile: 18.4secs
82.0bhp 61.1kW 83.1PS
@ 5200rpm
90.0lbft 122.0Nm @ 4400rpm
85.8bhp/ton 84.4bhp/tonne
56.3bhp/L 42.0kW/L 57.1PS/L
49.5ft/sec 15.1m/sec
36.0mpg 30.0mpgUS
7.8L/100km
Petrol 4-stroke piston
1457cc 88.9cu in
In-line 4 fuel injection
Compression ratio: 9.3:1
Bore: 73.0mm 2.9in
Stroke: 87.0mm 3.4in
Valve type/No: Overhead 12
Transmission: Manual
No. of forward speeds: 5
Wheels driven: Front
Springs F/R: Coil/Coil
Brake system: PA
Brakes F/R: Disc/Drum
Steering: Rack & pinion PA
Wheelbase: 238.0cm 93.7in
Track F: 140.0cm 55.1in
Track R: 143.0cm 56.3in
Length: 411.0cm 161.8in
Width: 166.1cm 65.4in
Height: 135.1cm 53.2in
Ground clearance: 14.0cm
5.5in
Kerb weight: 971.6kg 2140.0lb
Fuel: 45.0L 9.9gal 11.9galUS

909 Toyota
Paseo
1992 Japan
125.0mph 201.1kmh
0-60mph 96.5kmh: 10.3secs
0-1/4 mile: 17.6secs
100.0bhp 74.6kW 101.4PS
@ 6400rpm
91.0lbft 123.3Nm @ 3200rpm
102.5bhp/ton 100.8bhp/tonne
66.8bhp/L 49.8kW/L 67.7PS/L
61.0ft/sec 18.6m/sec
36.0mpg 30.0mpgUS
7.8L/100km
Petrol 4-stroke piston
1497cc 91.3cu in
In-line 4 fuel injection
Compression ratio: 9.4:1
Bore: 74.0mm 2.9in
Stroke: 87.0mm 3.4in
Valve type/No: Overhead 16
Transmission: Manual
No. of forward speeds: 5
Wheels driven: Front
Springs F/R: Coil/Coil
Brake system: PA
Brakes F/R: Disc/Drum

911 TVR
390 SE

Steering: Rack & pinion PA
Wheelbase: 238.0cm 93.7in
Track F: 140.5cm 55.3in
Track R: 139.4cm 54.9in
Length: 414.5cm 163.2in
Width: 165.6cm 65.2in
Height: 127.5cm 50.2in
Kerb weight: 992.0kg 2185.0lb
Fuel: 45.0L 9.9gal 11.9galUS

910 Treser
T1
1989 Germany
130.0mph 209.2kmh
0-60mph 96.5kmh: 8.5secs
130.0bhp 96.9kW 131.8PS
@ 5800rpm
168.0lbft 227.6Nm @ 4600rpm
128.6bhp/ton 126.4bhp/tonne
73.0bhp/L 54.4kW/L 74.0PS/L
54.8ft/sec 16.7m/sec
Petrol 4-stroke piston
1781cc 108.7cu in
In-line 4 fuel injection
Compression ratio: 10.0:1
Bore: 81.0mm 3.2in
Stroke: 86.4mm 3.4in
Valve type/No: Overhead 16
Transmission: Manual
No. of forward speeds: 5
Wheels driven: Rear
Springs F/R: Coil/Coil
Brake system: PA
Brakes F/R: Disc/Disc
Steering: Rack & pinion PA
Wheelbase: 249.9cm 98.4in
Track F: 145.5cm 57.3in
Track R: 148.8cm 58.6in
Length: 404.4cm 159.2in
Width: 173.0cm 68.1in
Height: 125.0cm 49.2in
Kerb weight: 1028.3kg 2265.0lb
Fuel: 71.9L 15.8gal 19.0galUS

911 TVR
390 SE
1987 UK
144.0mph 231.7kmh
0-50mph 80.5kmh: 4.4secs
0-60mph 96.5kmh: 5.7secs
0-1/4 mile: 14.2secs
0-1km: 25.6secs
275.0bhp 205.1kW 278.8PS
@ 5500rpm
270.0lbft 365.9Nm @ 3500rpm
230.7bhp/ton 226.9bhp/tonne
70.4bhp/L 52.5kW/L 71.4PS/L
46.4ft/sec 14.1m/sec
25.9mph 41.7kmh/1000rpm
16.6mpg 13.8mpgUS
17.0L/100km
Petrol 4-stroke piston
3905cc 238.0cu in
Vee 8 fuel injection
Compression ratio: 10.5:1
Bore: 93.5mm 3.7in
Stroke: 77.1mm 3.0in
Valve type/No: Overhead 16
Transmission: Manual
No. of forward speeds: 5
Wheels driven: Rear
Springs F/R: Coil/Coil
Brake system: PA
Brakes F/R: Disc/Disc
Steering: Rack & pinion PA
Wheelbase: 288.7cm 113.7in
Track F: 145.0cm 57.1in
Track R: 148.0cm 58.3in
Length: 401.3cm 158.0in
Width: 172.8cm 68.0in
Height: 120.5cm 47.4in
Ground clearance: 15.2cm
6.0in
Kerb weight: 1212.0kg 2669.6lb
Fuel: 61.4L 13.5gal
16.2galUS

912 TVR
S Convertible
1987 UK
130.0mph 209.2kmh
0-50mph 80.5kmh: 5.6secs
0-60mph 96.5kmh: 7.6secs
0-1/4 mile: 15.9secs
0-1km: 29.9secs
160.0bhp 119.3kW 162.2PS
@ 6000rpm
162.0lbft 219.5Nm @ 4300rpm
164.7bhp/ton 161.9bhp/tonne
57.3bhp/L 42.7kW/L 58.1PS/L
45.0ft/sec 13.7m/sec
23.3mph 37.5kmh/1000rpm
27.3mpg 22.7mpgUS
10.3L/100km
Petrol 4-stroke piston
2792cc 170.0cu in
Vee 6 fuel injection
Compression ratio: 9.2:1
Bore: 93.0mm 3.7in
Stroke: 68.5mm 2.7in
Valve type/No: Overhead 12
Transmission: Manual
No. of forward speeds: 5
Wheels driven: Rear
Springs F/R: Coil/Coil
Brake system: PA
Brakes F/R: Disc/Drum
Steering: Rack & pinion
Wheelbase: 228.6cm 90.0in
Track F: 139.8cm 55.0in
Track R: 139.8cm 55.0in
Length: 400.0cm 157.5in
Width: 145.0cm 57.1in
Height: 111.7cm 44.0in
Ground clearance: 14.3cm
5.6in
Kerb weight: 988.0kg 2176.2lb
Fuel: 54.6L 12.0gal
14.4galUS

913 TVR
420 SEAC
1989 UK
165.0mph 265.5kmh
0-60mph 96.5kmh: 5.0secs
300.0bhp 223.7kW 304.2PS
@ 5500rpm
290.0lbft 393.0Nm @ 4500rpm
305.4bhp/ton 300.4bhp/tonne
71.0bhp/L 52.9kW/L 71.9PS/L
46.3ft/sec 14.1m/sec
Petrol 4-stroke piston
4228cc 258.0cu in
Vee 8 fuel injection
Compression ratio: 10.5:1
Bore: 93.5mm 3.7in
Stroke: 77.0mm 3.0in
Valve type/No: Overhead 16
Transmission: Manual
No. of forward speeds: 5
Wheels driven: Rear
Springs F/R: Coil/Coil
Brake system: PA
Brakes F/R: Disc/Disc
Steering: Rack & pinion PA
Wheelbase: 238.8cm 94.0in
Track F: 145.0cm 57.1in
Track R: 148.1cm 58.3in
Length: 401.3cm 158.0in
Width: 172.7cm 68.0in
Height: 120.4cm 47.4in
Kerb weight: 998.8kg 2200.0lb
Fuel: 60.9L 13.4gal 16.1galUS

914 TVR
S Convertible
1989 UK
140.0mph 225.3kmh
0-60mph 96.5kmh: 7.0secs
0-1/4 mile: 14.5secs
160.0bhp 119.3kW 162.2PS
@ 6000rpm
162.0lbft 219.5Nm @ 4300rpm

164.8bhp/ton 162.0bhp/tonne
57.3bhp/L 42.7kW/L 58.1PS/L
45.0ft/sec 13.7m/sec
Petrol 4-stroke piston
2792cc 170.3cu in
Vee 6 fuel injection
Compression ratio: 9.2:1
Bore: 93.0mm 3.7in
Stroke: 68.5mm 2.7in
Valve type/No: Overhead 12
Transmission: Manual
No. of forward speeds: 5
Wheels driven: Rear
Springs F/R: Coil/Coil
Brake system: PA
Brakes F/R: Disc/Drum
Steering: Rack & pinion
Wheelbase: 228.6cm 90.0in
Track F: 139.7cm 55.0in
Track R: 139.7cm 55.0in
Length: 400.1cm 157.5in
Width: 145.0cm 57.1in
Height: 111.8cm 44.0in
Kerb weight: 987.4kg 2175.0lb
Fuel: 54.5L 12.0gal 14.4galUS

915 TVR
V8 S
1991 UK
150.0mph 241.4kmh
0-50mph 80.5kmh: 4.1secs
0-60mph 96.5kmh: 5.2secs
0-1/4 mile: 14.0secs
0-1km: 25.3secs
240.0bhp 179.0kW 243.3PS
@ 5750rpm
275.3lbft 373.0Nm @ 4200rpm
239.3bhp/ton 235.3bhp/tonne
60.8bhp/L 45.3kW/L 61.6PS/L
44.7ft/sec 13.6m/sec
26.9mph 43.3kmh/1000rpm
20.1mpg 16.7mpgUS
14.1L/100km
Petrol 4-stroke piston
3950cc 241.0cu in
Vee 8 fuel injection
Compression ratio: 10.5:1
Bore: 94.0mm 3.7in
Stroke: 71.0mm 2.8in
Valve type/No: Overhead 16
Transmission: Manual
No. of forward speeds: 5
Wheels driven: Rear
Springs F/R: Coil/Coil
Brake system: PA
Brakes F/R: Disc/Disc
Steering: Rack & pinion
Wheelbase: 228.6cm 90.0in
Track F: 143.8cm 56.6in
Track R: 143.8cm 56.6in
Length: 395.7cm 155.8in
Width: 166.4cm 65.5in
Height: 122.2cm 48.1in
Ground clearance: 14.0cm

5.5in
Kerb weight: 1020.0kg
2246.7lb
Fuel: 54.6L 12.0gal 14.4galUS

916 Vauxhall
Belmont 1.8 GLSi
1986 UK
120.0mph 193.1kmh
0-50mph 80.5kmh: 6.7secs
0-60mph 96.5kmh: 9.2secs
0-1/4 mile: 17.4secs
0-1km: 31.8secs
115.0bhp 85.8kW 116.6PS
@ 5800rpm
111.4lbft 150.9Nm @ 4800rpm
118.1bhp/ton 116.1bhp/tonne
64.0bhp/L 47.7kW/L 64.9PS/L
50.4ft/sec 15.4m/sec
24.8mph 39.9kmh/1000rpm
32.0mpg 26.6mpgUS
8.8L/100km
Petrol 4-stroke piston
1796cc 109.6cu in
In-line 4 fuel injection
Compression ratio: 9.5:1
Bore: 84.8mm 3.3in
Stroke: 79.5mm 3.1in
Valve type/No: Overhead 8
Transmission: Manual
No. of forward speeds: 5
Wheels driven: Front
Springs F/R: Coil/Coil
Brake system: PA
Brakes F/R: Disc/Drum
Steering: Rack & pinion
Wheelbase: 252.0cm 99.2in
Track F: 139.7cm 55.0in
Track R: 138.2cm 54.4in
Length: 421.9cm 166.1in
Width: 166.4cm 65.5in
Height: 139.7cm 55.0in
Ground clearance: 13.2cm
5.2in
Kerb weight: 990.6kg 2182.0lb
Fuel: 51.9L 11.4gal 13.7galUS

917 Vauxhall
Carlton CD 2.0i
1986 UK
123.0mph 197.9kmh
0-50mph 80.5kmh: 7.3secs
0-60mph 96.5kmh: 10.4secs
0-1/4 mile: 17.3secs
0-1km: 32.3secs
122.0bhp 91.0kW 123.7PS
@ 5400rpm
129.0lbft 174.8Nm @ 2600rpm
97.3bhp/ton 95.7bhp/tonne
61.1bhp/L 45.5kW/L 61.9PS/L
50.8ft/sec 15.5m/sec
22.5mph 36.2kmh/1000rpm
26.7mpg 22.2mpgUS
10.6L/100km

Petrol 4-stroke piston
1998cc 122.0cu in
In-line 4 fuel injection
Compression ratio: 10.0:1
Bore: 86.0mm 3.4in
Stroke: 86.0mm 3.4in
Valve type/No: Overhead 8
Transmission: Manual
No. of forward speeds: 5
Wheels driven: Rear
Springs F/R: Coil/Coil
Brake system: PA
Brakes F/R: Disc/Disc
Steering: Recirculating ball PA
Wheelbase: 273.0cm 107.5in
Track F: 145.0cm 57.1in
Track R: 146.8cm 57.8in
Length: 468.7cm 184.5in
Width: 177.2cm 69.8in
Height: 144.7cm 57.0in
Ground clearance: 17.9cm
7.0in
Kerb weight: 1275.0kg
2808.4lb
Fuel: 70.1L 15.4gal 18.5galUS

918 Vauxhall
Nova 1.3 GL 5-door
1986 UK
102.0mph 164.1kmh
0-50mph 80.5kmh: 8.0secs
0-60mph 96.5kmh: 11.8secs
0-1/4 mile: 18.6secs
0-1km: 34.8secs
70.0bhp 52.2kW 71.0PS
@ 5800rpm
74.5lbft 100.9Nm @ 3800rpm
84.4bhp/ton 83.0bhp/tonne
54.0bhp/L 40.2kW/L 54.7PS/L
46.6ft/sec 14.2m/sec
22.6mph 36.4kmh/1000rpm
35.1mpg 29.2mpgUS
8.0L/100km
Petrol 4-stroke piston
1297cc 79.0cu in
In-line 4 1 Carburettor
Compression ratio: 9.2:1
Bore: 75.0mm 2.9in
Stroke: 73.4mm 2.9in
Valve type/No: Overhead 8
Transmission: Manual
No. of forward speeds: 5
Wheels driven: Front
Springs F/R: Coil/Torsion bar
Brake system: PA
Brakes F/R: Disc/Drum
Steering: Rack & pinion
Wheelbase: 234.3cm 92.2in
Track F: 131.8cm 51.9in
Track R: 133.1cm 52.4in
Length: 362.2cm 142.6in
Width: 153.2cm 60.3in
Height: 136.5cm 53.7in
Ground clearance: 16.5cm

6.5in
Kerb weight: 843.0kg 1856.8lb
Fuel: 41.9L 9.2gal 11.1galUS

919 Vauxhall
Astra GTE 2.0i
1987 UK
125.0mph 201.1kmh
0-50mph 80.5kmh: 6.1secs
0-60mph 96.5kmh: 8.4secs
0-1/4 mile: 16.5secs
0-1km: 30.2secs
124.0bhp 92.5kW 125.7PS
@ 5600rpm
127.6lbft 172.9Nm @ 2600rpm
129.3bhp/ton 127.2bhp/tonne
62.1bhp/L 46.3kW/L 62.9PS/L
52.7ft/sec 16.0m/sec
21.6mph 34.8kmh/1000rpm
33.4mpg 27.8mpgUS
8.5L/100km
Petrol 4-stroke piston
1998cc 122.0cu in
In-line 4 fuel injection
Compression ratio: 10.0:1
Bore: 86.0mm 3.4in
Stroke: 86.0mm 3.4in
Valve type/No: Overhead 8
Transmission: Manual
No. of forward speeds: 5
Wheels driven: Front
Springs F/R: Coil/Coil
Brake system: PA
Brakes F/R: Disc/Drum
Steering: Rack & pinion PA
Wheelbase: 252.0cm 99.2in
Track F: 140.0cm 55.1in
Track R: 140.6cm 55.4in
Length: 399.8cm 157.4in
Width: 166.3cm 65.5in
Height: 139.5cm 54.9in
Ground clearance: 13.3cm
5.2in
Kerb weight: 975.0kg 2147.6lb
Fuel: 51.9L 11.4gal 13.7galUS

920 Vauxhall
Belmont 1.6 GL
1987 UK
112.0mph 180.2kmh
0-50mph 80.5kmh: 7.5secs
0-60mph 96.5kmh: 10.6secs
0-1/4 mile: 18.3secs
0-1km: 33.6secs
82.0bhp 61.1kW 83.1PS
@ 5400rpm
96.0lbft 130.1Nm @ 2600rpm
86.9bhp/ton 85.4bhp/tonne
51.3bhp/L 38.3kW/L 52.0PS/L
48.1ft/sec 14.7m/sec
23.5mph 37.8kmh/1000rpm
34.0mpg 28.3mpgUS
8.3L/100km
Petrol 4-stroke piston

1598cc 97.0cu in
In-line 4 1 Carburettor
Compression ratio: 10.0:1
Bore: 79.0mm 3.1in
Stroke: 81.5mm 3.2in
Valve type/No: Overhead 8
Transmission: Manual
No. of forward speeds: 5
Wheels driven: Front
Springs F/R: Coil/Coil
Brake system: PA
Brakes F/R: Disc/Drum
Steering: Rack & pinion
Wheelbase: 252.0cm 99.2in
Track F: 140.0cm 55.1in
Track R: 140.6cm 55.4in
Length: 421.8cm 166.1in
Width: 166.3cm 65.5in
Height: 140.0cm 55.1in
Ground clearance: 13.3cm
5.2in
Kerb weight: 960.0kg 2114.5lb
Fuel: 51.9L 11.4gal 13.7galUS

921 Vauxhall

Carlton 3000 GSi
1987 UK
135.0mph 217.2kmh
0-50mph 80.5kmh: 6.0secs
0-60mph 96.5kmh: 8.2secs
0-1/4 mile: 16.4secs
0-1km: 29.9secs
177.0bhp 132.0kW 179.4PS
@ 5600rpm
177.0lbft 239.8Nm @ 4400rpm
125.9bhp/ton 123.8bhp/tonne
59.6bhp/L 44.5kW/L 60.4PS/L
42.8ft/sec 13.0m/sec
23.2mph 37.3kmh/1000rpm
24.9mpg 20.7mpgUS
11.3L/100km
Petrol 4-stroke piston
2969cc 181.0cu in
In-line 6 fuel injection
Compression ratio: 9.4:1
Bore: 95.0mm 3.7in
Stroke: 69.8mm 2.7in
Valve type/No: Overhead 12
Transmission: Manual
No. of forward speeds: 5
Wheels driven: Rear
Springs F/R: Coil/Coil
Brake system: PA ABS
Brakes F/R: Disc/Disc
Steering: Recirculating ball PA
Wheelbase: 273.0cm 107.5in
Track F: 145.0cm 57.1in
Track R: 146.8cm 57.8in
Length: 468.7cm 184.5in
Width: 177.2cm 69.8in
Height: 144.7cm 57.0in
Ground clearance: 17.9cm
7.0in
Kerb weight: 1430.0kg

3149.8lb
Fuel: 75.1L 16.5gal 19.8galUS

922 Vauxhall

Cavalier 2.0i CD
1987 UK
115.0mph 185.0kmh
0-50mph 80.5kmh: 6.9secs
0-60mph 96.5kmh: 9.3secs
0-1/4 mile: 17.5secs
0-1km: 32.3secs
115.0bhp 85.8kW 116.6PS
@ 5600rpm
129.1lbft 174.9Nm @ 3000rpm
106.3bhp/ton 104.5bhp/tonne
57.6bhp/L 42.9kW/L 58.4PS/L
52.7ft/sec 16.0m/sec
26.9mph 43.3kmh/1000rpm
26.1mpg 21.7mpgUS
10.8L/100km
Petrol 4-stroke piston
1998cc 122.0cu in
In-line 4 fuel injection
Compression ratio: 9.2:1
Bore: 86.0mm 3.4in
Stroke: 86.0mm 3.4in
Valve type/No: Overhead 8
Transmission: Manual
No. of forward speeds: 5
Wheels driven: Front
Springs F/R: Coil/Coil
Brake system: PA
Brakes F/R: Disc/Drum
Steering: Rack & pinion PA
Wheelbase: 257.4cm 101.3in
Track F: 140.6cm 55.4in
Track R: 140.6cm 55.4in
Length: 426.4cm 167.9in
Width: 166.8cm 65.7in
Height: 138.5cm 54.5in
Ground clearance: 16.5cm
6.5in
Kerb weight: 1100.0kg
2422.9lb
Fuel: 61.0L 13.4gal 16.1galUS

923 Vauxhall

Cavalier SRi 130 4-door
1987 UK
117.0mph 188.3kmh
0-50mph 80.5kmh: 6.0secs
0-60mph 96.5kmh: 8.2secs
0-1/4 mile: 16.2secs
0-1km: 30.3secs
130.0bhp 96.9kW 131.8PS
@ 5650rpm
133.0lbft 180.2Nm @ 4600rpm
125.3bhp/ton 123.2bhp/tonne
65.1bhp/L 48.5kW/L 66.0PS/L
53.2ft/sec 16.2m/sec
21.3mph 34.3kmh/1000rpm
27.0mpg 22.5mpgUS
10.5L/100km
Petrol 4-stroke piston

1998cc 122.0cu in
In-line 4 fuel injection
Compression ratio: 10.0:1
Bore: 86.0mm 3.4in
Stroke: 86.0mm 3.4in
Valve type/No: Overhead 8
Transmission: Manual
No. of forward speeds: 5
Wheels driven: Front
Springs F/R: Coil/Coil
Brake system: PA
Brakes F/R: Disc/Drum
Steering: Rack & pinion
Wheelbase: 257.4cm 101.3in
Track F: 140.6cm 55.4in
Track R: 140.6cm 55.4in
Length: 436.6cm 171.9in
Width: 166.8cm 65.7in
Height: 139.5cm 54.9in
Ground clearance: 14.8cm
5.8in
Kerb weight: 1055.0kg
2323.8lb
Fuel: 61.0L 13.4gal 16.1galUS

924 Vauxhall

Senator 3.0i CD
1987 UK
126.0mph 202.7kmh
0-50mph 80.5kmh: 8.2secs
0-60mph 96.5kmh: 10.9secs
0-1/4 mile: 17.9secs
0-1km: 32.5secs
177.0bhp 132.0kW 179.4PS
@ 5600rpm
177.0lbft 239.8Nm @ 4400rpm
117.6bhp/ton 115.7bhp/tonne
59.6bhp/L 44.5kW/L 60.4PS/L
42.8ft/sec 13.0m/sec
30.9mph 49.7kmh/1000rpm
22.7mpg 18.9mpgUS
12.4L/100km
Petrol 4-stroke piston
2969cc 181.0cu in
In-line 6 fuel injection
Compression ratio: 9.4:1
Bore: 95.0mm 3.7in
Stroke: 69.8mm 2.7in
Valve type/No: Overhead 12
Transmission: Automatic
No. of forward speeds: 4
Wheels driven: Rear
Springs F/R: Coil/Coil
Brake system: PA
Brakes F/R: Disc/Disc
Steering: Recirculating ball PA
Wheelbase: 273.0cm 107.5in
Track F: 144.7cm 57.0in
Track R: 146.8cm 57.8in
Length: 484.3cm 190.7in
Width: 176.2cm 69.4in
Height: 144.7cm 57.0in
Ground clearance: 14.5cm
5.7in

Kerb weight: 1530.0kg
3370.0lb
Fuel: 75.1L 16.5gal 19.8galUS

925 Vauxhall

Carlton L 1.8i
1988 UK
122.0mph 196.3kmh
0-50mph 80.5kmh: 7.5secs
0-60mph 96.5kmh: 10.8secs
0-1/4 mile: 18.1secs
0-1km: 32.9secs
115.0bhp 85.8kW 116.6PS
@ 5600rpm
111.0lbft 150.4Nm @ 4600rpm
97.9bhp/ton 96.2bhp/tonne
64.0bhp/L 47.7kW/L 64.9PS/L
48.7ft/sec 14.8m/sec
22.8mph 36.7kmh/1000rpm
31.4mpg 26.1mpgUS
9.0L/100km
Petrol 4-stroke piston
1796cc 110.0cu in
In-line 4 fuel injection
Compression ratio: 10.0:1
Bore: 84.8mm 3.3in
Stroke: 79.5mm 3.1in
Valve type/No: Overhead 8
Transmission: Manual
No. of forward speeds: 5
Wheels driven: Rear
Springs F/R: Coil/Coil
Brake system: PA ABS
Brakes F/R: Disc/Disc
Steering: Recirculating ball PA
Wheelbase: 273.0cm 107.5in
Track F: 144.7cm 57.0in
Track R: 146.8cm 57.8in
Length: 468.7cm 184.5in
Width: 177.2cm 69.8in
Height: 144.7cm 57.0in
Ground clearance: 17.9cm
7.0in
Kerb weight: 1195.0kg
2632.2lb
Fuel: 75.1L 16.5gal 19.8galUS

926 Vauxhall

Carlton L 2.3D
1988 UK
105.0mph 168.9kmh
0-50mph 80.5kmh: 11.5secs
0-60mph 96.5kmh: 16.2secs
0-1/4 mile: 21.0secs
0-1km: 38.6secs
73.0bhp 54.4kW 74.0PS
@ 4400rpm
101.8lbft 138.0Nm @ 2400rpm
57.7bhp/ton 56.7bhp/tonne
32.3bhp/L 24.1kW/L 32.7PS/L
40.9ft/sec 12.5m/sec
23.7mph 38.1kmh/1000rpm
31.0mpg 25.8mpgUS
9.1L/100km

Diesel 4-stroke piston
2260cc 138.0cu in
In-line 4 fuel injection
Compression ratio: 23.0:1
Bore: 92.0mm 3.6in
Stroke: 85.0mm 3.3in
Valve type/No: Overhead 8
Transmission: Manual
No. of forward speeds: 5
Wheels driven: Rear
Springs F/R: Coil/Coil
Brake system: PA
Brakes F/R: Disc/Disc
Steering: Recirculating ball PA
Wheelbase: 273.0cm 107.5in
Track F: 144.7cm 57.0in
Track R: 146.8cm 57.8in
Length: 468.7cm 184.5in
Width: 177.2cm 69.8in
Height: 144.7cm 57.0in
Ground clearance: 17.9cm
7.0in
Kerb weight: 1287.0kg
2834.8lb
Fuel: 75.1L 16.5gal 19.8galUS

927 Vauxhall

Cavalier 4x4
1988 UK
124.0mph 199.5kmh
0-50mph 80.5kmh: 6.1secs
0-60mph 96.5kmh: 8.6secs
0-1/4 mile: 16.5secs
0-1km: 29.7secs
130.0bhp 96.9kW 131.8PS
@ 5600rpm
133.0lbft 180.2Nm @ 4600rpm
104.4bhp/ton 102.7bhp/tonne
65.1bhp/L 48.5kW/L 66.0PS/L
52.7ft/sec 16.0m/sec
20.3mph 32.7kmh/1000rpm
28.3mpg 23.6mpgUS
10.0L/100km
Petrol 4-stroke piston
1998cc 122.0cu in
In-line 4 fuel injection
Compression ratio: 10.0:1
Bore: 86.0mm 3.4in
Stroke: 86.0mm 3.4in
Valve type/No: Overhead 8
Transmission: Manual
No. of forward speeds: 5
Wheels driven: 4-wheel drive
Springs F/R: Coil/Coil
Brake system: PA
Brakes F/R: Disc/Disc
Steering: Rack & pinion PA
Wheelbase: 260.1cm 102.4in
Track F: 142.5cm 56.1in
Track R: 144.5cm 56.9in
Length: 443.0cm 174.4in
Width: 169.9cm 66.9in
Height: 140.0cm 55.1in
Ground clearance: 15.0cm

5.9in
Kerb weight: 1266.2kg
2789.0lb
Fuel: 61.0L 13.4gal 16.1galUS

928 Vauxhall

Nova GTE
1988 UK
119.0mph 191.5kmh
0-50mph 80.5kmh: 6.5secs
0-60mph 96.5kmh: 9.1secs
0-1/4 mile: 17.0secs
0-1km: 31.5secs
100.0bhp 74.6kW 101.4PS
@ 5600rpm
99.6lbft 135.0Nm @ 3400rpm
116.1bhp/ton 114.2bhp/tonne
62.6bhp/L 46.7kW/L 63.4PS/L
49.9ft/sec 15.2m/sec
19.9mph 32.0kmh/1000rpm
32.5mpg 27.1mpgUS
8.7L/100km
Petrol 4-stroke piston
1598cc 97.0cu in
In-line 4 fuel injection
Compression ratio: 10.0:1
Bore: 79.0mm 3.1in
Stroke: 81.5mm 3.2in
Valve type/No: Overhead 8
Transmission: Manual
No. of forward speeds: 5
Wheels driven: Front
Springs F/R: Coil/Coil
Brake system: PA
Brakes F/R: Disc/Drum
Steering: Rack & pinion
Wheelbase: 234.3cm 92.2in
Track F: 131.8cm 51.9in
Track R: 133.1cm 52.4in
Length: 362.2cm 142.6in
Width: 153.2cm 60.3in
Height: 136.5cm 53.7in
Ground clearance: 16.5cm
6.5in
Kerb weight: 876.0kg 1929.5lb
Fuel: 41.9L 9.2gal 11.1galUS

929 Vauxhall

Astra 1.7 DL
1989 UK
96.0mph 154.5kmh
0-50mph 80.5kmh: 10.6secs
0-60mph 96.5kmh: 15.1secs
0-1/4 mile: 19.8secs
0-1km: 37.6secs
57.0bhp 42.5kW 57.8PS
@ 4600rpm
77.5lbft 105.0Nm @ 2400rpm
58.3bhp/ton 57.3bhp/tonne
33.5bhp/L 25.0kW/L 34.0PS/L
40.0ft/sec 12.2m/sec
23.5mph 37.8kmh/1000rpm
44.2mpg 36.8mpgUS
6.4L/100km

Diesel 4-stroke piston
1699cc 104.0cu in
In-line 4 fuel injection
Compression ratio: 23.0:1
Bore: 82.5mm 3.2in
Stroke: 79.5mm 3.1in
Valve type/No: Overhead 8
Transmission: Manual
No. of forward speeds: 5
Wheels driven: Front
Springs F/R: Coil/Coil
Brake system: PA
Brakes F/R: Disc/Disc
Steering: Rack & pinion
Wheelbase: 252.0cm 99.2in
Track F: 140.0cm 55.1in
Track R: 140.6cm 55.4in
Length: 399.8cm 157.4in
Width: 166.3cm 65.5in
Height: 140.0cm 55.1in
Ground clearance: 13.5cm
5.3in
Kerb weight: 995.0kg 2191.6lb
Fuel: 51.9L 11.4gal 13.7galUS

930 Vauxhall

Cavalier 1.4L
1989 UK
107.0mph 172.2kmh
0-50mph 80.5kmh: 8.9secs
0-60mph 96.5kmh: 12.9secs
0-1/4 mile: 19.1secs
0-1km: 35.2secs
75.0bhp 55.9kW 76.0PS
@ 5600rpm
79.7lbft 108.0Nm @ 3000rpm
71.7bhp/ton 70.5bhp/tonne
53.7bhp/L 40.1kW/L 54.5PS/L
45.0ft/sec 13.7m/sec
22.3mph 35.9kmh/1000rpm
34.8mpg 29.0mpgUS
8.1L/100km
Petrol 4-stroke piston
1396cc 85.0cu in
In-line 4 1 Carburettor
Compression ratio: 9.4:1
Bore: 77.8mm 3.1in
Stroke: 73.4mm 2.9in
Valve type/No: Overhead 8
Transmission: Manual
No. of forward speeds: 5
Wheels driven: Front
Springs F/R: Coil/Torsion bar
Brake system: PA
Brakes F/R: Disc/Drum
Steering: Rack & pinion
Wheelbase: 260.0cm 102.4in
Track F: 142.0cm 55.9in
Track R: 142.6cm 56.1in
Length: 435.0cm 171.3in
Width: 170.0cm 66.9in
Height: 140.0cm 55.1in
Ground clearance: 17.9cm
7.0in

Kerb weight: 1064.0kg
2343.6lb
Fuel: 61.0L 13.4gal 16.1galUS

931 Vauxhall

Cavalier 1.6L
1989 UK
111.0mph 178.6kmh
0-50mph 80.5kmh: 8.7secs
0-60mph 96.5kmh: 12.1secs
0-1/4 mile: 25.0secs
0-1km: 35.0secs
82.0bhp 61.1kW 83.1PS
@ 5200rpm
93.7lbft 127.0Nm @ 2600rpm
78.7bhp/ton 77.4bhp/tonne
51.3bhp/L 38.3kW/L 52.0PS/L
46.4ft/sec 14.1m/sec
24.2mph 38.9kmh/1000rpm
34.1mpg 28.4mpgUS
8.3L/100km
Petrol 4-stroke piston
1598cc 97.0cu in
In-line 4 1 Carburettor
Compression ratio: 10.0:1
Bore: 79.0mm 3.1in
Stroke: 81.5mm 3.2in
Valve type/No: Overhead 8
Transmission: Manual
No. of forward speeds: 5
Wheels driven: Front
Springs F/R: Coil/Coil
Brake system: PA
Brakes F/R: Disc/Drum
Steering: Rack & pinion
Wheelbase: 260.0cm 102.4in
Track F: 142.0cm 55.9in
Track R: 142.6cm 56.1in
Length: 435.0cm 171.3in
Width: 170.0cm 66.9in
Height: 140.0cm 55.1in
Ground clearance: 17.9cm
7.0in
Kerb weight: 1059.0kg
2332.6lb
Fuel: 61.0L 13.4gal 16.1galUS

932 Vauxhall

Cavalier SRi
1989 UK
126.0mph 202.7kmh
0-50mph 80.5kmh: 6.5secs
0-60mph 96.5kmh: 8.9secs
0-1/4 mile: 16.4secs
0-1km: 30.2secs
130.0bhp 96.9kW 131.8PS
@ 5600rpm
132.8lbft 180.0Nm @ 4600rpm
124.7bhp/ton 122.6bhp/tonne
65.1bhp/L 48.5kW/L 66.0PS/L
52.7ft/sec 16.0m/sec
21.2mph 34.1kmh/1000rpm
25.3mpg 21.1mpgUS
11.2L/100km

Petrol 4-stroke piston
1998cc 122.0cu in
In-line 4 fuel injection
Compression ratio: 10.0:1
Bore: 86.0mm 3.4in
Stroke: 86.0mm 3.4in
Valve type/No: Overhead 8
Transmission: Manual
No. of forward speeds: 5
Wheels driven: Front
Springs F/R: Coil/Torsion bar
Brake system: PA ABS
Brakes F/R: Disc/Disc
Steering: Rack & pinion PA
Wheelbase: 260.0cm 102.4in
Track F: 142.6cm 56.1in
Track R: 144.4cm 56.9in
Length: 443.0cm 174.4in
Width: 170.0cm 66.9in
Height: 140.0cm 55.1in
Ground clearance: 14.9cm
5.9in
Kerb weight: 1060.0kg
2334.8lb
Fuel: 61.0L 13.4gal 16.1galUS

933 Vauxhall

Senator 2.5i
1989 UK
128.0mph 206.0kmh
0-50mph 80.5kmh: 6.8secs
0-60mph 96.5kmh: 9.3secs
0-1/4 mile: 17.2secs
0-1km: 31.7secs
140.0bhp 104.4kW 141.9PS
@ 5200rpm
151.3lbft 205.0Nm @ 4200rpm
97.2bhp/ton 95.6bhp/tonne
56.2bhp/L 41.9kW/L 57.0PS/L
39.3ft/sec 12.0m/sec
25.3mph 40.7kmh/1000rpm
22.2mpg 18.5mpgUS
12.7L/100km
Petrol 4-stroke piston
2490cc 152.0cu in
In-line 6 fuel injection
Compression ratio: 9.2:1
Bore: 87.0mm 3.4in
Stroke: 69.0mm 2.7in
Valve type/No: Overhead 12
Transmission: Manual
No. of forward speeds: 5
Wheels driven: Rear
Springs F/R: Coil/Coil
Brake system: PA ABS
Brakes F/R: Disc/Disc
Steering: Recirculating ball PA
Wheelbase: 273.0cm 107.5in
Track F: 145.0cm 57.1in
Track R: 146.8cm 57.8in
Length: 484.5cm 190.7in
Width: 176.3cm 69.4in
Height: 145.2cm 57.2in
Ground clearance: 14.6cm

5.7in
Kerb weight: 1465.0kg
3226.9lb
Fuel: 75.1L 16.5gal 19.8galUS

934 Vauxhall

Senator 3.0i CD
1989 UK
132.0mph 212.4kmh
0-50mph 80.5kmh: 7.0secs
0-60mph 96.5kmh: 9.3secs
0-1/4 mile: 17.2secs
0-1km: 30.2secs
177.0bhp 132.0kW 179.4PS
@ 5600rpm
177.1lbft 240.0Nm @ 4400rpm
117.6bhp/ton 115.7bhp/tonne
59.6bhp/L 44.5kW/L 60.4PS/L
42.9ft/sec 13.1m/sec
28.8mph 46.3kmh/1000rpm
19.3mpg 16.1mpgUS
14.6L/100km
Petrol 4-stroke piston
2969cc 181.0cu in
In-line 6 fuel injection
Compression ratio: 9.4:1
Bore: 95.0mm 3.7in
Stroke: 70.0mm 2.8in
Valve type/No: Overhead 12
Transmission: Automatic
No. of forward speeds: 4
Wheels driven: Rear
Springs F/R: Coil/Coil
Brake system: PA ABS
Brakes F/R: Disc/Disc
Steering: Recirculating ball PA
Wheelbase: 273.0cm 107.5in
Track F: 145.0cm 57.1in
Track R: 146.8cm 57.8in
Length: 484.5cm 190.7in
Width: 176.3cm 69.4in
Height: 145.2cm 57.2in
Ground clearance: 14.6cm
5.7in
Kerb weight: 1530.0kg
3370.0lb
Fuel: 75.1L 16.5gal 19.8galUS

935 Vauxhall

Calibra 2.0i 16v
1990 UK
139.0mph 223.7kmh
0-50mph 80.5kmh: 6.1secs
0-60mph 96.5kmh: 8.1secs
0-1/4 mile: 16.4secs
0-1km: 29.6secs
150.0bhp 111.9kW 152.1PS
@ 6000rpm
144.6lbft 196.0Nm @ 4800rpm
124.4bhp/ton 122.3bhp/tonne
75.1bhp/L 56.0kW/L 76.1PS/L
56.5ft/sec 17.2m/sec
22.1mph 35.6kmh/1000rpm
25.5mpg 21.2mpgUS

11.1L/100km
Petrol 4-stroke piston
1998cc 122.0cu in
In-line 4 fuel injection
Compression ratio: 10.5:1
Bore: 86.0mm 3.4in
Stroke: 86.0mm 3.4in
Valve type/No: Overhead 16
Transmission: Manual
No. of forward speeds: 5
Wheels driven: Front
Springs F/R: Coil/Coil
Brake system: PA ABS
Brakes F/R: Disc/Disc
Steering: Rack & pinion PA
Wheelbase: 260.1cm 102.4in
Track F: 142.5cm 56.1in
Track R: 144.5cm 56.9in
Length: 449.3cm 176.9in
Width: 168.9cm 66.5in
Height: 132.1cm 52.0in
Kerb weight: 1226.0kg
2700.4lb
Fuel: 63.7L 14.0gal 16.8galUS

936 Vauxhall

Carlton 3.0i CDX Estate
1990 UK
132.0mph 212.4kmh
0-50mph 80.5kmh: 6.0secs
0-60mph 96.5kmh: 8.4secs
0-1/4 mile: 16.4secs
0-1km: 30.5secs
177.0bhp 132.0kW 179.4PS
@ 5600rpm
177.1lbft 240.0Nm @ 4400rpm
120.0bhp/ton 118.0bhp/tonne
59.6bhp/L 44.5kW/L 60.4PS/L
42.8ft/sec 13.0m/sec
23.2mph 37.3kmh/1000rpm
21.0mpg 17.5mpgUS
13.5L/100km
Petrol 4-stroke piston
2969cc 181.0cu in
In-line 6 fuel injection
Compression ratio: 9.4:1
Bore: 95.0mm 3.7in
Stroke: 69.8mm 2.7in
Valve type/No: Overhead 12
Transmission: Manual
No. of forward speeds: 5
Wheels driven: Rear
Springs F/R: Coil/Coil
Brake system: PA ABS
Brakes F/R: Disc/Disc
Steering: Recirculating ball PA
Wheelbase: 273.1cm 107.5in
Track F: 144.8cm 57.0in
Track R: 147.3cm 58.0in
Length: 472.9cm 186.2in
Width: 193.3cm 76.1in
Height: 152.9cm 60.2in
Ground clearance: 17.8cm
7.0in

Kerb weight: 1500.0kg
3304.0lb
Fuel: 70.1L 15.4gal 18.5galUS

937 Vauxhall

Carlton GSi 3000 24v
1990 UK
148.0mph 238.1kmh
0-50mph 80.5kmh: 5.3secs
0-60mph 96.5kmh: 7.0secs
0-1/4 mile: 15.4secs
0-1km: 28.1secs
204.0bhp 152.1kW 206.8PS
@ 6000rpm
199.3lbft 270.0Nm @ 3600rpm
139.7bhp/ton 137.4bhp/tonne
68.7bhp/L 51.2kW/L 69.7PS/L
46.0ft/sec 14.0m/sec
24.4mph 39.3kmh/1000rpm
19.5mpg 16.2mpgUS
14.5L/100km
Petrol 4-stroke piston
2969cc 181.0cu in
In-line 6 fuel injection
Compression ratio: 10.1:1
Bore: 95.0mm 3.7in
Stroke: 70.0mm 2.8in
Valve type/No: Overhead 24
Transmission: Manual
No. of forward speeds: 5
Wheels driven: Rear
Springs F/R: Coil/Coil
Brake system: PA ABS
Brakes F/R: Disc/Disc
Steering: Recirculating ball PA
Wheelbase: 273.0cm 107.5in
Track F: 145.0cm 57.1in
Track R: 146.8cm 57.8in
Length: 468.7cm 184.5in
Width: 177.2cm 69.8in
Height: 144.7cm 57.0in
Ground clearance: 17.9cm
7.0in
Kerb weight: 1485.0kg
3270.9lb
Fuel: 75.1L 16.5gal 19.8galUS

938 Vauxhall

Cavalier GSi 2000 16v
1990 UK
134.0mph 215.6kmh
0-50mph 80.5kmh: 5.9secs
0-60mph 96.5kmh: 7.9secs
0-1/4 mile: 16.2secs
0-1km: 29.6secs
150.0bhp 111.9kW 152.1PS
@ 6000rpm
144.6lbft 196.0Nm @ 4800rpm
124.0bhp/ton 121.9bhp/tonne
75.1bhp/L 56.0kW/L 76.1PS/L
56.5ft/sec 17.2m/sec
21.1mph 33.9kmh/1000rpm
25.0mpg 20.8mpgUS
11.3L/100km

Petrol 4-stroke piston
1998cc 122.0cu in
In-line 4 fuel injection
Compression ratio: 10.5:1
Bore: 86.0mm 3.4in
Stroke: 86.0mm 3.4in
Valve type/No: Overhead 16
Transmission: Manual
No. of forward speeds: 5
Wheels driven: Front
Springs F/R: Coil/Coil
Brake system: PA ABS
Brakes F/R: Disc/Disc
Steering: Rack & pinion PA
Wheelbase: 260.0cm 102.4in
Track F: 142.6cm 56.1in
Track R: 144.4cm 56.9in
Length: 443.0cm 174.4in
Width: 170.0cm 66.9in
Height: 140.0cm 55.1in
Ground clearance: 14.9cm
5.9in
Kerb weight: 1230.0kg
2709.2lb
Fuel: 65.1L 14.3gal 17.2galUS

939 Vauxhall
Cavalier GSi 2000 16v 4x4
1990 UK
129.0mph 207.6kmh
0-50mph 80.5kmh: 6.1secs
0-60mph 96.5kmh: 8.5secs
0-1/4 mile: 16.2secs
0-1km: 30.2secs
150.0bhp 111.9kW 152.1PS
@ 6000rpm
144.6lbft 196.0Nm @ 4800rpm
113.5bhp/ton 111.6bhp/tonne
75.1bhp/L 56.0kW/L 76.1PS/L
56.5ft/sec 17.2m/sec
21.1mph 33.9kmh/1000rpm
24.0mpg 20.0mpgUS

11.8L/100km
Petrol 4-stroke piston
1998cc 122.0cu in
In-line 4 fuel injection
Compression ratio: 10.5:1
Bore: 86.0mm 3.4in
Stroke: 86.0mm 3.4in
Valve type/No: Overhead 16
Transmission: Manual
No. of forward speeds: 5
Wheels driven: 4-wheel drive
Springs F/R: Coil/Coil
Brake system: PA ABS
Brakes F/R: Disc/Disc
Steering: Rack & pinion PA
Wheelbase: 260.1cm 102.4in
Track F: 142.5cm 56.1in
Track R: 144.5cm 56.9in
Length: 443.0cm 174.4in
Width: 169.9cm 66.9in
Height: 140.0cm 55.1in
Ground clearance: 15.0cm
5.9in
Kerb weight: 1344.0kg
2960.3lb
Fuel: 65.1L 14.3gal 17.2galUS

940 Vauxhall
Nova 1.5TD Merit
1990 UK
102.0mph 164.1kmh
0-50mph 80.5kmh: 8.9secs
0-60mph 96.5kmh: 12.2secs
0-1/4 mile: 18.9secs
0-1km: 35.1secs
67.0bhp 50.0kW 67.9PS
@ 4600rpm
97.4lbft 132.0Nm @ 2600rpm
83.1bhp/ton 81.7bhp/tonne
45.0bhp/L 33.6kW/L 45.6PS/L
41.3ft/sec 12.6m/sec
24.2mph 38.9kmh/1000rpm

43.8mpg 36.5mpgUS
6.4L/100km
Diesel 4-stroke piston
1488cc 91.0cu in
turbocharged
In-line 4 fuel injection
Compression ratio: 22.5:1
Bore: 76.0mm 3.0in
Stroke: 82.0mm 3.2in
Valve type/No: Overhead 8
Transmission: Manual
No. of forward speeds: 5
Wheels driven: Front
Springs F/R: Coil/Coil
Brake system: PA
Brakes F/R: Disc/Drum
Steering: Rack & pinion
Wheelbase: 234.3cm 92.2in
Track F: 132.0cm 52.0in
Track R: 130.7cm 51.5in
Length: 362.2cm 142.6in
Width: 153.2cm 60.3in
Height: 136.5cm 53.7in
Ground clearance: 14.4cm
5.7in
Kerb weight: 820.0kg 1806.2lb
Fuel: 41.9L 9.2gal 11.1galUS

941 Vauxhall
Calibra 2.0i
1991 UK
128.0mph 206.0kmh
0-50mph 80.5kmh: 6.6secs
0-60mph 96.5kmh: 9.5secs
0-1/4 mile: 17.4secs
0-1km: 31.7secs
115.0bhp 85.8kW 116.6PS
@ 5200rpm
125.5lbft 170.0Nm @ 2600rpm
97.0bhp/ton 95.4bhp/tonne
57.6bhp/L 42.9kW/L 58.4PS/L
49.0ft/sec 14.9m/sec

21.2mph 34.1kmh/1000rpm
29.3mpg 24.4mpgUS
9.6L/100km
Petrol 4-stroke piston
1998cc 122.0cu in
In-line 4 fuel injection
Compression ratio: 9.2:1
Bore: 86.0mm 3.4in
Stroke: 86.0mm 3.4in
Valve type/No: Overhead 8
Transmission: Manual
No. of forward speeds: 5
Wheels driven: Front
Springs F/R: Coil/Coil
Brake system: PA ABS
Brakes F/R: Disc/Disc
Steering: Rack & pinion PA
Wheelbase: 260.1cm 102.4in
Track F: 142.5cm 56.1in
Track R: 144.5cm 56.9in
Length: 449.3cm 176.9in
Width: 168.9cm 66.5in
Height: 132.1cm 52.0in
Kerb weight: 1205.0kg
2654.2lb
Fuel: 63.7L 14.0gal 16.8galUS

942 Vauxhall
Calibra 4x4
1991 UK
132.0mph 212.4kmh
0-50mph 80.5kmh: 6.4secs
0-60mph 96.5kmh: 9.0secs
0-1/4 mile: 16.8secs
0-1km: 30.7secs
150.0bhp 111.9kW 152.1PS
@ 6000rpm
144.6lbft 196.0Nm @ 4800rpm
75.1bhp/L 56.0kW/L 76.1PS/L
112.3bhp/ton 110.5bhp/tonne
56.5ft/sec 17.2m/sec
22.1mph 35.6kmh/1000rpm

940 Vauxhall
Nova 1.5TD Merit

27.3mpg 22.7mpgUS
10.3L/100km
Petrol 4-stroke piston
1998cc 122.0cu in
In-line 4 fuel injection
Compression ratio: 10.5:1
Bore: 86.0mm 3.4in
Stroke: 86.0mm 3.4in
Valve type/No: Overhead 16
Transmission: Manual
No. of forward speeds: 5
Wheels driven: 4-wheel drive
Springs F/R: Coil/Coil
Brake system: PA ABS
Brakes F/R: Disc/Disc
Steering: Rack & pinion PA
Wheelbase: 260.1cm 102.4in
Track F: 142.5cm 56.1in
Track R: 144.5cm 56.9in
Length: 449.3cm 176.9in
Width: 168.9cm 66.5in
Height: 132.1cm 52.0in
Kerb weight: 1358.0kg
2991.2lb
Fuel: 63.7L 14.0gal 16.8galUS

943 Vauxhall
Nova 1.2 Luxe
1991 UK
94.0mph 151.2kmh
0-50mph 80.5kmh: 9.9secs
0-60mph 96.5kmh: 14.5secs
0-1/4 mile: 19.9secs
0-1km: 43.6secs
55.0bhp 41.0kW 55.8PS
@ 5600rpm
64.9lbft 88.0Nm @ 2200rpm
70.1bhp/ton 68.9bhp/tonne
46.0bhp/L 34.3kW/L 46.6PS/L
38.6ft/sec 11.8m/sec
23.0mph 37.0kmh/1000rpm
35.6mpg 29.6mpgUS
7.9L/100km
Petrol 4-stroke piston
1196cc 73.0cu in
In-line 4 1 Carburettor
Compression ratio: 9.2:1
Bore: 78.0mm 3.1in
Stroke: 63.0mm 2.5in
Valve type/No: Overhead 8
Transmission: Manual
No. of forward speeds: 5
Wheels driven: Front
Springs F/R: Coil/Coil
Brake system: PA
Brakes F/R: Disc/Drum
Steering: Rack & pinion
Wheelbase: 234.2cm 92.2in
Track F: 132.1cm 52.0in
Track R: 131.1cm 51.6in
Length: 362.2cm 142.6in
Width: 160.8cm 63.3in
Height: 136.4cm 53.7in
Ground clearance: 14.5cm

5.7in
Kerb weight: 798.0kg 1757.7lb
Fuel: 41.9L 9.2gal 11.1galUS

944 Vauxhall
Senator 3.0i 24v
1991 UK
140.0mph 225.3kmh
0-50mph 80.5kmh: 6.9secs
0-60mph 96.5kmh: 9.1secs
0-1/4 mile: 17.0secs
0-1km: 30.2secs
201.2bhp 150.0kW 204.0PS
@ 6000rpm
199.3lbft 270.0Nm @ 3600rpm
130.3bhp/ton 128.2bhp/tonne
67.8bhp/L 50.5kW/L 68.7PS/L
45.8ft/sec 14.0m/sec
27.6mph 44.4kmh/1000rpm
21.4mpg 17.8mpgUS
13.2L/100km
Petrol 4-stroke piston
2969cc 181.0cu in
In-line 6 fuel injection
Compression ratio: 10.0:1
Bore: 95.0mm 3.7in
Stroke: 69.8mm 2.7in
Valve type/No: Overhead 24
Transmission: Automatic
No. of forward speeds: 4
Wheels driven: Rear
Springs F/R: Coil/Coil
Brake system: PA ABS
Brakes F/R: Disc/Disc
Steering: Recirculating ball PA
Wheelbase: 273.0cm 107.5in
Track F: 145.0cm 57.1in
Track R: 146.8cm 57.8in
Length: 484.5cm 190.7in
Width: 176.3cm 69.4in
Height: 143.9cm 56.7in
Ground clearance: 14.6cm
5.7in
Kerb weight: 1570.0kg
3458.1lb
Fuel: 75.1L 16.5gal 19.8galUS

945 Vauxhall
1.4i GLS
1992 UK
108.0mph 173.8kmh
0-50mph 80.5kmh: 8.7secs
0-60mph 96.5kmh: 12.8secs
0-1/4 mile: 19.0secs
0-1km: 35.3secs
82.0bhp 61.1kW 83.1PS
@ 5800rpm
83.4lbft 113.0Nm @ 3400rpm
80.6bhp/ton 79.2bhp/tonne
58.7bhp/L 43.7kW/L 59.5PS/L
46.6ft/sec 14.2m/sec
22.2mph 35.7kmh/1000rpm
34.7mpg 28.9mpgUS
8.1L/100km

Petrol 4-stroke piston
1398cc 85.0cu in
In-line 4 fuel injection
Compression ratio: 9.8:1
Bore: 77.6mm 3.1in
Stroke: 73.4mm 2.9in
Valve type/No: Overhead 8
Transmission: Manual
No. of forward speeds: 5
Wheels driven: Front
Springs F/R: Coil/Coil
Brake system: PA ABS
Brakes F/R: Disc/Drum
Steering: Rack & pinion PA
Wheelbase: 251.7cm 99.1in
Track F: 142.5cm 56.1in
Track R: 143.0cm 56.3in
Length: 423.9cm 166.9in
Width: 168.9cm 66.5in
Height: 137.9cm 54.3in
Ground clearance: 13.7cm
5.4in
Kerb weight: 1035.0kg
2279.7lb
Fuel: 52.0L 11.4gal 13.7galUS

946 Vauxhall
Cavalier 1.7 GL TD
1992 UK
111.0mph 178.6kmh
0-50mph 80.5kmh: 9.2secs
0-60mph 96.5kmh: 12.8secs
0-1/4 mile: 19.2secs
0-1km: 35.3secs
82.0bhp 61.1kW 83.1PS
@ 4400rpm
124.0lbft 168.0Nm @ 2400rpm
71.0bhp/ton 69.8bhp/tonne
48.6bhp/L 36.3kW/L 49.3PS/L
41.4ft/sec 12.6m/sec
25.8mph 41.5kmh/1000rpm
37.1mpg 30.9mpgUS
7.6L/100km
Diesel 4-stroke piston
1686cc 102.9cu in
In-line 4 fuel injection
Compression ratio: 22.0:1
Bore: 79.0mm 3.1in
Stroke: 86.0mm 3.4in
Valve type/No: Overhead 8
Transmission: Manual
No. of forward speeds: 5
Wheels driven: Front
Springs F/R: Coil/Coil
Brake system: PA
Brakes F/R: Disc/Drum
Steering: Rack & pinion PA
Wheelbase: 260.0cm 102.4in
Track F: 143.0cm 56.3in
Track R: 143.0cm 56.3in
Length: 443.0cm 174.4in
Width: 170.0cm 66.9in
Height: 140.0cm 55.1in
Ground clearance: 17.9cm

7.0in
Kerb weight: 1175.0kg
2588.1lb
Fuel: 61.0L 13.4gal 16.1galUS

947 Vauxhall
Frontera 2.3 TD Estate
1992 UK
85.1mph 137.0kmh
0-50mph 80.5kmh: 12.2secs
0-60mph 96.5kmh: 18.1secs
0-1/4 mile: 20.9secs
0-1km: 39.9secs
100.0bhp 74.6kW 101.4PS
@ 4200rpm
158.7lbft 215.0Nm @ 2200rpm
54.1bhp/ton 53.2bhp/tonne
44.2bhp/L 33.0kW/L 44.9PS/L
39.1ft/sec 11.9m/sec
21.8mph 35.1kmh/1000rpm
22.8mpg 19.0mpgUS
12.4L/100km
Diesel 4-stroke piston
2260cc 138.0cu in
turbocharged
In-line 4 fuel injection
Compression ratio: 23.0:1
Bore: 92.0mm 3.6in
Stroke: 85.0mm 3.3in
Valve type/No: Overhead 8
Transmission: Manual
No. of forward speeds: 5
Wheels driven: 4-wheel
engageable
Springs F/R: Torsion bar/Leaf
Brake system: PA
Brakes F/R: Disc/Disc
Steering: Recirculating ball PA
Wheelbase: 276.0cm 108.7in
Track F: 144.0cm 56.7in
Track R: 144.5cm 56.9in
Length: 470.8cm 185.4in
Width: 172.8cm 68.0in
Height: 171.5cm 67.5in
Kerb weight: 1880.0kg
4141.0lb
Fuel: 80.0L 17.6gal 21.1galUS

948 Vector
W2
1990 USA
200.0mph 321.8kmh
0-60mph 96.5kmh: 4.0secs
0-1/4 mile: 11.8secs
600.0bhp 447.4kW 608.3PS
@ 5700rpm
580.0lbft 785.9Nm @ 4800rpm
483.4bhp/ton 475.4bhp/tonne
100.0bhp/L 74.6kW/L
101.4PS/L
57.5ft/sec 17.5m/sec
Petrol 4-stroke piston
6000cc 366.1cu in
turbocharged

948 Vector
W2

Vee 8 fuel injection
Compression ratio: 8.1:1
Bore: 101.6mm 4.0in
Stroke: 92.2mm 3.6in
Valve type/No: Overhead 16
Transmission: Automatic
No. of forward speeds: 3
Wheels driven: Rear
Springs F/R: Coil/Coil
Brakes F/R: Disc/Disc
Steering: Rack & pinion PA
Wheelbase: 261.6cm 103.0in
Track F: 160.0cm 63.0in
Track R: 165.1cm 65.0in
Length: 436.9cm 172.0in
Width: 193.0cm 76.0in
Height: 108.0cm 42.5in
Kerb weight: 1262.1kg
2780.0lb
Fuel: 106.0L 23.3gal
28.0galUS

949 Vector
W8 Twin Turbo
1991 USA
218.0mph 350.8kmh
0-50mph 80.5kmh: 3.3secs
0-60mph 96.5kmh: 4.2secs
0-1/4 mile: 12.0secs
625.0bhp 466.1kW 633.7PS
@ 5700rpm
630.0lbft 853.7Nm @ 4900rpm
421.7bhp/ton 414.6bhp/tonne
104.6bhp/L 78.0kW/L
106.1PS/L
57.5ft/sec 17.5m/sec
16.2mpg 13.5mpgUS
17.4L/100km
Petrol 4-stroke piston
5973cc 364.4cu in
turbocharged
Vee 8 fuel injection
Compression ratio: 8.0:1

Bore: 101.6mm 4.0in
Stroke: 92.1mm 3.6in
Valve type/No: Overhead 16
Transmission: Automatic
No. of forward speeds: 3
Wheels driven: Rear
Springs F/R: Coil/Coil
Brakes F/R: Disc/Disc
Steering: Rack & pinion PA
Wheelbase: 261.6cm 103.0in
Track F: 160.0cm 63.0in
Track R: 165.1cm 65.0in
Length: 436.9cm 172.0in
Width: 208.3cm 82.0in
Height: 108.0cm 42.5in
Ground clearance: 14.0cm
5.5in
Kerb weight: 1507.3kg
3320.0lb
Fuel: 106.0L 23.3gal
28.0galUS

950 Volkswagen
Caravelle Syncro
1986 Germany
82.0mph 131.9kmh
0-50mph 80.5kmh: 15.8secs
0-60mph 96.5kmh: 24.7secs
0-1/4 mile: 23.0secs
0-1km: 44.7secs
78.0bhp 58.2kW 79.1PS
@ 4600rpm
103.0lbft 139.6Nm @ 2600rpm
47.5bhp/ton 46.7bhp/tonne
40.8bhp/L 30.4kW/L 41.3PS/L
34.6ft/sec 10.6m/sec
18.0mph 29.0kmh/1000rpm
20.7mpg 17.2mpgUS
13.6L/100km
Petrol 4-stroke piston
1913cc 117.0cu in
Flat 4 1 Carburettor
Compression ratio: 8.6:1

Bore: 94.0mm 3.7in
Stroke: 68.9mm 2.7in
Valve type/No: Overhead 8
Transmission: Manual
No. of forward speeds: 4
Wheels driven: 4-wheel drive
Springs F/R: Coil/Coil
Brake system: PA
Brakes F/R: Disc/Drum
Steering: Rack & pinion PA
Wheelbase: 245.5cm 96.7in
Track F: 156.8cm 61.7in
Track R: 156.0cm 61.4in
Length: 457.0cm 179.9in
Width: 184.5cm 72.6in
Height: 199.0cm 78.3in
Ground clearance: 19.0cm
7.5in
Kerb weight: 1669.0kg
3676.2lb
Fuel: 68.2L 15.0gal 18.0galUS

951 Volkswagen
Golf GTi 16v
1986 Germany
125.0mph 201.1kmh
0-50mph 80.5kmh: 5.9secs
0-60mph 96.5kmh: 8.0secs
0-1/4 mile: 16.2secs
0-1km: 29.8secs
139.0bhp 103.6kW 140.9PS
@ 6100rpm
124.0lbft 168.0Nm @ 4600rpm
140.1bhp/ton 137.8bhp/tonne
78.0bhp/L 58.2kW/L 79.1PS/L
57.6ft/sec 17.6m/sec
19.8mph 31.9kmh/1000rpm
25.3mpg 21.1mpgUS
11.2L/100km
Petrol 4-stroke piston
1781cc 109.0cu in
In-line 4 fuel injection
Compression ratio: 10.0:1

Bore: 81.0mm 3.2in
Stroke: 86.4mm 3.4in
Valve type/No: Overhead 16
Transmission: Manual
No. of forward speeds: 5
Wheels driven: Front
Springs F/R: Coil/Coil
Brake system: PA
Brakes F/R: Disc/Disc
Steering: Rack & pinion PA
Wheelbase: 247.5cm 97.4in
Track F: 142.7cm 56.2in
Track R: 142.2cm 56.0in
Length: 398.8cm 157.0in
Width: 167.6cm 66.0in
Height: 142.2cm 56.0in
Ground clearance: 11.5cm
4.5in
Kerb weight: 1009.0kg
2222.5lb
Fuel: 55.1L 12.1gal 14.5galUS

952 Volkswagen
Jetta GT
1986 Germany
119.0mph 191.5kmh
0-50mph 80.5kmh: 6.0secs
0-60mph 96.5kmh: 8.5secs
0-1/4 mile: 16.9secs
0-1km: 31.5secs
112.0bhp 83.5kW 113.5PS
@ 5500rpm
114.0lbft 154.5Nm @ 3100rpm
110.0bhp/ton 108.2bhp/tonne
62.9bhp/L 46.9kW/L 63.8PS/L
51.9ft/sec 15.8m/sec
21.2mph 34.1kmh/1000rpm
25.6mpg 21.3mpgUS
11.0L/100km
Petrol 4-stroke piston
1781cc 109.0cu in
In-line 4 fuel injection
Compression ratio: 10.0:1

Bore: 81.0mm 3.2in
Stroke: 86.4mm 3.4in
Valve type/No: Overhead 8
Transmission: Manual
No. of forward speeds: 5
Wheels driven: Front
Springs F/R: Coil/Coil
Brake system: PA
Brakes F/R: Disc/Disc
Steering: Rack & pinion
Wheelbase: 246.4cm 97.0in
Track F: 142.2cm 56.0in
Track R: 142.2cm 56.0in
Length: 431.8cm 170.0in
Width: 167.6cm 66.0in
Height: 142.2cm 56.0in
Ground clearance: 11.5cm
4.5in
Kerb weight: 1035.0kg
2279.7lb
Fuel: 54.6L 12.0gal 14.4galUS

953 Volkswagen
Quantum Synchro Wagon
1986 Germany
113.0mph 181.8kmh
0-50mph 80.5kmh: 7.0secs
0-60mph 96.5kmh: 9.6secs
0-1/4 mile: 17.0secs
115.0bhp 85.8kW 116.6PS
@ 5500rpm
126.0lbft 170.7Nm @ 3000rpm
85.9bhp/ton 84.4bhp/tonne
51.7bhp/L 38.5kW/L 52.4PS/L
51.9ft/sec 15.8m/sec
23.4mpg 19.5mpgUS
12.1L/100km
Petrol 4-stroke piston
2226cc 135.8cu in
In-line 5 fuel injection
Compression ratio: 8.5:1
Bore: 81.0mm 3.2in
Stroke: 86.4mm 3.4in
Valve type/No: Overhead 10
Transmission: Manual
No. of forward speeds: 5
Wheels driven: 4-wheel drive
Springs F/R: Coil/Coil
Brake system: PA
Brakes F/R: Disc/Disc
Steering: Rack & pinion PA
Wheelbase: 255.0cm 100.4in
Track F: 141.2cm 55.6in
Track R: 143.5cm 56.5in
Length: 465.1cm 183.1in
Width: 169.4cm 66.7in
Height: 154.7cm 60.9in
Kerb weight: 1362.0kg
3000.0lb
Fuel: 70.0L 15.4gal 18.5galUS

954 Volkswagen
Scirocco 16v
1986 Germany

125.0mph 201.1kmh
0-50mph 80.5kmh: 5.8secs
0-60mph 96.5kmh: 7.7secs
0-1/4 mile: 16.1secs
123.0bhp 91.7kW 124.7PS
@ 5800rpm
120.0lbft 162.6Nm @ 4250rpm
116.5bhp/ton 114.6bhp/tonne
69.1bhp/L 51.5kW/L 70.0PS/L
54.8ft/sec 16.7m/sec
30.0mpg 25.0mpgUS
9.4L/100km
Petrol 4-stroke piston
1781cc 108.7cu in
In-line 4 fuel injection
Compression ratio: 10.0:1
Bore: 81.0mm 3.2in
Stroke: 86.4mm 3.4in
Valve type/No: Overhead 16
Transmission: Manual
No. of forward speeds: 5
Wheels driven: Front
Springs F/R: Coil/Coil
Brake system: PA
Brakes F/R: Disc/Disc
Steering: Rack & pinion
Wheelbase: 240.0cm 94.5in
Track F: 138.9cm 54.7in
Track R: 135.9cm 53.5in
Length: 420.9cm 165.7in
Width: 164.6cm 64.8in
Height: 130.6cm 51.4in
Kerb weight: 1073.7kg
2365.0lb
Fuel: 52.2L 11.5gal
13.8galUS

955 Volkswagen
Scirocco GTX 16v
1986 Germany
120.0mph 193.1kmh
0-60mph 96.5kmh: 8.0secs
120.0bhp 89.5kW 121.7PS
@ 6000rpm
118.0lbft 159.9Nm @ 3500rpm
122.2bhp/ton 120.1bhp/tonne
67.4bhp/L 50.2kW/L 68.3PS/L
Petrol 4-stroke piston
1781cc 108.7cu in
In-line 4 fuel injection
Valve type/No: Overhead 16
Transmission: Manual
No. of forward speeds: 5
Wheels driven: Front
Springs F/R: Coil/Coil
Brakes F/R: Disc/Disc
Steering: Rack & pinion PA
Wheelbase: 240.0cm 94.5in
Length: 420.9cm 165.7in
Width: 162.6cm 64.0in
Height: 130.6cm 51.4in
Kerb weight: 998.8kg 2200.0lb
Fuel: 52.2L 11.5gal
13.8galUS

956 Volkswagen
Fox GL
1987 Brazil
100.0mph 160.9kmh
0-50mph 80.5kmh: 7.7secs
0-60mph 96.5kmh: 10.8secs
0-1/4 mile: 18.0secs
81.0bhp 60.4kW 82.1PS
@ 5500rpm
93.0lbft 126.0Nm @ 3250rpm
78.7bhp/ton 77.4bhp/tonne
45.5bhp/L 33.9kW/L 46.1PS/L
51.9ft/sec 15.8m/sec
30.0mpg 25.0mpgUS
9.4L/100km
Petrol 4-stroke piston
1780cc 108.6cu in
In-line 4 fuel injection
Compression ratio: 9.0:1
Bore: 81.0mm 3.2in
Stroke: 86.4mm 3.4in
Valve type/No: Overhead 8
Transmission: Manual
No. of forward speeds: 4
Wheels driven: Front
Springs F/R: Coil/Coil
Brake system: PA
Brakes F/R: Disc/Drum
Steering: Rack & pinion
Wheelbase: 235.7cm 92.8in
Track F: 134.9cm 53.1in
Track R: 136.9cm 53.9in
Length: 415.0cm 163.4in
Width: 160.0cm 63.0in
Height: 136.1cm 53.6in
Ground clearance: 15.2cm
6.0in
Kerb weight: 1046.5kg
2305.0lb
Fuel: 48.1L 10.6gal 12.7galUS

957 Volkswagen
Golf Cabriolet Clipper
1987 Germany
106.0mph 170.6kmh
0-50mph 80.5kmh: 7.4secs
0-60mph 96.5kmh: 10.5secs
0-1/4 mile: 17.8secs
0-1km: 32.8secs
90.0bhp 67.1kW 91.2PS
@ 5200rpm
107.0lbft 145.0Nm @ 3300rpm
96.7bhp/ton 95.1bhp/tonne
50.5bhp/L 37.7kW/L 51.2PS/L
49.1ft/sec 15.0m/sec
20.1mph 32.3kmh/1000rpm
30.0mpg 25.0mpgUS
9.4L/100km
Petrol 4-stroke piston
1781cc 109.0cu in
In-line 4 1 Carburettor
Compression ratio: 10.0:1
Bore: 81.0mm 3.2in
Stroke: 86.4mm 3.4in

Valve type/No: Overhead 8
Transmission: Manual
No. of forward speeds: 5
Wheels driven: Front
Springs F/R: Coil/Torsion bar
Brake system: PA
Brakes F/R: Disc/Drum
Steering: Rack & pinion
Wheelbase: 240.0cm 94.5in
Track F: 140.4cm 55.3in
Track R: 137.2cm 54.0in
Length: 381.5cm 150.2in
Width: 163.0cm 64.2in
Height: 139.5cm 54.9in
Ground clearance: 11.7cm
4.6in
Kerb weight: 946.0kg 2083.7lb
Fuel: 54.6L 12.0gal 14.4galUS

958 Volkswagen
GTi 16v
1987 Germany
122.0mph 196.3kmh
0-50mph 80.5kmh: 6.0secs
0-60mph 96.5kmh: 8.5secs
0-1/4 mile: 16.5secs
123.0bhp 91.7kW 124.7PS
@ 5800rpm
120.0lbft 162.6Nm @ 4250rpm
114.1bhp/ton 112.2bhp/tonne
69.1bhp/L 51.5kW/L 70.1PS/L
54.8ft/sec 16.7m/sec
28.8mpg 24.0mpgUS
9.8L/100km
Petrol 4-stroke piston
1780cc 108.6cu in
In-line 4 fuel injection
Compression ratio: 10.0:1
Bore: 81.0mm 3.2in
Stroke: 86.4mm 3.4in
Valve type/No: Overhead 16
Transmission: Manual
No. of forward speeds: 5
Wheels driven: Front
Springs F/R: Coil/Coil
Brake system: PA
Brakes F/R: Disc/Disc
Steering: Rack & pinion PA
Wheelbase: 247.1cm 97.3in
Track F: 143.0cm 56.3in
Track R: 142.2cm 56.0in
Length: 401.3cm 158.0in
Width: 167.9cm 66.1in
Height: 141.5cm 55.7in
Kerb weight: 1096.4kg
2415.0lb
Fuel: 54.9L 12.1gal 14.5galUS

959 Volkswagen
Jetta GTi 16v
1987 Germany
130.0mph 209.2kmh
0-50mph 80.5kmh: 5.4secs
0-60mph 96.5kmh: 7.1secs

0-1/4 mile: 15.9secs
0-1km: 29.5secs
139.0bhp 103.6kW 140.9PS
@ 6100rpm
123.5lbft 167.3Nm @ 4600rpm
140.0bhp/ton 137.6bhp/tonne
78.0bhp/L 58.2kW/L 79.1PS/L
57.6ft/sec 17.6m/sec
19.8mph 31.9kmh/1000rpm
28.8mpg 24.0mpgUS
9.8L/100km
Petrol 4-stroke piston
1781cc 109.0cu in
In-line 4 fuel injection
Compression ratio: 10.0:1
Bore: 81.0mm 3.2in
Stroke: 86.4mm 3.4in
Valve type/No: Overhead 16
Transmission: Manual
No. of forward speeds: 5
Wheels driven: Front
Springs F/R: Coil/Coil
Brake system: PA
Brakes F/R: Disc/Disc
Steering: Rack & pinion
Wheelbase: 246.4cm 97.0in
Track F: 142.2cm 56.0in
Track R: 142.2cm 56.0in
Length: 431.8cm 170.0in
Width: 167.6cm 66.0in
Height: 142.2cm 56.0in
Ground clearance: 11.5cm
4.5in
Kerb weight: 1010.0kg
2224.7lb
Fuel: 55.1L 12.1gal 14.5galUS

960 Volkswagen

Passat GT 5-door
1987 Germany
104.0mph 167.3kmh
0-50mph 80.5kmh: 7.5secs
0-60mph 96.5kmh: 10.9secs
0-1/4 mile: 18.8secs
0-1km: 34.8secs
90.0bhp 67.1kW 91.2PS
@ 5200rpm
107.0lbft 145.0Nm @ 3300rpm
86.6bhp/ton 85.1bhp/tonne
50.5bhp/L 37.7kW/L 51.2PS/L
49.1ft/sec 15.0m/sec
24.7mph 39.7kmh/1000rpm
30.3mpg 25.2mpgUS
9.3L/100km
Petrol 4-stroke piston
1781cc 109.0cu in
In-line 4 1 Carburettor
Compression ratio: 10.0:1
Bore: 81.0mm 3.2in
Stroke: 86.4mm 3.4in
Valve type/No: Overhead 8
Transmission: Manual
No. of forward speeds: 5
Wheels driven: Front

Springs F/R: Coil/Coil
Brake system: PA
Brakes F/R: Disc/Drum
Steering: Rack & pinion
Wheelbase: 255.0cm 100.4in
Track F: 140.0cm 55.1in
Track R: 140.8cm 55.4in
Length: 443.5cm 174.6in
Width: 168.5cm 66.3in
Height: 138.5cm 54.5in
Ground clearance: 15.2cm
6.0in
Kerb weight: 1057.0kg
2328.2lb
Fuel: 60.1L 13.2gal 15.9galUS

961 Volkswagen

GTi 16v
1988 Germany
122.0mph 196.3kmh
0-50mph 80.5kmh: 6.0secs
0-60mph 96.5kmh: 8.5secs
0-1/4 mile: 16.5secs
123.0bhp 91.7kW 124.7PS
@ 5800rpm
120.0lbft 162.6Nm @ 4250rpm
114.1bhp/ton 112.2bhp/tonne
69.1bhp/L 51.5kW/L 70.1PS/L
54.8ft/sec 16.7m/sec
28.8mpg 24.0mpgUS
9.8L/100km
Petrol 4-stroke piston
1780cc 108.6cu in
In-line 4 fuel injection
Compression ratio: 10.0:1
Bore: 81.0mm 3.2in
Stroke: 86.4mm 3.4in
Valve type/No: Overhead 16
Transmission: Manual
No. of forward speeds: 5
Wheels driven: Front
Springs F/R: Coil/Coil
Brake system: PA
Brakes F/R: Disc/Disc
Steering: Rack & pinion PA
Wheelbase: 247.1cm 97.3in
Track F: 143.0cm 56.3in
Track R: 142.2cm 56.0in
Length: 401.3cm 158.0in
Width: 167.9cm 66.1in
Height: 141.5cm 55.7in
Kerb weight: 1096.4kg
2415.0lb
Fuel: 54.9L 12.1gal 14.5galUS

962 Volkswagen

Jetta GLi 16v
1988 Germany
121.0mph 194.7kmh
0-50mph 80.5kmh: 6.2secs
0-60mph 96.5kmh: 8.8secs
0-1/4 mile: 16.6secs
123.0bhp 91.7kW 124.7PS
@ 5800rpm

120.0lbft 162.6Nm @ 4250rpm
109.1bhp/ton 107.3bhp/tonne
69.1bhp/L 51.5kW/L 70.1PS/L
54.8ft/sec 16.7m/sec
28.8mpg 24.0mpgUS
9.8L/100km
Petrol 4-stroke piston
1780cc 108.6cu in
In-line 4 fuel injection
Compression ratio: 10.0:1
Bore: 81.0mm 3.2in
Stroke: 86.4mm 3.4in
Valve type/No: Overhead 16
Transmission: Manual
No. of forward speeds: 5
Wheels driven: Front
Springs F/R: Coil/Coil
Brake system: PA
Brakes F/R: Disc/Disc
Steering: Rack & pinion PA
Wheelbase: 247.1cm 97.3in
Track F: 143.0cm 56.3in
Track R: 142.2cm 56.0in
Length: 436.1cm 171.7in
Width: 167.9cm 66.1in
Height: 141.5cm 55.7in
Kerb weight: 1146.3kg
2525.0lb
Fuel: 54.9L 12.1gal 14.5galUS

963 Volkswagen

Jetta Turbo Diesel
1988 Germany
104.0mph 167.3kmh
0-50mph 80.5kmh: 9.5secs
0-60mph 96.5kmh: 12.9secs
0-1/4 mile: 19.3secs
0-1km: 35.7secs
70.0bhp 52.2kW 71.0PS
@ 4500rpm
98.2lbft 133.0Nm @ 2600rpm
71.1bhp/ton 69.9bhp/tonne
44.1bhp/L 32.9kW/L 44.7PS/L
42.5ft/sec 13.0m/sec
23.9mph 38.5kmh/1000rpm
39.3mpg 32.7mpgUS
7.2L/100km
Diesel 4-stroke piston
1588cc 97.0cu in
turbocharged
In-line 4 fuel injection
Compression ratio: 23.0:1
Bore: 76.5mm 3.0in
Stroke: 86.4mm 3.4in
Valve type/No: Overhead 8
Transmission: Manual
No. of forward speeds: 5
Wheels driven: Front
Springs F/R: Coil/Coil
Brake system: PA
Brakes F/R: Disc/Drum
Steering: Rack & pinion PA
Wheelbase: 246.4cm 97.0in
Track F: 142.2cm 56.0in

Track R: 142.2cm 56.0in
Length: 431.8cm 170.0in
Width: 167.6cm 66.0in
Height: 142.2cm 56.0in
Ground clearance: 11.5cm
4.5in
Kerb weight: 1001.0kg
2204.8lb
Fuel: 54.6L 12.0gal 14.4galUS

964 Volkswagen

Passat GT
1988 Germany
122.0mph 196.3kmh
0-50mph 80.5kmh: 7.6secs
0-60mph 96.5kmh: 10.3secs
0-1/4 mile: 18.1secs
0-1km: 30.0secs
112.0bhp 83.5kW 113.5PS
@ 5400rpm
117.3lbft 159.0Nm @ 4000rpm
98.5bhp/ton 96.9bhp/tonne
62.9bhp/L 46.9kW/L 63.8PS/L
51.0ft/sec 15.5m/sec
24.6mph 39.6kmh/1000rpm
29.2mpg 24.3mpgUS
9.7L/100km
Petrol 4-stroke piston
1781cc 109.0cu in
In-line 4 fuel injection
Compression ratio: 10.0:1
Bore: 81.0mm 3.2in
Stroke: 86.4mm 3.4in
Valve type/No: Overhead 8
Transmission: Manual
No. of forward speeds: 5
Wheels driven: Front
Springs F/R: Coil/Coil
Brake system: PA
Brakes F/R: Disc/Disc
Steering: Rack & pinion PA
Wheelbase: 262.5cm 103.3in
Track F: 146.0cm 57.5in
Track R: 142.2cm 56.0in
Length: 457.5cm 180.1in
Width: 170.5cm 67.1in
Height: 143.0cm 56.3in
Ground clearance: 10.3cm
4.1in
Kerb weight: 1156.0kg
2546.3lb
Fuel: 70.1L 15.4gal 18.5galUS

965 Volkswagen

Polo C Saloon
1988 Germany
89.0mph 143.2kmh
0-50mph 80.5kmh: 11.8secs
0-60mph 96.5kmh: 16.9secs
0-1/4 mile: 20.6secs
0-1km: 38.7secs
45.0bhp 33.6kW 45.6PS
@ 5600rpm
55.1lbft 74.6Nm @ 3600rpm

57.4bhp/ton 56.4bhp/tonne
43.1bhp/L 32.2kW/L 43.7PS/L
36.1ft/sec 11.0m/sec
16.3mph 26.2kmh/1000rpm
30.9mpg 25.7mpgUS
9.1L/100km
Petrol 4-stroke piston
1043cc 64.0cu in
In-line 4 1 Carburettor
Compression ratio: 9.5:1
Bore: 75.0mm 2.9in
Stroke: 59.0mm 2.3in
Valve type/No: Overhead 8
Transmission: Manual
No. of forward speeds: 5
Wheels driven: Front
Springs F/R: Coil/Coil
Brake system: PA
Brakes F/R: Disc/Drum
Steering: Rack & pinion
Wheelbase: 233.5cm 91.9in
Track F: 132.0cm 52.0in
Track R: 134.6cm 53.0in
Length: 397.5cm 156.5in
Width: 160.0cm 63.0in
Height: 135.5cm 53.3in
Ground clearance: 11.7cm
4.6in
Kerb weight: 797.4kg 1756.4lb
Fuel: 41.9L 9.2gal 11.1galUS

966 Volkswagen

Scirocco 16v
1988 Germany
125.0mph 201.1kmh
0-50mph 80.5kmh: 5.8secs
0-60mph 96.5kmh: 8.3secs
0-1/4 mile: 16.2secs
123.0bhp 91.7kW 124.7PS
@ 5800rpm
120.0lbft 162.6Nm @ 4250rpm
116.5bhp/ton 114.6bhp/tonne
69.1bhp/L 51.5kW/L 70.0PS/L
54.8ft/sec 16.7m/sec
30.0mpg 25.0mpgUS
9.4L/100km
Petrol 4-stroke piston
1781cc 108.7cu in
In-line 4 fuel injection
Compression ratio: 10.0:1
Bore: 81.0mm 3.2in
Stroke: 86.4mm 3.4in
Valve type/No: Overhead 16
Transmission: Manual
No. of forward speeds: 5
Wheels driven: Front
Springs F/R: Coil/Coil
Brake system: PA
Brakes F/R: Disc/Disc
Steering: Rack & pinion
Wheelbase: 240.0cm 94.5in
Track F: 138.9cm 54.7in
Track R: 135.9cm 53.5in
Length: 420.9cm 165.7in

Width: 164.6cm 64.8in
Height: 130.6cm 51.4in
Kerb weight: 1073.7kg
2365.0lb
Fuel: 52.2L 11.5gal 13.8galUS

967 Volkswagen

Caravelle Carat Automatic
1989 Germany
93.0mph 149.6kmh
0-50mph 80.5kmh: 12.0secs
0-60mph 96.5kmh: 16.9secs
0-1/4 mile: 20.6secs
0-1km: 39.0secs
112.0bhp 83.5kW 113.5PS
@ 4800rpm
128.4lbft 174.0Nm @ 2800rpm
65.8bhp/ton 64.7bhp/tonne
53.1bhp/L 39.6kW/L 53.8PS/L
39.9ft/sec 12.2m/sec
19.5mph 31.4kmh/1000rpm
18.1mpg 15.1mpgUS
15.6L/100km
Petrol 4-stroke piston
2109cc 129.0cu in
Flat 4 fuel injection
Compression ratio: 10.5:1
Bore: 94.0mm 3.7in
Stroke: 76.0mm 3.0in
Valve type/No: Overhead 8
Transmission: Automatic
No. of forward speeds: 3
Wheels driven: Rear
Springs F/R: Coil/Mini blocks
Brake system: PA ABS
Brakes F/R: Disc/Drum
Steering: Rack & pinion PA
Wheelbase: 245.5cm 96.7in
Track F: 156.8cm 61.7in
Track R: 156.0cm 61.4in
Length: 457.0cm 179.9in
Width: 184.5cm 72.6in
Height: 199.0cm 78.3in
Ground clearance: 19.0cm
7.5in
Kerb weight: 1730.0kg
3810.6lb
Fuel: 60.5L 13.3gal 16.0galUS

968 Volkswagen

Corrado
1989 Germany
140.0mph 225.3kmh
0-60mph 96.5kmh: 7.8secs
158.0bhp 117.8kW 160.2PS
@ 5600rpm
166.0lbft 224.9Nm @ 4000rpm
143.9bhp/ton 141.5bhp/tonne
88.7bhp/L 66.1kW/L 89.9PS/L
52.9ft/sec 16.1m/sec
Petrol 4-stroke piston
1781cc 108.7cu in
supercharged
In-line 4

Compression ratio: 8.0:1
Bore: 81.0mm 3.2in
Stroke: 86.4mm 3.4in
Valve type/No: Overhead 8
Transmission: Manual
No. of forward speeds: 5
Wheels driven: Front
Springs F/R: Coil/Coil
Brake system: PA
Brakes F/R: Disc/Disc
Steering: Rack & pinion PA
Wheelbase: 247.1cm 97.3in
Track F: 143.5cm 56.5in
Track R: 142.7cm 56.2in
Length: 404.9cm 159.4in
Width: 167.4cm 65.9in
Height: 131.8cm 51.9in
Kerb weight: 1116.8kg
2460.0lb
Fuel: 54.9L 12.1gal
14.5galUS

969 Volkswagen

Corrado 16v
1989 Germany
132.0mph 212.4kmh
0-50mph 80.5kmh: 6.4secs
0-60mph 96.5kmh: 8.7secs
0-1/4 mile: 16.5secs
136.0bhp 101.4kW 137.9PS
@ 6300rpm
118.8lbft 161.0Nm @ 4800rpm
119.7bhp/ton 117.7bhp/tonne
76.4bhp/L 56.9kW/L 77.4PS/L
59.3ft/sec 18.1m/sec
19.8mph 31.9kmh/1000rpm
25.9mpg 21.6mpgUS
10.9L/100km
Petrol 4-stroke piston
1781cc 109.0cu in
In-line 4 fuel injection
Compression ratio: 10.0:1
Bore: 81.0mm 3.2in
Stroke: 86.0mm 3.4in
Valve type/No: Overhead 16
Transmission: Manual
No. of forward speeds: 5
Wheels driven: Front
Springs F/R: Coil/Torsion bar
Brake system: PA
Brakes F/R: Disc/Disc
Steering: Rack & pinion PA
Wheelbase: 247.5cm 97.4in
Track F: 142.7cm 56.2in
Track R: 142.2cm 56.0in
Length: 404.8cm 159.4in
Width: 167.4cm 65.9in
Height: 131.8cm 51.9in
Ground clearance: 11.5cm
4.5in
Kerb weight: 1155.0kg
2544.0lb
Fuel: 54.6L 12.0gal
14.4galUS

970 Volkswagen

Golf CL Catalyst
1989 Germany
99.0mph 159.3kmh
0-50mph 80.5kmh: 8.0secs
0-60mph 96.5kmh: 11.5secs
0-1/4 mile: 18.4secs
0-1km: 34.5secs
72.0bhp 53.7kW 73.0PS
@ 5200rpm
92.0lbft 124.7Nm @ 2700rpm
77.9bhp/ton 76.6bhp/tonne
45.1bhp/L 33.7kW/L 45.8PS/L
44.1ft/sec 13.4m/sec
24.6mph 39.6kmh/1000rpm
38.4mpg 32.0mpgUS
7.4L/100km
Petrol 4-stroke piston
1595cc 97.0cu in
In-line 4 1 Carburettor
Compression ratio: 9.0:1
Bore: 81.0mm 3.2in
Stroke: 77.4mm 3.0in
Valve type/No: Overhead 8
Transmission: Manual
No. of forward speeds: 5
Wheels driven: Front
Springs F/R: Coil/Coil
Brake system: PA
Brakes F/R: Disc/Drum
Steering: Rack & pinion
Wheelbase: 247.5cm 97.4in
Track F: 142.7cm 56.2in
Track R: 142.2cm 56.0in
Length: 398.8cm 157.0in
Width: 167.6cm 66.0in
Height: 142.2cm 56.0in
Ground clearance: 11.5cm 4.5in
Kerb weight: 940.0kg 2070.5lb
Fuel: 54.6L 12.0gal 14.4galUS

971 Volkswagen

Jetta Syncro
1989 Germany
108.0mph 173.8kmh
0-50mph 80.5kmh: 8.0secs
0-60mph 96.5kmh: 11.3secs
0-1/4 mile: 18.5secs
0-1km: 34.2secs
90.0bhp 67.1kW 91.2PS
@ 5200rpm
107.0lbft 145.0Nm @ 3300rpm
80.3bhp/ton 78.9bhp/tonne
50.5bhp/L 37.7kW/L 51.2PS/L
49.1ft/sec 15.0m/sec
19.6mph 31.5kmh/1000rpm
22.3mpg 18.6mpgUS
12.7L/100km
Petrol 4-stroke piston
1781cc 109.0cu in
In-line 4 1 Carburettor
Compression ratio: 10.0:1
Bore: 81.0mm 3.2in
Stroke: 86.4mm 3.4in

**976
Volkswagen**
Corrado G60

Valve type/No: Overhead 8
Transmission: Manual
No. of forward speeds: 5
Wheels driven: 4-wheel drive
Springs F/R: Coil/Torsion bar
Brake system: PA
Brakes F/R: Disc/Drum
Steering: Rack & pinion PA
Wheelbase: 246.4cm 97.0in
Track F: 142.2cm 56.0in
Track R: 142.2cm 56.0in
Length: 431.8cm 170.0in
Width: 167.6cm 66.0in
Height: 142.2cm 56.0in
Ground clearance: 11.5cm
4.5in
Kerb weight: 1140.0kg
2511.0lb
Fuel: 54.6L 12.0gal 14.4galUS

972 Volkswagen
Passat GT 16v
1989 Germany
129.0mph 207.6kmh
0-50mph 80.5kmh: 6.4secs
0-60mph 96.5kmh: 8.8secs
0-1/4 mile: 17.3secs
0-1km: 31.7secs
136.0bhp 101.4kW 137.9PS
@ 6300rpm
119.6lbft 162.0Nm @ 4800rpm
108.5bhp/ton 106.7bhp/tonne
76.4bhp/L 56.9kW/L 77.4PS/L
59.5ft/sec 18.1m/sec
25.0mph 40.2kmh/1000rpm
27.4mpg 22.8mpgUS
10.3L/100km
Petrol 4-stroke piston
1781cc 108.0cu in
In-line 4 fuel injection
Compression ratio: 10.0:1
Bore: 81.0mm 3.2in
Stroke: 86.4mm 3.4in

Valve type/No: Overhead 16
Transmission: Manual
No. of forward speeds: 5
Wheels driven: Front
Springs F/R: Coil/Torsion bar
Brake system: PA ABS
Brakes F/R: Disc/Disc
Steering: Rack & pinion PA
Wheelbase: 262.5cm 103.3in
Track F: 146.0cm 57.5in
Track R: 142.2cm 56.0in
Length: 457.5cm 180.1in
Width: 170.5cm 67.1in
Height: 143.0cm 56.3in
Ground clearance: 10.3cm
4.1in
Kerb weight: 1275.0kg
2808.4lb
Fuel: 70.1L 15.4gal 18.5galUS

973 Volkswagen
Corrado G60
1990 Germany
137.0mph 220.4kmh
0-50mph 80.5kmh: 6.4secs
0-60mph 96.5kmh: 8.9secs
0-1/4 mile: 16.5secs
158.0bhp 117.8kW 160.2PS
@ 5600rpm
166.0lbft 224.9Nm @ 4000rpm
131.3bhp/ton 129.1bhp/tonne
88.7bhp/L 66.1kW/L 89.9PS/L
52.9ft/sec 16.1m/sec
25.8mpg 21.5mpgUS
10.9L/100km
Petrol 4-stroke piston
1781cc 108.7cu in
supercharged
In-line 4 fuel injection
Compression ratio: 8.0:1
Bore: 81.0mm 3.2in
Stroke: 86.4mm 3.4in
Valve type/No: Overhead 8

Transmission: Manual
No. of forward speeds: 5
Wheels driven: Front
Springs F/R: Coil/Coil
Brake system: PA ABS
Brakes F/R: Disc/Disc
Steering: Rack & pinion PA
Wheelbase: 247.1cm 97.3in
Track F: 143.5cm 56.5in
Track R: 142.7cm 56.2in
Length: 404.9cm 159.4in
Width: 167.4cm 65.9in
Height: 131.8cm 51.9in
Ground clearance: 12.7cm
5.0in
Kerb weight: 1223.5kg
2695.0lb
Fuel: 54.9L 12.1gal 14.5galUS

974 Volkswagen
Passat CL TD Estate
1990 Germany
103.0mph 165.7kmh
0-50mph 80.5kmh: 10.7secs
0-60mph 96.5kmh: 15.6secs
0-1/4 mile: 20.1secs
0-1km: 37.5secs
80.0bhp 59.7kW 81.1PS
@ 4500rpm
114.4lbft 155.0Nm @ 2600rpm
65.3bhp/ton 64.3bhp/tonne
50.4bhp/L 37.6kW/L 51.1PS/L
42.5ft/sec 13.0m/sec
24.2mph 38.9kmh/1000rpm
37.0mpg 30.8mpgUS
7.6L/100km
Diesel 4-stroke piston
1588cc 97.0cu in
turbocharged
In-line 4 fuel injection
Compression ratio: 23.0:1
Bore: 76.5mm 3.0in
Stroke: 86.4mm 3.4in

Valve type/No: Overhead 8
Transmission: Manual
No. of forward speeds: 5
Wheels driven: Front
Springs F/R: Coil/Coil
Brake system: PA
Brakes F/R: Disc/Disc
Steering: Rack & pinion PA
Wheelbase: 262.5cm 103.3in
Track F: 146.0cm 57.5in
Track R: 142.2cm 56.0in
Length: 457.5cm 180.1in
Width: 170.5cm 67.1in
Height: 143.0cm 56.3in
Ground clearance: 10.3cm
4.1in
Kerb weight: 1245.0kg
2742.3lb
Fuel: 70.1L 15.4gal 18.5galUS

975 Volkswagen
Passat GL
1990 Germany
125.0mph 201.1kmh
0-50mph 80.5kmh: 7.5secs
0-60mph 96.5kmh: 10.9secs
0-1/4 mile: 17.8secs
134.0bhp 99.9kW 135.9PS
@ 5800rpm
133.0lbft 180.2Nm @ 4400rpm
100.6bhp/ton 98.9bhp/tonne
67.5bhp/L 50.4kW/L 68.5PS/L
58.8ft/sec 17.9m/sec
30.0mpg 25.0mpgUS
9.4L/100km
Petrol 4-stroke piston
1984cc 121.0cu in
In-line 4 fuel injection
Compression ratio: 10.8:1
Bore: 82.5mm 3.2in
Stroke: 92.8mm 3.6in
Valve type/No: Overhead 16
Transmission: Manual

No. of forward speeds: 5
Wheels driven: Front
Springs F/R: Coil/Coil
Brake system: PA ABS
Brakes F/R: Disc/Disc
Steering: Rack & pinion PA
Wheelbase: 262.4cm 103.3in
Track F: 147.8cm 58.2in
Track R: 142.0cm 55.9in
Length: 457.2cm 180.0in
Width: 170.4cm 67.1in
Height: 142.7cm 56.2in
Ground clearance: 15.2cm
6.0in
Kerb weight: 1355.2kg 2985.0lb
Fuel: 70.0L 15.4gal 18.5galUS

976 Volkswagen

Corrado G60
1991 Germany
139.0mph 223.7kmh
0-50mph 80.5kmh: 6.4secs
0-60mph 96.5kmh: 8.9secs
0-1/4 mile: 16.8secs
0-1km: 30.3secs
160.0bhp 119.3kW 162.2PS
@ 5600rpm
166.1lbft 225.0Nm @ 4000rpm
141.5bhp/ton 139.1bhp/tonne
89.8bhp/L 67.0kW/L 91.1PS/L
52.9ft/sec 16.1m/sec
23.9mph 38.5kmh/1000rpm
22.0mpg 18.3mpgUS
12.8L/100km
Petrol 4-stroke piston
1781cc 109.0cu in
supercharged
In-line 4 fuel injection
Compression ratio: 8.0:1
Bore: 81.0mm 3.2in
Stroke: 86.4mm 3.4in
Valve type/No: Overhead 8
Transmission: Manual
No. of forward speeds: 5
Wheels driven: Front
Springs F/R: Coil/Coil
Brake system: PA ABS
Brakes F/R: Disc/Disc
Steering: Rack & pinion PA
Wheelbase: 247.4cm 97.4in
Track F: 147.1cm 57.9in
Track R: 142.0cm 55.9in
Length: 404.6cm 159.3in
Width: 167.4cm 65.9in
Height: 131.6cm 51.8in
Ground clearance: 11.4cm
4.5in
Kerb weight: 1150.0kg
2533.0lb
Fuel: 55.1L 12.1gal 14.5galUS

977 Volkswagen

GTi 16v
1991 Germany

122.0mph 196.3kmh
0-60mph 96.5kmh: 8.4secs
0-1/4 mile: 16.8secs
134.0bhp 99.9kW 135.9PS
@ 5800rpm
133.0lbft 180.2Nm @ 4400rpm
119.8bhp/ton 117.8bhp/tonne
67.5bhp/L 50.4kW/L 68.5PS/L
58.8ft/sec 17.9m/sec
30.0mpg 25.0mpgUS
9.4L/100km
Petrol 4-stroke piston
1984cc 121.0cu in
In-line 4 fuel injection
Compression ratio: 10.8:1
Bore: 82.5mm 3.2in
Stroke: 92.8mm 3.6in
Valve type/No: Overhead 16
Transmission: Manual
No. of forward speeds: 5
Wheels driven: Front
Springs F/R: Coil/Coil
Brake system: PA
Brakes F/R: Disc/Disc
Steering: Rack & pinion PA
Wheelbase: 247.1cm 97.3in
Track F: 143.0cm 56.3in
Track R: 142.2cm 56.0in
Length: 404.1cm 159.1in
Width: 167.9cm 66.1in
Height: 141.5cm 55.7in
Kerb weight: 1137.3kg
2505.0lb
Fuel: 54.9L 12.1gal 14.5galUS

978 Volkswagen

Jetta GLi 16v
1991 Germany
125.0mph 201.1kmh
0-50mph 80.5kmh: 6.5secs
0-60mph 96.5kmh: 9.1secs
0-1/4 mile: 16.8secs
134.0bhp 99.9kW 135.9PS
@ 5800rpm
133.0lbft 180.2Nm @ 4400rpm
123.0bhp/ton 121.0bhp/tonne
67.5bhp/L 50.4kW/L 68.5PS/L
58.8ft/sec 17.9m/sec
28.8mpg 24.0mpgUS
9.8L/100km
Petrol 4-stroke piston
1984cc 121.0cu in
In-line 4 fuel injection
Compression ratio: 10.8:1
Bore: 82.5mm 3.2in
Stroke: 92.8mm 3.6in
Valve type/No: Overhead 16
Transmission: Manual
No. of forward speeds: 5
Wheels driven: Front
Springs F/R: Coil/Coil
Brake system: PA ABS
Brakes F/R: Disc/Disc
Steering: Rack & pinion

Wheelbase: 247.1cm 97.3in
Track F: 143.0cm 56.3in
Track R: 142.2cm 56.0in
Length: 438.4cm 172.6in
Width: 167.9cm 66.1in
Height: 141.5cm 55.7in
Kerb weight: 1107.8kg
2440.0lb
Fuel: 54.9L 12.1gal 14.5galUS

979 Volkswagen

Polo 1.3 CL Coupe
1991 Germany
99.0mph 159.3kmh
0-50mph 80.5kmh: 9.5secs
0-60mph 96.5kmh: 13.5secs
0-1/4 mile: 19.4secs
0-1km: 32.6secs
55.0bhp 41.0kW 55.8PS
@ 5200rpm
71.6lbft 97.0Nm @ 3000rpm
72.6bhp/ton 71.4bhp/tonne
43.2bhp/L 32.2kW/L 43.8PS/L
40.9ft/sec 12.5m/sec
21.2mph 34.1kmh/1000rpm
35.2mpg 29.3mpgUS
8.0L/100km
Petrol 4-stroke piston
1272cc 78.0cu in
In-line 4 fuel injection
Compression ratio: 10.0:1
Bore: 75.0mm 2.9in
Stroke: 72.0mm 2.8in
Valve type/No: Overhead 8
Transmission: Manual
No. of forward speeds: 5
Wheels driven: Front
Springs F/R: Coil/Coil
Brake system: PA
Brakes F/R: Disc/Drum
Steering: Rack & pinion
Wheelbase: 253.7cm 99.9in
Track F: 130.6cm 51.4in
Track R: 133.1cm 52.4in
Length: 375.9cm 148.0in
Width: 157.0cm 61.8in
Height: 132.1cm 52.0in
Ground clearance: 10.4cm
4.1in
Kerb weight: 770.0kg 1696.0lb
Fuel: 41.9L 9.2gal 11.1galUS

980 Volkswagen

Polo G40
1991 Germany
120.0mph 193.1kmh
0-50mph 80.5kmh: 6.1secs
0-60mph 96.5kmh: 8.4secs
0-1/4 mile: 16.5secs
0-1km: 26.2secs
113.0bhp 84.3kW 114.6PS
@ 6000rpm
110.7lbft 150.0Nm @ 3600rpm
136.0bhp/ton 133.7bhp/tonne

88.8bhp/L 66.2kW/L 90.1PS/L
47.2ft/sec 14.4m/sec
21.7mph 34.9kmh/1000rpm
30.1mpg 25.1mpgUS
9.4L/100km
Petrol 4-stroke piston
1272cc 78.0cu in
supercharged
In-line 4 fuel injection
Compression ratio: 8.0:1
Bore: 75.0mm 2.9in
Stroke: 72.0mm 2.8in
Valve type/No: Overhead 8
Transmission: Manual
No. of forward speeds: 5
Wheels driven: Front
Springs F/R: Coil/Coil
Brake system: PA
Brakes F/R: Disc/Drum
Steering: Rack & pinion
Wheelbase: 233.5cm 91.9in
Track F: 130.6cm 51.4in
Track R: 133.2cm 52.4in
Length: 376.5cm 148.2in
Width: 157.0cm 61.8in
Height: 132.5cm 52.2in
Ground clearance: 10.4cm
4.1in
Kerb weight: 845.0kg 1861.2lb
Fuel: 42.0L 9.2gal 11.1galUS

981 Volkswagen

Polo GT Coupe
1991 Germany
104.0mph 167.3kmh
0-50mph 80.5kmh: 8.6secs
0-60mph 96.5kmh: 11.7secs
0-1/4 mile: 18.7secs
0-1km: 34.4secs
75.0bhp 55.9kW 76.0PS
@ 5900rpm
73.1lbft 99.0Nm @ 3600rpm
94.7bhp/ton 93.2bhp/tonne
59.0bhp/L 44.0kW/L 59.8PS/L
46.4ft/sec 14.2m/sec
17.8mph 28.6kmh/1000rpm
34.9mpg 29.1mpgUS
8.1L/100km
Petrol 4-stroke piston
1272cc 78.0cu in
In-line 4 fuel injection
Compression ratio: 10.0:1
Bore: 75.0mm 2.9in
Stroke: 72.0mm 2.8in
Valve type/No: Overhead 8
Transmission: Manual
No. of forward speeds: 5
Wheels driven: Front
Springs F/R: Coil/Coil
Brake system: PA
Brakes F/R: Disc/Disc
Steering: Rack & pinion
Wheelbase: 253.7cm 99.9in
Track F: 130.6cm 51.4in

Track R: 133.1cm 52.4in
Length: 375.9cm 148.0in
Width: 157.0cm 61.8in
Height: 132.1cm 52.0in
Ground clearance: 10.4cm
4.1in
Kerb weight: 805.0kg 1773.1lb
Fuel: 41.9L 9.2gal 11.1galUS

982 Volkswagen

Golf 1.8 GL
1992 Germany
107.0mph 172.2kmh
0-50mph 80.5kmh: 9.0secs
0-60mph 96.5kmh: 12.7secs
0-1/4 mile: 19.2secs
0-1km: 35.1secs
90.0bhp 67.1kW 91.2PS
@ 5500rpm
107.0lbft 145.0Nm @ 2500rpm
84.4bhp/ton 83.0bhp/tonne
50.5bhp/L 37.7kW/L 51.2PS/L
51.9ft/sec 15.8m/sec
21.1mph 33.9kmh/1000rpm
32.2mpg 26.8mpgUS
8.8L/100km
Petrol 4-stroke piston
1781cc 108.7cu in
In-line 4 fuel injection
Compression ratio: 10.0:1
Bore: 81.0mm 3.2in
Stroke: 86.4mm 3.4in
Valve type/No: Overhead 8
Transmission: Manual
No. of forward speeds: 5
Wheels driven: Front
Springs F/R: Coil/Coil
Brake system: PA
Brakes F/R: Disc/Drum
Steering: Rack & pinion PA
Wheelbase: 247.5cm 97.4in
Track F: 147.0cm 57.9in

Track R: 146.0cm 57.5in
Length: 402.0cm 158.3in
Width: 169.0cm 66.5in
Height: 142.5cm 56.1in
Kerb weight: 1084.0kg
2387.7lb
Fuel: 55.1L 12.1gal 14.5galUS

983 Volkswagen

Golf GTi
1992 Germany
121.0mph 194.7kmh
0-50mph 80.5kmh: 7.2secs
0-60mph 96.5kmh: 9.9secs
0-1/4 mile: 17.5secs
0-1km: 32.1secs
115.0bhp 85.8kW 116.6PS
@ 5400rpm
122.0lbft 165.3Nm @ 3200rpm
58.0bhp/L 43.2kW/L 58.8PS/L
54.7ft/sec 16.7m/sec
22.1mph 35.6kmh/1000rpm
28.8mpg 24.0mpgUS
9.8L/100km
Petrol 4-stroke piston
1984cc 121.0cu in
In-line 4 fuel injection
Compression ratio: 10.4:1
Bore: 82.5mm 3.2in
Stroke: 92.8mm 3.6in
Valve type/No: Overhead 8
Transmission: Manual
No. of forward speeds: 5
Wheels driven: Front
Springs F/R: Coil/Coil
Brake system: PA
Brakes F/R: Disc/Disc
Steering: Rack & pinion PA
Wheelbase: 247.5cm 97.4in
Track F: 146.2cm 57.6in
Track R: 144.4cm 56.9in
Length: 402.0cm 158.3in

Width: 171.0cm 67.3in
Height: 140.5cm 55.3in
Fuel: 55.1L 12.1gal 14.5galUS

984 Volvo

340 GLE
1986 Sweden
107.0mph 172.2kmh
0-50mph 80.5kmh: 9.5secs
0-60mph 96.5kmh: 13.4secs
0-1/4 mile: 18.8secs
0-1km: 35.5secs
80.0bhp 59.7kW 81.1PS
@ 5400rpm
96.0lbft 130.1Nm @ 3000rpm
79.0bhp/ton 77.7bhp/tonne
46.5bhp/L 34.7kW/L 47.1PS/L
49.3ft/sec 15.0m/sec
21.9mph 35.2kmh/1000rpm
29.1mpg 24.2mpgUS
9.7L/100km
Petrol 4-stroke piston
1721cc 105.0cu in
In-line 4 1 Carburettor
Compression ratio: 10.0:1
Bore: 81.0mm 3.2in
Stroke: 83.5mm 3.3in
Valve type/No: Overhead 8
Transmission: Manual
No. of forward speeds: 5
Wheels driven: Rear
Springs F/R: Coil/Leaf
Brake system: PA
Brakes F/R: Disc/Drum
Steering: Rack & pinion
Wheelbase: 239.5cm 94.3in
Track F: 136.9cm 53.9in
Track R: 139.7cm 55.0in
Length: 443.0cm 174.4in
Width: 165.9cm 65.3in
Height: 143.0cm 56.3in
Ground clearance: 15.2cm

6.0in
Kerb weight: 1029.7kg
2268.0lb
Fuel: 45.0L 9.9gal 11.9galUS

985 Volvo

480 ES
1987 Sweden
112.0mph 180.2kmh
0-50mph 80.5kmh: 7.3secs
0-60mph 96.5kmh: 10.3secs
0-1/4 mile: 17.7secs
0-1km: 32.8secs
109.0bhp 81.3kW 110.5PS
@ 5800rpm
103.0lbft 139.6Nm @ 4000rpm
108.7bhp/ton 106.9bhp/tonne
63.3bhp/L 47.2kW/L 64.2PS/L
53.0ft/sec 16.1m/sec
21.3mph 34.3kmh/1000rpm
28.8mpg 24.0mpgUS
9.8L/100km
Petrol 4-stroke piston
1721cc 105.0cu in
In-line 4 fuel injection
Compression ratio: 10.5:1
Bore: 81.0mm 3.2in
Stroke: 83.5mm 3.3in
Valve type/No: Overhead 8
Transmission: Manual
No. of forward speeds: 5
Wheels driven: Front
Springs F/R: Coil/Coil
Brake system: PA
Brakes F/R: Disc/Drum
Steering: Rack & pinion PA
Wheelbase: 250.2cm 98.5in
Track F: 141.0cm 55.5in
Track R: 141.0cm 55.5in
Length: 425.8cm 167.6in
Width: 171.0cm 67.3in
Height: 131.8cm 51.9in

**983
Volkswagen**
Golf GTi

988 Volvo
440 Turbo

Ground clearance: 10.8cm
4.3in
Kerb weight: 1020.0kg
2246.7lb
Fuel: 47.8L 10.5gal 12.6galUS

986 Volvo
740 Turbo
1987 Sweden
128.0mph 206.0kmh
0-50mph 80.5kmh: 5.9secs
0-60mph 96.5kmh: 8.3secs
0-1/4 mile: 16.4secs
0-1km: 29.8secs
182.0bhp 135.7kW 184.5PS
@ 5800rpm
192.0lbft 260.2Nm @ 3400rpm
135.1bhp/ton 132.8bhp/tonne
78.6bhp/L 58.6kW/L 79.7PS/L
50.7ft/sec 15.5m/sec
25.2mph 40.5kmh/1000rpm
21.8mpg 18.2mpgUS
13.0L/100km
Petrol 4-stroke piston
2316cc 141.0cu in
turbocharged
In-line 4 fuel injection
Compression ratio: 9.0:1
Bore: 96.0mm 3.8in
Stroke: 80.0mm 3.1in
Valve type/No: Overhead 8
Transmission: Manual
No. of forward speeds: 5
Wheels driven: Rear
Springs F/R: Coil/Coil
Brake system: PA
Brakes F/R: Disc/Disc
Steering: Rack & pinion PA
Wheelbase: 277.1cm 109.1in
Track F: 146.0cm 57.5in
Track R: 146.0cm 57.5in
Length: 478.7cm 188.5in
Width: 176.0cm 69.3in

Height: 143.0cm 56.3in
Ground clearance: 17.9cm
7.0in
Kerb weight: 1370.0kg
3017.6lb
Fuel: 60.1L 13.2gal 15.9galUS

987 Volvo
760 GLE
1987 Sweden
114.0mph 183.4kmh
0-50mph 80.5kmh: 7.2secs
0-60mph 96.5kmh: 9.6secs
0-1/4 mile: 17.5secs
0-1km: 32.1secs
170.0bhp 126.8kW 172.4PS
@ 5400rpm
177.0lbft 239.8Nm @ 4500rpm
118.6bhp/ton 116.6bhp/tonne
59.7bhp/L 44.5kW/L 60.5PS/L
43.0ft/sec 13.1m/sec
25.8mph 41.5kmh/1000rpm
19.5mpg 16.2mpgUS
14.5L/100km
Petrol 4-stroke piston
2849cc 174.0cu in
Vee 6 fuel injection
Compression ratio: 10.0:1
Bore: 91.0mm 3.6in
Stroke: 73.0mm 2.9in
Valve type/No: Overhead 12
Transmission: Automatic
No. of forward speeds: 4
Wheels driven: Rear
Springs F/R: Coil/Coil
Brake system: PA ABS
Brakes F/R: Disc/Disc
Steering: Rack & pinion PA
Wheelbase: 277.0cm 109.1in
Track F: 147.0cm 57.9in
Track R: 152.0cm 59.8in
Length: 479.0cm 188.6in
Width: 176.0cm 69.3in

Height: 141.0cm 55.5in
Kerb weight: 1458.0kg
3211.4lb
Fuel: 80.1L 17.6gal 21.1galUS

988 Volvo
440 Turbo
1989 Sweden
127.0mph 204.3kmh
0-50mph 80.5kmh: 6.8secs
0-60mph 96.5kmh: 9.6secs
0-1/4 mile: 17.1secs
0-1km: 31.5secs
120.0bhp 89.5kW 121.7PS
@ 5400rpm
129.2lbft 175.0Nm @ 1800rpm
114.0bhp/ton 112.1bhp/tonne
69.7bhp/L 52.0kW/L 70.7PS/L
49.3ft/sec 15.0m/sec
23.1mph 37.2kmh/1000rpm
22.5mpg 18.7mpgUS
12.6L/100km
Petrol 4-stroke piston
1721cc 105.0cu in
turbocharged
In-line 4 fuel injection
Compression ratio: 8.1:1
Bore: 81.0mm 3.2in
Stroke: 83.5mm 3.3in
Valve type/No: Overhead 8
Transmission: Manual
No. of forward speeds: 5
Wheels driven: Front
Springs F/R: Coil/Coil
Brake system: PA ABS
Brakes F/R: Disc/Disc
Steering: Rack & pinion PA
Wheelbase: 250.5cm 98.6in
Track F: 142.0cm 55.9in
Track R: 143.0cm 56.3in
Length: 431.2cm 169.8in
Width: 171.0cm 67.3in
Height: 140.0cm 55.1in

Ground clearance: 17.9cm
7.0in
Kerb weight: 1070.0kg
2356.8lb
Fuel: 48.2L 10.6gal 12.7galUS

989 Volvo
740 GLE
1989 Sweden
125.0mph 201.1kmh
0-50mph 80.5kmh: 7.3secs
0-60mph 96.5kmh: 9.6secs
0-1/4 mile: 17.2secs
0-1km: 31.5secs
153.0bhp 114.1kW 155.1PS
@ 5700rpm
150.0lbft 203.3Nm @ 4450rpm
113.7bhp/ton 111.8bhp/tonne
66.1bhp/L 49.3kW/L 67.0PS/L
49.9ft/sec 15.2m/sec
26.7mpg 22.2mpgUS
10.6L/100km
Petrol 4-stroke piston
2316cc 141.3cu in
In-line 4 fuel injection
Compression ratio: 10.0:1
Bore: 96.0mm 3 8in
Stroke: 80.0mm 3.1in
Valve type/No: Overhead 16
Transmission: Automatic
No. of forward speeds: 4
Wheels driven: Rear
Springs F/R: Coil/Coil
Brake system: PA ABS
Brakes F/R: Disc/Disc
Steering: Rack & pinion PA
Wheelbase: 277.1cm 109.1in
Track F: 147.1cm 57.9in
Track R: 146.1cm 57.5in
Length: 478.5cm 188.4in
Width: 176.0cm 69.3in
Height: 141.0cm 55.5in
Kerb weight: 1368.8kg 3015.0lb
Fuel: 59.8L 13.1gal 15.8galUS

990 Volvo
740 Turbo
1989 Sweden
125.0mph 201.1kmh
0-50mph 80.5kmh: 5.6secs
0-60mph 96.5kmh: 7.8secs
0-1/4 mile: 16.0secs
160.0bhp 119.3kW 162.2PS
@ 5300rpm
187.0lbft 253.4Nm @ 2900rpm
116.5bhp/ton 114.6bhp/tonne
69.1bhp/L 51.5kW/L 70.0PS/L
46.4ft/sec 14.1m/sec
22.6mpg 18.8mpgUS
12.5L/100km
Petrol 4-stroke piston
2316cc 141.3cu in
turbocharged
In-line 4 fuel injection
Compression ratio: 8.7:1
Bore: 96.0mm 3.8in
Stroke: 80.0mm 3.1in
Valve type/No: Overhead 8
Transmission: Manual with
overdrive
Wheels driven: Rear
Springs F/R: Coil/Coil
Brake system: PA ABS
Brakes F/R: Disc/Disc
Steering: Rack & pinion PA
Wheelbase: 277.1cm 109.1in
Track F: 147.1cm 57.9in
Track R: 146.1cm 57.5in
Length: 478.5cm 188.4in
Width: 176.0cm 69.3in
Height: 141.0cm 55.5in
Kerb weight: 1396.0kg
3075.0lb
Fuel: 59.8L 13.1gal 15.8galUS

991 Volvo
440 GLE
1990 Sweden
107.0mph 172.2kmh
0-50mph 80.5kmh: 8.6secs
0-60mph 96.5kmh: 12.4secs
0-1/4 mile: 18.9secs
0-1km: 34.8secs
90.0bhp 67.1kW 91.2PS
@ 5800rpm
96.7lbft 131.0Nm @ 3600rpm
89.3bhp/ton 87.8bhp/tonne
52.6bhp/L 39.2kW/L 53.3PS/L
53.0ft/sec 16.1m/sec
22.4mph 36.0kmh/1000rpm
26.9mpg 22.4mpgUS
10.5L/100km
Petrol 4-stroke piston
1712cc 104.0cu in
In-line 4 1 Carburettor
Compression ratio: 9.5:1
Bore: 81.0mm 3.2in
Stroke: 83.5mm 3.3in
Valve type/No: Overhead 8

Transmission: Manual
No. of forward speeds: 5
Wheels driven: Front
Springs F/R: Coil/Coil
Brake system: PA ABS
Brakes F/R: Disc/Drum
Steering: Rack & pinion PA
Wheelbase: 250.5cm 98.6in
Track F: 142.0cm 55.9in
Track R: 143.0cm 56.3in
Length: 431.2cm 169.8in
Width: 171.0cm 67.3in
Height: 140.0cm 55.1in
Ground clearance: 17.9cm
7.0in
Kerb weight: 1025.0kg
2257.7lb
Fuel: 48.2L 10.6gal 12.7galUS

992 Volvo
460 GLi
1990 Sweden
109.0mph 175.4kmh
0-50mph 80.5kmh: 7.9secs
0-60mph 96.5kmh: 11.1secs
0-1/4 mile: 18.1secs
0-1km: 33.2secs
102.0bhp 76.1kW 103.4PS
@ 5600rpm
104.8lbft 142.0Nm @ 3900rpm
102.2bhp/ton 100.5bhp/tonne
59.3bhp/L 44.2kW/L 60.1PS/L
51.2ft/sec 15.6m/sec
21.5mph 34.6kmh/1000rpm
26.4mpg 22.0mpgUS
10.7L/100km
Petrol 4-stroke piston
1721cc 105.0cu in
In-line 4 fuel injection
Compression ratio: 10.0:1
Bore: 81.0mm 3.2in
Stroke: 83.5mm 3.3in
Valve type/No: Overhead 8
Transmission: Manual
No. of forward speeds: 5
Wheels driven: Front
Springs F/R: Coil/Coil
Brake system: PA ABS
Brakes F/R: Disc/Disc
Steering: Rack & pinion
Wheelbase: 250.2cm 98.5in
Track F: 142.0cm 55.9in
Track R: 142.7cm 56.2in
Length: 440.4cm 173.4in
Width: 168.9cm 66.5in
Height: 137.9cm 54.3in
Ground clearance: 17.8cm
7.0in
Kerb weight: 1015.0kg 2235.7lb
Fuel: 48.2L 10.6gal 12.7galUS

993 Volvo
460 Turbo
1990 Sweden

126.0mph 202.7kmh
0-50mph 80.5kmh: 6.2secs
0-60mph 96.5kmh: 8.9secs
0-1/4 mile: 16.7secs
0-1km: 30.7secs
122.0bhp 91.0kW 123.7PS
@ 5400rpm
129.2lbft 175.0Nm @ 3500rpm
114.9bhp/ton 113.0bhp/tonne
70.9bhp/L 52.9kW/L 71.9PS/L
49.3ft/sec 15.0m/sec
23.1mph 37.2kmh/1000rpm
26.4mpg 22.0mpgUS
10.7L/100km
Petrol 4-stroke piston
1721cc 105.0cu in
turbocharged
In-line 4 fuel injection
Compression ratio: 8.1:1
Bore: 81.0mm 3.2in
Stroke: 83.5mm 3.3in
Valve type/No: Overhead 8
Transmission: Manual
No. of forward speeds: 5
Wheels driven: Front
Springs F/R: Coil/Coil
Brake system: PA ABS
Brakes F/R: Disc/Disc
Steering: Rack & pinion PA
Wheelbase: 250.2cm 98.5in
Track F: 142.0cm 55.9in
Track R: 142.7cm 56.2in
Length: 440.4cm 173.4in
Width: 168.9cm 66.5in
Height: 137.9cm 54.3in
Ground clearance: 17.8cm
7.0in
Kerb weight: 1080.0kg
2378.8lb
Fuel: 48.2L 10.6gal 12.7galUS

994 Volvo
740 GLT 16v Estate
1990 Sweden
119.0mph 191.5kmh
0-50mph 80.5kmh: 6.6secs
0-60mph 96.5kmh: 9.4secs
0-1/4 mile: 16.9secs
0-1km: 31.5secs
155.0bhp 115.6kW 157.1PS
@ 5800rpm
149.8lbft 203.0Nm @ 4450rpm
111.1bhp/ton 109.2bhp/tonne
66.9bhp/L 49.9kW/L 67.8PS/L
50.7ft/sec 15.5m/sec
24.0mph 38.6kmh/1000rpm
21.1mpg 17.6mpgUS
13.4L/100km
Petrol 4-stroke piston
2316cc 141.0cu in
In-line 4 fuel injection
Compression ratio: 10.0:1
Bore: 96.0mm 3.8in
Stroke: 80.0mm 3.1in

Valve type/No: Overhead 16
Transmission: Manual
No. of forward speeds: 5
Wheels driven: Rear
Springs F/R: Coil/Coil
Brake system: PA ABS
Brakes F/R: Disc/Disc
Steering: Rack & pinion PA
Wheelbase: 277.1cm 109.1in
Track F: 147.0cm 57.9in
Track R: 146.0cm 57.5in
Length: 479.0cm 188.6in
Width: 176.0cm 69.3in
Height: 141.0cm 55.5in
Ground clearance: 17.9cm
7.0in
Kerb weight: 1419.0kg
3125.5lb
Fuel: 60.1L 13.2gal 15.9galUS

995 Volvo
740 Turbo
1990 Sweden
120.0mph 193.1kmh
0-50mph 80.5kmh: 6.4secs
0-60mph 96.5kmh: 8.8secs
0-1/4 mile: 16.7secs
162.0bhp 120.8kW 164.2PS
@ 4800rpm
195.0lbft 264.2Nm @ 3450rpm
118.0bhp/ton 116.0bhp/tonne
69.9bhp/L 52.2kW/L 70.9PS/L
42.0ft/sec 12.8m/sec
31.8mpg 26.5mpgUS
8.9L/100km
Petrol 4-stroke piston
2316cc 141.3cu in
turbocharged
In-line 4 fuel injection
Compression ratio: 8.7:1
Bore: 96.0mm 3.8in
Stroke: 80.0mm 3.1in
Valve type/No: Overhead 8
Transmission: Automatic
No. of forward speeds: 4
Wheels driven: Rear
Springs F/R: Coil/Coil
Brake system: PA ABS
Brakes F/R: Disc/Disc
Steering: Rack & pinion PA
Wheelbase: 277.1cm 109.1in
Track F: 147.1cm 57.9in
Track R: 146.1cm 57.5in
Length: 478.5cm 188.4in
Width: 176.0cm 69.3in
Height: 141.0cm 55.5in
Kerb weight: 1396.0kg
3075.0lb
Fuel: 59.8L 13.1gal 15.8galUS

996 Volvo
940 SE
1991 Sweden
124.0mph 199.5kmh

0-60mph 96.5kmh: 9.2secs
0-1/4 mile: 17.0secs
162.0bhp 120.8kW 164.2PS
@ 4000rpm
195.0lbft 264.2Nm @ 3450rpm
107.0bhp/ton 105.3bhp/tonne
69.9bhp/L 52.2kW/L 70.9PS/L
35.1ft/sec 10.7m/sec
24.0mpg 20.0mpgUS
11.8L/100km
Petrol 4-stroke piston
2316cc 141.3cu in
turbocharged
In-line 4 fuel injection
Compression ratio: 8.7:1
Bore: 96.0mm 3.8in
Stroke: 80.2mm 3.2in
Valve type/No: Overhead 8
Transmission: Automatic
No. of forward speeds: 4
Wheels driven: Rear
Springs F/R: Coil/Coil
Brake system: PA ABS
Brakes F/R: Disc/Disc
Steering: Rack & pinion PA
Wheelbase: 277.1cm 109.1in
Track F: 147.1cm 57.9in
Track R: 151.9cm 59.8in
Length: 486.9cm 191.7in
Width: 176.0cm 69.3in
Height: 141.0cm 55.5in
Kerb weight: 1539.1kg
3390.0lb
Fuel: 79.5L 17.5gal 21.0galUS

997 Volvo
940 SE Turbo
1991 Sweden
124.0mph 199.5kmh
0-50mph 80.5kmh: 6.7secs
0-60mph 96.5kmh: 9.3secs
0-1/4 mile: 17.2secs

0-1km: 31.4secs
155.0bhp 115.6kW 157.1PS
@ 5600rpm
172.7lbft 234.0Nm @ 3600rpm
113.4bhp/ton 111.5bhp/tonne
78.0bhp/L 58.2kW/L 79.1PS/L
49.0ft/sec 14.9m/sec
25.3mph 40.7kmh/1000rpm
21.2mpg 17.7mpgUS
13.3L/100km
Petrol 4-stroke piston
1986cc 121.0cu in
turbocharged
In-line 4 fuel injection
Compression ratio: 8.5:1
Bore: 89.0mm 3.5in
Stroke: 80.0mm 3.1in
Valve type/No: Overhead 8
Transmission: Manual
No. of forward speeds: 5
Wheels driven: Rear
Springs F/R: Coil/Coil
Brake system: PA
Brakes F/R: Disc/Disc
Steering: Rack & pinion PA
Wheelbase: 276.9cm 109.0in
Track F: 146.8cm 57.8in
Track R: 151.9cm 59.8in
Length: 486.9cm 191.7in
Width: 174.8cm 68.8in
Height: 141.0cm 55.5in
Ground clearance: 20.3cm
8.0in
Kerb weight: 1390.0kg
3061.7lb
Fuel: 61.9L 13.6gal 16.3galUS

998 Volvo
960 24v
1991 Sweden
129.0mph 207.6kmh
0-50mph 80.5kmh: 7.0secs

0-60mph 96.5kmh: 9.3secs
0-1/4 mile: 16.6secs
0-1km: 30.4secs
204.0bhp 152.1kW 206.8PS
@ 6000rpm
197.0lbft 267.0Nm @ 4300rpm
133.0bhp/ton 130.8bhp/tonne
69.8bhp/L 52.1kW/L 70.8PS/L
59.0ft/sec 18.0m/sec
25.8mph 41.5kmh/1000rpm
18.4mpg 15.3mpgUS
15.4L/100km
Petrol 4-stroke piston
2922cc 178.0cu in
In-line 6 fuel injection
Compression ratio: 10.7:1
Bore: 83.0mm 3.3in
Stroke: 90.0mm 3.5in
Valve type/No: Overhead 24
Transmission: Automatic
No. of forward speeds: 4
Wheels driven: Rear
Springs F/R: Coil/Coil
Brake system: PA ABS
Brakes F/R: Disc/Disc
Steering: Rack & pinion PA
Wheelbase: 276.9cm 109.0in
Track F: 146.8cm 57.8in
Track R: 151.9cm 59.8in
Length: 486.9cm 191.7in
Width: 174.8cm 68.8in
Height: 141.0cm 55.5in
Ground clearance: 20.3cm
8.0in
Kerb weight: 1560.0kg
3436.1lb
Fuel: 80.1L 17.6gal 21.1galUS

999 Volvo
960
1992 Sweden
135.0mph 217.2kmh

0-60mph 96.5kmh: 10.1secs
0-1/4 mile: 17.4secs
201.0bhp 149.9kW 203.8PS
@ 6000rpm
197.0lbft 266.9Nm @ 4300rpm
128.1bhp/ton 125.9bhp/tonne
68.8bhp/L 51.3kW/L 69.7PS/L
59.0ft/sec 18.0m/sec
24.0mpg 20.0mpgUS
11.8L/100km
Petrol 4-stroke piston
2922cc 178.3cu in
In-line 6 fuel injection
Compression ratio: 10.7:1
Bore: 83.0mm 3.3in
Stroke: 90.0mm 3.5in
Valve type/No: Overhead 24
Transmission: Automatic
No. of forward speeds: 4
Wheels driven: Rear
Springs F/R: Coil/Coil
Brake system: PA ABS
Brakes F/R: Disc/Disc
Steering: Rack & pinion PA
Wheelbase: 277.1cm 109.1in
Track F: 147.1cm 57.9in
Track R: 151.9cm 59.8in
Length: 486.9cm 191.7in
Width: 176.0cm 69.3in
Height: 141.0cm 55.5in
Kerb weight: 1595.8kg
3515.0lb
Fuel: 82.5L 18.1gal 21.8galUS

1000 Westfield
SEight
1991 UK
140.0mph 225.3kmh
0-50mph 80.5kmh: 3.3secs
0-60mph 96.5kmh: 4.3secs
0-1/4 mile: 13.2secs
273.0bhp 203.6kW 276.8PS

999 Volvo
960

1000 Westfield
SEight

@ 5700rpm
260.0lbft 352.3Nm @ 4200rpm
396.6bhp/ton 390.0bhp/tonne
69.1bhp/L 51.5kW/L 70.1PS/L
44.8ft/sec 13.7m/sec
27.0mph 43.4kmh/1000rpm
20.0mpg 16.7mpgUS
14.1L/100km
Petrol 4-stroke piston
3949cc 241.0cu in
Vee 8 4 Carburettor
Compression ratio: 10.0:1
Bore: 94.0mm 3.7in
Stroke: 71.9mm 2.8in
Valve type/No: Overhead 16
Transmission: Manual
No. of forward speeds: 5
Wheels driven: Rear
Springs F/R: Coil/Coil
Brakes F/R: Disc/Disc
Steering: Rack & pinion
Wheelbase: 236.2cm 93.0in
Track F: 152.4cm 60.0in
Track R: 158.8cm 62.5in
Length: 369.6cm 145.5in
Width: 162.6cm 64.0in
Height: 106.7cm 42.0in
Ground clearance: 10.2cm
4.0in
Kerb weight: 700.0kg 1541.8lb
Fuel: 45.5L 10.0gal 12.0galUS

1001 Yugo
GV
1986 Yugoslavia
90.0mph 144.8kmh
0-60mph 96.5kmh: 13.9secs
0-1/4 mile: 19.5secs
55.0bhp 41.0kW 55.8PS
@ 6000rpm
52.0lbft 70.5Nm @ 4600rpm
64.8bhp/ton 63.8bhp/tonne
49.3bhp/L 36.7kW/L 50.0PS/L

36.5ft/sec 11.1m/sec
15.6mph 25.1kmh/1000rpm
36.0mpg 30.0mpgUS
7.8L/100km
Petrol 4-stroke piston
1116cc 68.1cu in
In-line 4 1 Carburettor
Compression ratio: 9.2:1
Bore: 80.0mm 3.1in
Stroke: 55.5mm 2.2in
Valve type/No: Overhead 8
Transmission: Manual
No. of forward speeds: 4
Wheels driven: Front
Springs F/R: Coil/Leaf
Brake system: PA
Brakes F/R: Disc/Drum
Steering: Rack & pinion
Wheelbase: 215.1cm 84.7in
Track F: 130.8cm 51.5in
Track R: 131.3cm 51.7in
Length: 353.1cm 139.0in
Width: 154.2cm 60.7in
Height: 138.9cm 54.7in
Kerb weight: 862.6kg 1900.0lb
Fuel: 32.2L 7.1gal 8.5galUS

1002 Yugo
45 GLS
1987 Yugoslavia
83.0mph 133.5kmh
0-50mph 80.5kmh: 15.0secs
0-60mph 96.5kmh: 21.6secs
0-1/4 mile: 22.1secs
0-1km: 41.9secs
45.0bhp 33.6kW 45.6PS
@ 5800rpm
46.3lbft 62.7Nm @ 3300rpm
55.1bhp/ton 54.2bhp/tonne
49.8bhp/L 37.2kW/L 50.5PS/L
43.2ft/sec 13.1m/sec
14.7mph 23.7kmh/1000rpm
34.7mpg 28.9mpgUS

8.1L/100km
Petrol 4-stroke piston
903cc 55.0cu in
In-line 4 1 Carburettor
Compression ratio: 9.0:1
Bore: 65.0mm 2.6in
Stroke: 68.0mm 2.7in
Valve type/No: Overhead 8
Transmission: Manual
No. of forward speeds: 4
Wheels driven: Front
Springs F/R: Coil/Leaf
Brakes F/R: Disc/Drum
Steering: Rack & pinion
Wheelbase: 215.0cm 84.6in
Track F: 130.8cm 51.5in
Track R: 131.2cm 51.7in
Length: 353.0cm 139.0in
Width: 154.2cm 60.7in
Height: 139.0cm 54.7in
Ground clearance: 17.9cm
7.0in
Kerb weight: 830.0kg 1828.2lb
Fuel: 30.0L 6.6gal 7.9galUS

1003 Yugo
65A GLX
1988 Yugoslavia
97.0mph 156.1kmh
0-50mph 80.5kmh: 8.2secs
0-60mph 96.5kmh: 11.7secs
0-1/4 mile: 18.1secs
0-1km: 36.4secs
65.0bhp 48.5kW 65.9PS
@ 5800rpm
72.3lbft 98.0Nm @ 3500rpm
82.0bhp/ton 80.6bhp/tonne
50.1bhp/L 37.3kW/L 50.8PS/L
35.3ft/sec 10.7m/sec
19.3mph 31.1kmh/1000rpm
31.1mpg 25.9mpgUS
9.1L/100km
Petrol 4-stroke piston

1298cc 79.0cu in
In-line 4 1 Carburettor
Compression ratio: 9.1:1
Bore: 86.4mm 3.4in
Stroke: 55.5mm 2.2in
Valve type/No: Overhead 8
Transmission: Manual
No. of forward speeds: 5
Wheels driven: Front
Springs F/R: Coil/Leaf
Brake system: PA
Brakes F/R: Disc/Drum
Steering: Rack & pinion
Wheelbase: 215.0cm 84.6in
Track F: 130.8cm 51.5in
Track R: 131.2cm 51.7in
Length: 353.0cm 139.0in
Width: 154.2cm 60.7in
Height: 139.0cm 54.7in
Ground clearance: 17.9cm
7.0in
Kerb weight: 806.0kg 1775.3lb
Fuel: 30.0L 6.6gal 7.9galUS

1004 Yugo
Sana 1.4
1990 Yugoslavia
99.0mph 159.3kmh
0-50mph 80.5kmh: 9.4secs
0-60mph 96.5kmh: 13.2secs
0-1/4 mile: 19.2secs
0-1km: 35.9secs
70.0bhp 52.2kW 71.0PS
@ 6000rpm
78.2lbft 106.0Nm @ 2900rpm
76.5bhp/ton 75.3bhp/tonne
51.0bhp/L 38.0kW/L 51.7PS/L
44.5ft/sec 13.5m/sec
19.9mph 32.0kmh/1000rpm
28.1mpg 23.4mpgUS
10.1L/100km
Petrol 4-stroke piston
1372cc 84.0cu in

1004 Yugo
Sana 1.4

In-line 4 1 Carburettor
Compression ratio: 9.2:1
Bore: 80.5mm 3.2in
Stroke: 67.7mm 2.7in
Valve type/No: Overhead 8
Transmission: Manual
No. of forward speeds: 5
Wheels driven: Front
Springs F/R: Coil/Coil
Brake system: PA
Brakes F/R: Disc/Drum
Steering: Rack & pinion
Wheelbase: 250.0cm 98.4in
Track F: 140.0cm 55.1in
Track R: 140.0cm 55.1in
Length: 393.2cm 154.8in
Width: 165.8cm 65.3in
Height: 141.0cm 55.5in
Ground clearance: 17.9cm
7.0in
Kerb weight: 930.0kg 2048.5lb
Fuel: 47.8L 10.5gal 12.6galUS

1005 Zender
Vision 3
1989 Germany
175.0mph 281.6kmh
0-60mph 96.5kmh: 6.0secs
300.0bhp 223.7kW 304.2PS
@ 5000rpm
335.0lbft 453.9Nm @ 3750rpm
225.9bhp/ton 222.1bhp/tonne
54.1bhp/L 40.3kW/L 54.8PS/L
51.8ft/sec 15.8m/sec
Petrol 4-stroke piston
5547cc 338.4cu in
Vee 8 fuel injection
Compression ratio: 9.0:1
Bore: 96.5mm 3.8in
Stroke: 94.8mm 3.7in
Valve type/No: Overhead 16
Transmission: Manual
No. of forward speeds: 5
Wheels driven: Rear
Springs F/R: Coil/Coil
Brake system: PA

Brakes F/R: Disc/Disc
Steering: Rack & pinion
Wheelbase: 249.9cm 98.4in
Track F: 161.0cm 63.4in
Track R: 164.8cm 64.9in
Length: 406.9cm 160.2in
Width: 197.9cm 77.9in
Height: 111.0cm 43.7in
Kerb weight: 1350.6kg
2975.0lb
Fuel: 86.3L 19.0gal 22.8galUS

1006 Zender
Fact 4
1990 Germany
186.0mph 299.3kmh
0-60mph 96.5kmh: 4.3secs
448.0bhp 334.1kW 454.2PS
@ 6500rpm
390.0lbft 528.5Nm @ 4000rpm
411.3bhp/ton 404.4bhp/tonne
125.8bhp/L 93.8kW/L
127.5PS/L

61.4ft/sec 18.7m/sec
Petrol 4-stroke piston
3562cc 217.3cu in
turbocharged
Vee 8 fuel injection
Compression ratio: 9.3:1
Bore: 81.0mm 3.2in
Stroke: 86.4mm 3.4in
Valve type/No: Overhead 32
Transmission: Manual
No. of forward speeds: 5
Wheels driven: Rear
Springs F/R: Coil/Coil
Brakes F/R: Disc/Disc
Steering: Rack & pinion
Wheelbase: 249.9cm 98.4in
Track F: 161.0cm 63.4in
Track R: 164.8cm 64.9in
Length: 407.9cm 160.6in
Width: 199.9cm 78.7in
Height: 112.0cm 44.1in
Kerb weight: 1107.8kg 2440.0lb
Fuel: 99.9L 22.0gal 26.4galUS

1006 Zender
Fact 4